LONDON MATHEMATICAL SOCIETY STUDENT TEXTS

Managing Editor: Ian J. Leary,
Mathematical Sciences, University of Southampton, UK

London Mathematical Society Student Texts 96

Tensor Products of C^*-Algebras and Operator Spaces

The Connes–Kirchberg Problem

GILLES PISIER

Texas A&M University

CAMBRIDGE
UNIVERSITY PRESS

University Printing House, Cambridge CB2 8BS, United Kingdom

One Liberty Plaza, 20th Floor, New York, NY 10006, USA

477 Williamstown Road, Port Melbourne, VIC 3207, Australia

314–321, 3rd Floor, Plot 3, Splendor Forum, Jasola District Centre,
New Delhi – 110025, India

79 Anson Road, #06–04/06, Singapore 079906

Cambridge University Press is part of the University of Cambridge.

It furthers the University's mission by disseminating knowledge in the pursuit of
education, learning, and research at the highest international levels of excellence.

www.cambridge.org
Information on this title: www.cambridge.org/9781108479011
DOI: 10.1017/9781108782081

© Gilles Pisier 2020

First published 2020

Printed in the United Kingdom by TJ International Ltd. Padstow Cornwall

A catalogue record for this publication is available from the British Library.

ISBN 978-1-108-47901-1 Hardback
ISBN 978-1-108-74911-4 Paperback

Cambridge University Press has no responsibility for the persistence or accuracy
of URLs for external or third-party internet websites referred to in this publication
and does not guarantee that any content on such websites is, or will remain,
accurate or appropriate.

Contents

Introduction

These lecture notes are centered around two open problems, one formulated by Alain Connes in his famous 1976 paper [61], the other one by Eberhard Kirchberg in his landmark 1993 paper [155]. At first glance, these two problems seem quite different and the proof of their equivalence described at the end of [155] is not so easy to follow. One of our main goals is to explain in detail the proof of this equivalence in an essentially self-contained way. The Connes problem asks roughly whether traces on "abstract" von Neumann algebras can always be approximated (in a suitable way) by ordinary matrix traces. The Kirchberg problem asks whether there is a unique C^*-norm on the algebraic tensor product $\mathscr{C} \otimes \mathscr{C}$ when $\mathscr{C}$ is the full C^*-algebra of the free group $\mathbb{F}_\infty$ with countably many generators.

In the remarkable paper where he proved the equivalence, Kirchberg studied more generally the pairs of C^*-algebras (A, B) for which there is only one C^*-norm on the algebraic tensor product $A \otimes B$. We call such pairs "nuclear pairs." A C^*-algebra A is traditionally called nuclear if this holds for any C^*-algebra B. Our exposition chooses as its cornerstone Kirchberg's theorem asserting the nuclearity of what is for us the "fundamental pair," namely the pair $(\mathscr{B}, \mathscr{C})$ where $\mathscr{B} = B(\ell_2)$ (see Theorem 9.6). Our presentation leads us to highlight two properties of C^*-algebras, the Weak Expectation Property (WEP) and the Local Lifting Property (LLP).

The first one is a weak sort of extension property (or injectivity) while the second one is a weak sort of lifting property. The connection with the fundamental pair is very clear: A has the WEP (resp. LLP) if and only if the pair $(A, \mathscr{C})$ (resp. $(A, \mathscr{B})$) is nuclear. With this terminology, the Kirchberg problem reduces to proving the implication LLP $\Rightarrow$ WEP, but there are many more interesting reformulations that deserve mention and we will present them in detail. For instance this problem is equivalent to the question whether *every* (unital) C^*-algebra is a quotient of one with the WEP, or equivalently, in short,

1

is QWEP. In passing, although the P stands for property, we will sometimes write for short that A is WEP (or A is LLP) instead of A has the WEP (resp. LLP).

Incidentally, since Kirchberg (unlike Connes) explicitly conjectured a positive answer to all these equivalent questions in [155], we often refer to them as his conjectures.

One originality of our treatment (although already present in [155]) is that we try to underline the structural properties (or their failure), such as injectivity or projectivity, in parallel for the minimal and the maximal tensor product of C^*-algebras. This preoccupation can be traced back to the "fundamental pair" itself: Indeed, we may view $\mathscr{B}$ as "injectively universal" and $\mathscr{C}$ as "projectively universal." The former because any separable C^*-algebra A is a subalgebra of $\mathscr{B}$, the latter because any such A is a quotient of $\mathscr{C}$ (see Proposition 3.39).

In particular, we will emphasize the fact that the minimal tensor product is injective but not projective, while the maximal one is projective but not injective (see §7.4 and 7.2). This is analogous to the situation that prevails for the Banach space tensor products in Grothendieck's classical work, but unlike Banach space morphisms (i.e. bounded linear maps) the C^*-algebraic morphisms are automatically isometric if they are injective (see Proposition A.24). The lack of injectivity of the max-norm is a rephrasing of the fact that if $B_1 \subset B_2$ is an isometric (or equivalently injective) $*$-homomorphism between C^*-algebras and A is another C^*-algebra, it is in general *not true* that the resulting $*$-homomorphism

$$A \otimes_{\max} B_1 \to A \otimes_{\max} B_2 \tag{1}$$

is isometric (or equivalently injective). This means that the norm induced by $A \otimes_{\max} B_2$ on the algebraic tensor product $A \otimes B_1$ is *not* equivalent to the max-norm on $A \otimes B_1$. In sharp contrast, this does not happen for the min-norm: $A \otimes_{\min} B_1 \to A \otimes_{\min} B_2$ is always injective (=isometric), and this is why one often says that the minimal tensor product is "injective."

This "defect" of the max-tensor product leads us to single out the class of inclusions, $B_1 \subset B_2$, for which this defect disappears (i.e. (1) is injective for any A). We choose to call them "max-injective." We will see that this holds if and only if there is a projection $P : B_2^{**} \to B_1^{**}$ with $\|P\| = 1$. We will also show that if (1) is injective for $A = \mathscr{C}$ then it is injective for all A.

It turns out that a C^*-algebra A is WEP if and only if the embedding $A \subset B(H)$ is max-injective or, equivalently, if and only if there is a projection $P : B(H)^{**} \to A^{**}$ with $\|P\| = 1$. All these facts have analogues for the min-tensor product, but now its "defect" is the failure of "projectivity," meant in the following sense: Let $q : B_1 \to B_2$ be a surjective $*$-homomorphism and let A be any C^*-algebra. Let $\mathcal{I} = \ker(q)$. Then, although the associated

*-homomorphism $q_A : A \otimes_{\min} B_1 \to A \otimes_{\min} B_2$ is clearly surjective (indeed, it suffices for that to have a dense range), its kernel may be strictly larger than $A \otimes_{\min} \mathcal{I}$. As a result, the min-norm on the algebraic tensor product $A \otimes B_2$ ($= A \otimes (B_1/\mathcal{I})$) may be much smaller than the norm induced on it by $(A \otimes_{\min} B_1)/(A \otimes_{\min} \mathcal{I})$. In sharp contrast, this "defect" does not happen for the max-norm and we always have an isometric identification

$$A \otimes_{\max} (B_1/\mathcal{I}) = (A \otimes_{\max} B_1)/(A \otimes_{\max} \mathcal{I}).$$

Again this defect of the min-norm leads us to single out the quotient maps (i.e. the surjective *-homomorphisms) $q : B_1 \to B_2$ for which the defect does not appear, i.e. the maps such that for any A we have an isometry

$$A \otimes_{\min} B_2 = (A \otimes_{\min} B_1)/(A \otimes_{\min} \mathcal{I}). \tag{2}$$

Here again, we can give a rather neat characterization of such maps, this time as a certain form of lifting property, see §7.5. It turns out that if (2) holds for $A = \mathscr{B}$ then it holds for all C^*-algebras A. We call such a map q a "min-projective surjection." The usual terminology to express that (2) holds for any A is that B_1 viewed as an extension of B_2 by $\mathcal{I}$ is a "locally split extension" (we prefer not to use this term). This notion is closely connected with the notion of exact C^*-algebra.

A C^*-algebra A is called exact if (2) holds for any surjective $q : B_1 \to B_2$. This "exact" terminology is motivated by the fact that (2) holds if and only if the sequence

$$0 \to A \otimes_{\min} \mathcal{I} \to A \otimes_{\min} B_1 \to A \otimes_{\min} B_2 \to 0$$

is exact. But actually, for C^*-algebras, the exactness of that sequence boils down to the fact that the natural *-homomorphism

$$\frac{A \otimes_{\min} B_1}{A \otimes_{\min} \mathcal{I}} \to A \otimes_{\min} B_2$$

is isometric (=injective).

Although our main interest is in C^*-algebras, it turns out that many results have better formulations (and sometimes better proofs) when phrased using linear subspaces of C^*-algebras (the so-called operator spaces) or unital self-adjoint subspaces (the so-called operator systems). It is thus natural to try to describe as best as we can the class of linear transformations that preserve the C^*-tensor products. For the minimal norm, it is well known that the associated class is that of completely bounded (c.b.) maps. More precisely, given a linear map $u : A \to B$ between C^*-algebras we have for any C^*-algebra C

$$\|Id_C \otimes u : C \otimes_{\min} A \to C \otimes_{\min} B\| \leq \|u\|_{cb} \tag{3}$$

where $\|u\|_{cb}$ is the c.b. norm of u. Moreover, the sup over all C of the left-hand side of (3) is equal to $\|u\|_{cb}$, and it remains unchanged when restricted to $C \in \{M_n \mid n \geq 1\}$. The space of such maps is denoted by $CB(A, B)$.

The mapping u is called completely positive (in short c.p.) if $Id_C \otimes u : C \otimes_{\min} A \rightarrow C \otimes_{\min} B$ is positive (=positivity preserving) for any C, and to verify this we may restrict to $C = M_n$ for any $n \geq 1$. The cone formed of all such maps is denoted by $CP(A, B)$.

For the max tensor product, there is an analogue of (3) but the corresponding class of mappings is smaller than $CB(A, B)$. These are the decomposable maps denoted by $D(A, B)$, defined as linear combinations of maps in $CP(A, B)$. More precisely, for any u as previously we have

$$\|Id_C \otimes u : C \otimes_{\max} A \rightarrow C \otimes_{\max} B\| \leq \|u\|_{dec}, \qquad (4)$$

where $\|u\|_{dec}$ is the norm in $D(A, B)$. Moreover, the supremum over all C of the left-hand side of (4) is equal to the dec-norm of u composed with the inclusion $B \subset B^{**}$. The dec-norm was introduced by Haagerup in [104]. We make crucial use of several of the properties established by him in the latter paper. See Chapter 6.

The third class of maps that we analyze are the maps $u : A \rightarrow B$ such that for any C

$$\|Id_C \otimes u : C \otimes_{\min} A \rightarrow C \otimes_{\max} B\| \leq 1.$$

This holds if and only if u is the pointwise limit of a net of finite rank maps with $\|u\|_{dec} \leq 1$ (see Proposition 6.13). When u is the identity on A this means that A has the c.p. approximation property (CPAP) which, as is by now well known, characterizes nuclear C^*-algebras (see Corollary 7.12).

More generally, suppose given two C^*-norms α and β, defined on $A \otimes B$ for any pair (A, B). We denote by $A \otimes_\alpha B$ (resp. $A \otimes_\beta B$) the C^*-algebra obtained after completion of $A \otimes B$ equipped with α (resp. β).

Then we say that a linear map $u : A \rightarrow B$ between C^*-algebras is $(\alpha \rightarrow \beta)$-tensorizing if for any C^*-algebra C

$$\|Id_C \otimes u : C \otimes_\alpha A \rightarrow C \otimes_\beta B\| \leq 1.$$

In §7.1 we describe the factorizations characterizing such maps in all the cases when α and β are either the minimal or the maximal C^*-norm. We also include the case when u is only defined on a subspace $E \subset A$ using the norm induced on $C \otimes E$ by $C \otimes_\alpha A$. The main cases of interest are min $\rightarrow$ max (nuclearity) and max $\rightarrow$ max (decomposability). For the former, we refer to Chapter 10, where we characterize nuclear C^*-algebras in parallel with exactness.

The bidual A^{**} of a C^*-algebra A is isomorphic to a von Neumann algebra. In Chapter 8 we study the relations between C^*-norms on A and on A^{**}

and we describe the biduals of certain C^*-tensor products. The notion of local reflexivity plays an important role in that respect. We prove in §8.3 the equivalence of the injectivity of A^{**} and the nuclearity of A. In Corollary 7.12 (proved in §10.2) we show that for C^*-algebras nuclearity is equivalent to the completely positive approximation property (CPAP). We also show in Theorem 8.12 that injective von Neumann algebras are characterized by a weak* analogue of the CPAP, which is sometimes called "semidiscreteness."

But our main emphasis is on *nuclear pairs*: in §9.1 we prove the nuclearity of the fundamental pair $(\mathscr{B}, \mathscr{C})$ and in the rest of Chapter 9 we give various equivalent characterizations of C^*-algebras with the properties WEP, LLP, and QWEP, that we choose to define using nuclear pairs. The main ones are formulated using the bidual A^{**} of a C^*-algebra A (see §8.1). Let $i_A : A \to A^{**}$ be the natural inclusion. For instance:

(i) A is nuclear if and only if for some (or any) embedding $A^{**} \subset B(H)$ there is a projection $P : B(H) \to A^{**}$ with $\|P\|_{cb} = 1$.

(ii) A is WEP if and only if for some (or any) embedding $A \subset B(H)$ there is a projection $P : B(H)^{**} \to A^{**}$ with $\|P\|_{cb} = 1$.

(iii) A is QWEP if and only if for some embedding $A^{**} \subset B(H)^{**}$ there is a projection $P : B(H)^{**} \to A^{**}$ with $\|P\|_{cb} = 1$.

We then come to the central part of these notes: the Connes embedding problem whether any tracial probability space embeds in an ultraproduct of matricial ones (Chapter 12) and the Kirchberg conjecture (Chapter 13) that $\mathscr{C}$ is WEP or that every C^*-algebra is QWEP. We show that they are equivalent in Chapter 14. We also show the equivalence with a well-known conjecture from Banach space theory (Chapter 15). The latter essentially asserts that every von Neumann algebra is isometric (as a Banach space) to a quotient of $B(H)$ for some H. In yet another direction we show in Chapter 16 that all these conjectures are equivalent to one formulated by Tsirelson in the context of quantum information theory.

In one of its many equivalent forms, Kirchberg's conjecture reduces to LLP $\Rightarrow$ WEP for C^*-algebras. Actually, he originally conjectured also the converse implication but in Chapter 18 we show that this fails, by producing tensors $t \in \mathscr{B} \otimes \mathscr{B}$ for which the min and max norms are different; in other words the pair $(\mathscr{B}, \mathscr{B})$ is not nuclear. The proof combines ideas from finite-dimensional operator space theory (indeed $t \in E \otimes F$ for some finite-dimensional subspaces E, F of $\mathscr{B}$) together with estimates of spectral gaps, that allow us to show that a certain constant $C(n)$ defined next is $<n$ for some n. The latter constant involves a sequence of integers N_m and a sequence $(u_1(m), \ldots, u_n(m))$ of n-tuples of unitary $N_m \times N_m$-matrices and their complex conjugates $(\overline{u_1(m)}, \ldots, \overline{u_n(m)})$. We then set

$$C(n) = \inf \left\{ \sup_{m \neq m'} \left\| \sum_1^n \overline{u_j(m)} \otimes u_j(m') \right\| \right\}, \tag{5}$$

where the last norm is meant in $M_{N_m N_{m'}}$ and the infimum runs over all possible sizes (N_m) and all possible sequences $(u_1(m), \ldots, u_n(m))$ of n-tuples of unitary $N_m \times N_m$-matrices.

Using unitary random matrices we will show that $C(n) = 2\sqrt{n-1}$ (see §18.2). Nevertheless other more explicit (deterministic) constructions of sequences $(u_1(m), \ldots, u_n(m))$ responsible for $C(n) < n$ are of much interest such as property (T) groups, expanders, quantum expanders, or quantum analogues of spherical coding sequences. In each case we obtain a tensor $t \in \mathcal{B} \otimes \mathcal{B}$ such that $\|t\|_{\min} < \|t\|_{\max}$. We describe these delicate ingredients in Chapter 19.

In Chapter 20, we gather several applications of the preceding ideas to the structure of the metric space of all finite-dimensional operator spaces equipped with the cb-analogue of the Banach–Mazur "distance," that is defined when $\dim(E) = \dim(F)$ by

$$d_{cb}(E, F) = \inf\{\|u\|_{cb}\|u^{-1}\|_{cb} \mid u : E \to F \text{ invertible}\}.$$

For instance, for any finite-dimensional operator space E, the dual space E^* admits a natural operator space structure (described in §2.4) so that we may view both E and E^* as subspaces of $\mathcal{B}$. Thus the identity operator on E defines a tensor $t_E \in \mathcal{B} \otimes \mathcal{B}$. We show that (see (20.6))

$$\|t_E\|_{\mathcal{B} \otimes_{\max} \mathcal{B}} = \inf\{d_{cb}(E, F) \mid F \subset \mathcal{C}\}$$

where the infimum (which is actually attained) runs over all possible subspaces $F \subset \mathcal{C}$ with $\dim(F) = \dim(E)$.

The fact that $(\mathcal{B}, \mathcal{B})$ is not a nuclear pair actually implies that for arbitrary von Neumann algebras (M, N) the pair (M, N) is nuclear only if either M or N is nuclear. This follows from the fact that a nonnuclear von Neumann algebra must contain as a subalgebra the direct sum in the sense of ℓ_∞ of the family $\{M_n \mid n \geq 1\}$ of all matrix algebras, and there is automatically a conditional expectation onto it. The latter is explained in §12.6.

In Chapter 23 we present in detail two unpublished characterizations of the WEP due to Haagerup. The first one says that a C^*-algebra A has the WEP if and only if for any n and any linear map $u : \ell_\infty^n \to A$ the dec-norm of u coincides with its c.b. norm (see §23.2). This naturally complements his earlier results from the 1980s in [104]. Haagerup claimed this theorem at some point in the 1990s but apparently did not circulate a detailed proof of it, as he did for the second (more delicate) one, that we give in §23.5.

There, to put it very roughly ℓ_∞^n is replaced by ℓ_2^n. More precisely, the second characterization says that A has the WEP if and only if for any n and any $(a_1, \ldots, a_n) \in A^n$ we have

$$\left\| \sum \overline{a_j} \otimes a_j \right\|_{\min}^{1/2} = \left\| \sum \overline{a_j} \otimes a_j \right\|_{\max}^{1/2}.$$

An important ingredient for its proof is the identification, for any C^*-algebra A, of the norm

$$A^n \ni (a_j) \mapsto \left\| \sum \overline{a_j} \otimes a_j \right\|_{\max}^{1/2}$$

as the norm obtained on A^n ($n \geq 1$) by the complex interpolation method of parameter $\theta = 1/2$ between the ("row and column") norms

$$(a_j) \mapsto \left\| \sum a_j{}^* a_j \right\|^{1/2} \text{ and } (a_j) \mapsto \left\| \sum a_j a_j{}^* \right\|^{1/2}.$$

In order to give a reasonably self-complete proof of the latter fact we give a brief basic description of complex interpolation in Chapter 22.

One important consequence of this particular characterization is the fact that the WEP is stable under complete isomorphisms. Explicitly, if two C^*-algebras A, B are completely isomorphic as operator spaces, then A WEP $\Rightarrow$ B WEP. In other words, if we forget the algebraic structure of a C^*-algebra, the WEP is "remembered" by its operator space structure.

In a similar flavor (see Chapter 23), let $M \subset \mathcal{M}$ be von Neumann algebras, if there is a completely bounded projection $P : \mathcal{M} \to M$ onto M (i.e. M is "completely complemented" in $\mathcal{M}$) then there is a projection $Q : \mathcal{M} \to M$ that is completely positive with $\|Q\|_{cb} = 1$. Thus when $\mathcal{M} = B(H)$ we conclude that M is injective.

In Chapter 24 we show that the tensor product $M \otimes_{\min} N$ of two nonnuclear von Neumann algebras M and N (for instance for $M = N = \mathcal{B}$) fails the WEP (see Corollary 24.23). The proof is reminiscent of the earlier proof that (M, N) is not a nuclear pair. It makes crucial use of the constant that we denote by $C_0(n)$, that is defined in the same way as $C(n)$ in (5), but using unitaries associated to permutations instead of plain unitary matrices and restricting them to the orthogonal of the constant vector. Again, the key point is that $C_0(n) < n$. We review the recent results that establish the latter. In analogy with the case of $C(n)$ we can show that $C_0(n) = 2\sqrt{n-1}$ using a very recent result on random permutation matrices, and also that $C_0(3) < 3$ by delicate deterministic arguments: we can use either Selberg's famous spectral bound or known results on expanders in permutation groups.

Lastly, in Chapter 25 we gather a collection of open questions related to our main topics.

Prerequisites. These notes are written in a rather detailed style and should be accessible to graduate students and nonspecialists. The prerequisite background is kept to a minimum. Of course basic functional analysis is needed, but for operator algebras, the fundamental theorems we use are the classical ones, such as the bicommutant theorem and Kaplansky's Theorem, as well as basic facts about states, *-homomorphisms and the GNS construction, and we review all those in this book's Appendix.

Sources. The main source for these notes is Kirchberg's fundamental paper [155]. However, we have made extensive use of Haagerup's treatment of decomposable maps in [104]. This allowed us to reformulate many results known for completely positive maps or *-homomorphisms for just *linear* maps. In addition, Ozawa's surveys [189, 191, 192] have been an invaluable help and inspiration, as well as the (highly recommended) book [39] by Brown and Ozawa.

Many of Kirchberg's results on exactness are already presented in detail in Simon Wassermann's excellent 1994 notes [258], the present volume can be viewed as a sequel and an updated complement to his.

Almost all chapters are followed by a Notes and Remarks section where we try to complement the references given in the text, and sometimes add some pointers to the literature.

About operator spaces. Some results already appear in our 2003 book on operator spaces [208]. When convenient, we used the presentation from [208]. We describe several applications of operator space theory when they are relevant for our topic, but our *main focus* being here on tensor products of C^*-algebras, we will refrain from developing operator space theory for its own sake, and we refer the reader instead to [208], or to [80, 196].

About operator systems. Following Arveson's pioneering papers [12], much work (notably by Choi, Effros, and Lance) on operator systems appeared already in the 1970s which marked a first period when much progress on tensor products of C^*-algebras was achieved. In particular, Choi and Effros introduced in [47] a notion of duality for operator systems that prefigured the one for operator spaces developed after Ruan's 1987 Ph.D. thesis. The emphasis then moved on to operator spaces in the 1990s, and C^*-tensor products were investigated (following Kirchberg's impulse and Haagerup's work) in the more general framework of operator space tensor products, by Effros, Ruan, Blecher, Paulsen, and others. Curiously, operator systems made a reappearance more recently and their tensor products were investigated thoroughly in a series of papers, notably [150, 151]. This led to several characterizations of the WEP (see [90–92, 149, 152, 153]), connected to the Connes–Kirchberg problem, but for lack of space (and energy) we chose not to cover this.

We also had to leave out the connections of the Connes–Kirchberg problem with noncommutative real algebraic geometry, for which we refer the reader to [40, 163] and to Ozawa's survey [191].

Basic notation and conventions. The letter H (or $\mathcal{H}$) always stands for a Hilbert space. Our Hilbert spaces all have an inner product

$$(y, x) \mapsto \langle y, x \rangle$$

that is linear in x and antilinear in y.

We denote by $B(H)$ (resp. $K(H)$) the Banach algebra formed of all the bounded (resp. compact) linear operators on H equipped with the operator norm.

Let K be another Hilbert space. We denote by $K \otimes_2 H$ the Hilbert space tensor product, obtained by completing $K \otimes H$ equipped with the classical scalar product characterized by

$$\langle k \otimes h, k' \otimes h' \rangle = \langle k, k' \rangle \langle h, h' \rangle.$$

We denote by $\overline{K}$ the complex conjugate Hilbert space, which is classically identified with the dual K^*. Then $\overline{K} \otimes_2 H$ can be identified with the space of all the Hilbert–Schmidt maps from K to H.

The unitary group of a unital C^*-algebra A is denoted by $U(A)$.

The identity map on a linear space X is denoted by Id_X.

The unit ball of a normed space X is denoted by B_X.

Let $1 \leq p \leq \infty$. Let I be an arbitrary index set. We denote by $\ell_p(I)$ the set of families of complex scalars $x = (x_i)_{i \in I}$ such that $\sum_{i \in I} |x_i|^p < \infty$ ($\sup_{i \in I} |x_i| < \infty$ when $p = \infty$) equipped with the norm $\|x\|_p = (\sum_{i \in I} |x_i|^p)^{1/p}$ ($\sup_{i \in I} |x_i|$ when $p = \infty$).

When $I = \mathbb{N}$, we denote $\ell_p(I)$ simply by ℓ_p.

Let $(X_i)_{i \in I}$ be a family of Banach spaces. We denote by

$$\left(\oplus \sum_{i \in I} X_i \right)_p$$

their direct sum "in the sense of ℓ_p," equipped with the norm $(x_i) \mapsto (\sum \|x_i\|^p)^{1/p}$.

When $X_i = X$ for all $i \in I$, we denote $\left(\oplus \sum_{i \in I} X_i \right)_p$ by $\ell_p(I; X)$.

When $X_i = \mathbb{C}$ for all $i \in I$ we recover $\ell_p(I)$.

In the particular case when $p = \infty$ the space $\mathcal{X} = \left(\oplus \sum_{i \in I} X_i \right)_\infty$ is the set of those $x = (x_i)$ with $x_i \in X_i$ ($\forall i \in I$) such that $\|x\| = \sup_{i \in I} \|x_i\| < \infty$.

The unit ball of this space $\mathcal{X}$ is just the product $B_{\mathcal{X}} = \prod_{i \in I} B_{X_i}$.

Let $n \geq 1$ be an integer. We denote by ℓ_p^n the space $\mathbb{C}^n$ equipped with the norm

$$x \mapsto \|x\| = \left(\sum_1^n |x_j|^p \right)^{1/p}.$$

Thus $\ell_p^n = \ell_p(I)$ for $I = \{1, \ldots, n\}$. When $p = 2$, the resulting space ℓ_2^n is the model for any n-dimensional Hilbert space.

When $p = \infty$, we set $\|x\| = \sup_j |x_j|$, the resulting space ℓ_∞^n is the model for any n-dimensional commutative C^*-algebra.

We denote by M_n (resp. $M_{n \times m}$) the space of $n \times n$ (resp. $n \times m$) matrices with complex entries. More generally, for any vector space E we will denote by $M_n(E)$ (resp. $M_{n \times m}(E)$) the space of $n \times n$ (resp. $n \times m$) matrices with entries in E. Thus $M_n = M_n(\mathbb{C})$ (resp. $M_{n \times m} = M_{n \times m}(\mathbb{C})$).

A linear mapping $u : X \to Y$ between Banach spaces with $\|u\| \leq 1$ is called "contractive" or "a contraction." We say that $u : X \to Y$ is a metric surjection if $u(X) = Y$ and the image of the open unit ball of X coincides with the open unit ball of Y. Then passing to the quotient by $\ker(u)$ produces an isometric isomorphism from $X/\ker(u)$ to Y.

A mapping $u : A \to B$ between C^*-algebras is called a $*$-homomorphism if it is a homomorphism of algebras such that $u(x^*) = u(x)^*$ for all $x \in A$. When $B = B(H)$ for some Hilbert space H the term "representation" is often used instead of $*$-homomorphism.

Some abbreviations frequently used. c.b. for completely bounded, c.p. for completely positive, u.c.p. for unital and completely positive, c.c. for completely contractive.

Acknowledgment. These lecture notes are partially based on the author's notes for topics courses given at Texas A&M University (Fall 2014 and 2016, Spring 2018), and a minicourse at the Winter School on Operator Spaces, Noncommutative Probability and Quantum Groups, held in Métabief in Dec. 2014, organized by the Laboratoire de Mathématiques de Besançon. I am indebted to Mateusz Wasilewski whose careful reading led to a number of corrections and improvements of a first draft. The author is very grateful to all the auditors and readers of the various drafts who by their questions and remarks helped improve several chapters, in particular to Roy Araiza, Li Gao, Kei Hasegawa, Guixiang Hong, Alexandre Nou, Mikael de la Salle, Andrew Swift, Simeng Wang, John Weeks, Hao Xing, and particularly Ignacio Vergara. Special thanks are due to Michiya Mori for his critical reading of a close to final version, which allowed me to correct numerous inadequacies.

1

Completely bounded and completely positive maps

Basics

In this opening chapter, we present the fundamental extension and factorization properties of c.b. and c.p. maps.

1.1 Completely bounded maps on operator spaces

We start with a few basic facts about operator spaces.

Definition 1.1 ("Noncommutative Banach spaces") An operator space E is a closed subspace of $B(H)$, i.e. we are given an inclusion mapping

$$E \subset B(H). \qquad (1.1)$$

Equipped with the norm induced by the one of $B(H)$, the space E is then a Banach space. Thus if we wish, we may think of an operator space as a Banach space given with the additional structure of an isometric embedding as in (1.1). Then for each $n \geq 1$ we have $M_n(E) \subset M_n(B(H))$. Since $M_n(B(H)) \simeq B(H \oplus \cdots \oplus H)$, ($n$-times) we can equip $M_n(E)$ with the norm induced on it by $M_n(B(H))$ or equivalently by $B(H \oplus \cdots \oplus H)$. Thus the data of the embedding (1.1) immediately leads to the sequence of norms

$$\{\|.\|_{M_n(E)} \mid n \geq 1\}.$$

In operator space theory, the usual norm from Banach space theory is replaced by that sequence of norms. The ordinary norm on E corresponds to $n = 1$.

Equivalently, we can think of an operator space E as a closed subspace of a C^*-algebra A itself embedded in $B(H)$ so we have

$$E \subset A \subset B(H).$$

For instance, we could use for A the C^*-algebra generated by E in $B(H)$ (i.e. the smallest C^*-subalgebra containing E) or we could take $A = B(H)$.

Actually, we can avoid reference to $B(H)$: a more abstract but still equivalent viewpoint is to define an operator space as a closed subspace of an (abstract) C^*-algebra A. Then the space $M_n(A)$ of $n \times n$ matrices with entries in A is also a C^*-algebra for the usual matrix operations, and hence (since it is complete) it has a unique C^*-norm, which we can again induce on the subspace $M_n(E)$. Thus the embedding $E \subset A$ automatically yields a sequence of norms $\{\|.\|_{M_n(E)} \mid n \geq 1\}$. It is easy to see (using Gelfand theory to embed A in some $B(H)$, see §A.11) that these two definitions of operator spaces are equivalent.

Occasionally, we will consider a linear subspace $E \subset B(H)$ that is not closed and treat it as an operator space. This simply means that we are referring to the norm closure of E in $B(H)$.

We will refer to the sequence of norms $\{\|.\|_{M_n(E)} \mid n \geq 1\}$ as the "operator space structure" of E. We extract from the preceding discussion that we can equip any concrete operator space $E \subset B(H)$ with a natural (somewhat "abstract") operator space structure.

"Operator space Theory" (see the books [80, 208]) took off after Ruan's 1987 thesis where he identified the abstract sequences of norms on $M_n(E)$ (E a vector space) that come from a concrete realization of E as a subspace of some $B(H)$. Operator space theory was then developed in the works of Effros–Ruan, Blecher–Paulsen, and others as a generalization of the Choi–Effros theory of operator systems developed in the 1970s in the series [45–48].

In this theory, the *bounded* linear maps (between Banach spaces) are replaced by the *completely bounded* linear ones (between operator spaces).

Let $u : E \to F$ be a linear map between operator spaces. For any given $n \geq 1$, we denote by $u_n : M_n(E) \to M_n(F)$ the linear map defined by

$$u_n([a_{ij}]) = [u(a_{ij})].$$

Definition 1.2 A map $u : E \to F$ is called completely bounded (in short c.b.) if

$$\sup_{n \geq 1} \|u_n\|_{M_n(E) \to M_n(F)} < \infty.$$

We define $\|u\|_{cb} = \sup_{n \geq 1} \|u_n\|_{M_n(E) \to M_n(F)}$ and we denote by $CB(E, F)$ the Banach space of all c.b. maps from E into F equipped with the c.b. norm.

In tensor product notation, using $M_n(E) \simeq M_n \otimes E$ we write $u_n = Id_{M_n} \otimes u : M_n \otimes E \to M_n \otimes F$ so that $u_n\left(\sum a_k \otimes x_k\right) = \sum a_k \otimes u(x_k)$. We have then

$$\|u\|_{cb} = \sup_{n \geq 1} \sup_{t \in B_{M_n(E)}} \|(Id_{M_n} \otimes u)(t)\|_{M_n(F)}. \tag{1.2}$$

If $G \subset B(\mathcal{H})$ is another operator space and if $v : F \to G$ is c.b., then the composition $vu : E \to G$ clearly remains c.b. and we have

$$\|vu\|_{cb} \leq \|v\|_{cb}\|u\|_{cb}.$$

Of course, when $n = 1$, 1×1 matrices are just elements of E, so that $u_1 : M_1(E) \to M_1(F)$ is nothing but u itself. In particular we have $\|u\| \leq \|u\|_{cb}$ and $CB(E, F) \subset B(E, F)$. In general this is a strict inclusion, but for linear forms or if $\mathrm{rk}(u) = 1$ we have $\|u\| = \|u\|_{cb}$. Indeed, assuming $E \subset B(H)$, for any $u : E \to F$ of the form $u(a) = \xi(a)b$, $\xi \in E^*, b \in F$ (note that $\|u\| = \|b\|\|\xi\|$) we have for any $a \in M_n(E)$

$$\|[u(a_{ij})]\| = \|b\|\|[\xi(a_{ij})]\|_{M_n} = \|b\| \sup_{x,y \in B_{\ell_2^n}} \left| \sum \xi(a_{ij})\overline{y_i}x_j \right|$$

$$\leq \|u\| \sup_{x,y \in B_{\ell_2^n}} \left\| \sum a_{ij}\overline{y_i}x_j \right\|_E$$

$$= \|u\| \sup \left\{ \left| \sum \langle y_i k, a_{ij}x_j h \rangle \right| \, \Big| \, x, y \in B_{\ell_2^n}, h, k \in B_H \right\}$$

and hence

$$\|[u(a_{ij})]\| \leq \|u\|\|a\|_{M_n(E)}.$$

Thus, whenever u has rank 1, we have

$$\|u\|_{cb} = \|u\|. \tag{1.3}$$

Proposition 1.3 (c.b. with commutative range) *Let $F \subset B(H)$ be an operator space. Let A_F be the C^*-algebra generated by F.*

(i) For any $n \geq 1$ and any x in $M_n(F)$ we have

$$\|x\|_{M_n(F)} \geq \sup \left\{ \left\| \sum \lambda_i \mu_j x_{ij} \right\|_F \, \Big| \, \lambda_i \in \mathbb{C}, \mu_j \in \mathbb{C}, \right.$$

$$\left. \sum |\lambda_i|^2 \leq 1, \sum |\mu_j|^2 \leq 1 \right\}.$$

(ii) If A_F is commutative we have equality in (i). Then, if E is an arbitrary operator space, any bounded map $u : E \to F$ is c.b. and satisfies $\|u\|_{cb} = \|u\|$.

Proof (i) is an easy exercise. As for (ii), when A_F is commutative (see §A.12), we can assume $A_F = C_0(\Omega)$ and also $M_n(A_F) = C_0(\Omega; M_n)$, for some locally compact space Ω. Then equality in (i) is very simple to check and (ii) is then immediate. $\qquad\square$

Remark 1.4 (c.b. with commutative domain) The preceding result is not valid if the domain is assumed commutative. For any $n > 2$ there is a map

$T : \ell_\infty^n \rightarrow B(\ell_2)$ with $\|T\| \leq 1$ and $\|T\|_{cb} \geq \sqrt{n/2} > 1$. Indeed, let $(u_j)_{1 \leq j \leq n}$ be a matricial spin system i.e. a system of unitary self-adjoint $N \times N$ matrices that are anticommuting i.e. satisfying $\forall i \neq j \quad u_i u_j + u_j u_i = 0$. Let $T : \ell_\infty^n \rightarrow M_N$ be defined by $T(e_j) = u_j/(2n)^{1/2}$. Then T satisfies the announced bounds. The proof that $\|T\|_{cb} \geq \sqrt{n/2}$ uses the elementary identity $\| \sum_1^n \overline{u_j} \otimes u_j \|_{\min} = n$, valid for any n-tuple of unitary matrices (see (18.5)). As for $\|T\| \leq 1$ we refer to [104, p. 209] for a proof. Haagerup also shows in [104] that $\|T\| = \|T\|_{cb}$ when $n = 2$, which we will prove in Remark 3.13.

When $\|u\|_{cb} \leq 1$, we say that u is "completely contractive" (or "a complete contraction").

The notion of isometry is replaced by that of "complete isometry": a linear map $u : E \rightarrow F$ is said to be a complete isometry (or completely isometric) if $u_n : M_n(E) \rightarrow M_n(F)$ is an isometry for all $n \geq 1$.

An invertible mapping $u : E \rightarrow F$ is said to be a complete isomorphism if both u and u^{-1} are c.b. Clearly, a completely isometric surjective map is a complete isomorphism.

Remark 1.5 For instance if $S : H \rightarrow \mathcal{H}$ is an isometry, then the linear map $u_S : B(\mathcal{H}) \rightarrow B(H)$ defined by $u_S(x) = S^* x S$ is completely isometric. This is easily checked by observing that S induces an isometry S_n from $\ell_2^n(H)$ to $\ell_2^n(\mathcal{H})$, such that $(u_S)_n(y) = S_n^* y S_n$ for any $y \in M_n(B(\mathcal{H})) = B(\ell_2^n(\mathcal{H}))$. Thus $(u_S)_n$ is of the same form as u_S. In particular if $S : H \rightarrow \mathcal{H}$ is a surjective isometry then u_S is completely isometric isomorphism.

Remark 1.6 If E, F are C^*-algebras and $u : E \rightarrow F$ is a $*$-homomorphism then $u_n : M_n(E) \rightarrow M_n(F)$ is also a $*$-homomorphism, and hence (see Proposition A.24) we have $\|u_n\| = 1$ for all n (unless $u = 0$), which shows that u is automatically a complete contraction.

Moreover, if u is injective then u_n is obviously also injective. Therefore (see Proposition A.24) u_n is isometric and u is automatically a complete isometry.

Definition 1.7 Let $E \subset B(H)$ and $G \subset B(K)$ be operator spaces. We have a natural embedding

$$G \otimes E \subset B(K \otimes_2 H)$$

that allows us to define

$$G \otimes_{\min} E = \overline{G \otimes E}^{\text{norm}} \subset B(K \otimes_2 H).$$

The space $G \otimes_{\min} E$ is then called the *minimal* tensor product of G and E.

In particular, in the case $G = B(\ell_2^n) = M_n$, we have an obvious completely isometric identification

$$M_n(E) = M_n \otimes_{\min} E. \tag{1.4}$$

Indeed, by Remark 1.5 this simply follows from the Hilbert space identification $\ell_2^n(H) \simeq \ell_2^n \otimes_2 H$.

Remark 1.8 (Associativity of the minimal tensor product) Let $E_j \subset B(H_j)$ be operator spaces ($1 \le j \le n$). We define similarly

$$E_1 \otimes_{\min} \cdots \otimes_{\min} E_n = \overline{E_1 \otimes \cdots \otimes E_n} \subset B(H_1 \otimes_2 \cdots \otimes_2 H_n).$$

Since we have $H_1 \otimes_2 H_2 \otimes_2 H_3 \simeq (H_1 \otimes_2 H_2) \otimes_2 H_3 \simeq H_1 \otimes_2 (H_2 \otimes_2 H_3)$, by Remark 1.5 we also have completely isometrically

$$E_1 \otimes_{\min} E_2 \otimes_{\min} E_3 \simeq (E_1 \otimes_{\min} E_2) \otimes_{\min} E_3 \simeq E_1 \otimes_{\min} (E_2 \otimes_{\min} E_3). \tag{1.5}$$

Thus we may also view $E_1 \otimes_{\min} \cdots \otimes_{\min} E_n = \overline{E_1 \otimes \cdots \otimes E_n}$ as obtained from successive minimal tensor products of suitable pairs, and we may suppress the parentheses since they become irrelevant.

Remark 1.9 (Commutativity of the minimal tensor product) Since we have $K \otimes_2 H \simeq H \otimes_2 K$, by Remark 1.5 we also have completely isometrically

$$G \otimes_{\min} E \simeq E \otimes_{\min} G, \tag{1.6}$$

via $x \otimes y \mapsto y \otimes x$.

Remark 1.10 (Injectivity of the minimal tensor product) From the preceding definition the following property is obvious: Let $E_1 \subset E_2 \subset B(H)$ and $G_1 \subset G_2 \subset B(K)$ be operator subspaces, so that $E_1 \otimes G_1 \subset E_2 \otimes G_2$. Then for any $t \in E_1 \otimes G_1$ we have

$$\|t\|_{E_1 \otimes_{\min} G_1} = \|t\|_{E_2 \otimes_{\min} G_2}. \tag{1.7}$$

Proposition 1.11 *Let $u : E \to F$ be a c.b. map between two operator spaces. Then for any other operator space G the mapping $Id_G \otimes u : G \otimes E \to G \otimes F$ extends to a bounded mapping $u_G : G \otimes_{\min} E \to G \otimes_{\min} F$ and we have*

$$\|u\|_{cb} = \sup_G \|u_G\| = \sup_G \|u_G\|_{cb}, \tag{1.8}$$

where the suprema run over all possible G's.

Proof First observe that the choice of $G = M_n$ shows that

$$\sup\{\|u_G\| : G \text{ an operator space}\} \ge \|u\|_{cb}.$$

To prove the converse assume $G \subset B(K)$. For notational simplicity, assume $K = \ell_2$ and let $t = \sum_{k=1}^r a_k \otimes b_k \in G \otimes E$. Consider the natural embeddings $\ell_n^2 \simeq [e_1, \ldots, e_n] \subset \ell_2$ ($n \ge 1$) with respect to some choice $\{e_j : j \ge 1\}$

of orthonormal basis for ℓ_2 and the corresponding orthogonal projections $P_n : K \otimes_2 H \to \overline{\ell_2^n \otimes_2 H}$. Then $\cup_n \ell_2^n \otimes H = K \otimes_2 H$ and hence

$$\|t\|_{\min} = \sup_n \| P_n \, t_{|\ell_2^n \otimes H} : \ell_2^n \otimes_2 H \to \ell_2^n \otimes_2 H \|.$$

Let $a_k(i,j) = \langle e_i, a_k e_j \rangle$. It is not hard to see that $P_n \, t_{|\ell_2^n \otimes H}$ can be identified with the matrix $t_n \in M_n(E)$ given by $t_n(i,j) = \sum_k a_k(i,j) b_k$. This shows that

$$\|t\|_{G \otimes_{\min} E} = \sup_n \|t_n\|_{M_n(E)}, \tag{1.9}$$

We have $u_n(t_n) = \left[\sum_k a_k(i,j) u(b_k) \right] \in M_n(F)$. Applying (1.9) to $u_G(t) = \sum a_k \otimes u(b_k)$ gives us $\|u_G(t)\|_{\min} = \sup_n \|u_n(t_n)\|_{M_n(F)}$ and hence $\|u_G(t)\|_{\min} \le \|u\|_{cb} \sup_n \|t_n\|_{M_n(E)} = \|u\|_{cb} \|t\|_{\min}$, which shows $\|u_G\| \le \|u\|_{cb}$. Thus $\|u\|_{cb} = \sup_G \|u_G\|$. Then, substituting $M_n(G)$ for G, we easily deduce that $\sup_G \|u_G\| = \sup_G \|u_G\|_{cb}$. $\qquad \square$

Corollary 1.12 *Let E_1, F_1, E_2, F_2 be operator spaces. Let $u_1 \in CB(E_1, F_1)$ and $u_2 \in CB(E_2, F_2)$. Then $u_1 \otimes u_2$ continuously extends by density to a c.b. map $u_1 \otimes u_2 : E_1 \otimes_{\min} E_2 \longrightarrow F_1 \otimes_{\min} F_2$ such that*

$$\|u_1 \otimes u_2\|_{cb} \le \|u_1\|_{cb} \|u_2\|_{cb}. \tag{1.10}$$

Proof The argument is based on the obvious identity $u_1 \otimes u_2 = (u_1 \otimes Id_{F_2})(Id_{E_1} \otimes u_2)$, which gives us $\|u_1 \otimes u_2\| \le \|u_1 \otimes Id_{F_2}\| \|Id_{E_1} \otimes u_2\|$. By (1.8) we have $\|Id_{E_1} \otimes u_2\| \le \|u_2\|_{cb}$ and using (1.6) we also find $\|u_1 \otimes Id_{F_2}\| \le \|u_1\|_{cb}$. This gives us $\|u_1 \otimes u_2\| \le \|u_1\|_{cb} \|u_2\|_{cb}$. Now replacing u_1 by $Id_{M_n} \otimes u_1$, by (1.5) and (1.4) we obtain the announced (1.10) after taking the sup over n. $\qquad \square$

It is an easy exercise to show that (1.10) is actually an equality but we do not use this in the sequel.

We will now generalize (1.9).

Proposition 1.13 *For any $t = \sum a_k \otimes b_k \in G \otimes E$ we have*

$$\|t\|_{\min} = \sup_{n \ge 1} \left\{ \left\| \sum v(a_k) \otimes b_k \right\|_{M_n(E)} \middle| v \in CB(G, M_n), \|v\|_{cb} \le 1 \right\}.$$
$$\tag{1.11}$$

Furthermore

$$\|t\|_{\min} = \sup \left\{ \left\| \sum v(a_i) \otimes w(b_i) \right\|_{M_{nm}} \right\} \tag{1.12}$$

where the supremum runs over $n, m \ge 1$ and all pairs $v : G \to M_n$, $: E \to M_m$ with $\|v\|_{cb} \le 1$ and $\|w\|_{cb} \le 1$. (We can of course restrict to $n = m$ if we wish.)

Proof By (1.8) and (1.6) we have $\|v \otimes Id_E\| \leq \|v\|_{cb}$, so the left-hand side of (1.11) is $\geq$ the right-hand side. But by (1.9) we see that equality holds: indeed just observe that $t_n = (v_n \otimes Id_E)(t)$ with $v_n(\cdot) = a_n^* \cdot a_n$ where $a_n : \ell_2^n \to H$ denote the inclusion. To check (1.12) we again invoke (1.6), which allows us to apply (1.11) one more time on the second factor. $\qquad\square$

Remark 1.14 The preceding proposition shows that the min-norm on $G \otimes E$ depends only on the sequences of norms on $M_n(G)$ and $M_n(E)$ and not on the particular embeddings $G \subset B(K)$ and $E \subset B(H)$. Indeed, the latter sequences suffice to determine the norms of the spaces $CB(G, M_n)$ and $CB(E, M_n)$ (see Proposition 1.19 for more precision).

More generally, the same remark holds for the norm in $M_n(G \otimes_{\min} E)$ and hence the whole sequence of the norms $\|\cdot\|_{M_n(G \otimes_{\min} E)}$ depends only on the sequences of norms on $M_n(G)$ and $M_n(E)$.

Corollary 1.15 *If an element* $t \in G \otimes_{\min} E$ *is such that* $(v \otimes w)(t) = 0$ *for any* $v \in G^*$ *and* $w \in E^*$, *then* $t = 0$.

Proof This is immediate from (1.12). Indeed, the assumption remains obviously true for any $v : G \to M_n$ and $w : E \to M_m$. $\qquad\square$

Warning: The reason we emphasize the rather simple fact in Corollary 1.15 is that the analogous fact for the maximal tensor product of two C^*-algebras fails in general.

Remark 1.16 (Direct sum of operator spaces) Let $E_i \subset B(H_i)$ ($i \in I$) be a family of operator spaces. Let $E = \left(\oplus \sum_{i \in I} E_i \right)_\infty$. Note that $E \subset \left(\oplus \sum_{i \in I} B(H_i) \right)_\infty$, and that $\left(\oplus \sum_{i \in I} B(H_i) \right)_\infty$ is a C^*-algebra naturally embedded in $B(H)$ with $H = \left(\oplus \sum_{i \in I} H_i \right)_2$. This allows us to equip E with an operator space structure as a subspace of $B(H)$. Let $n \geq 1$. Any matrix $a \in M_n(E)$ is determined by a family $(a_i)_{i \in I}$ with $a_i \in M_n(E_i)$ for all $i \in I$. It is then easy to check that for any $n \geq 1$ and any $a \in M_n(E)$ we have

$$\|a\|_{M_n(E)} = \sup_{i \in I} \|a_i\|_{M_n(E_i)}. \qquad (1.13)$$

More generally, for any operator space $F \subset B(K)$, we have a natural isometric embedding

$$F \otimes_{\min} \left(\oplus \sum_{i \in I} E_i \right)_\infty \subset \left(\oplus \sum_{i \in I} F \otimes_{\min} E_i \right)_\infty, \qquad (1.14)$$

which is an isomorphism if $\dim(F) < \infty$ since both sides are then setwise identical.

The equality (1.13) shows furthermore that for any operator space D, a linear map $u : D \to E$ is c.b. if and only if the coordinates $u_i : D \to E_i$ are c.b. with $\sup_{i \in I} \|u_i\|_{cb} < \infty$ and we have

$$\|u\|_{cb} = \sup_{i \in I} \|u_i\|_{cb}. \tag{1.15}$$

1.2 Extension property of $B(H)$

We first recall that the spaces L_∞ are the injective objects in the category of Banach spaces.

Theorem 1.17 (Nachbin's Hahn–Banach theorem) *Let (Ω, μ) be any measure space, and let $E \subset X$ be any subspace of a Banach space X. Then any $u \in B(E, L_\infty(\mu))$ admits an extension $\widetilde{u} \in B(X, L_\infty(\mu))$ such that $\|\widetilde{u}\| = \|u\|$.*

$$
\begin{array}{ccc}
X & & \\
\big\uparrow & \searrow{\widetilde{u}} & \\
E & \xrightarrow{\ u\ } & L_\infty(\mu)
\end{array}
$$

The proof of Nachbin's theorem relies on several identifications. First we note the elementary isometric isomorphisms

$$B(E, F^*) \cong B(F, E^*) \cong \mathrm{Bil}(E \times F)$$

where $\mathrm{Bil}(E \times F)$ is the Banach space of all bounded bilinear forms on $E \times F$. Then we have an isometric identification

$$B(E, F^*) \cong (E \overset{\wedge}{\otimes} F)^* \tag{1.16}$$

where $\overset{\wedge}{\otimes}$ is the projective tensor product, i.e. the completion of the algebraic tensor product $E \otimes F$ with respect to the so-called projective norm (see §A.1)

$$\|t\|_\wedge = \inf \left\{ \sum_1^n \|a_j\| \|b_j\| : t = \sum_1^n a_j \otimes b_j \right\}.$$

Note that $E \overset{\wedge}{\otimes} F$ and $F \overset{\wedge}{\otimes} E$ can obviously be (isometrically) identified. The duality between tensors $t \in E \overset{\wedge}{\otimes} F$ and operators $u \in B(E, F^*)$ is defined first on rank one tensors by setting

$$\langle u, a \otimes b \rangle = \langle u(a), b \rangle,$$

then this can be extended to unambiguously define $\langle u, t \rangle$ for $t \in E \otimes F$ by linearity. Then by density we define $\langle u, t \rangle$ for $t \in E \overset{\wedge}{\otimes} F$, and (1.16) holds for this duality.

By a classical result (due to Grothendieck) when $F = L_1(\Omega, \mu)$ (on some measure space (Ω, μ)), the space $E \overset{\wedge}{\otimes} F$ (or equivalently $F \overset{\wedge}{\otimes} E$) can be identified isometrically to the (Bochner sense) vector valued L_1-space $L_1(\mu; E)$.

Sketch of Proof of Nachbin's Theorem Taking $F = L_1(\mu)$ in the preceding, we find

$$B(E, L_\infty(\mu)) = L_1(\mu; E)^* \quad \text{and} \quad B(X, L_\infty(\mu)) = L_1(\mu; X)^*.$$

Then since we have an isometric inclusion

$$L_1(\mu; E) \subset L_1(\mu; X)$$

Nachbin's Theorem can be deduced from the classical Hahn–Banach theorem.
$\square$

We will follow the same approach to prove the noncommutative version of Nachbin's Theorem, due to Arveson.

Theorem 1.18 (Arveson's Hahn–Banach theorem) *Let H be any Hilbert space, and let $X \subset B(\mathcal{H})$ be any operator space and let $E \subset X$ be any subspace. Then any $u \in CB(E, B(H))$ admits an extension $\tilde{u} \in B(X, B(H))$ such that $\|\tilde{u}\|_{cb} = \|u\|_{cb}$.*

$$
\begin{array}{ccc}
X & & \\
\uparrow & \overset{\tilde{u}}{\searrow} & \\
\mathrel{\Big\downarrow} & & \\
E & \overset{u}{\longrightarrow} & B(H)
\end{array}
$$

The projective tensor norm $\| \ \|_\wedge$ on $L_1(\mu) \otimes E$ will be replaced by the following one on $\overline{K} \otimes E \otimes H$ where H, K are Hilbert spaces:

For any $t \in \overline{K} \otimes E \otimes H$, we define (recall $\|\overline{k}\| = \|k\|$ for all $k \in K$)

$$\gamma_E(t) = \inf \left\{ \left(\sum_{i=1}^m \|k_i\|^2 \right)^{1/2} \|[a_{ij}]\|_{M_{m \times n}(E)} \left(\sum_{j=1}^n \|h_j\|^2 \right)^{1/2} \right\}$$

where the infimum runs over all representations of t of the form

$$t = \sum_{i=1}^m \sum_{j=1}^n \overline{k_i} \otimes a_{ij} \otimes h_j.$$

In analogy with Nachbin's Theorem, we will show that this norm satisfies:

(i) The dual space $(\overline{K} \otimes E \otimes H, \gamma_E)^*$ can be identified with $CB(E, B(H, K))$.

(ii) The natural inclusion $(\overline{K} \otimes E \otimes H, \gamma_E) \subset (\overline{K} \otimes X \otimes H, \gamma_X)$ is isometric.

Proof of Arveson's Hahn–Banach theorem Using (i) and (ii) the proof of Theorem 1.18 can be completed exactly as in the case of Banach spaces: We simply take $K = H$ and apply the Hahn–Banach theorem to the subspace $(\overline{K} \otimes E \otimes H, \gamma_E) \subset (\overline{K} \otimes X \otimes H, \gamma_X)$.

Thus the proof now reduces to the verification of (i) and (ii). It is easy to check that γ_E is a norm by arguing as follows:

Let $t = \sum_{i=1}^{m} \sum_{j=1}^{n} \overline{k_i} \otimes a_{ij} \otimes h_j$ and $t' = \sum_{p=1}^{m'} \sum_{q=1}^{n'} \overline{k'_p} \otimes a'_{pq} \otimes h'_q$ be elements of $\overline{K} \otimes E \otimes H$. We have obviously (consider the block diagonal matrix with blocks a and a')

$$\gamma_E(t + t') \le \left(\sum \|k_i\|^2 + \sum \|k'_p\|^2\right)^{1/2} \max\{\|[a_{ij}]\|_{M_{m \times n}(E)}, \|[a'_{pq}]\|_{M_{m' \times n'}(E)}\}$$
$$\times \left(\sum \|h_j\|^2 + \sum \|h'_q\|^2\right)^{1/2}.$$

But by homogeneity, for any $\varepsilon > 0$ there are suitable representations of t, t' such that

$$\|[a_{ij}]\|_{M_{m \times n}(E)} = 1 \quad \text{and} \quad \sum_{i=1}^{m} \|k_i\|^2 = \sum_{j=1}^{n} \|h_j\|^2 < \gamma_E(t) + \varepsilon,$$

as well as

$$\|[a'_{pq}]\|_{M_{m' \times n'}(E)} = 1 \quad \text{and} \quad \sum_{p=1}^{m} \|k'_p\|^2 = \sum_{q=1}^{n} \|h'_q\|^2 < \gamma_E(t') + \varepsilon.$$

Then we find for any $\varepsilon > 0$

$$\gamma_E(t + t') \le \gamma_E(t) + \gamma_E(t') + 2\varepsilon, \tag{1.17}$$

which shows that γ_E is subadditive and hence (since $\gamma_E(t)$ dominates the norm of t as a bounded trilinear form on $\overline{K}^* \times E^* \times H^*$) it is a norm. Consider the $\mathbb{C}$-linear correspondence

$$CB(E, B(H, K)) \ni u \mapsto \varphi_u \in (\overline{K} \otimes E \otimes H)^*$$

defined by $\varphi_u(t) = \sum_{i,j} \langle k_i, u(a_{ij})h_j \rangle$, if $t = \sum_{i,j} \overline{k_i} \otimes a_{ij} \otimes h_j$. Note that

$$\|\varphi_u\|_{\gamma_E^*} = \sup_{\gamma_E(t) < 1} |\varphi_u(t)| = \sup \left\{ \left| \sum_{i,j} \langle k_i, u(a_{ij})h_j \rangle \right| \Big| \sum \|h_j\|^2 \right.$$
$$\left. = \sum \|k_i\|^2 < 1, \|[a_{ij}]\|_{M_n(E)} = 1 \right\}$$
$$= \sup\{\|u(a_{ij})\|_{M_n(B(H,K))} \mid \|[a_{ij}]\|_{M_n(E)} = 1\} = \|u\|_{cb}.$$

Hence $\|\varphi_u\|_{\gamma_E^*} = \|u\|_{cb}$ and (i) is proved.

To verify (ii) consider $t \in \overline{K} \otimes E \otimes H$. Note that obviously $\gamma_X(t) \le \gamma_E(t)$ since there are more representations allowed in the definition of γ_X. To establish (ii) it suffices to prove that conversely $\gamma_E(t) \le \gamma_X(t)$, or equivalently, by homogeneity, that $\gamma_X(t) < 1$ implies $\gamma_E(t) < 1$. The assumption $\gamma_X(t) < 1$ implies that there exists a decomposition (a priori with $a_{ij} \in X$)

$$t = \sum_{i=1}^{m} \sum_{j=1}^{n} \overline{k_i} \otimes a_{ij} \otimes h_j,$$

such that

$$\left(\sum_{i=1}^{m} \|k_i\|^2\right)^{1/2} \cdot \|[a_{ij}]\|_{M_{m \times n}(X)} \cdot \left(\sum_{j=1}^{n} \|h_j\|^2\right)^{1/2} < 1.$$

To conclude it suffices to show that there is a (possibly different) representation of t with the same bounds but with $a_{ij} \in E$. We will use the following simple linear algebraic fact.

Claim: Let $h_1, \ldots, h_n \in H$, and let L be their linear span. Let $r = \dim(L) \leq n$. Then there exist r linearly independent vectors $\{h'_q : 1 \leq q \leq r\}$ in L and a rectangular matrix $C = [c_{qj}] \in M_{r \times n}$ such that $\|C\|_{M_{r \times n}} \leq 1$, $h_j = \sum_{q=1}^{r} c_{qj} h'_q$ for $1 \leq j \leq n$, and $\sum_{j=1}^{n} \|h_j\|^2 = \sum_{q=1}^{r} \|h'_q\|^2$. We denote by ${}^t C \in M_{n \times r}$ the transposed matrix.

Let s be the dimension of the span of the vectors (k_i) in K. It follows from the claim applied to (k_i) that there also exists a linearly independent set $\{\overline{k'_p} : 1 \leq p \leq s\}$ and a matrix $D \in M_{s \times m}$ such that $\|D\| \leq 1$, $\overline{k_i} = \sum_{p=1}^{s} d_{pi} \overline{k'_p}$, and $\sum_{i=1}^{m} \|k_i\|^2 = \sum_{p=1}^{s} \|k'_p\|^2$. Then, note that

$$t = \sum_{i=1}^{m} \sum_{j=1}^{n} \overline{k_i} \otimes a_{ij} \otimes h_j = \sum_{p=1}^{s} \sum_{q=1}^{r} \overline{k'_p} \otimes a'_{pq} \otimes h'_q,$$

where we have denoted $a = [a_{ij}]$ and $a' = [a'_{pq}] = D a {}^t C$, so that $\|a'\|_{M_{s \times r}(X)} \leq \|a\|_{M_{m \times n}(X)}$. But now, since $\{k'_p : 1 \leq p \leq s\}$ and $\{h'_q : 1 \leq q \leq r\}$ are linearly independent sets, we have

$$t \in \overline{K} \otimes E \otimes H \Rightarrow a'_{pq} \in E \quad \forall p, q,$$

so that $a' \in M_{s \times r}(E)$ and $\|a'\|_{M_{s \times r}(E)} = \|a'\|_{M_{s \times r}(X)} \leq \|a\|_{M_{m \times n}(X)}$. Thus we may write

$$\gamma_E(t) \leq \left(\sum_{p=1}^{s} \|k'_p\|^2\right)^{1/2} \cdot \|a'\|_{M_{s \times r}(E)} \cdot \left(\sum_{q=1}^{r} \|h'_q\|^2\right)^{1/2}$$

and hence

$$\gamma_E(t) \leq \left(\sum_{i=1}^{m} \|k_i\|^2\right)^{1/2} \cdot \|a\|_{M_{m \times n}(X)} \cdot \left(\sum_{j=1}^{n} \|h_j\|^2\right)^{1/2} < 1,$$

which completes the proof of (ii).

Proof of the Claim: Consider the linear operator $T : \ell_2^n \to H$ defined by $Te_i = h_i$ for each $1 \leq i \leq n$, where $\{e_i : 1 \leq i \leq n\}$ is the canonical basis of ℓ_2^n. Let $\{f_q : 1 \leq q \leq r\}$ denote any orthonormal basis for $N = \ker(T)^\perp$

and let P denote the orthogonal projection of ℓ_2^n onto N. Set $h_q' = Tf_q$ for $1 \le q \le r$. Clearly, $\|T\|_{HS}^2 = \|TP\|_{HS}^2 = \sum_{q=1}^r \|Tf_q\|^2$ and hence

$$\sum_{i=1}^n \|h_i\|^2 = \sum_{q=1}^r \|h_q'\|^2.$$

Let $C \in M_{r \times n}$ be the matrix representing P with respect to the bases (e_j) and (f_q), so that $Pe_j = \sum_q c_{qj} f_q$. Then $h_j = T(e_j) = TP(e_j) = \sum_q c_{qj} h_q'$. This proves the claim. □

The following variant due to Roger Smith is often useful.

Proposition 1.19 *Consider operator spaces E, X with $E \subset X$ and $u : E \to M_n$. Then we have*

$$\|u\|_{cb} = \|u_n\|_{M_n(E) \to M_n(M_n)}. \tag{1.18}$$

Moreover, $u : E \to M_n$ admits an extension $\tilde{u} : X \to M_n$ such that $\|\tilde{u}_n\| = \|u_n\|$.

Proof For any $m \ge 1$ and any $x_1, \ldots, x_m$ in ℓ_2^n with $\sum_1^m \|x_i\|^2 \le 1$ there are an $m \times n$ scalar matrix $b = [b_{jk}]$ with $\|[b_{jk}]\| \le 1$ and vectors $\tilde{x}_1, \ldots, \tilde{x}_n$ in ℓ_2^n such that $\sum_1^n \|\tilde{x}_i\|^2 \le 1$ and

$$\forall j \le m \quad x_j = \sum_{k=1}^n b_{jk} \tilde{x}_k.$$

This follows from the claim in the preceding proof. Similarly, for any $y_1, \ldots, y_m$ in ℓ_2^n there are a scalar matrix $c = [c_{il}]$ with $\|[c_{il}]\| \le 1$ and $\tilde{y}_1, \ldots, \tilde{y}_n$ in ℓ_2^n such that $\sum_1^n \|\tilde{y}_i\|^2 \le 1$ and

$$\forall i \le m \quad y_i = \sum_{l=1}^n c_{il} \tilde{y}_l.$$

Hence for any $m \times m$ matrix $[a_{ij}]$ in $M_m(E)$ we have

$$\sum_{i,j=1}^m \langle y_i, u(a_{ij}) x_j \rangle = \sum_{k,l=1}^n \langle \tilde{y}_l, u(\alpha_{lk}) \tilde{x}_k \rangle$$

where $[\alpha_{lk}] \in M_n(E)$ is defined by $[\alpha_{lk}] = c^*.[a_{ij}].b$ (matrix product). Therefore:

$$\|[u(a_{ij})]\|_{M_m(M_n)} \le \|[u(\alpha_{kl})]\|_{M_n(M_n)} \le \|u_n\|_{M_n(E) \to M_n(M_n)} \|[\alpha_{lk}]\|_{M_n(E)}$$

$$\le \|u_n\|_{M_n(E) \to M_n(M_n)} \|[a_{ij}]\|_{M_n(E)}.$$

Thus we obtain $\|u\|_{cb} \le \|u_n\|$ and hence $\|u\|_{cb} = \|u_n\|$. The second assertion is then a corollary of Theorem 1.18. □

1.3 Completely positive maps

We start with the definition of complete positivity.

Definition 1.20 Let $E \subset A$, $F \subset B$ be linear subspaces of C^*-algebras A, B. We set $M_n(E)_+ = M_n(E) \cap M_n(A)_+$ and similarly for $M_n(F)_+$. We will say that a linear map $u : E \to F$ is completely positive (c.p. in short) if for any n we have $u_n(M_n(E)_+) \subset M_n(F)_+$. We will denote by $CP(E, F)$ the set of all such linear maps.

Actually, to get interesting examples this framework is too general and most of the time we need to work not only with operator subspaces, but either with C^*-algebras (the case $E = A, F = B$) or with operator *systems* as defined later on in Definition 1.36.

Remark 1.21 Let $v \in CP(E, B(H))$. Then for any bounded $V : K \to H$ (H, K Hilbert) the mapping $x \mapsto V^*v(x)V$ clearly belongs to $CP(E, B(K))$. Since any *-homomorphism $\pi : A \to B$ is clearly c.p., this shows that any u of the form $u(x) = V^*\pi(x)V$ is also c.p.

By a classical dilation result due to Stinespring, it turns out that the example we just described is actually the general form of a c.p. map.

Theorem 1.22 (Stinespring's theorem) *Let A be a C^*-algebra. Then a linear map $u : A \to B(H)$ is c.p. if and only if there is a Hilbert space $\widehat{H}$, an operator $V : H \to \widehat{H}$ and a *-homomorphism $\pi : A \to B(\widehat{H})$ (unital if A is unital) such that*

$$\forall a \in A \quad u(a) = V^*\pi(a)V. \tag{1.19}$$

If A is unital and $u(1) = 1$ then V is an isometry. Moreover, if u is c.p. and A is unital, we have

$$\|u\| = \|u\|_{cb} = \|u(1)\| \tag{1.20}$$

and in any case (unital or not)

$$\|u\| = \sup\{\|u(a)\| \mid a \geq 0 \ \|a\| \leq 1\}. \tag{1.21}$$

Let (x_i) be an increasing net in $B_A \cap A_+$ such that $x_i \geq 0$, $\|x_i\| \leq 1$ and $\|x_i x - x\| \to 0$ for any $x \in A$ (thus (x_i) is an approximate unit of A in the sense of §A.15), then

$$\|u\| = \|u\|_{cb} = \lim \|u(x_i)\|. \tag{1.22}$$

Proof Assume $u \in CP(A, B(H))$. We introduce a scalar product on $A \otimes H$ by setting

$$\forall t = \sum a_j \otimes b_j \in A \otimes H, \quad \langle t, t \rangle = \sum_{ij} \langle b_i, u(a_i^* a_j) b_j \rangle,$$

so that $t \to \langle s, t \rangle$ is linear and $s \to \langle s, t \rangle$ is antilinear. The complete positivity of u shows that $\langle t, t \rangle \geq 0$. Therefore after passing to the quotient and completing we obtain a Hilbert space $\widehat{H}$. The $*$-homomorphism π defined by $\pi(a)(t) = \sum aa_j \otimes b_j$ extends to a continuous $*$-homomorphism (unital if A is unital) from A to $B(\widehat{H})$. If A is unital, let $Vh = 1 \otimes h$. Then $\|V\| \leq 1$ and $\langle 1 \otimes h, \pi(a)(1 \otimes h) \rangle = \langle h, u(a)h \rangle$. Therefore $\langle Vh, \pi(a)Vh \rangle = \langle h, u(a)h \rangle$ or equivalently $V^* \pi(a) V = u(a)$. If $u(1) = 1$, then $V^*V = I$ and V is an isometry. If A is unital, we have clearly $\|u\| = \|V^*V\| = \|u(1)\|$. Moreover, it is easy to check that $\|u\|_{cb} \leq \|V\|^2 \|\pi\|_{cb}$ and since $\|\pi\|_{cb} = 1$ (indeed, π_n is a $*$-homomorphism for any n, and hence $\|\pi_n\| = 1$ by (A.17)), we find $\|u\|_{cb} \leq \|u(1)\|$, which proves (1.20).

If A is not unital, it admits an approximate unit (see §A.15). This will allow us to extend u to a c.p. map $\widehat{u}$ defined on a larger *unital* C^*-algebra $\widetilde{A}$ containing A. Then the already proved unital case applied to $\widehat{u}$ will give us the factorization (1.19). Assume A separable for simplicity. Then we have a sequence $0 \leq \cdots \leq \chi_n \leq \chi_{n+1} \leq \cdots$ in A_+ with $\|\chi_n\| < 1$ such that $\|\chi_n^{1/2} a \chi_n^{1/2} - a\| \to 0$ for any $a \in A$. Assume $A \subset B(\mathcal{H})$ and let $\widetilde{A} = A + \mathbb{C}I \subset B(\mathcal{H})$ be the unitization of A. We claim that there is a c.p. map $\widehat{u} : \widetilde{A} \to B(H)$ extending u. Note that, since u is positive, $u(\chi_n)$ is order increasing and bounded (which means the same for $\langle h, u(\chi_n)h \rangle, \forall h \in H$), and hence converges in the weak operator topology to an operator $T \in B(H)$ with $\|T\| \leq \|u\|$. We then define (assuming $I \notin A$)

$$\widehat{u}(a + \lambda I) = u(a) + \lambda T.$$

Since $\|u(\chi_n^{1/2} a \chi_n^{1/2} - a)\| \to 0 \quad \forall a \in A$, we have (in the weak operator topology)

$$\forall x \in \widetilde{A} \quad \widehat{u}(x) = \lim_{n \to \infty} u(\chi_n^{1/2} x \chi_n^{1/2}),$$

from which it is easy to see that $\widehat{u}$ is c.p. Moreover, we have

$$\|u\| \leq \|u\|_{cb} \leq \|\widehat{u}\|_{cb} = \|\widehat{u}(1)\| = \|T\| \leq \|u\|.$$

Thus we conclude

$$\|u\| = \|u\|_{cb} = \|\widehat{u}(1)\| = \sup\{\|u(\chi)\| \mid \chi \geq 0 \; \|\chi\| < 1\}.$$

For any approximate unit (x_i) in $B_A \cap A_+$ we have $\|x_i \chi x_i - \chi\| \to 0$ and hence

$$\|u(\chi)\| \leq \lim \|u(x_i \chi x_i)\|$$

and, if $\chi \geq 0, \|\chi\| < 1$, we have $0 \leq x_i \chi x_i \leq x_i^2 \leq x_i$ and hence $0 \leq u(x_i \chi x_i) \leq u(x_i^2) \leq u(x_i)$ so that we obtain (1.22). $\qquad\square$

Remark 1.23 In Theorem 1.22 we may clearly replace $\widehat{H}$ by $\widehat{H}_0 = \overline{\pi(A)(VH)}$ and $a \mapsto \pi(a)$ by $a \mapsto \pi(a)_{|\widehat{H}_0}$. Then we obtain (1.19) with in addition $\widehat{H} = \overline{\pi(A)(VH)}$. In the latter case π is called a minimal dilation of u. It is easy to check that it is unique up to conjugation by a unitary.

Remark 1.24 Any positive linear form $\varphi : E \to \mathbb{C}$ on a subspace $E \subset A$ of a C^*-algebra is c.p. This can be checked directly using $\varphi\left(\sum_{ij} \overline{x_i} a_{ij} x_j\right) = \sum_{ij} \overline{x_i} \varphi(a_{ij}) x_j$ $(x_i \in \mathbb{C})$ and the fact that for $b \in M_n(\mathbb{C})$ we have $[b_{ij}] \geq 0$ if and only if $\sum_{ij} \overline{x_i} b_{ij} x_j \geq 0$ $\forall n, \forall (x_i) \in \mathbb{C}^n$, together with the observation that $a \in M_n(E) \cap M_n(A)_+$ implies $\sum_{ij} \overline{x_i} a_{ij} x_j \in E \cap A_+$.

Remark 1.25 The classical GNS factorization (see §A.13) of a positive linear form $\varphi \in A^*$ rewrites φ as $\varphi(a) = \langle \xi, \pi(a)\xi \rangle$ for some Hilbert space $\widehat{H}$, some $*$-homomorphism $\pi : A \to B(\widehat{H})$ and some $\xi \in \widehat{H}$. This can be rewritten as (1.19) with $V : C \to \widehat{H}$ defined by $V(\lambda) = \lambda \xi$ $(\lambda \in \mathbb{C})$. Thus this can be viewed as a particular case (namely the case $B(H) = \mathbb{C}$) of Theorem 1.22.

Corollary 1.26 *Let A, B, G be C^*-algebras. For any $u \in CP(A, B)$ the mapping $u_G = Id_G \otimes u$ is a c.p. map from $G \otimes_{\min} A$ to $G \otimes_{\min} B$.*

Proof Assume $G \subset B(K)$. By Theorem 1.22 we have $u(\cdot) = V^*\pi(\cdot)V$. Therefore $u_G(\cdot) = (Id_K \otimes V)^* \pi_G(\cdot)(Id_K \otimes V)$ and $\pi_G : G \otimes_{\min} A \to G \otimes_{\min} B(\widehat{H})$ is a $*$-homomorphism. A more direct alternative proof can be given following the same idea as for Proposition 1.11. $\qquad\square$

By composition as for (1.10) we obtain as an immediate consequence:

Corollary 1.27 *Let A_j, B_j be C^*-algebras $(j = 1, 2)$. Let $u_j \in CP(A_j, B_j)$ $(j = 1, 2)$. Then $u_1 \otimes u_2$ extends to a c.p. map from $A_1 \otimes_{\min} A_2$ to $B_1 \otimes_{\min} B_2$.*

Remark 1.28 (Positivity with commutative range) Let A, B and $u : E \to F \subset B$ be as in Definition 1.20. Assume B commutative and unital. Then u positive $\Rightarrow u$ completely positive.

Indeed, $B \simeq C(T)$ for some compact set T (see §A.12), and $M_n(C(T))_+$ consists of the functions $a : T \to M_n$ such that $a(t) \in (M_n)_+$ for any $t \in T$. Therefore, u is c.p. if and only if $x \mapsto u(x)(t)$ is c.p. for any $t \in T$, or equivalently, by the preceding Remark 1.24, if and only if $x \mapsto u(x)(t)$

is a positive linear form on E for any $t \in T$. Thus, if u is positive, it is "automatically" c.p.

Remark 1.29 A simple example of a positive (unital) linear map that is not c.p. is the transposition on M_n for $n > 1$ or on $B(\ell_2)$ or $K(\ell_2)$ (see also Remark 2.2).

Remark 1.30 Consider a linear mapping $u : M_n \to A$ into a C^*-algebra A. Let $a \in M_n(A)$ be the matrix defined by $a_{ij} = u(e_{ij})$. Then $u \in CP(M_n, A)$ if and only if $a \in M_n(A)_+$.

Indeed, note that $a = u_n(\xi)$ with $\xi \in M_n(M_n)_+$ defined by $\xi_{ij} = e_{ij}$. From this the "only if" part follows. Conversely, if $a \in M_n(A)_+$ then we have $a = b^*b$ for some $b \in M_n(A)$, and hence $u(x) = \sum_k b^*_{ki}x_{ij}b_{kj}$ ($x \in M_n$) from which it is easy to deduce that u is of the form (1.19) and hence is c.p. We leave the details as an exercise. Note that if $u \in CP(M_n, A)$ we have $\|u\| = \|u(1)\| = \left\| \sum a_{ii} \right\| = \left\| \sum_{ik} b^*_{ik} b_{ik} \right\|$.

Remark 1.31 In the case when $u(e_{ij}) = 0$ for any $i \neq j$, u can be identified with a mapping from ℓ^n_∞ to A, and the associated matrix a is diagonal. Then $u : \ell^n_\infty \to A$ is c.p. if and only if u (or equivalently a) is positive.

Remark 1.32 (Positivity with commutative domain) Consider C^*-algebras A, B and assume now that A is commutative. Then any *positive* linear map $u : A \to B$ is c.p. A simple way to see this is to observe that the identity of A is the pointwise limit of a net of maps $u_i : A \to A$ of the form

$$u_i : A \xrightarrow{v_i} \ell^{n(i)}_\infty \xrightarrow{w_i} A$$

where $n(i)$ are integers and v_i, w_i are *positive* contractions (see Remark A.20). By Remark 1.28 v_i is c.p. Using this we are reduced to show that $uw_i : \ell^{n(i)}_\infty \to B$ is c.p., and this case is covered by the end of the preceding Remark 1.31.

Remark 1.33 (Positivity for unital forms) Let A a unital C^*-algebra and $f \in A^*$. Then

$$f \geq 0 \Leftrightarrow f(1) = \|f\|.$$

Indeed, if A is commutative, say $A = C(T)$ with T compact, this is a well-known characterization of positive measures on T. But if we fix $x \in A_+$, and let A_x denote the (commutative) unital C^*-algebra generated by x,

$$f(1) = \|f\| \Rightarrow f(1) = \|f_{|A_x}\| \Rightarrow f_{|A_x} \geq 0 \Rightarrow f(x) \geq 0,$$

which proves the implication from right to left. The converse direction can be proved easily using Cauchy–Schwarz for the inner product $\langle y, x \rangle = f(y^*x)$, applied with $y = 1$ and $x \in B_A$. We actually already proved a more general fact when we proved (1.20).

We note in passing that for any $a \in A$

$$a \geq 0 \Leftrightarrow \varphi(a) \geq 0 \ \forall \varphi \in A_+^*. \tag{1.23}$$

Remark 1.34 (Positivity for unital maps) More generally let $u : A \to B$ be a linear map with values in a C^*-algebra B. If $u(1) = 1$ and $\|u\| = 1$ then u is positive. Indeed, by the preceding remark, for any $F \in B_+^*$, the linear form $f : x \mapsto F(u(x))$ satisfies $f(1) = \|f\|$ and hence is positive. Conversely, if u is positive and unital then $\|u\| = 1$ by Corollary A.19.

The next statement, which is a sort of recapitulation, introduces an important "bridge" between c.p. and c.b. maps.

Theorem 1.35 *Let $E \subset A$ be a subspace of a unital C^*-algebra such that $1 \in E$. Let $u : E \to B(H)$ be a unital linear map (i.e. such that $u(1) = 1$). Then $\|u\|_{cb} = 1$ if and only if u extends to a c.p. map $\widehat{u} : A \to B(H)$.*

Proof By the injectivity of $B(H)$ (see Theorem 1.18), if $\|u\|_{cb} = 1$ there is an extension $\widehat{u} : A \to B(H)$ with $\|\widehat{u}\|_{cb} = \|u\|_{cb} = 1$. Since $\widehat{u}(1) = 1$, the preceding remark shows that $\widehat{u}$ is positive, but since the same can be applied to $\widehat{u}_n = Id_{M_n} \otimes \widehat{u}$, we conclude that $\widehat{u}$ is actually c.p. Conversely, if u admits a c.p. extension $\widehat{u}$ then we have $\|\widehat{u}\| = \|\widehat{u}(1)\| = 1$ and hence $\|u\| = 1$. Again applying this to $u_n : M_n(E) \to M_n(B(H))$ we obtain $\|u_n\| = 1 \ \forall n$ and hence $\|u\|_{cb} = 1$. $\qquad\square$

Definition 1.36 A (not necessarily closed) subspace $S \subset B(H)$ for some Hilbert space H is called an **operator system**, if it is self-adjoint and unital.

Clearly, if S is an operator system, $M_n(S) \subset M_n(B(H))$ also is one.

Let $S_+ = S \cap B(H)_+$. If $S \subset B(H)$ is an operator system, S coincides with the linear span of S_+. Indeed, if $a \in S$ and $a = a^*$, then $a = \|a\|1 - (\|a\|1 - a) \in S_+ - S_+$. This explains why c.p. maps can be used efficiently on operator systems.

The elementary proof of the next lemma is left to the reader, to whom we recall the arithmetic/geometric mean inequality $\sqrt{st} \leq (s + t)/2, \forall s, t \geq 0$.

Lemma 1.37

(i) *Let $s_1, s_2, a \in B(H)$. Assume $s_1, s_2 \geq 0$. Then*

$$\begin{pmatrix} s_1 & a \\ a^* & s_2 \end{pmatrix} \geq 0 \Leftrightarrow |\langle y, ax \rangle| \leq (\langle y, s_1 y \rangle \langle x, s_2 x \rangle)^{1/2}$$

$$\leq (\langle y, s_1 y \rangle + \langle x, s_2 x \rangle)/2 \ \forall x, y \in H. \tag{1.24}$$

When this holds, we have

$$\|a\| \le (\|s_1\|\|s_2\|)^{1/2} \le (\|s_1\| + \|s_2\|)/2. \qquad (1.25)$$

(ii) In particular,

$$\begin{pmatrix} 1 & a \\ a^* & 1 \end{pmatrix} \ge 0 \Leftrightarrow \|a\| \le 1.$$

Lemma 1.38 *Let $w : E \to B(H)$ be a map defined on an operator space $E \subset B(K)$. Let $S \subset M_2(B(K))$ be the operator system consisting of all matrices $\begin{pmatrix} \lambda 1 & a \\ b^* & \mu 1 \end{pmatrix}$ with $\lambda, \mu \in \mathbb{C}$, $a, b \in E$, and let $W : S \to M_2(B(H))$ be the (unital) mapping defined by*

$$W\left(\begin{pmatrix} \lambda 1 & a \\ b^* & \mu 1 \end{pmatrix}\right) = \begin{pmatrix} \lambda 1 & w(a) \\ w(b)^* & \mu 1 \end{pmatrix}.$$

Then $\|w\|_{cb} \le 1$ if and only if W is c.p.

Proof The easy direction is W c.p. $\Rightarrow \|w\|_{cb} \le 1$. Indeed, if W is c.p. we have $\|W\|_{cb} = \|W(1)\| = 1$, and a fortiori $\|w\|_{cb} \le 1$.

We now turn to the converse. Assume $\|w\|_{cb} \le 1$. Consider an element $s \in M_n(S)$, say $s = \begin{pmatrix} \lambda & a \\ b^* & \mu \end{pmatrix}$, $\lambda, \mu \in M_n(\mathbb{C}1)$, $a, b \in M_n(E)$. Assume $s \ge 0$, then necessarily $\lambda, \mu \in M_n(\mathbb{C}1)_+$ and $a = b$. Fix $\varepsilon > 0$ and let $\lambda_\varepsilon = \lambda + \varepsilon 1$ and $\mu_\varepsilon = \mu + \varepsilon 1$ (invertible perturbations of λ and μ). Let $s_\varepsilon = s + \varepsilon 1$. Let us denote $x_\varepsilon = \lambda_\varepsilon^{-1/2} a \mu_\varepsilon^{-1/2}$ and let $W_n = Id_{M_n} \otimes W$, $w_n = Id_{M_n} \otimes w$. We have

$$\begin{pmatrix} 1 & x_\varepsilon \\ x_\varepsilon^* & 1 \end{pmatrix} = \begin{pmatrix} \lambda_\varepsilon^{-1/2} & 0 \\ 0 & \mu_\varepsilon^{-1/2} \end{pmatrix} s_\varepsilon \begin{pmatrix} \lambda_\varepsilon^{-1/2} & 0 \\ 0 & \mu_\varepsilon^{-1/2} \end{pmatrix}, \qquad (1.26)$$

and hence the left-hand side of the preceding equation is ≥ 0, which implies by part (ii) in Lemma 1.37 that $\|x_\varepsilon\| \le 1$. Therefore if $\|w\|_{cb} \le 1$, we have $\|w_n(x_\varepsilon)\| \le 1$, which implies that

$$\begin{pmatrix} 1 & w_n(x_\varepsilon) \\ w_n(x_\varepsilon)^* & 1 \end{pmatrix} \ge 0.$$

But this last matrix is the same as $W_n \begin{pmatrix} 1 & x_\varepsilon \\ x_\varepsilon^* & 1 \end{pmatrix}$. Now, applying W_n to both sides of (1.26) and using the linearity of w, we find

$$W_n \begin{pmatrix} 1 & x_\varepsilon \\ x_\varepsilon^* & 1 \end{pmatrix} = \begin{pmatrix} \lambda_\varepsilon^{-1/2} & 0 \\ 0 & \mu_\varepsilon^{-1/2} \end{pmatrix} W_n(s_\varepsilon) \begin{pmatrix} \lambda_\varepsilon^{-1/2} & 0 \\ 0 & \mu_\varepsilon^{-1/2} \end{pmatrix}.$$

Thus

$$W_n(s_\varepsilon) = \begin{pmatrix} \lambda_\varepsilon^{1/2} & 0 \\ 0 & \mu_\varepsilon^{1/2} \end{pmatrix} \begin{pmatrix} 1 & w_n(x_\varepsilon) \\ w_n(x_\varepsilon)^* & 1 \end{pmatrix} \begin{pmatrix} \lambda_\varepsilon^{1/2} & 0 \\ 0 & \mu_\varepsilon^{1/2} \end{pmatrix}.$$

Since the right-hand side is ≥ 0, we have $W_n(s_\varepsilon) \geq 0$ and letting $\varepsilon \to 0$ we conclude that $W_n(s) \geq 0$, whence that W is c.p. □

We now give the c.p. version of Arveson's extension theorem:

Theorem 1.39 (Arveson's extension theorem/C.P. version) *Let $E \subset A \subset B(K)$ be an operator system included in a unital C^*-subalgebra of $B(K)$. Any c.p. map $u : E \to B(H)$ satisfies*

$$\|u\|_{cb} = \|u\| = \|u(1)\|. \tag{1.27}$$

Moreover, $u : E \to B(H)$ extends to a c.p. map $\widehat{u} : A \to B(H)$ such that $\|\widehat{u}\|_{cb} = \|u(1)\|$.

Proof We first establish (1.27). By part (ii) in Lemma 1.37, we see that $\|x\| \leq 1 \Rightarrow \begin{pmatrix} 1 & x \\ x^* & 1 \end{pmatrix} \geq 0 \Rightarrow \begin{pmatrix} u(1) & u(x) \\ u(x^*) & u(1) \end{pmatrix} \geq 0$. Hence by part (i) in Lemma 1.37, we find that $\|u(x)\| \leq \|u(1)\| \Rightarrow \|u\| = \|u(1)\|$.

Similarly, $x \in M_n(S), \|x\| \leq 1 \Rightarrow \begin{pmatrix} 1 & x \\ x^* & 1 \end{pmatrix} \geq 0 \Rightarrow \|u_n(x)\| \leq \|u_n(1)\| = \|u(1)\|$. Hence $\|u\|_{cb} = \sup_n \|u_n\| \leq \|u(1)\|$, i.e. $\|u\|_{cb} = \|u(1)\|$, proving (1.27).

Let $\varphi \in A^*$ be any state on A. Since u is positive, we know that $u(1) \geq 0$. We may assume that $\|u(1)\| = 1$, so that $0 \leq u(1) \leq 1$. Consider $E \oplus E \subset A \oplus A \subset B(K) \oplus B(K) \subset B(K \oplus K)$.

Let $v : E \oplus E \to B(H)$ be the mapping defined by

$$v(x \oplus y) = u(x) + (1 - u(1))\varphi(y).$$

Then v is clearly c.p. and $v(1 \oplus 1) = 1$. By (1.27) $\|v\|_{cb} = 1$. Therefore, by Theorem 1.35 v admits a c.p. extension $\widehat{v} : A \oplus A \to B(H)$ with $\|\widehat{v}\|_{cb} = \|v(1 \oplus 1)\| = 1$. Then the map $\widehat{u}$ defined by $\widehat{u}(x) = \widehat{v}(x \oplus 0)$, is a c.p. extension of u and $\|\widehat{u}\|_{cb} \leq \|\widehat{v}\|_{cb} = 1 = \|u(1)\|$. Thus (since $1 = \|u(1)\| = \|\widehat{u}(1)\| \leq \|\widehat{u}\|_{cb}$) we obtain $\|\widehat{u}\|_{cb} = \|u(1)\|$. □

Lemma 1.40 (Schur product of c.p. mappings) *Let B, C be C^*-algebras (or merely linear subspaces of C^*-algebras). Fix a number $n \geq 1$. Consider $u \in CP(A, M_n(B))$ and $v \in CP(B, M_n(C))$. Then the mapping $x \mapsto [v_{ij}(u_{ij}(x))]$ is in $CP(A, M_n(C))$.*

Proof We identify $M_n(B)$ with $M_n \otimes_{\min} B$, so that $u(x) = \sum e_{ij} \otimes u_{ij}(x)$. Consider the composition

$$w = (Id_{M_n} \otimes v) \circ u : A \to M_n \otimes_{\min} M_n \otimes_{\min} C.$$

Since complete positivity is obviously preserved under composition, $w \in CP(A, M_n \otimes_{\min} M_n \otimes_{\min} C)$. We have

$$w(x) = \sum_{ij} e_{ij} \otimes v \circ u_{ij}(x) = \sum_{ij} \sum_{k\ell} e_{ij} \otimes e_{k\ell} \otimes v_{k\ell} \circ u_{ij}(x).$$

Let $S : \ell_2^n \to \ell_2^n \otimes \ell_2^n$ be the isometry defined by $Se_j = e_j \otimes e_j$. We may assume $C \subset B(H)$. Then we have

$$(S \otimes Id_H)^* w(x)(S \otimes Id_H) = \sum e_{ij} \otimes v_{ij}(u_{ij}(x)),$$

and hence the latter mapping is c.p. $\square$

1.4 Normal c.p. maps on von Neumann algebras

Recall that a bounded linear map $u : M \to N$ between von Neumann algebras is called "normal" if it is continuous when M and N are both equipped with the weak* topology, or equivalently if $u^*(N_*) \subset M_*$. We wish to record here the following variant of the extension theorem.

Theorem 1.41 *Let $M \subset B(K)$ be a von Neumann algebra. Let $u : M \to B(H)$ be a normal c.p. map with $\|u\| = 1$. Then there is a normal c.p. map $\tilde{u} : B(K) \to B(H)$ extending u with $\|\tilde{u}\| = 1$. More precisely, there are $\widehat{H}$, a normal $*$-homomorphism $\widehat{\pi} : B(K) \to B(\widehat{H})$ and a contraction $W : H \to \widehat{H}$ such that $u(x) = W^*\widehat{\pi}(x)W$ for all $x \in M$. Moreover, we can obtain the latter with $\widehat{H}$ of the form $\widehat{H} = L \otimes_2 K$ and with $\widehat{\pi}(b) = Id_L \otimes b$ ($b \in B(K)$) for some Hilbert space L.*

Proof Consider a *minimal* dilation of u of the form $u(x) = V^*\pi(x)V$ for all $x \in M$ as in Remark 1.23. Let $\xi, \xi' \in \pi(M)(VH)$, say $\xi = \pi(m)(Vh)$, $\xi' = \pi(m')(Vh')$ ($m, m' \in M, h, h' \in H$). We have $\langle \xi', \pi(x)\xi \rangle = \langle h', u(m'^*xm)h \rangle$, which shows that $x \mapsto \langle \xi', \pi(x)\xi \rangle$ is normal on M. By the density of the linear span of $\pi(M)(VH)$ in $\widehat{H}$ this implies that $\pi : M \to B(\widehat{H})$ is normal (see Remark A.42). Then the result follows immediately by the special form of the *normal* $*$-homomorphisms described in Theorem A.61. Indeed, the latter says that we can find a Hilbert space L, a subspace $E \subset L \otimes_2 K$ (invariant under $Id_L \otimes M$) and a unitary $U : \widehat{H} \to E$ such that $\pi(x) = U^*P_E(Id_L \otimes x)_{|E}U$ for any $x \in M$. Let $j_E : E \to L \otimes_2 H$ be the inclusion, let $\widehat{\pi}(b) = Id_L \otimes b$ and $W = j_E UV : H \to L \otimes_2 K$. Then the mapping $\tilde{u}$ defined by $\tilde{u}(b) = W^*\widehat{\pi}(b)W$ for any $b \in B(H)$ is a c.p. extension of u with $\|\tilde{u}\| = 1$. $\square$

Remark 1.42 In the preceding situation if u is unital the extension $\tilde{u}$ is unital and W is an isometry.

In the case of the inclusion $A \subset A^{**}$ we also have a simple extension property, as follows.

Lemma 1.43 *Let A be a C^*-algebra, M a von Neumann algebra. Then for any $u \in CP(A, M)$ the mapping $\ddot{u} : A^{**} \to M$ is a normal c.p. map extending u with $\|\ddot{u}\| = \|u\|$.*

Proof We recall that, by density, $\ddot{u}$ can be viewed as the unique $(\sigma(A^{**}, A^*),$ $\sigma(M, M_*))$-continuous extension of u (see (A.32)). By the weak*-density of $B_A \cap A_+$ in $B_{A^{**}} \cap (A^{**})_+$ (see (A.36)), it follows that $\ddot{u}$ is positive. Applying that to u_n and observing (see Proposition A.58) that $M_n(A)^{**} = M_n(A^{**})$ shows that $\ddot{u}_n$ is positive, and hence $\ddot{u}$ is completely positive. $\qquad\square$

1.5 Injective operator algebras

Definition 1.44 A C^*-algebra (or an operator space) A is called injective if there exists a completely isometric embedding $A \subset B(H)$ and a projection $P : B(H) \to A$ with $\|P\|_{cb} = 1$.

Of course, $B(H)$ is the fundamental example of an injective C^*-algebra. We will focus on C^*-algebras (see Remark 1.49 for the case of operator spaces).

The next result is classical.

Theorem 1.45 (Tomiyama's theorem [245]) *If $A \subset B$ is a C^*-subalgebra of a C^*-algebra B, any linear projection $P : B \to A$ with $\|P\| = 1$ is automatically completely positive and completely bounded with $\|P\|_{cb} = 1$. Moreover, P is a conditional expectation in the sense that*

$$\forall a_1, a_2 \in A, \ \forall b \in B \quad P(a_1 b a_2) = a_1 P(b) a_2. \tag{1.28}$$

Proof The (nontrivial) proof can be found in Takesaki's book (see [240, p. 131]) or in [39, p. 13]. We skip it here. $\qquad\square$

Proposition 1.46 (Extension property) *Consider a C^*-subalgebra $A \subset B(H)$. The following are equivalent.*

(i) *A is injective.*

(ii) *For any completely isometric embedding $A \subset B$ into a C^*-algebra B there is a projection $Q : B \to A$ with $\|Q\|_{cb} = 1$.*

(iii) *For any pair of operator spaces with $X_1 \subset X_2$ (completely isometrically) any $u \in CB(X_1, A)$ admits an extension $\tilde{u} \in CB(X_2, A)$ with $\|\tilde{u}\|_{cb} = \|u\|_{cb}$.*

(iv) *For any pair B_1, B_2 of C^*-algebras with $B_1 \subset B_2$ (C^*-subalgebra), any u in $CB(B_1, A)$ admits an extension $\widetilde{u} \in CB(B_2, A)$ with $\|\widetilde{u}\|_{cb} = \|u\|_{cb}$.*

Proof Let $j_1 : A \to B(H)$ and $j_2 : A \to B$ denote completely isometric embeddings. By the extension Theorem 1.18, $\exists \widetilde{j}_1 \in CB(B, B(H))$ such that $\|\widetilde{j}_1\|_{cb} = \|j_1\|_{cb}$ and $\widetilde{j}_1 j_2 = j_1$. Assume (i). Let $P : B(H) \to j_1(A)$ be a projection such that $\|P\|_{cb} = 1$. Then $Q = j_2 j_1^{-1}{}_{|j_1(A)} P \widetilde{j}_1$ is a projection onto $j_2(A)$ with $\|Q\|_{cb} = 1$. This shows (i) $\Rightarrow$ (ii).

Assume (ii). Consider $A \subset B(H)$ and a projection $P : B(H) \to A$ with $\|P\|_{cb} = 1$. By the extension Theorem 1.18, any $u \in CB(X_1, A)$ admits an extension $v \in CB(X_2, B(H))$ with $\|v\|_{cb} = \|u\|_{cb}$. Then $\widetilde{u} = P\widetilde{v} \in CB(X_2, A)$ satisfies (iii). This shows (ii) $\Rightarrow$ (iii). (iii) $\Rightarrow$ (iv) is trivial.

Assume (iv). Consider $A \subset B(H)$. Let $B_1 = A$, $B_2 = B(H)$ and $u = Id_A$. Then $\widetilde{u} : B(H) \to A$ is a projection with cb norm 1, and hence (i) holds. $\square$

Corollary 1.47 *Let A be a C^*-algebra and $A_1 \subset A$ a C^*-subalgebra. Assume that there is a projection $P : A \to A_1$ with $\|P\| = 1$. If A is injective then A_1 is also injective.*

We end this section with a simple stability property of injective C^*-algebras.

Proposition 1.48 *Let $(A_i)_{i \in I}$ be a family of C^*-algebras. Then $\left(\oplus \sum_{i \in I} A_i \right)_\infty$ is injective if and only if A_i is injective for all $i \in I$.*

Proof If $A_i \subset B(H_i)$ and $P_i : B(H_i) \to A_i$ are projections with $\|P_i\|_{cb} = 1$ then the mapping $P : \left(\oplus \sum_{i \in I} B(H_i) \right)_\infty \to \left(\oplus \sum_{i \in I} A_i \right)_\infty$ taking (x_i) to $(P_i x_i)$ is clearly c.b. with $\|P\|_{cb} \leq 1$. Let $H = \left(\oplus \sum_{i \in I} H_i \right)_2$. If we denote by $V_i : H_i \to H$ the natural (isometric) inclusion, and if we define $Q_i : B(H) \to B(H_i)$ by $Q_i(T) = V_i^* T V_i$ and $Q : B(H) \to \left(\oplus \sum_{i \in I} B(H_i) \right)_\infty$ by

$$Q(T) = (Q_i(T))_{i \in I},$$

then Q is clearly c.b. with $\|Q\|_{cb} = \sup_i \|Q_i\|_{cb} = 1$. We may identify H_i with a subspace of H so that V_i becomes the inclusion. Then $\left(\oplus \sum_{i \in I} A_i \right)_\infty$ can be naturally identified with the C^*-subalgebra of $B(H)$ formed of all operators $T \in B(H)$ such that $TH_i \subset H_i$ and $T_{|H_i} \in A_i$ for all $i \in I$. With this convention, PQ is a projection from $B(H)$ onto $\left(\oplus \sum_{i \in I} A_i \right)_\infty$ with $\|PQ\|_{cb} = 1$. This proves the "if" part. The converse follows easily from the fact that the canonical projection from $\left(\oplus \sum_{i \in I} A_i \right)_\infty$ to A_i (which is a $*$-homomorphism) has cb-norm = 1 for all $i \in I$. $\square$

Remark 1.49 For operator spaces, Tomiyama's theorem does not hold. Therefore, it is more natural to say that an operator space $E \subset B(H)$ is c-injective if there is a projection $P : B(H) \to E$ with $\|P\|_{cb} \leq c$. With this terminology, we call E injective if it is 1-injective. The reader will easily check that Proposition 1.46, and Proposition 1.48 remain valid for injective (i.e. 1-injective) operator spaces.

See [185] for an example of a *bounded* linear map $u : A \to B(H)$ defined on a C^*-subalgebra $A \subset B(\mathcal{H})$ that does *not* extend to a bounded map on $B(\mathcal{H})$.

We return to injectivity for von Neumann algebras in §8.3. See [80, §6] for more information on injective operator spaces.

1.6 Factorization of completely bounded (c.b.) maps

We now turn to the factorization of c.b. maps. This important result, proved independently by Wittstock, Haagerup, and Paulsen in the early 1980s, can be viewed as the "linearization" of the Stinespring factorization Theorem 1.22.

Theorem 1.50 (Factorization of c.b. maps) *Let H, K be Hilbert spaces. Consider an operator space $E \subset B(K)$. Let $B \subset B(K)$ be a unital C^*-algebra such that $E \subset B \subset B(K)$. Consider a c.b. map*

$$
\begin{array}{c}
B \\
\cup \\
E \xrightarrow{\ u\ } B(H)
\end{array}
$$

Then there is a Hilbert space $\widehat{H}$, a unital $*$-homomorphism $\pi : B \longrightarrow B(\widehat{H})$ and operators $V_1, V_2 \in B(H, \widehat{H})$ such that $\|V_1\| \, \|V_2\| = \|u\|_{cb}$ and

$$
\forall x \in E \qquad u(x) = V_2^* \pi(x) V_1. \tag{1.29}
$$

Conversely, if (1.29) holds then u is c.b. and

$$
\|u\|_{cb} \leq \|V_1\| \, \|V_2\|. \tag{1.30}
$$

In addition, if $V_1 = V_2$, then u is completely positive.

The corresponding diagram is as follows:

$$
\begin{array}{ccc}
B & \xrightarrow{\ \pi\ } & B(\widehat{H}) \\
\uparrow & & \downarrow {\scriptstyle b \mapsto V_2^* b V_1} \\
E & \xrightarrow{\ u\ } & B(H)
\end{array}
$$

Proof It suffices to prove this when $B = B(K)$. We may assume $\|u\|_{cb} = 1$. Consider the operator system $S \subset M_2(B(K))$ defined in Lemma 1.38 and let $W \in CP(S, M_2(B(H)))$ be defined by

$$W\left(\begin{pmatrix} \lambda 1 & a \\ b^* & \mu 1 \end{pmatrix}\right) = \begin{pmatrix} \lambda 1 & u(a) \\ u(b)^* & \mu 1 \end{pmatrix}.$$

By (Arveson's extension) Theorem 1.39, there is $\widehat{W} \in CP(M_2(B(K)), M_2(B(H)))$ extending W, to which we may apply (Stinespring's) Theorem 1.22, with $A = M_2(B(K))$ and, setting $\mathcal{H} = H \oplus H$, we have $M_2(B(H)) = B(\mathcal{H})$. This gives us a Hilbert space $\widehat{\mathcal{H}}$, an operator $V : \mathcal{H} \to \widehat{\mathcal{H}}$ and a ∗-homomorphism $\sigma : M_2(B(K)) \to M_2(B(H))$ such that for any $x \in E$ we have

$$\begin{pmatrix} 0 & u(x) \\ 0 & 0 \end{pmatrix} = V^* \sigma \begin{pmatrix} 0 & x \\ 0 & 0 \end{pmatrix} V$$

or equivalently

$$\begin{pmatrix} 0 & u(x) \\ 0 & 0 \end{pmatrix} = V^* \sigma \begin{pmatrix} x & 0 \\ 0 & x \end{pmatrix} \sigma \begin{pmatrix} 0 & 1 \\ 0 & 0 \end{pmatrix} V$$

and hence if $P_1 : H \oplus H \to H$ (resp. $P_2 : H \oplus H \to H$) is the first (resp. second) coordinate projection, we have

$$u(x) = P_1 V^* \sigma \begin{pmatrix} x & 0 \\ 0 & x \end{pmatrix} \sigma \begin{pmatrix} 0 & 1 \\ 0 & 0 \end{pmatrix} V P_2^*$$

and we obtain (1.29) with $\pi(x) = \sigma \begin{pmatrix} x & 0 \\ 0 & x \end{pmatrix}$, $V_1 = \sigma \begin{pmatrix} 0 & 1 \\ 0 & 0 \end{pmatrix} V P_2^*$ and $V_2^* = P_1 V^*$. Note that $\|V_1\| \|V_2\| \le \|V\|^2 = 1$.

Conversely, if (1.29) holds we obviously have $1 = \|u\|_{cb} \le \|V_1\| \|V_2\|$. Moreover, if $V_1 = V_2$, u is clearly c.p.   □

Remark 1.51 In (1.29) we may without loss of generality replace $\widehat{H}$ by

$$\widehat{H}_0 = \overline{\operatorname{span}\{\pi(b)V_1 h \mid b \in B, h \in H\}} \subset \widehat{H}.$$

Indeed, since $\widehat{H}_0$ is an invariant subspace for all $\pi(b)$s for $b \in B$, the mapping $b \mapsto P_{\widehat{H}_0}\pi(b)_{|\widehat{H}_0}$ is a representation of B on $\widehat{H}_0$ and we may define $W_1, W_2 \in B(H, \widehat{H}_0)$ by $W_1 h = V_1 h$ and $W_2 h = P_{\widehat{H}_0} V_2 h$, so that $\|W_1\| \|W_2\| \le \|u\|_{cb}$ and

$$\forall x \in E \qquad u(x) = W_2^* \pi(x) W_1.$$

The point of this remark is that if H, B are both separable, we can ensure that $\widehat{H}$ also is. Since it is a trivial matter to enlarge $\widehat{H}$ if necessary, this shows that if $H = \ell_2$ and B is separable we can always take $\widehat{H} = \ell_2$. Note that if E is separable, we may replace B by the separable C^*-algebra generated by E. Thus it suffices to assume E separable.

Similarly, if $\dim(B) < \infty$ and $\dim(H) = \infty$, we can ensure that $\widehat{H}$ and H have the same Hilbertian dimension and hence by unitary equivalence we can take $\widehat{H} = H$.

For emphasis and for later reference, we state as separate corollaries parts of Theorem 1.50 that will be used frequently in the sequel. The first one repeats part of Theorem 1.35.

Corollary 1.52 *Let $E \subset B(H)$ be an operator space containing I. Consider a map $u : E \to B(K)$. If $u(I) = I$ and $\|u\|_{cb} = 1$, then there is a Hilbert space $\widehat{H}$ with $K \subset \widehat{H}$ and a unital representation $\pi : B(H) \to B(\widehat{H})$ such that*

$$u(x) = P_K \pi(x)_{|K}. \qquad\qquad \forall x \in E$$

In particular, u is completely positive.

Proof By Theorem 1.50, we have $u(\cdot) = V_2^* \pi(\cdot) V_1$. By homogeneity, we may assume $\|V_1\| = \|V_2\| = 1$. Since $I = u(I) = V_2^* \pi(I) V_1 = V_2^* V_1$, we have $\langle V_2 h, V_1 h \rangle = \|h\|^2$ for any $h \in H$. This forces $\|V_1 h\| = \|V_2 h\| = \|h\|$, and also $V_1 h = V_2 h$. Thus V_1 is an isometric embedding of K into $\widehat{H}$. Identifying K with $V_1(K)$, $u(\cdot) = V_2^* \pi(\cdot) V_1$ becomes $u(\cdot) = P_K \pi(\cdot)_{|K}$. $\qquad\square$

The second corollary is the decomposability of c.b. maps into $B(H)$ as linear combinations of c.p. maps. We should emphasize that while this holds for c.b. maps with range in $B(H)$, it usually fails for c.b. maps with range in a C^*-subalgebra $B \subset B(H)$: in general we cannot get the c.p. maps u_js to be B-valued. See Chapter 6 for more on the decomposability theme.

Corollary 1.53 *Any c.b. map $u : E \to B(K)$ can be decomposed as $u = u_1 - u_2 + i(u_3 - u_4)$ where u_1, u_2, u_3, u_4 are c.p. maps with $\|u_j\|_{cb} \le \|u\|_{cb}$. More precisely, we have $\|u_1 + u_2\| \le \|u\|_{cb}$ and $\|u_3 + u_4\| \le \|u\|_{cb}$.*

Proof By Theorem 1.50, we have $u(\cdot) = V_2^* \pi(\cdot) V_1$. By homogeneity, we may assume $\|V_1\| = \|V_2\| = \|u\|_{cb}^{1/2} = 1$. Let us denote $V = V_1$ and $V_2 = W$, so that $u(\cdot) = W^* \pi(\cdot) V$. Then the result simply follows from the polarization formula: we define u_1, u_2, u_3, u_4 by

$$u_1(\cdot) = 4^{-1}(V + W)^* \pi(\cdot)(V + W), \qquad u_2(\cdot) = 4^{-1}(V - W)^* \pi(\cdot)(V - W),$$

$$u_3(\cdot) = 4^{-1}(V + iW)^* \pi(\cdot)(V + iW), \qquad u_4(\cdot) = 4^{-1}(V - iW)^* \pi(\cdot)(V - iW).$$

Then, by (1.30), $\|u_j\|_{cb} \le 1$ for $j = 1, 2, 3, 4$ and $u = u_1 - u_2 + i(u_3 - u_4)$. Note that actually $(u_1 + u_2)(\cdot) = 2^{-1}(V^* \pi(\cdot) V + W^* \pi(\cdot) W)$ and hence again by (1.30) $\|u_1 + u_2\|_{cb} \le (\|V\|^2 + \|W\|^2)/2 \le 1$. Similarly $\|u_3 + u_4\|_{cb} \le 1$. $\qquad\square$

Remark 1.54 (GNS and Hahn decomposition) Let A be a C^*-algebra and let $f \in A^*$. Then it is well known that that there are a $*$-homomorphism $\pi : A \to B(H)$ and vectors $\eta, \xi \in H$ such that for any $x \in A$

$$f(x) = \langle \eta, \pi(x)\xi \rangle$$

and $\|\xi\|\|\eta\| = \|f\|_{A^*}$. This can be derived from the classical GNS factorization (see (A.16)).

We would like to point out to the reader that we have proved it in passing. Indeed, we can view this fact as a particular case of Theorem 1.50 applied with $E = A$ to the linear map $f : A \to \mathbb{C}$, since by Proposition 1.3 we have $\|f\|_{cb} = \|f\|$ in this case.

Moreover, if f is a self-adjoint form i.e. if $f(x) = \overline{f(x^*)}$ for any $x \in A$ then we recover the classical Hahn decomposition: there are $f_+, f_- \in A_+^*$ such that $f = f_+ - f_-$ and $\|f_+\| + \|f_-\| = \|f\|$. Indeed, we can take $f_{\pm}(x) = 4^{-1}\langle \eta \pm \xi, \pi(x)(\eta \pm \xi)\rangle$ and by homogeneity we may assume $\|\eta\| = \|\xi\| = \|f\|^{1/2}$. Then

$$\|f_+\| + \|f_-\| \le 4^{-1}(\|\eta + \xi\|^2 + \|\eta - \xi\|^2) = 2^{-1}(\|\eta\|^2 + \|\xi\|^2) = \|f\|.$$

We already saw in (1.3) that every finite rank map $u : E \to F$ (between arbitrary operator spaces) is c.b. Let $\alpha(n)$ be the best constant C such that, for any E, F, any map $u : E \to F$ of rank n satisfies

$$\|u\|_{cb} \le C\|u\|.$$

To majorize $\alpha(n)$, we will need the following classical lemma.

Lemma 1.55 (Auerbach's lemma) *Let E be an arbitrary n-dimensional normed space. There is a biorthogonal system $x_j \in E$, $\xi_j \in E^*$ ($j = 1, 2, \ldots, n$) such that $\|x_j\| = \|\xi_j\| = 1$ for all $j = 1, \ldots, n$.*

Proof Choose $x_1, \ldots, x_n$ in the unit sphere of E on which the function $x \to |\det(x_1, \ldots, x_n)|$ attains its maximum, supposed equal to $C > 0$. Then let $\xi_j(y) = C^{-1}\det(x_1, \ldots, x_{i-1}, y, x_{i+1}, \ldots, x_n)$. The desired properties are easy to check.   □

Remark 1.56 We may as well assume $\dim(F) = n$. Then by Auerbach's lemma we can write Id_F as the sum of n rank one maps of unit norm. This immediately implies

$$\alpha(n) \le n.$$

However, this is *not best possible*: It is known (due to Éric Ricard, see [208, p. 145]) that $\alpha(n) \le n/2^{1/4}$, but the exact value of $\alpha(n)$, or of $\limsup \alpha(n)/n$, does not seem to be known although it is known that the latter limit is $\ge 1/2$, by an argument due to Paulsen (see [196] or [208, p. 75]).

To illustrate the use of the factorization theorem, we end by a well-known characterization of complete boundedness for Schur mulipliers.

Let I be any set. We denote by $(e_i)_{i \in I}$ the canonical basis of $\ell_2(I)$ and by e_{ij} $(i, j \in I)$ the matrix units in $B(\ell_2(I))$. To any $a \in B(\ell_2(I))$ we associate the matrix $[a_{ij}]$ $(i, j \in I)$ defined as usual by $a_{ij} = \langle e_i, a(e_j) \rangle$. Let $\Phi : I \times I \to \mathbb{C}$ be any function. Any linear mapping that takes $[a_{ij}]$ to $[a_{ij} \Phi(i, j)]$ is commonly called a "Schur multiplier." The next result characterizes the Schur mulipliers that are c.b. linear maps from $B(\ell_2(I))$ to itself.

Proposition 1.57 (Schur multipliers on $B(\ell_2)$) *Let $C \geq 0$ be a constant. The following are equivalent:*

(i) *There is a c.b. map $u : B(\ell_2(I)) \to B(\ell_2(I))$ such that $u(e_{ij}) = \Phi(i, j)e_{ij}$ for any $i, j \in I \times I$ with $\|u\|_{cb} \leq C$.*

(ii) *There are a Hilbert space H and bounded functions $y : I \to H$, $x : I \to H$ such that $\sup_I \|y\|_H \sup_I \|x\|_H \leq C$ and*

$$\forall i, j \in I \times I \quad \Phi(i, j) = \langle y(i), x(j) \rangle.$$

(iii) *The Schur multiplier $u : B(\ell_2(I)) \to B(\ell_2(I))$ that takes $[a_{ij}]$ to $[a_{ij} \Phi(i, j)]$ is c.b. with $\|u\|_{cb} \leq C$.*

Proof Assume (i). By Theorem 1.50 there is $\pi : B(\ell_2(I)) \to B(\widehat{H})$ and $V, W : \ell_2(I) \to \widehat{H}$ with $\|V\| \|W\| \leq C$ such that $\Phi(i, j)e_{ij} = V^*\pi(e_{ij})W$. This implies $\Phi(i, j) = \langle e_i, (V^*\pi(e_{ij})W)e_j \rangle$. Fix an element $o \in I$. Note $e_{ij} = e_{io}e_{oj}$. Therefore we have $\Phi(i, j) = \langle y(i), x(j) \rangle$ where $x(j) = \pi(e_{oj})We_j$ and $y(i) = \pi(e_{io})^*Ve_i$, and (ii) follows.

Assume (ii). Define $\pi : B(\ell_2(I)) \to B(H)$ by $\pi(x) = x \otimes Id_H$ and let $V_x : \ell_2(I) \to \ell_2(I) \otimes_2 H$ be the map taking e_i to $e_i \otimes x(i)$ $(i \in I)$. Note $\|V_x\| = \sup_I \|x\|_H$. Let $u(\cdot) = V_y^*\pi(\cdot)V_x$. Then u coincides with the Schur multiplier in (iii) and $\|u\|_{cb} \leq C$. Thus (ii) $\Rightarrow$ (iii). (iii) $\Rightarrow$ (i) is trivial. □

Remark 1.58 Actually, it is known that (iii) $\Rightarrow$ (ii) holds even if we merely assume that u is a *bounded* Schur multiplier with $\|u\| \leq C$. In other words the c.b. norm and the norm of a Schur multiplier on $B(\ell_2(I))$ are equal. We prove this as a consequence of a more general phenomenon in Corollary 2.7.

1.7 Normal c.b. maps on von Neumann algebras

It is well known that the Banach space $X = L_1(\mu)$ has the property that there is a contractive projection from X^{**} to X. This corresponds to the decomposition into absolutely continuous and singular parts (with respect to μ) in the abstract L_1-space X^{**}.

The noncommutative analogue is also true, as follows. Let M be a von Neumann algebra with predual M_*. We will work with M^{**} that we assume realized as a von Neumann subalgebra of $B(K)$. Since this is a source of mistakes, as a preliminary precaution Remark A.60 is recommended reading.

For any linear map $u : M \rightarrow B(H)$, we will use the notation (see §A.16)

$$\ddot{u} = (u^*_{|B(H)_*})^* : M^{**} \rightarrow B(H).$$

Note that $\ddot{u}$ is normal and for any $f \in B(H)_*$ and any $z \in M^{**}$ we have

$$\langle f, \ddot{u}(z) \rangle = \langle u^*(f), z \rangle. \tag{1.31}$$

If u is a unital $*$-homomorphism, $\ddot{u}$ is also one. In particular taking $u = Id_M$ we find a normal unital $*$-homomorphism $\pi : M^{**} \rightarrow M$ that extends the identity on M.

Consider then the set $\mathcal{I}_0 \subset M^{**}$ that is the annihilator of $M_* \subset M^*$. It is immediate that $\mathcal{I}_0$ is a weak* closed two-sided ideal in M^{**} because $\mathcal{I}_0 = \ker(\pi)$. It follows (see Remark A.34) that there is a central projection $Q_0 \in M^{**}$ such that $\mathcal{I}_0 = Q_0 M^{**}$.

Lemma 1.59 *For any bounded linear map $u : M \rightarrow B(H)$ we define*

$$\forall x \in M \quad u_{\mathcal{N}}(x) = \ddot{u}((1 - Q_0)x) \text{ and } u_{\mathcal{S}}(x) = \ddot{u}(Q_0 x).$$

Then $u_{\mathcal{N}} : M \rightarrow B(H)$ is normal with $\|u_{\mathcal{N}}\| \leq \|u\|$ and in the c.b. case $\|u_{\mathcal{N}}\|_{cb} \leq \|u\|_{cb}$. If u is normal we have $u_{\mathcal{N}} = u$. Moreover if u is a $*$-homomorphism (resp. is c.p.) so is $u_{\mathcal{N}}$.

Proof If u is a $*$-homomorphism (resp. is c.p.) so is $\ddot{u}$, thus (Q_0 being central) the last assertion is obvious, as well as the norm inequalities. To show that $u_{\mathcal{N}}$ is normal it suffices to show that $u^*_{\mathcal{N}}(B(H)_*) \subset M_*$. A priori we only know $u^*_{\mathcal{N}}(B(H)_*) \subset M^*$. By the bipolar criterion (applied to the duality between M^* and M^{**}) it suffices to show that $u^*_{\mathcal{N}}(B(H)_*) \subset M_*^{\perp\perp} = \mathcal{I}_0^\perp$, or equivalently to show that $\langle u^*_{\mathcal{N}}(f), z \rangle = 0$ for any $f \in B(H)_*$ and any $z \in \mathcal{I}_0$. This can be checked as follows. Let (z_i) be a bounded net in M tending weak* to $z \in \mathcal{I}_0$. Then (see Remark A.37)

$$(1 - Q_0)z_i \overset{\sigma(M^{**}, M^*)}{\longrightarrow} (1 - Q_0)z = 0.$$

Since $u_{\mathcal{N}}(z_i)$ tends weak* to $u^{**}_{\mathcal{N}}(z)$ we have

$$\langle u^*_{\mathcal{N}}(f), z \rangle = \langle f, u^{**}_{\mathcal{N}}(z) \rangle = \lim \langle f, u_{\mathcal{N}}(z_i) \rangle = \lim \langle f, \ddot{u}((1 - Q_0)z_i) \rangle$$
$$= \lim \langle u^*(f), (1 - Q_0)z_i \rangle = 0,$$

where the last step uses (1.31). This completes the proof that $u_{\mathcal{N}}$ is normal.

Assume u normal. We will show that $\langle f, u_{\mathcal{S}}(x) \rangle = 0$ for any $f \in B(H)_*$ and $x \in M$. Indeed, we have $\langle f, u_{\mathcal{S}}(x) \rangle = \langle u^*(f), Q_0 x \rangle$ and $Q_0 x \in \mathcal{I}_0 = M_*^{\perp}$ while $u^*(f) \in M_*$ therefore $\langle f, u_{\mathcal{S}}(x) \rangle = 0$. Since $B(H)_*$ separates the points of $B(H)$ this implies $u_{\mathcal{S}} = 0$ and hence $u_{\mathcal{N}} = u$. $\qquad\square$

Theorem 1.60 *Let $M \subset B(K)$ be a von Neumann algebra. Let $u : M \to B(H)$ be a normal c.b. map. Then there is a normal c.b. map $\widetilde{u} : B(K) \to B(H)$ extending u with $\|\widetilde{u}\|_{cb} = \|u\|_{cb}$.*

There are $\widehat{H}$, a normal $*$-homomorphism $\pi : B(K) \to B(\widehat{H})$ and operators $V_1 : H \to \widehat{H}$, $V_2 : H \to \widehat{H}$ with $\|V_1\| \|V_2\| = \|u\|_{cb}$ such that $u(x) = V_2^* \pi(x) V_1$ for all $x \in M$. Moreover, we can obtain this with $\widehat{H}$ of the form $\widehat{H} = L \otimes_2 K$ and with $\pi(x) = Id_L \otimes x$ $(x \in B(K))$ for some Hilbert space L.

Proof By Theorem 1.50 we can find $\widehat{H}$, $\pi : M \to B(\widehat{H})$ and V_1, V_2 such that $u(\cdot) = V_2^* \pi(\cdot) V_1$. We have clearly $\ddot{u}(\cdot) = V_2^* \ddot{\pi}(\cdot) V_1$ and hence $u_{\mathcal{N}}(\cdot) = V_2^* \pi_{\mathcal{N}}(\cdot) V_1$ and since u is normal $u = u_{\mathcal{N}}$. Thus we may replace π by $\pi_{\mathcal{N}}$ and assume that π is normal. Then the proof can be completed by applying Theorem 1.41 to $u = \pi$ (or, more directly, by invoking Theorem A.61). $\qquad\square$

We end by a very simple observation for later use.

Lemma 1.61 *Let A be a C^*-algebra, M a von Neumann algebra. Then for any $u \in CB(A, M)$ the mapping $\ddot{u} : A^{**} \to M$ is a normal c.b. map extending u with $\|\ddot{u}\|_{cb} = \|u\|_{cb}$.*

Proof Recall that $\ddot{u}$ is the unique $(\sigma(A^{**}, A^*), \sigma(M, M_*))$-continuous extension of u. Since $M_n(A)^{**} = M_n(A^{**})$ (see Proposition A.58), it follows from the weak$*$-density of $B_{M_n(A)}$ in $B_{M_n(A^{**})}$ that $\|\ddot{u}_n\| = \|u_n\|$. $\qquad\square$

1.8 Notes and remarks

The history of complete positivity starts with Stinespring's 1955 paper where he proves his factorization theorem for c.p. maps on a C^*-algebra A. The case when A is commutative was already known due to Naimark's work on spectral measures. In two major, very influential *Acta Mathematica* papers in 1969 and 1972 Arveson [12] considerably expanded on Stinespring's breakthrough. He made the crucial step of considering complete positivity for maps defined only on operator systems, and he proved his extension theorem. Later on, Choi and Effros [47] made a deep study of injectivity for operator systems, that somehow opened the way for the later development of operator space theory. While it seems a bit surprising in retrospect, the factorization of completely *bounded* maps emerged only in the early 1980s through independent works by Wittstock, Haagerup (unpublished), and Paulsen. We refer the reader to

Paulsen's book [196] for details and more proper credit on the genesis of that important result, which is fundamental for operator space theory. The latter was ignited by Ruan's 1987 Ph.D. thesis where his abstract characterization of operator spaces is proved. This opened the way to the study of duality for operator spaces (see §2.4), which was thoroughly investigated independently by Effros–Ruan and Blecher–Paulsen. The books by Effros and Ruan [80], by Paulsen [196], by Blecher and Le Merdy [27], as well as our own [208], provide multiple complements to our presentation of complete positivity and complete boundedness. See Størmer's [235] for more on the comparison between positivity and complete positivity. Theorem 1.57 and Remark 1.58 about Schur multipliers have a long history: some essentially equivalent formulation (up to some factor 2) can be traced back to Grothendieck's [98]. Later on the result was rediscovered independently by J. Gilbert and U. Haagerup (see [210] or [207, p. 100] for details). The appendix of Haagerup's manuscript dating from 1986 but recently published as [108] contains more results on Schur multipliers. See also Corollary 2.7.

2

Completely bounded and completely positive maps

A tool kit

In this follow-up chapter on c.b. and c.p. maps, we include a variety of related topics that will later allow us to better illustrate the C^*-algebraic tensor product theory, and its significance for spaces of linear mappings between operator spaces or C^*-algebras. We advise the reader to browse through this chapter on first reading and return to each specific topic whenever needed.

2.1 Rows and columns: operator Cauchy–Schwarz inequality

Let (x_j) be an n-tuple in a C^*-algebra A. Let $r \in M_n(A)$ be the "row matrix" that has (x_j) on its first row and zero everywhere else. In other words, $r_{1j} = x_j$ and $r_{ij} = 0$ for any $i > 1$. Equivalently, in tensor product notation $r = \sum_{j=1}^n e_{1j} \otimes x_j \in M_n \otimes A$. Then

$$\|r\|_{M_n(A)} = \left\| \sum x_j x_j^* \right\|^{1/2}.$$

Indeed, this follows from $\|r\|_{M_n(A)} = \|rr^*\|_{M_n(A)}^{1/2}$. Motivated by this, we will frequently use the notation

$$\|x\|_R = \left\| \sum x_j x_j^* \right\|^{1/2}.$$

Analogously, if $c \in M_n(A)$ is the "column matrix" that has (x_j) on its first column and zero everywhere else, or in tensor product notation if $c = \sum_{j=1}^n e_{j1} \otimes x_j \in M_n \otimes A$, then

$$\|c\|_{M_n(A)} = \left\| \sum x_j^* x_j \right\|^{1/2}.$$

This follows either from $\|c\|_{M_n(A)} = \|c^*c\|_{M_n(A)}^{1/2}$ or from the fact that $r = c^*$ is a row matrix. We will use the notation

$$\|x\|_C = \left\|\sum x_j^* x_j\right\|^{1/2}.$$

These simple remarks lead us to the following useful test to check whether a map is c.b. or to evaluate its c.b. norm.

Proposition 2.1 *Let $E \subset B(H), F \subset B(K)$ be operator spaces. Then any $u \in CB(E, F)$ satisfies the following inequalities for any n and any n-tuple (x_j) in E:*

$$\left\|\sum u(x_j)u(x_j)^*\right\|^{1/2} \leq \|u\|_{cb} \left\|\sum x_j x_j^*\right\|^{1/2},$$

$$\left\|\sum u(x_j)^* u(x_j)\right\|^{1/2} \leq \|u\|_{cb} \left\|\sum x_j^* x_j\right\|^{1/2}.$$

Proof Just observe that u_n takes a row (resp. column) matrix to a row (resp. column) matrix. □

Remark 2.2 Let $R_n = \operatorname{span}[e_{1j}] \subset M_n$ and $C_n = \operatorname{span}[e_{j1}] \subset M_n$. Using the preceding test the reader can check as an easy exercise that the linear mapping $u : R_n \to C_n$ defined by $u(e_{1j}) = e_{j1}$ (which is isometric) satisfies $\|u\|_{cb} \geq \sqrt{n}$, and in fact $\|u\|_{cb} = \sqrt{n}$.

Another classical exercise consists in checking that the c.b. norm of transposition viewed as a linear map on M_n is equal to n (hint: compute $\left\|\sum_{ij} e_{ij} \otimes e_{ij}\right\|$ and $\left\|\sum_{ij} e_{ij} \otimes e_{ji}\right\|$).

It will be convenient to record here several simple consequences of the classical Cauchy–Schwarz inequality.

Lemma 2.3 *Let $(a_i)_{i \in I}$ and $(b_i)_{i \in I}$ be finitely supported families of operators in $B(H)$. We have for any bounded family (x_i) in $B(H)$*

$$\left\|\sum_{i \in I} a_i x_i b_i\right\| \leq \left\|\sum a_i a_i^*\right\|^{1/2} \sup \|x_i\| \left\|\sum b_i^* b_i\right\|^{1/2}. \tag{2.1}$$

In particular

$$\left\|\sum_{i \in I} a_i b_i\right\| \leq \left\|\sum a_i a_i^*\right\|^{1/2} \left\|\sum b_i^* b_i\right\|^{1/2}. \tag{2.2}$$

Proof See (A.12). □

More generally:

Lemma 2.4 *Let $A \subset B(H)$ be a C^*-algebra. Let $(a_i)_{1 \leq i \leq n}$ and $(b_j)_{1 \leq j \leq n}$ be operators in A, and let $x = [x_{ij}] \in M_n(A)$. Then*

$$\left\|\sum_{i,j=1}^n a_i x_{ij} b_j\right\| \leq \left\|\sum a_i a_i^*\right\|^{1/2} \|x\|_{M_n(A)} \left\|\sum b_j^* b_j\right\|^{1/2}. \tag{2.3}$$

Proof

$$\left\| \sum_{i,j=1}^{n} a_i x_{ij} b_j \right\| = \sup_{\xi,\eta \in B_H} \left| \sum \langle \eta, a_i x_{ij} b_j \xi \rangle \right| = \sup_{\xi,\eta \in B_H} \left| \sum \langle a_i^* \eta, x_{ij} b_j \xi \rangle \right|$$

$$\leq \|[x_{ij}]\|_{M_n(B(H))} \sup_{\xi \in B_H} \left(\sum \|b_j \xi\|^2 \right)^{1/2} \sup_{\eta \in B_H} \left(\sum \|a_i^* \eta\|^2 \right)^{1/2}$$

$$\leq \|x\|_{M_n(E)} \left\| \sum b_j^* b_j \right\|^{1/2} \left\| \sum a_i a_i^* \right\|^{1/2}.$$

$\square$

In particular, we quote the following variant for future reference.

Lemma 2.5 *Let* $A \subset B(H)$ *be a* C^**-algebra and let* (a_j), (b_j) *be finite sequences in* A *such that* $\left\| \sum a_j a_j^* \right\| \leq 1$ *and* $\left\| \sum b_j^* b_j \right\| \leq 1$. *For any* $\varphi \in A^*$ *let* φ_j *be defined by* $\varphi_j(x) = \varphi(a_j x b_j)$. *Then*

$$\sum \|\varphi_j\|_{A^*} \leq \|\varphi\|_{A^*}. \tag{2.4}$$

Proof We have $\sum \|\varphi_j\|_{A^*} = \sup\{|\sum \varphi_j(x_j)|\}$ where the sup runs over all $x_j \in B_A$. Then by (2.1) we have

$$\sup \left\{ \left| \sum \varphi_j(x_j) \right| \right\} = \left| \varphi \left(\sum a_j x_j b_j \right) \right| \leq \|\varphi\|_{A^*},$$

whence (2.4). $\square$

2.2 Automatic complete boundedness

Let $u : E \to B$ be a bounded linear map from an operator space E to a C^*-algebra B. We already saw (see Proposition 1.3) that if B is commutative then u is automatically c.b. and $\|u\|_{cb} = \|u\|$. Essentially, this phenomenon reduces to the case $B = \mathbb{C}$. We will now show a very useful generalization of the latter fact involving "cyclicity." We recall that a $*$-homomorphism $\pi : A \to B(H)$ on a C^*-algebra is called cyclic if there is a vector $\xi \in H$ (itself called cyclic) such that $\overline{\pi(A)\xi} = H$. Note that when $\dim(H) = 1$ any nonzero π is trivially cyclic.

Theorem 2.6 ([232]) *Let* $E \subset B(\mathcal{H})$ *be an operator space. Let* $u : E \to B(H)$ *be a bounded linear map. Assume that there are unital* C^**-subalgebras* $A_1, A_2 \subset B(\mathcal{H})$ *and* $*$*-homomorphisms* $\pi_1 : A_1 \to B(H)$ *and* $\pi_2 : A_2 \to B(H)$ *with respect to which* E *is a bimodule and* u *is bimodular, meaning that for all* $a_j \in A_j$ *and all* $x \in E$ *we have*

$$a_1 x a_2 \in E \text{ and } u(a_1 x a_2) = \pi_1(a_1) u(x) \pi_2(a_2).$$

If π_1 *and* π_2 *are cyclic, then* u *is c.b. and* $\|u\|_{cb} = \|u\|$.

Proof We may assume $\|u\| = 1$. Let ξ_j be a cyclic unit vector for π_j. Let $n \geq 1$ and $x = [x_{ij}] \in M_n(E)$ with $\|x\|_{M_n(E)} \leq 1$. To complete the proof it suffices to show that $\|[u(x_{ij})]\|_{M_n(B(H))} \leq 1$, or that $\left| \sum_{ij} \langle k_i, u(x_{ij}) h_j \rangle \right| \leq 1$ for any $(k_i), (h_j) \in H^n$ such that $\sum \|k_i\|^2 < 1$ and $\sum \|h_j\|^2 < 1$. By cyclicity, we may assume that $k_i = \pi_1(a_i)\xi_1$ and $h_j = \pi_2(b_j)\xi_2$ for some $a_i \in A_1, b_j \in A_2$. Assume for the moment that $a = \left(\sum a_i^* a_i \right)^{1/2}$ and $b = \left(\sum b_j^* b_j \right)^{1/2}$ are invertible. We may then factorize a_i, b_j as $a_i = a_i' a$ and $b_j = b_j' b$ where $a_i' = a_i a^{-1}$ and $b_j' = b_j b^{-1}$. Let $\xi_1' = \pi_1(a)\xi_1$ and $\xi_2' = \pi_2(b)\xi_2$. A simple verification shows that $\sum a_i'^* a_i' = 1$, $\sum b_j'^* b_j' = 1$ and also that $\|\xi_1'\|^2 = \sum \|k_i\|^2 < 1$ and $\|\xi_2'\|^2 = \sum \|h_j\|^2 < 1$. Therefore using the modular assumptions we have

$$\left| \sum_{ij} \langle k_i, u(x_{ij}) h_j \rangle \right| = \left| \sum \langle \xi_1', \pi_1(a_i'^*) u(x_{ij}) \pi_2(b_j') \xi_2' \rangle \right|$$

$$= \left| \langle \xi_1', u \left(\sum a_i'^* x_{ij} b_j' \right) \xi_2' \rangle \right| \leq \|u\| \left\| \sum a_i'^* x_{ij} b_j' \right\|$$

and hence by (2.3) $\leq \|u\|$. This proves the result assuming a, b invertible. The general case requires a minor adjustment: fixing $\varepsilon > 0$ we set $a = \left(\varepsilon 1 + \sum a_i^* a_i \right)^{1/2}$ and $b = \left(\varepsilon 1 + \sum b_j^* b_j \right)^{1/2}$. Then a, b are invertible and a simple modification of the preceding argument leads to the same conclusion. $\square$

Corollary 2.7 (Schur multipliers as module maps) *For any Schur multiplier $u : B(\ell_2(I)) \to B(\ell_2(I))$ as in Proposition 1.57 we have $\|u\|_{cb} = \|u\|$.*

Proof We apply Theorem 2.6 taking for A_1 and A_2 the algebra of diagonal operators on $\ell_2(I)$. We may reduce to the case when I is countable, in which case the latter algebra has a cyclic vector. $\square$

2.3 Complex conjugation

Let E be a Banach space. We will denote by $\overline{E}$ the complex conjugate of E, i.e. the vector space E with the same norm but with the conjugate multiplication by a complex scalar. We will denote by $x \to \overline{x}$ the identity map from E to $\overline{E}$. Thus, x and $\overline{x}$ are the same element but we "declare" that

$$\forall \lambda \in \mathbb{C} \quad \lambda \overline{x} = \overline{\bar{\lambda} x}.$$

The space $\overline{E}$ is anti-isometric to E. Perhaps, we should warn the reader that although this notion is very simple, it is easy to get confused by it.

Remark 2.8 For any Hilbert space H the dual H^* is a Hilbert space that can be *canonically* identified with $\overline{H}$ using the (sesquilinear) scalar product. Let (e_j)

(resp. (f_i)) be othonormal bases in a Hilbert space H (resp. K). The spaces H^* (resp. K^*) can be equipped with the biorthogonal orthonormal bases (that can be identified with $(\overline{e_j})$ and $(\overline{f_i})$ in $\overline{H}$ and $\overline{K}$). Let $a \in B(H, K)$. We define $\overline{a} : \overline{H} \to \overline{K}$ by setting $\overline{a}(\overline{h}) = \overline{a(h)}$ for any $h \in H$. It is easy to see that $a \mapsto \overline{a}$ is an isomorphism from $\overline{B(H, K)}$ to $B(\overline{H}, \overline{K})$.

We can associate a bi-infinite matrix $[a_{ij}]$ in the usual way so that $a_{ij} = \langle f_i, u(e_j) \rangle$. Then on one hand $[\overline{a_{ij}}]$ is the matrix associated to $\overline{a}$, and on the other hand, the Banach space sense adjoint operator from K^* to H^* admits the transposed matrix $[a_{ji}]$ as associated matrix. We choose to denote it by $^T a : K^* \to H^*$ to avoid the conflict with the usual Hilbert sense adjoint $a^* : K \to H$. Note that $a \mapsto {}^T a$ is linear while $a \mapsto a^*$ is antilinear. A moment of thought shows that the mapping $^T a : \overline{K^*} \to \overline{H^*}$ can be canonically identified with $a^* : K \to H$.

Incidentally, it is perhaps worthwhile to remind the reader that while $H^* \simeq \overline{H}$ is canonical, the identifications $H^* \simeq H$ and $\overline{H} \simeq H$ depend on the choice of an orthonormal basis.

In sharp contrast, for general Banach spaces $\overline{E}$ is not ($\mathbb{C}$-linearly) isomorphic to E (see [33]).

When A is a C^*-algebra, $\overline{A}$ is also a C^*-algebra for the same product and involution. This allows us to extend the notion of complex conjugate to operator spaces: for any operator space $E \subset A$, we define $\overline{E} \subset \overline{A}$ as the corresponding subspace of $\overline{A}$.

Equivalently, assuming $E \subset B(H)$, by what precedes $\overline{B(H)}$ can be canonically identified with $B(\overline{H})$, thus the embedding

$$\overline{E} \subset \overline{B(H)} = B(\overline{H})$$

allows us to equip $\overline{E}$ with an operator space structure. Moreover, if $E \subset B(H)$ is a C^*- (resp. von Neumann) subalgebra, so is $\overline{E} \subset B(\overline{H})$.

To be more "concrete," if an operator space E is given as a collection of bi-infinite matrices $\{[a_{ij}] \mid a \in E\}$ (representing operators acting on ℓ_2), then the space formed of all operators with complex conjugate matrices $\{[\overline{a_{ij}}] \mid a \in E\}$ is $\mathbb{C}$-linearly (completely) isometrically isomorphic to $\overline{E}$.

Remark 2.9 It is an easy exercise to check that the injectivity of E is equivalent to that of $\overline{E}$.

Remark 2.10 Let $(e_i)_{i \in I}$ be an orthonormal basis of H. While the isomorphism $\overline{B(H)} \simeq B(\overline{H})$ is canonical (which means independent of the choice of orthonormal basis), there is also a noncanonical isomorphism $\overline{B(H)} \simeq B(H)$, associated to the usual (basis dependent) isometric isomorphism $\overline{H} \simeq H$ that takes $\overline{e_i}$ to e_i ($i \in I$). In the case of $M_n = B(\ell_2^n)$, the (completely isometric) isomorphism $\pi : M_n \to \overline{M_n}$ is the linear map defined by $\pi([a_{ij}]) = \overline{[a_{ij}]}$ or in tensor product notation $\pi\left(\sum a_{ij}e_{ij}\right) = \sum \overline{a_{ij}}e_{ij} = \sum a_{ij}\overline{e_{ij}}$. The underlying map from M_n to itself is the complex conjugation.

Let E be an operator space. By the preceding definition of $\overline{E}$ the norm of $M_n(\overline{E}) = M_n \otimes_{\min} \overline{E}$ is characterized by the following identity

$$\forall a_j \in M_n \quad \forall x_j \in E \quad \left\|\sum_1^n a_j \otimes \overline{x_j}\right\|_{M_n \otimes_{\min} \overline{E}} = \left\|\sum_1^n \overline{a_j} \otimes x_j\right\|_{\overline{M_n} \otimes_{\min} E}.$$

In other words, the operator space structure of $\overline{E}$ is precisely defined so that $M_n \otimes_{\min} \overline{E}$ is naturally anti-isometric to $\overline{M_n} \otimes_{\min} E$. For each matrix $[a_{ij}]$ in $M_n(E)$, we simply have

$$\|[\overline{a_{ij}}]\|_{M_n(\overline{E})} = \|[a_{ij}]\|_{M_n(E)}.$$

In sharp contrast to $B(H)$, there are examples (see [60]) of von Neumann algebras E which fail to be C^*-isomorphic to $\overline{E}$.

Let H, K be Hilbert spaces. Assume $H = \ell_2(I_1)$, $K = \ell_2(I_2)$. Any $\xi \in H \otimes_2 K$ can be represented by a kernel $[\xi(i_1, i_2)]$ so that the series $\xi = \sum_{I_1 \times I_2} \xi(i_1, i_2)e_{i_1} \otimes e_{i_2}$ converges in $H \otimes_2 K \simeq \ell_2(I_1 \times I_2)$. To any pair $\xi, \eta \in H \otimes_2 K$ we associate a linear form on $B(H) \otimes_{\min} B(K) \subset B(H \otimes_2 K)$ defined by

$$\varphi_{\xi, \eta}(t) = \langle \xi, t\eta \rangle.$$

Let $h_\xi : \ell_2(I_2) \to \ell_2(I_1)$ be the (Hilbert–Schmidt) linear operator associated to ξ in the usual way, so that if $\xi = e_{i_1} \otimes e_{i_2}$ then (with the usual matrix conventions) $h_\xi = e_{i_1 i_2}$ $((i_1, i_2) \in I_1 \times I_2)$. As usual we associate to an operator $a \in B(H)$ the matrix $[a_{ij}]$ defined by $a_{ij} = \langle e_i, ae_j \rangle$ for $(i, j) \in I_1$ and similarly for b. We denote by ${}^t b \in B(K)$ the operator associated to the transposed matrix $[b_{ji}]$.

We observe that (note that $h_\xi^* a h_\eta {}^t b \in S_1(K, K)$)

$$\forall a \in B(H), \ b \in B(K), \quad \varphi_{\xi, \eta}(a \otimes b) = \operatorname{tr}(h_\xi^* a h_\eta {}^t b). \qquad (2.5)$$

Indeed, it suffices to check this when $\xi = e_{i_1} \otimes e_{i_2}$ and $\eta = e_{j_1} \otimes e_{j_2}$ and then both terms in (2.5) are easily seen to be equal to $a_{i_1, j_1} b_{i_2, j_2}$.

We now wish to apply the same formula to $B(H) \otimes \overline{B(K)}$. In that case, to any $\xi \in H \otimes_2 \overline{K}$ we associate the kernel $[\xi(i_1, i_2)]$ defined by $\xi = \sum_{I_1 \times I_2} \xi(i_1, i_2)e_{i_1} \otimes \overline{e_{i_2}}$ so that $h_\xi : K \to H$ remains the same. The formula becomes

$$\forall a \in B(H), \ b \in B(K), \quad \varphi_{\xi, \eta}(a \otimes \overline{b}) = \operatorname{tr}(h_\xi^* a h_\eta b^*). \qquad (2.6)$$

Indeed, ${}^t \overline{b}$ can be identified with the adjoint operator $b^* \in B(H)$, with associated matrix $[\overline{b_{ji}}]$.

Let $t \in B(H) \otimes \overline{B(K)}$, be a finite sum $t = \sum a_j \otimes \overline{b_j}$. Then $\|t\|_{B(H) \otimes_{\min} \overline{B(K)}} = \sup\{|\varphi_{\xi,\eta}(t)|\}$ where the sup runs over ξ, η in the unit ball of $H \otimes_2 \overline{K}$. This gives us

$$\left\| \sum a_j \otimes \overline{b_j} \right\|_{B(H) \otimes_{\min} \overline{B(K)}} = \sup_{y, x \in B_2} \left| \sum_j \operatorname{tr}(y^* a_j x b_j^*) \right|, \qquad (2.7)$$

where B_2 denotes the unit ball in $S_2(K, H)$. Let us denote by $\| \ \|_2$ the norm in $S_2(K, H)$ or $S_2(H, K)$. Then (2.7) can be rewritten as

$$\left\| \sum a_j \otimes \overline{b_j} \right\|_{B(H) \otimes_{\min} \overline{B(K)}} = \sup_{x \in B_2} \left\| \sum a_j x b_j^* \right\|_2 = \sup_{y \in B_2} \left\| \sum b_j^* y^* a_j \right\|_2.$$
$$(2.8)$$

We need to record all variants of this for future reference:

Proposition 2.11 *Consider finite sequences (a_j) in $B(H)$ and (b_j) in $B(K)$. Then we have*

$$\left\| \sum \overline{a_j} \otimes b_j \right\|_{\overline{B(H)} \otimes_{\min} B(K)} = \sup_{x \in B_2} \left\| \sum b_j x^* a_j^* \right\|_2 = \sup_{y \in B_2} \left\| \sum a_j^* y b_j \right\|_2$$
$$(2.9)$$

$$= \sup_{x \in B_2} \left\| \sum a_j x b_j^* \right\|_2 = \sup_{y \in B_2} \left\| \sum b_j^* y^* a_j \right\|_2.$$
$$(2.10)$$

Proof Since $\left\| \sum \overline{a_j} \otimes b_j \right\|_{\overline{B(H)} \otimes_{\min} B(K)} = \left\| \sum b_j \otimes \overline{a_j} \right\|_{B(K) \otimes_{\min} \overline{B(H)}}$ by Remark 1.9, it is easy to derive (2.9) from (2.8), exchanging the roles of $(a_j), B(H)$ and $(b_j), B(K)$. Then (2.10) follows using $\|T\|_2 = \|T^*\|_2$ for all T in $S_2(H, K)$ or $S_2(K, H)$. $\qquad \square$

When $H = K$ and $a_j = b_j$ we can make these formulae more precise:

Proposition 2.12 *Consider a finite sequence (a_j) in $B(H)$. Then we have*

$$\left\| \sum \overline{a_j} \otimes a_j \right\|_{\overline{B(H)} \otimes_{\min} B(H)} = \sup \left\{ \left| \sum_j \operatorname{tr}(x a_j^* y^* b_j) \right| \ \middle| \ x, y \in B_{S_2(H)}, x, y \geq 0 \right\}.$$
$$(2.11)$$

Moreover, if $\sum \overline{a_j} \otimes a_j$ is self-adjoint, then the preceding supremum is unchanged if we restrict it to $x = y \geq 0$ in $B_{S_2(H)}$.

Proof Every x in S_2 can be written as $x_1 - x_2 + i(x_3 - x_4)$ with $x_1, \ldots, x_4$ all ≥ 0 such that $\|x_1\|_2^2 + \|x_2\|_2^2 + \|x_3\|_2^2 + \|x_4\|_2^2 = \|x\|_2^2$. From this fact (applied to both x and y) it is easy to deduce (2.11) from (2.7). Moreover, when $\sum \overline{a_j} \otimes a_j$ is self-adjoint, the sesquilinear form $(y, x) \mapsto \sum_j \operatorname{tr}(x a_j^* y^* a_j)$ is symmetric, and hence (by polarization) the supremum in (2.11) remains the same if we restrict it to $x = y$. $\qquad \square$

Remark 2.13 (Opposite C^*-Algebra) Let A be a C^*-algebra. We define the opposite C^*-algebra, that we denote by A^{op}, as the same C^*-algebra but with the reverse product, so that the product of a, b in A^{op} is defined as ba. The involution remains unchanged. It turns out that this is nothing but another way to consider $\overline{A}$. Indeed, as is easy to check, the mapping $\overline{a} \mapsto a^*$ is a ($\mathbb{C}$-linear) isomorphism from $\overline{A}$ to A^{op}. Therefore for any C^*-algebra B and any $a_j \in A$, $b_j \in B$ ($1 \le j \le n$) we have

$$\left\| \sum \overline{a_j} \otimes b_j \right\|_{\overline{A} \otimes_{\min} B} = \left\| \sum a_j^* \otimes b_j \right\|_{A^{op} \otimes_{\min} B}. \qquad (2.12)$$

If H, K are Hilbert spaces, we have clearly $\overline{H \otimes_2 K} \simeq \overline{H} \otimes_2 \overline{K}$, and hence $\overline{B(H)} \otimes_{\min} \overline{B(K)} \simeq \overline{B(H) \otimes_{\min} B(K)}$. Therefore:

$$\left\| \sum \overline{a_j} \otimes b_j \right\|_{\overline{A} \otimes_{\min} B} = \left\| \sum \overline{a_j \otimes b_j} \right\|_{\overline{A \otimes_{\min} B}} = \left\| \sum a_j \otimes \overline{b_j} \right\|_{A \otimes_{\min} \overline{B}}.$$
$$(2.13)$$

Remark 2.14 (Conjugate of group representation) We will also use complex conjugation for group representations. Given a group G and a unitary group representation $\pi : G \to B(H)$, let $\overline{\pi}(t) = \overline{\pi(t)}$ for any $t \in G$. This defines a unitary representation $\overline{\pi} : G \to B(\overline{H})$.

2.4 Operator space dual

The existence of the operator space dual is a consequence of Ruan's fundamental theorem, describing for any vector space E, the sequences of norms on the spaces $\{M_n(E) \mid n \ge 1\}$ that come from an embedding of E in $B(H)$, in other words that are associated to an operator space structure on E. But actually we prefer to give a direct proof avoiding the use of Ruan's theorem.

Theorem 2.15 *Let E be an operator space. There is a Hilbert space $\mathcal{H}$ and an isometric embedding*

$$J : E^* \to B(\mathcal{H})$$

such that, for all $n \ge 1$ and all $\xi = [\xi_{ij}] \in M_n(E^*)$, we have

$$\| [J(\xi_{ij})] \|_{M_n(B(\mathcal{H}))} = \| u_\xi \|_{cb}$$

where $u_\xi : E \to M_n$ is the linear map naturally associated to ξ.

Proof Consider the set $\mathcal{D}$ that is the disjoint union of the unit balls in $M_n(E)$, i.e. we have

$$\mathcal{D} = \bigcup_{n \ge 1} \overset{\cdot}{B}_{M_n(E)}.$$

Then for any $t \in \mathcal{D}$, we have $t \in B_{M_n(E)}$ for some $n = n(t)$ and we denote by

$$v_t : E^* \to M_{n(t)}$$

the linear map associated to t. We then define

$$\forall \xi \in E^* \quad J(\xi) = \bigoplus_{t \in \mathcal{D}} v_t(\xi) \in \left(\oplus \sum M_{n(t)} \right)_\infty.$$

Since, by (1.3), $\|\xi\|_{cb} = \|\xi\|$ and $v_t(\xi) = (Id_{M_{n(t)}} \otimes \xi)(t) \in M_{n(t)}$, we have

$$\|v_t(\xi)\|_{M_{n(t)}} \le \|\xi\|$$

and hence $\|J(\xi)\| \le \|\xi\|$, but using only $t \in B_{M_1(E)} = B_E$ we find $\|J(\xi)\| \ge \|\xi\|$. Thus J is isometric.

Consider now $\xi = [\xi_{ij}] \in M_n(E^*)$ with associated linear map $u_\xi : E \to M_n$. We have then by (1.2)

$$\|u_\xi\|_{cb} = \sup_{t \in \mathcal{D}} \|(Id_{M_{n(t)}} \otimes u_\xi)(t)\|_{M_{n(t)}(M_n)}$$

but since (modulo permutation of factors)

$$M_{n(t)}(M_n) \ni (Id_{M_{n(t)}} \otimes u_\xi)(t) \simeq [v_t(\xi_{ij})] \in M_n(M_{n(t)})$$

and since such a permutation clearly preserves the norms we have

$$\|u_\xi\|_{cb} = \sup_{t \in \mathcal{D}} \|[v_t(\xi_{ij})]\|_{M_n(M_{n(t)})} = \|[J(\xi_{ij})]\|_{M_n(B(\mathcal{H}))},$$

where at the last step we used (1.13). $\qquad \square$

The operator space dual E^* is defined as the one obtained by equipping E^* with the sequence of the norms on $M_n(E^*)$ derived from the embedding J in the preceding statement. Thus we have isometrically

$$M_n(E^*) = M_n \otimes_{\min} E^* = CB(E, M_n). \tag{2.14}$$

More generally, we have for any operator space G an isometric inclusion

$$G \otimes_{\min} E^* \subset CB(E, G) \text{ (or equivalently } E^* \otimes_{\min} G \subset CB(E, G)). \tag{2.15}$$

Indeed, this is easy to deduce from the case $G = M_n$ using (1.9) or (1.11).

The norm in $M_n(E^*)$ is described in a more suggestive way by the formula

$$\forall \xi \in M_n(E^*) \| \xi \|_{M_n(E^*)} = \sup\{\|\xi \cdot x\|_{M_n(M_m)} \mid x \in B_{M_m(E)}, m \ge 1\} \tag{2.16}$$

where

$$\xi \cdot x = \sum_{1 \le ij \le n, 1 \le k,l \le m} e_{ij} \otimes e_{kl} \, \xi_{ij}(x_{kl}).$$

By (1.18) the equality (2.16) still holds if we restrict the sup to $m = n$.

The resulting formula then appears as an extension of $\|\xi\|_{E^*} = \sup\{|\xi(x)| \mid x \in B_E\}$ (case $n = 1$).

We can reverse the roles of E and E^*, as follows. We will show that for any G we also have isometric embeddings

$$E \otimes_{\min} G \subset CB(E^*, G) \text{ (or equivalently } G \otimes_{\min} E \subset CB(E^*, G)). \quad (2.17)$$

Let $t \in E \otimes G$ and let $\tilde{t} : E^* \to G$ be the associated linear map. For any $v \in CB(E, M_n)$ let $t_v \in M_n(E^*)$ be the associated tensor. By (2.14), the correspondence $v \mapsto t_v$ is a bijection from $B_{CB(E, M_n)}$ to $B_{M_n(E^*)}$. Observe that $(v \otimes Id_G)(t) = (Id_{M_n} \otimes \tilde{t})(t_v)$. By (1.9) (with G and E interchanged) we have $\|t\|_{\min} = \sup\{\|(v \otimes Id_G)(t)\|_{M_n(G)} \mid n \geq 1, v \in B_{CB(E, M_n)}\}$, and hence

$$\|t\|_{\min} = \sup\{\|(Id_{M_n} \otimes \tilde{t})(t_v)\|_{M_n(G)} \mid n \geq 1, t_v \in B_{M_n(E^*)}\} = \|\tilde{t}\|_{cb},$$

which proves that (2.17) is isometric.

Taking $G = M_n$ in (2.17) and recalling (2.14) we find that the inclusion $M_n(E) \subset M_n((E^*)^*)$ is isometric. This shows that the inclusion $E \subset (E^*)^* = E^{**}$ is completely isometric. In particular, if E is reflexive as a Banach space, then $(E^*)^* = E$ completely isometrically.

From these remarks, the following statement emerges naturally:

Lemma 2.16 *Let E, F be operator spaces. For any $u \in CB(E, F)$ the adjoint $u^* : F^* \to E^*$ is c.b. and*

$$\|u^*\|_{cb} = \|u\|_{cb}.$$

Proof Composition by u on the left gives us a contraction from $CB(F, M_n)$ to $CB(E, M_n)$ which can be restated as $\|(u^*)_n : M_n(F^*) \to M_n(E^*)\| \leq \|u\|_{cb}$. This implies $\|u^*\|_{cb} \leq \|u\|_{cb}$. Iterating we find $\|(u^*)^*\|_{cb} \leq \|u^*\|_{cb}$, and by the preceding remark $\|u\|_{cb} \leq \|(u^*)^*\|_{cb}$. $\square$

2.5 Bi-infinite matrices with operator entries

In the completely bounded context, it is natural to wonder whether one can replace $M_n(E)$ by the space $M_\infty(E)$ of bi-infinite matrices with entries in E. Let us first clarify what we mean by $M_\infty(E)$. Let $E \subset B(H)$ be an operator space. Recall the notation $\ell_2(H) = H \oplus H \oplus \cdots$. We may clearly represent an operator $a \in B(\ell_2(H))$ by a matrix $[a_{ij}]$ with entries in $B(H)$. We denote by $M_\infty(B(H))$ the space of such matrices equipped with the norm transplanted from $B(\ell_2(H))$ (so that $M_\infty(B(H)) \simeq B(\ell_2(H))$ isometrically).

Then we define $M_\infty(E) \subset M_\infty(B(H))$ as the subspace formed of those $a \in M_\infty(B(H))$ such that $a_{ij} \in E$ for all i, j.

The norm of $a \in M_\infty(E)$ is easy to compute in terms of the truncated matrices: we have (assuming the indices i, j run over $\{1, 2, \ldots\}$)

$$\|a\|_{M_\infty(E)} = \sup_n \|[a_{ij}]_{1 \leq i, j \leq n}\|_{M_n(E)}.$$

Moreover, if we are given a priori the entries $a_{ij} \in E$, then there is $a \in M_\infty(E)$ admitting these as its entries if and only if

$$\sup_n \|[a_{ij}]_{1 \leq i, j \leq n}\|_{M_n(E)} < \infty.$$

Therefore it is evident that for any operator spaces E, F and any $u \in CB(E, F)$ we have

$$\|u\|_{cb} = \|u_\infty : M_\infty(E) \to M_\infty(F)\|$$

where $u_\infty : M_\infty(E) \to M_\infty(F)$ is defined by

$$u_\infty([a_{ij}]) = [u(a_{ij})].$$

While we used ℓ_2 for simplicity, we could just as well use an arbitrary Hilbert space $\mathcal{H}$ instead: We define $M_\mathcal{H}(B(H)) = B(\mathcal{H} \otimes_2 H)$, and

$$M_\mathcal{H}(E) = \{x \in M_\mathcal{H}(B(H)) \mid x(\xi, \eta) \in E \ \forall \xi, \eta \in \mathcal{H}\}$$

where this time, for any $\xi, \eta \in \mathcal{H}$, we have denoted by $x(\xi, \eta)$ the element of $B(H)$ obtained under the natural action of the functional $f_{\xi, \eta} \in B(\mathcal{H})^*$ defined by $f_{\xi, \eta}(y) = \langle \xi, y\eta \rangle$. Equivalently, for any $\eta \in \mathcal{H}$, we let $v_\eta : H \to \mathcal{H} \otimes_2 H$ be defined by $v_\eta(h) = \eta \otimes h$ for any $h \in H$, and we set

$$x(\xi, \eta) = v_\xi^* x v_\eta.$$

The collection of all the finite-dimensional subspaces of $\mathcal{H}$ directed by inclusion gives us a simple formula for the norm of $a \in M_\mathcal{H}(E)$. We have (the easy verification is left to the reader)

$$\|a\|_{M_\mathcal{H}(E)} = \sup \|(v^* \otimes I)a(v \otimes I)\|_{M_n(E)} \tag{2.18}$$

where the supremum runs over all n and all isometric embeddings $v : \ell_2^n \to \mathcal{H}$.

Again if $\mathcal{H}$ is infinite dimensional, we have

$$\|u\|_{cb} = \|u_\mathcal{H} : M_\mathcal{H}(E) \to M_\mathcal{H}(F)\| \tag{2.19}$$

where $u_\mathcal{H} : M_\mathcal{H}(E) \to M_\mathcal{H}(F)$ is defined as the only mapping such that $[u_\mathcal{H}(a)](\xi, \eta) = u(a(\xi, \eta))$ for any $\xi, \eta \in \mathcal{H}$. Moreover, if we are given a priori a sesquilinear map $(\xi, \eta) \mapsto a(\xi, \eta)$ (linear in η, antilinear in ξ) from $\mathcal{H} \times \mathcal{H}$ to E, then this will come from an element of $M_\mathcal{H}(E)$ if and only if the matrices $[a_{ij}^v]$ defined by $[a_{ij}^v] = a(ve_i, ve_j)$ satisfy

$$\sup \|[a_{ij}^v]\|_{M_n(E)} < \infty \tag{2.20}$$

with the sup as before.

Using the definition of the dual operator space E^* given in §2.4, we find that we have an isometric identity

$$CB(E, B(\mathcal{H})) = M_{\mathcal{H}}(E^*). \qquad (2.21)$$

Indeed, if $\dim(\mathcal{H}) < \infty$ this is the very definition of $M_{\mathcal{H}}(E^*)$. Now, if $\mathcal{H}$ is arbitrary, for any $u \in CB(E, B(\mathcal{H}))$ we have $\|u\|_{cb} = \sup\{\|x \mapsto v^*u(x)v\|_{cb}\}$ where the sup is as before, and moreover if $u : E \to B(\mathcal{H})$ is any linear map then $u \in CB(E, B(\mathcal{H}))$ iff $\sup\{\|x \mapsto v^*u(x)v\|_{cb}\} < \infty$ where the sup is as before. Comparing this with (2.18) and (2.20), we obtain (2.21).

Note that the algebraic tensor product $B(\mathcal{H}) \otimes B(H)$ is weak*-dense in $B(\mathcal{H} \otimes_2 H)$. When the subspaces $F \subset B(\mathcal{H})$ and $E \subset B(H)$ are weak* closed in $B(H)$, we denote by $F \bar{\otimes} E \subset B(\mathcal{H} \otimes_2 H)$ the weak* closure of $E \otimes F$. With this notation we have $B(\mathcal{H} \otimes_2 H) = B(\mathcal{H}) \bar{\otimes} B(H)$. We observe for later use:

Proposition 2.17 *If $E \subset B(H)$ is weak* closed in $B(H)$, then for any $\mathcal{H}$ we have*

$$M_{\mathcal{H}}(E) = B(\mathcal{H}) \bar{\otimes} E. \qquad (2.22)$$

Proof It is easy to check that if E is weak* closed in $B(H)$ then $M_{\mathcal{H}}(E)$ is weak* closed in $B(\mathcal{H} \otimes_2 H)$, whence the inclusion $B(\mathcal{H}) \bar{\otimes} E \subset M_{\mathcal{H}}(E)$. To show the reverse assume $\mathcal{H} = \ell_2$ for notational simplicity. Then for any $a \in M_{\infty}(E)$ it is easy to see that the truncated matrices $[a_{ij}]_{1 \le i, j \le n}$ (which are obviously in $B(\mathcal{H}) \otimes E$) tend weak* to a in $M_{\infty}(B(H)) = B(\mathcal{H} \otimes_2 H)$ when $n \to \infty$. This proves the reverse inclusion. The case of a general $\mathcal{H}$ is similar. We skip the details. $\qquad \square$

Since we use the spaces $M_n(E)$ throughout these notes it is worthwhile to observe that the representations of M_n are very special:

Proposition 2.18 *Let $\pi : M_n \to A \subset B(H)$ be an isometric unital $*$-homomorphism embedding M_n in a C^*-algebra A. Let $A_1 = A \cap \pi(M_n)'$ be the relative commutant of $\pi(M_n)$ in A. Then A is isomorphic to $M_n(A_1)$, and with this isomorphism π can be identified with $x \mapsto x \otimes 1$.*

Remark 2.19 The case $A = M_N$ shows that the existence of a *unital* embedding $M_n \subset M_N$ requires that n divides N, while for a nonunital one $n \le N$ obviously suffices (we may just add zero entries).

We leave the proof of Proposition 2.18 to the reader, but the proof is an easy modification of the following one, which is a von Neumann variant of it, to be used toward the end of these notes.

Proposition 2.20 *Let $\mathcal{H}$ be any Hilbert space. Let $\pi : B(\mathcal{H}) \to \mathcal{M} \subset B(H)$ be a normal (isometric and unital) $*$-homomorphism embedding $B(\mathcal{H})$ in a von Neumann algebra $\mathcal{M}$. Let $\mathcal{M}_1 = \mathcal{M} \cap \pi(B(\mathcal{H}))'$. Then $\mathcal{M}$ is isomorphic as a von Neumann algebra to $B(\mathcal{H})\bar{\otimes}\mathcal{M}_1$, and modulo this isomorphism π can be identified with $B(\mathcal{H}) \ni b \mapsto b \otimes 1$.*

Proof We may assume $\mathcal{H} = \ell_2(I)$ for some set I. For notational simplicity, we assume $I = \{1, 2, \ldots\}$. Let $E_{ij} = \pi(e_{ij}) \in \mathcal{M}$. We will use the standard identities

$$E_{ik}E_{k'j} = E_{ij} \text{ if } k = k' \text{ and } E_{ik}E_{k'j} = 0 \text{ otherwise.} \qquad (2.23)$$

Note that $\mathcal{M}_1 = \mathcal{M} \cap \{E_{ij} \mid i, j \geq 1\}'$. For any $x \in B(\mathcal{H})$ let $x(n) = P_n x P_n$ where P_n is the canonical projection onto the span of the first n basis vectors in $\mathcal{H}$. Let $Q_n = \pi(P_n) = \sum_1^n E_{ii}$. Then $x(n) \to x$, in particular $P_n \to I$ and hence $\pi(x(n)) \to \pi(x)$ and $Q_n \to I$; these limits are meant when $n \to \infty$ and with respect to the weak* topology, as all the limits in the rest of the proof. For any $a \in \mathcal{M}$ we have $Q_n a Q_n \to a$. Let $a_{ij} = \lim_n \sum_{k=1}^n E_{ki} a E_{jk}$. We have $\sum_{1 \leq i, j \leq n} E_{ij}a_{ij} = \sum_{1 \leq i, j \leq n} E_{ii} a E_{jj} = Q_n a Q_n$. Using (2.23) one checks easily that $a_{ij} \in \mathcal{M}_1$. We define a mapping $\sigma : B(\mathcal{H}) \bar{\otimes} \mathcal{M}_1 \to \mathcal{M}$ by setting

$$\sigma\left(\lim \sum_{i, j=1}^n e_{ij} \otimes y_{ij}\right) = \lim \sum_{i, j=1}^n E_{ij}y_{ij}.$$

It is easy to check with (2.23) that σ is a $*$-homomorphism, and since $\sigma([a_{ij}]) = a$ for any $a \in \mathcal{M}$, σ is onto $\mathcal{M}$. Moreover, we have $\|a\| = \lim \|Q_n a Q_n\| = \lim \left\| \sum_{1 \leq i, j \leq n} E_{ij}a_{ij} \right\| = \|\sigma([a_{ij}])\|$. Therefore, σ is an isometric $*$-isomorphism from $B(\mathcal{H}) \bar{\otimes} \mathcal{M}_1$ to $\mathcal{M}$. Since the latter are both von Neumann algebras, we know (see Remark A.38) that σ and its inverse are automatically normal (actually it is easy to see this directly in the present situation). Lastly, we have $\sigma(b \otimes 1) = \lim \sum_{i, j=1}^n E_{ij}b_{ij} = \lim \pi\left(\sum_{i, j=1}^n e_{ij}b_{ij} \right)$, and hence (since π is normal) $\sigma(b \otimes 1) = \pi(b)$ for any $b \in B(\mathcal{H})$. $\qquad \square$

For more information on all questions involving operator spaces or algebras and dual topologies, we refer the reader to [27].

2.6 Free products of C^*-algebras

We start by recalling the definition of the free product of algebras: let $(A_i)_{i \in I}$ be a family of algebras (resp. unital algebras). We will denote by $\dot{\mathcal{A}}$ (resp. $\mathcal{A}$) their free product in the category of algebras (unital algebras). This object is characterized as the unique algebra (resp. unital algebra) A containing each A_i as a (resp. unital) subalgebra and such that if we are given another object B

and morphisms $\varphi_i : A_i \rightarrow B$ $(i \in I)$, there is a unique morphism $\varphi : A \rightarrow B$ such that $\varphi_{|A_i} = \varphi_i$ for all i. Moreover A is generated by the union of the A_i's viewed as subalgebras (resp. unital subalgebras) of A. Similarly, if all the A_i's are $*$-algebras, we can equip A with a $*$-algebra structure (meaning with an involution) so that for any $*$-algebra B and $*$-morphisms $\varphi_i : A_i \rightarrow B$, there is a unique $*$-morphism such that $\varphi_{|A_i} = \varphi_i$ for all i, and the A_i's can be viewed as $*$-subalgebras of A.

One can think of a typical element a of $\dot{A}$ (resp. $\mathcal{A}$) as a sum of products of elements of the union of the A_i's. More precisely, $a = \sum a_k$ with each a_k of the form $a_k = a_k(i_1) \ldots a_k(i_{\ell(k)})$ with $a_k(i_j) \in A_{i_j}$ $(1 \leq j \leq \ell(k))$; when two consecutive terms are in the same algebra, say when $i_j = i_{j+1}$ we may replace $a_k(i_j)a_k(i_{j+1})$ by the single term $(a_k(i_j)a_k(i_{j+1})) \in A_{i_j}$, thus reducing the product. When all such reductions have been done, we obtain a product of the same form but for which $i_1 \neq i_2 \neq \cdots \neq i_{\ell(k)}$, the product is then called reduced. For any scalar $\lambda \in \mathbb{C}$ we have $\lambda a = \sum (\lambda a_k(i_1)) \ldots a_k(i_{\ell(k)})$. The product in $\dot{A}$ (resp. $\mathcal{A}$) is defined by $(\sum_k a_k)(\sum_m b_m) = \sum_{k,m} a_k b_m$ and by concatenation of the product terms $a_k b_m$ i.e.

$$[a_k(i_1) \ldots a_k(i_{\ell(k)})][b_m(j_1) \ldots b_m(j_{\ell(m)})]$$

$$= a_k(i_1) \ldots a_k(i_{\ell(k)})b_m(j_1) \ldots b_m(j_{\ell(m)}).$$

If we now assume that $(A_i)_{i \in I}$ is a family of C^*-algebras (resp. unital ones), then we can equip $\dot{A}$ (resp. $\mathcal{A}$) with a (resp. unital) C^*-algebra structure in the following way.

Let $\mathcal{F}$ be either $\mathcal{A}$ or $\dot{A}$. Let C be the collection of all $*$-homomorphisms $\pi : \mathcal{F} \rightarrow B(H_\pi)$ (automatically such that $\|\pi_{|A_i}\| \leq 1$ for all i in I). Let $j : \mathcal{F} \rightarrow \bigoplus_{\pi \in C} B(H_\pi)$ be the embedding defined by $j(x) = \bigoplus_{\pi \in C} \pi(x)$ for all x in $\mathcal{F}$. Clearly j is a $*$-homomorphism and (by standard algebraic facts) it is injective. This allows us to equip $\mathcal{F}$ with the noncomplete C^*-algebra structure associated to j and, after completion with respect to the norm

$$\forall x \in \mathcal{F} \quad \|x\| = \sup_{\pi \in C} \|\pi(x)\| \tag{2.24}$$

we obtain a C^*-algebra (resp. a unital one), admitting $\mathcal{F}$ as a dense subalgebra. We will denote by $\dot{*}_{i \in I} A_i$ (resp. $*_{i \in I} A_i$) the resulting (resp. unital) C^*-algebra, which we call the free product of the family of (resp. unital) C^*-algebras $(A_i)_{i \in I}$. See [15] and [254] for basic facts on free products.

Let A be any of these two free products. Let $\sigma_i : A_i \rightarrow A$ be the natural embedding. Then for any family of morphisms $\pi_i : A_i \rightarrow B(H)$ there is a unique morphism $\pi : A \rightarrow B(H)$ such that $\pi \sigma_i = \pi_i$ for all i.

Remark 2.21 The constructions of $*_{i \in I} A_i$ or $\dot{*}_{i \in I} A_i$ that we just sketched deliberately avoids a few points. For instance it is true but not obvious that the canonical morphism from $\mathcal{A}$ to $*_{i \in I} A_i$ is injective. This point is addressed in [23] and [171, p. 37]. See also [254, p. 4].

Using this it is easy to check that the norm we just defined in (2.24) on either $\mathcal{A}$ or $\dot{\mathcal{A}}$ is the *maximal C*-norm*.

Let A_1, A_2 be unital C*-algebras with subspaces $E_j \subset A_j$ ($j = 1, 2$). By suitably restricting the product map, we have a natural embedding $E_1 \otimes E_2 \subset A_1 * A_2$ with range equal to the linear span of the set $E_1 E_2 \subset A_1 * A_2$. The induced operator space structure on $E_1 \otimes E_2$ is that of the so-called Haagerup tensor product $E_1 \otimes_h E_2$, which, despite its importance, we chose not to present in detail in the present volume. The reader will find a full treatment in our previous book [208, ch. 5] (or in [80, 196]). In the present one, the main result we need is stated as the next lemma.

Let A_1, A_2 be unital C*-algebras with subspaces $E_j \subset A_j$ ($j = 1, 2$). Let $E \subset A_1 * A_2$ be the linear span of the set $E_1 E_2 \subset A_1 * A_2$. We wish to describe the operator space structure on E induced by this embedding, i.e. the norm in $M_n(E)$ for all n. Actually we can do this directly for $n = \infty$:

Lemma 2.22 *Assume* $\dim(H) = \infty$. *Let* $t \in B(H) \otimes E$. *Then* $\|t\|_{\min} \le 1$ *if and only if there are* $t_j \in B(H) \otimes E_j$ *with* $\|t_j\|_{\min} \le 1$ *such that*

$$t = t_1 \odot t_2, \tag{2.25}$$

where the sign $\odot$ *means the bilinear mapping defined from*

$$(B(H) \otimes E_1) \times (B(H) \otimes E_2) \to B(H) \otimes E \text{ by } (b_1 \otimes e_1) \odot (b_2 \otimes e_2)$$
$$= (b_1 b_2 \otimes e_1 e_2) \ (b_j \in B(H), e_j \in E_j).$$

Equivalently, $\odot$ *is the restriction of the usual product on* $B(H) \otimes (A_1 * A_2)$.

About the proof Since t_1, t_2 are tensors of finite rank, the statement reduces immediately to the case when E_1, E_2 are finite dimensional. Using mainly results from [50] it was proved in [204] that E can be identified completely isometrically with the Haagerup tensor product $E_1 \otimes_h E_2$. Then the result follows from [208, cor. 5.9, p. 95]. □

Actually Lemma 2.22 is valid for any number n of factors $E_j \subset A_j$ ($1 \le j \le n$).

More generally, one can describe very efficiently the (operator space) structure of a free product $*_{i \in I} A_i$ of C*-algebras, as follows.

Theorem 2.23 (Blecher–Paulsen Factorization [28]) *Let* $(A_i)_{i \in I}$ *be a family of unital C*-algebras, let* $A = *_{i \in I} A_i$ *be the unital free product, and let*

$\mathcal{A} \subset A$ be the algebraic free product. Let $n \geq 1$ and consider $x \in M_n(\mathcal{A})$. Then $\|x\| < 1$ if and only if there are m, $i_1 \neq i_2 \neq \cdots \neq i_m$ in I and rectangular matrices $x_k \in M_{p_k \times q_k}(A_{i_k})$ with $\|x_k\| < 1$ such that $p_1 = q_m = n$ and

$$x = x_1 \ldots x_m. \tag{2.26}$$

About the proof This beautiful result is a consequence of the Blecher–Ruan–Sinclair (BRS in short) characterization of operator algebras from [29]. One equips $M_n(\mathcal{A})$ with the norm

$$\|x\|_n = \inf \prod \|x_k\| \tag{2.27}$$

where the infimum runs over all factorizations $x = x_1 \ldots x_m$ of the kind just described. By the BRS characterization there is for some H a unital homomorphism $\pi : \mathcal{A} \to B(H)$ such that

$$\|x\|_n = \|(Id_{M_n} \otimes \pi)(x)\|_{M_n(B(H))}. \tag{2.28}$$

Let $\pi_i = \pi_{|A_i}$. Clearly $\|x\|_n \leq \|x\|_{M_n(A)}$ whenever $x \in M_n(A_i)$. Therefore by (2.28) $\|\pi_i\|_{cb} \leq 1$. Since $\pi_i(1) = 1$, π_i is c.p. by Theorem 1.35 and a fortiori self-adjoint. Consequently π_i must be a $*$-homomorphism for any i. This shows that $\pi = *\pi_i$ on $\mathcal{A}$. By definition of $A = *_{i \in I} A_i$, π extends to a (contractive) $*$-homomorphism on A. But now by (2.27) we clearly have $\|x\|_{M_n(A)} \leq \|x\|_n$ for any n, in particular $\|x\|_A \leq \|\pi(x)\|$ for any $x \in A$. By the maximality of $\|x\|_A$ (as reflected in (2.24)) equality must hold, so that π is the restriction of an isometric, and hence completely isometric, $*$-homomorphism from A to $B(H)$. Thus we conclude that $\|x\|_{M_n(A)} = \|x\|_n$ for any $x \in \mathcal{A}$. $\qquad\square$

The following very useful result is due to Boca [30].

Theorem 2.24 (Boca's theorem) *Let* $(A_i)_{i \in I}$ *be a family of unital C^*-algebras and let* $A = *_{i \in I} A_i$ *be the unital free product. For each* $i \in I$ *let* f_i *be a state on* A_i. *Let* $u_i : A_i \to B$ *be unital c.p. maps with values in another unital C^*-algebra B. There is a unital c.p. map $u : A \to B$ such that for any m, any $i_1 \neq i_2 \neq \cdots \neq i_m$ in I and any $a_k \in A_{i_k}$ such that $f_{i_k}(a_k) = 0$ for all k we have*

$$u(a_1 a_2 \ldots a_m) = u_{i_1}(a_1) u_{i_2}(a_2) \ldots u_{i_m}(a_m). \tag{2.29}$$

See [68] for a recent proof of Boca's theorem using the Stinespring dilation Theorem 1.22.

2.7 Universal C^*-algebra of an operator space

Let E be an operator space. The universal C^*-algebra generated by E will be denoted by $C^*\langle E \rangle$. Its definition will be given after the next statement.

Theorem 2.25 *Let E be an operator space. There is a (resp. unital) C^*-algebra A and a completely isometric embedding $j : E \to A$ with the following properties:*

(i) *For any (resp. unital) C^*-algebra B and any completely contractive map $u : E \to B$ there is a (resp. unital) representation $\pi : A \to B$ extending u, i.e. such that $\pi j = u$.*
(ii) *The (resp. unital) algebra generated by $j(E)$ is dense in A.*

Moreover, (ii) ensures that the (resp. unital) representation π in (i) is unique.

Proof The proof is immediate. Let I be the "collection" of all u as in (i) with range generating a (resp. unital) C^*-algebra denoted by B_u, so that $u : E \to B_u$ satisfies $\|u\|_{cb} \le 1$. Let

$$\mathcal{B}_E = \left(\bigoplus \sum\nolimits_{u \in I} B_u \right)_\infty.$$

We define $j : E \to \mathcal{B}_E$ by

$$j(x) = (u(x))_{u \in I}.$$

By (1.13) it is easy to check that j is completely isometric. Then, if we define A to be the (resp. unital) C^*-algebra generated by $j(E)$ in $\mathcal{B}_E$, the announced universal property of A is immediate. $\qquad\square$

Notation: We will denote by $C^*\langle E \rangle$ (resp. $C_u^*\langle E \rangle$) the (resp. unital) C^*-algebra A appearing in the preceding statement.

Note that $C^*\langle E \rangle$ is essentially unique. Indeed, if $j_1 : E \to A_1$ is another completely isometric embedding into a C^*-algebra A_1 with the property in Theorem 2.25, then the universal property of A (resp. A_1) implies the existence of a representation $\pi : A \to A_1$ (resp. $\pi_1 : A_1 \to A$) such that $\pi j = j_1$ (resp. $\pi_1 j_1 = j$). Since C^*-representations are automatically contractive we have $\|\pi\| \le 1$, $\|\pi_1\| \le 1$ and $\pi_1 = \pi^{-1}$ on $j_1(E)$ hence on the $*$-algebra generated by $j_1(E)$, which is dense in A_1 by assumption. This implies that π is an isometric isomorphism from A onto A_1.

Similarly, $C_u^*\langle E \rangle$ is characterized as the unique unital C^*-algebra C containing E completely isometrically in such a way that, for any unital C^*-algebra B (actually we may restrict to $B = B(H)$ with H arbitrary), any c.c. map $u : E \to B$ *uniquely* extends to a unital representation (i.e. $*$-homomorphism) from C to B.

It is easy to see that $C^*\langle E \rangle$ can be identified to the C^*-algebra generated by E in $C^*_u\langle E \rangle$. Indeed, the latter has the property in Theorem 2.25.

Remark 2.26 If two operator spaces E, F are completely isometrically isomorphic, then E and F can be realized as "concrete" operator subspaces $E \subset A$ and $F \subset B$ of two isomorphic C^*-algebras A and B, for which there is an isometric $*$-homomorphism $\pi : A \to B$ such that $\pi(E) = F$.

Indeed, let $A = C^*\langle E \rangle$ and $B = C^*\langle F \rangle$, let $u : E \to F$ be a completely isometric isomorphism, let $\pi : C^*\langle E \rangle \to C^*\langle F \rangle$ be the (unique) extension of u (as in Theorem 2.25) and let $\sigma : C^*\langle F \rangle \to C^*\langle E \rangle$ be the (unique) extension of u^{-1}. Then clearly we must have (by unicity again) $\sigma\pi = Id_{C^*\langle E \rangle}$ and $\pi\sigma = Id_{C^*\langle F \rangle}$ so that $\sigma = \pi^{-1}$.

Remark 2.27 (Universal C^*-algebra of a contraction) Consider for example the simplest choice of E, namely $E = \mathbb{C}$. Then $C^*\langle \mathbb{C} \rangle$ can be described more explicitly as the completion of the space of polynomials P in the formal variables X, X^* equipped with the C^*-norm $\|P\| = \sup\{\|P(x, x^*)\|\}$ where the sup runs over all H and all $x \in B(H)$ with $\|x\| \leq 1$. Indeed, the linear mapping $u_x : \mathbb{C} \to B(H)$ taking 1 to x satisfies trivially $\|u_x\|_{cb} = \|x\|$, so that $\|x\| \leq 1$ if and only if u_x extends to a $*$-homomorphism $\pi_x : C^*\langle \mathbb{C} \rangle \to B(H)$ with $\|\pi_x\| = 1$.

The analogue for the unital case requires that we consider polynomials "with a constant term" of the form $\lambda 1 + P(X, X^*)$ with $\lambda \in \mathbb{C}$. We then equip the resulting unital algebra of polynomials with the norm $\sup\{\|\lambda Id_H + P(x, x^*)\|\}$ over all H's and all contractions $x \in B(H)$ as before. The completion can now be identified with $C^*_u\langle \mathbb{C} \rangle$.

2.8 Completely positive perturbations of completely bounded maps

Warning: this section is devoted to a rather technical point. To avoid interrupting the flow of our presentation, we advise the reader to skip it until it becomes needed (in §9.4).

Notation (Order defined by the cone of c.p. maps). Let $u, v : E \to B$ be linear maps from an operator system E to a C^*-algebra. If $v - u \in CP(E, B)$, we will write

$$u \preccurlyeq v. \tag{2.30}$$

This is the partial order on $CB(E, B)$ for which $CP(E, B)$ is the positive cone.

As we saw previously (see Theorem 1.35), in analogy with a well-known fact in measure theory, if a complete contraction on an operator system is unital, then it is automatically c.p. In this section, we will prove quantitative versions of this: if a unital map has a cb-norm close to 1, then the map is close to a c.p. map. In the first result the range is an arbitrary C^*-algebra, while in the second one it is $B(H)$. The first case is restricted to a finite-dimensional domain but the second one is not.

Theorem 2.28 ([77]) *Let E be an n-dimensional operator system, B a unital C^*-algebra.*

(i) *For any $\varphi \in CP(E, B)$ there is a linear form $f \in E^*$ such that $f \geq 0$, $\|f\| \leq 2n\|\varphi\|$ and the mapping $\Psi : E \to B$ defined by $\Psi(x) = f(x)1 - \varphi(x)$ is c.p. (equivalently we have $0 \preccurlyeq \varphi \preccurlyeq f(\cdot)1$).*

(ii) *For any self-adjoint unital linear map $\varphi : E \to B$ such that $\|\varphi\|_{cb} \leq 1 + \delta$, where $0 < \delta < 1/2n$, there is a u.c.p. map $\psi : E \to B$ such that $\|\varphi - \psi\|_{cb} \leq 8n\delta$.*

Proof We reproduce the proofs from [77] with a cosmetic change.
(i) We first consider the case of a self-adjoint (not necessarily c.p.) linear map $\varphi : E \to B$ of rank 1 with $\|\varphi\| \leq 1$ of the form $\varphi(x) = f(x)b$ for some $f \in B_{E^*}$ and $b \in B_B$ both self-adjoint. We will denote this map by $f \otimes b$.

We claim that there is $F \in B_{E^*}$ with $F \geq 0$ and $\|F\| \leq 1$ such that, with respect to the natural ordering (2.30) of $CB(E, B)$, we have

$$-F \otimes 1 \preccurlyeq f \otimes b \preccurlyeq F \otimes 1.$$

Assuming $E \subset B(H)$, f can be extended to a self-adjoint form $f' \in B(H)^*$ with $\|f'\| \leq \|f\|$ (indeed, we may take for f' the real part of the Hahn–Banach extension of f). By the classical Hahn decomposition (see Remark 1.54), we have $f' = f'_+ - f'_-$ with $f'_\pm \geq 0$ such that $\|f'_+\| + \|f'_-\| \leq \|f'\|$. By restricting this to E we find a decomposition $f = f_+ - f_-$ with $f_\pm \in E^*$ positive and such that $\|f_+\| + \|f_-\| \leq \|f\|$.

In parallel, any $b \in B_B$ with $b = b^*$ can be decomposed as $b = b_+ - b_-$ with $b_\pm \geq 0$ in B such that $\|b_+ + b_-\| \leq 1$. But it is easy to check that for the c.p. order for any $0 \leq f_1 \leq f_2$ ($f_1, f_2 \in E^*$) and $0 \leq b_1 \leq b_2$ ($b_1, b_2 \in B$) we have $0 \preccurlyeq f_1 \otimes b_1 \preccurlyeq f_2 \otimes b_2$. Therefore we have

$$-T \preccurlyeq (f_+ - f_-) \otimes (b_+ - b_-) \preccurlyeq T$$

with $T = (f_+ + f_-) \otimes (b_+ + b_-)$ and a fortiori also with $T = (f_+ + f_-) \otimes 1$ since $0 \leq b_+ + b_- \leq 1$. Thus our claim follows with $F = f_+ + f_-$.

Now, since $\dim(E) = n$, by Auerbach's lemma 1.55, there is a biorthogonal system (ξ_j, x_j) in $B_{E^*} \times B_E$ such that $Id_E = \sum_1^n \xi_j \otimes x_j$. Note that for any $x \in E$ we have $x^* = \left(\sum \xi_j(x)x_j \right)^*$ and also (since $x^* \in E$) $x^* = \sum_1^n \xi_j(x^*)x_j$. Thus (setting $\xi_{j_*} = \xi_j(x^*)$) we have $\sum \xi_j \otimes x_j = \sum \xi_{j_*} \otimes x_j^*$, and hence

$$\sum \xi_j \otimes x_j = \sum (\xi_j + \xi_{j_*})/2 \otimes (x_j + x_j^*)/2$$
$$- \sum (\xi_j - \xi_{j_*})/(2i) \otimes (x_j - x_j^*)/(2i) = \sum_1^{2n} f_j \otimes y_j,$$

with f_j, y_j self-adjoint in $B_{E^*} \times B_E$. This gives us

$$\forall x \in E \quad x = \sum_1^{2n} f_j(x)y_j. \tag{2.31}$$

Now let $\varphi \in CP(E, B)$. Using (2.31), we may write φ as $\varphi = \sum f_j \otimes b_j$ with $b_j = \varphi(y_j)$. We may assume $\|\varphi\| = 1$ (by homogeneity). Note $b_j = b_j^*$ and $\|b_j\| \le \|\varphi\|\|y_j\| \le 1$. Therefore, using the first part of the proof we obtain $F_j \ge 0$ with $\|F_j\| \le 1$, and hence $\|\sum F_j\| \le 2n$, such that

$$-(\Sigma F_j) \otimes 1 \preccurlyeq \varphi \preccurlyeq (\Sigma F_j) \otimes 1.$$

In particular, $0 \preccurlyeq \varphi \preccurlyeq f(\cdot)1$ with $f = \Sigma F_j \ge 0$ and $\|f\| \le 2n$. In other words $x \mapsto \Psi(x) = f(x)1 - \varphi(x)$ is in $CP(E, B)$.

(ii) Assume $B \subset B(H)$. By Corollary 1.53 we can write $\varphi = \varphi_1 - \varphi_2$ with $\varphi_1, \varphi_2 \in CP(E, B(H))$ such that $\|\varphi_1 + \varphi_2\| \le \|\varphi\|_{cb}$. Let $a_j = \varphi_j(1)$. Then $a_1, a_2 \ge 0$, $a_1 - a_2 = 1$ and $\|a_1\| \le \|a_1 + a_2\| \le 1 + \delta$. Thus, for any unit vector ξ in H we have

$$\langle \xi, a_2\xi \rangle = \langle \xi, (a_1 - 1)\xi \rangle = \langle \xi, a_1\xi \rangle - 1 \le \delta,$$

and hence by (1.27)

$$\|\varphi_2\| = \|a_2\| \le \delta.$$

By part (i) there is $f \in E^*$ with $f \ge 0$ and $\|f\| \le 2n\delta$ such that the mapping $\psi_2 : E \to B(H)$ defined by $\psi_2(x) = f(x)1 - \varphi_2(x)$ is c.p. We then set

$$\psi_0 = \varphi_1 + \psi_2.$$

Note that, since $\psi_0(x) - \varphi(x) = f(x)1 \in \mathbb{C}1$, we have $\psi_0(E) \subset B$, $\psi_0 \in CP(E, B)$ and

$$\|\psi_0 - \varphi\|_{cb} \le 2n\delta. \tag{2.32}$$

To obtain a *unital* ψ we need to work a bit more. Let $b = \psi_0(1)$. Then $b \ge 0$ and $\|b - 1\| = \|\psi_0(1) - \varphi(1)\| \le 2n\delta < 1$, and hence b is invertible. A fortiori, by functional calculus (since $|\sqrt{b} - 1| \le |b - 1| \ \forall b \in [0, 2]$), we have $\|\sqrt{b} - 1\| \le \|b - 1\| \le 2n\delta$. For any $y \in B$ we have

$$\|b^{\frac{1}{2}}yb^{\frac{1}{2}} - y\| = \|(b^{\frac{1}{2}} - 1)yb^{\frac{1}{2}} + y(b^{\frac{1}{2}} - 1)\|$$

$$\leq (\|b^{\frac{1}{2}} - 1\| \|b^{\frac{1}{2}}\| + \|b^{\frac{1}{2}} - 1\|)\|y\| \leq 6n\delta\|y\|.$$

A similar estimate holds for any k and any $y = [y_{ij}] \in M_k(B)$. Thus if we now set

$$\psi(x) = b^{-\frac{1}{2}}\psi_0(x)b^{-\frac{1}{2}},$$

so that $\psi_0(\cdot) = b^{\frac{1}{2}}\psi(\cdot)b^{\frac{1}{2}}$, we find $\|\psi_0 - \psi\|_{cb} = \|b^{\frac{1}{2}}\psi b^{\frac{1}{2}} - \psi\|_{cb} \leq 6n\delta\|\psi\|_{cb} = 6n\delta$. Thus we obtain by (2.32)

$$\|\varphi - \psi\|_{cb} \leq \|\varphi - \psi_0\|_{cb} + \|\psi_0 - \psi\|_{cb} \leq 8n\delta,$$

and $\psi \in CP(E, B)$ is unital with $\|\psi\|_{cb} = \|\psi(1)\| = 1$ (by (1.27)). $\square$

The case when $B = B(H)$ is much simpler:

Lemma 2.29 (Kirchberg) *Let A be a unital C^*-algebra and let $\varphi : A \to B(H)$ be a completely bounded self-adjoint unital map. Then there is a u.c.p. map $\psi : A \to B(H)$ such that $\|\varphi - \psi\|_{cb} \leq \|\varphi\|_{cb} - 1$.*

Proof Let $\varepsilon = \|\varphi\|_{cb} - 1$. By the factorization Theorem 1.50 there are $\widehat{H}, \pi : A \to B(\widehat{H})$ and $V, W \in B(H, \widehat{H})$ such that $\|V\|\|W\| = 1 + \varepsilon$ and $\varphi(a) = V^*\pi(a)W$ for all a in A. By homogeneity we may assume $\|V\| = \|W\| = (1 + \varepsilon)^{\frac{1}{2}}$. Since φ is unital we have $V^*W = Id_H$, and since it is self-adjoint $V^*\pi(a)W = W^*\pi(a)V$. Note $W^*V = Id_H$. Let $\psi(a) = (V^*\pi(a)V + W^*\pi(a)W)/2$. We have then

$$\psi(a) - \varphi(a) = \frac{1}{2}(V^*\pi(a)V + W^*\pi(a)W - V^*\pi(a)W - W^*\pi(a)V)$$

$$= \frac{1}{2}(V - W)^*\pi(a)(V - W).$$

Note that $\psi - \varphi$ (as well as ψ) is c.p. Moreover $(V - W)^*(V - W) = V^*V + W^*W - 2Id_H \leq 2\varepsilon Id_H$. Therefore, $\|\psi - \varphi\| = \|\psi - \varphi\|_{cb} \leq \varepsilon$. $\square$

2.9 Notes and remarks

For Schur multipliers as in Remark 1.58 the equality $\|u\| = \|u\|_{cb}$ is due to Haagerup, and for "module maps" as in Theorem 2.6 it was proved by Roger Smith in [232]. The theory of free products of C^*-algebras can be traced back to Avitsur's seminal paper [15], but it really took off with Voiculescu's "free probability" (see [254]). Note, however, that in the latter theory, it is the *reduced* free product of two C^*-algebras equipped with states that plays the central role, while for the present volume we are mostly concerned with the maximal free product of unital C^*-algebras (i.e. amalgamated over $\mathbb{C}1$). For this kind of free product it is rather striking that unital c.p. maps are "admissible" morphisms as described by Boca's theorem from [30]. As already mentioned the duality theory for operator spaces goes back to Effros–Ruan and Blecher–Paulsen. The results in §2.4 and §2.5 are due to them, except for Propositions 2.20 and 2.18, both well-known facts. The remarkable factorization in (2.26) (due to Blecher and Paulsen [28]) is a consequence of the no-less remarkable Blecher–Ruan–Sinclair characterization of operator algebras; the central underlying concept is the Haagerup tensor product of operator spaces. These important topics are treated in detail in our previous book [208] which explains our reluctance to expand on that same theme in the present one. We describe these results in Theorem 2.23 and, making here an exception, refer the reader to [208] or [196] for detailed proofs.

See Loring's book [171] for a discussion of amalgamated full free products as solutions of a universal problem. Loring also considers in [171] free products of unital algebras amalgamated over $\mathbb{C}1$ but relative to *nonunital* embeddings into the free product, which can be a quite different object. Concerning §2.8 and the completely positive ordering, we should mention the following Radon–Nikodym theorem due to Arveson [12]. Let A be a C^*-algebra, let $u, v \in CP(A, B(H))$. Let $v(\cdot) = V^*\pi(\cdot)V$ be the minimal Stinespring factorization of v (which means that $V \in B(H, \widehat{H})$ with $\overline{V\pi(A)(H)} = \widehat{H}$). Then $0 \preccurlyeq u \preccurlyeq v$ if and only if there is $T \in \pi(A)' \subset B(H)$ with $0 \le T \le 1$ such that $u(\cdot) = V^*T\pi(\cdot)V$. Moreover, such a T is unique. See [197] for a sort of completely bounded analogue.

3

C^*-algebras of discrete groups

Group representations are one of the main sources of examples of C^*-algebras. The universal representation of a group G gives rise to the full or maximal C^*-algebra $C^*(G)$, while the left regular representation leads to the reduced C^*-algebra $C^*_\lambda(G)$. In this chapter we review some of their main properties when G is a discrete group.

3.1 Full (=Maximal) group C^*-algebras

We first recall some classical notation from noncommutative Abstract Harmonic Analysis on an arbitrary discrete group G.

We denote by e (and sometimes by e_G) the unit element. Let $\pi : G \to B(\mathcal{H})$ be a unitary representation of G. We denote by $C^*_\pi(G)$ the C^*-algebra generated by the range of π.

Equivalently, $C^*_\pi(G)$ is the closed linear span of $\pi(G)$.

In particular, this applies to the so-called universal representation of G, a notion that we now recall. Let $(\pi_j)_{j \in I}$ be a family of unitary representations of G, say

$$\pi_j : G \to B(H_j)$$

in which every equivalence class of a cyclic unitary representation of G has an equivalent copy. Now one can define the "universal" representation $U_G : G \to B(\mathcal{H})$ of G by setting

$$U_G = \oplus_{j \in I} \pi_j \quad \text{on} \quad \mathcal{H} = \oplus_{j \in I} H_j.$$

Then the associated C^*-algebra $C^*_{U_G}(G)$ is simply denoted by $C^*(G)$ and is called the "full" (or the "maximal") C^*-algebra of the group G, to distinguish it from the "reduced" one that is described in the sequel. Note that

$$C^*(G) = \overline{\text{span}}\{U_G(t) \mid t \in G\}.$$

Let π be any unitary representation of G. By a classical argument, π is unitarily equivalent to a direct sum of cyclic representations, hence for any finitely supported function $x : G \to \mathbb{C}$ we have

$$\left\| \sum x(t)\pi(t) \right\| \leq \left\| \sum x(t)U_G(t) \right\|. \tag{3.1}$$

In particular, if π is the trivial representation

$$\left| \sum x(t) \right| \leq \left\| \sum x(t)U_G(t) \right\|. \tag{3.2}$$

Equivalently (3.1) means

$$\left\| \sum x(t)U_G(t) \right\|_{B(\mathcal{H})} = \sup \left\{ \left\| \sum x(t)\pi(t) \right\|_{B(H_\pi)} \right\}$$

where the supremum runs over all possible unitary representations $\pi : G \to B(H_\pi)$ on an arbitrary Hilbert space H_π. More generally, for any Hilbert space K and any finitely supported function $x : G \to B(K)$ we have

$$\left\| \sum x(t) \otimes U_G(t) \right\|_{B(K \otimes_2 \mathcal{H})} = \sup \left\{ \left\| \sum x(t) \otimes \pi(t) \right\|_{B(K \otimes_2 H_\pi)} \right\}$$

where the sup is the same as before.

There is an equivalent description in terms of the group algebra $\mathbb{C}[G]$, the elements of which are simply the formal linear combinations of the elements of G, equipped with the obvious natural $*$-algebra structure. One equips $\mathbb{C}[G]$ with the norm (actually a C^*-norm)

$$\sum_{t \in G} x(t)t \mapsto \sup \left\{ \left\| \sum x(t)\pi(t) \right\| \right\}$$

where the supremum runs over all possible unitary representations π of G. One can then define $C^*(G)$ as the completion of $\mathbb{C}[G]$ with respect to the latter norm.

These formulae show that the norm of $C^*(G)$ is the largest possible C^*-norm on $\mathbb{C}[G]$. Whence the term "maximal" C^*-algebra of G.

Remark 3.1 (A recapitulation) By (3.1) there is a $1-1$ correspondence between the unitary representations $\pi : G \to B(H)$ and the $*$-homomorphisms $\psi : C^*(G) \to B(H)$. More precisely, for any π there is a unique $\psi : C^*(G) \to B(H)$ such that $\forall g \in G \quad \psi(U_G(g)) = \pi(g)$, or if we view G as a subset of $\mathbb{C}[G] \subset C^*(G)$ (which means we identify g and $U_G(g)$), we have

$$\forall g \in G \quad \pi(g) = \psi(g).$$

Remark 3.2 (c.b. and c.p. maps on $C^*(G)$) A linear map $u : C^*(G) \to B(K)$ is c.b. if and only if there exists a unitary group representation $\pi : G \to B(H_\pi)$ and operators $V, W : K \to H_\pi$ such that

$$\forall t \in G \quad u(U_G(t)) = W^* \pi(t) V.$$

Moreover, we have $\|u\|_{cb} = \inf\{\|W\|\|V\|\}$ and the infimum is attained. Indeed, in view of the preceding remark this follows immediately from Theorem 1.50. The c.p. case is characterized similarly but with $V = W$. When $K = \mathbb{C}$ and hence $B(K) = \mathbb{C}$, this gives us a description of the dual of $C^*(G)$, as well as a characterization of states on $C^*(G)$.

The next result (in which we illustrate the preceding remark in the case of multipliers) is classical, and fairly easy to check.

Proposition 3.3 (Multipliers on $C^*(G)$) *Let $\varphi : G \rightarrow \mathbb{C}$. Consider the associated linear operator M_φ (a so-called multiplier, see §3.4) defined on $\operatorname{span}\{U_G(t) \mid t \in G\}$ by $M_\varphi\left(\sum x(t) U_G(t)\right) = \sum x(t)\varphi(t) U_G(t)$. Then M_φ extends to a bounded operator on $C^*(G)$ if and only if there are a unitary representation $\pi : G \rightarrow B(H_\pi)$ and ξ, η in H_π such that*

$$\forall t \in G \quad \varphi(t) = \langle \eta, \pi(t)\xi \rangle. \tag{3.3}$$

Moreover we have for the resulting bounded operator (still denoted by M_φ)

$$\|M_\varphi\| = \|M_\varphi\|_{cb} = \inf\{\|\xi\|\|\eta\|\} \tag{3.4}$$

where the infimum (which is attained) runs over all possible π, ξ, η for which this holds. Lastly, if M_φ is positive (3.3) holds with $\xi = \eta$, and then M_φ is completely positive on $C^*(G)$.

Proof If $\|M_\varphi : C^*(G) \rightarrow C^*(G)\| \leq 1$, let $f(x) = \sum_{t \in G} \varphi(t) x(t)$. Then by (3.2) $f \in C^*(G)^*$ with $\|f\| \leq 1$. Note $f(U_G(t)) = \varphi(t)$. By Remark 1.54 there are π, ξ, and η with $\|\xi\|\|\eta\| \leq \|f\| \leq 1$ such that (3.3) holds. If M_φ (and hence f) is positive we find this with $\xi = \eta$. For the converse, since (like any unitary group representation) the mapping $U_G(t) \mapsto U_G(t) \otimes \pi(t)$ extends to a continuous $*$-homomorphism $\sigma : C^*(G) \rightarrow B(\mathcal{H} \otimes_2 H_\pi)$, we have $M_\varphi(\cdot) = V_2^* \sigma(\cdot) V_1$, with $V_1 h = h \otimes \xi$ and $V_2 h = h \otimes \eta$ ($h \in \mathcal{H}$) from which we deduce by (1.30) $\|M_\varphi\|_{cb} \leq \|\xi\|\|\eta\|$. If $\xi = \eta$ then $V_1 = V_2$ and hence M_φ is c.p. on $C^*(G)$. □

Remark 3.4 By Remark 3.2 and (3.4) the space of bounded multipliers on $C^*(G)$ can be identified isometrically with $C^*(G)^*$. If f_φ is the linear form on $C^*(G)$ taking $U_G(t)$ to $\varphi(t)$ ($t \in G$) we have

$$\|M_\varphi\| = \|f_\varphi\|_{C^*(G)^*}.$$

Proposition 3.5 *Let G be a discrete group and let $\Gamma \subset G$ be a subgroup. Then the correspondence $U_\Gamma(t) \rightarrow U_G(t)$, ($t \in \Gamma$) extends to an isometric*

(C*-algebraic) embedding J of C*(Γ) into C*(G). Moreover, there is a completely contractive and completely positive projection P from C*(G) onto the range of this embedding, defined by $P(U_G(t)) = U_G(t)$ for any $t \in \Gamma$ and $P(U_G(t)) = 0$ otherwise.

Proof By the universal property of $C^*(\Gamma)$ the unitary representation $\Gamma \supset \gamma \mapsto U_G(\gamma)$ extends to a *-homomorphism $J : C^*(\Gamma) \to C^*(G)$ with $\|J\| = 1$. Let $\varphi = 1_\Gamma$. The projection P described in Proposition 3.5 coincides with the multiplier M_φ acting on $C^*(G)$. Thus, by Proposition 3.3 it suffices to show that there is a unitary representation $\pi : G \to B(H_\pi)$ of G and a unit vector $\xi \in H_\pi$ such that $\varphi(t) = \langle \xi, \pi(t)\xi \rangle$. Let $G = \bigcup_{s \in G/\Gamma} s\Gamma$ be the disjoint partition of G into left cosets. For any $t \in G$ the mapping $s\Gamma \mapsto ts\Gamma$ defines a permutation $\sigma(t)$ of the set G/Γ, and $t \mapsto \sigma(t)$ is a homomorphism. Let $H_\pi = \ell_2(G/\Gamma)$ and let $\pi : G \to B(H_\pi)$ be the unitary representation defined on the unit vector basis by $\pi(t)(\delta_s) = \delta_{\sigma(t)(s)}$ for any $s \in G/\Gamma$. Let $[[\Gamma]] \in G/\Gamma$ denote the coset Γ (i.e. $s\Gamma$ for $s = 1_G$) and let $\xi = \delta_{[[\Gamma]]}$. Then it is immediate that $\varphi(t) = \langle \xi, \pi(t)\xi \rangle$ for any $t \in G$. $\qquad\square$

Remark 3.6 Let G be a discrete group and let $E \subset C^*(G)$ be any separable subspace. We claim that there is a countable subgroup $\Gamma \subset G$ such that with the notation of Proposition 3.5 we have $E \subset J(C^*(\Gamma))$. Indeed, since $C^*(G) \subset \overline{\text{span}}[U_G(t) \mid t \in G]$ for any fixed $x \in C^*(G)$ there is clearly a countable subgroup $\Gamma_x \subset G$ and an analogous J_x such that $x \subset J_x(C^*(\Gamma_x))$. Arguing like this for each x in a dense countable sequence in E and taking the group generated by all the resulting Γ_x's gives us the claim.

By Proposition 3.5 this shows that there is a separable C^*-subalgebra $C \subset C^*(G)$ with $E \subset C$ for which there is a c.p. projection $P : C^*(G) \to C$.

Remark 3.7 Let G be any discrete group, let $A = C^*(G)$. Then $\bar{A} \simeq A$. Indeed, since for any unitary representation π on G, the complex conjugate $\bar{\pi}$ (as in Remark 2.14) is also a unitary representation, the correspondence $\pi \mapsto \bar{\pi}$ is a bijection on the set of unitary representations, from which the $\mathbb{C}$-linear isomorphism $\Phi : C^*(G) \to \overline{C^*(G)}$ follows immediately. Denoting by U_G the universal representation of G, this isomorphism takes $U_G(t)$ to $\overline{U_G(t)}$. Note that $\bar{A} \simeq A$ is in general not true (see [60]).

3.2 Full *C**-algebras for free groups

In this section, we start by comparing the C^*-algebras of free groups of different cardinals. Our goal is to make clear that we can restrict to $\mathscr{C} = C^*(\mathbb{F}_\infty)$ (or if we wish to $C^*(\mathbb{F}_2)$) for the various properties of interest to us in the sequel. Then we describe the operator space structure of the span of the free generators in $C^*(\mathbb{F})$ when $\mathbb{F}$ is any free group. The following simple lemma will be often invoked when we wish to replace $C^*(\mathbb{F})$ by $C^*(\mathbb{F}_\infty)$.

Lemma 3.8 *Let $\mathbb{F}$ be a free group with generators $(g_i)_{i \in I}$. Let $E \subset C^*(\mathbb{F})$ be any separable subspace. Then the inclusion $E \subset C^*(\mathbb{F})$ admits an extension $T_E : C^*(\mathbb{F}) \to C^*(\mathbb{F})$ that can be factorized as*

$$T_E : C^*(\mathbb{F}) \xrightarrow{\ w\ } C^*(\mathbb{F}_\infty) \xrightarrow{\ v\ } C^*(\mathbb{F})$$

where v, w are contractive c.p. maps.

For any C^*-algebra D and any $x \in D \otimes E$ we have

$$\|x\|_{D \otimes_{\max} C^*(\mathbb{F})} = \|(Id_D \otimes w)(x)\|_{D \otimes_{\max} C^*(\mathbb{F}_\infty)}. \tag{3.5}$$

In particular, $E \subset C^*(\mathbb{F})$ is completely isometric to $w(E) \subset C^*(\mathbb{F}_\infty)$.

Proof For any $x \in C^*(\mathbb{F})$ there is clearly a *countable* subgroup $\Gamma_x \subset \mathbb{F}$ such that

$$x \in \overline{\mathrm{span}}[U_\mathbb{F}(t) \mid t \in \Gamma_x].$$

By the separability of E, we can find a countable subgroup Γ such that $E \subset \overline{\mathrm{span}}[U_\mathbb{F}(t) \mid t \in \Gamma]$. Since any element of $t \in \Gamma$ can be written using only finitely many "letters" in $\{g_i \mid i \in I\}$, we may assume that Γ is the free subgroup generated by $(g_i)_{i \in I'}$ for some countable subset $I' \subset I$. Then, identifying $\overline{\mathrm{span}}[U_\mathbb{F}(t) \mid t \in \Gamma]$ with $C^*(\Gamma)$, Proposition 3.5 yields a mapping $T = JP : C^*(\mathbb{F}) \to C^*(\mathbb{F})$ with the required factorization through $C^*(\Gamma) = C^*(\mathbb{F}_{I'})$ that is the identity when restricted to E. If I' is infinite the proof is complete: since $C^*(\mathbb{F}_{I'}) = C^*(\mathbb{F}_\infty)$ we may take $T_E = T$.

Otherwise, we note that $\mathbb{F}_{I'} \subset \mathbb{F}_\infty$ as a subgroup and hence by Proposition 3.5 again we have a factorization of the same type $C^*(\mathbb{F}_{I'}) \xrightarrow{J'} C^*(\mathbb{F}_\infty) \xrightarrow{P'} C^*(\mathbb{F}_{I'})$ from which it is easy to conclude.

Note

$$\|x\|_{D \otimes_{\max} C^*(\mathbb{F})} = \|(Id_D \otimes T_E)(x)\|_{D \otimes_{\max} C^*(\mathbb{F})} = \|(Id_D \otimes vw)(x)\|_{D \otimes_{\max} C^*(\mathbb{F})}.$$

By Corollary 4.18 since v, w are c.p. contractions we have

$$\|x\|_{D \otimes_{\max} C^*(\mathbb{F})} \leq \|(Id_D \otimes w)(x)\|_{D \otimes_{\max} C^*(\mathbb{F}_\infty)} \quad \text{and}$$

$$\|(Id_D \otimes w)(x)\|_{D \otimes_{\max} C^*(\mathbb{F}_\infty)} \leq \|x\|_{D \otimes_{\max} C^*(\mathbb{F})},$$

from which (3.5) follows. $\qquad\square$

Let $\mathbb{F}$ be a free group with generators $(g_i)_{i \in I}$. We start with a basic property of the span of the free generators in $C^*(\mathbb{F})$.

Lemma 3.9 *Let $\mathbb{F}$ be a free group with generators $(g_i)_{i \in I}$. Let $U_i = U_{\mathbb{F}}(g_i) \in C^*(\mathbb{F})$. Let $E = \mathrm{span}[(U_i)_{i \in I}, 1] \subset C^*(\mathbb{F})$ and $E_I = \mathrm{span}[(U_i)_{i \in I}] \subset C^*(\mathbb{F})$. Then for any linear map $u : E \to B(H)$ and any $v : E_I \to B(H)$ we have*

$$\|u\|_{cb} = \|u\| = \max\{\sup_{i \in I} \|u(U_i)\|, \|u(1)\|\} \quad and$$

$$\|v\|_{cb} = \|v\| = \max\{\sup_{i \in I} \|v(U_i)\|\}. \tag{3.6}$$

Proof It clearly suffices to show that $\max\{\sup_{i \in I} \|u(U_i)\|, \|u(1)\|\} \leq 1$ implies $\|u\|_{cb} \leq 1$. When $u(1) = 1$ and all $u(U_i)$ are unitaries this is easy: indeed there is a (unique) group representation $\sigma : \mathbb{F} \to B(H)$ such that $\sigma(g_i) = u(U_i)$ and the associated linear extension $u_\sigma : C^*(\mathbb{F}) \to B(H)$ is a $*$-homomorphism automatically satisfying $\|u_\sigma\|_{cb} = 1$, and hence $\|u\|_{cb} = 1$. This same argument works if we merely assume that $u(1)$ is unitary. Indeed, we may replace u by $x \mapsto u(1)^{-1}u$, which takes us back to the previous easy case. Since the general case is easy to reduce to that of a finite set, we assume that I is finite. Then the Russo–Dye Theorem A.18 shows us that any u such that $\max\{\sup_{i \in I} \|u(U_i)\|, \|u(1)\|\} \leq 1$ lies in the closed convex hull of u's for which $u(1)$ and all the $u(U_i)$s are unitaries, and hence $\|u\|_{cb} \leq 1$ in that case also. $\square$

The first part of the next result is based on the classical observation that a unitary representation $\pi : \mathbb{F} \to B(H)$ is entirely determined by its values $u_i = \pi(g_i)$ on the generators, and if we let π run over all possible unitary representations, then we obtain all possible families (u_i) of unitary operators. The second part is also well known.

Lemma 3.10 *Let $A \subset B(H)$ be a C^*-algebra. Let $\mathbb{F}$ be a free group with generators $(g_i)_{i \in I}$. Let $U_i = U_{\mathbb{F}}(g_i) \in C^*(\mathbb{F})$. Let $(x_i)_{i \in I}$ be a family in A with only finitely many nonzero terms. Consider the linear map $T : \ell_\infty(I) \to A$ defined by $T((\alpha_i)_{i \in I}) = \sum_{i \in I} \alpha_i x_i$. Then we have*

$$\left\|\sum_{i \in I} U_i \otimes x_i\right\|_{C^*(\mathbb{F}) \otimes_{\min} A} = \|T\|_{cb} = \sup\left\{\left\|\sum u_i \otimes x_i\right\|_{\min}\right\} \tag{3.7}$$

where the sup runs over all possible Hilbert spaces K and all families (u_i) of unitaries on K. Actually, the latter supremum remains the same if we restrict it to finite-dimensional Hilbert spaces K. Moreover, in the case when $A = B(H)$ with $\dim(H) = \infty$, we have

$$\left\|\sum_{i \in I} U_i \otimes x_i\right\|_{C^*(\mathbb{F}) \otimes_{\min} B(H)} = \inf\left\{\left\|\sum y_i y_i^*\right\|^{1/2} \left\|\sum z_i^* z_i\right\|^{1/2}\right\} \tag{3.8}$$

where the infimum, which runs over all possible factorizations $x_i = y_i z_i$ with y_i, z_i in $B(H)$, is actually attained.

Moreover, all this remains true if we enlarge the family $(U_i)_{i \in I}$ by including the unit element of $C^*(\mathbb{F})$.

Proof It is easy to check going back to the definitions that on one hand

$$\left\| \sum U_i \otimes x_i \right\|_{\min} = \sup \left\{ \left\| \sum u_i \otimes x_i \right\|_{\min} \right\},$$

where the sup runs over all possible families of unitaries (u_i), and on the other hand that

$$\|T\|_{cb} = \sup \left\{ \left\| \sum t_i \otimes x_i \right\|_{\min} \right\},$$

where the sup runs over all possible families of contractions (t_i). By the Russo–Dye Theorem A.18, any contraction is a norm limit of convex combinations of unitaries, so (3.7) follows by convexity. Actually, the preceding sup obviously remains unchanged if we let it run only over all possible families of contractions (t_i) on a *finite-dimensional* Hilbert space. Thus it remains unchanged when restricted to families of *finite-dimensional* unitaries (u_i).

Now assume $\|T\|_{cb} = 1$. By the factorization of *c.b.* maps we can write $T(\alpha) = V^* \pi(\alpha) W$ where $\pi : \ell_\infty(I) \to B(\widehat{H})$ is a representation and where V, W are in $B(H, \widehat{H})$ with $\|V\| \, \|W\| = \|T\|_{cb}$. Since we assume $\dim(H) = \infty$ and may assume I finite (because $i \mapsto x_i$ is finitely supported), by Remark 1.51 we may as well take $\widehat{H} = H$. Let $(e_i)_{i \in I}$ be the canonical basis of $\ell_\infty(I)$, we set

$$y_i = V^* \pi(e_i) \quad \text{and} \quad z_i = \pi(e_i) W.$$

It is then easy to check $\left\| \sum y_i y_i^* \right\|^{1/2} \left\| \sum z_i^* z_i \right\|^{1/2} \le \|V\| \, \|W\| = \|T\|_{cb}$. Thus we obtain one direction of (3.8). The converse follows from (2.2) (easy consequence of Cauchy–Schwarz) applied to $a_i = U_i \otimes y_i$ and $b_i = 1 \otimes z_i$. Finally, the last assertion follows from the forthcoming Remark 3.12. $\qquad\square$

Remark 3.11 (Russo–Dye) The Russo–Dye Theorem A.18 shows that the sup of any continuous convex function on the unit ball of a unital C^*-algebra coincides with its sup over all its unitary elements.

Remark 3.12 Let $\{0\}$ be a singleton disjoint from the set I and let $\dot{I} = \{0\} \cup I$. Then for any finitely supported family $\{x_j \mid j \in \dot{I}\}$ in $B(H)$ (H arbitrary) we have

$$\left\| I \otimes x_0 + \sum_{i \in I} U_i \otimes x_i \right\|_{\min} = \sup \left\{ \left\| \sum_{j \in \dot{i}} u_j \otimes x_j \right\|_{\min} \right\} \tag{3.9}$$

where the supremum runs over all possible families $(u_j)_{j \in \dot{i}}$ of unitaries.
    Indeed, since

$$\left\| \sum_{j \in \dot{i}} u_j \otimes x_j \right\|_{\min} = \left\| I \otimes x_0 + \sum_{i \in I} u_0^{-1} u_i \otimes x_i \right\|_{\min},$$

the right-hand side of (3.9) is the same as the supremum of

$$\left\| I \otimes x_0 + \sum_{i \in I} u_i \otimes x_i \right\|_{\min} \tag{3.10}$$

over all possible families of unitaries $(u_i)_{i \in I}$. Therefore (recalling $U(g_i) = U_i$) we find

$$\left\| I \otimes x_0 + \sum_{i \in I} U_i \otimes x_i \right\|_{\min} = \sup \left\{ \left\| I \otimes x_0 \right. \right.$$
$$\left. \left. + \sum_{i \in I} u_i \otimes x_i \right\|_{\min} \right| u_i \text{ unitary} \right\},$$

where the sup runs over all Hilbert spaces $\mathcal{H}$ and all families (u_i) of unitaries in $B(\mathcal{H})$.

Moreover, by the same argument we used for Lemma 3.10, we can restrict to finite-dimensional $\mathcal{H}$'s:

$$\left\| I \otimes x_0 + \sum_{i \in I} U_i \otimes x_i \right\|_{\min} = \sup_{n \geq 1} \left\{ \left\| I \otimes x_0 \right. \right.$$
$$\left. \left. + \sum_{i \in I} u_i \otimes x_i \right\|_{\min} \right| u_i \, n \times n \text{ unitaries} \right\}$$
$$\tag{3.11}$$

so that the supremum on the right-hand side is restricted to families of *finite-dimensional* unitaries. Indeed, by Russo–Dye (Remark 3.11) the suprema of (3.10) taken over u_i's in the unit ball of $B(H)$ and over unitary u_i's are the same. Replacing u_i by $P_E u_{i|E}$ with $E \subset H$, $\dim(E) < \infty$ shows that the supremum of (3.10) is the same if we restrict it to u_i's in the unit ball of $B(E)$ with $\dim(E) < \infty$. Then, invoking Russo–Dye (Remark 3.11) again, we obtain (3.11).

Remark 3.13 Using (3.11) when I is a singleton and the fact that a single unitary generates a commutative unital C^*-algebra, it is easy to check that $\|T\| = \|T\|_{cb}$ for any $T : \ell_\infty^2 \to B(H)$.

Remark 3.14 ($\ell_1(I)$ as operator space) In the particular case $A = \mathbb{C}$, (3.7) becomes

$$\left\| \sum_{i \in I} U_i x_i \right\|_{C^*(\mathbb{F})} = \sum_{i \in I} |x_i|, \tag{3.12}$$

which shows that $E_I = \overline{\text{span}}[U_i, i \in I] \simeq \ell_1(I)$ isometrically.

Note that (3.8) generalizes the classical fact that $B_{\ell_1} = B_{\ell_2} B_{\ell_2}$ for the pointwise product.

More generally, Lemma 3.9 shows that the dual operator space E_I^* can be identified with the von Neumann algebra $\ell_\infty(I)$ equipped with its natural operator space structure as a C^*-algebra, i.e. the one such that we have $M_n(\ell_\infty(I)) = \ell_\infty(I; M_n)$ isometrically for all n. Lemma 3.10 describes the dual operator space of the operator space (actually a C^*-subalgebra) $c_0(I) \subset \ell_\infty(I)$ that is the closed span of the canonical basis in $\ell_\infty(I)$. We obtain $c_0(I)^* = E_I$ completely isometrically, which is the operator space analogue of the isometric identity $c_0(I)^* = \ell_1(I)$. Indeed, together with Lemma 3.9, (3.7) tells us that $CB(c_0(I), M_n) = M_n(E_I)$ isometrically for all n.

3.3 Reduced group C^*-algebras: Fell's absorption principle

We denote by $C_\lambda^*(G)$ (resp. $C_\rho^*(G)$) the so-called reduced C^*-algebra generated in $B(\ell_2(G))$ by λ_G (resp. ρ_G). Equivalently, $C_\lambda^*(G) = \overline{\text{span}}\{\lambda_G(t) \mid t \in G\}$ and $C_\rho^*(G) = \overline{\text{span}}\{\rho_G(t) \mid t \in G\}$. Note that $\lambda_G(t)$ and $\rho_G(s)$ commute for all t, s in G.

We denote λ_G and ρ_G simply by λ and ρ (and U_G by U) when there is no ambiguity.

The following very useful result is known as Fell's "absorption principle."

Proposition 3.15 *For any unitary representation* $\pi : G \to B(H)$, *we have*

$$\lambda_G \otimes \pi \simeq \lambda_G \otimes I \quad (\text{unitary equivalence}).$$

Here I stands for the trivial representation of G in B(H) (i.e. I(t) = Id_H $\forall t \in G$). *In particular, for any finitely supported functions* $a : G \to \mathbb{C}$ *and* $b : G \to B(\ell_2)$, *we have*

$$\left\| \sum a(t)\lambda_G(t) \otimes \pi(t) \right\|_{C_\lambda^*(G) \otimes_{\min} B(H)} = \left\| \sum a(t)\lambda_G(t) \right\|, \tag{3.13}$$

$$\left\| \sum b(t) \otimes \lambda_G(t) \otimes \pi(t) \right\|_{B(\ell_2) \otimes_{\min} C_\lambda^*(G) \otimes_{\min} B(H)} = \left\| \sum b(t) \otimes \lambda_G(t) \right\|_{B(\ell_2) \otimes_{\min} C_\lambda^*(G)}.$$

Proof Note that $\lambda_G \otimes \pi$ acts on the Hilbert space $K = \ell_2(G) \otimes_2 H \simeq \ell_2(G; H)$. Let $V : K \to K$ be the unitary operator taking $x = (x(t))_{t \in G}$ to $(\pi(t^{-1})x(t))_{t \in G}$. A simple calculation shows that

$$V^{-1}(\lambda_G(t) \otimes Id_H)V = \lambda_G(t) \otimes \pi(t).$$

$\square$

We will often use the following immediate consequence:

Corollary 3.16 *For any unitary representation* $\pi : G \to B(H)$, *the linear map*

$$\sigma_\pi : \text{span}[\lambda_G(G)] \to B(\ell_2(G) \otimes_2 H) \text{ defined by}$$
$$\sigma_\pi(\lambda_G(g)) = \lambda_G(t) \otimes \pi(t) \ (\forall t \in G)$$

*extends to a (contractive) *-homomorphism from* $C_\lambda^*(G)$ *to* $B(\ell_2(G) \otimes_2 H)$.

Remark 3.17 Let $\mathbb{F}$ be a free group with free generators (g_j). Then for any finitely supported sequence of scalars (a_j), for any H and for any family (u_j) of unitary operators in $B(H)$ we have

$$\left\| \sum a_j \lambda(g_j) \otimes u_j \right\|_{\min} = \left\| \sum a_j \lambda(g_j) \right\|.$$

Indeed, this follows from (3.13) applied to the function a defined by $a(g_j) = a_j$ and $= 0$ elsewhere, and to the unique unitary representation π of F such that $\pi(g_j) = u_j$.

Proposition 3.18 *Let G be a discrete group and let $\Gamma \subset G$ be a subgroup. Then the correspondence $\lambda_\Gamma(t) \to \lambda_G(t)$, $(t \in \Gamma)$ extends to an isometric $(C*-algebraic)$ embedding $J_\lambda : C_\lambda^*(\Gamma) \to C_\lambda^*(G)$. Moreover there is a completely contractive and completely positive projection P_λ from $C_\lambda^*(G)$ onto the range of this embedding, taking $\lambda_G(t)$ to 0 for any $t \notin \Gamma$.*

Proof Let $Q = G/\Gamma$ and let $G = \bigcup_{q \in Q} \Gamma g_q$ be the partition of G into (disjoint) right cosets. For convenience, let us denote by 1 the equivalence class of the unit element of G. Since $G \simeq \Gamma \times Q$, we have an identification

$$\ell_2(G) \simeq \ell_2(\Gamma) \otimes_2 \ell_2(Q)$$

such that

$$\forall t \in \Gamma \quad \lambda_G(t) = \lambda_\Gamma(t) \otimes I.$$

This shows of course that J_λ is an isometric embedding. Moreover, we have a natural (linear) isometric embedding $V : \ell_2(\Gamma) \to \ell_2(G)$ (note that the range of V coincides with $\ell_2(\Gamma) \otimes \delta_1$ in the preceding identification), such that $\lambda_\Gamma(t) = V^* \lambda_G(t) V$ for all $t \in \Gamma$. Let $u(x) = V^* x V$. Clearly for any $t \in G$ we have $u(\lambda_G(t)) = \lambda_\Gamma(t)$ if $t \in \Gamma$ and $u(\lambda_G(t)) = 0$ if $t \notin \Gamma$. Therefore $P_\lambda = J_\lambda u$ is the announced completely positive and completely contractive projection from $C_\lambda^*(G)$ onto $J_\lambda C_\lambda^*(\Gamma)$. $\qquad\square$

As an immediate application, we state for further use the following particular case:

Corollary 3.19 (The diagonal subgroup in $G \times G$) *Let $\Delta = \{(g,g) \mid g \in G\} \subset G \times G$ be the diagonal subgroup. There are:*

- *a complete isometry $J^\Delta : C_\lambda^*(G) \to C_\lambda^*(G) \otimes_{\min} C_\lambda^*(G)$ such that $J^\Delta(\lambda_G(t)) = \lambda_G(t) \otimes \lambda_G(t)$, and*
- *a c.p. map $Q^\Delta : C_\lambda^*(G) \otimes_{\min} C_\lambda^*(G) \to C_\lambda^*(G)$ with $\|Q^\Delta\| = 1$ such that $Q^\Delta(\lambda_G(t) \otimes \lambda_G(s)) = 0$ whenever $s \neq t$ and $Q^\Delta(\lambda_G(t) \otimes \lambda_G(t)) = \lambda_G(t)$.*

Proof We apply Proposition 3.18 to the subgroup Δ and we use the identification

$$C_\lambda^*(G) \otimes_{\min} C_\lambda^*(G) \simeq C_\lambda^*(G \times G)$$

which follows easily from the definition of both sides (see §4.3 for more such identifications). $\qquad \square$

The projection P_λ in the preceding proposition is an example of mapping associated to a "multiplier."

3.4 Multipliers

Let $\varphi : G \to \mathbb{C}$ be a (bounded) function and let π be a unitary representation of G. Let M_φ be the linear mapping defined on the linear span of $\{\pi(t) \mid t \in G\}$ by

$$\forall t \in G \quad M_\varphi(\pi(t)) = \varphi(t)\pi(t).$$

As anticipated in Proposition 3.3, we say that φ is a bounded (resp. c.b. rresp. c.p.) multiplier on $C_\pi^*(G)$ if M_φ extends to a bounded (resp. c.b. rresp. c.p.) linear map on $C_\pi^*(G)$.

We will be mainly interested in the cases when $\pi = \lambda_G$ or $\pi = U_G$.

In the commutative case (or when G is amenable) the bounded or c.b. multipliers of $C_\lambda^*(G)$ coincide with the linear combinations of positive definite functions, and the latter, as we explain next, are the c.p. multipliers. However, in general the situation is more complicated. The next statement characterizes the c.b. case. We may even include $B(H)$-valued multipliers.

Theorem 3.20 ([35, 136]) *Let G be a discrete group, H a Hilbert space. The following properties of a function $\varphi : G \to B(H)$ are equivalent:*

(i) *The linear mapping defined on* $\mathrm{span}[\lambda(t) \mid t \in G]$ *by*

$$M_\varphi(\lambda(t)) = \lambda(t) \otimes \varphi(t)$$

extends to a c.b. map
$$M_\varphi : C_\lambda^*(G) \to C_\lambda^*(G) \otimes_{\min} B(H) \subset B(\ell_2(G) \otimes_2 H) \text{ with } \|M_\varphi\|_{cb} \leq 1.$$

(ii) *There is a Hilbert space $\widehat{H}$ and bounded functions $x : G \to B(H, \widehat{H})$ and $y : G \to B(H, \widehat{H})$ with $\sup_{t \in G} \|x(t)\| \leq 1$ and $\sup_{s \in G} \|y(s)\| \leq 1$ such that*

$$\varphi(s^{-1}t) = y(s)^*x(t). \qquad \forall s, t \in G$$

Proof Assume (i). Then by Theorem 1.50 there are a Hilbert space $\widehat{H}$, a representation $\pi : C^*_\lambda(G) \to B(\widehat{H})$ and operators $V_j : \ell_2(G) \otimes_2 H \to \widehat{H}$ ($j = 1, 2$) with $\|V_1\|\|V_2\| \le 1$ such that

$$\forall \theta \in G \quad \lambda(\theta) \otimes \varphi(\theta) = M_\varphi(\lambda(\theta)) = V_2^* \pi(\lambda(\theta)) V_1. \tag{3.14}$$

We will use this for $\theta = s^{-1}t$, in which case we have $\langle \delta_{s^{-1}}, \lambda(\theta)\delta_{t^{-1}} \rangle = 1$. We define $x(t) \in B(H, \widehat{H})$ and $y(s) \in B(H, \widehat{H})$ by $x(t)h = \pi(\lambda(t)) V_1(\delta_{t^{-1}} \otimes h)$ and $y(s)k = \pi(\lambda(s))V_2(\delta_{s^{-1}} \otimes k)$. Note that when $\theta = s^{-1}t$

$$\langle \delta_{s^{-1}} \otimes k, (\lambda(\theta) \otimes \varphi(\theta))(\delta_{t^{-1}} \otimes h) \rangle = \langle k, \varphi(s^{-1}t)h \rangle,$$

and hence (3.14) implies

$$\langle k, \varphi(s^{-1}t)h \rangle = \langle k, y(s)^*x(t)h \rangle,$$

and we obtain (ii).

Conversely assume (ii). Define $\pi : C^*_\lambda(G) \to B(\ell_2(G) \otimes_2 \widehat{H})$ by $\pi(x) = x \otimes Id_{\widehat{H}}$. Let

$$V_j : \ell_2(G) \otimes_2 H \to \ell_2(G) \otimes_2 \widehat{H}$$

be defined by $V_1(\delta_t \otimes h) = \delta_t \otimes x(t)h$ and $V_2(\delta_s \otimes k) = \delta_s \otimes y(s)k$. Note that $\|V_1\| = \sup_{t \in G} \|x(t)\| \le 1$ and $\|V_2\| = \sup_{s \in G} \|y(s)\| \le 1$. Then for any θ, t, s, h, k we have

$$\langle \delta_s \otimes k, V_2^* \pi(\lambda(\theta)) V_1(\delta_t \otimes h) \rangle = \langle \delta_s, \lambda(\theta)\delta_t \rangle \langle k, y(s)^*x(t)h \rangle$$

$$= \langle \delta_s \otimes k, (\lambda(\theta) \otimes \varphi(\theta))(\delta_t \otimes h) \rangle,$$

equivalently $V_2^* \pi(\lambda(\theta)) V_1 = M_\varphi(\lambda(\theta))$, so the converse part of Theorem 1.50 yields (ii) $\Rightarrow$ (i). $\square$

In the particular case $\mathbb{C} = B(H)$ the preceding result yields:

Corollary 3.21 (Characterization of c.b. multipliers on $C^*_\lambda(G)$) *Consider a complex-valued function $\varphi : G \to \mathbb{C}$. Then $\|M_\varphi : C^*_\lambda(G) \to C^*_\lambda(G)\|_{cb} \le 1$ if and only if there are Hilbert space valued functions x, y with $\sup_t \|x(t)\| \le 1$ and $\sup_s \|y(s)\| \le 1$ such that*

$$\varphi(s^{-1}t) = \langle y(s), x(t) \rangle. \qquad \forall s, t \in G$$

Remark 3.22 (On positive definiteness) A function $\varphi : G \to \mathbb{C}$ is called positive definite if for any n and any $t_1, \ldots, t_n \in G$ the $n \times n$-matrix $[\varphi(t_i^{-1}t_j)]$ is positive (semi)definite, i.e. we have

$$\forall x \in \mathbb{C}^n \quad \sum \overline{x_i}x_j\varphi(t_i^{-1}t_j) \ge 0.$$

Equivalently

$$\forall x \in \mathbb{C}[G] \quad \sum \overline{x(s)} x(t) \varphi(s^{-1}t) \geq 0.$$

Using the scalar product defined by the latter condition, we find, after passing to the quotient and completing in the usual way, a Hilbert space H_φ and a mapping $\mathbb{C}[G] \to H_\varphi$ denoted by $x \mapsto \dot{x}$ with dense range (so that $\|\dot{x}\|^2_{H_\varphi} = \sum \overline{x(s)} x(t) \varphi(s^{-1}t)$ for all $x \in \mathbb{C}[G]$) and a unitary representation π_φ of G extending left translation on $\mathbb{C}[G]$. Let $\delta_e \in \mathbb{C}[G]$ denote the indicator function of the unit element of G. We have

$$\langle \dot{\delta}_e, \pi_\varphi(g)\dot{\delta}_e \rangle_{H_\varphi} = \langle \dot{\delta}_e, \dot{\delta}_g \rangle_{H_\varphi} = \varphi(g). \tag{3.15}$$

Thus φ is a (diagonal) matrix coefficient of π.

Conversely, any φ of the form $\varphi(g) = \langle \xi, \pi(g)\xi \rangle$ (with π unitary and $\xi \in H_\pi$) is positive definite.

Proposition 3.23 *Let $\varphi : G \to \mathbb{C}$. The following are equivalent:*

(i) *φ is a c.p. multiplier of $C^*_\lambda(G)$.*
(ii) *φ is positive definite.*

Moreover, in that case we have $\|M_\varphi\| = \|M_\varphi\|_{cb} = \varphi(e)$ where e is the unit of G.

Proof Assume (i). Let $t_1, \ldots, t_n \in G$. Consider the matrix a defined by $a_{ij} = \lambda_G(t_i)^{-1} \lambda_G(t_j)$. Clearly $a \in M_n(C^*_\lambda(G))_+$. Then $(Id_{M_n} \otimes M_\varphi)(a) = [\varphi(t_i^{-1} t_j) a_{ij}] \in M_n(C^*_\lambda(G))_+$. Therefore, for any $x_1, \ldots, x_n \in \ell_2(G)$ we have $\sum \varphi(t_i^{-1} t_j) \langle x_i, a_{ij} x_j \rangle \geq 0$. Choosing $x_j = \lambda_j \delta_{t_j^{-1}}$ ($\lambda_j \in \mathbb{C}$) we find $\langle x_i, a_{ij} x_j \rangle = \overline{\lambda_i} \lambda_j$ for all i, j, and we conclude that φ is positive definite.

Assume (ii). By (3.15) we have for any $g \in G$

$$M_\varphi(\lambda_G(g)) = \varphi(g)\lambda_G(g) = V^*([\lambda_G \otimes \pi_\varphi](g))V$$

where $V : \ell_2(G) \to \ell_2(G) \otimes_2 H_\varphi$ is defined by $V(h) = h \otimes \dot{\delta}_e$. By Corollary 3.16 we have $M_\varphi(\cdot) = V^*(\sigma_\pi(\cdot))V$, and hence M_φ is c.p. on $C^*_\lambda(G)$. Moreover $M_\varphi(1) = \varphi(e)1$, so $\|M_\varphi(1)\| = \varphi(e)$. $\square$

Remark 3.24 The reader can easily check that the preceding statement remains valid for $B(H)$-valued functions, in analogy with Theorem 3.20, for the natural extension of positive definiteness, defined by requesting that $[\varphi(t_i^{-1} t_j)] \in M_n(B(H))_+$ for all n. Such functions are sometimes called *completely* positive definite.

In the preceding construction, we associated a linear mapping M_φ to a function φ. We now go conversely. We will associate to a c.b. mapping a multiplier. In other words, we will describe a linear projection from the set of c.b. maps to the subspace formed by those associated to multipliers.

Proposition 3.25 (Haagerup) *Let $u : C_\lambda^*(G) \to C_\lambda^*(G)$ be a c.b. map. Then the function φ_u defined by (recall e is the unit of G)*

$$\varphi_u(t) = \langle \delta_t, u(\lambda_G(t))\delta_e \rangle$$

is a c.b. multiplier on $C_\lambda^*(G)$ with $\|M_{\varphi_u}\|_{cb} \le \|u\|_{cb}$. If u is c.p. then the multiplier is also c.p.

If u has finite rank then $\varphi_u \in \ell_2(G)$.

Moreover, if $u = M_\varphi$ then $\varphi_u = \varphi$.

Proof We have

$$\|Id_{C_\lambda^*(G)} \otimes u : C_\lambda^*(G) \otimes_{\min} C_\lambda^*(G) \to C_\lambda^*(G) \otimes_{\min} C_\lambda^*(G)\| \le \|u\|_{cb}.$$

It is easy to see that $C_\lambda^*(G) \otimes_{\min} C_\lambda^*(G)$ can be identified with $C_\lambda^*(G \times G)$. With this identification, we have, for the mappings J^Δ, Q^Δ in Corollary 3.19, for any $t \in G$

$$\varphi_u(t)\lambda_G(t) = Q^\Delta[Id_{C^*(G)} \otimes u]J^\Delta(\lambda_G(t)).$$

In other words, $M_{\varphi_u} = Q^\Delta[Id_{C^*(G)} \otimes u]J^\Delta$. All the assertions are now evident. We just note that if u has rank 1, say $u(x) = f(x)y$ with $f \in C_\lambda^*(G)^*$ and $y \in C_\lambda^*(G)$, then $\varphi_u(t) = f(\lambda_G(t))y(t)$, and $t \mapsto f(\lambda_G(t))$ is bounded while $y(t) = \langle \delta_t, y\delta_e \rangle$ is in $\ell_2(G)$; this shows $\varphi_u \in \ell_2(G)$. $\qquad\square$

Remark 3.26 With the notation of the next section we have

$$\varphi_u(t) = \tau_G(\lambda_G(t)^* u(\lambda_G(t))),$$

while with that of §11.2 it becomes $\varphi_u(t) = \langle \lambda_G(t), u(\lambda_G(t)) \rangle_{L_2(\tau_G)}$.

The preceding two statements combined show that if u is decomposable as a linear combination of c.p. maps on $C_\lambda^*(G)$ (as in Chapter 6) then φ_u is a linear combination of positive definite functions. In particular:

Corollary 3.27 *Let $\varphi : G \to \mathbb{C}$. The associated mapping M_φ is decomposable on $C_\lambda^*(G)$ if and only if φ is a linear combination of positive definite functions.*

We will now complete the description started in Proposition 3.3 of multipliers on the full algebra $C^*(G)$. In this case the picture is simpler.

Proposition 3.28 *Let $\varphi : G \to \mathbb{C}$. The following are equivalent:*

 (i) *φ is a bounded multiplier on $C^*(G)$.*

 (ii) *φ is a linear combination of positive definite functions.*

(iii) *φ is c.b. multiplier on $C^*(G)$.*

Moreover, φ is positive definite if and only if M_φ is c.p. on $C^*(G)$.

Proof We already know (i) ⇔ (iii) from Proposition 3.3. Assume (i). Then by Proposition 3.3 φ satisfies (3.3) for some π, η, ξ. By the polarization formula, we can rewrite φ as a linear combination of four functions of the form $t \mapsto \langle \xi, \pi(t)\xi \rangle$ with $\eta = \xi$. But the latter are clearly positive definite. This shows (i) ⇒ (ii). Assume φ positive definite. By (3.15) and by the case $\xi = \eta$ in Proposition 3.3 M_φ is c.p. and hence a fortiori c.b. Now (ii) ⇒ (iii) is clear. □

3.5 Group von Neumann Algebra

We denote by $M_G \subset B(\ell_2(G))$ the von Neumann algebra generated by λ_G. This means that $M_G = \lambda_G(G)''$. Equivalently M_G is the weak* closure of the linear span of $\lambda_G(G)$, and also the weak* closure of $C_\lambda(G)$. See §A.16 for some background on von Neumann algebras (in particular on the bicommutant Theorem A.46).

Let $f \in \ell_2(G)$. Note that a priori, the operator of left convolution by f, $T_f : x \mapsto f * x$ is only bounded from $\ell_2(G)$ to $\ell_\infty(G)$. An operator $T \in B(\ell_2(G))$ belongs to M_G if and only if there is a (uniquely determined by $f = T_f(\delta_e)$) function $f \in \ell_2(G)$ such that $x \mapsto f * x$ defines a bounded operator on $\ell_2(G)$ such that $T = T_f$.

We have

$$M_G' = \lambda_G(G)' = \rho_G(G)'' \text{ and } \rho_G(G)' = M_G.$$

Let $\Gamma \subset G$ be a subgroup. Since the embedding $J_\lambda : C_\lambda^*(\Gamma) \to C_\lambda^*(G)$ in Proposition 3.18 is clearly bicontinuous with respect to the weak* topologies of $B(\ell_2(\Gamma))$ and of $B(\ell_2(G))$, it extends to an embedding

$$M_\Gamma \subset M_G,$$

with which we may identify M_Γ to a von Neumann subalgebra of M_G.

Let $\{\delta_t \mid t \in G\}$ denote the canonical basis of $\ell_2(G)$. There is a distinguished tracial state τ_G defined on M_G by

$$\tau_G(T) = \langle \delta_e, T(\delta_e) \rangle.$$

Of course this makes sense on the whole of $B(\ell_2(G))$, but it is tracial only if we restrict to M_G:

$$\forall S, T \in M_G \quad \tau_G(TS) = \tau_G(ST).$$

Clearly τ_G is "normal" (meaning continuous for the weak* topology of $B(\ell_2(G))$) and faithful (meaning $\tau_G(T^*T) = 0 \Rightarrow T = 0$) and $\tau_G(1) = 1$. Thus (M_G, τ_G) is the basic example of a "tracial (or noncommutative) probability space" that we will consider in Chapter 12 when we discuss the Connes embedding problem.

Remark 3.29 Let φ be as in Corollary 3.21. Let $\Phi(s,t) = \varphi(s^{-1}t)$. Then the Schur multiplier $u_\Phi : B(\ell_2(G)) \to B(\ell_2(G))$ associated to Φ according to (iii) in Theorem 1.57 is completely contractive on $B(\ell_2(G))$ if and only if φ satisfies the equivalent conditions in Corollary 3.21. Moreover, the latter Schur multiplier is weak* continuous, meaning continuous from $B(\ell_2(G))$ to $B(\ell_2(G))$ when both spaces are equipped with the weak* topology. Therefore, if we restrict to M_G we obtain a weak* continuous (also called normal) complete contraction from M_G to M_G that extends the multiplier $M_\varphi : C_\lambda^*(G) \to C_\lambda^*(G)$. We will call the resulting maps weak* continuous multipliers on M_G.

A similar argument, based on Proposition 3.23, shows that φ is positive definite if and only if M_φ extends to a weak* continuous c.p. multiplier on M_G.

Lastly, the conclusion of Proposition 3.25 holds with the same proof for any c.b. map $u : M_G \to M_G$. The resulting multiplier M_{φ_u} is weak* continuous on M_G, with $\|M_{\varphi_u}\|_{cb} \leq \|u\|_{cb}$. Moreover, if u is c.p. on M_G, so is M_{φ_u}.

3.6 Amenable groups

We review some basic facts on amenability.

A discrete group G is called amenable if it admits an invariant mean, i.e. a functional φ in $\ell_\infty(G)_+^*$ with $\varphi(1) = 1$ such that $\varphi(\delta_t * f) = \varphi(f)$ for any f in $\ell_\infty(G)$ and any t in G.

Theorem 3.30 *The following are equivalent:*

   (i) *G is amenable.*
   (i)' *There is a net (h_i) in the unit sphere of $\ell_2(G)$ that is approximately translation invariant, i.e. such that $\|\lambda_G(t)h_i - h_i\|_2 \to 0$ for any $t \in G$.*
   (ii) *$C^*(G) = C_\lambda^*(G)$.*
   (iii) *For any finitely supported function $f : G \to \mathbb{C}$ we have*
         $\left| \sum f(t) \right| \leq \left\| \sum f(t)\lambda_G(t) \right\|$.
   (iii)' *For any finite subset $E \subset G$, we have $|E| = \left\| \sum_{t \in E} \lambda_G(t) \right\|$.*
   (iv) *There is a generating subset $S \subset G$ with $e \in S$ such that, for any finite subset $E \subset S$, we have $|E| = \left\| \sum_{t \in E} \lambda_G(t) \right\|$.*
   (v) *M_G is injective.*

Proof Assume (i). Let φ be the invariant mean. Note that φ is in the unit ball of $\ell_1(G)_+^{**}$. Therefore, there is a net (φ_i) in the unit ball of $\ell_1(G)_+$ tending in the sense of $\sigma(\ell_1(G)^{**}, \ell_1(G)^*)$ to φ. Let **1** be the constant function equal to 1 on G. Since $\varphi_i(1) \to 1$, we may assume after renormalization that $\varphi_i(1) = \|\varphi_i\|_{\ell_1(G)} = 1$. Fix $t \in G$. Since $\delta_t * \varphi = \varphi$, we have $\delta_t * \varphi_i - \varphi_i \to 0$ when $i \to \infty$. But since $\delta_t * \varphi_i - \varphi_i$ lies in $\ell_1(G)$ this means that $\lim_{i \to \infty}(\delta_t * \varphi_i - \varphi_i) = 0$ for the weak topology of $\ell_1(G)$. By (Mazur's) Theorem A.9, passing to convex combinations of elements of a subnet (here we leave some details to the reader, see Remark A.10) we may assume that $\lim_{i \to \infty} \|\delta_t * \varphi_i - \varphi_i\|_{\ell_1(G)} = 0$. A priori, this was obtained for each fixed t, but, by suitably refining the argument (here again we skip some details), we can obtain the same for each finite subset $T \subset G$. Let $h_i = \sqrt{\varphi_i}$. We claim that $\|\delta_t * h_i - h_i\|_2 \to 0$ for any $t \in T$. This claim clearly implies (i)'. To check the claim, using $|x^{1/2} - y^{1/2}| \le |x - y|^{1/2}$ for any $x, y \in \mathbb{R}_+$, we observe that $|\delta_t * h_i(s) - h_i(s)| \le |\delta_t * \varphi_i(s) - \varphi_i(s)|^{1/2}$ and hence $\|\delta_t * h_i - h_i\|_2 \to 0$ for any $t \in T$. This shows (i) $\Rightarrow$ (i)'.

Assume (i)'. Let $x = \sum x(t)\lambda_G(t) \in \text{span}[\lambda_G(t) \mid t \in G]$. Let $\pi : G \to B(H)$ be any unitary representation. By the absorption principle (3.13) $\|\sum x(t)\lambda_G(t)\| = \|\sum x(t)\pi(t) \otimes \lambda_G(t)\|$. We claim that $\|\sum x(t)\pi(t) \otimes \lambda_G(t)\| \ge \|\sum x(t)\pi(t)\|$. Indeed, let f_i be the state on $B(\ell_2(G))$ defined by $f_i(T) = \langle h_i, T h_i \rangle$. Then we have clearly

$$\left\| [Id \otimes f_i] \left(\sum x(t)\pi(t) \otimes \lambda_G(t) \right) \right\| \le \left\| \sum x(t)\pi(t) \otimes \lambda_G(t) \right\|$$

but

$$[Id \otimes f_i] \left(\sum x(t)\pi(t) \otimes \lambda_G(t) \right) = \sum x(t)\pi(t) f_i(\lambda_G(t)) \to \sum x(t)\pi(t),$$

where at the last step we use $f_i(\lambda_G(t)) = \langle h_i, \delta_t * h_i \rangle \to 1$. This implies the claim and hence $\|\sum x(t)\lambda_G(t)\| \ge \|\sum x(t)\pi(t)\|$. Taking the sup over the πs we obtain (by "maximality" of U_G) $\|\sum x(t)\lambda_G(t)\| = \|\sum x(t)U_G(t)\|$. This shows (i)' $\Rightarrow$ (ii).

Assume (ii). Then (iii) holds by (3.2), and (iii) $\Rightarrow$ (iii)' $\Rightarrow$ (iv) are trivial.

Assume (iv). We will show (i)'. Fix E as in (iv). Let $M_E = |E|^{-1} \sum_{t \in E} \lambda_G(t)$ so that $\|M_E\| = 1$. There is a net (x_i) in the unit sphere of $\ell_2(G)$ such that $\|M_E(x_i)\| \to 1$. By the uniform convexity of $\ell_2(G)$ (see §A.3), this implies $\delta_t * x_i - x_i \to 0$ in $\ell_2(G)$ for any $t \in E$. Rearranging the net (here again we leave the details to the reader) we find a net (h_i) in the unit sphere of $\ell_2(G)$ such that the same holds for any $t \in S$, and since S generates G, still the same for any $t \in G$. This shows (iv) $\Rightarrow$ (i)'.

Assume (i)'. We will show (i). Let $\varphi \in \ell_\infty(G)^*$ be defined by

$$\forall x \in \ell_\infty(G) \quad \varphi(x) = \lim_{\mathcal{U}} \sum x(t)|h_i(t)|^2,$$

where $\mathcal{U}$ is an ultrafilter refining the net (see Remark A.6). Let $D_x \in B(\ell_2(G))$ be the diagonal operator associated to x. Note that $\varphi(x) = \lim_{\mathcal{U}} \langle h_i, D_x h_i \rangle$, and also

$$\lambda_G(t)D_x\lambda_G(t)^{-1} = D_{\delta_t * x}. \tag{3.16}$$

Therefore

$$\varphi(\delta_t * x) = \lim_{\mathcal{U}}\langle h_i, D_{\delta_t * x}h_i\rangle = \lim_{\mathcal{U}}\langle \lambda_G(t)h_i, D_x\lambda_G(t)h_i\rangle$$
$$= \lim_{\mathcal{U}}\langle h_i, D_x h_i\rangle = \varphi(x).$$

Thus φ is an invariant mean, so (i) holds. This proves the equivalence of (i)–(iv), (i)', and (iii)'. It remains to show that (i) and (v) are equivalent.

Assume (i). We will show that there is a c.p. projection $P : B(\ell_2(G)) \to M_G$ with $\|P\| = 1$. Let $T \in B(\ell_2(G))$. We define $\Phi_T : G \to B(\ell_2(G))$ by $\Phi_T(g) = \rho_G(g)T\rho_G(g)^{-1}$. We will define $P(T)$ as the "integral" with respect to φ of the function Φ_T, but some care is needed since φ is not really a measure on G. Let $[T(s,t)]$ be the "matrix" associated to T defined by $T(s,t) = \langle \delta_s, T\delta_t\rangle$ $(s,t \in G)$. Observe that $g \mapsto \Phi_T(g)(s,t)$ is in $\ell_\infty(G)$. Then we set

$$P(T)(s,t) = \varphi(\Phi_T(\cdot)(s,t)).$$

This defines a matrix and it is easy to see that the associated linear operator on span$[\delta_t \mid t \in G]$ extends to a bounded one (still denoted by $P(T)$) on $\ell_2(G)$ such that $\|P(T)\| \le \|T\|$. We have $\Phi_T(g)(s,t) = T(sg,tg)$ and hence, by the left invariance of φ, $P(T)(s,t) = P(T)(st^{-1},e)$. This shows that $P(T)$ acts on $\ell_2(G)$ as a left convolution bounded operator, in other words $P(T) \in M_G$. Moreover, if $T \in M_G$ then T commutes with ρ_G so we have $P(T) = T$. This proves that $P : B(\ell_2(G)) \to M_G$ is a contractive projection. A simple verification left to the reader shows that it is c.p. (but this is automatic by Tomiyama's Theorem 1.45). This shows (i) $\Rightarrow$ (v).

Assume (v). Let $P : B(\ell_2(G)) \to M_G$ be a projection with $\|P\| = 1$. Invoking Theorem 1.45 again, we know that P is a c.p. conditional expectation. We define

$$\forall x \in \ell_\infty(G) \quad \varphi(x) = \tau_G(P(D_x)) = \langle \delta_e, P(D_x)\delta_e\rangle.$$

Clearly $\varphi \in \ell_\infty(G)^*_+$, $\varphi(1) = 1$ and by (3.16), (1.28) and the trace property of τ_G

$$\forall t \in G \quad \varphi(\delta_t * x) = \tau_G[P(\lambda_G(t)D_x\lambda_G(t)^{-1})] = \tau_G[\lambda_G(t)P(D_x)\lambda_G(t)^{-1}]$$
$$= \tau_G[P(D_x)] = \varphi(x).$$

Thus φ is an invariant mean on G. This shows (v) $\Rightarrow$ (i). $\qquad\square$

Remark 3.31 If the generating set S is finite, the condition (iv) obviously reduces (by the triangle inequality) to

$$|S| = \left\|\sum\nolimits_{t \in S}\lambda_G(t)\right\|.$$

Remark 3.32 The net (h_i) in (i)' is sometimes called asymptotically left invariant. By density (and after renormalization) when it exists, it can always be found in the group algebra $\mathbb{C}[G]$.

Remark 3.33 (On Følner sequences) It is well known (see e.g. [194]) that for any amenable discrete group G the net (h_i) appearing in (i)' in Theorem 3.30 can be chosen of the form $h_i = 1_{B_i}|B_i|^{-1/2}$ for some family (B_i) of finite subsets of G. For (h_i) of the latter form, (i)' boils down to the assertion that the symmetric differences $(tB_i)\triangle B_i$ satisfy

$$\forall t \in G \quad |tB_i \triangle B_i||B_i|^{-1} \to 0.$$

A net of finite subsets (B_i) satisfying this is called a Følner net, and a Følner sequence when the index set is $\mathbb{N}$. Thus a (resp. countable) group G is amenable if and only if it admits a Følner net (resp. sequence). For instance, for $G = \mathbb{Z}^d$ ($1 \le d < \infty$), the sequence $B_n = [-n, n]^d$ is a Følner sequence.

This gives us the following special property of the *reduced C^*-algebra*, called the CPAP in the sequel (see Definition 4.8):

Lemma 3.34 *If G is amenable, there is a net of finite rank maps $u_i \in CP(C^*_\lambda(G), C^*_\lambda(G))$ (resp. $u_i \in CP(C^*(G), C^*(G))$) with $\|u_i\| = 1$ that tends pointwise to the identity on $C^*_\lambda(G)$ (resp. $C^*(G)$). Moreover, in both cases the u_i's are multiplier operators.*

Proof By Remark 3.32, there is a net (h_i) in $\mathbb{C}[G]$ in the unit sphere of $\ell_2(G)$ such that $\|\lambda_G(t)h_i - h_i\|_2 \to 0$ for any t in G. Let $h_i^*(t) = \overline{h_i(t^{-1})}$ ($t \in G$). A simple verification show that $\varphi_i = h_i^* * h_i$ is a positive definite function on G such that $\varphi_i(e) = \|h_i\|_2^2 = 1$. Moreover, φ_i is finitely supported and tends pointwise to the constant function 1 on G. Let u_i be the associated multiplier operator on $C^*_\lambda(G)$ (resp. $C^*(G)$). Its rank being equal to the cardinality of the support of φ_i is finite. By Proposition 3.23 (resp. Proposition 3.28), u_i is c.p. and since $u_i(1) = \varphi_i(e)1 = 1$, we have $\|u_i\| = 1$ by (1.20). For any $x = \sum x(t)\lambda_G(t)$ (resp. $x = \sum x(t)U_G(t)$) with x finitely supported, $u_i(x)$ obviously tends to x in the norm of $C^*_\lambda(G)$ (resp. $C^*(G)$). Since such finite sums are dense in $C^*_\lambda(G)$ (resp. $C^*(G)$) and $\sup_i \|u_i\| < \infty$, we conclude that $u_i(x) \to x$ for any $x \in C^*_\lambda(G)$ (resp. $C^*(G)$). $\square$

Remark 3.35 (Examples of amenable groups) All commutative groups are amenable. If G is commutative (and discrete), its dual $\widehat{G}$ is defined as the group formed of all homomorphisms $\gamma : G \to \mathbb{T}$, which is compact for the pointwise convergence topology. For any finitely supported function $f : G \to \mathbb{C}$ we define its "Fourier transform" by $\widehat{f}(\gamma) = \sum f(g)\overline{\gamma(g)}$. (This is the usual convention but we could remove the bar from $\overline{\gamma(g)}$ if we wished). As is entirely classical $f \mapsto \widehat{f}$ extends to an isometric isomorphism from $\ell_2(G)$ to $L_2(\widehat{G}, m)$, where m is the normalized Haar measure on $\widehat{G}$, and convolution of two functions on G is transformed into the pointwise product of their Fourier transforms. Using the latter fact one shows that the correspondence $f \mapsto \widehat{f}$ extends to an isometric isomorphism from $C_\lambda^*(G)$ to the C^*-algebra $C(\widehat{G})$ of all continuous functions on $\widehat{G}$. Thus in the commutative case we have

$$C^*(G) = C_\lambda^*(G) \simeq C(\widehat{G}). \tag{3.17}$$

All finitely generated groups of polynomial growth are amenable. The growth is defined using the length. If G is generated by a symmetric set S the smallest number of elements of S needed to write an element $g \in G$ (as a word in letters in S) is denoted by $\ell_S(g)$. The growth function is the function $\Phi(R) = |\{g \in G \mid \ell_S(g) \le R\}|$. The group G is called of polynomial growth if $\Phi(R)$ grows less than a power of R when $R \to \infty$. For instance $G = \mathbb{Z}^n$ is of polynomial growth (but $\mathbb{F}_n$ is not whenever $n \ge 2$).

Remark 3.36 By Kesten's famous work on the spectral radius of random walks on the free group $\mathbb{F}_n$ with n generators, the set $S_1 \subset \mathbb{F}_n$ formed of the $2n$ elements of length 1 (i.e. these are either generators or their inverses), satisfies

$$\left\| \sum\nolimits_{s \in S_1} \lambda_{\mathbb{F}_n}(s) \right\| = 2\sqrt{2n-1}. \tag{3.18}$$

Kesten also observed that it is not difficult to deduce from this that for any group G and any symmetric subset $S \subset G$ with $|S| = k$ we have

$$\left\| \sum\nolimits_{s \in S} \lambda_G(s) \right\| \ge 2\sqrt{k-1}.$$

Akemann and Ostrand [2] proved that any $S \subset S_1$ in $\mathbb{F}_n$ with $|S| = k$ satisfies

$$\left\| \sum\nolimits_{s \in S} \lambda_{\mathbb{F}_n}(s) \right\| = 2\sqrt{k-1}. \tag{3.19}$$

In particular

$$\left\| \sum\nolimits_{j=1}^{n} \lambda_{\mathbb{F}_n}(g_j) \right\| = 2\sqrt{n-1}. \tag{3.20}$$

The subsets S of a discrete group for which (3.19) holds have been characterized by Franz Lehner in [166], as the translates of the union of a free set and the unit.

Let $S \subset \mathbb{F}_n$ be the set formed of the unit and the n free generators, so that $|S| = n+1$. Then a variant of what precedes is that for $G = \mathbb{F}_n$

$$\left\| \sum\nolimits_{s \in S} \lambda_G(s) \right\| = 2\sqrt{n}. \tag{3.21}$$

When $n \ge 2$, this is $< n+1$, and hence (iii) or (iv) in Theorem 3.30 fails. This shows that $\mathbb{F}_n$ is not amenable for $n \ge 2$.

Since amenability passes to subgroups (by Proposition 3.18 and (i)$\Leftrightarrow$ (iv) in Theorem 3.30), any group containing a copy of $\mathbb{F}_2$ as a subgroup is nonamenable. The converse, whether nonamenable groups must contain $\mathbb{F}_2$, remained a major open question for a long time but was disproved by A. Olshanskii, see [126] for details. See Monod's [178] for what seems to be currently the simplest construction of nonamenable groups not containing $\mathbb{F}_2$ as a subgroup.

3.7 Operator space spanned by the free generators in $C^*_\lambda(\mathbb{F}_n)$

The next statement gives us a description up to complete isomorphism of the span of the generators in $C^*_\lambda(\mathbb{F}_n)$ (and also implicitly in $C^*_\lambda(\mathbb{F}_\infty)$). See [168] for a more precise (completely isometric) description.

Theorem 3.37 *Let $(g_j)_{1 \le j \le n}$ be the generators in $\mathbb{F}_n(n \ge 1)$. Then for any Hilbert space H and any $a_j \in B(H)(1 \le j \le n)$ we have*

$$\max\left\{\left\|\sum a_j^* a_j\right\|^{1/2}, \left\|\sum a_j a_j^*\right\|^{1/2}\right\} \le \left\|\sum a_j \otimes \lambda_{\mathbb{F}_n}(g_j)\right\|_{\min}$$
$$\le \left\|\sum a_j^* a_j\right\|^{1/2} + \left\|\sum a_j a_j^*\right\|^{1/2}.$$
$$(3.22)$$

In particular for any $\alpha_j \in \mathbb{C}$ we have

$$\left(\sum |\alpha_j|^2\right)^{1/2} \le \left\|\sum \alpha_j \lambda_{\mathbb{F}_n}(g_j)\right\| \le 2\left(\sum |\alpha_j|^2\right)^{1/2}.$$

Proof We will first prove the upper bound in (3.22). Let $C_i^+ \subset \mathbb{F}_n$ (resp. $C_i^- \subset \mathbb{F}_n$) be the subset formed by all the reduced words which start with g_i (resp. g_i^{-1}). Note: except for the empty word e, every element of G can be written as a reduced word in the generators admitting a well-defined "first" and "last" letter (where we read from left to right). Let P_i^+ (resp. P_i^-) be the orthogonal projection on $\ell_2(\mathbb{F}_n)$ with range $\overline{\text{span}}[\delta_t \mid t \in C_i^+]$ (resp. $\overline{\text{span}}[\delta_t \mid t \in C_i^-]$). The $2n$ projections $\{P_i^+, P_i^- \mid 1 \le i \le n\}$ are mutually orthogonal. Then it is easy to check that

$$\lambda_{\mathbb{F}_n}(g_j) = \lambda_{\mathbb{F}_n}(g_j)P_j^- + \lambda_{\mathbb{F}_n}(g_j)(1 - P_j^-)$$
$$= \lambda_{\mathbb{F}_n}(g_j)P_j^- + P_j^+ \lambda_{\mathbb{F}_n}(g_j)(1 - P_j^-)$$
$$= \lambda_{\mathbb{F}_n}(g_j)P_j^- + P_j^+ \lambda_{\mathbb{F}_n}(g_j)$$

so that setting $\lambda_{\mathbb{F}_n}(g_j) = x_j + y_j$ with $x_j = \lambda_{\mathbb{F}_n}(g_j)P_j^-$ and $y_j = P_j^+ \lambda_{\mathbb{F}_n}(g_j)$ we find

$$\left\| \sum x_j^* x_j \right\| = \left\| \sum P_j^- \right\| \le 1 \quad \text{and} \quad \left\| \sum y_j y_j^* \right\| = \left\| \sum P_j^+ \right\| \le 1.$$

Therefore for any finite sequence (a_j) in $B(H)$ we have by (1.11) (note $a_j \otimes x_j = (a_j \otimes 1)(1 \otimes x_j)$ and similarly for $a_j \otimes y_j$)

$$\left\| \sum a_j \otimes \lambda_{\mathbb{F}_n}(g_j) \right\| \le \left\| \sum a_j \otimes x_j \right\| + \left\| \sum a_j \otimes y_j \right\|$$

$$\le \left\| \sum a_j a_j^* \right\|^{1/2} + \left\| \sum a_j^* a_j \right\|^{1/2}.$$

The inverse inequality follows from a more general one valid for any discrete group G: for any finitely supported function $a : G \to B(H)$ we have

$$\max \left\{ \left\| \sum a(t)^* a(t) \right\|^{1/2}, \left\| \sum a(t) a(t)^* \right\|^{1/2} \right\} \le \left\| \sum a(t) \otimes \lambda_G(t) \right\|_{\min}.$$

$$(3.23)$$

To check this, let $T = \sum a(t) \otimes \lambda_G(t)$. For any h in B_H we have $T(h \otimes \delta_e) = \sum a(t)h \otimes \delta_t$ so that $\|T(h \otimes \delta_e)\| = \left(\sum_t \|a(t)h\|^2 \right)^{1/2}$ and hence

$$\left\| \sum a(t)^* a(t) \right\|^{1/2} = \sup_{h \in B_H} \left(\sum \|a(t)h\|^2 \right)^{1/2} \le \|T\|.$$

Similarly since $T^* = \sum a(t^{-1})^* \otimes \lambda_G(t)$ we find

$$\left\| \sum a(t) a(t)^* \right\|^{1/2} \le \|T^*\| = \|T\|$$

and we obtain (3.23). In the case $G = \mathbb{F}_n$, (3.23) implies the left-hand side of (3.22). The second inequality follows by taking $a_j = \alpha_j 1$. □

Corollary 3.38 *For any $n, N \ge 1$ and any unitaries $a \in \mathbb{U}_n$, $x_1, \ldots, x_n \in \mathbb{U}_N$ we have*

$$\left\| \sum_{i,j=1}^n a_{ij} x_i \otimes \lambda_{\mathbb{F}_n}(g_j) \right\| \le 2\sqrt{n}.$$

Proof Let $a_j = \sum_i a_{ij} x_i$. Since a is unitary a simple verification (using (A.13)) shows that we have $\left\| \sum a_j^* a_j \right\|^{1/2} \le \left\| \sum x_j^* x_j \right\|^{1/2} = \sqrt{n}$ and $\left\| \sum a_j a_j^* \right\|^{1/2} \le \left\| \sum x_j x_j^* \right\|^{1/2} = \sqrt{n}$. □

3.8 Free products of groups

Let $(G_i)_{i \in I}$ be a family of groups. The free product $G = *_{i \in I} G_i$ is a group containing each G_i as a subgroup and possessing the following universal property that characterizes it: for any group G' and any family of homomorphisms $f_i : G_i \to G'$, there is a *unique* homomorphism $f : G \to G'$ extending each f_i.

When $I = \{1, 2\}$ we denote $G_1 * G_2$ the free product $*_{i \in I} G_i$.

When $I = \{1, \ldots, n\}$ and $G_1 = \cdots = G_n = \mathbb{Z}$ it is easy to see that $G = *_{i \in I} G_i$ can be identified with $\mathbb{F}_n$.

More generally, any free group $\mathbb{F}$ that is generated by a family of free elements $(g_i)_{i \in I}$ can be identified with the free product $*_{i \in I} G_i$ relative to $G_i = \mathbb{Z}$ for all $i \in I$. We denote that group by $\mathbb{F}_I$.

It is well known that any group G is a quotient of some free group. Indeed, if G is generated by a family $(t_i)_{i \in I}$, let $f : \mathbb{F}_I \to G$ be the (unique) homomorphism such that $f(g_i) = t_i$ for all $i \in I$. Then f is onto G. Thus $G \simeq \mathbb{F}_I / \ker(f)$. The analogous fact for C^*-algebras is the next statement.

Proposition 3.39 *Any unital C^*-algebra A is a quotient of $C^*(\mathbb{F}_I)$ for some set I. If A is separable (resp. is generated by n unitaries) then we can take $I = \mathbb{N}$ (resp. $I = \{1, \ldots, n\}$).*

Proof Let G be the unitary group of A. Let $f : \mathbb{F}_I \to G$ be a surjective homomorphism. Let $\pi : C^*(\mathbb{F}_I) \to A$ be the associated $*$-homomorphism, as in Remark 3.1. By the Russo–Dye Theorem A.18, the range of π is dense in A, but since it is closed (see §A.14), π must be surjective. Thus A is a quotient of $C^*(\mathbb{F}_I)$. If A is generated as a C^*-algebra by a family of unitaries $(u_i)_{i \in I}$, we can replace G in the preceding argument by the group generated by $(u_i)_{i \in I}$. This settles the remaining assertions. $\qquad\square$

Remark 3.40 As we saw in Remark 3.36, $\mathbb{F}_2 = \mathbb{Z} * \mathbb{Z}$ is not amenable. More generally it can be shown that $\mathbb{Z}_n * \mathbb{Z}_m$ is not amenable if $n \geq 2$ and $m \geq 3$, and in fact contains a subgroup isomorphic to $\mathbb{F}_\infty$. The group $\mathbb{Z}_2 * \mathbb{Z}_2$ is a slightly surprising exception, it is amenable because it happens to have polynomial growth (an exercise left to the reader).

3.9 Notes and remarks

The main results of this section are by now well known, and sometimes for general locally compact groups (for instance Proposition 3.5 is proved in greater generality in [225]), but we choose to focus on the discrete ones. Section 3.2 on free groups is just a reformulation of operator space duality illustrated on the pair (ℓ_1, ℓ_∞). Lemmas 3.9 and 3.10 are elementary facts from operator space theory (see [80, 208]). The classical reference that exploited C^*-algebra theory in noncommutative harmonic analysis is Eymard's thesis [85]. The name of Fell is attached to the notions of weak containment and weak equivalence of group representations, which apparently led him to the principle enunciated in Theorem 3.15. Concerning multipliers, those considered in Theorem 3.20 are sometimes called Herz–Schur multipliers (in honor of Carl Herz). The characterization in Theorem 3.20 and its Corollary is due to Jolissaint [136], but the simple proof we give is due to Bożejko and Fendler [35]. Our treatment is inspired by Haagerup's unpublished (but widely circulated) notes on multipliers, where in particular he proves Proposition 3.25. There are many known characterizations of amenability, the main one going back to Kesten, with variants due to Hulanicki and many authors. We refer the reader to [194] (or [199]) for details and references. Theorem 3.37 appears in [118]. In [168] Lehner gives an exact computation of the norm of $\sum a_j \otimes \lambda_{\mathbb{F}_n}(g_j)$ when the coefficients a_j are matricial or equivalently when $\dim(H) < \infty$.

4

C*-tensor products

A norm on a *-algebra $\mathcal{A}$ is called a C^*-norm if it satisfies

$$\|x\| = \|x^*\|, \qquad \|xy\| \le \|x\|\,\|y\| \quad \text{and} \quad \|x^*x\| = \|x\|^2 \qquad (4.1)$$

for any x, y in $\mathcal{A}$.

The completion A of $(\mathcal{A}, \|.\|)$ then becomes a C^*-algebra. It is useful to point out that, after completion, the norm is unique: there is only one C^*-norm on a C^*-algebra. In particular if two C^*-norms on $\mathcal{A}$ are distinct then they are not equivalent, since otherwise they would produce the same completion, where the C^*-norm is unique.

In particular any *-isomorphism between C^*-algebras must be isometric.

More generally, it is useful to record here that any injective *-homomorphism between C^*-algebras is automatically isometric. Consequently, a *-homomorphism between C^*-algebras must have a closed range, and the range is isometric to a quotient C^*-algebra of the source of the map. Indeed, the kernel of any *-homomorphism $u: A \to B$ is a (closed two-sided and self-adjoint) ideal $\mathcal{I} \subset A$. Passing to the quotient gives us an injective (and hence isometric) *-homomorphism $A/\mathcal{I} \to B$, which must have a closed range. When u is surjective we have $B \cong A/\mathcal{I}$. See Proposition A.24 for more details.

4.1 C*-norms on tensor products

Let A_1, A_2 be two C^*-algebras. Their algebraic tensor product $A_1 \otimes A_2$ is a *-algebra for the natural operations defined by

$$(a_1 \otimes a_2) \cdot (b_1 \otimes b_2) = a_1 b_1 \otimes a_2 b_2$$

and

$$(a_1 \otimes a_2)^* = a_1^* \otimes a_2^*.$$

Thus a norm $\| \ \|$ on $A_1 \otimes A_2$ is a C^*-norm if it satisfies (4.1) for any x, y in $A_1 \otimes A_2$.

This subject was initiated in the 1950s by Turumaru in Japan. Later work by Takesaki and Guichardet leads to the following result.

Theorem 4.1 *There is a minimal C^*-norm $\| \ \|_{min}$ and a maximal one $\| \ \|_{max}$, so that any C^*-norm $\| \cdot \|$ on $A_1 \otimes A_2$ must satisfy*

$$\|x\|_{min} \le \|x\| \le \|x\|_{max} \quad \forall x \in A_1 \otimes A_2. \tag{4.2}$$

We denote by $A_1 \otimes_{min} A_2$ (resp. $A_1 \otimes_{max} A_2$) the completion of $A_1 \otimes A_2$ for the norm $\| \ \|_{min}$ (resp. $\| \ \|_{max}$).

The maximal C^*-norm is easy to describe. We simply write

$$\|x\|_{max} = \sup \|\pi(x)\|_{B(H)} \tag{4.3}$$

where the supremum runs over all possible Hilbert spaces H and all possible *-homomorphisms $\pi : A_1 \otimes A_2 \to B(H)$.

The minimal (or spatial) norm can be described as follows: embed A_1 and A_2 as C^*-subalgebras of $B(H_1)$ and $B(H_2)$ respectively, then for any $x = \sum a_i^1 \otimes a_i^2$ in $A_1 \otimes A_2$, $\|x\|_{min}$ coincides with the norm induced by the space $B(H_1 \otimes_2 H_2)$, i.e. we have an embedding (i.e. an isometric *-homomorphism) of the completion, denoted by $A_1 \otimes_{min} A_2$, into $B(H_1 \otimes_2 H_2)$. The resulting C^*-algebra does not depend on the particular embeddings $A_1 \subset B(H_1)$ and $A_2 \subset B(H_2)$.

More generally, even if we allow completely isometric linear embeddings $A_1 \subset B(H_1)$ and $A_2 \subset B(H_2)$, we obtain the same norm (i.e. the min-norm) induced on $A_1 \otimes A_2$. So that, actually, the minimal tensor product of operator spaces, introduced in §1.1, coincides with the minimal C^*-tensor product when restricted to two C^*-algebras. See §1.1 for more on this.

Proof of Theorem 4.1 Let $x \mapsto \|x\|_\alpha$ be a C^*-norm on $A_1 \otimes A_2$. After completion we find a C^*-algebra $A_1 \otimes_\alpha A_2$ and a (Gelfand–Naimark) embedding $\pi : A_1 \otimes_\alpha A_2 \subset B(H)$. For any $x \in A_1 \otimes A_2$ we have $\|x\|_\alpha = \|\pi(x)\|$, which shows $\|x\|_\alpha \le \|x\|_{max}$. This proves the second inequality in (4.2).

In particular, the minimal norm must satisfy $\| \ \|_{min} \le \| \ \|_{max}$. This goes back to Guichardet. The lower bound $\| \ \|_{min} \le \| \ \|_\alpha$ is due to Takesaki and is much more delicate. For a proof, see either [240] or [146]. $\qquad\square$

It is easy to see (at least in the unital case) that for any *-homomorphism $\pi : A_1 \otimes A_2 \to B(H)$ there is a pair of (necessarily contractive) *-homomorphisms $\pi_i : A_i \to B(H)$ $(i = 1, 2)$ *with commuting ranges* such that

$$\pi(a_1 \otimes a_2) = \pi_1(a_1)\pi_2(a_2) \qquad \forall a_1 \in A_1 \quad \forall a_2 \in A_2. \qquad (4.4)$$

Indeed, in the unital case we just set $\pi_1(a_1) = \pi(a_1 \otimes 1)$ and $\pi_2(a_2) = \pi(1 \otimes a_2)$. In the general case, the same idea works with approximate units (see Remark 4.2).

Conversely any such pair $\pi_j : A_j \to B(H)$ $(j = 1, 2)$ of $*$-homomorphisms with commuting ranges determines uniquely a $*$-homomorphism $\pi : A_1 \otimes A_2 \to B(H)$ by setting $\pi(a_1 \otimes a_2) = \pi_1(a_1)\pi_2(a_2)$.

For any finite sum $x = \sum a_k^1 \otimes a_k^2$ in $A_1 \otimes A_2$ we will use the notation

$$(\pi_1 \cdot \pi_2)(x) = \sum \pi_1(a_k^1) \otimes \pi_2(a_k^2), \qquad (4.5)$$

with which $\pi = \pi_1 \cdot \pi_2$. Then, we can rewrite (4.3) as:

$$\|x\|_{\max} = \sup \left\{ \left\| \sum \pi_1(a_k^1)\pi_2(a_k^2) \right\| \right\} = \sup \{ \|(\pi_1 \cdot \pi_2)(x)\| \}, \qquad (4.6)$$

where the supremum runs over all possible such pairs (π_1, π_2).

Since $\|\pi_1\| \le 1$ and $\|\pi_2\| \le 1$ we have $\left\| \sum a_k^1 \otimes a_k^2 \right\|_{\max} \le \sum \|a_k^1\| \|a_k^2\|$ and hence (see §A.1)

$$\|x\|_{\max} \le \|x\|_{\wedge} = \inf \left\{ \sum \|a_k^1\| \|a_k^2\| \right\} \qquad (4.7)$$

where the infimum runs over all possible ways to write x as a finite sum of tensors of rank 1. Incidentally, this ensures that (4.6) or (4.3) is finite.

Remark 4.2 ((4.4) still holds in the nonunital case) In the general a priori nonunital case, we claim that it is still true that any $*$-homomorphism $\pi : A_1 \otimes A_2 \to B(H)$ that is "nondegenerate" (meaning here such that $V = \pi(A_1 \otimes A_2)(H)$ is dense in H) must be of the form (4.4). Let $t = \sum a_1^j \otimes a_2^j \in A_1 \otimes A_2$. We denote for $x_1 \in A_1$ (resp. $x_2 \in A_2$) $x_1 \cdot t = \sum x_1 a_1^j \otimes a_2^j$ (resp. $x_2 \cdot \cdot t = \sum a_1^j \otimes x_2 a_2^j$). Let $\xi = \sum_1^n \pi(t_k)h_k \in V$ $(t_k \in A_1 \otimes A_2)$. We define $\pi_1(x_1)\xi = \sum_1^n \pi(x_1 \cdot t_k)h_k$ and similarly $\pi_2(x_2)\xi = \sum_1^n \pi(x_2 \cdot \cdot t_k)h_k$. Then $\pi(x_1 \otimes x_2)\xi = (\pi_1 \cdot \pi_2)(x_1 \otimes x_2)(\xi)$ and hence $\pi(t)\xi = (\pi_1 \cdot \pi_2)(t)(\xi)$ for any $t \in A_1 \otimes A_2$. Moreover π_j extends to a bounded $*$-homomorphism $\pi_j : A_j \to B(H)$. Indeed, a simple verification shows that $\|\pi_1(x_1)\xi\|^2 \le \|x_1\|^2 \|\xi\|^2$ and similarly $\|\pi_2(x_2)\xi\|^2 \le \|x_2\|^2 \|\xi\|^2$. This proves the claim.

Remark 4.3 ((4.6) still holds in the nonunital case) Now if V is not assumed dense in H, the existence of approximate units shows that $\|\pi(t)\| = \|\pi(t)_{|\overline{V}}\|$ so that since we can always replace H by $\overline{V}$ we conclude that (4.6) still holds in the general a priori nonunital case.

Remark 4.4 The norm $\|x\|_{\wedge}$ appearing in (4.7) is called the projective norm. It is the largest among the reasonable tensor norms on tensor products of Banach spaces in Grothendieck's sense (see §A.1), but it is not adapted to our context because it is not a C^*-norm.

Remark 4.5 It follows that any C^*-norm $\|\cdot\|$ on $A_1 \otimes A_2$ automatically satisfies

$$\|a_1 \otimes a_2\| = \|a_1\| \|a_2\| \quad \forall a_1 \in A_1, \forall a_2 \in A_2.$$

Indeed, it is easy to show that $\|a_1 \otimes a_2\|_{\max} \leq \|a_1\| \|a_2\|$ and $\|a_1 \otimes a_2\|_{\min} \geq \|a_1\| \|a_2\|$.

Note that if α is either min or max we have canonically

$$A_1 \otimes_\alpha A_2 \simeq A_2 \otimes_\alpha A_1. \tag{4.8}$$

Indeed, in both cases the flip $a_1 \otimes a_2 \mapsto a_2 \otimes a_1$ extends to an isomorphism.

The basic definitions extend to tensor products of n-tuples $A_1, A_2, \ldots, A_n$ of C^*-algebras, but when $\alpha =$ either min or max the resulting tensor products are "associative" so that we may reduce consideration if we wish to the case $n = 2$. Indeed, "associative" means here the identity

$$(A_1 \otimes_\alpha A_2) \otimes_\alpha A_3 = A_1 \otimes_\alpha A_2 \otimes_\alpha A_3 = A_1 \otimes_\alpha (A_2 \otimes_\alpha A_3), \tag{4.9}$$

which is easy to check and shows that the theory of multiple products reduces, by iteration, to that of products of pairs.

Remark 4.6 Let $A = \left(\oplus \sum_1^n A_j \right)_\infty$ be the direct sum of a *finite* family of C^*-algebras. It is easy to check that for any representation $\pi : A \to B(H)$ that is nondegenerate (that is such that $\overline{\pi(A)H} = H$) there is an orthogonal decomposition $H = \oplus \sum_1^n H_j$ and nondegenerate representations $\pi_j : A_j \to B(H_j)$ such that π can be unitarily identified with $\pi_1 \oplus \cdots \oplus \pi_n$. Using this it is easy to check that, when α is either min or max, for any C^*-algebra B we have

$$\left(\oplus \sum_1^n A_j \right)_\infty \otimes_\alpha B \simeq \left(\oplus \sum_1^n A_j \otimes_\alpha B \right)_\infty. \tag{4.10}$$

For $\alpha = \min$, a more general identity holds, see (1.14).

Let (B_1, B_2) be another pair of C^*-algebras and let $\pi_i : A_i \to B_i$ $(i = 1, 2)$ be $*$-homomorphisms. Then it is immediate from the definition that $\|(\pi_1 \otimes \pi_2)(t)\|_{\max} \leq \|t\|_{\max}$ for any $t \in A_1 \otimes A_2$ and hence $\pi_1 \otimes \pi_2$ defines a $*$-homomorphism from $A_1 \otimes_{\max} A_2$ to $B_1 \otimes_{\max} B_2$. For the minimal tensor product, this is also true because $*$-homomorphisms are automatically complete contractions (see Remark 1.6). Indeed, consider c.b. maps $u_i : A_i \to B_i$ $(i = 1, 2)$. Then (see §1.1) $u_1 \otimes u_2$ defines a c.b. map from $A_1 \otimes_{\min} A_2$ to $B_1 \otimes_{\min} B_2$ with $\|u_1 \otimes u_2\|_{cb} = \|u_1\|_{cb} \|u_2\|_{cb}$.

In sharp contrast, the analogous property does *not* hold for the max-tensor products. However, it does hold if we moreover assume that u_1 and u_2 are *completely positive* (*resp. decomposable*) and then (see the forthcoming Corollary 6.12 and §7.1) the resulting map $u_1 \otimes u_2$ also is completely positive (resp. decomposable) from $A_1 \otimes_{\max} A_2$ to $B_1 \otimes_{\max} B_2$, and we have

$$\forall x \in A_1 \otimes A_2 \quad \|(u_1 \otimes u_2)(x)\|_{B_1 \otimes_{\max} B_2} \leq \|u_1\| \|u_2\| \|x\|_{A_1 \otimes_{\max} A_2},$$

$$(\text{resp. } \|u_1 \otimes u_2\|_{dec} \leq \|u_1\|_{dec} \|u_2\|_{dec}).$$

As we will see in Theorem 7.6 decomposable maps are the "right" analogue of c.b. maps when one replaces the minimal tensor products by the maximal ones.

If $B_1 = B(H_1)$ and $B_2 = B(H_2)$ (or merely if both B_1 and B_2 are assumed injective) then $\|u_1\|_{dec} = \|u_1\|_{cb}$ and $\|u_2\|_{dec} = \|u_2\|_{cb}$ (see Proposition 6.7), so in this particular case there is no problem, tensor products of c.b. maps are bounded both on the minimal and maximal tensor products.

We have obviously a bounded $*$-homomorphism $q : A_1 \otimes_{\max} A_2 \to A_1 \otimes_{\min} A_2$, which (as all C^*-representations) has a closed range, hence $A_1 \otimes_{\min} A_2$ is C^*-isomorphic to the quotient $(A_1 \otimes_{\max} A_2)/\ker(q)$. The observation that in general q is not injective is at the basis of the theory of nuclear C^*-algebras.

4.2 Nuclear C^*-algebras (a brief preliminary introduction)

In these notes, we will emphasize the notion of nuclear pair rather than that of nuclear C^*-algebra (which is by now well known), i.e. we will focus attention on specific pairs (A, B) of C^*-algebras such that the min and max C^*-norms coincide on $A \otimes B$. See Chapter 9. Thus in our presentation the theory of nuclear C^*-algebras becomes embedded in that of nuclear pairs. Nevertheless, it seems more convenient to give here first a brief overview of nuclear C^*-algebras.

Definition 4.7 A C^*-algebra A is called nuclear if for any C^*-algebra B we have $\| \ \|_{\min} = \| \ \|_{\max}$ on $A \otimes B$ or in short if $A \otimes_{\min} B = A \otimes_{\max} B$. In that case, there is only one C^*-norm on $A \otimes B$.

This notion was introduced (under a different name) by Takesaki and was especially investigated by Lance [165], who saw the connection with the following property:

Definition 4.8 A C^*-algebra A has the completely positive approximation property (in short CPAP) if the identity on A is the pointwise limit of a net of finite rank c.p. maps.

We will see in Corollary 7.12 that the CPAP is actually equivalent to nuclearity. Since we place this result in a much broader context (see Theorem 7.10), we delay the full details of its proof till §10.2.

Remark 4.9 If A and B are nuclear then $A \otimes_{\min} B$ is also nuclear. This is easy to check using (4.9). By (4.10), $A \oplus B$ is nuclear, as well as the direct sum of finitely many nuclear C^*-algebras.

Remark 4.10 (Examples of nuclear C^*-algebras) For example, if $\dim(A) < \infty$, A is nuclear, because $A \otimes B$ (there is no need to complete it!) is already a C^*-algebra, hence it admits a unique C^*-norm. All commutative C^*-algebras are nuclear.

Indeed, any such algebra A is isometric to the space $C_0(T)$ of continuous functions vanishing at ∞ on a locally compact space T (that can be taken compact in the unital case). It is a well-known fact from Banach space theory that all such spaces have the metric approximation property, meaning that the identity is the pointwise limit of a net of finite rank maps of norm at most 1. Moreover, in the particular case of $A = C_0(T)$ we can arrange the latter net to be formed of positive maps. Since the latter maps are automatically c.p. (see Remark 1.28) this shows that A has the CPAP, and hence by Corollary 7.12 that A is nuclear.

It is an easy exercise to show that $K(H)$ has the CPAP, and hence is nuclear. For the same reason (although it is not so immediate) the Cuntz algebras are nuclear. The Cuntz algebra O_n for $n \in \mathbb{N} \cup \{\infty\}$ is the C^*-subalgebra of $B(\ell_2)$ generated by an n-tuple (a sequence if $n = \infty$) of isometries (S_j) on ℓ_2 such that $\sum S_j S_j^* = 1$. It can be shown that, given any fixed n, all such algebras are isomorphic, regardless of the choice of (S_j).

In sharp contrast, $B(\ell_2)$ is *not* nuclear and $C^*(\mathbb{F}_2)$ does not embed in a nuclear C^*-algebra (due to Simon Wasserman [255, 257]); we will prove these facts in the sequel (see Theorem 12.29 or Corollary 18.12 and Proposition 7.34).

Other examples or counterexamples can be given among group C^*-algebras. For any discrete group G, the full C^*-algebra $C^*(G)$ or the reduced one $C_\lambda^*(G)$ (as defined in §3.1 and §3.3) is nuclear if and only if G is amenable. See the subsequent Corollary 7.13 for details. So for instance if $G = \mathbb{F}_I$ with $|I| \geq 2$ then $C^*(G)$ and $C_\lambda^*(G)$ are *not* nuclear. (Note that for continuous groups the situation is quite different: Connes [61] proved that, for any separable connected locally compact group G, $C^*(G)$ and $C_\lambda^*(G)$ are nuclear.)

4.3 Tensor products of group C^*-algebras

The following results are easy exercises:

Let G_1, G_2 be two discrete groups. Then

$$C^*(G_1) \otimes_{\max} C^*(G_2) \simeq C^*(G_1 \times G_2), \tag{4.11}$$

$$C_\lambda^*(G_1) \otimes_{\min} C_\lambda^*(G_2) \simeq C_\lambda^*(G_1 \times G_2), \tag{4.12}$$

and similarly for the free product $G_1 * G_2$:

$$C^*(G_1) * C^*(G_2) \simeq C^*(G_1 * G_2). \tag{4.13}$$

These identities can be extended to arbitrary families $(G_i)_{i \in I}$ in place of the pair (G_1, G_2). In particular, we have

$$\underset{i \in I}{*} \, C^*(G_i) \simeq C^* \left(\underset{i \in I}{*} \, G_i \right).$$

The next result (essentially from [208, p.150]) illustrates the usefulness of the Fell principle (see Proposition 3.15).

Theorem 4.11 *Let $\Psi_G : C^*_\lambda(G) \otimes_{\max} C^*_\lambda(G) \to M_G \otimes_{\max} M_G$ be the extension of the natural inclusion $C^*_\lambda(G) \otimes C^*_\lambda(G) \subset M_G \otimes M_G$. We have an isometric *-homomorphism*

$$J_G : C^*(G) \to C^*_\lambda(G) \otimes_{\max} C^*_\lambda(G)$$

taking $U_G(t)$ to $\lambda_G(t) \otimes \lambda_G(t)$ $(t \in G)$, and a completely contractive c.p. mapping

$$P_G : M_G \otimes_{\max} M_G \to C^*(G)$$

such that $Id_{C^*(G)} = P_G \Psi_G J_G$ as in the diagram

$$Id_{C^*(G)} : C^*(G) \xrightarrow{J_G} C^*_\lambda(G) \otimes_{\max} C^*_\lambda(G) \xrightarrow{\Psi_G} M_G \otimes_{\max} M_G \xrightarrow{P_G} C^*(G).$$

Moreover, for all $a, b \in M_G$, such that $a(\delta_e) = \sum_{t \in G} a(t)\delta_t$, $b(\delta_e) = \sum_{t \in G} b(t)\delta_t$, we have (absolutely convergent series)

$$P_G(a \otimes b) = \sum_{t \in G} a(t)b(t)U_G(t).$$

We can express the preceding as a commuting diagram:

$$
\begin{array}{ccc}
C^*_\lambda(G) \otimes_{\max} C^*_\lambda(G) & \xrightarrow{\ \Psi_G\ } & M_G \otimes_{\max} M_G \\
{\scriptstyle J_G} \big\uparrow & & \big\downarrow {\scriptstyle P_G} \\
C^*(G) & \xrightarrow[\ Id_{C^*(G)}\ ]{} & C^*(G)
\end{array}
$$

Proof Let $x \in M_G \otimes M_G$ (algebraic tensor product). For $s \in G$ let $f_s \in M_G^*$ be the natural linear form defined by $f_s(a) = \langle \delta_s, a\delta_e \rangle$, so that $a = \sum f_s(a)\lambda_G(s)$ (convergence in $L_2(\tau_G)$) for any $a \in M_G$. Clearly $x(s,t) = (f_s \otimes f_t)(x)$ is well defined. Note that $(a \otimes b)(s,t) = f_s(a)f_t(b) = a(s)b(t)$ and $\left(\sum_s |f_s(a)|^2 \right)^{1/2} \le \|a\|_{M_G}$ $(a, b \in M_G)$. Thus (Cauchy–Schwarz) $\sum_t |(a \otimes b)(t,t)| \le \|a\|_{M_G}\|b\|_{M_G}$. This shows that $\sum_t |x(t,t)| < \infty$ for any $x \in M_G \otimes M_G$.

We will show the following claim:

$$\forall x \in M_G \otimes M_G \quad \left\| \sum_t x(t,t) U_G(t) \right\|_{C^*(G)} \leq \|x\|_{M_G \otimes_{\max} M_G}. \quad (4.14)$$

Then we set $P_G(x) = \sum_t x(t,t) U_G(t)$. This implies the result. Indeed, in the converse direction we have obviously by maximality

$$\left\| \sum x(t,t) \lambda_G(t) \otimes \lambda_G(t) \right\|_{\max} \leq \left\| \sum x(t,t) U_G(t) \otimes U_G(t) \right\|_{\max}$$

$$\leq \left\| \sum x(t,t) U_G(t) \right\|.$$

Therefore (4.14) implies at the same time that the map J_G (and also $\Psi_G J_G$) defines an isometric $*$-homomorphism and that $P_G \Psi_G$ is a contractive map onto $C^*(G)$. The proof of the claim will actually show that P_G is c.p. Incidentally, $J_G P_G$ is a "conditional expectation" onto $J_G(C^*(G)) \subset C_\lambda^*(G) \otimes_{\max} C_\lambda^*(G)$, in the sense of Theorem 1.45.

We now prove the claim. Let $\pi : G \to B(H)$ be a unitary representation of G. We introduce a pair of commuting representations (π_1, π_2) as follows:

$$\pi_1(\lambda_G(t)) = \lambda_G(t) \otimes \pi(t) \quad \text{and} \quad \pi_2(\lambda_G(t)) = \rho_G(t) \otimes I.$$

Note that both π_1 and π_2 extend to normal isometric representations of M_G. For π_1 this follows from the Fell absorption principle. For π_2, it follows from the fact that $\rho_G \simeq \lambda_G$ (indeed if $W : \ell_2(G) \to \ell_2(G)$ is the unitary taking δ_t to $\delta_{t^{-1}}$, then $W \lambda_G(\cdot) W^* = \rho_G(\cdot)$).

Since π_1 and π_2 have commuting ranges, we have (recall the notation (4.5))

$$\|(\pi_1.\pi_2)(x)\|_{B(\ell_2(G) \otimes_2 H)} \leq \|x\|_{M_G \otimes_{\max} M_G}, \quad (4.15)$$

and hence if we restrict the left-hand side to $K = \delta_e \otimes H \subset \ell_2(G) \otimes_2 H$, we obtain (note that $\langle \delta_e, \lambda_G(s) \rho_G(t) \delta_e \rangle = 1$ if $s = t$ and zero otherwise) $Id \otimes \sum_t x(t,t) \pi(t) = P_K(\pi_1 \cdot \pi_2)(x)_{|K}$ and hence

$$\left\| \sum_t x(t,t) \pi(t) \right\|_{B(H)} \leq \|x\|_{M_G \otimes_{\max} M_G}. \quad (4.16)$$

Finally, taking the supremum over π, we obtain the announced claim (4.14). This argument shows that P_G is c.p. and $\|P_G\|_{cb} \leq 1$. $\qquad \square$

Remark 4.12 By (4.16) applied when π is the trivial representation, we have for any $x \in M_G \otimes M_G$

$$\left| \sum x(t,t) \right| \leq \|x\|_{M_G \otimes_{\max} M_G}.$$

Note that since the matrix of $\lambda_G(t)$ has real entries (equal to 0 or 1) $\lambda_G(t) = \overline{\lambda_G(t)}$. Therefore the correspondence $\sum f(t) \lambda_G(t) \mapsto \sum f(t) \overline{\lambda_G(t)}$ is an isomorphism from $C_\lambda^*(G)$ to $\overline{C_\lambda^*(G)}$, that extends to an isomorphism from M_G to $\overline{M_G}$. Thus we also have (say assuming $(s,t) \mapsto x(s,t)$ finitely supported)

$$
\begin{aligned}
\left| \sum x(t,t) \right| &\le \left\| \sum_{s,t} x(s,t) \lambda_G(s) \otimes \overline{\lambda_G(t)} \right\|_{M_G \otimes_{\max} \overline{M_G}} \\
&\le \left\| \sum_{s,t} x(s,t) \lambda_G(s) \otimes \overline{\lambda_G(t)} \right\|_{C_\lambda^*(G) \otimes_{\max} \overline{C_\lambda^*(G)}}.
\end{aligned}
\tag{4.17}
$$

Remark 4.13 It will be convenient to record here the following fact similar to (4.16) (when π is the trivial representation). Consider $x \in \mathbb{C}[G] \otimes \mathbb{C}[G]$. Then, via the maps $C^*(G) \to C_\lambda^*(G) \subset M_G$, x determines an element $x_U \in C^*(G) \otimes_{\max} C^*(G)$, an element $x_\lambda \in C_\lambda^*(G) \otimes_{\max} C_\lambda^*(G)$, and lastly $x_m \in M_G \otimes_{\max} M_G$ (we will not use this notation later on). Since λ_G and ρ_G obviously extend to representations on M_G with commuting ranges, we have

$$
\begin{aligned}
\|(\lambda_G . \rho_G)(x_m)\|_{B(\ell_2(G))} &\le \|x_m\|_{M_G \otimes_{\max} M_G} \le \|x_\lambda\|_{C_\lambda^*(G) \otimes_{\max} C_\lambda^*(G)} \\
&\le \|x_U\|_{C^*(G) \otimes_{\max} C^*(G)}.
\end{aligned}
\tag{4.18}
$$

As earlier, for any $x \in M_G \otimes M_G$, let $x(s,t) = \langle \delta_s \otimes \delta_t, x(\delta_e \otimes \delta_e) \rangle$ $(s,t \in G)$. Since $\sum x(t,t) = \langle \delta_e, (\lambda_G \cdot \rho_G)(x)\delta_e \rangle$, we have

$$
\left| \sum_{t \in G} x(t,t) \right| \le \|(\lambda_G \cdot \rho_G)(x)\|_{B(\ell_2(G))}.
\tag{4.19}
$$

4.4 A brief repertoire of examples from group C^*-algebras

It is often hard to calculate norms of operators, and hence also of tensors in C^*-tensor products. The group case provides us with many instances where there are nice formulae. For convenience we recapitulate them here.

Proposition 4.14 *Let G be a discrete group, and $\pi : G \to B(H)$ a unitary representation. Let $f : G \to \mathbb{C}$ be any finitely supported function, then*

$$
\begin{aligned}
&\left\| \sum f(t) U_G(t) \otimes U_G(t) \right\|_{C^*(G) \otimes_{\min} C^*(G)} \\
&= \left\| \sum f(t) U_G(t) \otimes U_G(t) \right\|_{C^*(G) \otimes_{\max} C^*(G)} = \left\| \sum f(t) U_G(t) \right\|.
\end{aligned}
\tag{4.20}
$$

$$
\forall f \ge 0 \quad \left\| \sum f(t) U_G(t) \right\| = \sum f(t).
\tag{4.21}
$$

$$
\left\| \sum f(t) U_G(t) \otimes \lambda_G(t) \right\|_{C^*(G) \otimes_{\min} C_\lambda^*(G)} = \left\| \sum f(t) \lambda_G(t) \right\|_{C_\lambda^*(G)}.
\tag{4.22}
$$

$$
\left\| \sum f(t) \pi(t) \otimes \lambda_G(t) \right\|_{C_\pi^*(G) \otimes_{\min} C_\lambda^*(G)} = \left\| \sum f(t) \lambda_G(t) \right\|_{C_\lambda^*(G)}.
\tag{4.23}
$$

$$\left\| \sum f(t)\lambda_G(t) \otimes \lambda_G(t) \right\|_{C^*_\lambda(G) \otimes_{\max} C^*_\lambda(G)} \geq \left| \sum f(t) \right|. \tag{4.24}$$

$$\left\| \sum f(t)\lambda_G(t) \otimes \lambda_G(t) \right\|_{C^*_\lambda(G) \otimes_{\max} M_G} \geq \left| \sum f(t) \right|. \tag{4.25}$$

$$\forall f \geq 0 \quad \left\| \sum f(t)\lambda_G(t) \otimes \lambda_G(t) \right\|_{C^*_\lambda(G) \otimes_{\max} C^*_\lambda(G)}$$
$$= \left\| \sum f(t)\lambda_G(t) \otimes \lambda_G(t) \right\|_{M_G \otimes_{\max} M_G} = \sum f(t). \tag{4.26}$$

$$\forall f \geq 0 \quad \left\| \sum f(t)U_G(t) \otimes \lambda_G(t) \right\|_{C^*(G) \otimes_{\max} C^*_\lambda(G)}$$
$$= \left\| \sum f(t)U_G(t) \otimes \lambda_G(t) \right\|_{C^*(G) \otimes_{\max} M_G} = \sum f(t). \tag{4.27}$$

Proof (4.20): It is easy to show that $U_G \otimes U_G$ dominates U_G since U_G contains the trivial representation, and the converse is obvious by maximality.

(4.21): Indeed, $\geq \sum f(t)$ holds because of the presence of the trivial representation in U_G, and $\leq \sum |f(t)|$ follows from the triangle inequality.

Both (4.22) and (4.23) follow from Fell's principle (see (3.13)).

Let $\|x\|$ (resp. $\|y\|$) be the (common) left-hand side of (4.24) and (4.26) (resp. of (4.27)). Then

$$\|y\| \geq \|x\| \geq \left\| \sum f(t)\lambda_G(t)\rho_G(t) \right\| \geq \left\| \sum f(t)\lambda_G(t)\rho_G(t)\delta_e \right\| = \left| \sum f(t) \right|$$

where at the last step we use $\lambda_G(t)\rho_G(t)\delta_e = \delta_e$. When $f \geq 0$ the triangle inequality gives the converse.

The same argument is valid for (4.25) and for the terms involving M_G in (4.26) and (4.27) since λ_G and ρ_G obviously extend to mutually commuting $*$-homomorphisms on M_G. □

4.5 States on the maximal tensor product

By definition, a state on a C^*-algebra A is a positive linear form φ of unit norm. If A is unital, a functional $\varphi \in A^*$ is a state if and only if $\|\varphi\| = \varphi(1) = 1$ (see Remark 1.33). If A is not unital, a positive functional φ on A is a state if and only if $\varphi(x_i) \to 1$ when (x_i) is any fixed approximate unit in A as in (1.22). Indeed, this follows from Remark 1.24 and (1.22). When this holds we say that φ is approximately unital.

Remark 4.15 Let $\| \ \|_\alpha$ be any C^*-norm on the algebraic tensor product $A_1 \otimes A_2$ of two C^*-algebras. The set of elements of the form $\{x^*x \mid x \in A_1 \otimes A_2\}$ is clearly α-dense in the set $\{x^*x \mid x \in A_1 \otimes_\alpha A_2\}$, or equivalently α-dense in $(A_1 \otimes_\alpha A_2)_+$. This shows that when $\varphi \in (A_1 \otimes_\alpha A_2)^*$, if φ is positive on $A_1 \otimes A_2$ then it is positive on $A_1 \otimes_\alpha A_2$. If both algebras are unital then $1 \otimes 1 \in A_1 \otimes A_2$, and hence, by (1.22) and Remark 1.24, the set of states on $A_1 \otimes_\alpha A_2$ is simply formed of the set of positive unital functionals on $A_1 \otimes A_2$ that are α-continuous. In the nonunital case, if $x_i \geq 0$ (resp. $y_j \geq 0$) is any approximate unit in B_{A_1} (resp. B_{A_2}) as in (1.22), then $x_i \otimes y_j$ is also one in the unit ball of $A_1 \otimes_\alpha A_2$. Thus, in any case, the set of states on $A_1 \otimes_\alpha A_2$ is simply formed of the set of approximately unital positive functionals on the *algebraic* tensor product that happen to be continuous for the norm $\| \ \|_\alpha$.

The following statement describes the states on $A_1 \otimes_{\max} A_2$. Not surprisingly this is the largest possibility: the set of all normalized positive functionals on $A_1 \otimes A_2$.

Theorem 4.16 *Let* $\varphi : A_1 \otimes A_2 \to \mathbb{C}$ *be a linear form and let* $u_\varphi : A_1 \to A_2^*$ *be the corresponding linear map defined by* $u_\varphi(a_1)(a_2) = \varphi(a_1 \otimes a_2)$. *The following are equivalent:*

(i) φ *extends to a positive linear form in* $(A_1 \otimes_{\max} A_2)^*$.

(ii) $u_\varphi : A_1 \to A_2^*$ *is completely positive in the following sense:*

$$\sum_{i,j} u_\varphi(x_{ij})(y_{ij}) \geq 0 \quad \forall n \ \forall x \in M_n(A_1)_+ \ \forall y \in M_n(A_2)_+. \quad (4.28)$$

(iii) φ *is a positive linear form on* $A_1 \otimes A_2$, *in the sense that* $\varphi(t^*t) \geq 0$ *for any* $t \in A_1 \otimes A_2$.

When this holds

$$\|\varphi\|_{(A_1 \otimes_{\max} A_2)^*} = \|u_\varphi\|. \quad (4.29)$$

Thus $\|\varphi\|_{(A_1 \otimes_{\max} A_2)^*} = 1$ (*i.e.* φ *is a state on* $A_1 \otimes_{\max} A_2$) *if and only if* $\|u_\varphi\| = 1$.

Proof Assume (i) with φ of norm 1. By the GNS construction (see §A.13), there are a representation $\pi : A_1 \otimes_{\max} A_2 \to B(H)$ and ξ in the unit ball of H such that $\varphi(\cdot) = \langle \xi, \pi(\cdot)\xi \rangle$. We may assume that $\pi = \pi_1 \cdot \pi_2$ as in (4.4). Let x, y be as in (ii). Let $z = y^{1/2}$ so that $y_{ij} = \sum_k z_{ki}^* z_{kj}$ (note $z_{ik} = z_{ki}^*$) we claim that the matrix $[\pi_1(x_{ij})\pi_2(y_{ij})]$ is positive. Indeed, for each *fixed* k the matrix $[a_{ij}^k]$ defined by

$$a_{ij}^k = \pi_2(z_{ki})^* \pi_1(x_{ij})\pi_2(z_{kj})$$

can be rewritten as a product $C_k^*[\pi_1(x_{ij})]C_k$ showing that it is positive and since π_1, π_2 have commuting ranges we have

$$\pi_1(x_{ij})\pi_2(y_{ij}) = \sum_k \pi_2(z_{ki})^* \pi_1(x_{ij})\pi_2(z_{kj}) = \sum_k a_{ij}^k.$$

This proves our claim. Let $\widetilde{\xi} \in H \oplus \cdots \oplus H$ (n times) be defined by $\widetilde{\xi} = \xi \oplus \cdots \oplus \xi$. Then we have

$$\sum u_\varphi(x_{ij})(y_{ij}) = \langle \widetilde{\xi}, [\pi_1(x_{ij})\pi_2(y_{ij})]\widetilde{\xi} \rangle \geq 0,$$

which (by homogeneity) shows that (i) $\Rightarrow$ (ii).

Assume (ii). Consider $t = \sum a_j \otimes b_j$ in $A_1 \otimes A_2$. Then $t^*t = \sum_{i,j} a_i^* a_j \otimes b_i^* b_j$ hence by (ii)

$$\varphi(t^*t) = \sum_{i,j} u_\varphi(a_i^* a_j)(b_i^* b_j) \geq 0,$$

which shows that (iii) holds.

Assume (iii). By the GNS construction applied to $A_1 \otimes A_2$, there are a $*$-homomorphism $\pi : A_1 \otimes A_2 \to B(H)$ and ξ in H such that $\varphi(\cdot) = \langle \xi, \pi(\cdot)\xi \rangle$. But any $*$-homomorphism $\pi : A_1 \otimes A_2 \to B(H)$ extends to one on the whole of $A_1 \otimes_{\max} A_2$. Thus φ also extends to $A_1 \otimes_{\max} A_2$ and satisfies (i). Moreover, if we assume A_1, A_2 and π unital we have $\|\varphi\|_{\max}^* = \varphi(1 \otimes 1) = u_\varphi(1)(1) \leq \|u_\varphi\|$, but also $|u_\varphi(a_1)(a_2)| \leq \|\varphi\|_{\max}^* \|a_1 \otimes a_2\|_{\max} = \|\varphi\|_{\max}^*$ for any unit vectors a_1, a_2, and hence $\|\varphi\|_{\max}^* = \|u_\varphi\|$. In the nonunital case it is easy to modify this argument using an approximate unit (we leave the details to the reader). $\qquad\square$

Remark 4.17 Let φ be as in (4.28). Consider C^*-algebras B_j and $u_j \in CP$ (B_j, A_j) $(j = 1, 2)$. Then the composition $(u_2)^* u_\varphi u_1$ is obviously c.p. from B_1 to B_2^* in the sense of (4.28).

By definition of the maximal tensor product, the following inequality (4.30) is clear when u_1, u_2 are $*$-homomorphisms, but we will crucially use the following more general fact.

Corollary 4.18 *Let* $u_i : A_i \to B_i$ $(i = 1, 2)$ *be c.p. maps between C^*-algebras. Then the linear map* $u_1 \otimes u_2 : A_1 \otimes A_2 \to B_1 \otimes B_2$ *extends to a c.p. map from* $A_1 \otimes_{\max} A_2$ *to* $B_1 \otimes_{\max} B_2$ *and*

$$\forall x \in A_1 \otimes A_2 \quad \|(u_1 \otimes u_2)(x)\|_{B_1 \otimes_{\max} B_2} \leq \|u_1\|\|u_2\| \, \|x\|_{A_1 \otimes_{\max} A_2}. \quad (4.30)$$

Proof We may assume $\|u_i\| \leq 1$. Consider any φ in the unit ball of $(B_1 \otimes_{\max} B_2)_+^*$, define ψ by $\psi(t) = \varphi((u_1 \otimes u_2)(t))$ $(t \in A_1 \otimes A_2)$. We claim that $\psi \in (A_1 \otimes_{\max} A_2)_+^*$. Indeed, $u_\psi : A_1 \to A_2^*$ is given by $u_\psi = (u_2)^* u_\varphi u_1$, and the latter map is c.p. by Remark 4.17. By Theorem 4.16 $\psi \in (A_1 \otimes_{\max} A_2)_+^*$ and $\|\psi\|_{(A_1 \otimes_{\max} A_2)^*} = \|u_\psi\| \leq \|u_\varphi\|\|u_1\|\|u_2\| \leq 1$. Note

$$\|(u_1 \otimes u_2)(t)\|_{\max} = \sup\{|\varphi((u_1 \otimes u_2)(t))| \mid \|\varphi\|_{(B_1 \otimes_{\max} B_2)^*} \leq 1\}.$$

Since any element φ in the unit ball of the dual of a C^*-algebra such as $(B_1 \otimes_{\max} B_2)^*$ decomposes as a linear combination of positive elements

$\varphi = \varphi_1 - \varphi_2 + i(\varphi_3 - \varphi_4)$ all in the unit ball, it is clear that $u_1 \otimes u_2$ must be bounded (say by 4) from $(A_1 \otimes A_2, \| \ \|_{\max})$ to $(B_1 \otimes B_2, \| \ \|_{\max})$ and hence extends to a bounded map $u : A_1 \otimes_{\max} A_2 \to B_1 \otimes_{\max} B_2$. To complete the proof we show that the latter extension u is positive, i.e. that $t \in (A_1 \otimes_{\max} A_2)_1 \Rightarrow u(t) \in (B_1 \otimes_{\max} B_2)_+$. Indeed, it suffices to check that $\varphi(u(t)) \geq 0$ for all φ in $(B_1 \otimes_{\max} B_2)^*_+$, but with the preceding notation we have $\varphi(u(t)) = \psi(t)$ with $\psi \in (A_1 \otimes_{\max} A_2)^*_+$, so the positivity of u is clear. Replacing A_1 by $M_n(A_1)$, we obtain its complete positivity. By (1.20), if A_1, A_2 are unital we have

$$\|u_1 \otimes u_2 : A_1 \otimes_{\max} A_2 \to B_1 \otimes_{\max} B_2\| \leq \|(u_1 \otimes u_2)(1 \otimes 1)\|$$

$$= \|u_1(1)\| \cdot \|u_2(1)\| \leq \|u_1\| \cdot \|u_2\|.$$

In the nonunital case, we obtain the same conclusion using approximate units. $\square$

Corollary 4.19 *Let $u_i : A_i \to B(H)$ ($i = 1, 2$) be c.p. maps with commuting ranges. Then the linear map*

$$u_1.u_2 : A_1 \otimes A_2 \to B(H)$$

defined by $(u_1.u_2)(x_1 \otimes x_2) = u_1(x_1)u_2(x_2)$ extends to a c.p. map from $A_1 \otimes_{\max} A_2$ to $B(H)$ with norm $\leq \|u_1\|\|u_2\|$.

Proof Let $B_i \subset B(H)$ be the C^*-subalgebra generated by $u_i(A_i)$. Since the u_i's are self-adjoint so are their ranges. Therefore B_1 and B_2 mutually commute. Let $\pi : B_1 \otimes_{\max} B_2 \to B(H)$ be the $*$-homomorphism defined by $\pi(b_1 \otimes b_2) = b_1 b_2$. Since $u_1 \cdot u_2 = \pi(u_1 \otimes u_2)$, by the preceding corollary $u_1 \cdot u_2$ is a composition of c.p. maps with $\|u_1 \cdot u_2\| \leq \|\pi\|\|u_1 \otimes u_2\| \leq \|u_1\|\|u_2\|$. $\square$

4.6 States on the minimal tensor product

We clearly have a surjective $*$-homomorphism $Q : A_1 \otimes_{\max} A_2 \to A_1 \otimes_{\min} A_2$. Let $\mathcal{I} = \ker(Q)$ so that

$$A_1 \otimes_{\min} A_2 = (A_1 \otimes_{\max} A_2)/\mathcal{I}.$$

Therefore

$$(A_1 \otimes_{\min} A_2)^* = \mathcal{I}^\perp \subset (A_1 \otimes_{\max} A_2)^*.$$

Note that by (1.3) and (1.10) we have a natural canonical inclusion

$$A_1^* \otimes A_2^* \subset (A_1 \otimes_{\min} A_2)^*.$$

By Corollary 1.15, we have

$$\ker(Q) = \{t \in A_1 \otimes_{\max} A_2 \mid \langle t, \xi_1 \otimes \xi_2 \rangle = 0 \ \forall \xi_1 \in A_1^*, \xi_2 \in A_2^*\}. \quad (4.31)$$

Lemma 4.20

$$(A_1 \otimes_{\min} A_2)^* = \overline{A_1^* \otimes A_2^*}$$

where the closure is with respect to pointwise convergence on $A_1 \otimes_{\min} A_2$. Moreover, when viewed as a subset of $(A_1 \otimes_{\max} A_2)^*$, the set $(A_1 \otimes_{\min} A_2)^*$ is the closure of $A_1^* \otimes A_2^*$ with respect to pointwise convergence on $A_1 \otimes_{\max} A_2$.

Proof By Hahn–Banach, to check the second assertion it suffices to show that any $t \in A_1 \otimes_{\max} A_2$ that vanishes on $A_1^* \otimes A_2^*$ belongs to $\mathcal{I} = \ker(Q)$, i.e. satisfies $Q(t) = 0$. But by (4.31) this is clear. The first assertion is then clear.   □

Let $A_j \subset B(H_j)$ be C^*-algebras. Then, by definition, $A_1 \otimes_{\min} A_2 \subset B(H_1 \otimes_2 H_2)$ isometrically. Therefore any state on $A_1 \otimes_{\min} A_2$ is the restriction of a state on $B(H_1 \otimes_2 H_2)$. Let $H = H_1 \otimes_2 H_2$. We should first recall how to tackle states on $B(H)$. First note that the unit ball of $B(H)^*$ is the weak* closure of the convex hull of the elements which come from rank one operators on H (see §A.10), i.e. the functionals of the form

$$\varphi_{\xi,\eta}(T) = \langle \xi, T\eta \rangle$$

for some $\xi, \eta \in B_H$. Similarly the positive part of the unit ball of $B(H)^*$ is the weak* closure of the convex hull of the set of functionals of the form $\varphi_{\xi,\xi}$ for some $\xi \in B_H$. Lastly, the set of states on $B(H)$ is the weak* closure of the convex hull of $\{\varphi_{\xi,\xi} \mid \|\xi\| = 1\}$. Moreover, if $F \subset H$ is a dense linear subspace, the latter set is the same as the weak* closure of the convex hull of $\{\varphi_{\xi,\xi} \mid \|\xi\| = 1, \xi \in F\}$.

Let $A \subset B(H)$ be a C^*-algebra. We will say that a map $u \in CP(A, M_n)$ is obtained by compression if there is an isometry $V : \ell_2^n \to H$ such that $u(x) = V^*xV \in B(\ell_2^n) \simeq M_n$ for any $x \in A$. We will say that a linear form φ on $A_1 \otimes A_2$ comes from a matricial state (resp. obtained by compression) if there are integers $n(j)$, maps $u_j \in CP(A_j, M_{n(j)})$ with $\|u_j\| \leq 1$ ($j = 1, 2$) (resp. both obtained by compression) and a state ψ on $M_{n(1)} \otimes_{\min} M_{n(2)}$ such that

$$\forall x \in A_1 \otimes A_2 \quad \varphi(x) = \psi((u_1 \otimes u_2)(x)). \quad (4.32)$$

Note that the notion of map $u \in CP(A, M_n)$ or of state "obtained by compression" depends on the embedding $A \subset B(H)$, while that of state coming from a matricial state does not.

We can now refine the description of the states on $A_1 \otimes_{\min} A_2$ given in the preceding lemma.

Theorem 4.21 *Let $A_j \subset B(H_j)$ be C^*-algebras. Let $\varphi : A_1 \otimes A_2 \to \mathbb{C}$ be a linear form and let $u_\varphi : A_1 \to A_2^*$ be the corresponding linear map. We first assume A_1, A_2 unital and $\varphi(1 \otimes 1) = 1$. The following are equivalent:*

(i) *The functional φ extends to a state on $A_1 \otimes_{\min} A_2$.*

(i)' *The functional φ satisfies $|\varphi(x)| \leq \|x\|_{\min}$ for any $x \in A_1 \otimes A_2$.*

(ii) *The functional φ is the pointwise limit on $A_1 \otimes A_2$ of a net of functionals that come from matricial states obtained by compression.*

(iii) *The functional φ is the pointwise limit on $A_1 \otimes A_2$ of a net of functionals that come from matricial states.*

(iv) *The map $u_\varphi : A_1 \to A_2^*$ is the pointwise limit with respect to the weak* topology on A_2^* of a net of finite rank c.p. maps of unit norm from A_1 to A_2^*.*

Moreover, in the nonunital case, if we replace our assumption on φ by $\sup\{\varphi(x \otimes y) \mid 0 \leq x, 0 \leq y, \|x\| < 1, \|y\| < 1\} = 1$, the equivalence still holds.

Proof By our normalization assumption, (i) and (i)' are equivalent (see the discussion at the beginning of §4.5). Assume (i) and $A_j \subset B(H_j)$, $j = 1, 2$. Let $H = H_1 \otimes_2 H_2$. Then φ is the restriction of a state on $B(H_1 \otimes_2 H_2)$. By the remarks before Theorem 4.21, φ is the pointwise limit of states on $B(H)$ of the form

$$T \mapsto \sum_1^N \lambda_k \varphi_{\xi_k, \xi_k}(T) \tag{4.33}$$

where $\lambda_k > 0$, $\sum_1^N \lambda_k = 1$ and the ξ_k's are unit vectors in H. We may assume (by density) that in each case there are finite-dimensional subspaces $K_j \subset H_j$ such that $\xi_k \in K_1 \otimes_2 K_2$ for all $k \leq N$. Then the resulting states come from matricial states. Indeed, letting $u_j(x) = P_{K_j} x_{|K_j} \in B(K_j)$, and denoting by ψ_k the state on $B(K_1) \otimes_{\min} B(K_2) = B(K_1 \otimes_2 K_2)$ defined by φ_{ξ_k, ξ_k}, we may write $\varphi_{\xi_k, \xi_k}(T) = \psi_k((u_1 \otimes u_2)(T))$. Then the state $\sum \lambda_k \varphi_{\xi_k, \xi_k}$ satisfies (4.32) with $n(j) = \dim(K_j)$ and $\psi = \sum \lambda_k \psi_k$. This proves (i) $\Rightarrow$ (ii) and (ii) $\Rightarrow$ (iii) is trivial.

Note that (4.32) implies $u_\varphi = (u_2)^* u_\psi u_1$, and hence by Remark 4.17 u_φ is a c.p. map of finite rank when φ comes from matricial states. Moreover, $\|u_\varphi\| = \varphi(1 \otimes 1)$ by (4.29).

Assume (iii). Let φ_i be the pointwise approximating functionals coming from matricial states. Since we can replace them by $\varphi_i / \varphi_i(1 \otimes 1)$ and $\varphi_i(1 \otimes 1) \to 1$ we obtain (iv).

Assume (iv). Then by Theorem 4.16, φ is the pointwise limit on the set $\{a_1 \otimes a_2 \mid a_j \in A_j\}$ (or equivalently, by linearity, on $A_1 \otimes A_2$) of a net $\varphi_i \in A_1^* \otimes A_2^*$, formed of states on $A_1 \otimes_{\max} A_2$. By density, since this net is equicontinuous on $A_1 \otimes_{\max} A_2$, we have pointwise convergence on the whole of $A_1 \otimes_{\max} A_2$, and hence by Lemma 4.20 we obtain (i). $\qquad\square$

Remark 4.22 Consider a C^*-algebra $A \subset B(H)$. Let $\mathcal{H} = H \oplus H \oplus \cdots$. Then $\pi : a \mapsto a \oplus a \oplus \cdots$ is an embedding of A in $B(\mathcal{H})$. We will say that any such embedding $A \subset B(\mathcal{H})$ has infinite multiplicity. When $A \subset B(\mathcal{H})$ has infinite multiplicity, any state φ on A of the form $a \mapsto \sum \lambda_k \langle \xi_k, a\xi_k \rangle$ (with unit vectors $\xi_k \in H$) can be rewritten as a vector state on $B(\mathcal{H})$. More precisely, if we define $\xi' = (\lambda_k^{1/2} \xi_k)$ then ξ' is a unit vector in $\mathcal{H}$ and we have $\varphi(T) = \langle \xi', \pi(T)\xi' \rangle$. Thus, any state on A is a pointwise limit of vector states (relative to $\mathcal{H}$).

We will need an obvious generalization of this trick for $A = A_1 \otimes_{\min} A_2$ with $H = H_1 \otimes_2 H_2$ and $A_j \subset B(H_j)$. Let $\mathcal{H}_j = H_j \oplus H_j \oplus \cdots$ and $\pi_j : B(H_j) \to B(\mathcal{H}_j)$ be again such that $\pi_j(a) = a \oplus a \oplus \cdots$. For notational simplicity we give ourselves fixed orthonormal bases in H_1 and H_2 which allow us to define unambiguously the transpose ${}^t a$ of $a \in B(H_j)$ simply as the operator associated to the transposed matrix. We then define ${}^t \pi_j(a) = \pi_j({}^t a) = {}^t a \oplus {}^t a \oplus \cdots$ for any $a \in B(H_j)$.

Thus we obtain:

Proposition 4.23 *Let φ be a state on $A_1 \otimes_{\min} A_2$. With the preceding notation, there is a net of finite rank operators $z_i : \mathcal{H}_2 \to \mathcal{H}_1$ with Hilbert–Schmidt norm 1, i.e. $\operatorname{tr}(z_i^* z_i) = 1$ such that*

$$\forall (a, b) \in A_1 \times A_2 \quad \varphi(a \otimes b) = \lim \operatorname{tr}(z_i^* \pi_1(a) z_i \, {}^t\pi_2(b)).$$

Proof The state φ is the limit of states of the form (4.33). Each state φ_{ξ_k, ξ_k} can be written as described in (2.5) (with unit vectors $\xi = \eta$). Thus it is easy to complete the proof using the same idea as in Remark 4.22 (with z_i acting diagonally). $\qquad\square$

Remark 4.24 The preceding proposition shows that a state φ on $A_1 \otimes_{\min} A_2$ is the pointwise limit of states that come from matricial *vector* states (that is for which ψ in (4.32) is a vector state on $M_n \otimes_{\min} M_m \simeq B(\ell_2^n \otimes_2 \ell_2^m)$).

Actually, we will more often use the following variant for $\overline{A_1} \otimes_{\min} A_2$ (see also Proposition 2.11):

Proposition 4.25 *In the preceding situation, for any state φ on $A_1 \otimes_{\min} \overline{A_2}$ there is a net of finite rank operators $z_i : \mathcal{H}_2 \to \mathcal{H}_1$ with $\operatorname{tr}(z_i^* z_i) = 1$ such that*

$$\forall (a, b) \in A_1 \times A_2 \quad \varphi(a \otimes \bar{b}) = \lim \operatorname{tr}(z_i^* \pi_1(a) z_i \pi_2(b)^*).$$

Similarly, for any state φ on $\overline{A_1} \otimes_{\min} A_2$ there is a net of finite rank operators $h_i : \mathcal{H}_2 \to \mathcal{H}_1$ with $\operatorname{tr}(h_i^* h_i) = 1$ such that

$$\varphi(\bar{a} \otimes b) = \lim \operatorname{tr}(h_i^* \pi_1(a)^* h_i \pi_2(b)).$$

Proof The first part is just a rewriting of the preceding Proposition with (2.6) in place of (2.5). For the second part we observe that $b \otimes \bar{a} \mapsto \varphi(\bar{a} \otimes b)$ defines a state on $A_2 \otimes_{\min} \overline{A_1}$. $\qquad\square$

4.7 Tensor product with a quotient C^*-algebra

We will need the following basic fact on the behavior of C^*-tensor products with respect to quotient C^*-algebras. We will return to this topic specifically for the minimal tensor product more extensively in §7.5.

Lemma 4.26 *Let A, B be C^*-algebras and let $\mathcal{I} \subset A$ be a closed ideal so that $A/\mathcal{I}$ is a C^*-algebra. Let $\| \ \|_\alpha$ be any C^*-norm on $A \otimes B$. Let $E \subset B$ be an arbitrary subspace. We denote by $A \overline{\otimes}_\alpha E$ (resp. $\mathcal{I} \overline{\otimes}_\alpha E$) the closure of $A \otimes E$ (resp. $\mathcal{I} \otimes E$) in $A \otimes_\alpha B$. Let*

$$Q_\alpha[E] = \frac{A \overline{\otimes}_\alpha E}{\mathcal{I} \overline{\otimes}_\alpha E}.$$

Then, if $E \subset F \subset B$ are arbitrary subspaces, we have a natural isometric embedding

$$Q_\alpha[E] \subset Q_\alpha[F].$$

Proof The proof uses the classical fact that the ideal $\mathcal{I}$ has a two-sided approximate unit formed of elements a_i with $0 \le a_i$ and $\|a_i\| \le 1$ (see §A.15).

Let $T_i : A \overline{\otimes}_\alpha F \to \mathcal{I} \overline{\otimes}_\alpha F$ be the operator defined by $T_i(x \otimes y) = a_i x \otimes y$. If B is unital, this is just the left multiplication by $a_i \otimes 1$. Note that $\|T_i\| \le 1$ and $\|I - T_i\| \le 1$. Moreover, $T_i(\varphi) \to \varphi$ for any φ in $\mathcal{I} \otimes F$.

Let us denote by $d(\cdot, \cdot)$ the distance in the norm of $A \otimes_\alpha B$. Note that by density we have for any $x \in A \otimes B$

$$d(x, \mathcal{I} \overline{\otimes}_\alpha F) = d(x, \mathcal{I} \otimes F),$$

and similarly for E. We claim that for any $x \in A \otimes F$

$$d(x, \mathcal{I} \overline{\otimes}_\alpha F) = \limsup_i \|(1 - T_i)(x)\|.$$

Let $y \in \mathcal{I} \otimes F$. Since $\|1 - T_i\| \le 1$ we have

$$\|x - T_i x\|_\alpha \le \|(I - T_i)(x - y)\|_\alpha + \|(I - T_i)(y)\|_\alpha$$

$$\le \|x - y\|_\alpha + \|(I - T_i)(y)\|_\alpha,$$

and $\|(I - T_i)(y)\|_\alpha \to 0$ for any $y \in \mathcal{I} \otimes F$. Thus we obtain $\limsup_i \|x - T_i x\|_\alpha \le \|x - y\|_\alpha$ and hence $\limsup_i \|x - T_i x\|_\alpha \le d(x, \mathcal{I} \overline{\otimes}_\alpha F)$. Since

$T_i(x) \in \mathcal{I} \otimes F$, we have $\liminf_i \|x - T_i x\|_\alpha \geq d(x, \mathcal{I}\overline{\otimes}_\alpha F)$ which proves the claim (and the convergence of $\|x - T_i x\|_\alpha$). Now to show that for any x in $A \otimes E$ we have

$$d(x, \mathcal{I} \otimes F) = d(x, \mathcal{I} \otimes E),$$

it suffices to observe that $T_i x \in \mathcal{I} \otimes E$ for any $x \in A \otimes E$. This gives us

$$d(x, \mathcal{I} \otimes E) \leq \liminf_i \|x - T_i x\|_\alpha = d(x, \mathcal{I} \otimes F)$$

and the converse is obvious since $\mathcal{I} \otimes E \subset \mathcal{I} \otimes F$. $\qquad\square$

The following simple fact will be invoked several times.

Lemma 4.27 *Let A, B be C^*-algebras and let $\mathcal{I} \subset A$ be a closed (two-sided, self-adjoint) ideal. Let $\|\ \|_\alpha$ be any C^*-norm on $A \otimes B$. Then*

$$(A \otimes B) \cap \overline{\mathcal{I}\otimes B}^\alpha = \mathcal{I} \otimes B.$$

Moreover, if we denote by $Q : A \otimes B \to (A \otimes B)/(\mathcal{I} \otimes B)$ the quotient map, then for any $t \in (A \otimes B)/(\mathcal{I} \otimes B)$ we have

$$\|t\|_{(A\otimes_\alpha B)/\overline{\mathcal{I}\otimes B}^\alpha} = \inf\{\|\hat{t}\|_{A\otimes_\alpha B} \mid \hat{t} \in A \otimes B,\ Q(\hat{t}) = t\}. \qquad (4.34)$$

Proof Let $t = \sum_1^n a_k \otimes b_k \in A \otimes B$. Let (x_i) be a (bounded) approximate unit of $\mathcal{I}$ and (y_j) one for B in the sense of §A.15. Clearly for any $z \in \mathcal{I} \otimes B$ we have $\|z - z(x_i \otimes y_j)\|_\alpha \to 0$, and by equicontinuity this remains true for any $z \in \overline{\mathcal{I} \otimes B}^\alpha$. Therefore, if $t \in \overline{\mathcal{I} \otimes B}^\alpha$ then $\|\sum_{k=1}^n a_k \otimes b_k - \sum_1^n a_k x_i \otimes b_k y_j\|_\alpha \to 0$ and hence also (since $\|b_k y_j - b_k\| \to 0$) $\|\sum_{k=1}^n (a_k - a_k x_i) \otimes b_k\|_\alpha \to 0$. We may assume the b_k is linearly independent. Then it follows that $a_k x_i \to a_k$ and hence $a_k \in \bar{\mathcal{I}} = \mathcal{I}$ for any k. This proves $(A \otimes B) \cap (\mathcal{I}\overline{\otimes}_\alpha B) \subset \mathcal{I} \otimes B$ and the converse is trivial.

Let $s \in A \otimes B$ be a representative of t modulo $\mathcal{I} \otimes B$. Then $\|t\|_{(A\otimes_\alpha B)/\overline{\mathcal{I}\otimes B}^\alpha} = \inf\{\|s + \eta\|_\alpha \mid \eta \in \overline{\mathcal{I}\otimes B}^\alpha\}$ and by density this is $= \inf\{\|s + \eta\|_\alpha \mid \eta \in \mathcal{I}\otimes B\}$, which is the same as (4.34). $\qquad\square$

4.8 Notes and remarks

The study of tensor products of C^*-algebras was initiated in the 1950s by Turumaru in Japan, and continued by Takesaki [238] and Guichardet [100] (see also [101]). Much work was then done on nuclear C^*-algebras, (to which we return in §10.2). In the process, this clarified what we know on C^*-tensor products. For instance, Lance's paper [165] contains a lot of information on the latter, in particular (4.11) and (4.12), and the fact that $C^*_\lambda(G)$ is nuclear if and only if G is amenable. Lance [165] showed that the CPAP implies nuclearity.

Choi–Effros and Kirchberg [45, 154] independently proved the converse (see §10.2). In some variant, Theorem 4.11 appears in [208, p. 150]). As an example of application this shows that exactness is not stable by the max-tensor product (see Remark 10.5), which was left open in Kirchberg's early works on exactness (see [156, p. 75 (P3)]). Lance [165] found the description of the states of the maximal tensor product that appears in §4.5. The corresponding result for states on the minimal one in §4.6 is used in Kirchberg's work [155] but had probably long been known to experts.

Section 4.7 is probably well known, but our treatment is influenced by Arveson's [13].

5

Multiplicative domains of c.p. maps

When dealing with a contraction $u : A \to B$ between C^*-algebras it is often interesting to identify the largest C^*-subalgebra of A on which u behaves as a $*$-homomorphism. For c.p. maps there is a useful description of the largest such C^*-subalgebra, called the multiplicative domain of u.

5.1 Multiplicative domains

The unreasonable effectiveness of completely positive contractions in C^*-algebra theory is partially elucidated by the next statement.

Theorem 5.1 *Let* $u : A \to B$ *be a c.p. map between* C^*-algebras with $\|u\| \leq 1$.

(i) *Then if* $a \in A$ *satisfies* $u(a^*a) = u(a)^*u(a)$, *we have necessarily*

$$u(xa) = u(x)u(a), \forall x \in A$$

and the set of such a's forms an algebra.

(ii) *Let* $D_u = \{a \in A \mid u(a^*a) = u(a)^*u(a) \text{ and } u(aa^*) = u(a)u(a)^*\}$. *Then* D_u *is a* C^*-subalgebra of A *(called the multiplicative domain of* u*) and* $u_{|D_u}$ *is a* $*$-homomorphism. Moreover, we have

$$\forall a, b \in D_u \ \forall x \in A \quad u(ax) = u(a)u(x), \ u(xb) = u(x)u(b)$$
$$\text{and } u(axb) = u(a)u(x)u(b). \tag{5.1}$$

Proof First recall a classical inequality for $x \in L_2(m)$ when m is a probability

$$\int |x|^2 dm \geq |\int x \, dm|^2.$$

We will show that u satisfies a similar Cauchy–Schwarz inequality (first used by Choi for 2-positive maps, but see also [143] for earlier similar results for positive ones), as follows.

$$\forall x \in A \quad u(x^*x) \geq u(x)^*u(x). \tag{5.2}$$

This is easy for c.p. maps. Indeed, by Theorem 1.22, we can write u as $u(\cdot) = V^*\pi(\cdot)V$ for some representation $\pi : A \to B(\widehat{H})$ and $V : H \to \widehat{H}$ with $B \subset B(H)$. Then we have for all T with $0 \leq T \leq 1$

$$u(x^*x) = V^*\pi(x)^*\pi(x)V \geq V^*\pi(x)^*T\pi(x)V,$$

and hence choosing $T = VV^*$ we obtain (5.2). This implies that the "defect"

$$\varphi(x, y) = u(x^*y) - u(x)^*u(y)$$

behaves like a B-valued scalar product. In particular, we clearly have (by Cauchy–Schwarz)

$$|\langle \xi, \varphi(x, y)\xi \rangle| \leq \langle \xi, \varphi(x, x)\xi \rangle^{1/2} \langle \xi, \varphi(y, y)\xi \rangle^{1/2} \quad \forall \xi \in H \quad \forall x, y \in A.$$

This shows (taking $y = a$) that if $\varphi(a, a) = 0$ we have $\langle \xi, \varphi(x, a)\xi \rangle = 0$ for all ξ, hence $\varphi(x, a) = 0$ for all x in A. Changing x to x^* (and recalling that a c.p. map is self-adjoint) we obtain

$$\forall x \in A \quad u(xa) = u(x)u(a).$$

Thus we have proved

$$\{a \in A \mid u(a^*a) = u(a)^*u(a)\} = \{a \in A \mid u(xa) = u(x)u(a) \quad \forall x \in A\}. \tag{5.3}$$

It is easy to see that the right-hand side of this equality is an algebra (using associativity). This proves (i). To check (ii), we note that reversing the roles of a and a^* in (i) we have $u(aa^*) = u(a)u(a)^*$ if and only if $u(ay) = u(a)u(y)$ for all y in A. Note that $a \in D_u$ if and only if both a and a^* belong to the set (5.3). Therefore D_u is a C^*-algebra, we have $u(ab) = u(a)u(b)$ for any a, b in D_u which proves (ii) and (5.1) holds. $\qquad\square$

Remark 5.2 In the situation of Theorem 5.1, let $\pi = u_{|D_u} : D_u \to B$. Then $\ker(\pi)$ is a hereditary C^*-subalgebra of A (and of course an ideal of D_u). Indeed, if $0 \leq y \leq x$ with $y \in A$ and $x \in D_u$ then $\pi(x) = 0 \Rightarrow u(y) = 0$ and $0 \leq u(y^2) \leq \|y\|u(y) = 0$ so that $y \in D_u$ and hence $y \in \ker(\pi)$.

As a consequence, we have

Corollary 5.3 (On bimodular maps) *Let $C \subset B$ be a C^*-subalgebra of a C^*-algebra B. Let $\pi : C \to \pi(C) \subset B(H)$ be a representation. Then*

*any contractive c.p. map (in particular any unital c.p. map) $u : B \to B(H)$
extending π must satisfy*

$$u(c_1 x c_2) = \pi(c_1) u(x) \pi(c_2),$$

*for all $x \in B$ and all $c_1, c_2 \in C$ i.e. u must be a C-bimodule map (for the
action defined by π).*

In particular (taking for π the identity on C):

Corollary 5.4 (On conditional expectations) *Let $C \subset B$ be a C^*-subalgebra
of a C^*-algebra B. Then any contractive c.p. projection $P : B \to C$ is a
conditional expectation, i.e.,*

$$P(c_1 x c_2) = c_1 P(x) c_2, \ \forall x \in B \ \forall c_1, c_2 \in C.$$

5.2 Jordan multiplicative domains

In some situations, we will have to deal with a contractive mapping $u : A \to B$
that is merely positive (and hence preserving self-adjointness). In that case,
there is an analogue of Theorem 5.1 where the product in A and B is replaced
by the Jordan product defined by $x \circ y = (xy + yx)/2$. A linear subspace Δ
of a C^*-algebra is called a Jordan subalgebra if it is stable under the Jordan
product. A linear map $u : \Delta \to B(H)$ is called a Jordan morphism if

$$\forall a, b \in \Delta \quad u(a \circ b) = u(a) \circ u(b).$$

If in addition Δ and u are self-adjoint (meaning $a^* \in \Delta$ and $u(a^*) = u(a)^*$
for any $a \in \Delta$), we will say that u is a Jordan $*$-morphism.

Theorem 5.5 *Let $u : A \to B$ be a positive unital map between unital
C^*-algebras with $\|u\| \le 1$.*
　　If (and only if) $a \in A$ satisfies $u(a^* \circ a) = u(a)^* \circ u(a)$, we have

$$\forall x \in A \quad u(x \circ a) = u(x) \circ u(a).$$

*The set Δ_u of such a's forms a closed self-adjoint Jordan subalgebra (called
the Jordan multiplicative domain of u) and $u_{|\Delta_u}$ is a Jordan $*$-morphism.*

Proof Since u preserves positivity it also preserves self-adjointness. Note that
the commutativity of $\circ$ dispenses us from distinguishing left and right products.
In particular, Δ_u is self-adjoint. Let $a \in A$ with $a = a^*$. We claim that

$$u(a^2) \ge u(a)^2. \tag{5.4}$$

Let C_a be the C^*-algebra generated by a. Since C_a is commutative, the
restriction of u to C_a is c.p. by Remark 1.32. Therefore, the claim follows

from (5.2) applied to the latter restriction. Note for later use that if $u(a^2) = u(a)^2$, Theorem 5.1 implies that u is multiplicative on C_a, and in particular

$$u(a^4) = u(a^2)^2. \tag{5.5}$$

Now for $x \in A$ of the form $x = a + ib$ with a, b self-adjoint we have $x^* \circ x = a^2 + b^2$ and $u(x) = u(a) + iu(b)$. Therefore, (5.4) implies

$$u(x^* \circ x) \geq u(x)^* \circ u(x). \tag{5.6}$$

From that point on, the proof can be completed like for Theorem 5.1. Note that

$$\Delta_u = \{a \in A \mid u(x \circ a) = u(x) \circ u(a) \ \forall x \in A\}. \tag{5.7}$$

However, some extra care is needed to show that Δ_u (which is clearly self-adjoint) is a Jordan algebra, because the Jordan product is *not associative*. Since Δ_u is a self-adjoint subspace, it suffices to show that $a \circ b \in \Delta_u$ for any pair a, b of self-adjoint elements of Δ_u. Since $a \circ b = ((a+b)^2 - (a-b)^2)/4$, it suffices to show that $a = a^*$ and $a \in \Delta_u$ implies $a^2 \in \Delta_u$, or equivalently that $a = a^*$ and $u(a^2) = u(a)^2$ implies $u(a^4) = u(a^2)^2$, but we already observed this in (5.5). $\qquad\square$

At some point in the sequel we will crucially need the following result due to Størmer (see [123]).

Theorem 5.6 *Let A be a C^*-algebra. Let $r : A \to B(H)$ be a Jordan $*$-morphism. There is a projection p in $r(A)'' \cap r(A)'$ such that the decomposition*

$$\forall a \in A \quad r(a) = pr(a) + (1-p)r(a)$$

decomposes r as the sum of a $*$-homomorphism $a \mapsto pr(a)$ $(= pr(a)p)$ and $*$-antihomomorphism $a \mapsto (1-p)r(a)$ $(= (1-p)r(a)(1-p))$.

We will use this (without proof) via the following consequence:

Theorem 5.7 *Let M, N be von Neumann algebras. Let $\varphi : N \to M$ be a normal (surjective) linear map such that $\varphi(B_N) = B_M$. Then there are mutually orthogonal projections p, q in N such that M embeds in $pNp \oplus (qNq)^{op}$ as a von Neumann subalgebra admitting a contractive conditional expectation onto it. More precisely, there is an injective normal $*$-homomorphism $r : M \to pNp \oplus (qNq)^{op}$ and a normal contractive (and c.p.) projection $P : pNp \oplus (qNq)^{op} \to r(M)$.*

To prepare for the proof we first need some background on support projections.

Remark 5.8 Let $N \subset B(H)$ be a von Neumann algebra. Let $\mathcal{P}_N$ denote the set of (self-adjoint) projections in N. Let $(p_i)_{i \in I}$ be any family in $\mathcal{P}_N$. Then there is a unique element denoted by $\vee\{p_i \mid i \in I\}$ in $\mathcal{P}_N$ that is minimal among all projections $q \in \mathcal{P}_N$ such that $q \geq p_i$ for all $i \subset I$. To verify this just observe that by the bicommutant Theorem A.46 a (self-adjoint) projection $p \in B(H)$ is in $\mathcal{P}_N$ if and only if p commutes with N' or equivalently if $p(H) \subset H$ is an invariant subspace for N'. By this criterion if $E = \overline{\operatorname{span}}[\cup_{i \in I} p_i(H)]$ the projection P_E is in N and hence we may define simply $\vee\{p_i \mid i \in I\} = P_E$.

Remark 5.9 Let $\varphi : N \to B(H)$ be a positive map that is weak* to weak* continuous (in other words φ is "normal"). The support projection s_φ of φ in N is defined as

$$s_\varphi = 1 - \vee\{p \mid p \in \mathcal{P}_N, \varphi(p) = 0\}.$$

We claim that $\varphi(1 - s_\varphi) = 0$. Let $\mathcal{J} = \{x \in N \mid \varphi(x^*x) = 0\}$. Recall $x^*y^*yx \leq \|y\|^2 x^*x$ for any $x, y \in N$. Thus $\mathcal{J}$ is a weak* closed left ideal in N. By Remark A.36 there is a projection $P \in \mathcal{J}$ such that $\mathcal{J} = NP$. Now for $p \in \mathcal{P}_N, \varphi(p) = 0$ implies $p \in NP$ and hence $\ker(P) \subset \ker(p)$ or equivalently $p(H) \subset P(H)$ therefore also $\overline{\operatorname{span}}[p(H) \mid \varphi(p) = 0] \subset P(H)$. It follows that $1 - s_\varphi \leq P$ and hence $\varphi(1 - s_\varphi) = 0$, which proves our claim.

This implies that $\varphi(x) = \varphi(s_\varphi x s_\varphi)$ for any $x \in N$. Indeed, we have $x - s_\varphi x s_\varphi = x(1 - s_\varphi) + (1 - s_\varphi)x s_\varphi$ and it is easy to check (hint: compose with a state and use Cauchy–Schwarz) that $\varphi(x(1 - s_\varphi)) = \varphi((1 - s_\varphi)y) = 0$ for any $x, y \in N$.

Lastly, $\varphi(x) \neq 0$ for any nonzero $x \in (s_\varphi N s_\varphi)_+$. Otherwise in the commutative von Neumann algebra generated by x in $s_\varphi N s_\varphi$ we would find a nonzero projection p such that $\varepsilon p \leq x$ for some $\varepsilon > 0$ and hence $\varphi(p) = 0$, so that $p \leq 1 - s_\varphi$ which is absurd for $0 \neq p \in s_\varphi N s_\varphi$. Moreover, s_φ is the largest projection in $\mathcal{P}_N$ with the latter property.

Proof of Theorem 5.7 We follow closely Ozawa's presentation in [189].

Our first goal is to show that our assumption implies the existence of a normal *positive unital* surjective linear map $\varphi' : N \to M$ such that $\varphi'(B_N) = B_M$. Let $C = \{x \in N \mid \|x\| \leq 1, \varphi(x) = 1\}$. Observe that C is a (nonvoid) convex subset (actually, as will soon become apparent, a "face") of the unit ball of N. Since φ is normal, C is $\sigma(N, N_*)$ compact and hence has extreme points by the Krein–Milman theorem. Let U be an extreme point of C. We claim that U is necessarily an extreme point of B_N. Indeed, if U is the midpoint of a segment $[a, b]$ in B_N, then $\varphi(U) = 1$ is the midpoint of the segment $[\varphi(a), \varphi(b)]$ in the unit ball of M. But it is easy to see (by the uniform convexity of the Hilbert space on which M is realized, see §A.3) that this forces $\varphi(a) = \varphi(b)$, and hence $\varphi(a) = \varphi(b) = 1$. The latter means that $a, b \in C$, and since

$U \in \text{ext}(C)$, we must have $a = b = U$. This proves the claim that $U \in \text{ext}(B_N)$. By a well-known characterization of $\text{ext}(B_N)$ (see e.g. [240, p. 48]) U is a partial isometry. Since $\varphi(U) = 1$ we have

$$\forall x \in N \quad \varphi(x) = \varphi(UU^*x). \tag{5.8}$$

Indeed, for any normal state g on M we have $g(\varphi(U)) = 1$ so that the functional $f \in N_*$ defined by $f(x) = g(\varphi(x))$ satisfies $\|f\|_{N_*} = 1 = f(U)$, so by Lemma A.41 we have $f(x) = f(UU^*x)$ or $g(\varphi(x)) = g(\varphi(UU^*x))$ and since this holds for any g, we obtain (5.8).

Let us define $\varphi' : N \to M$ by $\varphi'(x) = \varphi(Ux)$. Then $\varphi'(1) = 1 - \|\varphi'\|$ and, by Remark 1.34, φ' is automatically positive. By (5.8) φ' still takes the closed unit ball of N onto that of M. Thus we reach our first goal.

Thus, replacing φ by φ' we may assume that φ is in addition positive and unital. Let $\mathcal{P}$ denote the set of projections in N. Let $e \in \mathcal{P}$ denote the support projection s_φ of φ (see Remark 5.9). Then $\varphi(a) \neq 0$ for any nonzero $a \in (eNe)_+$. Replacing N by eNe (which has e as its unit) and φ by its restriction to eNe, we may assume that $e = 1$. Since $\varphi(x) = \varphi(exe)$ for any $x \in N$, after this change from N to eNe our assumption $\varphi(B_N) = B_M$ still holds. Let Δ_φ be the Jordan multiplicative domain. By Theorem 5.5, the latter is a self-adjoint Jordan subalgebra, which is weak* closed as can be easily deduced from (5.7), and $\varphi_{|\Delta_\varphi} : \Delta_\varphi \to M$ is a normal Jordan $*$-morphism. We first claim that $\varphi_{|\Delta_\varphi}$ is injective. Indeed, for $x \in \Delta_\varphi$, if $\varphi(x) = 0$ then $\varphi(x)^* \circ \varphi(x) = \varphi(x^* \circ x) = 0$ and since $1 = e$ is the support of φ we have $x^* \circ x = 0$ and hence $x = 0$.

Secondly, we claim that $\varphi_{|\Delta_\varphi}$ is surjective. Let $v \in U(M)$. We will show that there is $w \in U(N)$ such that $\varphi(w) = v$. Since $\varphi(B_N) = B_M$ we know there is $w \in B_N$ satisfying this. But now the positivity of φ implies $\varphi(1 - w^* \circ w) \geq 0$ and by (5.6) $\varphi(1 - w^* \circ w) \leq 0$. Therefore we must have $1 - w^* \circ w = 0$, so that w is necessarily unitary. Lastly, since w and $\varphi(w)$ are both unitary and $\varphi(1) = 1$, we must have $w \in \Delta_\varphi$. This shows that the range of $\varphi_{|\Delta_\varphi}$ contains $U(M)$, and since M is linearly spanned by $U(M)$, this proves the surjectivity of $\varphi_{|\Delta_\varphi}$.

Thus $\varphi_{|\Delta_\varphi}$ is an invertible Jordan $*$-morphism from Δ_φ onto M. We now apply Theorem 5.6 to the inverse Jordan $*$-morphism $\psi : M \to \Delta_\varphi \subset N$: there are mutually orthogonal projections $p, q \in \mathcal{P}$ with $p + q = 1$ in $\Delta_\varphi'' \cap \Delta_\varphi' \subset N$ such that the mapping $r : M \to pNp \oplus (qNq)^{op}$ defined by

$$r(x) = p\psi(x) \oplus q\psi(x)$$

is an injective $*$-homomorphism. We may write just as well $r(x) = p\psi(x)p \oplus q\psi(x)q$ since p, q commute with the range of ψ. Moreover, since $\psi(M) = \Delta_\varphi$ is weak* closed in N and commutes with p (and q), $r(M)$ is also weak* closed in $pNp \oplus (qNq)^{op}$. Thus $r(M)$ is a von Neumann algebra, $r : M \rightarrow r(M)$ is a *-isomorphism and hence (recall Remark A.38) r is automatically normal. Consider $y = y_1 \oplus y_2 \in pNp \oplus (qNq)^{op}$. Let $t(y) = y_1 + y_2 \in N$. Then $t : pNp \oplus (qNq)^{op} \rightarrow N$ is isometric (positive but in general *not* c.p.). The mapping $P : pNp \oplus (qNq)^{op} \rightarrow r(M)$ defined by $P(y) = (r\varphi t)(y)$ is a contractive (normal) projection onto $r(M)$ (because $tr = \psi$ and $\varphi\psi$ is the identity on M). By Tomiyama's theorem 1.45, it is a c.p. projection.   □

Remark 5.10 Conversely, if r, p, q, P are as in the conclusion of Theorem 5.7, then the map φ of the form $\varphi(x) = r^{-1}P((pxp, qxq))$, is a normal (positive unital) map onto M such that $\varphi(B_N) = B_M$.

5.3 Notes and remarks

The theory of multiplicative domains for c.p. (or merely 2-positive) maps is due to Choi [44]. It was preceded by important work on positive maps by Kadison and Størmer; see [235] for references and information on the latter maps. See [123] for information on Jordan algebras. More recent results on Jordan multiplicative domains appear in Størmer's paper [234].

6

Decomposable maps

This chapter is devoted to linear maps that are decomposable as linear combinations of c.p. maps and to the appropriate norm denoted by $\| \cdot \|_{dec}$. As will soon be clear, these maps and the dec-norm play the same role for the max-tensor product as cb-maps and the cb-norm with respect to the min-tensor product.

6.1 The dec-norm

Let $A \subset B(H)$ be a closed subspace forming an *operator system* and B a C^*-algebra. We will denote by $D(A, B)$ the set of all "decomposable" maps $u : A \to B$, i.e. the maps that are in the linear span of $CP(A, B)$. This means that $u \in D(A, B)$ if and only if there are $u_j \in CP(A, B)$ $(j = 1, 2, 3, 4)$ such that

$$u = u_1 - u_2 + i(u_3 - u_4).$$

A simple minded choice of norm would be to take $\|u\| = \inf \sum_1^4 \|u_j\|$, but this is not the optimal choice. In many respects, the "right" norm on $D(A, B)$ is the following one, introduced by Haagerup in [104]. We denote

$$\|u\|_{dec} = \inf\{\max\{\|S_1\|, \|S_2\|\}\} \tag{6.1}$$

where the infimum runs over all maps $S_1, S_2 \in CP(A, B)$ such that the map

$$V : x \to \begin{pmatrix} S_1(x) & u(x) \\ u(x^*)^* & S_2(x) \end{pmatrix} \tag{6.2}$$

is in $CP(A, M_2(B))$.

We will use the notation

$$u_*(x) = u(x^*)^*.$$

Note that $u = u_*$ if and only if u takes self-adjoint elements of A to self-adjoint elements of B. This holds in particular for any c.p. map u.

With this notation, we can write

$$V = \begin{pmatrix} S_1 & u \\ u_* & S_2 \end{pmatrix}.$$

Then $D(A, B)$ equipped with the norm $\| \ \|_{dec}$ is a Banach space. To clarify this, let us denote by $D'(A, B)$ the set of those linear mappings $u : A \to B$ such that there are $S_1, S_2 \in CP(A, B)$ for which the preceding map V is in $CP(A, M_2(B))$.

Remark 6.1 Let $\lambda \in \mathbb{C}$. Consider the matrices $a = \begin{pmatrix} 0 & 1 \\ 1 & 0 \end{pmatrix}$ and $b = \begin{pmatrix} 1 & 0 \\ 0 & \lambda \end{pmatrix}$.
Let V be as in (6.2). Then $V' : x \mapsto a^* V(x) a$ and $V'' : x \mapsto b^* V(x) b$ are also c.p. Note

$$V' = \begin{pmatrix} S_2 & u_* \\ u & S_1 \end{pmatrix} \quad V'' = \begin{pmatrix} S_1 & \lambda u \\ \bar{\lambda} u_* & |\lambda|^2 S_2 \end{pmatrix}. \tag{6.3}$$

We will show that actually:

Lemma 6.2 $D(A, B) = D'(A, B)$ *and* $D(A, B)$ *is a Banach space for the norm* $\| \ \|_{dec}$.

Proof Let $u \in D'(A, B)$ with V as in (6.2). Note that we have $u(x) = \begin{pmatrix} 1 & 0 \end{pmatrix} V(x) \begin{pmatrix} 0 \\ 1 \end{pmatrix}$. Therefore by the polarization formula this implies $u \in D(A, B)$. Thus $D'(A, B) \subset D(A, B)$.

We now claim that $D'(A, B)$ is a vector space. Clearly it is stable by addition. A look at V'' in (6.3) shows that $u \in D'(A, B) \Rightarrow \lambda u \in D'(A, B)$, proving the claim. Now if $u \in CP(A, B)$, and if we denote $\chi = \sum e_{1i}$, the mapping

$$x \mapsto \begin{pmatrix} u(x) & u(x) \\ u(x) & u(x) \end{pmatrix} = \left(\sum_{1 \le i, j \le 2} e_{ij} \right) \otimes u(x) = \chi^* \chi \otimes u(x) \tag{6.4}$$

is clearly c.p. Therefore $CP(A, B) \subset D'(A, B)$. But since $D'(A, B)$ is a vector space, this implies $D(A, B) \subset D'(A, B)$.

The easy verification that $\| \ \|_{dec}$ is a norm for which $D(A, B)$ is complete is left to the reader. □

Remark 6.3 It is easy to show that the infimum in the definition (6.1) of the dec-norm is a minimum (i.e. this infimum is attained) when the range B is a von Neumann algebra, or when there is a contractive c.p. projection from B^{**} to B. Haagerup raises in [104] the (apparently still open) question whether it is always a minimum.

Lemma 6.4 *Let* $u : A \to B$ *be "self-adjoint" i.e. such that* $u_* = u$, *and let* $S_1, S_2 \in CP(A, B)$.

If $V = \begin{pmatrix} S_1 & u \\ u & S_2 \end{pmatrix} \in CP(A, M_2(B))$ *in other words if* $0 \preccurlyeq V$ (*see* (2.30))
then

$$-(S_1 + S_2)/2 \preccurlyeq u \preccurlyeq (S_1 + S_2)/2.$$

Proof For any $a \in A_+$ we have $V(a) \geq 0$, and hence $\pm u(a) \leq (S_1(a) + S_2(a))/2$ by (1.24). Therefore, the two mappings $a \mapsto (S_1(a) + S_2(a))/2 \mp u(a)$ are positive mappings (i.e. positivity preserving). But since $V_n = \begin{pmatrix} (S_1)_n & u_n \\ u_n & (S_2)_n \end{pmatrix}$ is assumed positive for any n, we conclude that the same two mappings are c.p., which proves the lemma. $\qquad\square$

Lemma 6.5 *The following simple properties hold:*

(i) *If* $u \in CP(A, B)$, *then* $\|u\|_{dec} = \|u\|_{cb} = \|u\|$.
(ii) *If* $u(x) = u(x^*)^*$ (*i.e.* u *is "self-adjoint"*) *then*

$$\|u\|_{dec} = \inf\{\|u_1 + u_2\| \mid u_1, u_2 \in CP(A, B), u = u_1 - u_2\}. \quad (6.5)$$

(iii) *To any* $u : A \to B$ *we associate the self-adjoint mapping* $\tilde{u} = \begin{pmatrix} 0 & u \\ u_* & 0 \end{pmatrix}$.
Then $u \in D(A, B)$ *if and only if* $\tilde{u} \in D(A, M_2(B))$ *and* $\|u\|_{dec} = \|\tilde{u}\|_{dec}$.

Proof (i) If $u \in CP(A, B)$, then $\begin{pmatrix} u & u \\ u & u \end{pmatrix} \in CP(A, M_2(B))$ and hence $\|u\|_{dec} \leq \|u\|$. Conversely, for any $x \geq 0$ in the unit ball of A, if V is as in (6.2), then $\begin{pmatrix} S_1(x) & u(x) \\ u(x) & S_2(x) \end{pmatrix} \geq 0$ and hence, by Lemma 1.37, $\|u(x)\| \leq \max\{\|S_1(x)\|, \|S_2(x)\|\}$. Therefore $\|u\| \leq \max\{\|S_1\|, \|S_2\|\}$ by (1.21) and hence $\|u\| \leq \|u\|_{dec}$. Since u is c.p. we already know (see (1.22)) that $\|u\| = \|u\|_{cb}$.

(ii) Assume $u = u_1 - u_2$ with $u_1, u_2 \in CP(A, B)$. Then $V_1 = \begin{pmatrix} u_1 & u_1 \\ u_1 & u_1 \end{pmatrix} \in CP(A, M_2(B))$ and (use (6.3) with $\lambda = -1$) $V_2 = \begin{pmatrix} u_2 & -u_2 \\ -u_2 & u_2 \end{pmatrix} \in CP(A, M_2(B))$, and hence $V_1 + V_2 \in CP(A, M_2(B))$. This shows $\|u\|_{dec} = \|u_1 - u_2\|_{dec} \leq \|u_1 + u_2\|$, and hence $\|u\|_{dec} \leq \inf\{\|u_1 + u_2\| \mid u = u_1 - u_2\}$.

Conversely, if $u = u_*$ and if $\begin{pmatrix} S_1 & u \\ u & S_2 \end{pmatrix} \in CP(A, M_2(B))$, let $T = (S_1 + S_2)/2$. Then by Lemma 6.4 we have $-T \preccurlyeq u \preccurlyeq T$ and hence we can write $u = u_1 - u_2$ with $u_1 = (u + T)/2$ and $u_2 = (-u + T)/2$. Then $u_1, u_2 \in CP(A, B)$

and $u_1 + u_2 = T$. Thus $\|u_1 + u_2\| = \|(S_1 + S_2)/2\| \le \max\{\|S_1\|, \|S_2\|\}$. So we find $\inf\{\|u_1 + u_2\| \mid u = u_1 - u_2\} \le \|u\|_{dec}$.

(iii) Assume $\tilde{u} = U_1 - U_2$ with $U_1, U_2 \in CP(A, M_2(B))$. Note that U_1, U_2 coincide on the diagonal and are self-adjoint. Let (S_1, S_2) be their diagonal coefficients, which are clearly in $CP(A, B)$. We have then mappings $u_1 : A \to B$ and $u_2 : A \to B$ such that

$$U_1 = \begin{pmatrix} S_1 & u_1 \\ u_{1*} & S_2 \end{pmatrix} \text{ and } U_2 = \begin{pmatrix} S_1 & u_2 \\ u_{2*} & S_2 \end{pmatrix}.$$

This implies $\|u_1\|_{dec} \le \max\{\|S_1\|, \|S_2\|\}$ and $\|u_2\|_{dec} \le \max\{\|S_1\|, \|S_2\|\}$. Therefore $\|u\|_{dec} \le \|u_1\|_{dec} + \|u_2\|_{dec} \le 2\max\{\|S_1\|, \|S_2\|\} \le \|U_1 + U_2\|$, where for the last inequality we used the classical inequality

$$\max\{\|a\|, \|d\|\} = \left\| \begin{pmatrix} a & 0 \\ 0 & d \end{pmatrix} \right\| \le \left\| \begin{pmatrix} a & b \\ c & d \end{pmatrix} \right\|.$$

This shows that $\|u\|_{dec} \le \inf\{\|U_1 + U_2\| \mid \tilde{u} = U_1 - U_2\} = \|\tilde{u}\|_{dec}$, where for the last $=$ we use (ii) for $\tilde{u}$.

Conversely, if $\|u\|_{dec} < 1$ there are c.p. maps S_1, S_2 with $\begin{pmatrix} S_1 & u \\ u_* & S_2 \end{pmatrix}$ c.p. and $\max\{\|S_1\|, \|S_2\|\} < 1$. Let $S = \begin{pmatrix} S_1 & 0 \\ 0 & S_2 \end{pmatrix}$. Then (recall (6.3) with $\lambda = -1$) $S \pm \tilde{u}$ is c.p. and hence $\tilde{u} = U_1 - U_2$ with $U_1 = (S + \tilde{u})/2$ and $U_2 = (S - \tilde{u})/2$, and $\|U_1 + U_2\| = \|S\| < 1$. This shows that $\|\tilde{u}\|_{dec} \le \|u\|_{dec}$. $\square$

Proposition 6.6 *The following additional properties hold:*

(i) *We have* $D(A, B) \subset CB(A, B)$ *and*

$$\forall u \in D(A, B) \quad \|u\|_{cb} \le \|u\|_{dec}. \tag{6.6}$$

(ii) *If* $u \in D(A, B)$ *and* $v \in D(B, C)$ *then* $vu \in D(A, C)$ *and*

$$\|vu\|_{dec} \le \|v\|_{dec}\|u\|_{dec}. \tag{6.7}$$

Proof (i) Assume first that u is self-adjoint, i.e. $u = u_*$ and $\|u\|_{dec} < 1$. Then, by part (ii) in Lemma 6.5, $u = u_1 - u_2$ with u_1, u_2 c.p. and $\|u_1 + u_2\| < 1$. We claim that

$$\sup\{\|u(x)\| \mid x = x^*, \|x\| \le 1\} \le 1. \tag{6.8}$$

Indeed, first consider $x \ge 0$ in the unit ball of A. We have then $\pm u(x) \le (u_1 + u_2)(x)$ and hence $\|u(x)\| \le 1$. But now if $x = x^*$ and $\|x\| \le 1$, we have $\pm x \le |x| = x^+ + x^-$, and hence $\pm u(x) = \pm(u_1 - u_2)(x) \le (u_1 + u_2)(|x|)$, which implies $\|u(x)\| \le \|\,|x|\,\| \le 1$, proving the claim.

But (6.8) is valid also for $Id_{M_n} \otimes u = Id_{M_n} \otimes u_1 - Id_{M_n} \otimes u_2$ for any n. In particular, using $n = 2$ since the matrix $\begin{pmatrix} 0 & x \\ x^* & 0 \end{pmatrix}$ is self-adjoint in $M_2(A)$, we have for any $x \in A$

$$\|u(x)\| \leq \left\| \begin{pmatrix} 0 & u(x) \\ u(x^*) & 0 \end{pmatrix} \right\| \leq \left\| \begin{pmatrix} 0 & x \\ x^* & 0 \end{pmatrix} \right\| = \|x\|,$$

which implies $\|u\| \leq 1$. Since we may replace u by $Id_{M_n} \otimes u$ for any n, we conclude $\|u\|_{cb} \leq 1$. By homogeneity, this proves (i) for self-adjoint u's. But by part (iii) in Lemma 6.5, we have $\|u\|_{dec} = \|\tilde{u}\|_{dec}$, and by what we just proved $\|\tilde{u}\|_{cb} \leq \|\tilde{u}\|_{dec}$. Since we have obviously $\|u\|_{cb} \leq \|\tilde{u}\|_{cb}$, we obtain (i).

(ii) Assume that $\begin{pmatrix} S_1 & u \\ u_* & S_2 \end{pmatrix} \in CP(A, M_2(B))$ and $\begin{pmatrix} T_1 & v \\ v_* & T_2 \end{pmatrix} \in CP(B,$ $M_2(C))$. Then by Lemma 1.40 we have $\begin{pmatrix} T_1 S_1 & vu \\ v_* u_* & T_2 S_2 \end{pmatrix} \in CP(A, M_2(C))$. Therefore, observing that $v_* u_* = (vu)_*$, we have

$$\|vu\|_{dec} \leq \max\{\|T_1 S_1\|, \|T_2 S_2\|\} \leq \max\{\|T_1\|, \|T_2\|\} \max\{\|S_1\|, \|S_2\|\},$$

and (ii) follows.

Note that for self-adjoint mappings there is a very direct argument: if $u = u_1 - u_2$ and $v = v_1 - v_2$ we have $vu = (v_1 u_1 + v_2 u_2) - (v_1 u_2 + v_2 u_1)$ (a difference of two c.p. maps) and hence $\|vu\|_{dec} \leq \|(v_1 u_1 + v_2 u_2) + (v_1 u_2 + v_2 u_1)\| = \|(v_1 + v_2)(u_1 + u_2)\| \leq \|v_1 + v_2\| \|u_1 + u_2\|$, and recalling (6.5), this yields (ii) for self-adjoint maps u, v. $\qquad \square$

The preceding results are valid with an arbitrary range. However, the special case when the range is $B(H)$ (or is injective) is quite important:

Proposition 6.7 *If $B = B(H)$ or if B is an injective C^*-algebra, then*

$$D(A, B) = CB(A, B)$$

and for any $u \in CB(A, B)$ we have

$$\|u\|_{dec} = \|u\|_{cb}. \tag{6.9}$$

Proof Assume $B = B(H)$. By the factorization of c.b. maps (see Theorem 1.50) we can write $u(x) = V^* \pi(x) W$ ($x \in A$) with $\|V\| = \|W\| = \|u\|_{cb}^{1/2}$. Let $S_1(x) = V^* \pi(x) V$ and $S_2(x) = W^* \pi(x) W$. Note $u_*(x) = W^* \pi(x) V$. Then the map

$$\begin{pmatrix} S_1 & u \\ u_* & S_2 \end{pmatrix} = \begin{pmatrix} V^* & 0 \\ W^* & 0 \end{pmatrix} \begin{pmatrix} \pi & 0 \\ 0 & \pi \end{pmatrix} \begin{pmatrix} V & W \\ 0 & 0 \end{pmatrix}$$

is c.p. (by Remark 1.21) and hence $\|u\|_{dec} \leq \max\{\|V\|^2, \|W\|^2\} = \|u\|_{cb}$. Equality holds by (6.6).

If $B \subset B(H)$ is injective there is a contractive c.p. projection $P : B(H) \to B$. Note that by (i) in Lemma 6.5 $\|P\|_{dec} - 1$. Then by (ii) in Proposition 6.6

$$\|u : A \to B\|_{dec} \leq \|u : A \to B(H)\|_{dec}\|P : B(H) \to B\|_{dec} = \|u\|_{cb}.$$

Again equality holds by (6.6). $\qquad\qquad\qquad\qquad\qquad\qquad\qquad\qquad\qquad\square$

In analogy with (1.15) we have:

Lemma 6.8 (Decomposable maps into a direct sum) *Let A and $(B_i)_{i \in I}$ be C^*-algebras and let $B = \left(\oplus \sum_{i \in I} B_i\right)_\infty$. Let $u : A \to B$. We denote $u_i = p_i u : A \to B_i$. Then $u \in D(A, B)$ if only if all the u_i's are decomposable with $\sup_{i \in I} \|u_i\|_{dec} < \infty$ and we have*

$$\|u\|_{dec} = \sup_{i \in I} \|u_i\|_{dec}. \tag{6.10}$$

Proof Assume $\sup_{i \in I} \|u_i\|_{dec} < 1$. We then have c.p. maps $V_i : A \to M_2(B_i)$ such that $V_{i\,12} = u_i$ and such that $\max\{\|V_{i\,11}\|, \|V_{i\,22}\|\} < 1$. By (6.6) the mapping $V : A \to \left(\oplus \sum_{i \in I} M_2(B_i)\right)_\infty$ associated to $(V_i)_{i \in I}$ is well defined and clearly c.p. Since we may identify $\left(\oplus \sum_{i \in I} M_2(B_i)\right)_\infty$ with $M_2(B)$ so that $u = V_{12}$, we obtain $\|u\|_{dec} \leq 1$. By homogeneity this proves $\|u\|_{dec} \leq \sup_{i \in I} \|u_i\|_{dec}$. The converse follows from (6.7). $\qquad\qquad\square$

In the von Neumann algebra setting, the next lemma will be useful.

Lemma 6.9 (Decomposability extends to the bidual) *Let $u : A \to M$ be a linear map from a C^*-algebra A to a von Neumann algebra M. Then $u \in D(A, M) \Rightarrow \ddot{u} \in D(A^{**}, M)$ and $\|\ddot{u}\|_{dec} = \|u\|_{dec}$.*

Proof By Lemma 1.43 we know $u \in CP(A, M) \Rightarrow \ddot{u} \in CP(A^{**}, M)$. Assume $\|u\|_{dec} < 1$. Then let $S_1, S_2 \in CP(A, B)$ with $\|S_1\| < 1, \|S_2\| < 1$ be such that $V = \begin{pmatrix} S_1 & u \\ u_* & S_2 \end{pmatrix}$ is c.p. and note that $\ddot{V} = \begin{pmatrix} \ddot{S}_1 & \ddot{u} \\ \ddot{u}_* & \ddot{S}_2 \end{pmatrix}$. This implies $\|\ddot{u}\|_{dec} < 1$, and hence by homogeneity, $\|\ddot{u}\|_{dec} \leq \|u\|_{dec}$. The converse is obvious (say by (ii) in Proposition 6.6). $\qquad\qquad\qquad\qquad\qquad\qquad\qquad\qquad\qquad\qquad\square$

The next statement provides us with simple examples of decomposable maps (note that we will show in Lemma 6.24 that (6.12) is somewhat optimal).

Proposition 6.10 *Let A be a C^*-algebra.*

(i) *Fix a, b in A. Let $u : A \to A$ be defined by $u(x) = a^*xb$, then*

$$\|u\|_{dec} \leq \|a\|\,\|b\|.$$

(ii) *Let $u : M_n \to A$ be a linear mapping into a C^*-algebra. Assume that $u(e_{ij}) = a_i^* b_j$ with a_i, b_j in A. Let $\|a\|_C = \left\| \sum a_i^* a_i \right\|^{1/2}$. Then*

$$\|u\|_{dec} \leq \|a\|_C \|b\|_C. \tag{6.11}$$

(iii) *More generally, if $u(e_{ij}) = \sum_{1 \leq k \leq m} a_{ki}^* b_{kj}$ with a_{ki}, b_{kj} in A then*

$$\|u\|_{dec} \leq \left\| \sum_{ki} a_{ki}^* a_{ki} \right\|^{1/2} \left\| \sum_{kj} b_{kj}^* b_{kj} \right\|^{1/2}. \tag{6.12}$$

Proof (i) Let $V : A \to M_2(A)$ be the mapping defined by

$$V(x) = \begin{pmatrix} a^* x a & a^* x b \\ b^* x a & b^* x b \end{pmatrix}.$$

An elementary verification shows that $V(x) = t^* \begin{pmatrix} x & 0 \\ 0 & x \end{pmatrix} t$ where $t = 2^{-1/2}$ $\begin{pmatrix} a & b \\ a & b \end{pmatrix}$. Clearly this shows that V is c.p. hence by definition of the dec-norm we have

$$\|u\|_{dec} \leq \max\{\|V_{11}\|, \|V_{22}\|\}$$

where $V_{11}(x) = a^* x a$ and $V_{22}(x) = b^* x b$. Thus we obtain $\|u\|_{dec} \leq \max\{\|a\|^2, \|b\|^2\}$. Applying this to the mapping $x \to u(x)\|a\|^{-1}\|b\|^{-1}$ we find $\|u\|_{dec} \leq \|a\| \|b\|$.

(ii) Let $a^* = (a_1^*, a_2^*, \ldots, a_n^*), b^* = (b_1^*, b_2^*, \ldots, b_n^*)$ viewed as row matrices with entries in A (so that a and b are column matrices). Then, for any x in M_n, $u(x)$ can be written as a matrix product:

$$u(x) = a^* x b.$$

We again introduce the mapping $V : M_n \to M_2(A)$ defined by $V(x) = \begin{pmatrix} a^* x a & a^* x b \\ b^* x a & b^* x b \end{pmatrix}$. Again we note $V(x) = t^* \begin{pmatrix} x & 0 \\ 0 & x \end{pmatrix} t$ where $t = 2^{-1/2}$ $\begin{pmatrix} a & b \\ a & b \end{pmatrix} \in M_{2n \times 2}(A)$ which shows that V is c.p. so we obtain $\|u\|_{dec} \leq \max\{\|V_{11}\|, \|V_{22}\|\} \leq \max\{\|b\|^2, \|a\|^2\} = \max\{\|\sum b_j^* b_j\|, \|\sum a_i^* a_i\|\}$ and by homogeneity this yields (6.11).

(iii) We have $u = \sum u_k$ where $u_k : M_n \to A$ is defined by $u_k(e_{ij}) = a_{ki}^* b_{kj}$. Let $V_k : M_n \to M_2(A)$ be associated to u_k as in (ii). Let $\mathcal{V} = \sum V_k$. Clearly $\mathcal{V}$ is c.p. and hence

$$\|u\|_{dec} \leq \max\{\|\mathcal{V}_{11}\|, \|\mathcal{V}_{22}\|\} = \max\{\|\mathcal{V}_{11}(1)\|, \|\mathcal{V}_{22}(1)\|\},$$

which yields (6.12). $\qquad\square$

Proposition 6.11 *Let A, B, C be C^*-algebras. For any $u \in D(A, B)$*

$$\forall x \in C \otimes A \quad \|(Id_C \otimes u)(x)\|_{C \otimes_{\max} B} \leq \|u\|_{dec} \|x\|_{C \otimes_{\max} A}. \tag{6.13}$$

Moreover, the mapping $Id_C \otimes u : C \otimes_{\max} A \to C \otimes_{\max} B$ is decomposable and its norm satisfies

$$\|Id_C \otimes u\|_{D(C \otimes_{\max} A, C \otimes_{\max} B)} \leq \|u\|_{dec}. \tag{6.14}$$

Proof Since by (6.6) $\|Id_C \otimes u\| \leq \|Id_C \otimes u\|_{dec}$, it suffices to prove (6.14). We already saw in Corollary 4.18 the analogous property for c.p. maps. Let $u \in D(A, B)$. Assume that the map $V = \begin{pmatrix} S_1 & u \\ u_* & S_2 \end{pmatrix}$ is in $CP(A, M_2(B))$. Then by Corollary 4.18 the mapping

$$Id_C \otimes V : C \otimes_{\max} A \to C \otimes_{\max} M_2(B)$$

is c.p. and using $C \otimes_{\max} M_2(B) \cong M_2(C \otimes_{\max} B)$, we may view it as $\begin{pmatrix} Id_C \otimes S_1 & Id_C \otimes u \\ Id_C \otimes u_* & Id_C \otimes S_2 \end{pmatrix}$. Since $Id_C \otimes u_* = (Id_C \otimes u)_*$ this implies that $Id_C \otimes u \in D(C \otimes_{\max} A, C \otimes_{\max} B)$ with dec-norm

$$\leq \max\{\|Id_C \otimes S_1 : C \otimes_{\max} A \to C \otimes_{\max} B\|,$$
$$\|Id_C \otimes S_2 : C \otimes_{\max} A \to C \otimes_{\max} B\|\}$$

but, by (4.30) since S_1, S_2 are c.p. , this is the same as $\max\{\|S_1\|, \|S_2\|\}$. Taking the infimum over all possible S_1, S_2, we obtain (6.14). □

Corollary 6.12 *Let $u_j \in D(A_j, B_j)$ ($j = 1, 2$) be decomposable mappings between C^*-algebras. Then $u_1 \otimes u_2$ extends to a decomposable mapping in $D(A_1 \otimes_{\max} A_2, B_1 \otimes_{\max} B_2)$ such that*

$$\|u_1 \otimes u_2\|_{D(A_1 \otimes_{\max} A_2, B_1 \otimes_{\max} B_2)} \leq \|u_1\|_{dec} \|u_2\|_{dec}. \tag{6.15}$$

Proof Just write $u_1 \otimes u_2 = (u_1 \otimes Id)(Id \otimes u_2)$. □

When the mapping u has finite rank then a stronger result holds. We can go min $\to$ max:

Proposition 6.13 *Let $u \in D(A, B)$ be a* finite rank *map between C^*-algebras. For any C^*-algebra C we have*

$$\forall x \in C \otimes A \quad \|(Id_C \otimes u)(x)\|_{C \otimes_{\max} B} \leq \|u\|_{dec} \|x\|_{C \otimes_{\min} A}. \tag{6.16}$$

Proof For any finite-dimensional subspace $F \subset B$, the min and max norms are clearly equivalent on $C \otimes F$. Thus since its rank is finite u defines a bounded map $Id_C \otimes u : C \otimes_{\min} A \to C \otimes_{\max} B$. That same map has norm at most $\|u\|_{dec}$ as a map from $C \otimes_{\max} A$ to $C \otimes_{\max} B$. But since we have a metric surjection

$q : C \otimes_{\max} A \to C \otimes_{\min} A$ taking the open unit ball onto the open unit ball, it follows automatically that

$$\|Id_C \otimes u : C \otimes_{\min} A \to C \otimes_{\max} B\| = \|Id_C \otimes u : C \otimes_{\max} A \to C \otimes_{\max} B\|$$
$$\leq \|u\|_{dec}.$$

$\square$

6.2 The δ-norm

In this section, we introduce a "hybrid" tensor product $E \otimes_\delta A$ of an operator space E and a C^*-algebra A, which is convenient to compute the dec-norms of certain important examples, as we do in the next section. We will use it again in §10.3 to relate nuclearity and c.p. approximation properties. The main point is Theorem 6.15, which provides us with a useful factorization for the elements in the unit ball of $(C \otimes_{\max} A) \cap (E \otimes A)$ when C is the universal C^* algebra of E.

We need the obvious generalization of the notation (4.5): given operator spaces E, F and linear mappings $\theta : E \to B(H)$ and $\pi : F \to B(H)$, we denote by $\theta \cdot \pi : E \otimes F \to B(H)$ the linear mapping defined by

$$(\theta \cdot \pi)\left(\sum x_j \otimes y_j\right) = \sum \theta(x_j)\pi(y_j) \quad (x_j \in E, y_j \in F).$$

Let A be a C^*-algebra. For any $y \in E \otimes A$ we define

$$\Delta(y) = \sup\{\|\theta \cdot \pi(y)\|_{B(H)}\} \tag{6.17}$$

where the supremum runs over all Hilbert spaces H and all pairs (θ, π) where $\pi : A \to B(H)$ is a *-homomorphism and $\theta : E \to \pi(A)'$ is a complete contraction.

When A is unital, we claim that (6.17) remains unchanged if we restrict to *unital* *-homomorphisms. Indeed, if π is not unital, let $p = \pi(1)$, then p is a projection on H and it is immediate that $\theta \cdot \pi(y) = \theta' \cdot \pi'(y)$ where $\theta' = p\theta p$ and $\pi' = p\pi p$; thus if we replace π by π', which we view as a unital *-homomorphism into $B(p(H))$, and θ by θ', we find $\|\theta \cdot \pi(y)\| = \|\theta' \cdot \pi'(y)\|$ whence the claim.

Lemma 6.14 *We view E as embedded into $C^*\langle E \rangle$ (as defined in §2.7). Then*

$$\Delta(y) = \|y\|_{C^*\langle E \rangle \otimes_{\max} A}. \tag{6.18}$$

Proof Since $\pi(A)'$ is a C^*-algebra, any completely contractive map $\theta : E \to \pi(A)'$ extends to a representation $\widehat{\theta} : C^*\langle E \rangle \to \pi(A)'$. Hence we clearly have $\forall y \in E \otimes A$

$$\Delta(y) \leq \sup \|\widehat{\theta} \cdot \pi(y)\| \leq \|y\|_{C^*\langle E \rangle \otimes_{\max} A}.$$

The reverse inequality is clear (using an embedding $C^*\langle E \rangle \otimes_{\max} A \subset B(H)$).

$\square$

The main motivation for introducing Δ as in (6.17) is the following.

Theorem 6.15 *Let $A \subset B(H)$ be a unital C^*-algebra and let E be an operator space. Consider an element y in $E \otimes A$. Let*

$$\delta(y) = \inf \left\{ \|x\|_{M_n(E)} \left\| \sum a_i a_i^* \right\|^{1/2} \left\| \sum b_j^* b_j \right\|^{1/2} \right\} \qquad (6.19)$$

where the infimum runs over all possible n and all possible representations of y of the form

$$y = \sum_{i,j=1}^{n} x_{ij} \otimes a_i b_j. \qquad (6.20)$$

Then

$$\Delta(y) = \delta(y). \qquad (6.21)$$

Proof We adapt a proof from [208] that actually is valid even if the subalgebra A is not assumed self-adjoint (in that case the π's are assumed to be unital completely contractive homomorphisms).

We first show the easy inequality $\Delta(y) \leq \delta(y)$. Let (θ, π) be as in the definition of $\Delta(y)$. Then we have, assuming (6.20)

$$(\theta \cdot \pi)(y) = \sum \pi(a_i)\theta(x_{ij})\pi(b_j)$$

and hence by (2.3) $\Delta(y) \leq \delta(y)$.

To show the converse, since δ and Δ are norms (for δ this can be proved by the same idea as for γ_E in (1.17)), it suffices to show that

$$\Delta^* \leq \delta^*.$$

So let $\varphi : E \otimes A \to \mathbf{C}$ be a linear form such that $\delta^*(\varphi) \leq 1$, or equivalently such that

$$|\varphi(y)| \leq \delta(y). \qquad \forall y \in E \otimes A$$

The following Lemma 6.16 is the heart of the proof. With the notation from that lemma, we have

$$\varphi(y) = \langle \eta_1, (\widetilde{v} \cdot \pi)(y)\eta_2 \rangle$$

therefore

$$|\varphi(y)| \leq \|\widetilde{v} \cdot \pi(y)\| \leq \Delta(y).$$

This completes the proof that $\Delta^* \leq \delta^*$, and hence that $\delta \leq \Delta$. $\square$

Lemma 6.16 *Let $\varphi : E \otimes A \to \mathbb{C}$ be a linear form with $\delta^*(\varphi) \leq 1$. Then there are a Hilbert space $\widetilde{H}$ and a representation $\pi : A \to B(\widetilde{H})$ together with a completely contractive map $\widetilde{v} : E \to B(\widetilde{H})$ and unit vectors $\eta_1 \in \widetilde{H}$ and $\eta_2 \in \widetilde{H}$ such that for any a, b, c in A and any x in E we have*

$$\varphi(x \otimes ab) = \langle \eta_1, \pi(a)\widetilde{v}(x)\pi(b)\eta_2 \rangle \text{ and } \widetilde{v}(x)\pi(c) = \pi(c)\widetilde{v}(x).$$

Proof We first claim that there are two representations $\pi_1 : A \to B(H_1)$ and $\pi_2 : A \to B(H_2)$ together with a completely contractive map $v : E \to B(H_2, H_1)$ and unit vectors $\xi_1 \in H_1$ and $\xi_2 \in H_2$ such that for any a, b in A and any x in E we have

$$\varphi(x \otimes ab) = \langle \xi_1, \pi_1(a)v(x)\pi_2(b)\xi_2 \rangle. \tag{6.22}$$

Recall the classical arithmetic/geometric mean inequality:

$$\forall \alpha, \beta \geq 0 \quad (\alpha\beta)^{1/2} = \inf_{s>0}(\alpha/s + s\beta)/2. \tag{6.23}$$

By the definition of δ, this implies that if $x \in B_{M_n(E)}$

$$\left| \varphi \left(\sum x_{ij} \otimes a_i b_j \right) \right| \leq (1/2) \left(\left\| \sum a_i a_i^* \right\| + \left\| \sum b_j^* b_j \right\| \right).$$

Let S be the set of pairs of states (f_1, f_2) on A and let $F : S \to \mathbb{R}$ be the function defined by

$$F(f_1, f_2) = (1/2)\left(f_1\left(\sum a_i a_i^* \right) + f_2\left(\sum b_j^* b_j \right) \right) - \Re\left(\varphi\left(\sum x_{ij} \otimes a_i b_j \right) \right). \tag{6.24}$$

Then

$$\sup_{t \in S} F(t) \geq 0.$$

Let $\mathcal{F} \subset \ell_\infty(S, \mathbb{R})$ be the set of all such functions. It is easy to check that $\mathcal{F}$ is a *convex* cone. Indeed, if F' is the function associated to x', a_i', b_j' with $x' \in B_{M_{n'}(E)}$ then $F + F'$ is associated to $\begin{pmatrix} x & 0 \\ 0 & x' \end{pmatrix}$, with the sequences (a_i, a_i') and (b_j, b_j') obtained by concatenation. Moreover, S is a weak* compact convex subset of $A^* \oplus A^*$ and each $F \in \mathcal{F}$ is affine and weak* continuous. By the variant of Hahn–Banach described in Lemma A.16 there is (f_1, f_2) in S such that for any $x \in B_{M_n(E)}$ and $a_i, b_j \in A$

$$\Re\left(\varphi\left(\sum x_{ij} \otimes a_i b_j \right) \right) \leq (1/2)\left(f_1\left(\sum a_i a_i^* \right) + f_2\left(\sum b_j^* b_j \right) \right).$$

By (6.23) (since the left-hand side is unchanged when we replace (a_i, b_j) by $(a_i/s, sb_j)$ for any $s > 0$) we find automatically

$$\Re\left(\varphi\left(\sum x_{ij} \otimes a_i b_j \right) \right) \leq \left(f_1\left(\sum a_i a_i^* \right) f_2\left(\sum b_j^* b_j \right) \right)^{1/2}.$$

Changing a_i to za_i with $z \in \mathbb{C}, |z| = 1$ does not change the right-hand side either, thus we also find

$$\left| \varphi \left(\sum x_{ij} \otimes a_i b_j \right) \right| \le \left(f_1 \left(\sum a_i a_i^* \right) f_2 \left(\sum b_j^* b_j \right) \right)^{1/2}. \qquad (6.25)$$

In particular, for any $x \in B_E$ and $a, b \in A$

$$|\varphi(x \otimes ab)| \le (f_1(aa^*) f_2(b^*b))^{1/2}.$$

Let $\pi_1 : A \to B(H_1)$ (resp. $\pi_2 : A \to B(H_2)$) be the GNS representation of the state f_1 (resp. f_2) with cyclic unit vector $\xi_1 \in H_1$ (resp. $\xi_2 \in H_2$). This gives us

$$|\varphi(x \otimes ab)| \le \|\pi_1(a^*)\xi_1\| \|\pi_2(b)\xi_2\|.$$

Recall that since we use cyclic vectors, the sets $\{\pi_2(b)\xi_2 \mid b \in A\}$ and $\{\pi_1(a^*)\xi_1 \mid a \in A\}$ are dense in H_2 and H_1 respectively. Thus we can unambiguously define a linear mapping $v : E \to B(H_2, H_1)$ with $\|v\| \le 1$ such that for any $x \in E, a, b \in A$

$$\varphi(x \otimes ab) = \langle \pi_1(a^*)\xi_1, v(x)\pi_2(b)\xi_2 \rangle.$$

But going back to (6.25) this shows us that actually for any $x \in B_{M_n(E)}$ we have

$$\left| \sum \langle \pi_1(a_i^*)\xi_1, v(x_{ij})\pi_2(b_j)\xi_2 \rangle \right| \le \left(\sum \|\pi_1(a_i^*)\xi_1\|^2 \right)^{1/2} \left(\sum \|\pi_2(b_j)\xi_2\|^2 \right)^{1/2}.$$

In other words, we actually have $\|v_n\| \le 1$ for any $n \ge 1$ and hence $\|v\|_{cb} \le 1$, which proves our claim.

Then, writing $(ac)b = a(cb)$ into (6.22) and using the density just mentioned, we find for all c in A

$$\forall x \in E \quad v(x)\pi_2(c) = \pi_1(c)v(x). \qquad (6.26)$$

Let $\pi : A \to B(H_1 \oplus H_2)$ and $\tilde{v} : E \to B(H_1 \oplus H_2)$ be defined by

$$\pi(a) = \begin{pmatrix} \pi_1(a) & 0 \\ 0 & \pi_2(a) \end{pmatrix} \quad \text{and} \quad \tilde{v}(x) = \begin{pmatrix} 0 & v(x) \\ 0 & 0 \end{pmatrix}.$$

Then (6.26) implies $\tilde{v}(x)\pi(c) = \pi(c)\tilde{v}(x)$, for $x \in E$, $c \in A$, and π is a unital representation on A. Lastly, letting $\eta_2 = \begin{pmatrix} 0 \\ \xi_2 \end{pmatrix}$ and $\eta_1 = \begin{pmatrix} \xi_1 \\ 0 \end{pmatrix}$ we find

$$\varphi(x \otimes ab) = \langle \eta_1, \pi(a)\tilde{v}(x)\pi(b)\eta_2 \rangle. \qquad \square$$

Remark 6.17 Let E be an operator space and A a unital C^*-algebra. Consider $y \in E^* \otimes A$. Let $u : E \to A$ be the associated linear map. If $\delta(y) \le 1$ then for any $\varepsilon > 0$ u admits a factorization of the form

$$E \xrightarrow{v} M_n \xrightarrow{w} A,$$

with $\|v\|_{cb} \leq 1 + \varepsilon$ and $\|w\|_{dec} \leq 1$. Indeed, we can write y as $y = \sum x_{ij} \otimes a_i^* b_j$ with $\|x\|_{M_n(E^*)} \leq 1 + \varepsilon$ and $\|a\|_C \|b\|_C \leq 1$ (recall the notation $\|a\|_C = \left\| \sum a_i^* a_i \right\|^{1/2}$). Define $w : M_n \to A$ by $w(e_{ij}) = a_i^* b_j$, and let $v : E \to M_n$ be the map associated to x, so that $\|v\|_{cb} = \|x\|_{M_n(E^*)} \leq 1 + \varepsilon$. Then by (6.11) we have $\|w\|_{dec} \leq 1$. $\qquad\square$

Remark 6.18 The space $E \otimes_\delta A$ can be described as a quotient of the Haagerup tensor product $A \otimes_h E \otimes_h A$ via the map $a \otimes e \otimes b \mapsto e \otimes ab$, see [208, p. 241].

6.3 Decomposable extension property

We previously described (see §1.2) an analogue of the Hahn–Banach extension theorem for c.b. maps. In this section we give an analogue for the maximal tensor product, where decomposable maps replace the c.b. ones.

At this point, we advise the reader to review Lemma 1.38 and Corollary 5.3 on bimodule maps. We will use a generalization of Lemma 1.38 for bimodule maps on "operator modules," as follows (this result appears in [236], see also [197]).

Lemma 6.19 *Let* $C \subset B(K)$ *be a unital C^*-algebra given with a representation* $\pi : C \to B(H)$. *Let* $\mathcal{E} \subset B(K)$ *be a C-bimodule, i.e. an operator space stable by (left and right) multiplication by any element of C. Consider a bimodule map* $w : \mathcal{E} \to B(H)$, *i.e. a map satisfying* $w(c_1 x c_2) = \pi(c_1) w(x) \pi(c_2)$, $(c_1, c_2 \in C, x \in \mathcal{E})$. *Let* $S \subset M_2(B(K))$ *be the operator system consisting of all matrices of the form* $\begin{pmatrix} \lambda & a \\ b^* & \mu \end{pmatrix}$ *with* $\lambda, \mu \in C, a, b \in \mathcal{E}$. *Let* $W : S \to M_2(B(H))$ *be defined by*

$$W\left(\begin{pmatrix} \lambda & a \\ b^* & \mu \end{pmatrix} \right) = \begin{pmatrix} \pi(\lambda) & w(a) \\ w(b)^* & \pi(\mu) \end{pmatrix}.$$

Then $\|w\|_{cb} \leq 1$ *if and only if W is c.p.*

Proof The proof of Lemma 1.38 can be easily modified to yield this. $\qquad\square$

We now give an extension property (one more version of Hahn–Banach) for maps defined on a subspace of the maximal tensor product. We will repeat the same trick later on in §8.3.

Theorem 6.20 *Let A be a unital C^*-algebra, $E \subset A$ an operator space and $M \subset B(H)$ a von Neumann algebra. Let* $u : E \to M$ *be a bounded linear map. Let* $\widehat{u} : M' \otimes E \to B(H)$ *be the linear map defined when* $x' \in M', x \in E$ *by* $\widehat{u}(x' \otimes x) = x' u(x)$.

Let

$$M' \otimes_{\overline{\max}} E = \overline{M' \otimes E} \subset M' \otimes_{\max} A$$

denote the closure of $M' \otimes E$ in $M' \otimes_{\max} A$ equipped with the norm induced by $M' \otimes_{\max} A$.

Then $\widehat{u}$ *extends to a c.b. map on* $M' \otimes_{\overline{\max}} E$ *with* $\|\widehat{u} : M' \otimes_{\overline{\max}} E \to B(H)\|_{cb} \leq 1$ *if and only if there is* $\widetilde{u} \in D(A, M)$ *with* $\|\widetilde{u}\|_{dec} \leq 1$ *extending u. In other words*

$$\|\widehat{u} : M' \otimes_{\overline{\max}} E \to B(H)\|_{cb} = \inf\{\|\widetilde{u} : A \to M\|_{dec} \mid \widetilde{u}_{|E} = u\}$$

and the infimum is attained.

Proof Assume there is an extension $\widetilde{u}$ with $\|\widetilde{u} : A \to M\|_{dec} \leq 1$. By (6.14) and (6.6) $\|Id_{M'} \otimes \widetilde{u} : M' \otimes_{\max} A \to M' \otimes_{\max} M\|_{cb} \leq 1$ and hence a fortiori $\|Id_{M'} \otimes u : M' \otimes_{\overline{\max}} E \to M' \otimes_{\max} M\|_{cb} \leq 1$. Since the product map $p : M' \otimes M \to B(H)$ defines trivially a $*$-homomorphism on $M' \otimes_{\max} M$ we have $\|p : M' \otimes_{\max} M \to B(H)\|_{cb} \leq 1$, and hence $\|\widehat{u} : M' \otimes_{\overline{\max}} E \to B(H)\|_{cb} \leq 1$ since $\widehat{u} = p(Id_{M'} \otimes u)$.

Conversely, assume $\|\widehat{u} : M' \otimes_{\overline{\max}} E \to B(H)\|_{cb} \leq 1$. Let K be a suitable Hilbert space so that $M' \otimes_{\max} A \subset B(K)$, let $C = M' \otimes 1 \subset B(K)$, let $\pi : C \to B(H)$ be the natural identification $M' \otimes 1 \simeq M'$ and let $\mathcal{E} = M' \otimes_{\overline{\max}} E$. Note that $\mathcal{E}$ is a C-bimodule, and that $\widehat{u} : cl\mathcal{E} \to B(H)$ is a C-bimodule map. Hence, with the same notation as in Lemma 6.19 that we apply here to $w = \widehat{u}$ the mapping $W : S \to M_2(B(H))$ must be completely positive (and unital). Recall that by Theorem 1.39 any unital c.p. map $V : A_1 \to B(H)$ on a (unital) operator system $A_1 \subset A_2$ admits a (unital) c.p. extension $\widetilde{V} : A_2 \to B(H)$. Thus, let

$$\widetilde{W} : M_2(M' \otimes_{\max} A) \to M_2(B(H))$$

be a completely positive extension of $W : S \to M_2(B(H))$, and let T be the restriction of $\widetilde{W}$ to $M_2(1 \otimes A)$. Identifying A to $1 \otimes A \subset B(K)$, we may view T as a mapping from $M_2(A)$ to $M_2(B(H))$. Note $T(1) = \widetilde{W}(1) = 1$. In the present case of modular maps we claim that T has the following special form

$$\forall x \in M_2(A) \quad T(x) = \begin{pmatrix} T_{11}(x_{11}) & T_{12}(x_{12}) \\ T_{21}(x_{21}) & T_{22}(x_{22}) \end{pmatrix} \tag{6.27}$$

with $T_{12|E} = u$ and moreover that $T(M_2(A)) \subset M_2(M)$.

Taking this claim temporarily for granted, we will now conclude the proof. Since T is unital and c.p. we have $\max\{\|T_{11}\|, \|T_{22}\|\} \leq \|T\| = 1$, the maps T_{11} and T_{22} are c.p. and moreover the mapping $R : a \mapsto T\left(\begin{pmatrix} a & a \\ a & a \end{pmatrix}\right)$ is clearly c.p. on A, and a fortiori self-adjoint. Let $\widetilde{u} = T_{12}$ (and as usual $\widetilde{u}_*(a) = \widetilde{u}(a^*)^*$ for any $a \in A$). Then we have

$$\forall a \in A \quad R(a) = \begin{pmatrix} T_{11}(a) & \tilde{u}(a) \\ \tilde{u}_*(a) & T_{22}(a) \end{pmatrix}.$$

Therefore, by definition of the dec-norm, we have

$$\|\tilde{u}\|_{dec} \le \max\{\|T_{11}\|, \|T_{22}\|\} = 1.$$

Thus it only remains to prove the claim. For this purpose, observe that, by its definition, W is a $*$-homomorphism on the algebra of matrices $\begin{pmatrix} c_1 & 0 \\ 0 & c_2 \end{pmatrix}$ with c_1, c_2 in $C = M' \otimes 1$. Let $\mathcal{D}$ denote the set of all such matrices. By Corollary 5.3 the map $\tilde{W}$ must be $\mathcal{D}$-bimodular, i.e. we have

$$\tilde{W}(y_1 x y_2) = W(y_1)\tilde{W}(x)W(y_2), \quad \forall y_1, y_2 \in \mathcal{D} \quad \forall x \in M_2(B(H)). \quad (6.28)$$

Applying this with y_1, y_2 equal to either $\begin{pmatrix} 1 & 0 \\ 0 & 0 \end{pmatrix}$ or $\begin{pmatrix} 0 & 0 \\ 0 & 1 \end{pmatrix}$, we find that $\tilde{W}$ is necessarily such that $\tilde{W}(x)_{ij}$ depends only on x_{ij}. A fortiori the same is true for $T = \tilde{W}_{|M_2(A)}$. Thus we can write a priori T in the form (6.27). Moreover, since $\tilde{W}$ extends W, we know $\tilde{W}_{12}$ extends $W_{12} = w = \hat{u}$, and hence restricting this to $A \simeq 1 \otimes A$ (on which $\hat{u} = u$) we see that $\tilde{u}$ extends u.

Lastly, it remains to check that all T_{ij}s take their values in M. Equivalently it suffices to check that all the terms $T_{ij}(x_{ij})$ $(i, j = 1, 2, x_{ij} \in A)$ commute with M'. But this is an easy consequence of (6.28). Indeed, since $1 \otimes A$ trivially commutes with $M' \otimes 1$, any $x \in M_2(A)$ commutes with any $y \in \mathcal{D}$ of the form $y = \begin{pmatrix} m' \otimes 1 & 0 \\ 0 & m' \otimes 1 \end{pmatrix} \in \mathcal{D}$ with $m' \in M'$, and hence by (6.28) $W(y)\tilde{W}(x) = \tilde{W}(yx) = \tilde{W}(xy) = \tilde{W}(x)W(y)$. Equivalently $W(y)T(x) = T(x)W(y)$ and since $W(y) = \begin{pmatrix} m' & 0 \\ 0 & m' \end{pmatrix}$, this implies that $T_{ij}(x_{ij})$ all commute with any $m' \in M'$, and hence take their values in $M'' = M$. This completes the proof of the claim, and of Theorem 6.20. $\square$

Corollary 6.21 (Infinite multiplicity) *In the situation of Theorem 6.20, if the embedding $M \subset B(H)$ has infinite multiplicity, which means we assume $H = \ell_2 \otimes_2 \mathcal{H}$ and there is an isomorphism $\pi : M \to \mathcal{M}$ with $\mathcal{M} \subset B(\mathcal{H})$ such that the embedding $M \subset B(H)$ is of the form $x \mapsto Id_{\ell_2} \otimes \pi(x)$ $(x \in M)$, then*

$$\|\hat{u} : M' \otimes_{\overline{\max}} E \to B(H)\| = \|\hat{u} : M' \otimes_{\overline{\max}} E \to B(H)\|_{cb}. \quad (6.29)$$

Proof We have $M' = B(\ell_2)\bar{\otimes}\mathcal{M}'$. Let $v = \pi u : E \to \mathcal{M}$. A simple verification shows that $\hat{u}$ restricted to $[B(\ell_2) \otimes \mathcal{M}'] \otimes E \subset M' \otimes E$ can be identified with $Id_{B(\ell_2)} \otimes \hat{v}$. Moreover, we may use this idenfication with

$M_n \subset B(\ell_2)$ in place of $B(\ell_2)$ and the isomorphism $M_n(M') \otimes_{\overline{\max}} E = M_n(M' \otimes_{\overline{\max}} E)$ to show that $\|\widehat{u}\| = \|\widehat{v}\|_{cb}$. But since π is an isomorphism we also have

$$\inf\{\|\widetilde{u} : A \to M\|_{dec} \mid \widetilde{u}_{|E} = u\} = \inf\{\|\widetilde{v} : A \to \mathcal{M}\|_{dec} \mid \widetilde{v}_{|E} = v\}.$$

By Theorem 6.20 this last equality implies $\|\widehat{u}\|_{cb} = \|\widehat{v}\|_{cb}$, so that we obtain $\|\widehat{u}\| = \|\widehat{v}\|_{cb} = \|\widehat{u}\|_{cb}$.   □

Corollary 6.22 (Case $E = \ell_\infty^n$) *In the situation of Theorem 6.20, assume either that $E = A$ or that $E \subset A$ is a C^*-subalgebra for which there is a contractive c.p. projection $P : A \to E$. Then*

$$\|u\|_{D(E,M)} = \|\widehat{u}\|_{cb}. \tag{6.30}$$

In particular, this holds for $E = \ell_\infty^n$, $E = M_n$ or when E is an injective C^*-algebra.

Proof We claim $\|u\|_{dec} = \inf\{\|\widetilde{u} : A \to M\|_{dec} \mid \widetilde{u}_{|E} = u\}$. The case $E = A$ is trivial, and the other one very simple. Indeed, by (6.7) for any extension $\widetilde{u}$ we have $\|u : E \to M\|_{dec} = \|\widetilde{u}_{|E}\|_{dec} \le \|\widetilde{u}\|_{dec}$ and if we choose $\widetilde{u} = uP$ we have $\|\widetilde{u}\|_{dec} \le \|u\|_{dec}\|P\|_{dec} = \|u\|_{dec}$.   □

If M' admits a cyclic vector, or equivalently if M admits a separating vector (see Lemma A.62) then the mere boundedness of $\widehat{u}$ on $M' \otimes_{\overline{\max}} E$ ensures that it is c.b. Actually this is a general phenomenon, as the next lemma shows.

Corollary 6.23 (Cyclic case) *In the situation of Theorem 6.20, assume that $M' \subset B(H)$ admits a cyclic vector. Let $\|\cdot\|_\alpha$ be a C^*-norm on $M' \otimes A$ and let $M' \otimes_\alpha A$ be the resulting C^*-algebra (after completion). Let*

$$M' \otimes_{\overline{\alpha}} E = \overline{M' \otimes E}^\alpha \subset M' \otimes_\alpha A$$

equipped with the norm induced by $M' \otimes_\alpha A$.
    Then if $\widehat{u}$ defines (by density) a bounded map on $M' \otimes_{\overline{\alpha}} E$, the latter is (automatically) c.b. and its c.b. norm is equal to its norm.

Proof We may assume $M' \otimes_\alpha A \subset B(\mathcal{H})$. The operator space $\mathcal{E} = M' \otimes_{\overline{\alpha}} E \subset M' \otimes_\alpha A \subset B(\mathcal{H})$ is clearly a bimodule with respect to $M' \otimes 1$ viewed as a C^*-subalgebra of $B(\mathcal{H})$. Let $\pi : M' \otimes 1 \to B(H)$ be the natural embedding. If $\widehat{u}$ defines (by density) a bounded map on $M' \otimes_{\overline{\alpha}} E$, the latter is (like $\widehat{u}$ itself) bimodular with respect to $\pi : M' \otimes 1 \to B(H)$. Therefore, by Theorem 2.6 the resulting map $\widehat{u} : M' \otimes_{\overline{\alpha}} E \to B(H)$ is c.b. with equality of the norm and the c.b. norm.   □

6.4 Examples of decomposable maps

We end this section with several important examples. First we invite the reader to recall Remark 1.30. We will now refine Proposition 6.10.

Lemma 6.24 *Consider a linear mapping* $u : M_n \to A$ *into a* C^*-*algebra* A. *Let* $\mathfrak{a} \in M_n(A)$ *be the matrix defined by* $\mathfrak{a}_{ij} = u(e_{ij})$. *Then*

$$\|u\|_{dec} = \inf \left\{ \left\| \sum_{k,j} a_{kj}^* a_{kj} \right\|^{1/2} \left\| \sum_{k,j} b_{kj}^* b_{kj} \right\|^{1/2} \,\middle|\, a,b \in M_n(A), \ \mathfrak{a} = a^*b \right\}. \tag{6.31}$$

Proof By (iii) in Proposition 6.10 we know that $\|u\|_{dec}$ is $\le$ the right-hand side of (6.31). Conversely, assume $\|u\|_{dec} < 1$. Let $V \in CP(M_n, M_2(A))$ be such that $V_{12} = u$ and $\|V_{jj}\| < 1$ for $j = 1, 2$. Let $\alpha_{ij} = V(e_{ij}) \in M_2(A)$. We denote its entries by $\alpha_{ij_{11}}, \alpha_{ij_{12}}, \ldots \in A$. By Remark 1.30, $\alpha \in M_n(M_2(A))_+$, and hence $\alpha = \beta^*\beta$ for some $\beta \in M_n(M_2(A))$. Then

$$\mathfrak{a}_{ij} = \alpha_{ij_{12}} = \sum_k \beta_{ki_{11}}^* \beta_{kj_{12}} + \beta_{ki_{21}}^* \beta_{kj_{22}}.$$

Moreover,

$$\sum_{k,j} \beta_{kj_{11}}^* \beta_{kj_{11}} + \beta_{kj_{21}}^* \beta_{kj_{21}} = \sum_j \alpha_{jj_{11}} = V(1)_{11}$$

and hence

$$\left\| \sum_{k,j} \beta_{kj_{11}}^* \beta_{kj_{11}} + \beta_{kj_{21}}^* \beta_{kj_{21}} \right\| = \|V(1)_{11}\| = \|V_{11}\| < 1.$$

Similarly

$$\left\| \sum_{k,j} \beta_{kj_{12}}^* \beta_{kj_{12}} + \beta_{kj_{22}}^* \beta_{kj_{22}} \right\| \le \|V_{22}\| < 1.$$

Thus we "almost" conclude as desired that the right-hand side of (6.31) is < 1. The only trouble is that we obtain a representation of the form $\mathfrak{a} = a^*b$ with matrices a, b of size $2n \times n$ instead of $n \times n$. This is easy to fix using the following elementary factorization (essentially the polar decomposition): assuming A unital for simplicity, for any $\varepsilon > 0$ we set $\gamma = (a^*a + \varepsilon 1)^{-1/2} a^*b (b^*b + \varepsilon 1)^{-1/2}$. Let $x = (a^*a + \varepsilon 1)^{-1/2} a^*$ and $y = b(b^*b + \varepsilon 1)^{-1/2}$. Then clearly $\|x\| = \|xx^*\|^{1/2} \le 1$, $\|y\| = \|y^*y\|^{1/2} \le 1$ and hence $\|\gamma\| = \|xy\| \le 1$. Then we have

$$a^*b = (a^*a + \varepsilon 1)^{1/2} \gamma (b^*b + \varepsilon 1)^{1/2} = a'^*b',$$

where $a' = (a^*a + \varepsilon 1)^{1/2}$ and $b' = \gamma (b^*b + \varepsilon 1)^{1/2}$. But now a', b' are both in $M_n(A)$ and such that $a'^*a' \le a^*a + \varepsilon 1$ and $b'^*b' \le b^*b + \varepsilon 1$. It follows that $\|\sum_{kj} a_{kj}'^* a_{kj}'\| \le \|\sum_{kj} a_{kj}^* a_{kj}\| + n^2\varepsilon$ and $\|\sum_{kj} b_{kj}'^* b_{kj}'\| \le \|\sum_{kj} b_{kj}^* b_{kj}\| + n^2\varepsilon$. Since $\varepsilon > 0$ is arbitrary this proves that the right-hand side of (6.31) is < 1. By homogeneity, this completes the proof. $\square$

For emphasis, we single out the next example, which will play an important role in the sequel. The reader should compare this to the earlier description of the unit ball of $CB(\ell_\infty^n, A)$ in (3.7).

Lemma 6.25 *Consider a linear mapping* $T : \ell_\infty^n \to A$ *into a* C^*-*algebra* A. *Let* $x_j = T(e_j)$ $(1 \le j \le n)$. *Then*

$$\|T\|_{dec} = \inf\left\{ \left\|\sum_j a_j^* a_j\right\|^{1/2} \left\|\sum_j b_j^* b_j\right\|^{1/2} \ \Big| \ a_j, b_j \in A, \ x_j = a_j^* b_j \right\}.$$
(6.32)

Proof We may identify ℓ_∞^n with the C^*-subalgebra of diagonal matrices in M_n. We know there is a contractive c.p. projection $P : M_n \to \ell_\infty^n$. By (i) in Lemma 6.5 both P and the inclusion $\ell_\infty^n \to M_n$ have dec-norm $= 1$. By (6.7) we have

$$\|T\|_{D(\ell_\infty^n, A)} = \|TP\|_{D(M_n, A)}.$$
(6.33)

Using this it is easy to deduce (6.32) from (6.31). We leave the details to the reader. □

Lemma 6.26 *In the situation of Theorem 6.15, assume that* A *is a unital* C^*-*algebra, let* F *be another operator space and* B *another* C^*-*algebra. Consider* $u_1 \in CB(E, F)$ *and* $u_2 \in D(A, B)$. *Then for all* y *in* $E \otimes A$, *we have*

$$\delta((u_1 \otimes u_2)(y)) \le \|u_1\|_{cb} \|u_2\|_{dec} \delta(y).$$
(6.34)

Proof Assume $\|u_1\|_{cb} = 1$. Note that $u_1 : E \to F$ extends to a C^*-representation from $C^*\langle E \rangle$ to $C^*\langle F \rangle$. Then (6.34) is an immediate consequence of (6.15) and (6.18). □

In the particular cases when $E = M_n^*$ or $E = \ell_\infty^{n*}$, Theorem 6.15 becomes:

Corollary 6.27 *In Theorem 6.15, assume* A *is a unital* C^*-*algebra and let* $E = M_n^*$ *(resp.* $E = \ell_\infty^{n*}$*) for some* $n \ge 1$, *viewed as dual operator spaces in the sense of §2.4. Then for all* y *in* $E \otimes A$, *with associated linear map* $\tilde{y} : M_n \to A$ *(resp.* $\tilde{y} : \ell_\infty^n \to A$*), we have*

$$\delta(y) = \|\tilde{y}\|_{D(M_n, A)} \quad (resp. \ \delta(y) = \|\tilde{y}\|_{D(\ell_\infty^n, A)}).$$
(6.35)

Proof Let $y \in E \otimes A$ when $E = M_n^*$. Let $\tilde{y}(e_{ij}) = \mathfrak{a}_{ij}$. Let (ξ_{ij}) be biorthogonal to the standard basis (e_{ij}) in M_n, so that $y = \sum \xi_{ij} \otimes \mathfrak{a}_{ij}$. Assume $\mathfrak{a} = a^* b$ as in (6.31). Then

$$y = \sum_{i,k,j} \xi_{ij} \otimes a_{ki}^* b_{kj} = \sum_{i,k,l,j} 1_{k=l} \xi_{ij} \otimes a_{ki}^* b_{lj}$$

$$= \sum_{i,k,l,j} z((i,k),(j,l)) a_{ki}^* b_{lj}$$

with $z((i,k),(j,l)) = 1_{k=l}\xi_{ij}$. By definition of $\delta(y)$ we have

$$\delta(y) \leq \left\|\sum\nolimits_{i,k,l,j} e_{ij} \otimes e_{kl} \otimes z((i,k),(j,l))\right\|_{M_n \otimes_{\min} M_n \otimes_{\min} E^*} \|(a_{ki})\|_C \|(b_{lj})\|_C$$

and

$$\left\|\sum\sum\nolimits_{i,k,l,j} e_{ij} \otimes e_{kl} \otimes z((i,k),(j,l))\right\|_{M_n \otimes_{\min} M_n \otimes_{\min} E^*}$$

$$= \left\|\sum\nolimits_{i,j} e_{ij} \otimes \left(\sum\nolimits_k e_{kk}\right) \otimes \xi_{ij}\right\|_{M_n \otimes_{\min} M_n \otimes_{\min} E^*}$$

$$= \left\|\sum\nolimits_{i,j} e_{ij} \otimes \xi_{ij}\right\|_{M_n \otimes_{\min} E^*} = \|Id_E\|_{cb} = 1.$$

Thus we obtain $\delta(y) \leq \|\tilde{y}\|_{dec}$ by (6.31). We now turn to the reverse inequality (which incidentally is valid for any E). If we have $y = \sum_{i,j=1}^N x_{ij} a_i b_j$ for some $N \geq 1$, then let $v : E^* \to M_N$ be the map defined by $v(\xi) = (\xi(x_{ij}))$ ($\xi \in E^*$). Note $\|v\|_{cb} = \|x\|_{M_N(E)}$ by (2.14). Let $w : M_N \to A$ be defined by $w(e_{ij}) = a_i b_j$. Then by (6.11)

$$\|w\|_{dec} \leq \left\|\sum a_i a_i^*\right\|^{1/2} \left\|\sum b_j^* b_j\right\|^{1/2}. \tag{6.36}$$

Whence since $\tilde{y} = wv$ by part (ii) in Proposition 6.6 and Proposition 6.7:

$$\|\tilde{y}\|_{dec} \leq \|v\|_{dec} \|w\|_{dec} = \|v\|_{cb} \|w\|_{dec}$$

$$\leq \|x\|_{M_N(E)} \left\|\sum a_i a_i^*\right\|^{1/2} \left\|\sum b_j^* b_j\right\|^{1/2}.$$

Taking the infimum over all N and all possible (a_i) and (b_j), we obtain $\|\tilde{y}\|_{dec} \leq \delta(y)$.

The case when $E = \ell_\infty^n$ is proved similarly but using (6.32) instead of (6.31). $\qquad\square$

We invite the reader to compare the following fact with Lemma 3.10.

Lemma 6.28 *Let $\mathbb{F}$ be a free group with (free) generators $(g_i)_{i \in I}$ and let $U_i = U_{\mathbb{F}}(g_i) \in C^*(\mathbb{F})(i \in I)$. We augment I by one element by setting formally $\hat{I} = I \cup \{0\}$, and we set g_0 equal to the unit in $\mathbb{F}$ so that $U_0 = U_{\mathbb{F}}(g_0) = 1$. Let $(x_i)_{i \in \hat{I}}$ be a finitely supported family in a C^*-algebra A and let $T : \ell_\infty(\hat{I}) \to A$ be the mapping defined by $T((\alpha_i)_{i \in \hat{I}}) = \sum_{i \in \hat{I}} \alpha_i x_i$. Then we have*

$$\left\|\sum\nolimits_{i \in \hat{I}} U_i \otimes x_i\right\|_{C^*(\mathbb{F}) \otimes_{\max} A} = \|T\|_{dec}. \tag{6.37}$$

Proof Let $E = \overline{span}[U_i, \mid i \in \hat{I}]$. Let $y = \sum U_i \otimes x_i \in E \otimes A$. Recall that, by Theorem 6.15, $\delta(y) = \Delta(y)$, and hence by Corollary 6.27, we have $\|T\|_{dec} = \Delta(y)$, and by definition (recalling Lemma 3.9)

$$\Delta(y) = \sup \left\| \sum v_i \pi(x_i) \right\|$$

where the supremum runs over all representations $\pi : A \to B(H)$ and all families (v_i) of contractions in $\pi(A)' \subset B(H)$. By Remark 3.11 (Russo–Dye), the latter supremum remains unchanged if we let it run only over all the families (v_i) of *unitaries* in $\pi(A)'$. Equivalently, by (4.6) this means:

$$\Delta(y) = \left\| \sum_{i \in I} U_i \otimes x_i \right\|_{C^*(\mathbb{F}) \otimes_{\max} A},$$

which establishes (6.37). $\qquad\square$

When A is a von Neumann algebra with a separating vector the preceding lemma can be significantly reinforced and it gives us, in the case $|I| = n$, a rather pretty formula for the dec-norm of a mapping $T : \ell_\infty^n \to M$.

Theorem 6.29 *In the situation of the preceding lemma, assume that $A = M$ where M is a von Neumann algebra with either infinite multiplicity (in the sense of Corollary 6.21) or with a separating vector. Then*

$$\left\| \sum_{i \in I} U_i \otimes x_i \right\|_{C^*(\mathbb{F}) \otimes_{\max} M} = \|T\|_{dec} = \sup \left\{ \left\| \sum_{i \in I} u_i x_i \right\| \ \Big| \ u_i \in U(M') \right\}.$$
$$(6.38)$$

Proof Since (x_i) is assumed finitely supported, we may assume $|I| < \infty$, say $|I| = n$. Let $\widehat{T} : M' \otimes \ell_\infty(I) \to B(H)$ be the mapping defined as in Theorem 6.20 by $\widehat{T}\left(\sum y_i \otimes e_i \right) = \sum y_i x_i$. By (6.29) and (6.30) and Corollary 6.23 we have $\|T\|_{dec} = \|\widehat{T}\|$, which is the second equality in (6.38). The first one just repeats (6.37). $\qquad\square$

Remark 6.30 Note that the second equality in (6.38) does not hold in general. For instance if $M = B(\ell_2)$ then $M' = \mathbb{C}1$, and the third term in (6.38) is equal to $\|T\|$, which in general is $< \|T\|_{cb}$, and a fortiori $< \|T\|_{dec}$ (see Remark 1.4).

Remark 6.31 Let $C = M_n * C^*(\mathbb{Z})$. By Proposition 2.18, we have $C \simeq M_n \otimes_{\min} B_n = M_n(B_n)$ where B_n is a unital C^*-algebra called the Brown algebra for Larry Brown who introduced it in [36] (see [54] for more on B_n). In the isomorphism $C \simeq M_n \otimes_{\min} B_n = M_n(B_n)$, the embedding $M_n \subset C$ becomes $x \mapsto x \otimes 1$. Let $U = \sum e_{ij} \otimes U_{ij} \in U(M_n(B_n))$ be the unitary corresponding to the (single) generator of $\mathbb{Z}$ in $C^*(\mathbb{Z}) \subset C$. Let A be a C^*-algebra. As before, let $E = M_n^*$ (operator space dual), let $\xi_{ij} \in E$ be biorthogonal to the usual basis (e_{ij}) of M_n, let $a_{ij} \in A$ and $y = \sum_{i,j=1}^n \xi_{ij} \otimes a_{ij} \in E \otimes A$. Then

$$\|\widetilde{y}\|_{D(M_n, A)} = \left\| \sum \xi_{ij} \otimes a_{ij} \right\|_{C^*\langle E \rangle \otimes_{\max} A} = \left\| \sum U_{ij} \otimes a_{ij} \right\|_{B_n \otimes_{\max} A}. \quad (6.39)$$

The first equality is the same as (6.35). We will prove the second one. For any unital C^*-algebra $D \subset B(H)$, there is a 1–1 correspondence $\pi \mapsto \sum e_{ij} \otimes \pi(U_{ij})$ between the set of unital $*$-homomorphisms $\pi : B_n \to D$ and $U(M_n(D))$. Indeed, $\pi \leftrightarrow Id_{M_n} \otimes \pi : M_n(B_n) \to M_n(D)$ and since $M_n(B_n) \simeq M_n * C^*(\mathbb{Z})$, each $Id_{M_n} \otimes \pi$ is determined by a $*$-homomorphism $\rho : C^*(\mathbb{Z}) \to M_n(D)$ coupled with $x \mapsto x \otimes 1$ on M_n, and of course ρ is determined by its value on the single generator of $\mathbb{Z}$, and hence by a single element of $U(M_n(D))$. Therefore for any $x \in M_n(B(H))$

$$\sup\left\{\left\|\sum \pi(U_{ij})x_{ij}\right\| \,\Big|\, \pi : B_n \to D\right\} = \sup\left\{\left\|\sum u_{ij}x_{ij}\right\| \,\Big|\, u \in U(M_n(D))\right\}.$$

By Remark 3.11 (Russo–Dye) applied to $M_n(D)$, we have

$$\sup\left\{\left\|\sum u_{ij}x_{ij}\right\| \,\Big|\, u \in U(M_n(D))\right\} = \sup\left\{\left\|\sum z_{ij}x_{ij}\right\| \,\Big|\, z \in B_{M_n(D)}\right\}.$$

Let $\sigma : A \to B(H)$ be a $*$-homomorphism. Applying this to $D = \sigma(A)'$ and $x_{ij} = \sigma(a_{ij})$, we find

$$\sup\left\{\left\|\sum \pi(U_{ij})\sigma(a_{ij})\right\| \,\Big|\, \pi : B_n \to \sigma(A)'\right\}$$
$$= \sup\left\{\left\|\sum z_{ij}\sigma(a_{ij})\right\| \,\Big|\, z \in B_{M_n(\sigma(A)')}\right\},$$

and since $M_n(D) = E^* \otimes_{\min} D = CB(E, D)$ isometrically when $E = M_n^*$ (see (2.17)) the last term is the same as $\|\sum \xi_{ij} \otimes a_{ij}\|_{C^*\langle E\rangle \otimes_{\max} A}$. This proves (6.39).

In the situation of Theorem 6.29 (with $\Lambda = M$) the norms in (6.39) are equal to

$$\sup\left\{\left\|\sum u_{ij}a_{ij}\right\| \,\Big|\, u \in U(M_n(M'))\right\}.$$

This follows from Theorem 6.20 just like for (6.38).

Remark 6.32 (Computing some dec-norms) Assume $|I| = n$ (and hence $|\dot{I}| = n + 1$).

In the preceding theorem, consider the case when $A = C_\lambda^*(\mathbb{F}_I)$ and $x_i = \lambda_{\mathbb{F}_I}(g_i)$. Then by (4.27) we have $\|T\|_{dec} = n+1$, and by (3.7),(4.22) and (3.21) $\|T\|_{cb} = 2\sqrt{n}$, so that $\|T\|_{dec} \neq \|T\|_{cb}$ when $n > 1$. The same equalities clearly hold if $C_\lambda^*(\mathbb{F}_I)$ is replaced by $M_{\mathbb{F}_I}$.

More generally, assume that $A = M$ is a *finite* von Neumann algebra, as defined in §11.2. Then if (x_i) is *any* family of unitaries in M we have

$$\|T\|_{dec} = n + 1.$$

Indeed, we have clearly $\left\| \sum_{i \in \dot{I}} U_i \otimes x_i \right\|_{C^*(\mathbb{F}) \otimes_{\max} M} \geq \left\| \sum_{i \in \dot{I}} x_i^* \otimes x_i \right\|_{M^{op} \otimes_{\max} M}$
and using the left and right multiplications L and R on $L_2(\tau)$ we find

$$\left\| \sum_{i \in \dot{I}} x_i^* \otimes x_i \right\|_{M^{op} \otimes_{\max} M} \geq \sum_{l \in \dot{I}} \langle 1, R(x_i^*) L(x_i) 1 \rangle = n + 1.$$

Actually, the same reasoning shows that for any family of scalars $(\alpha_i)_{i \in \dot{I}}$ we have

$$\left\| \sum_{i \in \dot{I}} \alpha_i U_i \otimes x_i \right\|_{C^*(\mathbb{F}) \otimes_{\max} M} \geq \left\| \sum_{i \in \dot{I}} \alpha_i x_i^* \otimes x_i \right\|_{M^{op} \otimes_{\max} M} \geq \sum_{i \in \dot{I}} \alpha_i,$$

and by the triangle inequality this becomes an equality when $\alpha_i \geq 0$ for all $i \in \dot{I}$.

Remark 6.33 (The exceptional case when $|\dot{I}| = 2$) The case $n = 2$ is in sharp contrast with the preceding remark : *Any linear map $T : \ell_\infty^2 \to A$ into an arbitrary C^*-algebra satisfies*

$$\|T\|_{dec} = \|T\|_{cb} = \|T\|.$$

We already observed $\|T\|_{cb} = \|T\|$ in Remark 3.13. Since the C^*-algebra generated by the unit and a single unitary is commutative and hence nuclear, we can replace the max-norm by the min-norm in (6.37), then the first equality in (3.7) shows that $\|T\|_{dec} = \|T\|_{cb}$.

Remark 6.34 By Proposition 6.7 if M is injective then $\|T\|_{dec} = \|T\|_{cb}$ for any n and any $T : \ell_\infty^n \to M$. At the end of [104] Haagerup asks whether the converse holds, and even simply for $n = 3$. We return to this open problem later on in Corollary 23.5 and Remark 23.4.

For the record, we now turn to decomposable multipliers on $C^*(G)$. They turn out to be the same as the bounded ones, as the next remark shows.

Remark 6.35 (Decomposable multipliers on $C^*(G)$) Recall that a bounded linear map $u : C^*(G) \to C^*(G)$ is called a multiplier if there is a (necessarily bounded) function $\varphi : G \to \mathbb{C}$ such that $u(U_G(t)) = \varphi(t) U_G(t)$. In that case, we claim that $u \in D(C^*(G), C^*(G))$ and $\|u\|_{dec} = \|u\|$. Indeed, assuming $\|u\| = 1$, by Proposition 3.3 we can write $\varphi(t) = \langle \eta, \pi(t) \xi \rangle$ where π is a unitary representation on G and $\|\eta\| = \|\xi\| = 1$. Let S_1 (resp. S_2) be the bounded multiplier mapping on $C^*(G)$ associated to the function $\varphi_1(t) = \langle \eta, \pi(t) \eta \rangle$ (resp. $\varphi_2(t) = \langle \xi, \pi(t) \xi \rangle$). By Proposition 3.3 $\|S_1\| = \|S_2\| = 1$. Let $A = C^*(G)$. It is easy to check using (6.4) that the linear map from A to $M_2(A)$ that takes $a = U_G(t)$ to $\begin{pmatrix} S_1(a) & u(a) \\ u(a^*)^* & S_2(a) \end{pmatrix} = (\eta\ \xi) \begin{pmatrix} \pi(a) & \pi(a) \\ \pi(a) & \pi(a) \end{pmatrix} \begin{pmatrix} \eta \\ \xi \end{pmatrix} \otimes a$ is c.p. and hence we have $\|u\|_{dec} \leq \max_{j=1,2} \{\|S_j\|\} = 1$, which proves the claim.

The next two results are the analogue of Proposition 3.25 but with $C^*(G)$ in place of $C^*_\lambda(G)$.

Proposition 6.36 *Let G be a discrete group. Let $T \in D(C^*(G), M_G)$. Define $\psi_T : G \to \mathbb{C}$ by*

$$\psi_T(t) = \langle \delta_t, T(U_G(t))(\delta_e) \rangle.$$

Then the linear mapping that takes $U_G(t)$ to $\psi_T(t)U_G(t)$ extends to a decomposable multiplier $\widetilde{T} \in D(C^*(G), C^*(G))$ with $\|\widetilde{T}\| = \|\widetilde{T}\|_{cb} = \|\widetilde{T}\|_{dec} \le \|T\|_{dec}$.

Proof Let $\dot{Q}_G : C^*(G) \to M_G$ be the natural $*$-homomorphism taking $U_G(t)$ to $\lambda_G(t)$. Let

$$T_G \in D(C^*(G) \otimes_{\max} C^*(G), M_G \otimes_{\max} M_G)$$

denote the extension of $T \otimes \dot{Q}_G$ given by (6.15) so that $\|T_G\|_{dec} \le \|T\|_{dec}$. Let $\widetilde{T} = P_G T_G J_G$ where P_G and J_G are as in Theorem 4.11. The definitions of P_G and J_G show that $\widetilde{T}$ is the multiplier corresponding to ψ_T and $\|\widetilde{T}\|_{dec} \le \|T\|_{dec}$ by (6.7). By (3.4) we have $\|\widetilde{T}\| = \|\widetilde{T}\|_{cb}$, and $\|\widetilde{T}\| = \|\widetilde{T}\|_{dec}$ by Remark 6.35. $\qquad\square$

Corollary 6.37 *In the preceding situation, there is a contractive projection Q from $D(C^*(G), C^*(G))$ onto the subspace formed of all the multipliers in $D(C^*(G), C^*(G))$.*

Proof Let $u \in D(C^*(G), C^*(G))$, then $T = \dot{Q}_G u \in D(C^*(G), M_G)$ by (6.7). Let $Q(u) = \widetilde{T}$. Then $\|Q(u)\|_{dec} \le \|u\|_{dec}$ by (6.7) and $Q(u) = u$ if u is a multiplier. $\qquad\square$

6.5 Notes and remarks

The results of §6.1 on the dec-norm come mainly from [104]. Those of §6.2 come from [208]. The δ-norm was developed there to present the equivalence nuclear $\Leftrightarrow$ CPAP (due to Choi–Effros and Kirchberg) in a framework better suited for linear maps on operator spaces. We exploit this in the sequel in §10.2. Lemma 6.28 which appeared in [204] was directly inspired by a previous result of Haagerup in [104, lemma 3.5], which was essentially the second equality in (6.38). The examples of §6.4 are variations suggested by the δ-norm with roots in Haagerup's ideas in [104]. Haagerup in [104, proposition 3.4] states and proves Remark 6.33 for maps on ℓ^2_∞ with values in a von Neumann algebra.

7

Tensorizing maps and functorial properties

In this chapter, we introduce a major tool to study tensor products. We will try to identify the linear mappings $u : A \to B$ between C^*-algebras that are "tensorizing" meaning by this that, for any other C^*-algebra C, $Id_C \otimes u$ gives rise to a bounded map that is bounded from $C \otimes A$ to $C \otimes B$ when the domain and the range are equipped with given C^*-norms, respectively $\| \cdot \|_\alpha$ and $\| \cdot \|_\beta$. In order for this to make sense, of course we need $\| \cdot \|_\alpha$ and $\| \cdot \|_\beta$ defined on $C \otimes A$ and $C \otimes B$ for any C. More generally, we will consider the case of maps u that are defined only on a subspace $E \subset A$ (with $E \otimes C$ equipped with the induced norm), but for simplicity we will restrict ourselves to the minimal and maximal C^*-norms.

7.1 $(\alpha \to \beta)$-tensorizing linear maps

This is meant as a quick preliminary overview of the topic. Some of the main statements are proved in detail later on in this volume in §10.2.

Definition 7.1 Let A, B be C^*-algebras. Let α, β be one of the symbols min or max. Let $E \subset A$ be an operator subspace of a C^*-algebra A. We will say that a linear map $u : E \to B$ is $(\alpha \to \beta)$-tensorizing if for any C^*-algebra C we have

$$\forall x \in C \otimes E \quad \|Id_C \otimes u(x)\|_\beta \leq \|x\|_\alpha.$$

When this holds $Id_C \otimes u$ extends to a contraction as indicated in the following diagram.

$$
\begin{array}{c}
C \otimes_\alpha A \\
\cup \\
\overline{C \otimes E}^{\|\cdot\|_\alpha} \xrightarrow{\ Id_C \otimes u\ } C \otimes_\beta B.
\end{array}
$$

Remark It should be emphasized that this notion, when $\alpha = \max$ depends on the particular embedding $E \subset A$ under consideration.

The preceding definition equivalently means that:

$$\sup \|Id_C \otimes u : \overline{C \otimes E}^{\|\cdot\|_\alpha} \to C \otimes_\beta B\| \le 1 \tag{7.1}$$

where the sup runs over all possible C's. This can be automatically strengthened to

$$\sup \|Id_C \otimes u : \overline{C \otimes E}^{\|\cdot\|_\alpha} \to C \otimes_\beta B\|_{cb} \le 1. \tag{7.2}$$

Indeed, replacing C by $M_n(C)$ gives control of the cb-norm, and $M_n(C) \otimes_\alpha A = M_n(C \otimes_\alpha A)$ when either $\alpha = \min$ or $\alpha = \max$, and similarly for β.

By Proposition 1.11, the case $(\min, \min)$ is clear:

Proposition 7.2 *The map u is $(\min \to \min)$-tensorizing if and only if* $\|u\|_{cb} \le 1$.

Note that for $u : E \to B$ to be $(\min \to \min)$-tensorizing, by Proposition 1.11, it suffices to consider the min-tensor product with the C^*-algebra $C = \mathscr{B}$.

The case of $(\max \to \max)$ is given by the following remarkable Theorem 7.6 due to Kirchberg [161].

Let us denote by $i_B : B \to B^{**}$ the canonical inclusion map of B into B^{**} viewed as a von Neumann algebra as usual. See §A.16 for background on this. Anticipating a little on the subsequent Proposition 7.26 and Corollary 7.27, we will need the following preliminary fact on the bidual.

Lemma 7.3 *Let B be any C^*-algebra. Then for any C^*-algebra C we have*

$$\forall x \in C \otimes B \quad \|x\|_{C \otimes_{\max} B} = \|x\|_{C \otimes_{\max} B^{**}}.$$

Proof Let (π_1, π_2) be a pair of representations with commuting ranges taking values in some $B(H)$ with π_1 (resp. π_2) defined on C (resp. B). Let $\ddot{\pi}_2 : B^{**} \to \pi_1(C)'$ is as in §A.16. By (4.6), we have $\|(\pi_1 \cdot \pi_2)(x)\|_{B(H)} = \|(\pi_1 \cdot \ddot{\pi}_2)(x)\|_{B(H)} \le \|x\|_{C \otimes_{\max} B^{**}}$, whence $\|x\|_{C \otimes_{\max} B} \le \|x\|_{C \otimes_{\max} B^{**}}$, and the converse is obvious by (4.6). $\qquad\square$

Theorem 7.4 *The map u is $(\max \to \max)$-tensorizing if and only if u admits a decomposable extension $\tilde{u} : A \to B^{**}$ with $\|\tilde{u}\|_{dec} \le 1$, as in the following commutative diagram.*

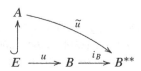

Remark 7.5 We will see in Corollary 7.16 that for $u : E \to B$ to be (max $\to$ max)-tensorizing it suffices to consider the max-tensor product with the C^*-algebra $C = \mathscr{C}$.

Proof of Theorem 7.4 The idea is to apply Theorem 6.20 to the map $i_B u$. To prove the if part, assume we have an extension $\tilde{u} : A \to B^{**}$ with $\|\tilde{u}\|_{dec} \le 1$. By (6.13), $\tilde{u}$ and a fortiori its restriction $i_B u$ is (max $\to$ max)-tensorizing. By Lemma 7.3, the map u itself is (max $\to$ max)-tensorizing. Conversely, assume u is (max $\to$ max)-tensorizing. We will use $C = M'$ in order to apply Theorem 6.20. Let $M = B^{**}$ viewed as a von Neumann algebra embedded in $B(H)$ for some H, so that $B \subset B^{**} \subset B(H)$. Let $\hat{u} : M' \otimes_{\max} E \to B(H)$ be as defined in Theorem 6.20. Since (7.1) implies (7.2) we have

$$\|Id_C \otimes u : \overline{C \otimes E}^{\|\cdot\|_{\max}} \to C \otimes_{\max} B\|_{cb} \le 1$$

for all C. Then, choosing $C = M'$, we observe that $\hat{u}$ is the composition of the map $Id_{M'} \otimes u$ with the $*$-homomorphism $\sigma : M' \otimes_{\max} B \to B(H)$ defined by $\sigma(c \otimes b) = cb = bc$. This shows that $\|\hat{u} : M' \otimes_{\overline{\max}} E \to B(H)\|_{cb} \le 1$. Then Theorem 6.20 applied to the map $i_B u : E \to M = B^{**}$ shows that there is an extension $\tilde{u} : A \to B^{**}$ with $\|\tilde{u}\|_{dec} \le 1$. $\square$

In the particular case when $E = A$, we must have $\tilde{u} = i_B u$, whence:

Theorem 7.6 ([161]) *Let A, B be C^*-algebras and let $u : A \to B$ be a linear map. Then u is (max $\to$ max)-tensorizing if and only if $i_B u : A \to B^{**}$ is decomposable with $\|i_B u\|_{dec} \le 1$.*

Remark 7.7 It is easy to check that this holds if and only if $\|u^{**}\|_{D(A^{**}, B^{**})} \le 1$. Note also that $u^{**} = i_{\ddot{B}} u$ with the notation in §A.16.

We state here for future reference a consequence of (6.13):

Corollary 7.8 *Any c.p. map u with $\|u\| \le 1$ between C^*-algebras is (max $\to$ max)-tensorizing.*

Proof This follows from Corollary 4.18 (and $\|u\|_{dec} = \|u\|$ by (ii) in Proposition 6.6). $\square$

Remark 7.9 Let A, B, G be C^*-algebras. Let $u : A \to B$ be an ($\alpha \to \beta$)-tensorizing linear map (where α or β can be either min or max). If u is c.p. then the mapping $Id_G \otimes u$ extends to a c.p. map from $G \otimes_\alpha A$ to $G \otimes_\beta B$. This extends Corollary 1.26.

Indeed, by Corollary 4.18 for any $t \in G \otimes A$ that is of the form $t = a^* a$ with $a \in G \otimes A$ we have $(Id_G \otimes u)(t) \in (G \otimes_{\max} B)_+$ and a fortiori $(Id_G \otimes u)(t) \in (G \otimes_\beta B)_+$. But since $Id_G \otimes u$ is ($\alpha \to \beta$)-bounded and the set of such t's is clearly dense in $G \otimes_\alpha A$, it follows that $Id_G \otimes u$ extends to a positive map from $G \otimes_\alpha A$ to $G \otimes_\beta B$. Replacing G by $M_n(G)$ the assertion follows.

The case (min $\to$ max) is closely related to the notion of nuclearity: a C^*-algebra A is nuclear if and only if the identity map on A is (min $\to$ max)-tensorizing. Kirchberg and Choi–Effros [45, 154] independently showed that this implies a strong approximation property for A called the CPAP (see Definition 4.8). The following statement for more general linear mappings (in place of Id_A) originates in their work. We postpone its proof to §10.2 (see Theorem 10.14).

Theorem 7.10 *Let $u : E \to B$ be a linear mapping from an operator space to a C^*-algebra. The following assertions are equivalent.*

(i) *The map u is (min $\to$ max)-tensorizing.*
(ii) *There is a net of finite rank maps $u_i : E \to B$ admitting factorizations through matrix algebras of the form*

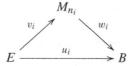

with $\|v_i\|_{cb}\|w_i\|_{dec} \le 1$ such that $u_i = w_i v_i$ converges pointwise to u.
(iii) *There is a net $u_i : A \to B$ of finite rank maps with $\sup \|u_i\|_{dec} \le 1$ that tends pointwise to u when restricted to E.*

In particular when $E = A$ we get (see Corollary 10.16 for a detailed proof):

Theorem 7.11 *Let $u : A \to B$ be a completely positive and unital linear mapping between two unital C^*-algebras. The following assertions are equivalent.*

(i) *u is (min $\to$ max)-tensorizing.*
(ii) *There is a net of finite rank maps (u_i) admitting factorizations through matrix algebras of the form*

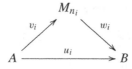

where v_i, w_i are c.p. maps with $\|v_i\|\|w_i\| \le 1$ such that $u_i = w_i v_i$ converges pointwise to u.
(iii) *There is a net $u_i : A \to B$ of finite rank c.p. maps that tends pointwise to u.*

Lastly, when $E = A = B$, we obtain the classical characterization of nuclear C^*-algebras:

Corollary 7.12 *The following properties of a C^*-algebra A are equivalent.*

(i) *A is nuclear,*
(ii) *A has the CPAP, i.e. the identity on A is the pointwise limit of a net of finite rank c.p. maps.*

As for the remaining case max $\to$ min, we leave it as an exercise for the reader. The answer is the same as for min $\to$ min (hint: take $C = M_n$).

Let us illustrate this with group C^*-algebras:

Corollary 7.13 (Nuclearity versus amenability) *The following properties of a discrete group G are equivalent.*

(i) *G is amenable.*
(ii) *$C_\lambda^*(G)$ is nuclear.*
(iii) *$C^*(G)$ is nuclear.*
(iv) *The canonical quotient map $Q_G : C^*(G) \to C_\lambda^*(G)$ is* (min $\to$ max)-*tensorizing.*
(v) *The natural $*$-homomorphism $\dot{Q}_G : C^*(G) \to M_G$ is* (min $\to$ max)-*tensorizing.*

Note that the group G is amenable if and only if $C_\lambda^*(G) = C^*(G)$, by Theorem 3.30.

Proof (i) $\Rightarrow$ (ii) (resp. (i) $\Rightarrow$ (iii)) follow from the preceding corollary since by Lemma 3.34 $C_\lambda^*(G)$ (resp. $C^*(G)$) has the CPAP when G is amenable. Assume (ii). Then we claim $Q_G : C^*(G) \to C_\lambda^*(G)$ is (min $\to$ max)-tensorizing; indeed, being a $*$-homomorphism, it is clearly (min $\to$ min)-tensorizing (and also (max $\to$ max)-tensorizing), so composing Q_G with $Id_{C_\lambda^*(G)}$ which is (min $\to$ max)-tensorizing gives the claim. Thus (ii) $\Rightarrow$ (iv).

A similar argument using $Id_{C^*(G)}$ (and (max $\to$ max) for Q_G) shows (ii) $\Rightarrow$ (iv).

(iv) $\Rightarrow$ (v) is trivial. Assume (v). Then for any $f \in \mathbb{C}[G]$, since $\dot{Q}_G(U_G(t)) = \lambda_G(t)$ we have

$$\left\| \sum f(t)\lambda_G(t) \otimes \lambda_G(t) \right\|_{C_\lambda^*(G) \otimes_{\max} M_G} \leq \left\| \sum f(t)\lambda_G(t) \otimes U_G(t) \right\|_{C_\lambda^*(G) \otimes_{\min} C^*(G)}$$

and hence by (4.22) and (4.25) we have

$$\left| \sum f(t) \right| \leq \left\| \sum f(t)\lambda_G(t) \right\|$$

and G is amenable by Theorem 3.30. This shows (v) $\Rightarrow$ (i) which completes the proof. $\qquad\square$

7.2 ‖ ‖max is projective (i.e. exact) but not injective

We first observe that for the *algebraic* tensor product we have no problem. It is both injective and projective. In fact if A, B and $\mathcal{I} \subset A$ are merely vector spaces we have a linear isomorphism

$$(A/\mathcal{I}) \otimes B = (A \otimes B)/(\mathcal{I} \otimes B). \tag{7.3}$$

When A, B are $*$-algebras and $\mathcal{I} \subset A$ is a self-adjoint ideal, this isomorphism is also a $*$-homomorphism.

We start with a basic fact, which will be generalized in Proposition 7.19.

Lemma 7.14 *Let* $\mathcal{I} \subset A$ *be a* (*closed, self-adjoint, two-sided*) *ideal in a* C^*-*algebra. We have then* (*isometrically*) *for any* C^*-*algebra* B

$$\mathcal{I} \otimes_{\max} B \subset A \otimes_{\max} B. \tag{7.4}$$

Equivalently, $\mathcal{I} \otimes_{\max} B$ *can be identified with the closure of* $\mathcal{I} \otimes B$ *in* $A \otimes_{\max} B$.

Proof Let (x_i) denote the net formed by all $x_i \in \mathcal{I}_+$ such that $\|x_i\| < 1$. We view these as a generalized sequence ($i \le j$ means $x_i \le x_j$). It is well known (see §A.15) that $\|xx_i - x\| \to 0$ and $\|x_i x - x\| \to 0$ for any $x \in \mathcal{I}$ when $i \to \infty$ (along the net). Since $x_i \le \sqrt{x_i}$ (and $\sqrt{x_i} \in \mathcal{I}_+$ with $\|\sqrt{x_i}\| < 1$), we also have $\|x\sqrt{x_i} - x\| + \|\sqrt{x_i}x - x\| \to 0$, and by the triangle inequality $\|\sqrt{x_i}x\sqrt{x_i} - x\| \le \|\sqrt{x_i}x\sqrt{x_i} - x\sqrt{x_i}\| + \|x\sqrt{x_i} - x\| \le \|\sqrt{x_i}x - x\| + \|x\sqrt{x_i} - x\|$. Therefore

$$\forall x \in \mathcal{I} \quad \|\sqrt{x_i}x\sqrt{x_i} - x\| \to 0. \tag{7.5}$$

Let $t \in \mathcal{I} \otimes B$. Assume $\|t\|_{A \otimes_{\max} B} < 1$. Let $P_i : A \to \mathcal{I}$ be the c.p. map defined by $P_i(x) = \sqrt{x_i}x\sqrt{x_i}$ for $x \in A$. By (4.30) we have $\|(P_i \otimes Id_B)(t)\|_{\mathcal{I} \otimes_{\max} B} \le \|t\|_{A \otimes_{\max} B} < 1$. By (7.5)

$$\|(P_i \otimes Id_B)(t) - t\|_{\mathcal{I} \otimes_{\max} B} \to 0$$

and hence $\|t\|_{\mathcal{I} \otimes_{\max} B} \le 1$. By homogeneity this shows that $\|t\|_{\mathcal{I} \otimes_{\max} B} \le \|t\|_{A \otimes_{\max} B}$, and the reverse inequality is trivial, proving (7.4). □

Proposition 7.15 (Exactness of the max-tensor product) *Let* A, B *be* C^*-*algebras and let* $\mathcal{I} \subset A$ *be a* (*closed, self-adjoint, two-sided*) *ideal. We have then* (*isometrically*)

$$(A/\mathcal{I}) \otimes_{\max} B = (A \otimes_{\max} B)/(\mathcal{I} \otimes_{\max} B). \tag{7.6}$$

In other words the sequence

$$\{0\} \to \mathcal{I} \otimes_{\max} B \to A \otimes_{\max} B \to (A/\mathcal{I}) \otimes_{\max} B \to \{0\}$$

is exact. More precisely, any x in $(A/\mathcal{I}) \otimes B$ with $\|x\|_{\max} < 1$ admits a lifting $\widehat{x}$ in $A \otimes B$ such that $\|\widehat{x}\|_{\max} < 1$.

Proof To verify (7.6), we use (7.3). Let $\rho : A \otimes_{\max} B \to (A/\mathcal{I}) \otimes_{\max} B$ denote the natural representation (obtained from $A \to A/\mathcal{I}$ after tensoring with the identity of B). Obviously, ρ vanishes on $\mathcal{I} \otimes_{\max} B$. Hence denoting by $Q : A \otimes_{\max} B \to (A \otimes_{\max} B)/(\mathcal{I} \otimes_{\max} B)$ the quotient map, we have a factorization of ρ of the form $\rho = \pi Q$ where $\pi : (A \otimes_{\max} B)/(\mathcal{I} \otimes_{\max} B) \to (A/\mathcal{I}) \otimes_{\max} B$ is a $*$-homomorphism such that

$$\|\pi : (A \otimes_{\max} B)/(\mathcal{I} \otimes_{\max} B) \to (A/\mathcal{I}) \otimes_{\max} B\| \le 1. \qquad (7.7)$$

By Lemma 4.27 we have an injective $*$-homomorphism

$$(A \otimes B)/(\mathcal{I} \otimes B) \subset (A \otimes_{\max} B)/(\mathcal{I} \otimes_{\max} B).$$

Thus the norm of $(A \otimes_{\max} B)/(\mathcal{I} \otimes_{\max} B)$ induces a C^*-norm on $(A \otimes B)/(\mathcal{I} \otimes B)$, but by (7.3) we may view it as a C^*-norm on $(A/\mathcal{I}) \otimes B$. By (7.7) the latter C^*-norm dominates the maximal C^*-norm on $(A/\mathcal{I}) \otimes B$, and hence it must coincide with it.

Since we know (Lemma 7.14)) that $\mathcal{I} \otimes_{\max} B$ is the closure of $\mathcal{I} \otimes B$ in $A \otimes_{\max} B$, the last assertion follows from (4.34). $\qquad\square$

Corollary 7.16 *Let $E \subset A$ be an operator subspace of a C^*-algebra A and let $u : E \to B$ be a linear map into another C^*-algebra B. The following are equivalent.*

(i) *The map u is (max $\to$ max)-tensorizing i.e. for any C^*-algebra C we have*

$$\forall x \in C \otimes E \quad \|(Id_C \otimes u)(x)\|_{C \otimes_{\max} B} \le \|x\|_{C \otimes_{\max} A}.$$

(ii) *The same as* (i) *holds but restricted to $C = \mathscr{C}$.*

Proof If (ii) holds, we claim that it holds when $C = C^*(\mathbb{F})$ for any free group $\mathbb{F}$. Indeed, since we may assume $x \in E_1 \otimes E$ for some separable subspace $E_1 \subset C^*(\mathbb{F})$, this is an immediate consequence of Lemma 3.8. Now, since any unital C is a quotient of $C^*(\mathbb{F})$ for a suitable $\mathbb{F}$ (see Proposition 3.39), (i) for such C's follows from the preceding proposition and Lemma 4.26 applied with $\alpha = \max$. When C is not unital, we may view it as an ideal in its unitization $\widetilde{C}$ and by Lemma 7.14 we have $C \otimes_{\max} D \subset \widetilde{C} \otimes_{\max} D$ for all D in particular for both $D = A$ and $D = B$. Thus (i) for $\widetilde{C}$ implies (i) for C. This shows (ii) $\Rightarrow$ (i). $\qquad\square$

Proposition 7.15 shows that the maximal tensor product behaves well with respect to quotients. However, in sharp contrast with the minimal norm, it does not behave well at all with respect to subalgebras, i.e. when $D \subset A$ is a C^*-subalgebra of a C^*-algebra and C is another C^*-algebra, the $*$-homomorphism $C \otimes_{\max} D \to C \otimes_{\max} A$ is in general *NOT* injective.

Indeed, if this holds and if the pair (A, C) is nuclear, it follows that the pair (D, C) is nuclear. Taking $C = \mathscr{C}$, this would imply that the WEP is inherited by subalgebras, and hence that any C^*-subalgebra $A \subset B(H)$ is WEP, which is of course absurd.

We will now give an explicit example showing that this fails for the inclusion $C_\lambda^*(G) \subset B(\ell_2(G))$ with $C = C^*(G)$ whenever G is a nonamenable discrete group such that $(C^*(G), B(H))$ is a nuclear pair. With the terminology of §9.4 this means that $C = C^*(G)$ has the LLP. In particular this holds when $G = \mathbb{F}_n$ for $n \geq 2$.

Let G be a discrete group and let $S \subset G$ be a finite subset and let $f : S \to \mathbb{R}_+$ be any function. We will draw this from our repertoire in §4.4. Let

$$x = \sum\nolimits_{t \in S} f(t) U_G(t) \otimes \lambda_G(t) \in C^*(G) \otimes C_\lambda^*(G) \subset C^*(G) \otimes B(\ell_2(G)).$$

By (4.27)

$$\|x\|_{C^*(G) \otimes_{\max} C_\lambda^*(G)} = \sum\nolimits_{t \in S} f(t),$$

but since the min and max norms coincide by assumption on $C^*(G) \otimes B(\ell_2(G))$ we have

$$\|x\|_{C^*(G) \otimes_{\max} B(\ell_2(G))} = \|x\|_{C^*(G) \otimes_{\min} B(\ell_2(G))} = \|x\|_{C^*(G) \otimes_{\min} C_\lambda^*(G)},$$

and hence by (4.22) (i.e. by Fell's absorption principle)

$$= \left\| \sum\nolimits_{t \in S} f(t) \lambda_G(t) \right\|_{B(\ell_2(G))}.$$

By Kesten's criterion (see Theorem 3.30), if G is not amenable, there is a finite subset $S \subset G$ such that $|S| > \left\| \sum_{t \in S} \lambda_G(t) \right\|_{B(\ell_2(G))}$, thus we obtain as announced that if $D = C_\lambda^*(G)$, $A = B(\ell_2(G))$ and $C = C^*(G)$, the $*$-homomorphism $C \otimes_{\max} D \to C \otimes_{\max} A$ is not injective (since it is not isometric). The same proof works for the inclusion of $D = M_G$ in $B(\ell_2(G))$.

More explicitly, if we consider the inclusion $C_\lambda^*(\mathbb{F}_n) \subset B(\ell_2(\mathbb{F}_n))$ then we have:

$$\left\| I \otimes I + \sum\nolimits_1^n U_{\mathbb{F}_n}(g_j) \otimes \lambda_{\mathbb{F}_n}(g_j) \right\|_{C^*(\mathbb{F}_n) \otimes_{\max} B(\ell_2(\mathbb{F}_n))} = 2\sqrt{n} \qquad (7.8)$$

but

$$\left\| I \otimes I + \sum\nolimits_1^n U_{\mathbb{F}_n}(g_j) \otimes \lambda_{\mathbb{F}_n}(g_j) \right\|_{C^*(\mathbb{F}_n) \otimes_{\max} C_\lambda^*(\mathbb{F}_n)} = n + 1 \qquad (7.9)$$

and these are different for $n > 1$.

More generally, using a different reasoning, we have:

Proposition 7.17 *Consider* $D = C^*_\lambda(G)$ *and* $A = B(\ell_2(G))$. *If* G *is not amenable, then the inclusion* $D \subset A$ *is not* max-*injective (in the sense of Definition 7.18). In particular, this holds when* $G = \mathbb{F}_n$ *for* $n \geq 2$.

Proof Indeed, assume that $D \subset A$ is max-injective. Then by iteration, $D \otimes_{\max} \bar{D} \to A \otimes_{\max} \bar{D} \to A \otimes_{\max} \bar{A}$ is isometric. Anticipating a bit, we will see in Theorem 23.7 that, when $A = B(\ell_2(G))$, the norms of $A \otimes_{\max} \bar{A}$ and $A \otimes_{\min} \bar{A}$ coincide on the "positive definite cone" formed of the tensors of the form $\sum_1^n a_j \otimes \overline{a_j}$, and hence by Fell's principle $\| \sum_{s \in S} \lambda(s) \otimes \overline{\lambda(s)} \|_{A \otimes_{\max} \bar{A}} = \| \sum_{s \in S} \lambda(s) \otimes \overline{\lambda(s)} \|_{A \otimes_{\min} \bar{A}} = \| \sum_{s \in S} \lambda(s) \|$. But by (4.17) we have $\| \sum_{s \in S} \lambda(s) \otimes \overline{\lambda(s)} \|_{D \otimes_{\max} \bar{D}} = |S|$. Therefore, by Kesten's criterion (see Theorem 3.30) G is amenable. $\square$

7.3 max-injective inclusions

Motivated by the last section, we are naturally led to the following notion.

Definition 7.18 Let A be a C^*-algebra and let $D \subset A$ be a C^*-subalgebra. We will say that the inclusion $D \subset A$ is max-injective if for any C^*-algebra C, the max-norm on $C \otimes D$ coincides with the norm induced by $C \otimes_{\max} A$. Equivalently, this means that the map $C \otimes_{\max} D \to C \otimes_{\max} A$ is isometric (or equivalently injective).

As already observed in Proposition 7.17, this does not always hold.
    Here are some examples of max-injective inclusions $D \subset A$:

Proposition 7.19 *Let* A *be a* C^*-*algebra and let* $D \subset A$ *be a* C^*-*subalgebra. The inclusion* $D \subset A$ *is* max-*injective in each of the following cases.*

   (i) *If there is a c.p. projection* $P : A \to D$ *with* $\|P\| = 1$.
  (ii) *More generally, if we have "approximate projections," meaning by this that there is a net of c.p. mappings* $P_i : A \to D$ *with* $\|P_i\| \leq 1$ *such that* $P_i(x) \to x \quad \forall x \in D$.
 (iii) *If* D *is a (closed two-sided self-adjoint) ideal of* A, *we already saw in Lemma 7.14 that* $D \subset A$ *is* max-*injective. Actually this still holds if* D *is merely a hereditary subalgebra of* A *(i.e. we have* $x \in A, y \in D \quad 0 \leq x \leq y \Rightarrow x \in D$*).*

Proof (i) clearly suffices, since P will be (max $\to$ max)-tensorizing by Corollary 7.8. (ii) suffices for essentially the same reason: the P_i's are all (max $\to$ max)-tensorizing. We claim that case (ii) implies case (iii). Indeed, let (P_i) be as in the proof of Lemma 7.14. Note that if $x \in D_+$ we have $0 \leq P_i(x) \leq \|x\| x_i$. Thus in both cases (ideal or hereditary) we have $P_i(x) \in D$ and we obtain the desired "approximate projections," proving the claim. $\square$

Remark 7.20 The notion of max-injectivity is relevant to the study of pairs (A, B) of C^*-algebras such that $A \otimes_{\min} B = A \otimes_{\max} B$, that we call nuclear pairs in the sequel.

For instance, if we assume that $D \subset A$ is a max-injective inclusion, then clearly:

$$(A, B) \text{ nuclear} \Rightarrow (D, B) \text{ nuclear}. \tag{7.10}$$

Remark 7.21 Let D, C be C^*-algebras. Let $A = D \otimes_{\min} C$ and B another C^*-algebra. Then:

$$(A, B) \text{ nuclear} \Rightarrow (D, B) \text{ nuclear}. \tag{7.11}$$

Indeed, if C is unital we have an embedding $D \simeq D \otimes 1 \subset A$ and a unital c.p. projection $P : A \to D \otimes 1$ defined by $P(d \otimes c) = d \otimes (\varphi(c)1)$, where φ is any state on C (recall Remark 1.24). By (i) in Proposition 7.19, the inclusion $D \subset A$ is max-injective. If C is not unital, the same conclusion can be obtained using an approximate unit and (ii) in Proposition 7.19. Therefore (7.11) is a special case of (7.10).

Remark 7.22 By Proposition 3.5, for any subgroup $\Gamma \subset G$ of a discrete group G, the inclusion $C^*(\Gamma) \subset C^*(G)$ is max-injective. Thus for any C^*-algebra B we have isometrically $C^*(\Gamma) \otimes_{\max} B \subset C^*(G) \otimes_{\max} B$. Let $\mathbb{F}$ be any free group. For any fixed $t \in C^*(\mathbb{F}) \otimes B$ there is an at most countably generated free subgroup $\Gamma \subset \mathbb{F}$ (that may depend on t) such that (viewing $C^*(\Gamma) \subset C^*(\mathbb{F})$) we have $t \in C^*(\Gamma) \otimes B$. The group Γ is isomorphic to $\mathbb{F}_n$ for some $1 \leq n \leq \infty$. Thus the norm of t in $C^*(\mathbb{F}) \otimes_{\max} B$ can be computed in $C^*(\mathbb{F}_n) \otimes_{\max} B$. In fact if there is a copy of $\mathbb{F}_\infty$ such that $\Gamma \subset \mathbb{F}_\infty \subset \mathbb{F}$, we may compute the latter norm simply in $C^*(\mathbb{F}_\infty) \otimes_{\max} B$.

It is natural to wonder whether in the nonseparable case one can compute the norm of the max-tensor product using separable C^*-subalgebras. The next two statements address this question.

Lemma 7.23 *Let A, B be C^*-algebras. Let $t \in A \otimes B$. Then for any $\varepsilon > 0$ there are separable C^*-subalgebras $A_1 \subset A$, $B_1 \subset B$ such that $t \in A_1 \otimes B_1$ and*

$$\|t\|_{A_1 \otimes_{\max} B_1} \leq (1 + \varepsilon) \|t\|_{A \otimes_{\max} B}. \tag{7.12}$$

A fortiori $t \in A_1 \otimes B$ *and we have*

$$\|t\|_{A_1 \otimes_{\max} B} \leq (1 + \varepsilon) \|t\|_{A \otimes_{\max} B}. \qquad (7.13)$$

Proof A moment of thought shows that (7.13) implies (7.12) by iteration (recall (4.8)). Thus it suffices to find A_1 separable for which (7.13) holds. Assume A unital. Then by Proposition 3.39 for a suitable free group $\mathbb{F}$ there is an onto $*$-homomorphism $q : C^*(\mathbb{F}) \to A$. By (7.6) and (4.34) for any $\varepsilon > 0$ there is $\hat{t} \in C^*(\mathbb{F}) \otimes B$ with $\|\hat{t}\|_{C^*(\mathbb{F}) \otimes_{\max} B} \leq (1 + \varepsilon) \|t\|_{A \otimes_{\max} B}$ such that $(q \otimes Id_B)(\hat{t}) = t$. We may assume $\hat{t} \in E \otimes B$ for some separable (and even finite-dimensional) subspace $E \subset C^*(\mathbb{F})$. By Remark 3.6 there is a separable C^*-subalgebra $C \subset C^*(\mathbb{F})$ containing E and admitting a contractive c.p. projection $P : C^*(\mathbb{F}) \to C$. By (i) in Proposition 7.19 we have $\|\hat{t}\|_{C^*(\mathbb{F}) \otimes_{\max} B} = \|\hat{t}\|_{C \otimes_{\max} B}$. Let $A_1 = q(C)$. Then A_1 is separable, $t = (q \otimes Id_B)(\hat{t}) \in A_1 \otimes B$ and by (say) (4.30)

$$\|t\|_{A_1 \otimes_{\max} B} \leq \|\hat{t}\|_{C \otimes_{\max} B} = \|\hat{t}\|_{C^*(\mathbb{F}) \otimes_{\max} B} \leq (1 + \varepsilon) \|t\|_{A \otimes_{\max} B}.$$

This proves the unital case. The nonunital one follows by a unitization argument. $\qquad \square$

Proposition 7.24 *Let A, B be C^*-algebras. Let $E \subset A$ be a separable subspace. Assume that B is separable. There is a separable C^*-subalgebra A_1 such that $E \subset A_1 \subset A$ such that the $*$-homomorphism $A_1 \otimes_{\max} B \to A \otimes_{\max} B$ is isometric.*

Proof Assume A unital. We may clearly assume that E is a C^*-subalgebra. We first claim that there is a separable unital C^*-subalgebra E_1 such that $E \subset E_1 \subset A$ and such that for any $t \in E \otimes B$ we have $\|t\|_{E_1 \otimes_{\max} B} = \|t\|_{A \otimes_{\max} B}$. Indeed, let $\{t_n\}$ be a dense sequence in $E \otimes B$ with respect to the norm in $E \otimes_{\max} B$. A fortiori $\{t_n\}$ is dense in $E \otimes B$ for the (smaller) norm induced by $E_1 \otimes_{\max} B$ for any $E_1 \supset E$. Moreover, we may and do assume that each element in $\{t_n\}$ appears infinitely many times. By Lemma 7.23 for each n there is a separable unital C^*-subalgebra $D_n \subset A$ such that $t_n \in D_n \otimes B$ and $\|t_n\|_{D_n \otimes_{\max} B} \leq (1 + 1/n)\|t_n\|_{A \otimes_{\max} B}$. Let $E_1 \subset A$ be the C^*-subalgebra generated by $\cup_{n \geq 1} D_n$. Since $D_n \subset E_1$ we have obviously $\|t_n\|_{E_1 \otimes_{\max} B} \leq \|t_n\|_{D_n \otimes_{\max} B} \leq (1 + 1/n)\|t_n\|_{A \otimes_{\max} B}$ for all $n \geq 1$, and hence (since each element t_n appears with n arbitrary large) we obtain $\|t_n\|_{E_1 \otimes_{\max} B} = \|t_n\|_{A \otimes_{\max} B}$ for all $n \geq 1$, and by the density of $\{t_n\}$ this proves the claim. By iteration, this claim gives us a sequence of separable unital C^*-subalgebra E_n with $E_0 = E$ such that $E_{n-1} \subset E_n \subset A$ and such that for any $t \in E_{n-1} \otimes B$ we have $\|t\|_{E_n \otimes_{\max} B} = \|t\|_{A \otimes_{\max} B}$ for all $n \geq 1$. Then let $A_1 = \overline{\cup_{n \geq 1} E_n}$. Let $t \in A_1 \otimes B$. To show that $\|t\|_{A_1 \otimes_{\max} B} = \|t\|_{A \otimes_{\max} B}$ we may assume by density that $t \in \cup_{n \geq 1} E_n \otimes B$ or equivalently that $t \in E_n \otimes B$ for some

$n \geq 1$. Then $\|t\|_{E_{n+1} \otimes_{\max} B} = \|t\|_{A \otimes_{\max} B}$ and again since $E_{n+1} \subset A_1$, we have $\|t\|_{A_1 \otimes_{\max} B} \leq \|t\|_{E_{n+1} \otimes_{\max} B}$. Thus we conclude $\|t\|_{A_1 \otimes_{\max} B} \leq \|t\|_{A \otimes_{\max} B}$ which completes the proof since the reverse inequality is obvious from the start. □

Remark 7.25 In the situation of Proposition 7.24 if $A \otimes_{\min} B = A \otimes_{\max} B$ then $A_1 \otimes_{\min} B = A_1 \otimes_{\max} B$

We now turn to a very simple but quite useful example: for any C^*-algebra A the inclusion $A \subset A^{**}$ is max-injective (see §A.17 for background on A^{**}).

Proposition 7.26

(i) *For any C^*-algebras A_1, A_2 we have an isometric embedding*

$$A_1 \otimes_{\max} A_2 \to A_1^{**} \otimes_{\max} A_2.$$

(ii) *More generally, we have an isometric embedding*

$$A_1 \otimes_{\max} A_2 \to A_1^{**} \otimes_{\max} A_2^{**}.$$

Proof We start with a preliminary observation.

If $\sigma_1 : A_1 \to B(H)$ and $\sigma_2 : A_2 \to B(H)$ are $*$-homomorphisms with commuting ranges then $\ddot{\sigma}_1 : A_1^{**} \to B(H)$ and $\ddot{\sigma}_2 : A_2^{**} \to B(H)$ still have commuting ranges. Indeed, let $M_1 = \sigma_1(A_1)''$ and $M_2 = \sigma_2(A_2)''$. The latter are mutually commuting von Neumann algebras in $B(H)$. Since $\ddot{\sigma}_1$ is the unique $(\sigma(A_1^{**}, A_1^*), \sigma(M_1, M_{1*}))$-continuous extension of σ_1 and $M_1 = \overline{\sigma_1(A_1)}^{\text{weak}*}$ (by the bicommutant Theorem A.46) we have $\ddot{\sigma}_1(A_1) \subset M_1$ and similarly $\ddot{\sigma}_2(A_2) \subset M_2$.

For (i) (resp. for (ii)), it clearly suffices to show that if $\sigma_1 : A_1 \to B(H)$ and $\sigma_2 : A_2 \to B(H)$ are $*$-homomorphisms with commuting ranges, then $\ddot{\sigma}_1$ and σ_2 (resp. $\ddot{\sigma}_1$ and $\ddot{\sigma}_2$) still have commuting ranges, which is what the preceding observation says. Actually we can also deduce (ii) from (i) by iteration. □

In other words (see Definition 7.18), (i) means:

Corollary 7.27 *For any C^*-algebra A the inclusion $i_A : A \to A^{**}$ is max-injective.*

Remark 7.28 For (min $\to$ max) and (max $\to$ max)-tensorizing maps in the sense of §7.1, it is worthwhile to record here the following observation: Let $u : E \to B$ be a linear map. Then u is (min $\to$ max)-tensorizing (resp. (max $\to$ max)-tensorizing) if and only if the same is true for $i_B u : E \to B^{**}$.

Indeed, this is immediate, given the preceding Corollary (applied to B instead of A).

Theorem 7.29 *Consider an inclusion $D \subset A$ between C^*-algebras, and the bitransposed inclusion $D^{**} \subset A^{**}$. Let $i_D : D \to D^{**}$ denote as before the canonical inclusion. The following properties are equivalent:*

(i) *The inclusion $D \subset A$ is max-injective.*

(i)' *The map $C \otimes_{\max} D \to C \otimes_{\max} A$ is isometric when $C = \mathscr{C}$.*

(ii) *There is a contractive map $T : A \to D^{**}$ such that $T_{|D} = i_D$, or equivalently such that the following diagram commutes.*

$$
\begin{array}{ccc}
A & & \\
\uparrow & \searrow^{T} & \\
\big\downarrow & & \\
D & \xrightarrow{\ i_D\ } & D^{**}
\end{array}
$$

(iii) *There is a contractive and normal c.p. projection $P : A^{**} \to D^{**}$.*

(iii)' *There is a contractive projection $P : A^{**} \to D^{**}$.*

Proof The equivalence of (i) and (i)' follows from Corollary 7.16.

The implication (i) $\Rightarrow$ (ii) can be deduced from Theorem 7.4 (applied with $E = D$ and $u = Id_D$): (i) holds if and only if there is $T : A \to D^{**}$ with $\|T\|_{dec} = 1$ extending the inclusion $D \subset D^{**}$. When this holds, a fortiori (ii) holds.

Assume (ii). We will use $\ddot{T}$ as defined in §A.16. Then (see Proposition A.50) $P = \ddot{T} : A^{**} \to D^{**}$ is a normal projection onto D^{**} with $\|P\| = 1$. By Theorem 1.45, P is automatically c.p. so that (iii) holds, and (iii) $\Rightarrow$ (iii)' is trivial.

Assume (iii)'. By Theorem 1.45 again, P is automatically c.p. so that by part (i) in Proposition 7.19 the inclusion $D^{**} \subset A^{**}$ is max-injective. Let C be a C^*-algebra. Let $t \in C \otimes D$ such that $\|t\|_{C \otimes_{\max} A} \leq 1$. A fortiori we have clearly $\|t\|_{C \otimes_{\max} A^{**}} \leq 1$ and hence $\|t\|_{C \otimes_{\max} D^{**}} \leq 1$ by our assumption. By Corollary 7.27 (applied to D) we have $\|t\|_{C \otimes_{\max} D} = \|t\|_{C \otimes_{\max} D^{**}} \leq 1$, and hence by homogeneity $\|t\|_{C \otimes_{\max} D} \leq \|t\|_{C \otimes_{\max} A}$ for all $t \in C \otimes D$. The converse being obvious, this completes the proof that (iii)' $\Rightarrow$ (i).

The last step can be rephrased more abstractly like this: since the inclusions $D \subset D^{**}$ and $A \subset A^{**}$ are max-injective (see Corollary 7.27), it is formally immediate that the max-injectivity of $D^{**} \subset A^{**}$ implies that of $D \subset A$, as can be read on this diagram:

$$
\begin{array}{ccc}
D^{**} \otimes_{\max} C & \longrightarrow & A^{**} \otimes_{\max} C \\
\uparrow & & \uparrow \\
\big\downarrow & & \big\downarrow \\
D \otimes_{\max} C & \longrightarrow & A \otimes_{\max} C
\end{array}
$$

$\square$

Corollary 7.30 *The properties in Theorem 7.29 are also equivalent to:*

(iv) *Any c.p. contraction $u : D \to M$ into a von Neumann algebra M extends to a c.p. (complete) contraction $\widetilde{u} : A \to M$.*

(iv)' *For any von Neumann algebra M, any $u \in D(D, M)$ extends to a mapping $\widetilde{u} \in D(A, M)$ with $\|\widetilde{u}\|_{dec} = \|u\|_{dec}$, as in the following diagram.*

$$
\begin{array}{ccc}
 & A & \\
 \Big\uparrow & \diagdown\ \widetilde{u} & \\
 & & \diagdown \\
 D & \xrightarrow{\ u\ } & M
\end{array}
$$

Proof Assume (iii) and let u be as in (iv)'. By Lemma 6.9, $\ddot{u} \in D(D^{**}, M)$ and $\|\ddot{u}\|_{dec} = \|u\|_{dec}$. Consider $\ddot{u} P : A^{**} \to M$ and let $\widetilde{u} = (\ddot{u} P)_{|A} : A \to M$. By (6.7) and by part (i) from Lemma 6.5, we have $\|\widetilde{u}\|_{dec} \le \|\ddot{u}\|_{dec}\|P\|_{dec} = \|\ddot{u}\|_{dec}\|P\| = \|u\|_{dec}$. Thus (iii) $\Rightarrow$ (iv)'. Using Lemma 1.43 the same argument yields (iii) $\Rightarrow$ (iv). Conversely if we assume either (iv) or (iv)' and apply it to $u = i_D : D \to D^{**}$ we obtain (ii). $\qquad\square$

Remark 7.31 Any $T : A \to D^{**}$ satisfying (ii) in Theorem 7.29 is automatically c.p. and completely contractive. This follows from Tomiyama's Theorem 1.45 since $\ddot{T}$ is a projection onto D^{**} and $T = \ddot{T}_{|A}$.

Remark 7.32 If there is a net of maps $T_i : A \to D$ with $\|T_i\| \le 1$ such that $\|T_i(x) - x\| \to 0$ for any $x \in D$, then the inclusion $D \subset A$ is max-injective. Indeed, $\mathcal{U}$ being an ultrafilter refining the net (see §A.4), let $T(x) = \lim_{\mathcal{U}} T_i(x)$ for all x in A with respect to $\sigma(D^{**}, D^*)$, then $T : A \to D^{**}$ satisfies (ii) in Theorem 7.29.

The next statement will be very useful when dealing with QWEP C^*-algebras in §9.7.

Theorem 7.33 *Let A, B be C^*-algebras with B unital. Let $\varphi : A \to B$ be a c.p. map such that $\varphi(B_A) = B_B$. Then the restriction to the multiplicative domain*

$$\varphi_{|D_\varphi} : D_\varphi \to B$$

is a surjective $*$-homomorphism, and moreover the inclusion

$$D_\varphi \subset A$$

is max-*injective.*

Proof Note that for any unitary y in B there is $x \in B_A$ such that $\varphi(x) = y$. By the Choi–Cauchy–Schwarz inequality (5.2), we have

$$1 = \varphi(x)^*\varphi(x) \le \varphi(x^*x) \le 1$$

and hence $\varphi(x)^*\varphi(x) = \varphi(x^*x)$. Similarly, $\varphi(x)\varphi(x)^* = \varphi(xx^*)$, so that $x \in D_\varphi$ and hence

$$D_\varphi \supset \{x \in B_A \mid \varphi(x) \in U(B)\},$$

where $U(B)$ denotes as usual the set of unitaries in B. It follows that $\varphi(D_\varphi) \supset U(B)$, so that the restriction

$$\pi = \varphi_{|D_\varphi} : D_\varphi \to B$$

is a surjective $*$-homomorphism. Let C be another C^*-algebra. Let $j : D_\varphi \to A$ be the inclusion. We will now show that the $*$-homomorphism

$$j^C : C \otimes_{\max} D_\varphi \to C \otimes_{\max} A$$

that extends $Id_C \otimes j$ is injective.

Let $x \in C \otimes_{\max} D_\varphi$ be such that $j^C(x) = 0$. We will show that $x = 0$.

Let $D^0 = \ker(\pi) \subset D_\varphi$. By Remark 5.2 D^0 is a *hereditary* C^*-subalgebra of A (and an ideal in D_φ), and hence $D^0 \subset A$ is max-injective by Proposition 7.19 (iii). Let us denote

$$\varphi^C : C \otimes_{\max} A \to C \otimes_{\max} B$$

and

$$\pi^C : C \otimes_{\max} D_\varphi \to C \otimes_{\max} B$$

the natural extensions. Clearly, $\varphi^C j^C = \pi^C$. Therefore $\pi^C(x) = 0$. But by the projectivity (or exactness) of max (see (7.6)), we have $\ker(\pi^C) = C \otimes_{\max} D^0$ and hence $x \in C \otimes_{\max} D^0 \subset C \otimes_{\max} D_\varphi$. But since $D^0 \subset A$ is max-injective, the composition

$$C \otimes_{\max} D^0 \to C \otimes_{\max} D_\varphi \to C \otimes_{\max} A$$

is injective, and we conclude that $x = 0$ in $C \otimes_{\max} D_\varphi$. $\qquad\qquad \square$

7.4 $\| \ \|_{\min}$ is injective but not projective (i.e. not exact)

In this section, in analogy with what we saw for the maximal tensor products in §7.2, we investigate if or when the canonical identification

$$(A/\mathcal{I}) \otimes B = (A \otimes B)/(\mathcal{I} \otimes B) \qquad (7.14)$$

remains valid, after completion, for the minimal tensor products.

By its very definition, the minimal tensor product is obviously injective, even in the operator space setting (see Remark 1.10). In the C^*-case, if E_j, G_j ($j = 1, 2$) are C^*-algebras and $E_j \subset G_j$ ($j = 1, 2$) are injective

∗-homomorphisms, then $E_1 \otimes_{\min} E_2 \subset G_1 \otimes_{\min} G_2$ is also an injective ∗-homomorphism.

In particular, if $\mathcal{I} \subset A$ is an ideal so that $A/\mathcal{I}$ is a C^*-algebra, we have for any B an embedding $\mathcal{I} \otimes_{\min} B \subset A \otimes_{\min} B$, and $\mathcal{I} \otimes_{\min} B$ is an ideal in $A \otimes_{\min} B$. It is somewhat natural, in analogy with (7.14), to expect that the quotient $(A \otimes_{\min} B)/(\mathcal{I} \otimes_{\min} B)$ should be identifiable with $(A/\mathcal{I}) \otimes_{\min} B$. However, although it is true for many B's, it is not so for all B.

We will now give an explicit example showing that this fails for $A = C^*(\mathbb{F}_n)$ whenever $n \geq 2$. See §3.1 for notation and background on operator algebras such as $C^*(G)$ and M_G associated to a discrete group G. The same argument works for $A = C^*(G)$ assuming G nonamenable, but assuming also that $C^*(G)$ has the LLP and that G is approximately linear (so-called hyperlinear), i.e. that the von Neumann algebra of G, namely $M_G = \lambda_G(G)''$ is QWEP (see Remark 9.68). When $G = \mathbb{F}_n$ we will see later on in these notes that the latter properties hold (see (9.5) and Theorem 12.21).

In analogy with Proposition 7.17, we will prove (the traditional terminology for this would be that for $G = \mathbb{F}_n$, the algebra $C^*(G)$, viewed as an extension of $C^*_\lambda(G)$, is not "locally split"):

Proposition 7.34 *Recall that $Q_G : C^*(G) \to C^*_\lambda(G)$ denotes the canonical quotient map. For $G = \mathbb{F}_n$ with $n > 1$, if $A = B = C^*(G)$ and $\mathcal{I} = \ker(Q_G)$, the natural ∗-homomorphism*

$$(A \otimes_{\min} B)/(\mathcal{I} \otimes_{\min} B) \to (A/\mathcal{I}) \otimes_{\min} B \qquad (7.15)$$

is not injective. Equivalently (just exchanging A and B), the homomorphism

$$(B \otimes_{\min} A)/(\mathcal{I} \otimes_{\min} A) \to (B/\mathcal{I}) \otimes_{\min} A \qquad (7.16)$$

is not injective.

To prove this we need to anticipate slightly: we will review a few facts that will be proved later on in these notes.

For a group G, we will consider the following property:

Property 7.35 *The map $\dot{Q}_G : C^*(G) \to M_G$ that is the same as Q_G but viewed as taking values in M_G admits a factorization $\dot{Q}_G : C^*(G) \overset{w}{\to} B \overset{v}{\to} M_G$ where v, w are c.p. maps with $\|v\|_{cb} \leq 1$ and $\|w\|_{cb} \leq 1$, and where B is a C^*-algebra such that the pair $(B, C^*(G))$ is nuclear.*

For further reference, we introduce a notion due to Kirchberg [155].

Definition 7.36 A group G is said to have the factorization property (or simply property (F)) if for any $x \in C^*(G) \otimes C^*(G)$ we have

$$\|[\lambda_G \cdot \rho_G](x)\|_{B(\ell_2(G))} \leq \|x\|_{C^*(G) \otimes_{\min} C^*(G)}, \qquad (7.17)$$

where $[\lambda_G \cdot \rho_G]: C^*(G) \otimes C^*(G) \to B(\ell_2(G))$ denotes the $*$-homomorphism that takes $U_G(s) \otimes U_G(t)$ to $\lambda_G(s)\rho_G(t)$ $(s, t \in G)$.

Remark 7.37 This definition should be compared with (4.18). In particular, (4.18) shows that the factorization property holds if G is amenable, because $C^*(G)$ is then nuclear.

Lemma 7.38 *If G satisfies Property 7.35 then G has the factorization property* (7.17).

Proof Since c.p. maps with cb-norm ≤ 1 tensorize both the minimal and maximal tensor products, the following maps are all of norm 1:

$$C^*(G) \otimes_{\min} C^*(G) \to B \otimes_{\min} C^*(G)$$
$$= B \otimes_{\max} C^*(G) \to M_G \otimes_{\max} C^*(G) \to M_G \otimes_{\max} M_G$$

and the composition is equal to $\dot{Q}_G \otimes \dot{Q}_G$ on the algebraic tensor product. Thus

$$\|\dot{Q}_G \otimes \dot{Q}_G : C^*(G) \otimes_{\min} C^*(G) \to M_G \otimes_{\max} M_G\| \leq 1.$$

Then (4.18) yields the conclusion. $\qquad\square$

Lemma 7.39 *Assume that G satisfies the factorization property* (7.17). *If* (7.15) *is injective (or equivalently isometric) with $A = B = C^*(G)$ and $\mathcal{I} = \ker(Q_G)$, then G is amenable.*

Proof By (7.17) for any $x \in C^*(G) \otimes C^*(G)$ we have

$$\|[\lambda_G \cdot \rho_G](x)\|_{B(\ell_2(G))} \leq \|x\|_{C^*(G) \otimes_{\min} C^*(G)}.$$

Let $\mathcal{I} = \ker(Q_G)$. Clearly the corresponding mapping $x \mapsto [\lambda_G \cdot \rho_G](x)$ vanishes on $\mathcal{I} \otimes_{\min} C^*(G)$. Therefore, we can write

$$\|[\lambda_G \cdot \rho_G](x)\|_{B(\ell_2(G))} \leq \|(Q_G \otimes Id)x\|_{\frac{C^*(G) \otimes_{\min} C^*(G)}{\mathcal{I} \otimes_{\min} C^*(G)}}.$$

Therefore, if (7.15) is isometric

$$\|[\lambda_G \cdot \rho_G](x)\|_{B(\ell_2(G))} \leq \|(Q_G \otimes Id)x\|_{C_\lambda^*(G) \otimes_{\min} C^*(G)}$$
$$= \left\|\sum x(s, t)\lambda_G(s) \otimes U_G(t)\right\|_{C_\lambda^*(G) \otimes_{\min} C^*(G)}.$$

Let $S \subset G$ be any finite set. Applying this to $x = \sum_{s \in S} U_G(s) \otimes U_G(s)$ so that $(Q_G \otimes Id)x = \sum_{s \in S} \lambda_G(s) \otimes U_G(s)$, and using (4.19) we find

$$|S| \leq \left\|\sum_{s \in S} \lambda_G(s) \otimes U_G(s)\right\|$$

and hence by (4.22) (Fell's absorption principle)

$$|S| \le \left\| \sum_{s \in S} \lambda_G(s) \right\|.$$

Thus G is amenable by Theorem 3.30. ☐

Proof of Proposition 7.34 We will show in the sequel (see Corollary 12.22) that $G = \mathbb{F}_n$ verifies the factorization appearing in 7.35 for some B with the WEP. We will also show (see Theorem 9.6 or rather Corollary 9.40) that $(B, C^*(G))$ is a nuclear pair. Thus $G = \mathbb{F}_n$ satisfies Property 7.35 and consequently also the factorization property (7.17) but is not amenable. By Lemma 7.39, we obtain Proposition 7.34. ☐

More explicitly, viewing $C_\lambda^*(\mathbb{F}_n) = C^*(\mathbb{F}_n)/\mathcal{I}$ we have by (4.22) and the preceding proof

$$\left\| 1 \otimes 1 + \sum_1^n \lambda_{\mathbb{F}_n}(g_j) \otimes U_{\mathbb{F}_n}(g_j) \right\|_{C_\lambda^*(\mathbb{F}_n) \otimes_{\min} C^*(\mathbb{F}_n)} = 2\sqrt{n}$$

but

$$\left\| 1 \otimes 1 + \sum_1^n \lambda_{\mathbb{F}_n}(g_j) \otimes U_{\mathbb{F}_n}(g_j) \right\|_{\frac{C^*(\mathbb{F}_n) \otimes_{\min} C^*(\mathbb{F}_n)}{\mathcal{I} \otimes_{\min} C^*(\mathbb{F}_n)}} = n + 1$$

and these are different for $n > 1$.

7.5 min-projective surjections

Motivated by the last section, we are naturally led to the following notion.

Definition 7.40 Let $q : A \to C$ be a surjective $*$-homomorphism. Let $\mathcal{I} = \ker(q)$ so that $C \cong A/\mathcal{I}$. We will say that the surjection q is min-projective if for any C^*-algebra B, the min-norm on $B \otimes C = B \otimes (A/\mathcal{I})$ coincides with the norm induced by $(B \otimes_{\min} A)/(B \otimes_{\min} \mathcal{I})$. Equivalently, this means that the canonical map $q_B : B \otimes_{\min} A \to B \otimes_{\min} C$ (that extends $Id_B \otimes q$) satisfies

$$\ker(q_B) = B \otimes_{\min} \ker(q). \tag{7.18}$$

Thus $q : A \to A/\mathcal{I}$ is min-projective if for any B

$$B \otimes_{\min} (A/\mathcal{I}) = (B \otimes_{\min} A)/(B \otimes_{\min} \mathcal{I}),$$

or equivalently (see Remark 1.9)

$$(A/\mathcal{I}) \otimes_{\min} B = (A \otimes_{\min} B)/(\mathcal{I} \otimes_{\min} B). \tag{7.19}$$

Remark 7.41 In analogy with Remark 7.20, let B be another C^*-algebra. If $q : A \to C$ is min-projective, then (7.6) (with the roles of B, A interchanged) shows that

$$(A, B) \text{ nuclear } \Rightarrow (C, B) \text{ nuclear.} \tag{7.20}$$

Remark 7.42 Obviously $q_B(B \otimes_{\min} \mathcal{I}) = 0$. Thus we have

$$\|(B \otimes_{\min} A)/(B \otimes_{\min} \mathcal{I}) \to B \otimes_{\min} C\| \leq 1.$$

Therefore, q is min-projective if and only if for any B and any $t \in B \otimes (A/\mathcal{I})$ we have conversely

$$\|t\|_{(B \otimes_{\min} A)/(B \otimes_{\min} \mathcal{I})} \leq \|t\|_{B \otimes_{\min} (A/\mathcal{I})}. \tag{7.21}$$

Or equivalently, for any $t \in (A/\mathcal{I}) \otimes B$ we have

$$\|t\|_{(A \otimes_{\min} B)/(\mathcal{I} \otimes_{\min} B)} \leq \|t\|_{(A/\mathcal{I}) \otimes_{\min} B}. \tag{7.22}$$

As observed in the preceding section, the latter does not always hold. In the subsequent §10.1 we will study the C^* algebras B (these are called "exact") such that (7.18) (or (7.19)) holds for any quotient map q. But for the moment, we content ourselves with a simple characterization of the quotient maps for which (7.18) holds for any B. We will need the following useful lemma which requires a specific notation. Let $\mathcal{I} \subset A$ be a (closed two-sided) ideal in a C^*-algebra A. Let E be an operator space. As in Lemma 4.26 we denote for simplicity

$$Q[E] = \frac{A \otimes_{\min} E}{\mathcal{I} \otimes_{\min} E}.$$

Then if F is another operator space and if $u : E \to F$ is a c.b. map we clearly have a bounded linear map

$$u_{[Q]} : Q[E] \to Q[F]$$

naturally associated to $Id_A \otimes u$ such that

$$\|u_{[Q]}\| \leq \|u\|.$$

Lemma 7.43 *If u is an isometry, then $u_{[Q]}$ also is one. In particular, if $E \subset F$ then $Q[E] \subset Q[F]$ (isometrically).*

Proof For simplicity we assume that $E \subset F$ and u is the inclusion map. Then the result is a particular case of Lemma 4.26 applied for $\alpha = \min$. ☐

Next we show that, just like $A \to A/\mathcal{I}$, the quotient map $E \otimes_{\min} A \to (E \otimes_{\min} A)/(E \otimes_{\min} \mathcal{I})$ takes the closed unit ball to the closed unit ball.

Lemma 7.44 *Let $\mathcal{I} \subset A$ be an ideal in a C^*-algebra A. Consider an operator space E and let $q_{[E]} : A \otimes_{\min} E \to Q[E]$ denote the quotient map. Then for* *any $\widehat{y}$ in $Q[E]$, there is an element y in $A \otimes_{\min} E$ that lifts $\widehat{y}$ (i.e. $q_{[E]}(y) = \widehat{y}$) such that $\|y\|_{\min} = \|\widehat{y}\|_{Q[E]}$.*

Proof Choose y_0 in $A \otimes_{\min} E$ such that $q_{[E]}(y_0) = \widehat{y}$. It is easy to check that $q_{[E]}$ has the properties appearing in Lemmas A.31 and A.32 suitably modified. Therefore, repeating the argument appearing before Lemma A.33, we obtain a Cauchy sequence $y_0, y_1, \ldots, y_n, \ldots$ in $A \otimes_{\min} E$ such that $q_{[E]}(y_n) = \widehat{y}$ for all n and $\|y_n\|_{\min} \to \|\widehat{y}\|_{Q[E]}$ when $n \to \infty$. Thus $y = \lim y_n$ is a lifting with the same norm as $\widehat{y}$. $\qquad\square$

Remark 7.45 (Description of $\ker(q_B)$) Let $q_B : B \otimes_{\min} A \to B \otimes_{\min} A/\mathcal{I}$ be as in (7.18). Let $t \in B \otimes_{\min} A$. Then $t \in \ker(q_B)$ if and only if $(\xi \otimes Id_A)(t) \in \mathcal{I}$ for any $\xi \in B^*$. Indeed, $q_B(t) = 0$ is the same as $(\xi \otimes Id_{A/\mathcal{I}})(q_B(t)) = 0$ for any $\xi \in B^*$ (see Corollary 1.15) and it is immediate that $\xi \otimes Id_{A/\mathcal{I}} = (Id_{\mathbb{C}} \otimes q)(\xi \otimes Id_A)(t)$ and of course we may identify $Id_{\mathbb{C}} \otimes q$ with q.

Thus when (7.18) fails there are ts satisfying this which fail to be in the min-closure of $B \otimes \mathcal{I}$ (although they are in that of $B \otimes A$). This shows that the failure of (7.18) is closely related to nontrivial approximation problems. In [246] Tomiyama introduced the related notion of Fubini product. When given operator subspaces $Y \subset B$ and $X \subset A$ in C^*-algebras, we may consider the subspace $F(Y, X) \subset B \otimes_{\min} A$ (called the Fubini product) formed of all those $t \in B \otimes_{\min} A$ such that $(\xi \otimes Id_A) \in X$ and $(Id_B \otimes \eta) \in Y$ for any $(\xi, \eta) \in B^* \times A^*$. Obviously $Y \otimes_{\min} X \subset F(Y, X)$ but in general the latter is larger. For instance when (7.18) fails, what precedes shows us that $B \otimes_{\min} \mathcal{I} \neq F(B, \mathcal{I})$. See [246] for more information on this interesting notion that we will not use.

Definition 7.46 Fix a constant $c \geq 0$. Let C be a C^*-algebra (or merely an operator space). Let $u : C \to A/\mathcal{I}$ be a linear map into a quotient C^*-algebra. Let $q : A \to A/\mathcal{I}$ denote the quotient map. We will say that u is *c-liftable* if there is $v : C \to A$ with $\|v\|_{cb} \leq c$ that lifts u in the sense that $qv = u$.

We will say that u is *locally c-liftable* (or admits a local c-lifting) if, for any finite-dimensional subspace $E \subset C$, the restriction $u_{|E}$ is c-liftable, or more explicitly for any finite dimensional $E \subset C$ there is $v^E : E \to A$ with $\|v^E\|_{cb} \leq c$ such that $qv^E = u_{|E}$.

Remark 7.47 Assume that $\mathcal{I}$ is completely complemented in A, by which we mean that there is a c.b. projection P from A onto $\mathcal{I}$. Then it is easy to see that the identity of $A/\mathcal{I}$ is c-liftable for $c = 1 + \|P\|_{cb}$. We just define $v : A/\mathcal{I} \to A$ by $v(x) = (I - P)(\widehat{x})$ where $\widehat{x} \in A$ is any lifting of $x \in A/\mathcal{I}$.

See Corollary 9.47 for more on unital c.p. maps that are locally 1-liftable. In that case a unital c.p. v^E can be found.

Proposition 7.48 *Let C be any operator space. Fix a constant $c \geq 0$. Let $u : C \to A/\mathcal{I}$ be a linear map into a quotient C^*-algebra. Let $q : A \to A/\mathcal{I}$ denote the quotient map. The following are equivalent.*

(i) *For any C^*-algebra B, u defines a map $Id_B \otimes u : B \otimes C \to (B \otimes A)/(B \otimes \mathcal{I})$ such that*

$$\|Id_B \otimes u : B \otimes_{\min} C \to (B \otimes_{\min} A)/(B \otimes_{\min} \mathcal{I})\| \leq c.$$

(ii) *Same as (i) with $B = \mathcal{B}$.*

(iii) *The map u is locally c-liftable.*

(iii)' *The map u is locally $(c + \varepsilon)$-liftable for any $\varepsilon > 0$.*

Proof (i) $\Rightarrow$ (ii) is trivial. Assume (ii). Let $s_E \in E^* \otimes C$ be the tensor associated to the inclusion map $j_E : E \subset C$ of a finite-dimensional subspace. We view $E^* \subset \mathcal{B}$ completely isometrically (see Theorem 2.15). Then $(Id_{\mathcal{B}} \otimes u)(s_E) \in E^* \otimes (A/\mathcal{I})$ is the tensor associated to $u_{|E}$, and $\|s_E\|_{\min} = \|j_E\|_{cb} = 1$. By (ii) we have

$$\|(Id_{\mathcal{B}} \otimes u)(s_E)\|_{(\mathcal{B} \otimes_{\min} A)/(\mathcal{B} \otimes_{\min} \mathcal{I})} \leq c\|s_E\|_{\min} = c.$$

By Lemma 7.43 we have

$$\|(Id_{\mathcal{B}} \otimes u)(s_E)\|_{(\mathcal{B} \otimes_{\min} A)/(\mathcal{B} \otimes_{\min} \mathcal{I})} = \|(Id_{\mathcal{B}} \otimes u)(s_E)\|_{(E^* \otimes_{\min} A)/(E^* \otimes_{\min} \mathcal{I})},$$

and since $E^* \otimes_{\min} F = CB(E, F)$ isometrically, we find an isometric identity

$$(E^* \otimes_{\min} A)/(E^* \otimes_{\min} \mathcal{I}) = CB(E, A)/CB(E, \mathcal{I})$$

and taking Lemma 7.44 into account, we find that there is $v \in CB(E, A)$ with $\|v\|_{cb} \leq c$ such that $qv = u_{|E}$. This shows (ii) $\Rightarrow$ (iii) and (iii) $\Rightarrow$ (iii)' is trivial. To complete the proof it clearly suffices to show (iii) $\Rightarrow$ (i). Assume (iii). Let $t \in B \otimes C$. Let $E \subset C$ be finite dimensional and such that $t \in B \otimes E$. Let $v^E : E \to A$ with $\|v^E\|_{cb} \leq c$ lifting $u_{|E} : E \to A/\mathcal{I}$. Let $\hat{t} = (Id_B \otimes v^E)(t) \in B \otimes A$. We have $\|\hat{t}\|_{B \otimes_{\min} A} \leq \|v^E\|_{cb}\|t\|_{\min} \leq c\|t\|_{\min}$. Moreover, $(Id_B \otimes q)(\hat{t}) = (Id_B \otimes u)(t)$. By (4.34) applied with $\alpha = \min$, this shows

$$\|(Id_B \otimes u)(t)\|_{(B \otimes_{\min} A)/(B \otimes_{\min} \mathcal{I})} \leq c\|t\|_{B \otimes_{\min} C},$$

and the latter means that (i) holds. □

Remark 7.49 In the situation of Proposition 7.48, assume that $\mathcal{B} \otimes_{\min} A = \mathcal{B} \otimes_{\max} A$ (this is what we take as definition of the LLP for A in the sequel). In that case the conditions in Proposition 7.48 are equivalent to

(ii)'

$$\|Id_{\mathcal{B}} \otimes u : \mathcal{B} \otimes_{\min} C \to \mathcal{B} \otimes_{\max} (A/\mathcal{I})\| \leq c.$$

Indeed, this follows from (7.6) (but one has to exchange the roles of the letters A and B there).

When $u = Id_{A/\mathcal{I}}$ and $c = 1$, Proposition 7.48 becomes:

Corollary 7.50 *Let* $q : A \rightarrow C$ *be a surjective* *-*homomorphism. Let* $\mathcal{I} = \ker(q)$ *so that* $C \cong A/\mathcal{I}$. *The following are equivalent*

(i) q *is* min-*projective (i.e.* (7.18) *holds for any* B).
(ii) *We have* (7.18) *for* $B = \mathscr{B}$ *(i.e. for* $B = B(\ell_2)$).
(iii) *For any* $\varepsilon > 0$, *the identity map* $Id_{A/\mathcal{I}} : A/\mathcal{I} \rightarrow A/\mathcal{I}$ *is locally* $(1 + \varepsilon)$-*liftable.*
(iii)' *Same as* (iii) *with* $\varepsilon = 0$.

Remark 7.51 Here to emphasize the analogy injective/projective, we did not conform to the existing terminology: One usually says, using the exact sequence $0 \rightarrow \mathcal{I} \rightarrow A \rightarrow A/\mathcal{I} \rightarrow 0$, that A is a "locally split extension" (of $A/\mathcal{I}$ by $\mathcal{I}$) to express the property (iii) in Corollary 7.50, which equivalently means that $A \rightarrow A/\mathcal{I}$ is min-projective.

By Remark 7.47, Corollary 7.50 implies:

Corollary 7.52 *If the ideal* $\mathcal{I}$ *is completely complemented in* A *the quotient map* $A \rightarrow A/\mathcal{I}$ *is* min-*projective.*

Remark 7.53 Note that in the subsequent Corollary 9.47, we show that, in the unital case (when A, C and q are unital), the properties in Corollary 7.50 are equivalent to

(iv) For any finite-dimensional operator system $E \subset C$ there is a unital c.p. map $u^E : E \rightarrow A$ (with $\|u^E\|_{cb} = 1$) that lifts q in the sense that qu^E coincides with the inclusion map $E \subset C$.

7.6 Generating new C^*-norms from old ones

Following the path Grothendieck opened up for Banach space tensor products in [98], we now derive several additional C^*-tensor products from the minimal and maximal ones using the injective and projective universality of $B(H)$ and $C^*(\mathbb{F})$ respectively. Kirchberg already followed that same road first in [155], then more systematically in [158].

Let $A_1, A_2, B_1, B_2, C_1, C_2$ be C^*-algebras. The basic idea is two-fold:

(i) If we are given an embedding $A_1 \subset B_1$ we have a linear embedding $A_1 \otimes A_2 \subset B_1 \otimes A_2$. Then given a C^*-norm α on $B_1 \otimes A_2$ its restriction to $A_1 \otimes A_2$ defines an a priori new C^*-norm on $A_1 \otimes A_2$, denoted by α_1. If α is the min-norm then so is α_1 (see Remark 1.10), but when α is the max-norm the induced norm α_1 is in general not the max-norm on $A_1 \otimes A_2$, since by §7.2 the max-norm is not injective. Thus this produces an a priori new C^*-norm on $A_1 \otimes A_2$.

(ii) If we are given a surjective $*$-homomorphism $q_1 : C_1 \to A_1$ so that $A_1 = C_1 / \mathcal{I}_1$ ($\mathcal{I}_1 = \ker(q_1)$), we have a surjection $q_1 \otimes Id_{A_2} : C_1 \otimes A_2 \to A_1 \otimes A_2$. Then given a C^*-norm α on $C_1 \otimes A_2$ we can define a new C^*-norm on $A_1 \otimes A_2$ as the norm induced on $A_1 \otimes A_2$ by the natural C^*-norm in $(C_1 \otimes_\alpha A_2)/\overline{\mathcal{I}_1 \otimes A_2}^\alpha$. The resulting C^*-norm on $A_1 \otimes A_2$ is denoted by α^1. If α is the max-norm then by (7.6) so is α^1, but when α is the min-norm, α^1 is in general not the min-norm on $A_1 \otimes A_2$, since by §7.4 the min-norm is not projective. Thus this produces an a priori new C^*-norm on $A_1 \otimes A_2$.

One can also apply these constructions to the second factor, and produce in this way another pair of a priori new C^*-norms α_2, α^2 on $A_1 \otimes A_2$.

In [98] (see also [71]) Grothendieck applied these constructions in the Banach space category starting from the minimal and maximal norms among what he called the reasonable tensor norms. In the Banach space setting, he denoted by $/\alpha$ and $\backslash\alpha$ (resp. $\alpha\backslash$ and $\alpha/$) the analogues of α_1 and α^1 (resp. α_2 and α^2). We will now describe a couple of possibilities that this idea offers us when we apply it to C^*-norms.

We start with the C^*-norms derived from the (injective) universality of $B(H)$. Assume (as we may) $A_j \subset B(H_j)$ ($j = 1, 2$). Then the norm induced on $A_1 \otimes A_2$ by $B(H_1) \otimes_{\max} A_2$ is a C^*-norm on $A_1 \otimes A_2$, that we denote by $\max_1$. Let $u_j : A_j \to B_j$ ($j = 1, 2$) be c.b. maps. Consider $u_1 \otimes u_2 : A_1 \otimes A_2 \to B_1 \otimes B_2$. Then

$$\|u_1 \otimes u_2 : A_1 \otimes_{\max_1} A_2 \to B_1 \otimes_{\max_1} B_2\| \le \|u_1\|_{cb}\|u_2\|_{dec}. \qquad (7.23)$$

Indeed, this follows from the extension property of c.b. maps (Theorem 1.18) together with (6.15) and (6.9).

Using again the extension property, one shows that the $\max_1$-norm does not depend on the embedding $A_1 \subset B(H_1)$. Moreover, we have

$$\forall t \in A_1 \otimes A_2 \quad \|t\|_{\max_1} = \inf \|t\|_{B_1 \otimes_{\max} A_2}$$

where the inf runs over all C^*-algebras B_1 containing A_1 as a C^*-subalgebra.

The same construction applied to the second factor leads to another C^*-norm on $A_1 \otimes A_2$, that we denote by $\max_2$. By (4.8), we have $A_1 \otimes_{\max_2} A_2 \simeq A_2 \otimes_{\max_1} A_1$, so this is not really new.

The most interesting case is when we do this operation on both factors: continuing, we are led to denote by $\max_{12}$ the norm induced on $A_1 \otimes A_2$ by $B(H_1) \otimes_{\max} B(H_2)$. Doing the two operations in reverse order would lead us to define $\max_{21}$ but this is clearly the same as $\max_{12}$. Thus, to lighten the notation, we denote it in the sequel by $\| \ \|_M$ and the completed tensor product by $A_1 \otimes_M A_2$ (see Definition 20.11). It has the supplementary advantage that, like the minimal norm, it yields at the same time a tensor product of operator spaces. Indeed, by the same argument as for (7.23), this time we have

$$\|u_1 \otimes u_2 : A_1 \otimes_M A_2 \to B_1 \otimes_M B_2\| \leq \|u_1\|_{cb}\|u_2\|_{cb}, \qquad (7.24)$$

and the definition of $\| \ \|_M$ makes sense even when A_1, A_2 are just operator spaces. Moreover, we have

$$\forall t \in A_1 \otimes A_2 \quad \|t\|_M = \inf \|t\|_{B_1 \otimes_{\max} B_2}$$

where the inf runs over all C^*-algebras B_j containing A_j as a C^*-subalgebra ($j = 1, 2$), or even just completely isometrically.

We now turn to the C^*-norms derived from the (projective) universality of $C^*(\mathbb{F})$.

Assume that $A_j = C_j/\mathcal{I}_j$ with each C_j of the form $C_j = C^*(G_j)$ for some free group G_j. Let $q_j : C_j \to C_j/\mathcal{I}_j$ be the quotient map. Let $t \in A_1 \otimes A_2$. We define

$$\|t\|_{\min^1} = \|t\|_{(C_1 \otimes_{\min} A_2)/(\mathcal{I}_1 \otimes_{\min} A_2)}.$$

Clearly this is a C^*-norm on $A_1 \otimes A_2$. Let $u_t : A_2^* \to A_1$ be the finite rank linear map associated to t. Using the property (2.15) of the dual operator space together with (4.34) we immediately obtain the following reinterpretation of $\|t\|_{\min^1}$ in terms of c.b. liftings (here w* means weak*):

$$\|t\|_{\min^1} = \inf\{\|v\|_{cb} \mid v : A_2^* \to C_1, \text{ w* continuous, } \mathrm{rk}(v) < \infty, \ q_1 v = u_t\}. \qquad (7.25)$$

Obviously we could do the same on the right-hand side to define $\otimes_{\min^2}$ but by (4.8) this boils down to the same notion. However, if we do it twice, then we obtain a genuinely new tensor product. More precisely, using $\otimes_L$ for what should be denoted $\otimes_{\min^{12}}$, we set

$$\|t\|_{A_1 \otimes_L A_2} = \|t\|_{(C_1 \otimes_{\min} C_2)/(\mathcal{I}_1 \otimes_{\min} C_2 + C_1 \otimes_{\min} \mathcal{I}_2)}.$$

Note $A_2^* \subset C_2^*$. Again we have

$$\|t\|_{A_1 \otimes_L A_2} = \inf\{\|v\|_{cb} \mid v : C_2^* \to C_1, \text{ w* continuous, } \mathrm{rk}(v) < \infty, \ q_1 v_{|A_2^*} = u_t\}. \qquad (7.26)$$

We will show later on in Remark 9.45 when discussing the LLP that (7.25) and (7.26) are independent of the choice of C_1, C_2 as long as they are of the required form $C^*(G)$ with G a free group (or simply as long as they have the LLP).

7.7 Notes and remarks

This chapter is inspired mainly from Kirchberg's ideas, but we introduce special terms (such as "max-injective" and "min-projective") in order to emphasize as much as possible properties of linear maps in the spirit of operator space theory. We feel some features become much clearer. There are analogies with the situation in Banach space theory according to Grothendieck's viewpoint in [98]. We explain this in §7.6. For more in this direction, see Kirchberg's presentation of his works in [158], where he systematically adopts a category theory standpoint. Kirchberg communicated Theorem 7.6 to the author with permission to include it in [208]. What we call min-projective surjection is very much the center of attention in Effros and Haagerup's paper [77], but they do not give it a name. Proposition 7.48 is essentially there (see [77, th. 3.2]), except they use finite-dimensional operator systems and approximate c.p. liftings, but, by part (ii) in Proposition 9.42, this is equivalent to our formulation with locally liftable maps. Incidentally, Effros and Haagerup mention a 1971 paper by Douglas and Howe [73] on Toeplitz operators as an early source for the fact that the existence of liftings implies that the quotient map is min-projective as in Proposition 7.48. The latter is the C^*-algebraic analogue of a well-known principle in homological algebra. It is amusing to observe, as a historical curiosity, that in [73, prop. 2] Douglas and Howe only assume that the lifting is bounded while their proof uses its *complete* boundedness; this defect is observed in the later paper [20] (also quoted in [77]) and repaired by invoking [5, th. 7].

8

Biduals, injective von Neumann algebras, and C^*-norms

In this chapter, we review the main results involving C^*-algebras and their biduals proved in the aftermath of Connes's breakthroughs [61] on injective von Neumann algebras. In particular, we show in Theorem 8.16 that the nuclearity of a C^*-algebra A is equivalent to the injectivity of A^{**}.

8.1 Biduals of C^*-algebras

We will use here the basic facts and notation introduced in §A.16: when A is a C^*-algebra and M a von Neumann one, for all $u : A \to M$ we recall that

$$\ddot{u} = (u^*_{|M_*})^* : A^{**} \to M.$$

The following statement is merely a recapitulation.

Theorem 8.1 *Let $u : A \to M$ be a linear map from a C^*-algebra to a von Neumann algebra.*

  (i) *If u is a $*$-homomorphism then $\ddot{u} : A^{**} \to M$ is a normal*
     $*$-homomorphism.
  (ii) *$u \in CP(A, M) \Rightarrow \ddot{u} \in CP(A^{**}, M)$ and $\|\ddot{u}\| = \|u\|$.*
  (iii) *$u \in CB(A, M) \Rightarrow \ddot{u} \in CB(A^{**}, M)$ and $\|\ddot{u}\|_{cb} = \|u\|_{cb}$.*
  (iv) *$u \in D(A, M) \Rightarrow \ddot{u} \in D(A^{**}, M)$ and $\|\ddot{u}\|_{dec} = \|u\|_{dec}$.*

Proof We recall that, by density, $\ddot{u}$ can be viewed as the unique $(\sigma(A^{**}, A^*)$, $\sigma(M, M_*))$-continuous extension of u. (i) is a well-known consequence of the very definition of A^{**} (see Theorem A.55). (ii) (resp. (iii)) was proved in Lemma 1.43 (resp. Lemma 1.61) and (iv) in Lemma 6.9. $\qquad\square$

8.2 The nor-norm and the bin-norm

Let A be a C^*-algebra and M a von Neumann algebra. In this "hybrid" situation, one defines a C^*-norm on $A \otimes M$ as follows. For any $t \in A \otimes M$ we set

$$\|t\|_{\text{nor}} = \sup \|(\sigma \cdot \pi)(t)\| \qquad (8.1)$$

where the sup runs over all H's and all commuting pairs of $*$-homomorphisms $\sigma : A \to B(H), \pi : M \to B(H)$ with π assumed normal (i.e. continuous with respect to both weak* topologies on M and $B(H)$). It is easy to check that this is indeed a C^*-norm intermediate between the minimal and maximal norms. We denote by $A \otimes_{\text{nor}} M$ the corresponding completion.

Remark 8.2 There is obviously a similar definition of the nor-norm on $M \otimes A$. In some situation (say if A happens to be a von Neumann algebra too) this may lead to some confusion, so that we should use a notation that distinguishes both cases (such as nor_1 and nor_2), but for simplicity we prefer not to do that. In the cases we consider in the sequel there is no risk of confusion.

Remark 8.3 Let A^1, M^1 be respectively a C^*-algebra and a von Neumann algebra. Let $\sigma : A \to A^1$ and $\pi : M \to M^1$ be $*$-homomorphisms (resp. isomorphisms). Then if π is normal, the mapping $\sigma \otimes \pi$ obviously defines (by density) a $*$-homomorphism (resp. an isomorphism) from $A \otimes_{\text{nor}} M$ to $A^1 \otimes_{\text{nor}} M^1$. See Remark A.38 for clarification.

We now turn to a more symmetric situation. Let M, N be von Neumann algebras. On $M \otimes N$ one defines the "binormal" norm of $t \in M \otimes N$ by

$$\|t\|_{\text{bin}} = \sup\{\|(\pi . \sigma)(t)\|\} \qquad (8.2)$$

where the sup runs over all pairs of *normal* $*$-homomorphism $\pi : M \to B(H)$ and $\sigma : N \to B(H)$ with commuting ranges.

This is clearly a C^*-norm. We denote by $M \otimes_{\text{bin}} N$ the completion.

Using normal embeddings $M \subset B(H_1)$, $N \subset B(H_2)$, $H = H_1 \otimes_2 H_2$ and the usual pair $\pi(x) = x \otimes 1$, $\sigma(y) = 1 \otimes y$, we find for any $t \in M \otimes N$

$$\|t\|_{\text{min}} \le \|t\|_{\text{bin}}. \qquad (8.3)$$

When $M = A^{**}$ and $N = B^{**}$ are biduals, then we may clearly write (by the extension property of biduals, see Theorem 8.1 (i))

$$\|t\|_{\text{bin}} = \sup\{\|(\ddot{\pi} . \ddot{\sigma})(t)\|\} \qquad (8.4)$$

where the sup runs over all $*$-homomorphisms $\pi : A \to B(H)$ and $\sigma : B \to B(H)$ with commuting ranges.

Remark 8.4 Let M be a von Neumann algebra and A a C^*-algebra. Then the norm induced on $A \otimes M$ by the bin-norm on $A^{**} \otimes M$ coincides with the nor-norm on $A \otimes M$ as defined in (8.1). This is easy to check using again part (i) in Theorem 8.1.

Moreover, for any $t \in A^{**} \otimes M$, we have clearly $\|t\|_{\text{bin}} \leq \|t\|_{\text{nor}}$ (where for the nor-norm we view A^{**} as a C^*-algebra).

8.3 Nuclearity and injective von Neumann algebras

We will need the following basic fact:

Proposition 8.5 *A von Neumann algebra $M \subset B(H)$ is injective if and only if its commutant M' is injective.*

The (not so simple) proof is based on the following

Lemma 8.6 *Let $M \subset B(H)$ and $N \subset B(\mathcal{H})$ be isomorphic von Neumann algebras. If N' is injective then M' is injective.*

Proof By Theorem A.61 we may write the isomorphism $T : M \rightarrow N$ as the product of three isomorphisms of three different kinds: amplification, compression and spatial. Clearly it suffices to check the lemma for each of the three kinds, and this turns out to be an easy exercise that we leave to the reader. $\qquad\square$

Proof of Proposition 8.5 By a very well-known fact there is a realization of M say $\psi : M \subset B(\mathcal{H})$ for which $\psi(M)$ and $\psi(M)'$ are anti-isomorphic. This is part of what is called the standard form of M (see [102] or [241, p. 151] and the proof of the subsequent Theorem 23.30). Then clearly $\psi(M)$ injective $\Leftrightarrow \psi(M)'$ injective (recall Remarks 2.9 and 2.13). Recall that by Definition 1.44 injectivity is stable by completely isometric isomorphisms. Thus if M is injective, so is $\psi(M)$ and hence $\psi(M)'$ is injective. By Lemma 8.6 it follows that M' is injective. Reversing the roles of M and M' gives the converse. $\qquad\square$

The fact that M injective $\Leftrightarrow M'$ injective allows us to prove a simple, but important stability property of injective von Neumann algebras:

Proposition 8.7 *If $\{N_i \mid i \in I\}$ is a family, directed by inclusion, of injective von Neumann algebras in $B(H)$ then their weak* closure $M = \overline{\cup_{i \in I} N_i} \subset B(H)$ is injective.*

Proof Note that $N_i \subset N_j$ is equivalent to $N'_j \subset N'_i$. Therefore the commutants (N'_i) form a decreasing directed family. Let $P_i : B(H) \rightarrow N'_i$ be a (completely) contractive projection. Let $\mathcal{U}$ be an ultrafilter refining the underlying net formed by (N_i') (see Remark A.6). Then the mapping P defined by $P(x) = \lim_{\mathcal{U}} P_i(x)$ (the limit being in the weak* topology of $B(H)$) is clearly a (completely) contractive projection onto $M' = \cap_{i \in I} N_i'$. Thus M', and hence M itself, is injective. $\square$

The von Neumann algebras M that can be written as the (weak*) closure $M = \overline{\cup_{i \in I} N_i}$ of an ascending union (directed by inclusion) of finite-dimensional von Neumann algebras N_i are sometimes called "hyperfinite," but, as already emphasized by many authors (including Connes himself in [61, p. 113]) the term "approximately finite dimensional" (AFD) is more appropriate. Thus the last statement shows that AFD $\Rightarrow$ injective. The converse (say, when H is separable) is a celebrated deep result of Connes [61], that we state without proof:

Theorem 8.8 (Finite-dimensional approximation of injective von Neumann algebras) *Any injective von Neumann algebra M is approximately finite dimensional (AFD).*

This major advance led to a number of deep characterizations of injective von Neumann algebras and nuclear C^*-algebras. Although we do not include the proof of Theorem 8.8, we will try in this section to include complete proofs of the latter characterizations, for which (unlike for Theorem 8.8) reasonably simple proofs are now available. In particular we will now show that injectivity is equivalent to the weak* CPAP, which is a rather natural analogue of the CPAP for von Neumann algebras. This is often called "semidiscreteness" but we prefer to use a term that emphasizes the analogy with the CPAP.

Definition 8.9 A von Neumann algebra M has the weak* CPAP (in other words is "semidiscrete") if the identity on M is the pointwise weak* limit of a net of finite rank normal c.p. maps, i.e. there is a net of weak* continuous unital c.p. maps $u_i : M \to M$ of finite rank such that $u_i(x) \to x$ in the weak* topology for any $x \in M$.

Remark 8.10 When this holds, we may assume that $u_i = v_i^*$ for some $v_i : M_* \to M_*$, and the net (v_i) converges pointwise to the identity of M_* with respect to $\sigma(M_*, M)$ (i.e. the weak topology of M_*). By Mazur's Theorem A.9 after passing to convex combinations we may assume that (v_i) converges pointwise to the identity of M_* for the norm topology.

We first give a characterization of injectivity obtained by a simple but very important trick involving extensions of maps on the max tensor product; the latter is called "The Trick" in [39]. We already used a variant of this idea previously for the extension property in Theorem 6.20, and again for the characterization of max-injective inclusions in Theorem 7.29. Actually the main point of the next statement (i.e. the if part) can be deduced from Theorem 6.20 (with $E = A$) but we prefer to repeat the argument in the present situation, thus avoiding the use of operator modules and Lemma 6.19.

Proposition 8.11 *Let A be a C^*-algebra, $\pi : A \to B(H)$ a $*$-homomorphism and let $M = \pi(A)'' \subset B(H)$ be the von Neumann algebra it generates. Let $\widehat{\pi} : M' \otimes A \to B(H)$ be the $*$-homomorphism associated to the product, so that $\widehat{\pi}(x' \otimes a) = x'\pi(a)$ for all $a \in A, x' \in M'$. Let*

$$M' \otimes_{\overline{\max}} A = \overline{M' \otimes A}^{\max} \subset B(H) \otimes_{\max} A$$

be (as in Theorem 6.20) the closure of $M' \otimes A$ in $B(H) \otimes_{\max} A$. Then M' is injective if and only if $\widehat{\pi}$ extends to a continuous $*$-homomorphism from $M' \otimes_{\overline{\max}} A$ to $B(H)$ (equivalently $\widehat{\pi}$ is bounded with respect to the norm induced by $B(H) \otimes_{\max} A$).

Proof We first treat the case when π and hence $\widehat{\pi}$ are unital. Assume $\widehat{\pi}$ continuous on $M' \otimes_{\overline{\max}} A$, so that automatically (see Remark 1.6)

$$\|\widehat{\pi} : M' \otimes_{\overline{\max}} A \to B(H)\|_{cb} = 1.$$

Let $\varphi : B(H) \otimes_{\max} A \to B(H)$ be a complete contraction extending $\widehat{\pi}$ according to Theorem 1.18. Since φ is unital it is c.p. and its multiplicative domain D obviously includes $M' \otimes A$. Let $P : B(H) \to B(H)$ be defined by $P(b) = \varphi(b \otimes 1)$ for any $b \in B(H)$. Since $1 \otimes a \in D$ for any $a \in A$, we have (see Corollary 5.3) $P(b)\pi(a) = \varphi((b \otimes 1)(1 \otimes a)) = \varphi((1 \otimes a)(b \otimes 1)) = \pi(a)P(b)$. This shows $P(b) \in \pi(A)' = M'$. Moreover, we have $P(m') = m'$ for any $m' \in M'$ since φ extends $\widehat{\pi}$. It follows that P is a completely contractive and c.p. projection from $B(H)$ onto M'; in other words M' is injective. This proves the "if part" in the unital case.

If A is not unital, let (x_i) be an approximate unit of A as in §A.15 and let Q be the weak* (or w.o.t.) limit of $\pi(x_i)$ in $B(H)$. Then $Q\pi(a) = \pi(a)Q = \pi(a)$ for any $a \in A$. It follows that Q is a (self-adjoint) projection in the center $M \cap M'$ of M. A moment of thought shows that $M'' = QMQ \oplus B(K)$ with $K = (I - Q)(H)$. Thus we are reduced to show that QMQ is injective. Then we may as well replace H by $Q(H)$ and Q by I. In that case, if we define $P(b)$ as the w.o.t.-limit of $\varphi(b \otimes x_i)$, the same reasoning as for the unital case leads to the desired result.

Conversely, if M' is injective, there is a (completely) contractive c.p. projection $P' : B(H) \to M'$. By (4.30) $\|P' \otimes Id_A : B(H) \otimes_{\max} A \to M' \otimes_{\max} A\| = 1$, and by definition of the maximal tensor product $\|\widehat{\pi} : M' \otimes_{\max} A \to B(H)\| = 1$. Therefore, for any $t \in M' \otimes A$, since $t = (P' \otimes Id_A)(t)$ we have

$$\|\widehat{\pi}(t)\|_{B(H)} = \|\widehat{\pi}(P' \otimes Id_A)(t)\|_{B(H)} \leq \|(P' \otimes Id_A)(t)\|_{M' \otimes_{\max} A}$$

$$\leq \|t\|_{B(H) \otimes_{\max} A} = \|t\|_{M' \otimes_{\overline{\max}} A}.$$

$\square$

Theorem 8.12 *All injective von Neumann algebras have the weak* CPAP.*

Proof Here we are facing an embarrassing situation. The ingredients for a complete proof are scattered in the sequel in a more general framework emphasizing the WEP instead of injectivity. Rather than move the proof to a much later chapter we choose to give it here with references to the text coming ahead, hoping for the reader's indulgence. We believe our (exceptional) choice gives a more focused global picture of injectivity.

The first part of the proof is simply a reduction to the case when M admits a finite faithful normal tracial state τ. Indeed, if this case is settled it is very easy to deduce from it the case when M is semifinite, because semifiniteness of M implies (see Remark 11.1) the existence of a net of normal unital c.p. maps on M tending weak* to the identity and each with range in a finite von Neumann subalgebra of M, which inherits the injectivity of M. Once the semifinite case is settled, Takesaki's duality theorem 11.3 comes to our rescue and produces the general case, taking Remark 11.4 into account.

Thus it suffices to consider a tracial probability space (M, τ) (in the sense of Definition 11.6). If M is injective, then by (22.15) for any finite set (x_j) in M we have

$$\left\| \sum \overline{x_j} \otimes x_j \right\|_{\overline{M} \otimes_{\max} M} = \left\| \sum \overline{x_j} \otimes x_j \right\|_{\overline{M} \otimes_{\min} M}. \tag{8.5}$$

Since left and right multiplication are commuting representations of M on $L_2(\tau)$ (see §11.3) we always have $\sum \|x_j\|_2^2 \leq \left\| \sum \overline{x_j} \otimes x_j \right\|_{\overline{M} \otimes_{\max} M}$ but injectivity implies by (8.5)

$$\sum \|x_j\|_2^2 \leq \left\| \sum \overline{x_j} \otimes x_j \right\|_{\overline{M} \otimes_{\min} M}.$$

We assume $M \subset B(L_2(\tau))$. By Theorem 11.38 (with $\pi = Id_M$ and $E = M$) and (i) $\Rightarrow$ (ii) in Theorem 4.21 there is a net of *finite-dimensional* subspaces $H_i \subset L_2(\tau)$ and $K_i \subset L_2(\tau)$ and states f_i on $\overline{B(H_i)} \otimes_{\min} B(K_i)$ such that

$$\forall y, x \in M \quad \tau(y^* x) = \lim_i f_i(\overline{P_{H_i} y_{|H_i}} \otimes P_{K_i} x_{|K_i}). \tag{8.6}$$

Replacing both H_i and K_i by $H_i + K_i \subset L_2(\tau)$ we may assume for simplicity that $K_i = H_i$. Since M is dense in $L_2(\tau)$ by perturbation we may assume that $H_i \subset M$. This gives us the advantage (purely for convenience in the sequel) that there is $c_i < \infty$ such that

$$\forall y \in M \quad \|P_{H_i} y_{|H_i}\| \leq c_i \tau(|y|), \tag{8.7}$$

which is easy to deduce from (11.2). We will now describe the state $\varphi_i \in (\overline{M} \otimes_{\max} M)^*$ defined by

$$\forall x, y \in M \quad \varphi_i(\overline{y} \otimes x) = f_i(\overline{P_{H_i} y_{|H_i}} \otimes P_{H_i} x_{|H_i}).$$

By Theorem 4.16 and the discussion of $M_* \simeq L_1(\tau)$ in §11.2 there is a finite rank map $\eta_i : M \to L_1(\tau)$ that is c.p. in the sense of (4.28) such that

$$\forall x, y \in M \quad \varphi_i(\overline{y} \otimes x) = \tau(y^* \eta_i(x)).$$

The fact that η_i takes its values in M_* rather that M^* is due to the fact that φ_i is separately normal when considered as a bilinear form on $\overline{M} \times M$. Its rank is finite because $\dim(H_i) < \infty$ ensures that the latter bilinear form is of finite rank. The condition (8.7) gives us $|\tau(y^* \eta_i(x))| \leq c_i \tau(|y|) \|x\|$ from which $\|\eta_i(x)\| \leq c_i \|x\|$ follows by (11.4). Thus $\eta_i(M) \subset M \subset L_1(\tau)$. Since $\eta_i : M \to L_1(\tau)$ satisfies (4.28), one can check using (11.9) for $M_n(M)$ that $\eta_i \in CP(M, M)$. We now wish to modify η_i to make it unital. Observe that by (8.6) $\eta_i(1) \to 1$ for $\sigma(M_*, M)$. Using Mazur's Theorem A.9 we may pass to convex combinations and assume that $\eta_i(1) \to 1$ in norm in M_*. Assume for a moment that $\eta_i(1)$ is invertible in M, then we define for $x \in M$

$$u_i(x) = \eta_i(1)^{-1/2} \eta_i(x) \eta_i(1)^{-1/2}.$$

We have for all $x, y \in M$

$$\tau(y^* x) = \lim_i \tau(\eta_i(1)^{1/2} y^* \eta_i(1)^{1/2} u_i(x))$$

but since $\eta_i(1)^{1/2} \to 1$ in $L_2(\tau)$ by the Powers–Størmer inequality (11.38) (as extended to general traces by Araki, Connes, and Haagerup, see [241, (9) p. 143], see also [134, appendix]) we must have as well

$$\tau(y^* x) = \lim_i \tau(y^* u_i(x)).$$

Thus we have obtained a net of finite rank unital c.p. maps (u_i) tending to the identity in the $\sigma(M, M)$-sense but since $\|u_i\| = \|u_i(1)\| = 1$, the net being equicontinuous and M dense in M_*, we conclude that $u_i(x) \to x$ for $\sigma(M, M_*)$, whence the weak* CPAP. The only drawback is that we assumed $\eta_i(1)$ invertible. This can be fixed easily by replacing η_i by $\eta_{i,\varepsilon}$ defined by $\eta_{i,\varepsilon}(x) = \eta_i(x) + \varepsilon \tau(x) 1$ $(x \in M)$ in the preceding reasoning and letting $\varepsilon > 0$ tend to zero as part of our net. We skip the details. $\square$

Theorem 8.13 *The following properties of a von Neumann algebra $M \subset B(H)$ are equivalent:*

(i) *M is injective.*
(ii) *M has the weak* CPAP.*
(iii) *For any C^*-algebra A and any $t \in A \otimes M$ we have $\|t\|_{\text{nor}} = \|t\|_{\text{min}}$.*
(iv) *For any von Neumann algebra N and any $t \in N \otimes M$ we have*
$$\|t\|_{\text{bin}} = \|t\|_{\text{min}}.$$

Proof We already know (i) $\Rightarrow$ (ii) by Theorem 8.12. Assume (ii). Let (u_i) be as in Definition 8.9. Let $t \in A \otimes M$. By (6.16) we have $\|(Id_A \otimes u_i)(t)\|_{\text{max}} \leq \|t\|_{\text{min}}$ and hence for any commuting pair of $*$-homomorphisms $\sigma : A \to B(H)$ and $\pi : M \to B(H)$ we have $\|(\sigma \cdot \pi)((Id_A \otimes u_i)(t))\| \leq \|t\|_{\text{min}}$. Now if π is assumed normal it is easy to check that $(\sigma \cdot \pi)((Id_A \otimes u_i)(t))$ tends weak* to $(\sigma \cdot \pi)(t)$ and hence $\|(\sigma \cdot \pi)(t)\| \leq \|t\|_{\text{min}}$. This shows $\|t\|_{\text{nor}} \leq \|t\|_{\text{min}}$. The converse inequality is obvious.

(iii) $\Rightarrow$ (iv) is trivial since $\|\cdot\|_{\text{bin}} \leq \|\cdot\|_{\text{nor}}$.

Assume (iv) with $N = M'$. We claim that M' is injective and hence (i) holds by Proposition 8.5. Indeed, (iv) implies that the map $\widehat{\pi}$ in Proposition 8.11 (with $A = M$ and $\pi : M \to B(H)$ the embedding) is continuous on $M' \otimes M$, with respect to the min-norm, i.e. the norm induced by $B(H) \otimes_{\text{min}} M$; a fortiori it is continuous with respect to the norm induced by $B(H) \otimes_{\text{max}} M$, so M' is injective by Proposition 8.11. $\qquad\square$

The equivalence between (ii) and (iii) in the next result is somewhat surprising. We state this for emphasis.

Theorem 8.14 (A consequence of the weak* CPAP) *The following properties of a von Neumann algebra $M \subset B(H)$ are equivalent:*

(i) *M is injective.*
(ii) *The product mapping p defined on $M' \otimes M$ by $p(x' \otimes x) = x'x$ defines a contractive $*$-homomorphism from $M' \otimes_{\text{min}} M$ to $B(H)$.*
(iii) *The product mapping p defines a contractive $*$-homomorphism from $M' \otimes M$ equipped with the norm induced by $B(H) \otimes_{\text{max}} B(H)$.*

Proof Assume (i). Note that p is clearly contractive on $M' \otimes_{\text{bin}} M$, so that (ii) holds by Theorem 8.13 applied with $N = M'$.

(ii) $\Rightarrow$ (iii) is clear since the norm induced on $M' \otimes M$ by $B(H) \otimes_{\text{max}} B(H)$ dominates the minimal C^*-norm, i.e. the one of $M' \otimes_{\text{min}} M$.

Assume (iii). We will apply Proposition 8.11 with $A = M$. Note that the norm induced on $M' \otimes M$ by $B(H) \otimes_{\text{max}} B(H)$ is clearly majorized by the norm induced by $B(H) \otimes_{\text{max}} M$. Thus Proposition 8.11 shows that M' is injective, but since we may exchange the roles of M and M', M is injective. $\qquad\square$

We now derive the consequences of the preceding (major) theorems for nuclear C^*-algebras.

Corollary 8.15 *Let* A *be a nuclear* C^*-*algebra. Then for any* $*$-*homomorphism* $\pi : A \to B(H)$ *the von Neumann algebra* $M = \pi(A)''$ *generated by* π *is injective. In particular,* A^{**} *is injective.*

Proof Since A is nuclear, we have an isometric embedding

$$M' \otimes_{\max} A \subset B(H) \otimes_{\max} A,$$

and by definition of $M' \otimes_{\max} A$ the map $\widehat{\pi}$ in Proposition 8.11 clearly satisfies $\|\widehat{\pi} : M' \otimes_{\max} A \to B(H)\| \leq 1$. Thus M' and also M by Proposition 8.5 are injective. $\qquad\square$

Actually the converse of the preceding corollary also holds, as the next statement shows.

Theorem 8.16 *A* C^*-*algebra* A *is nuclear if and only if its bidual* A^{**} *is injective.*

Proof Assume A^{**} injective. By Theorem 8.12 it has the weak* CPAP. By Remark 8.10 there is a net of finite rank maps (v_i) on A^* that are preadjoints of unital c.p. maps and tend pointwise to the identity on A^*. Let B be any C^*-algebra. Let $u : B \to A^*$ be a c.p. map. Composing with u_i gives us a net of c.p. maps tending pointwise to u. It follows from the description of the set of states on $B \otimes_{\max} A$ and $B \otimes_{\min} A$ given in §4.5 and §4.6 that they must coincide. Thus we conclude that $B \otimes_{\max} A = B \otimes_{\min} A$, which means A is nuclear. The converse was already part of the preceding statement. $\qquad\square$

Corollary 8.17 ([46]) *Nuclearity is preserved under quotients.*

Proof Let $A/\mathcal{I}$ be a quotient C^*-algebra. By (A.37) we have $A^{**} \simeq (A/\mathcal{I})^{**} \oplus \mathcal{I}^{**}$. Therefore, A^{**} injective implies $(A/\mathcal{I})^{**}$ injective. $\qquad\square$

Note that there is no known really simple and direct proof of Corollary 8.17.

Corollary 8.18 ([46]) *Nuclearity is preserved under "extensions." This means that if* $\mathcal{I} \subset A$ *is an ideal in a* C^*-*algebra, and if both* $\mathcal{I}$ *and* $A/\mathcal{I}$ *are nuclear, then* A *is nuclear.*

Proof This is a corollary of Proposition 7.15 (on the exactness of the max-tensor product).

This assertion can also be seen as an easy consequence of Theorem 8.16 and the fact that for any ideal $\mathcal{I} \subset A$, we have by (A.37) a C^*-isomorphism

$$A^{**} \simeq (A/\mathcal{I})^{**} \oplus \mathcal{I}^{**}.$$

Indeed, the latter isomorphism shows that A^{**} is injective if and only if both $(A/\mathcal{I})^{**}$ and $\mathcal{I}^{**}$ are injective. $\qquad\square$

As we will see in the next chapter, injectivity is equivalent to the WEP for von Neumann algebras (see Corollary 9.26). Thus the reader will find more conditions equivalent to injectivity there, as well as in §11.7 on hypertraces.

8.4 Local reflexivity of the maximal tensor product

In the next section we will study local reflexivity. A C^*-algebra A is locally reflexive if for any B and any $t \in A^{**} \otimes B$ we have $\|t\|_{(A \otimes_{\min} B)^{**}} \leq \|t\|_{A^{**} \otimes_{\min} B}$ (see what follows for clarification). Equivalently this means that for any finite-dimensional operator space E we have $CB(E, A)^{**} = CB(E, A^{**})$ isometrically. We will soon show (see Remarks 8.32 and 8.33) that this does not always hold, even when $E = \ell_\infty^n$ ($n > 2$). In sharp contrast, we show in the present section that a property analogous to local reflexivity does hold for the max-tensor product, and moreover we always have $D(E, A)^{**} = D(E, A^{**})$ isometrically when $E = M_n$ or $E = \ell_\infty^n$ ($n \geq 1$).

Let A, B be unital C^*-algebras. We first need to clarify how we embed $A^{**} \otimes B^{**}$ into the biduals $(A \otimes_{\max} B)^{**}$ and $(A \otimes_{\min} B)^{**}$.

Let us denote simply by $i_0 : A \otimes_{\max} B \to (A \otimes_{\max} B)^{**}$ (resp. $i_1 : A \otimes_{\min} B \to (A \otimes_{\min} B)^{**}$) the natural inclusion. Define $\pi_0 : A \to (A \otimes_{\max} B)^{**}$ (resp. $\pi_1 : A \to (A \otimes_{\min} B)^{**}$) and $\sigma_0 : B \to (A \otimes_{\max} B)^{**}$ (resp. $\sigma_1 : B \to (A \otimes_{\min} B)^{**}$) by $\pi_0(a) = i_0(a \otimes 1)$ (resp. $\pi_1(a) = i_1(a \otimes 1)$) and $\sigma_0(b) = i_0(1 \otimes b)$ (resp. $\sigma_1(b) = i_1(1 \otimes b)$). Then π_0, σ_0 (resp. π_1, σ_1) are $*$-homomorphisms with commuting ranges such that $i_0 = \pi_0 \cdot \sigma_0$ (resp. $i_1 = \pi_1 \cdot \sigma_1$). Actually we can define similar pairs (π_0, σ_0) (resp. (π_1, σ_1)) in the nonunital case using Remark 4.2 and the observation that the universal representations of $A \otimes_{\max} B$ and $A \otimes_{\min} B$, being direct sums of cyclic ones, are nondegenerate.

Let $q : A \otimes_{\max} B \to A \otimes_{\min} B$ be the quotient map. Then $q^{**} : (A \otimes_{\max} B)^{**} \to (A \otimes_{\min} B)^{**}$ is a normal $*$-homomorphism onto $(A \otimes_{\min} B)^{**}$.

Note that we have canonical linear embeddings

$$A^* \otimes B^* \subset (A \otimes_{\max} B)^* \quad (\text{resp. } A^* \otimes B^* \subset (A \otimes_{\min} B)^*)$$

that, for any $(f, g) \in A^* \times B^*$, take $f \otimes g$ to the linear map $f \otimes g : A \otimes_{\max} B \to \mathbb{C}$ (resp. $f \otimes g : A \otimes_{\min} B \to \mathbb{C}$).

Proposition 8.19 *There are natural inclusions* $J_{\max} : A^{**} \otimes B^{**} \to (A \otimes_{\max} B)^{**}$ *and* $J_{\min} : A^{**} \otimes B^{**} \to (A \otimes_{\min} B)^{**}$ *such that for all* $(a'', b'') \in A^{**} \times B^{**}$ *and* $(f, g) \in A^* \times B^*$ *we have*

$$\langle J_{\max}(a'' \otimes b''), f \otimes g \rangle = a''(f)b''(g) = \langle J_{\min}(a'' \otimes b''), f \otimes g \rangle.$$

Moreover $J_{\min} = q^{**} J_{\max}$.

Proof Let $\ddot{\pi}_0 : A^{**} \to (A \otimes_{\max} B)^{**}$ (resp. $\ddot{\pi}_1 : A^{**} \to (A \otimes_{\min} B)^{**}$) and $\ddot{\sigma}_0 : B^{**} \to (A \otimes_{\max} B)^{**}$ (resp. $\ddot{\sigma}_1 : B^{**} \to (A \otimes_{\min} B)^{**}$) be the normal $*$-homomorphisms extending π_0 and σ_0 (resp. π_1 and σ_1), still with commuting ranges. This gives us a $*$-homomorphism $J_{\max} : A^{**} \otimes B^{**} \to (A \otimes_{\max} B)^{**}$ (resp. $J_{\min} : A^{**} \otimes B^{**} \to (A \otimes_{\min} B)^{**}$) defined by

$$J_{\max} = \ddot{\pi}_0 \cdot \ddot{\sigma}_0 \text{ and } J_{\min} = \ddot{\pi}_1 \cdot \ddot{\sigma}_1.$$

Claim: For any $t \in A^{**} \otimes B^{**}$ and any $F \in (A \otimes_{\min} B)^*$ we have

$$\langle J_{\max}(t), q^*(F) \rangle = \langle J_{\min}(t), F \rangle. \tag{8.8}$$

Moreover if $F = f \otimes g$ with $(f, g) \in A^* \times B^*$, then

$$\langle J_{\min}(t), F \rangle = \langle t, f \otimes g \rangle \tag{8.9}$$

where the last pairing is the canonical one between $A^{**} \otimes B^{**}$ and $A^* \otimes B^*$. *Proof of the Claim:* It suffices to prove (8.8) for any t of the form $t = a'' \otimes b''$ with $(a'', b'') \in A^{**} \times B^{**}$. Equivalently, it suffices to prove

$$\langle J_{\max}(a'' \otimes b''), q^*(F) \rangle = \langle J_{\min}(a'' \otimes b''), F \rangle. \tag{8.10}$$

It is easy to verify going back to the definitions that, for any fixed $F \in (A \otimes_{\min} B)^*$, both sides of (8.10) are separately weak* continuous bilinear forms on $A^{**} \times B^{**}$, which coincide (and are equal to F) on $A \times B$. Therefore they coincide on $A^{**} \times B^{**}$. This proves (8.8) and hence $J_{\min} = q^* J_{\max}$. Now if $F = f \otimes g$ with $(f, g) \in A^* \times B^*$, then $(a'', b'') \mapsto \langle t, F \rangle = f(a'') g(b'')$ is also a separately weak* continuous bilinear form on $A^{**} \times B^{**}$ coinciding with the preceding two on $A \times B$. This implies (8.9), completing the proof of the claim. By the second part of Remark A.1, (8.9) shows that $J_{\min}$ is injective, and since $J_{\min} = q^{**} J_{\max}$ so is $J_{\max}$. $\square$

It will be convenient to record here a simple observation.

Lemma 8.20 *Let A, B be C^*-algebras. Let $\pi : A \to B(H)$ and $\sigma : B \to B(H)$ be representations with commuting ranges such that $\pi \cdot \sigma : A \otimes B \to B(H)$ extends to a contractive $*$-homomorphism $T : A \otimes_{\min} B \to B(H)$. Then $\ddot{T} : (A \otimes_{\min} B)^{**} \to B(H)$ satisfies*

$$\forall a'' \in A^{**}, b'' \in B^{**} \quad \ddot{T}(a'' \otimes b'') = \ddot{\pi}(a'') \ddot{\sigma}(b''), \tag{8.11}$$

where the embedding $A^{**} \otimes B^{**} \subset (A \otimes_{\min} B)^{**}$ is implicitly meant to be $J_{\min}$.

Proof Note that by definition the mapping $(a'', b'') \mapsto J_{\min}(a'' \otimes b'')$ is separately normal. Therefore both sides of (8.11) are separately normal bilinear maps on $A^{**} \times B^{**}$. Since they clearly coincide on $A \times B$, and A (resp. B) is weak* dense in A^{**} (resp. B^{**}), they must coincide on $A^{**} \times B^{**}$. $\square$

We now come to the version of local reflexivity satisfied by the maximal tensor product. In the next two statements, the embedding $A^{**} \otimes B^{**} \subset (A \otimes_{\max} B)^{**}$ is implicitly meant to be $J_{\max}$.

Theorem 8.21 *For any B and any $t \in A^{**} \otimes B$*

$$\|t\|_{(A \otimes_{\max} B)^{**}} \leq \|t\|_{A^{**} \otimes_{\max} B}. \tag{8.12}$$

More precisely, we have

$$\|t\|_{(A \otimes_{\max} B)^{**}} = \|t\|_{A^{**} \otimes_{\mathrm{bin}} B^{**}} \leq \|t\|_{A^{**} \otimes_{\max} B^{**}} = \|t\|_{A^{**} \otimes_{\max} B}. \tag{8.13}$$

We start by proving first a more precise version of (8.13):

Theorem 8.22 *The norm induced by $(A \otimes_{\max} B)^{**}$ on $A^{**} \otimes B^{**}$ coincides with the bin-norm.*

Proof By definition $J_{\max} : A^{**} \otimes B^{**} \to (A \otimes_{\max} B)^{**}$ is of the form $J_{\max} = \ddot{\pi}_0.\ddot{\sigma}_0$. From this follows, by definition of the bin-norm that for any $t \in A^{**} \otimes B^{**}$ we have

$$\|J_{\max}(t)\| \leq \|t\|_{\mathrm{bin}}.$$

Actually, the reverse inequality also holds. To show this consider an isometric embedding

$$\varphi : A^{**} \otimes_{\mathrm{bin}} B^{**} \subset B(H)$$

such that the restriction to each factor is normal. (This obviously exists, just consider the direct sum of all $\ddot{\pi} \cdot \ddot{\sigma}$ as in (8.4).) Note that, by definition, the bin norm of $A^{**} \otimes B^{**}$ restricted to $A \otimes B$ coincides with the max-norm of $A \otimes B$. Thus we have a *-homomorphism

$$\psi : A \otimes_{\max} B \to B(H)$$

obtained by restricting φ to $A \otimes B$. But now we have a normal extension $\ddot{\psi} : (A \otimes_{\max} B)^{**} \to B(H)$ (with $\|\ddot{\psi}\| = 1$ of course). We claim that

$$\forall a \in A^{**}, \quad \forall b \in B^{**}, \quad \ddot{\psi} J_{\max}(a \otimes b) = \varphi(a \otimes b), \tag{8.14}$$

and hence

$$\forall t \in A^{**} \otimes B^{**}, \quad \ddot{\psi} J_{\max}(t) = \varphi(t).$$

Indeed, both sides of (8.14) are separately normal bilinear maps on $A^{**} \times B^{**}$, which coincide on $A \times B$. By weak* density again, (8.14) follows. From this we deduce

$$\forall t \in A^{**} \otimes B^{**}, \quad \|t\|_{A^{**} \otimes_{\mathrm{bin}} B^{**}} = \|\varphi(t)\| \leq \|J_{\max}(t)\|_{(A \otimes_{\max} B)^{**}}.$$

$\square$

Remark 8.23 We cannot replace the bin-norm by the max-norm in Theorem 8.22. Indeed, consider $A = B = K(H)$ (compact operators). Now A, B are nuclear so $A \otimes_{max} B = A \otimes_{min} B = K(H \otimes_2 H)$. Thus $(A \otimes_{max} B)^{**} = B(H \otimes_2 H)$, and $A^{**} = B^{**} = B(H)$. The norm induced by $(A \otimes_{max} B)^{**}$ on $B(H) \otimes B(H)$ is now the min norm. But it is known (see §18.1) that the min and max norm are not equivalent on $B(H) \otimes B(H)$.

Proof of Theorem 8.21 By the maximality of the max-norm, (8.12) is clear. Moreover, since the inclusion $B \subset B^{**}$ is max-injective (see Corollary 7.27) we have $\|t\|_{A^{**} \otimes_{max} B} = \|t\|_{A^{**} \otimes_{max} B^{**}}$ for any $t \in A^{**} \otimes B$ and the rest follows by Theorem 8.22. □

Lemma 8.24 *Let C, A be C^*-algebras. Let $u : C \to A^{**}$ and let (u_i) be a net in the unit ball of $D(C, A)$ such that $u_i(x) \to u(x)$ with respect to $\sigma(A^{**}, A^*)$ for any $x \in C$. Then $u \in D(C, A^{**})$ with $\|u\|_{dec} \le 1$.*

Proof By definition of $\|u_i\|_{dec}$ there are $V_i \in CP(C, M_2(A))$ of the form

$$V_i : x \to \begin{pmatrix} S_1^i(x) & u_i(x) \\ u_i(x^*)^* & S_2^i(x) \end{pmatrix} \text{ with } \|S_1^i\| \le 1 \quad \|S_2^i\| \le 1.$$

Passing to a subnet we may assume that $S_1^i(x)$ and $S_2^i(x)$ are $\sigma(A^{**}, A^*)$-convergent for any $x \in C$ to $S_1(x) \in A^{**}$ and $S_2(x) \in A^{**}$, so that $\|S_1\| \le 1, \|S_2\| \le 1$. Then the limit V of (V_i) is clearly in $CP(C, M_2(A^{**}))$. Since $V = \begin{pmatrix} S_1(x) & u(x) \\ u(x^*)^* & S_2(x) \end{pmatrix}$, we have $u \in D(C, A^{**})$ and $\|u\|_{dec} \le 1$. □

In sharp contrast with (8.20), we have

Theorem 8.25 *For any n and any C^*-algebra A, we have natural isometric identifications*

$$D(M_n, A^{**}) = D(M_n, A)^{**} \text{ and } D(\ell_\infty^n, A^{**}) = D(\ell_\infty^n, A)^{**}.$$

Proof Note that the spaces $D(M_n, A^{**})$ and $D(M_n, A)^{**}$ are setwise identical. The inclusion $D(M_n, A)^{**} \to D(M_n, A^{**})$ has norm ≤ 1 by Lemma 8.24. For the reverse inclusion, we use the description of the unit ball of $D(M_n, A)$ given in (6.31). Let u be in the open unit ball of $D(M_n, A^{**})$. Define $\mathfrak{a} \in M_n(A^{**})$ by $\mathfrak{a}_{ij} = u(e_{ij})$. We can find $a, b \in M_n(A^{**})$ such that $\mathfrak{a} = a^*b$ and (by homogeneity) $\max \left\{ \left\| \sum_{kj} a_{kj}^* a_{kj} \right\|^{1/2}, \left\| \sum_{kj} b_{kj}^* b_{kj} \right\|^{1/2} \right\} < 1$. Since $M_n(A^{**}) = M_n(A)^{**}$ (see Proposition A.58), there are nets (a^γ) and (b^δ) in $M_n(A)$ that are $\sigma(A^{**}, A^*)$-convergent to a and b and such that $\max \left\{ \left\| \sum_{kj} a_{kj}^{\gamma *} a_{kj}^\gamma \right\|^{1/2}, \left\| \sum_{kj} b_{kj}^{\delta *} b_{kj}^\delta \right\|^{1/2} \right\} \le 1$. Let $\mathfrak{a}_{ij}^{\gamma,\delta} = \sum_k a_{ki}^{\gamma *} b_{kj}^\delta$. Let us now assume that $A^{**} \subset B(H)$ (as a von Neumann subalgebra). We then have for all $h', h \in H$

$$\lim_\gamma \lim_\delta \langle h', \mathfrak{a}_{ij}^{\gamma,\delta} h \rangle = \sum_k \lim_\gamma \lim_\delta \langle a_{ki}^\gamma h', b_{kj}^\delta h \rangle$$

$$= \sum_k \langle a_{ki} h', b_{kj} h \rangle = \langle h', \mathfrak{a}_{ij} h \rangle.$$

Thus $\lim_\gamma \lim_\delta \mathfrak{a}_{ij}^{\gamma,\delta} = \mathfrak{a}_{ij}$ in the w.o.t. of $B(H)$. Then since the adjoint of the embedding $A^{**} \subset B(H)$ takes $B(H)_*$ onto A^*, the set of functionals on A of the form $x \mapsto \langle h', xh \rangle$ $(h', h \in H)$ is total in A^* and hence since the $\mathfrak{a}^{\gamma,\delta}$s are uniformly bounded we must have $\lim_\gamma \lim_\delta \mathfrak{a}_{ij}^{\gamma,\delta} = \mathfrak{a}_{ij}$ for $\sigma(A^{**}, A^*)$. This shows $\|u\|_{D(M_n,A)^{**}} \leq 1$. Thus we conclude that the reverse inclusion: $D(M_n, A^{**}) \to D(M_n, A)^{**}$ has norm ≤ 1.

If we replace M_n by ℓ_∞^n the same proof works using (6.32). Alternatively we can use the realization of ℓ_∞^n as diagonal matrices in M_n and (6.33) to deduce the case of ℓ_∞^n from that of M_n. $\square$

8.5 Local reflexivity

Following [77] a C^*-algebra A is called locally reflexive if for any B and any $t \in A^{**} \otimes B$ we have

$$\|t\|_{(A \otimes_{\min} B)^{**}} \leq \|t\|_{A^{**} \otimes_{\min} B}, \qquad (8.15)$$

or equivalently (see Remark 1.9) for any $t \in B \otimes A^{**}$ we have

$$\|t\|_{(B \otimes_{\min} A)^{**}} \leq \|t\|_{B \otimes_{\min} A^{**}}. \qquad (8.16)$$

In (8.15) and throughout this section, we (implicitly) use the map $J_{\min} : A^{**} \otimes B^{**} \to (A \otimes_{\min} B)^{**}$ from Proposition 8.19 to view $A^{**} \otimes B^{**}$ as included in $(A \otimes_{\min} B)^{**}$.

In sharp contrast with the Banach space analogue of local reflexivity, briefly described in §A.8 (from which the terminology comes) this property does not always hold (see Remark 8.33). It is implied by "exactness" but the converse is an open problem.

Remark 8.26 (Reversing (8.16) and (8.15)) Let A be an arbitrary C^*-algebra. Then the reverse inequalities to (8.16) or (8.15) hold: for any $t \in B \otimes A^{**}$ we have

$$\|t\|_{B \otimes_{\min} A^{**}} \leq \|t\|_{(B \otimes_{\min} A)^{**}}. \qquad (8.17)$$

This is immediate by the minimality of the min-norm among C^*-norms on $B \otimes A^{**}$. However, we find it instructive to include a direct "hands on" proof, as follows. Let E be an n-dimensional operator space such that $t \in E \otimes A^{**}$. The space $E \otimes_{\min} A^{**}$ is isomorphic to $[A^{**}]^n$. The unit ball of the space $E \otimes_{\min} A^{**}$ is closed for the weak* topology (i.e. the topology induced by $\sigma(A^{**}, A^*)$)

(we leave this as an exercise). Therefore $E \otimes_{\min} A^{**}$ is isometrically a dual Banach space. Let $J : E \otimes_{\min} A \to E \otimes_{\min} A^{**}$ denote the isometric inclusion. Equivalently $J = Id_E \otimes i_A$. Then, with the notation in (A.32), we have $\|\ddot{J} : [E \otimes_{\min} A]^{**} \to E \otimes_{\min} A^{**}\| \leq 1$, and $\ddot{J}$ is the identity on $[E \otimes_{\min} A]^{**} \simeq E \otimes_{\min} A^{**} \simeq [A^{**}]^n$.

Remark 8.27 Let $t \in B \otimes A^{**}$ and let $E \subset B$ be finite dimensional such that $t \in E \otimes A^{**}$. Clearly, since $E \otimes_{\min} A^{**} \subset B \otimes_{\min} A^{**}$ and $[E \otimes_{\min} A]^{**} \subset [B \otimes_{\min} A]^{**}$ are isometric inclusions, (8.16) holds if and only if for any finite-dimensional $E \subset B$ and any such $t \in E \otimes A^{**}$ we have

$$\|t\|_{[E \otimes_{\min} A]^{**}} \leq \|t\|_{E \otimes_{\min} A^{**}}. \tag{8.18}$$

Using the identification $CB(E^*, A) = E \otimes_{\min} A$ (see §2.4), in which we may exchange the roles of E and E^*, and the preceding remark we see that A is locally reflexive if and only if for any finite-dimensional operator space E we have $CB(E, A)^{**} = CB(E, A^{**})$ isometrically.

Let us record this important fact.

Proposition 8.28 *A C^*-algebra A is locally reflexive if and only for any $u \in CB(E, A^{**})$*

$$\|u\|_{CB(E, A)^{**}} \leq \|u\|_{CB(E, A^{**})}, \tag{8.19}$$

or more explicitly, any $u \in B_{CB(E, A^{**})}$ is the pointwise-weak* limit of a net in $B_{CB(E, A)}$.

The latter reformulation explains the analogy with (A.10). Actually the reverse of (8.19) always holds by Remark 8.27, and hence we have equality in (8.19). This shows that when A is locally reflexive, we have an isometric embedding $A^{**} \otimes_{\min} B \subset [A \otimes_{\min} B]^{**}$ for any B.

Theorem 8.29 *Any nuclear C^*-algebra is locally reflexive.*

Proof Let A, B be C^*-algebras. Assume A nuclear. Then

$$A \otimes_{\min} B = A \otimes_{\max} B \Rightarrow (A \otimes_{\min} B)^{**} = (A \otimes_{\max} B)^{**}.$$

By Theorem 8.22 we get $\|A^{**} \otimes_{\mathrm{bin}} B^{**} \to (A \otimes_{\max} B)^{**}\| \leq 1$. By Corollary 8.15 the algebra A^{**} is injective and hence $A^{**} \otimes_{\mathrm{bin}} B^{**} = A^{**} \otimes_{\min} B^{**}$. Thus we obtain $\|A^{**} \otimes_{\min} B^{**} \to (A \otimes_{\min} B)^{**}\| \leq 1$ (this is called property (C) in Remark 8.34), which implies a fortiori the local reflexivity of A. $\square$

In the opposite direction, the next statement will allow us to produce explicit examples failing local reflexivity.

Proposition 8.30 *If a C^*-algebra A is locally reflexive then for any ideal $\mathcal{I} \subset A$ and any C^*-algebra B we have*

$$B \otimes_{\min} (A/\mathcal{I}) = (B \otimes_{\min} A)/(B \otimes_{\min} \mathcal{I}).$$

In other words, the quotient map $A \to A/\mathcal{I}$ is min-*projective in the sense of Definition 7.40.*

Proof Recall that the canonical map $[(B \otimes_{\min} A)/(B \otimes_{\min} \mathcal{I})] \to B \otimes_{\min} (A/\mathcal{I})$ has unit norm. Thus it suffices to prove the same for its inverse. Since $B \otimes_{\min} \mathcal{I}$ is an ideal in $B \otimes_{\min} A$ we have canonically (see (A.37))

$$(B \otimes_{\min} A)^{**} \simeq [(B \otimes_{\min} A)/(B \otimes_{\min} \mathcal{I})]^{**} \oplus (B \otimes_{\min} \mathcal{I})^{**}.$$

Moreover since $A^{**} \simeq (A/\mathcal{I})^{**} \oplus \mathcal{I}^{**}$ (again by (A.37)) we have an embedding

$$B \otimes_{\min} (A/\mathcal{I})^{**} \subset B \otimes_{\min} A^{**}.$$

By the local reflexivity of A we may write (with maps of unit norm)

$$B \otimes_{\min} (A/\mathcal{I})^{**} \subset B \otimes_{\min} A^{**} \to (B \otimes_{\min} A)^{**}$$
$$\to [(B \otimes_{\min} A)/(B \otimes_{\min} \mathcal{I})]^{**},$$

and a fortiori $B \otimes_{\min} (A/\mathcal{I}) \to [(B \otimes_{\min} A)/(B \otimes_{\min} \mathcal{I})]^{**}$ has unit norm. But the range of the latter $*$-homomorphism is included in $(B \otimes_{\min} A)/(B \otimes_{\min} \mathcal{I})$, therefore we find that the map

$$B \otimes_{\min} (A/\mathcal{I}) \to [(B \otimes_{\min} A)/(B \otimes_{\min} \mathcal{I})]$$

also has unit norm.   □

Remark 8.31 (Local reflexivity and injectivity) Let $E \subset F$ be an inclusion of operator spaces with E finite dimensional. Assume that a C^*-algebra A has the following extension property: for any $u : E \to A$ there is $\widetilde{u} : F \to A$ extending u with $\|\widetilde{u}\|_{cb} = \|u\|_{cb}$.

If A is locally reflexive then A^{**} has the same property.

$$
\begin{array}{ccc}
F & & \\
\uparrow & \searrow^{\widetilde{u}} & \\
\Big\uparrow & & \searrow \\
E & \xrightarrow{\ u\ } & A \subset A^{**}
\end{array}
$$

Indeed, given $u \in CB(E, A^{**})$ we have a net $u_i \in CB(E, A)$ with $\|u_i\|_{cb} \leq \|u\|_{cb}$ tending weak* to u. Then the map $\widehat{u} \in CB(F, A^{**})$ equal to a pointwise-weak* cluster point of the net $(\widetilde{u}_i)$ is the desired extension of u.

When A is a von Neumann algebra the preceding property holds for all inclusions $E \subset F$ with $\dim(E) < \infty$ if and only if A is injective. Indeed, this follows by a simple weak* limit argument.

Remark 8.32 (Local reflexivity is inherited by subalgebras and quotients) Local reflexivity passes to C^*-subalgebras. Indeed, this follows directly from the definition, once one recalls that for any closed subspace $Y \subset X$ of a Banach space X we have an isometric canonical embedding $Y^{**} \subset X^{**}$ (see Remark A.54). When $A_1 \subset A$ is a C^*-subalgebra, we have $A_1^{**} \subset A^{**}$ (completely) isometrically, and the latter embedding realizes A_1^{**} as a von Neumann subalgebra of A^{**}. We also have similarly $(A_1 \otimes_{\min} B)^{**} \subset (A \otimes_{\min} B)^{**}$ isometrically. Therefore, if A is locally reflexive, for any $t \in A_1^{**} \otimes B \subset A^{**} \otimes B$ we have

$$\|t\|_{(A_1 \otimes_{\min} B)^{**}} = \|t\|_{(A \otimes_{\min} B)^{**}} \leq \|t\|_{A^{**} \otimes_{\min} B} = \|t\|_{A_1^{**} \otimes_{\min} B},$$

and we conclude that A_1 is locally reflexive.

Thus $B(H)$ must fail local reflexivity, otherwise any C^*-algebra would be locally reflexive. One quick way to see that $B(H)$ fails local reflexivity is to observe that if it were locally reflexive then $B(H)^{**}$ would be injective, and hence $B(H)$ would be nuclear (see Remark 8.31).

Local reflexivity passes to quotient C^*-algebras. We briefly sketch the easy argument for this. Let $q : A \to A/\mathcal{I}$ be the quotient map. By (A.37) we have a *-homomorphism $r : (A/\mathcal{I})^{**} \to A^{**}$ such that $q^{**}r = Id_{(A/\mathcal{I})^{**}}$. Let B be another C^*-algebra, we have $\|r \otimes Id_B : (A/\mathcal{I})^{**} \otimes_{\min} B \to A^{**} \otimes_{\min} B\| = 1$ and if A is locally reflexive $\|A^{**} \otimes_{\min} B \to (A \otimes_{\min} B)^{**}\| = 1$, while clearly $\|(q \otimes Id_B)^{**} : (A \otimes_{\min} B)^{**} \to ((A/\mathcal{I}) \otimes_{\min} B)^{**}\| = 1$. By composition it follows that $\|(A/\mathcal{I})^{**} \otimes_{\min} B \to ((A/\mathcal{I}) \otimes_{\min} B)^{**}\| = 1$, which means that $A/\mathcal{I}$ is locally reflexive. Thus, for some free group $\mathbb{F}$ (resp. for $\mathbb{F} = \mathbb{F}_\infty$) $C^*(\mathbb{F})$ (resp. $\mathscr{C}$) must fail local reflexivity, otherwise any (resp. any separable) C^*-algebra would be locally reflexive.

Remark 8.33 ($\mathscr{C}$ and $\mathscr{B}$ fail local reflexivity (quantitative estimate)) By Proposition 7.34 the quotient map $C^*(\mathbb{F}_n) \to C_\lambda^*(\mathbb{F}_n)$ is not min-projective when $1 < n \leq \infty$. By Proposition 8.30 this means that $C^*(\mathbb{F}_n)$ is not locally reflexive, and since the latter embeds in $\mathscr{B}$, a fortiori $\mathscr{B}$ is not locally reflexive.

More explicitly, let $G = \mathbb{F}_n$, $A = C^*(G)$, $H = \ell_2(G)$ and $M = M_G \subset B(H)$. Then the extension of λ_G defines a *-homomorphism $\pi : A \to M$, such that $\ddot{\pi} : A^{**} \to M$ is a surjective normal *-homomorphism (see Remark A.49). Let $\mathcal{I} = \ker(\ddot{\pi})$. Then $A^{**} \simeq M \oplus \mathcal{I}$ by (A.24), so that we have a natural embedding $M \subset A^{**}$, which we denote by $\Phi : M \to A^{**}$, that lifts $\ddot{\pi}$ so that $\ddot{\pi}\Phi = Id_M$. Let $U_j = U_G(g_j)$ as usual and $U_0 = 1$. Consider the tensor

$$t = \sum_0^n U_j \otimes \Phi(\pi U_j) \in A \otimes A^{**}.$$

Then

$$\|t\|_{A \otimes_{\min} A^{**}} = 2\sqrt{n} \text{ and } \|t\|_{(A \otimes_{\min} A)^{**}} = n + 1.$$

Indeed, $\|t\|_{A \otimes_{\min} A^{**}} = \left\| \sum_0^n U_j \otimes \pi U_j \right\|_{A \otimes_{\min} M} = 2\sqrt{n}$ by (4.27). By Lemma 8.20, the inequality $\|t\|_{(A \otimes_{\min} A)^{**}} \geq n + 1$ follows from the factorization property of the free groups that will be proved later on in Corollary 12.23. Indeed, let $\sigma : A \to B(H)$ be the representation associated to ρ_G. By the factorization property (see Definition 7.36) we know that $T = \sigma \cdot \pi : A \otimes_{\min} A \to B(H)$ is contractive. By Lemma 8.20 we have

$$\ddot{T}(t) = \sum \sigma(U_j) \ddot{\pi}(\Phi(\pi(U_j)))$$

$$= \sum \sigma(U_j) \pi(U_j), \text{ and } \|\ddot{T} : (A \otimes_{\min} A)^{**} \to B(H)\| \leq 1.$$

This gives us $n + 1 = \left\| \sum \sigma(U_j) \pi(U_j) \right\| \leq \|\ddot{T}(t)\| \leq \|t\|_{(A \otimes_{\min} A)^{**}}$.

Using (2.14) this can be reformulated using the linear operator $u : \ell_\infty^{n+1} \to A^{**}$ (corresponding to t) defined by $u(e_j) = \Phi(U_j)$, as follows

$$\|u\|_{CB(\ell_\infty^{n+1}, A^{**})} = 2\sqrt{n} \text{ and } \|u\|_{CB(\ell_\infty^{n+1}, A)^{**}} = n + 1. \qquad (8.20)$$

This argument shows that if a group G has the factorization property then G is amenable if and only if $C^*(G)$ is locally reflexive.

Remark 8.34 (Properties C and C') The origin of local reflexivity for C^*-algebras lies in [10]. There Archbold and Batty introduced two properties that they named C and C'.

A has property C if for any C^*-algebra B we have an isometric embedding

$$A^{**} \otimes_{\min} B^{**} \subset (A \otimes_{\min} B)^{**}.$$

A has property C' if for any C^*-algebra B we have an isometric embedding

$$A \otimes_{\min} B^{**} \subset (A \otimes_{\min} B)^{**}.$$

They showed in [10] that property C implies exactness, as defined in the sequel in §10.1, and it is obvious that $C \Rightarrow C'$. Kirchberg showed later on that C and C' are actually equivalent properties, and each is equivalent to exactness. Clearly property C implies local reflexivity but (as we already mentioned) the converse remains open. We refer the reader to [208, ch. 18] for more comments and to [39, ch. 9] for a complete proof of the equivalence of C and C', which is much more delicate than the one of C' with exactness that we prove later on in Proposition 10.12.

Remark 8.35 The definition of local reflexivity makes sense equally well for operator spaces. Surprisingly, it turns out that any predual of a von Neumann algebra is locally reflexive when viewed as an operator space. See [78].

8.6 Notes and remarks

The results of §8.1 on biduals are all well-known facts, while those on decomposable maps are based on [104]. Equations (6.14) and (6.15) appear in [141].

Concerning injectivity in §8.3 again the main references are Connes's [61], Lance's [165] and the Choi–Effros papers [45–48]. The original proof that injective factors on a separable Hilbert space are approximately finite dimensional (i.e. "hyperfinite") is an outstanding achievement of Connes [61].

The case of general von Neumann algebras was deduced from Connes's results by Elliott. See [82–84, 261] for clarifications on that question.

Later on, simpler proofs of Connes's result that injective implies AFD were given by Uffe Haagerup [105] and Sorin Popa [217]. See also chapter XVI in [241], [39, p. 333] and [4, ch. 11] for more recent detailed expositions.

The proof that injective ⇒ semidiscrete is more accessible. A simpler proof of that implication appears in S. Wassermann's [256]. Before Connes's work and the Choi–Effros papers a number of implications between injectivity and semidiscreteness (or in other words the weak* CPAP) appeared in the Effros–Lance paper [79] which was already circulating as a preprint around 1974. In particular, they proved the equivalence of injectivity and semidiscreteness for the von Neumann algebra of a discrete group. They also proved the equivalence of semidiscreteness with either (iii) or (iv) in Theorem 8.13 and also with (ii) in Theorem 8.14.

Local reflexivity originates in Archbold–Batty's [10] (see Remark 8.34) where they prove Theorem 8.29, but the subject owes a lot to a subsequent paper by Effros and Haagerup [77], and to Simon Wassermann's work in connection with exactness (see [258] for more references).

See [78] for more recent important work on operator space local reflexivity.

9

Nuclear pairs, WEP, LLP, QWEP

We start with a few general remarks around nuclearity for pairs.

Definition 9.1 A pair of C^* algebras (A, B) will be called a nuclear pair if

$$A \otimes_{\min} B = A \otimes_{\max} B,$$

or equivalently if the min- and max-norm are equal on the algebraic tensor product $A \otimes B$.

Remark 9.2 If the min- and max-norm are equivalent on $A \otimes B$, then they automatically are equal by Corollary A.26.

Remark 9.3 Let $A_1 \subset A$ and $B_1 \subset B$ be C^*-subalgebras. In general, the nuclearity of the pair (A, B) does *not* imply that of (A_1, B_1). As the sequel will demonstrate, this "defect" is a major feature of the notion of nuclearity. However, if (A_1, B_1) admit contractive c.p. projections (conditional expectations) $P : A \to A_1$ and $Q : B \to B_1$ then (A_1, B_1) inherits the nuclearity of (A, B). This is an immediate application of Corollary 7.8 (see also Remark 7.20) and Proposition 7.19.

More generally, the following holds:

Lemma 9.4 *Let A, D be C^*-algebras. We assume that Id_D factors through A in a certain "local" sense as follows: For any finite-dimensional subspace $E \subset D$ and any $\varepsilon > 0$ there is a factorization $E \xrightarrow{v} A \xrightarrow{w} D$ of the inclusion map with $\|v\|_{cb} \|w\|_{dec} \leq 1 + \varepsilon$.*

Let B be another C^*-algebra. Then, if (A, B) is nuclear, the same is true for (D, B).

Proof Let $x \in D \otimes B$. We may assume $x \in E \otimes B$ with $\dim(E) < \infty$. Then, since $x = (w \otimes Id_B)(v \otimes Id_B)(x)$ we have by (6.13)

$$\|x\|_{D\otimes_{\max}B} \leq \|w\|_{dec}\|(v \otimes Id_B)(x)\|_{A\otimes_{\max}B} = \|w\|_{dec}\|(v \otimes Id_B)(x)\|_{A\otimes_{\min}B}$$
$$\leq \|v\|_{cb}\|w\|_{dec}\|x\|_{E\otimes_{\min}B}.$$

Thus, since $\|x\|_{E\otimes_{\min}B} = \|x\|_{D\otimes_{\min}B}$, we obtain $\|x\|_{D\otimes_{\max}B} = \|x\|_{D\otimes_{\min}B}$.

$\square$

Remark 9.5 Since $i_D : D \to D^{**}$ is max-injective, the preceding argument works equally well if we only assume that i_D (instead of Id_D) factors through A in the same local sense as described in Lemma 9.4.

Recall that A is called nuclear if (A, B) is nuclear for all B.

The basic examples of nuclear C^*-algebras (see §4.2) include all commutative ones, the algebra $K(H)$ of all compact operators on an arbitrary Hilbert space H, $C^*(G)$ for all amenable discrete groups G and the Cuntz algebras.

While the meaning of nuclearity for a C^*-algebra seems by now fairly well understood, it is not so for pairs, as reflected by Kirchberg's fundamental conjecture from [155] that we discuss in detail in §13.

9.1 The fundamental nuclear pair $(C^*(\mathbb{F}_\infty), B(\ell_2))$

A large part of the sequel revolves around the two fundamental examples

$$\mathscr{B} = B(\ell_2) \quad \text{and} \quad \mathscr{C} = C^*(\mathbb{F}_\infty).$$

Note that these are both universal but in two different ways, injectively for $\mathscr{B} = B(\ell_2)$ (by this we mean that every separable C^*-algebra embeds in $\mathscr{B}$), projectively for $\mathscr{C} = C^*(\mathbb{F}_\infty)$ (by this we mean that every separable unital C^*-algebra is a quotient of $\mathscr{C}$, see Proposition 3.39 for details).

Kirchberg's conjecture is simply that the pair $(\mathscr{C}, \mathscr{C})$ is nuclear. The main goal of these notes is to introduce the reader to the state of the art related to this conjecture, and in particular its equivalence with a problem posed back in 1976 by Alain Connes [61].

In this context, Kirchberg [155] proved the following striking result. We will give the simpler proof from [204].

Theorem 9.6 (The fundamental pair) *The pair* $(C^*(\mathbb{F}_\infty), B(\ell_2)) = (\mathscr{C}, \mathscr{B})$ *is a nuclear pair, as well as* $(C^*(\mathbb{F}_\infty), B(H))$ *for any* H.

The following simple fact is essential for our argument.

Proposition 9.7 *Let* A, B *be two unital* C^*-algebras. *Let* $(u_i)_{i\in I}$ *be a family of unitary elements of* A *generating* A *as a unital* C^*-algebra (*i.e. the smallest unital* C^*-subalgebra of A containing them is A itself). *Let* $E \subset A$ *be the* *linear span of* $(u_i)_{i \in I}$ *and* 1_A. *Let* $T : E \to B$ *be a linear operator such that* $T(1_A) = 1_B$ *and taking each* u_i *to a unitary in* B. *Then* $\|T\|_{cb} \leq 1$ *suffices to ensure that* T *extends to a (completely) contractive* *-*homomorphism* $\widetilde{T} : A \to B$.

Moreover, $\widetilde{T} : A \to B$ *is the unique completely contractive (or equivalently the unique unital c.p.) map extending* T.

Lastly, if T *is completely isometric and* $T(E)$ *generates* B *(as a* C^**-algebra) then* $\widetilde{T}$ *is a* *-*isomorphism from* A *to* B.

Proof The first variant uses multiplicative domains (see §5.1). Consider B as embedded in $B(\mathcal{H})$. By Arveson's extension Theorem 1.18, T extends to a complete contraction $\widetilde{T} : A \to B(\mathcal{H})$. Since T is assumed unital, $\widetilde{T}$ is unital, and hence completely positive by Corollary 1.52. Now for any unitary U in the family $(u_i)_{i \in I}$, since $\widetilde{T}(U) = T(U)$ is unitary by assumption, we have

$$\{u_i \mid i \in I\} \subset D_{\widetilde{T}},$$

and since $D_{\widetilde{T}}$ is a C^*-algebra (see Theorem 5.1) this implies automatically $A = D_{\widetilde{T}}$, so that $\widetilde{T}$ is actually a (contractive) *-homomorphism into $B(\mathcal{H})$. Since $\widetilde{T}(u_i) = T(u_i)$ and the u_i's generate A, we must have $\widetilde{T}(A) \subset B$. Moreover, any other complete contraction $T' : A \to B$ extending T must be a *-homomorphism equal to $\widetilde{T}$ on E, and hence must be equal to $\widetilde{T}$.

Lastly, let $F = T(E) \subset B$. Note that $\widetilde{T}(A)$ is the C^*-algebra generated by F. Assume T completely isometric and $\widetilde{T}(A) = B$. Then we can apply the first part of the proof to $T^{-1} : F \to A$. This gives us a *-homomorphism $\sigma : B \to A$ with $\|\sigma\| \leq 1$ that is inverse to $\widetilde{T}$ and proves the last assertion. This completes the proof.

Alternate argument: The reader who so wishes can avoid the use of multiplicative domains by arguing like this: By Theorem 1.22 we can find an embedding $\mathcal{H} \subset K$ and a unital *-homomorphism $\pi : A \to B(K)$ such that $\widetilde{T}(a) = P_{\mathcal{H}} \pi(a)_{|\mathcal{H}}$. Then an elementary argument shows that if a unitary U on K is such that $P_{\mathcal{H}} U_{|\mathcal{H}}$ is still unitary, then U must commute with $P_{\mathcal{H}}$. Thus, by our assumption this commutation is true for $\pi(u_i)$ and hence for $\pi(A)$ since the $\pi(u_i)$'s generate it. This shows that $a \mapsto P_{\mathcal{H}} \pi(a)_{|\mathcal{H}}$ (which is the same as $\widetilde{T}$) is a *-homomorphism, necessarily with range in B.   □

The main idea of our proof of Kirchberg's Theorem 9.6 is that if E is the linear span of 1 and the free unitary generators of $C^*(\mathbb{F}_\infty)$, then it suffices to check that the min- and max-norms coincide on $E \otimes B(H)$. More generally, we will prove

Theorem 9.8 *Let* A_1, A_2 *be unital* C^**-algebras. Let* $(u_i)_{i \in I}$ *(resp.* $(v_j)_{j \in J}$*) be a family of unitary operators that generate* A_1 *(resp.* A_2*). Let* E_1 *(resp.* E_2*) be*

the closed span of $(u_i)_{i\in I}$ *(resp.* $(v_j)_{j\in J}$*). Assume* $1 \in E_1$ *and* $1 \in E_2$*. Then the following assertions are equivalent:*

(i) *The inclusion map* $E_1 \otimes_{\min} E_2 \to A_1 \otimes_{\max} A_2$ *is completely isometric.*

(ii) $A_1 \otimes_{\min} A_2 = A_1 \otimes_{\max} A_2$.

Proof The implication (ii) $\Rightarrow$ (i) is trivial (since $*$-homomorphisms are completely contractive), so we prove only the converse. Assume (i). Let $E = E_1 \otimes_{\min} E_2$. We view E as a subspace of $A = A_1 \otimes_{\min} A_2$. By (i), we have an inclusion map $T : E_1 \otimes_{\min} E_2 \to A_1 \otimes_{\max} A_2$ with $\|T\|_{cb} \leq 1$. By Proposition 9.7, T extends to a (contractive) $*$-homomorphism $\widetilde{T}$ from $A_1 \otimes_{\min} A_2$ to $A_1 \otimes_{\max} A_2$. Clearly $\widetilde{T}$ must preserve the algebraic tensor products $A_1 \otimes 1$ and $1 \otimes A_2$, hence also $A_1 \otimes A_2$. Thus we obtain (ii). $\square$

Remark 9.9 Let us denote by $E_1 \otimes 1 + 1 \otimes E_2$ the linear subspace $\{e_1 \otimes 1 + 1 \otimes e_2 \mid e_1 \in E_1,\ e_2 \in E_2\}$. Then, in the situation of Theorem 9.8, $E_1 \otimes 1 + 1 \otimes E_2$ generates $A_1 \otimes_{\min} A_2$, so that it suffices for the conclusion of Theorem 9.8 to assume that the operator space structures induced on $E_1 \otimes 1 + 1 \otimes E_2$ by the min and max norms coincide.

Proof of Kirchberg's Theorem 9.6 Let $A_1 = C^*(\mathbb{F}_\infty)$, $A_2 = B(H)$. We may clearly assume $\dim(H) = \infty$. Let $U_0 = 1$ and let $(U_i)_{i\geq 1}$ denote the free unitary generators of $\mathscr{C} = C^*(\mathbb{F}_\infty)$. We take $E_2 = B(H)$ and let E_1 be the linear span of $(U_i)_{i\geq 0}$.

Consider $x \in E_1 \otimes E_2$, with $\|x\|_{\min} < 1$. By Lemma 3.10 we can write $x = \sum_{i\geq 0} U_i \otimes x_i$ with $x_i \in B(H)$, $(x_i)_{i\geq 0}$ finitely supported, admitting a decomposition as $x_i = a_i b_i$ with $\left\|\sum a_i a_i^*\right\| < 1$, $\left\|\sum b_i^* b_i\right\| < 1$, $a_i, b_i \in B(H)$. Now, let $\pi : A_1 \otimes_{\max} A_2 \to B(\mathcal{H})$ be any faithful $*$-homomorphism. Let $\pi_1 = \pi_{|A_1 \otimes 1}$ and $\pi_2 = \pi_{|1 \otimes A_2}$. We have

$$\pi(x) = \sum_{i\geq 0} \pi_1(U_i)\pi_2(x_i) = \sum_{i\geq 0} \pi_1(U_i)\pi_2(a_i)\pi_2(b_i).$$

Since π_1 and π_2 have commuting ranges we have $\pi(x) = y$ with

$$y = \sum_{i\geq 0} \pi_2(a_i)\pi_1(U_i)\pi_2(b_i).$$

Now by (2.1) from Lemma 2.3 we have

$$\|y\| \leq \left\|\sum_{i\geq 0} \pi_2(a_i)\pi_2(a_i)^*\right\|^{1/2} \left\|\sum_{i\geq 0} \pi_2(b_i)^*\pi_2(b_i)\right\|^{1/2} < 1.$$

Thus we conclude that

$$\|x\|_{\max} = \|\pi(x)\| < 1.$$

This shows that the min and max norms coincide on $E_1 \otimes B(H)$. But since $\dim(H) = \infty$ we have $M_n(B(H)) \simeq B(H)$ for any n, and hence

$$M_n(E_1 \otimes_{\min} B(H)) = E_1 \otimes_{\min} M_n(B(H)) \simeq E_1 \otimes_{\min} B(H),$$

and also

$$M_n(A_1 \otimes_{\max} B(H)) - A_1 \otimes_{\max} M_n(B(H)) \sim A_1 \otimes_{\max} B(H),$$

therefore the latter coincidence of norms "automatically" implies that the inclusion

$$E_1 \otimes_{\min} B(H) \to A_1 \otimes_{\max} B(H)$$

is completely isometric. In other words, the operator space structures associated to the min and max norms coincide. We may clearly replace E_1 by its closure. Thus, the proof is concluded by Theorem 9.8 (here $E_2 = A_2 = B(H)$). $\qquad \square$

As a complement to his fundamental Theorem 9.6 Kirchberg observed the following general phenomenon.

Theorem 9.10 *Let $\mathbb{F}$ be any free group and M any von Neumann algebra. Then $\| \cdot \|_{\max} = \| \cdot \|_{\mathrm{nor}}$ on $C^*(\mathbb{F}) \otimes M$.*

Proof Assume $\mathbb{F} = \mathbb{F}_I$. Let $E = \overline{\mathrm{span}}[U_i \mid i \in \dot{I}]$ with $\{U_i \mid i \in \dot{I}\}$ as in Lemma 6.28. Let us denote by $E \otimes_{\overline{\max}} M$ (resp. $E \otimes_{\overline{\mathrm{nor}}} M)$) the operator space generated by $E \otimes M$ in $C^*(\mathbb{F}) \otimes_{\max} M$ (resp. $C^*(\mathbb{F}) \otimes_{\mathrm{nor}} M$). By Proposition 9.7 it suffices to prove that the natural mapping $E \otimes_{\overline{\max}} M \to E \otimes_{\overline{\mathrm{nor}}} M$ is completely isometric. Since we may replace M by $M_n(M)$ to pass from isometric to completely isometric (we skip the easy details for this point), it suffices to show that $\|t\|_{\max} = \|t\|_{\mathrm{nor}}$ for any $t \in E \otimes M$ (the max- and nor-norms being the ones induced on $E \otimes M$ by those of $C^*(\mathbb{F}) \otimes M$). Let $\pi : M \to B(H)$ be a normal embedding with infinite multiplicity (i.e. $H = \ell_2 \otimes_2 K$ and $\pi(\cdot) = Id_{\ell_2} \otimes \sigma(\cdot)$ where $\sigma : M \to B(K)$ embeds M as a von Neumann subalgebra). Then (6.38) means that for any $t \in E \otimes M$

$$\|t\|_{\max} = \sup_{\sigma} \|(\sigma . \pi)(t)\| \qquad (9.1)$$

where the sup runs over all $*$-homomorphisms $\sigma : C^*(\mathbb{F}) \to M'$, and hence $\|t\|_{\max} \leq \|t\|_{\mathrm{nor}}$. By maximality $\|t\|_{\max} = \|t\|_{\mathrm{nor}}$. The proof actually shows that (9.1) holds for all $t \in C^*(\mathbb{F}) \otimes M$. $\qquad \square$

In the same direction, the following variant will be useful.

Theorem 9.11 *Let $(A_i)_{i \in I}$ be a family of C^*-algebras and let $A = \left(\oplus \sum_{i \in I} A_i \right)_{\infty}$. Let $t \in \mathscr{C} \otimes A$. For each $i \in I$ let $t_i = (Id_{\mathscr{C}} \otimes p_i)(t) \in \mathscr{C} \otimes A_i$, where $p_i : A \to A_i$ is the coordinate projection. Then*

$$\|t\|_{\mathscr{C} \otimes_{\max} A} = \sup_{i \in I} \|t_i\|_{\mathscr{C} \otimes_{\max} A_i}. \qquad (9.2)$$

Proof By Proposition 9.7 it suffices to check that this holds for any $t \in E \otimes A$ where E is the linear span of the unitary generators (U_j) and the unit. (Indeed, one can replace A by $M_n(A)$ and observe that $M_n(A) \simeq \left(\oplus \sum_{i \in I} M_n(A_i) \right)_\infty$.) Then we may as well assume that $t \in E \otimes A$ with $E = \mathrm{span}[I, U_1, \ldots, U_{N-1}]$. Let $u : \ell_\infty^N \to A$ (resp. $u_i : \ell_\infty^N \to A_i$) be the linear map associated to t (resp. t_i). Then (6.37) shows us that (9.2) is equivalent to

$$\|u\|_{dec} = \sup_{i \in I} \|u_i\|_{dec},$$

which we observed in (6.10). $\square$

For our exposition, it will be convenient to adopt the following definitions (equivalent to the more standard ones by [155]).

Definition 9.12 Let A be a C^*-algebra.

We say that A has the WEP (or is WEP) if $(A, \mathscr{C})$ is a nuclear pair.

We say that A has the LLP (or is LLP) if $(A, \mathscr{B})$ is a nuclear pair.

We say that A is QWEP if it is a quotient (by a closed, self-adjoint, two-sided ideal) of a WEP C^*-algebra.

Here WEP stands for weak expectation property (and LLP for local lifting property).

Remark 9.13 If A has the WEP (resp. LLP) then any C^*-subalgebra $D \subset A$ such that the inclusion $D \subset A$ is max-injective has the WEP (resp. LLP) by (7.10).

Remark 9.14 By Proposition 3.5, for any subgroup $\Gamma \subset G$ of a discrete group G, the inclusion $C^*(\Gamma) \subset C^*(G)$ is max-injective. Therefore, for any given B, if the pair $(C^*(G), B)$ is nuclear, then the same is true for the pair $(C^*(\Gamma), B)$.

For instance, it is well known that $\mathbb{F}_\infty$ embeds as a subgroup in $\mathbb{F}_n$ for any $1 < n < \infty$. Therefore:

$$(C^*(\mathbb{F}_\infty), B) \text{ is nuclear} \Leftrightarrow (C^*(\mathbb{F}_n), B) \text{ is nuclear.} \tag{9.3}$$

In particular, $C^*(\mathbb{F}_n)$ is LLP for any $1 \le n \le \infty$.

Let $\mathbb{F}$ be any free group. By Lemmas 3.8 and 9.4 we can replace $\mathbb{F}_n$ by $\mathbb{F}$ in (9.3):

$$(C^*(\mathbb{F}_\infty), B) \text{ is nuclear} \Leftrightarrow (C^*(\mathbb{F}_n), B) \text{ is nuclear}$$
$$\Leftrightarrow (C^*(\mathbb{F}), B) \text{ is nuclear for any } \mathbb{F}. \tag{9.4}$$

In particular:

$$C^*(\mathbb{F}) \text{ is LLP for any free group } \mathbb{F}. \tag{9.5}$$

Remark 9.15 Let A, B, C be C^*-algebras. Assume that A is nuclear. If the pair (B, C) is nuclear then the pair $(A \otimes_{\min} B, C)$ is also nuclear. We leave the proof as an (easy) exercise (see (4.9)). In particular, if B has the WEP (resp. LLP) then the same is true of $A \otimes_{\min} B$.

Remark 9.16 If A is WEP, LLP (or QWEP) then the same is true of A^{op} or $\overline{A}$. This can be deduced easily from the fact that $A^{op} \simeq A$ (or $\overline{A} \simeq A$) when $A = \mathscr{C}$ and $A = \mathscr{B}$ (see Remarks 3.7 and 2.10). Moreover, by (4.10) the properties WEP, LLP (or QWEP) are stable under direct sum.

9.2 $C^*(\mathbb{F})$ is residually finite dimensional

A C^*-algebra is called residually finite dimensional (RFD) if for any $x \in A$ with $x \neq 0$ there is a $*$-homomorphism $\pi : A \to B(H)$ with $\dim(H) < \infty$ such that $\pi(x) \neq 0$.

It is easy to see that this holds if and only if there is a family of finite-dimensional $*$-homomorphisms

$$\pi_i : A \to M_{n(i)} \quad i \in I, \, n(i) < \infty$$

such that their direct sum

$$\oplus_{i \in I} \pi_i : A \to \left(\oplus \sum\nolimits_{i \in I} M_{n(i)} \right)_\infty \tag{9.6}$$

is an embedding. Equivalently A is RFD if and only if the $*$-homomorphism that is the direct sum, over all $n \geq 1$, of all $*$-homomorphisms $\pi : A \to M_n$ is injective.

Remark 9.17 If A is separable (and RFD), since there is a countable subset dense in the unit sphere, we can always find a countable family $(M_{n(i)})_{i \in I}$ for which (9.6) is an embedding.

In this section we prove the following result (due to Choi).

Theorem 9.18 *Let G be any free group with free generators $\{g_i \mid i \in I\}$. Then $C^*(G)$ is residually finite dimensional.*

Proof Let $\{\pi \mid \pi \in \widehat{G}_0\}$ denote the collection of all the finite-dimensional unitary representations of G (without repetitions). We define

$$\sigma(t) = \oplus\{\pi(t) \mid \pi \in \widehat{G}_0\}. \qquad \forall t \in G$$

Clearly, σ extends to a (contractive) $*$-homomorphism

$$u : C^*(G) \to C^*_\sigma(G) \subset \left(\oplus \sum\nolimits_{\pi \in \widehat{G}_0} B(H_\pi) \right)_\infty,$$

taking $U_G(t)$ to $\sigma(t)$. Let $E \subset C^*(G)$ be the linear span of 1 and $\{U_G(g_i) \mid i \in I\}$. To prove the theorem, we will show that u is isometric. By Proposition 9.7 it suffices to show that $u_{|E} : E \to C^*_\sigma(G)$ is completely isometric. Let $\dot{I} = \{0\} \cup I$ (disjoint union) and let $\{x_i \mid i \in \dot{I}\}$ be any finitely supported family in $B(H)$ (with H arbitrary). Then a typical element of $B(H) \otimes E$ is of the form

$$x = x_0 \otimes I + \sum_{i \in I} x_i \otimes U_G(g_i),$$

and we have

$$\|x\|_{B(H) \otimes_{\min} E} = \sup \left\| x_0 \otimes I + \sum_{i \in I} x_i \otimes u_i \right\|,$$

where the sup runs over all possible families $\{u_i \mid i \in I\}$ of unitaries (including infinite-dimensional ones). But we saw in (3.11) that this supremum remains the same if we let it run over all possible families of *finite-dimensional* unitaries. Thus (3.11) tells us that

$$\|x\|_{B(H) \otimes_{\min} E} = \|[Id_{B(H)} \otimes u](x)\|_{B(H) \otimes_{\min} C^*_\sigma(G)},$$

which shows that the restriction of u to E is completely isometric. By Proposition 9.7, u gives us an embedding of $C^*(G)$ into $\left(\oplus \sum_{\pi \in \widehat{G}_0} B(H_\pi) \right)_\infty$. □

Remark 9.19 If A, B are residually finite dimensional C^*-algebras, then $A \otimes_{\min} B$ is residually finite dimensional. Indeed, if $A \subset \left(\oplus \sum_{i \in I} M_{n(i)} \right)_\infty$ and $B \subset \left(\oplus \sum_{j \in J} M_{m(j)} \right)_\infty$ then

$$A \otimes_{\min} B \subset \left(\oplus \sum_{(i,j) \in I \times J} M_{n(i)} \otimes_{\min} M_{m(j)} \right)_\infty$$

$$= \left(\oplus \sum_{(i,j) \in I \times J} M_{n(i)m(j)} \right)_\infty.$$

Let us say for short that a state f on a C^*-algebra A is "a finite-dimensional state" (resp. "a finite-dimensional vector state") if there is a $*$-homomorphism $\pi : A \to B(H)$ on a *finite-dimensional* Hilbert space H and a state (resp. a vector state) F on $B(H)$ such that $f(a) = F(\pi(a))$ for all $a \in A$.

Using multiplicity, we observe that any finite-dimensional state is actually a finite-dimensional vector state: indeed, any state f on $B(H)$ with $\dim(H) < \infty$ is of the form $f(x) = \text{tr}(a^*xa) = \langle a, xa \rangle_{S_2(H)}$ for some unit vector $a \in S_2(H)$, and hence it becomes a vector state when we embed $B(H)$ in $B(S_2(H))$ by left multiplication. Thus, for later reference, we state the next result only for vector states.

Proposition 9.20 *A C^*-algebra A is residually finite dimensional if and only if any state on A is the pointwise limit of a net of finite-dimensional vector states.*

In particular this holds for the states on $\mathscr{C} \otimes_{\min} \mathscr{C}$.

Proof Assume A RFD, so that (9.6) is an embedding. Let $H_i = \ell_2^{n(i)}$ and $H = \oplus_{i \in I} H_i$. Any state f on A extends to a state on $B(H)$, which is a limit of normal states on $B(H)$. The latter states are limits of states f_γ in the convex hull of those of the form $a \mapsto \langle \xi, \pi(a)\xi \rangle$ where ξ is a unit vector in H, which, after truncation and renormalization, we may assume to be all in $\oplus_{i \in I(\gamma)} H_i$ for some *finite* subset $I(\gamma) \subset I$. Then f_γ is in the convex hull of states of the form

$$a \mapsto \langle \xi, (\oplus_{i \in I(\gamma)} \pi_i(a))\xi \rangle.$$

Since $\dim(\oplus_{i \in I(\gamma)} H_i) < \infty$, each f_γ is a finite-dimensional state. By the preceding observation, this proves the only if part.

Conversely, for any $a \in A$ we have $\|a\|^2 = \|a^*a\| = \sup f(a^*a)$ where the sup is over all states. If any such f is the pointwise limit of finite-dimensional states we can restrict the sup to the latter, and then we find $\|a\|^2 \le \sup \|\pi(a^*a)\|$ where the sup runs over all $*$-homomorphisms $\pi : A \to B(H)$ with H finite dimensional. Since the reverse inequality is obvious, we conclude that A is RFD.

The last assertion follows from Theorem 9.18 and Remark 9.19. $\square$

9.3 WEP (Weak Expectation Property)

We defined the WEP for A by the equality $A \otimes_{\min} \mathscr{C} = A \otimes_{\max} \mathscr{C}$. We will now see that it is equivalent to a weak form of extension property (a sort of weakening of injectivity), which is the original and more traditional definition of the WEP. Let $A \subset B(H)$ be a C^*-subalgebra. If A is injective, there is a completely contractive projection $P : B(H) \to A$, satisfying the properties of a conditional expectation by Theorem 1.45. Recall that the weak* closure $\overline{A}^{\mathrm{weak}*}$ of A in $B(H)$, is equal to A'' by Theorem A.46. A unital c.p. mapping $T : B(H) \to \overline{A}^{\mathrm{weak}*}$ is called a weak expectation if $T(a) = a$ for any $a \in A$. This concept goes back to Lance [165]. We will show that the WEP is equivalent to the existence of a weak expectation $T : B(H) \to \overline{\pi(A)}^{\mathrm{weak}*} = \pi(A)''$ for any H and any embedding $\pi : A \to B(H)$ (see Remark 9.23). But for our broader framework, it will be convenient to enlarge Lance's concept, as follows.

Definition 9.21 Let $A \subset B$ be a C^*-subalgebra of another one. A linear mapping $V : B \to A^{**}$ will be called a generalized weak expectation if $\|V\| \le 1$ and $V(a) = a$ for any $a \in A$.

If $V(B) \subset A$ then V is a conditional expectation in the usual sense as in Theorem 1.45.

In Theorem 7.29 we already gave an important characterization of the inclusions that admit a generalized weak expectation, as those such that $A \subset B$ is max-injective.

We will show that the WEP of A is equivalent to the existence of a generalized weak expectation $V : B \rightarrow A^{**}$ whenever A embeds in B. Indeed, this reduces to the case $B = B(H)$ treated in Theorem 9.31. Note that if the embedding $A \subset B(H)$ is the universal representation of A then $\overline{\pi(A)}^{\text{weak}*} = A^{**}$ (see §A.16) and hence the generalized notion of weak expectation coincides in this case with Lance's original one. See Remark 9.32 for more on generalized weak expectations.

Theorem 9.22 *Let $A \subset B(H)$ be a C^*-algebra. The following are equivalent.*

(i) *A has the WEP (i.e. $(A, \mathscr{C})$ is a nuclear pair).*

(ii) *The inclusion $A \subset B(H)$ is* max-*injective.*

(iii) *Any $*$-homomorphism $u : A \rightarrow M$ into a von Neumann algebra M extends to a completely positive and (completely) contractive mapping from $B(H)$ to M.*

(iii)' *Any $*$-homomorphism $u : A \rightarrow M$ into a von Neumann algebra M factors completely positively and (completely) contractively through $B(\mathcal{H})$ for some $\mathcal{H}$.*

(iv) *The inclusion $i_A : A \rightarrow A^{**}$ factors completely positively and (completely) contractively through $B(\mathcal{H})$ for some $\mathcal{H}$.*

Proof (i) $\Rightarrow$ (ii). Assume (i). Then

$$A \otimes_{\max} \mathscr{C} = A \otimes_{\min} \mathscr{C} \subset B(H) \otimes_{\min} \mathscr{C} = B(H) \otimes_{\max} \mathscr{C}$$

where the last equality is from Theorem 9.6. By (i) $\Leftrightarrow$ (i)' in Theorem 7.29, (ii) holds.

Assume (ii). Then (iii) holds by Corollary 7.30.

(iii) $\Rightarrow$ (iii)' is obvious, and using $u = i_A$ we see that (iii)'$\Rightarrow$(iv).

Assume (iv). Consider a completely positive and (completely) contractive factorization

$$i_A : A \rightarrow B(\mathcal{H}) \rightarrow A^{**}.$$

By Theorem 9.6 (recalling that by Corollary 7.8 contractive c.p. maps are (max $\rightarrow$ max)-tensorizing) we find a contractive factorization

$$i_A \otimes Id_{\mathscr{C}} : A \otimes_{\min} \mathscr{C} \rightarrow B(\mathcal{H}) \otimes_{\min} \mathscr{C} = B(\mathcal{H}) \otimes_{\max} \mathscr{C} \rightarrow A^{**} \otimes_{\max} \mathscr{C}$$

and since i_A is max-injective we conclude that $A \otimes_{\min} \mathscr{C} = A \otimes_{\max} \mathscr{C}$. In other words we obtain (i). $\qquad \square$

Remark 9.23 (On Lance's WEP) Lance's definition of the WEP for a C^*-algebra is different but easily seen to be equivalent to ours. Lance [165] says that a $*$-homomorphism $\pi : A \rightarrow B(H)$ has the WEP if the von Neumann algebra it generates, i.e. the weak$*$ closure $\overline{\pi(A)}^{\text{weak}*}$, admits a weak expectation. He then says that A has the WEP if every *faithful* π has the WEP. We claim that Lance's WEP is the same as our WEP in Theorem 9.22 (i). Indeed, by (i) $\Rightarrow$ (iii) in Theorem 9.22, A has Lance's WEP if it has our WEP (because if A has our WEP so does $\pi(A)$ for any faithful π). Conversely, Lance's WEP applied to the embedding $\pi : A \subset A^{**} \subset B(H)$ (assuming A^{**} embedded as a von Neumann algebra in $B(H)$) implies the existence of a weak expectation from $B(H)$ to A^{**} that is equal to π on A and hence by (iv) $\Rightarrow$ (i) in Theorem 9.22 that A has our WEP.

Using the extension properties of $B(\mathcal{H})$ described in Theorems 1.18 and 1.39, the following is an immediate consequence of (iv):

Corollary 9.24 *If A is WEP then it has the following weak forms of injectivity into the bidual:*

For any operator space X and any subspace $E \subset X$, any $u \in CB(E, A)$ admits an "extension" $\widetilde{u} \in CB(X, A^{**})$ with $\|\widetilde{u}\|_{cb} = \|u\|_{cb}$ such that $\widetilde{u}_{|E} = i_A u$.

If E, X are operator systems, then any $u \in CP(E, A)$ admits an "extension" $\widetilde{u} \in CP(X, A^{**})$ with $\|\widetilde{u}\| = \|u\|$ such that $\widetilde{u}_{|E} = i_A u$.

Remark 9.25 It is obvious that the second property in Corollary 9.24 characterizes WEP: just substitute to $E \subset X$ the inclusion $j_A : A \rightarrow B(H)$. That the first one also does is less obvious, but it will be shown by Theorem 23.7.

Corollary 9.26 *A von Neumann algebra M is injective if and only if it has the WEP.*

Proof If M has the WEP, then the identity of M factors through some $B(\mathcal{H})$ as in (iii) in Theorem 9.22 (take $A = M$ and $u = Id_M$). Therefore M is injective. Conversely, if M is injective the identity of M factors completely positively and (completely) contractively through $B(\mathcal{H})$, so (iv) in Theorem 9.22 (with $A = M$) follows immediately. $\qquad\square$

Recall that it is obvious by our definition that nuclear implies WEP, thus

$$\{\text{nuclear}\} \cup \{\text{injective}\} \subset \{WEP\}.$$

By Theorem 8.16 we can deduce from Corollary 9.26:

Corollary 9.27 *Let $\mathscr{C} = C^*(\mathbb{F}_\infty)$. A C^*-algebra A is nuclear if and only if*

$$\mathscr{C} \otimes_{\min} A^{**} = \mathscr{C} \otimes_{\max} A^{**},$$

i.e. if and only if the pair $(\mathscr{C}, A^{**})$ is nuclear.

Corollary 9.28 *A C^*-algebra A is both WEP and locally reflexive if and only if it is nuclear.*

Proof Assume that A has the WEP. By our definition of the WEP, $\mathscr{C} \otimes_{\min} A = \mathscr{C} \otimes_{\max} A$ and hence $(\mathscr{C} \otimes_{\min} A)^{**} = (\mathscr{C} \otimes_{\max} A)^{**}$. If A is locally reflexive, then $\|\mathscr{C} \otimes_{\min} A^{**} \to (\mathscr{C} \otimes_{\min} A)^{**}\| = 1$, and hence $\|\mathscr{C} \otimes_{\min} A^{**} \to (\mathscr{C} \otimes_{\max} A)^{**}\| = 1$. By Theorem 8.22 we have $\|\mathscr{C} \otimes_{\min} A^{**} \to \mathscr{C}^{**} \otimes_{\mathrm{bin}} A^{**}\| = 1$. But the norm induced on $\mathscr{C} \otimes A^{**}$ by the bin-norm on $\mathscr{C}^{**} \otimes A^{**}$ coincides with the nor-norm (see Remark 8.4). Therefore we have $\|\mathscr{C} \otimes_{\min} A^{**} \to \mathscr{C} \otimes_{\mathrm{nor}} A^{**}\| = 1$. Since by Theorem 9.10 $\mathscr{C} \otimes_{\mathrm{nor}} A^{**} = \mathscr{C} \otimes_{\max} A^{**}$, we conclude that $\|\mathscr{C} \otimes_{\min} A^{**} \to \mathscr{C} \otimes_{\max} A^{**}\| = 1$, and hence A is nuclear by Corollary 9.27. The converse is immediate since nuclear implies locally reflexive by Theorem 8.29. □

Corollary 9.29 *If the reduced C^*-algebra $C_\lambda^*(G)$ of a discrete group G has the WEP then G is amenable.*

Proof Assume that $C_\lambda^*(G)$ has the WEP. Let $M \subset B(\ell_2(G))$ be as usual the von Neumann algebra generated by $C_\lambda^*(G)$. Let $T : B(\ell_2(G)) \to M$ be the completely positive contraction extending the inclusion $C_\lambda^*(G) \to M$, given by (iii) in Theorem 9.22. By Corollary 5.3, T is $C_\lambda^*(G)$-bimodular. This implies in particular that $T(\lambda(g)x\lambda(g)^*) = \lambda(g)T(x)\lambda(g)^*$ for any $x \in B(\ell_2(G))$ and any $g \in G$. For any $x \in \ell_\infty(G)$, let $D_x \in B(\ell_2(G))$ denote the (diagonal) operator of multiplication by x. Note that

$$\lambda(g)D_x\lambda(g)^* = D_{\delta_g * x}.$$

Let $\tau_G : M \to \mathbb{C}$ be defined by $\tau_G(a) = \langle \delta_e, a\delta_e \rangle$. It is easy to check that $\tau_G(\lambda(g)a\lambda(g)^*) = \tau_G(a)$ for any $a \in M$ and $g \in G$ (in other words τ_G is a trace on M in the sense of §11.1). Let $\varphi \in \ell_\infty(G)_+^*$ be the functional defined by

$$\varphi(x) = \tau_G(T(D_x)) = \langle \delta_e, T(D_x)\delta_e \rangle.$$

Then we have

$$\varphi(\delta_g * x) = \tau_G(T(\lambda(g)D_x\lambda(g)^*)) = \tau_G(\lambda(g)T(D_x)\lambda(g)^*)$$
$$= \tau_G(T(D_x)) = \varphi(x).$$

Thus φ is an invariant mean on G. □

Proposition 9.30 *For any separable C^*-algebra $A \subset B(H)$ there is a separable unital C^*-algebra $A_1 \subset B(H)$ with the WEP such that $A \subset A_1$.*

Proof This follows directly from Proposition 7.24 and Remark 7.25 applied with $B = \mathscr{C}$ and $A = B(H)$.   □

Now that we know by Theorem 9.22 that A has the WEP if and only if the inclusion $A \subset B(H)$ is max-injective, let us review what Theorem 7.29 tells us about it.

Theorem 9.31 (On weak expectations) *Let $A \subset B(H)$ be a C^*-algebra and let $A^{**} \subset B(H)^{**}$ be the embedding obtained by bitransposition. The following are equivalent.*

   (i) *A has the WEP.*
  (ii) *There is a contractive linear map $V : B(H) \to A^{**}$ such that $V(x) = x$ for any $x \in A$ (in other words, V is a generalized weak expectation for $A \subset B(H)$).*
 (iii) *There is a projection $P : B(H)^{**} \to A^{**}$ with $\|P\| = 1$.*

Remark 9.32 (Comparing weak expectations and projections) We repeatedly use the observation recorded in Proposition A.50 that, assuming $A \subset B$, A, B being here merely Banach spaces, a linear map $V : B \to A^{**}$ is a generalized weak expectation if and only if $\ddot{V} : B^{**} \to A^{**}$ is a contractive projection. It can be shown (by a fairly easy application of the Hahn–Banach theorem) that such a V exists if and only the natural map $A \hat{\otimes} C \to B \hat{\otimes} C$ between the projective tensor products is isometric for any Banach space C, and actually it suffices to have this for $C = A^*$. This kind of duality argument can be generalized to treat the case when V is not necessarily a contraction. It can also be checked easily that the natural map $A \hat{\otimes} C \to A^{**} \hat{\otimes} C$ is isometric for any Banach space C. In analogy with the latter facts, in the C^*-algebra case, we showed in Theorem 7.29 that a generalized weak expectation exists if and only if the inclusion $A \subset B$ is max-injective, and in Corollary 7.27 that $i_A : A \to A^{**}$ is always max-injective.

Remark 9.33 (Warning on a trap) To avoid possible errors, we emphasize that (in sharp contrast with the analogue for injectivity) the existence of an embedding $u : A^{**} \subset B(H)^{**}$ admitting a c.p. contractive projection $P : B(H)^{**} \to A^{**}$ does *not* in general imply the WEP for A. Indeed, we will show in Theorem 9.72 that this holds if and only if A is QWEP. For the WEP to hold it is essential to assume in addition that $u = v^{**}$ for some $v : A \to B(H)$. Then we effectively can conclude that v is max-injective so A has the WEP by Theorem 9.22.

We now turn to the stability of the WEP under infinite direct sums in the sense of ℓ_∞.

Proposition 9.34 *For any family* $\{A_i \mid i \in I\}$ *of WEP C^*-algebras the direct sum* $\left(\oplus \sum_{i \in I} A_i\right)_\infty$ *also has the WEP.*

Proof With the notation in (9.2), for any $t \in \mathscr{C} \otimes A$ we have by (9.2) and (1.14)

$$\|t\|_{\mathscr{C} \otimes_{\max} A} = \sup_{i \in I} \|t_i\|_{\mathscr{C} \otimes_{\max} A_i} = \sup_{i \in I} \|t_i\|_{\mathscr{C} \otimes_{\min} A_i} = \|t\|_{\mathscr{C} \otimes_{\min} A},$$

which means that A has the WEP. $\qquad\square$

Remark 9.35 Thus if $\mathscr{C}$ denotes as usual the full C^*-algebra of $\mathbb{F}_\infty$, this means that if $(A_i, \mathscr{C})$ is nuclear for any i then $(A, \mathscr{C})$ is nuclear where $A = \left(\oplus \sum_{i \in I} A_i\right)_\infty$. However, this is not true if we replace $\mathscr{C}$ by an arbitrary C^* algebra. Indeed, nuclearity is not preserved by infinite direct sums of the type $A = \left(\oplus \sum_{i \in I} A_i\right)_\infty$. For example $\mathbb{B} = (\oplus \sum_{n \geq 1} M_n)_\infty$ is not nuclear (see Corollary 18.11).

We refer to [189, Lemma 3.2] for a different proof of Proposition 9.34 based on the following purely Banach space result, which as its corollary, is of independent interest.

Theorem 9.36 *Let B be any Banach space and let $A \subset B$ be a closed subspace. Then the following are equivalent:*

(i) *There is a projection $P : B^{**} \to A^{**}$ with $\|P\| = 1$.*
(ii) *For any $\varepsilon > 0$ and any finite-dimensional subspace $E \subset B$ there is a linear map $\varphi : E \to A$ with $\|\varphi\| \leq 1 + \varepsilon$ such that*

$$\forall a \in E \cap A \quad \varphi(a) = a.$$

Corollary 9.37 *Let $\{B_i \mid i \in I\}$ be a family of Banach spaces. Let $\{A_i \mid i \in I\}$ be a family of subspaces, with $A_i \subset B_i$ for each $i \in I$, such that there is a projection $P_i : B_i^{**} \to A_i^{**}$ with $\|P_i\| = 1$. Then there is a projection*

$$P : \left(\oplus \sum_{i \in I} B_i\right)_\infty^{**} \to \left(\oplus \sum_{i \in I} A_i\right)_\infty^{**}$$

with $\|P\| \leq 1$.

9.4 LLP (Local Lifting Property)

In Banach space theory, the "lifting property" of ℓ_1 is classical: for any bounded linear map u from ℓ_1 into a quotient Banach space X/Y and for any $\varepsilon > 0$, there is a lifting $\tilde{u} : \ell_1 \to X$ with $\|\tilde{u}\| \leq (1 + \varepsilon)\|u\|$. Moreover, the so-called $\mathcal{L}_1$-spaces satisfy a local variant of this. There are analogues of this lifting property for C^*-algebras and operator spaces (see [208, ch. 16]). We return to this in §9.5 and §21.2. But in this chapter we concentrate on the local variant called LLP, which is better understood. We show next that indeed the LLP as defined previously in Definition 9.12 is equivalent to a certain "local" lifting property. Recall that linear maps with values in a quotient $A/\mathcal{I}$ (of a C^*-algebra A by an ideal $\mathcal{I}$) that locally c-lift were defined in Definition 7.46.

Theorem 9.38 (Local lifting property) *The following properties of a C^*-algebra C are equivalent:*

(i) *C has the LLP i.e. the pair $(C, \mathcal{B})$ is nuclear.*
(ii) *Any u in the unit ball of $D(C, A/\mathcal{I})$ is locally 1-liftable for any quotient $A/\mathcal{I}$.*
(iii) *Any $u \in CP(C, A/\mathcal{I})$ with $\|u\| \leq 1$ is locally 1-liftable for any quotient $A/\mathcal{I}$.*
(iv) *Any $*$-homomorphism $u : C \to A/\mathcal{I}$ is locally 1-liftable for any quotient $A/\mathcal{I}$.*

Proof Assume (i). Let $u \in D(C, A/\mathcal{I})$ with $\|u\|_{dec} \leq 1$. By (6.13) for any C^*-algebra B we have

$$\|Id_B \otimes u : B \otimes_{\max} C \to B \otimes_{\max} (A/\mathcal{I})\| \leq 1 \qquad (9.7)$$

and by (7.6) (with the roles of A, B interchanged)

$$B \otimes_{\max} (A/\mathcal{I}) = (B \otimes_{\max} A)/(B \otimes_{\max} \mathcal{I}),$$

and hence a fortiori we have a "canonical" $*$-homomorphism of norm ≤ 1

$$B \otimes_{\max} (A/\mathcal{I}) \to (B \otimes_{\min} A)/(B \otimes_{\min} \mathcal{I}).$$

By definition of the LLP, we have $\mathcal{B} \otimes_{\max} C = \mathcal{B} \otimes_{\min} C$. Thus taking $B = \mathcal{B}$ in (9.7), we find

$$\|Id_{\mathcal{B}} \otimes u : \mathcal{B} \otimes_{\min} C \to (\mathcal{B} \otimes_{\min} A)/(\mathcal{B} \otimes_{\min} \mathcal{I})\| \leq 1$$

which means, by Proposition 7.48, that u is locally 1-liftable. Thus (i) $\Rightarrow$ (ii).

(ii) $\Rightarrow$ (iii) $\Rightarrow$ (iv) are trivial. It remains to show that (iv) $\Rightarrow$ (i).

Assume (iv). Consider $t \in C \otimes \mathcal{B}$. We can assume $t \in E \otimes \mathcal{B}$ with $E \subset C$ finite dimensional. Let $\mathbb{F}$ be a free group such that $C \simeq C^*(\mathbb{F})/\mathcal{I}$ for some ideal $\mathcal{I} \subset C^*(\mathbb{F})$ (see Proposition 3.39). Let us denote $A = C^*(\mathbb{F})$, let $q : A \to A/\mathcal{I} \simeq C$ be the quotient map, and let $u : C \to A/\mathcal{I}$ be the identity mapping (here we identify C with $A/\mathcal{I}$) so that $u_{|E} : E \to A/\mathcal{I}$ is the natural inclusion. By assumption (iv) $u_{|E}$ admits a completely contractive lifting $v : E \to A$, so that $qv = u_{|E}$. We have then $t = (q \otimes Id_{\mathcal{B}})(v \otimes Id_{\mathcal{B}})(t)$ and hence

$$\|t\|_{C\otimes_{\max}\mathscr{B}} \le \|(v \otimes Id_{\mathscr{B}})(t)\|_{A\otimes_{\max}\mathscr{B}}$$

$$= \|(v \otimes Id_{\mathscr{B}})(t)\|_{A\otimes_{\min}\mathscr{B}} \text{ (by (9.5))}$$

$$\le \|t\|_{E\otimes_{\min}\mathscr{B}} = \|t\|_{C\otimes_{\min}\mathscr{B}} \text{ (by (1.8))}.$$

Hence we have proved (iv) $\Rightarrow$ (i).

Alternative proof: Combine Remark 7.41 and (iii)' $\Leftrightarrow$ (i) in Corollary 7.50.
□

See Corollary 9.47 for more on the local liftings in points (iii) and (iv) of Theorem 9.38.

Remark 9.39 (On the OLLP) The reader is probably wondering why we do not include c.b. maps in the list of mappings from C to $A/\mathcal{I}$ appearing in Theorem 9.38. The reason is the corresponding property of C is a much stronger one called the OLLP. An operator space X is said to have the OLLP if its universal C^*-algebra $C_u^*\langle X \rangle$ (in the sense of §2.7) has the LLP. The OLLP is studied in detail by Ozawa in [186]. See also [208, p. 278]. The spaces ℓ_∞^3 or M_n for $n \ge 3$ are the simplest examples of C^*-algebras with the LLP but failing the OLLP. Thus, when C is one of these, it is *not* true that any u in the unit ball of $CB(C, A/\mathcal{I})$ is locally liftable.

We now come to the general form of Kirchberg's Theorem 9.6:

Corollary 9.40 (Generalized Kirchberg Theorem) *For any LLP C^*-algebra C and any WEP C^*-algebra B, the pair (C, B) is nuclear.*

Proof Assume C LLP and B WEP, to prove that (C, B) is nuclear we simply invoke (9.4) and repeat the reasoning for (iv) $\Rightarrow$ (i) in Theorem 9.38 but with B in place of $\mathscr{B}$.
□

Corollary 9.41 *Let A, B, C be C^*-algebras with $C = A/\mathcal{I}$. Assume that C has the LLP. If (A, B) is nuclear, then (C, B) is nuclear.*
In particular, if a QWEP C^*-algebra C has the LLP then it has the WEP.

Proof Since the identity on $C = A/\mathcal{I}$ is locally 1-liftable (with respect to $A \to A/\mathcal{I}$), Proposition 7.48 implies that $B \otimes_{\min} C \to (B \otimes_{\min} A)/(B \otimes_{\min} \mathcal{I})$ is well defined and of norm 1 (and hence is an isomorphism). If (A, B) is nuclear

$$(B \otimes_{\min} A)/(B \otimes_{\min} \mathcal{I}) = (B \otimes_{\max} A)/(B \otimes_{\max} \mathcal{I}) = B \otimes_{\max} C$$

where at the last step we used (7.6). Thus (C, B) is nuclear.
The second assertion corresponds to the case when A has the WEP and $B = \mathscr{C}$.
□

It is important to emphasize that the local liftings considered in Theorem 9.38 are all for maps defined on a whole C^*-algebra C. In the next statement the maps that we want to lift are only defined on a finite-dimensional operator subspace or system $E \subset C$ and in general they do not contractively extend to C.

Proposition 9.42 (On approximate liftings) *Let C be a unital C^*-algebra C, $E \subset C$ a finite-dimensional linear subspace and $u : E \to A/\mathcal{I}$ a linear operator. Let $c \geq 0$ be a constant.*

(i) *Assume that $u : E \to A/\mathcal{I}$ admits a lifting $v_1 : E \to A$ with $\|v_1\| \leq c$ and another one $v_2 : E \to A$ that is c.p. Then, for any $\varepsilon > 0$, there is $v \in CP(E, A)$ with $\|v\| \leq c$ such that $\|qv - u\| \leq \varepsilon$.*

(ii) *Assume that E is a (finite-dimensional) operator system, that A is unital, that $u \in CP(E, A/\mathcal{I})$ is unital and admits a lifting $v_1 : E \to A$ with $\|v_1\|_{cb} \leq 1$. Then, for any $\varepsilon > 0$, there is a unital $v \in CP(E, A)$ such that $\|qv - u\| \leq \varepsilon$.*

Proof (i) Let σ_i be a quasi-central approximate unit as in §A.15. For any $w : E \to A$ we denote $w^i(x) = (1 - \sigma_i)^{1/2} w(x)(1 - \sigma_i)^{1/2}$. By (A.19) we have $(1 - \sigma_i)w - w_i \to 0$ pointwise on E, and hence $q(w - w^i) \to 0$ pointwise on E. This gives us that

$$u - qv_2^i = q(v_2 - v_2^i) \to 0 \quad \text{pointwise on } E.$$

Since $\dim(E) < \infty$ pointwise convergence implies norm convergence, and hence

$$\|u - qv_2^i\| \to 0. \tag{9.8}$$

Moreover, if w takes all its values in the ideal $\mathcal{I}$, then $w^i \to 0$ pointwise on E, and hence $\|w^i\| \to 0$. In particular this holds for the mapping $w = v_2 - v_1 : E \to \mathcal{I}$. Then $v_2^i = v_1^i + w^i$, $\|v_2^i\| \leq c + \|w^i\|$, and v_2^i is c.p. Let $v = v_2^i c(c + \|w^i\|)^{-1}$ so that $\|v\| \leq c$ and $\|v - v_2^i\| \leq \|w^i\|$. Since $u = qv_2$ we have

$$\|qv - u\| \leq \|qv - qv_2^i\| + \|qv_2^i - u\| \leq \|w^i\| + \|qv_2^i - u\|.$$

Thus we can choose i large enough so that $\|qv - u\| \leq \varepsilon$.

(ii) We first observe that by replacing v_1 by $(v_1 + v_{1*})/2$ we may assume that v_1 is self-adjoint. Choose any $f \in C_+^*$ so that $f(1) = 1$ (i.e. f is a state on C). Let $w_i : E \to A$ be defined by $w_i(x) = v_1^i(x) + f(x)\sigma_i$, so that $qw_i = qv_1^i$ and $w_i(1) - 1 = (1 - \sigma_i)^{1/2}(v_1(1) - 1)(1 - \sigma_i)^{1/2}$. Since $qv_1(1) = u(1) = 1$, we know $1 - v_1(1) \in \mathcal{I}$, and hence $1 - w_i(1) \to 0$. Note that

$$w_i(x) = ((1 - \sigma_i)^{1/2} \ \sigma_i^{1/2}) \begin{pmatrix} v_1(x) & 0 \\ 0 & f(x)1 \end{pmatrix} \begin{pmatrix} (1 - \sigma_i)^{1/2} \\ \sigma_i^{1/2} \end{pmatrix}$$

and hence $\|w_i\|_{cb} \le \max\{\|v_1\|_{cb}, \|f\|\} \le 1$.

Fix $\delta > 0$ (to be specified). By (9.8) we can choose i far enough so that $\|qv_1^i - u\| \le \delta$, and also $\|1 - w_i(1)\| \le \delta$. We now set

$$\varphi(x) = w_i(x) + f(x)(1 - w_i(1))$$

so that φ is self-adjoint, $\varphi(1) = 1$ and $\|\varphi - w_i\|_{cb} \le \delta$, and hence $\|\varphi\|_{cb} \le \|w_i\|_{cb} + \delta \le 1 + \delta$. Let $n = \dim(E)$. By Theorem 2.28, there is a unital $v \in CP(E, A)$ such that $\|\varphi - v\|_{cb} \le 8n\delta$. Note that $q(w_i - v_1^i) = 0$ and hence $\|q(\varphi - v_1^i)\| = \|q(\varphi - w_i)\| \le \|\varphi - w_i\| \le \delta$. Therefore

$$\|qv - u\| \le \|q(v - \varphi)\| + \|q(\varphi - v_1^i)\| + \|qv_1^i - u\| \le 8n\delta + 2\delta.$$

Choosing δ so that $8n\delta + 2\delta < \varepsilon$, we obtain v with the desired property. $\quad\square$

Remark 9.43 Actually the preceding proof in part (ii) works as well if we merely assume that u admits a family of liftings $v_i \in CB(E, A)$ such that $\inf_{i \in I} \|v_i\|_{cb} = 1$.

To end this section we sketch a proof of the stability of the LLP under free products, generalizing the fact that $C^*(\mathbb{F}_2) = C^*(\mathbb{Z}) * C^*(\mathbb{Z})$ has the LLP, that we saw in Theorem 9.6.

Theorem 9.44 *Let C_1, C_2 be unital C^*-algebras with the LLP. Then $C_1 * C_2$ has the LLP.*

Proof Let $E \subset C_1 * C_2$ be the linear span of $\{a_1 a_2 \mid (a_1, a_2) \in C_1 \times C_2\}$. By Theorem 9.8 applied with $E_1 = A_1 = \mathcal{B}$ and $E_2 = E$ with $A_2 = C_1 * C_2$, it suffices to prove that for any $n \ge 1$ and $t \in M_n(\mathcal{B} \otimes E)$ with $\|t\|_{M_n(\mathcal{B} \otimes_{\min} E)} \le 1$ we have $\|t\|_{M_n(\mathcal{B} \otimes_{\max}(C_1 * C_2))} \le 1$.

Since $M_n(\mathcal{B}) \simeq \mathcal{B}$, it suffices to prove this for $n = 1$, i.e. for $t \in \mathcal{B} \otimes E$ with $\|t\|_{\min} \le 1$. By Lemma 2.22 we can factorize t as $t = t_1 \odot t_2$ with $t_j \in \mathcal{B} \otimes C_j$ in the unit ball of $\mathcal{B} \otimes_{\min} C_j$ $(j = 1, 2)$. Let $\pi : \mathcal{B} \to B(H)$ and $\sigma : C_1 * C_2 \to B(H)$ be $*$-homomorphisms with commuting ranges. Equivalently, the restrictions $\sigma_j = \sigma_{|C_j} : C_j \to B(H)$ $(j = 1, 2)$ have range in $\pi(\mathcal{B})'$. Let $s_j = (\pi . \sigma_j)(t_j) \in B(H)$. Then $\|s_j\| \le \|t_j\|_{\max}$ $(j = 1, 2)$, and by the assumed commutations we have $(\pi . \sigma)(t) = (\pi . \sigma)(t_1 \odot t_2) = s_1 s_2$, and hence

$$\|(\pi . \sigma)(t)\| \le \|s_1\| \|s_2\| \le \|t_1\|_{\max} \|t_2\|_{\max}.$$

If C_1 and C_2 have the LLP this implies $\|(\pi . \sigma)(t)\| \le 1$ and hence $\|t\|_{\max} \le 1$. $\quad\square$

Remark 9.45 We can now show that (7.25) and (7.26) are independent of the choice of C_1, C_2 as long as they have the LLP.

Let D_1 be another C^*-algebra with LLP admitting a surjective morphism $r_1 : D_1 \to A_1 = D_1/\ker(r_1)$. Consider $v : A_2^* \to C_1$ as in (7.25). Let $E \subset C_1$ be the (finite-dimensional) range of v. By Theorem 9.38 the LLP of C_1 implies that the map from C_1 to $D_1/\ker(r_1)$ is locally 1-liftable. So there is a lifting $w : E \to D_1$ such that $r_1 w(e) = e$ for all $e \in E$ and $\|w\|_{cb} = 1$. Then $s = wv : A_2^* \to D_1$ is a (weak* continuous with finite rank) lifting of u_t, with $\|s\|_{cb} \leq \|v\|_{cb}$. Since we can reverse the roles of D_1 and C_1, this shows that the norm in (7.25) is the same if we compute it using C_1, q_1 or using D_1, r_1.

Reasoning as we just did for (7.25) one can show that (7.26) does not depend on the choice of either C_1 or C_2, as long as both have the LLP.

9.5 To lift or not to lift (global lifting)

In this section we will prove several global lifting theorems, notably the well-known Choi–Effros lifting Theorem [49], for c.p. maps defined on nuclear C^*-algebras taking values in a quotient C^*-algebra. We return to this theme in the later §21.2 where we will formally introduce and discuss briefly the lifting property (LP). We will also explain there why if Kirchberg's conjecture holds then LLP $\Rightarrow$ LP in the separable case. This is based on Theorem 9.46 (which heads this section) and the fact that WEP implies a certain restricted form of extension property (to be proved in §21.2): for any finite-dimensional subspace $E \subset C$ of an LLP C^*-algebra C, and any $\varepsilon > 0$ any $u \in CB(E, W)$ from E to a space W with WEP admits an extension $\widehat{u} \in CB(C, W)$ with $\|\widehat{u}\|_{cb} \leq (1 + \varepsilon)\|u\|_{cb}$.

Our first statement, due to Arveson, that says that nicely liftable maps on separable spaces are stable by pointwise limits is a priori surprising: one would expect a stronger limit to be required for this to hold. Roughly, the idea of the proof is similar to that of Lemma A.33.

Theorem 9.46 (Pointwise limits of liftables are liftable) *Let E be a separable operator space. Let $\mathcal{I} \subset A$ be an ideal in a C^*-algebra and let $q : A \to A/\mathcal{I}$ be the quotient map. Consider a bounded linear map $u : E \to A/\mathcal{I}$. Assume that there is a net of complete contractions $v_\gamma : E \to A$ such that $q v_\gamma \to u$ pointwise on E. Then:*

(i) *The map u admits a completely contractive lifting $v : E \to A$, i.e. we have $\|v\|_{cb} \leq 1$ and $qv = u$.*

(ii) *If E is a (separable) operator system and if u and all the v_γ's are c.p. (resp. unital and c.p.) then we can find a c.p. (resp. unital and c.p.) lifting $v : E \to A$ with $\|v\|_{cb} \leq 1$.*

Proof Let $\{x_k\}$ be a dense sequence in the unit ball of E. Assume given a complete contraction (c.c. in short) w_n such that

$$\|qw_n x_k - ux_k\| < 2^{-n} \qquad \forall k = 1, \ldots, n. \tag{9.9}$$

Moreover, if u is c.p. we assume that w_n is c.p.

We claim there is a map $w_{n+1} : E \to A$ (c.p. if u and w_n are c.p.) with $\|w_{n+1}\|_{cb} \leq 1$ such that

$$\|qw_{n+1} x_k - ux_k\| < 2^{-n-1} \qquad \forall k = 1, \ldots, n+1 \tag{9.10}$$

and

$$\|(w_{n+1} - w_n)x_k\| < 2^{-n+1} \qquad \forall k = 1, \ldots, n. \tag{9.11}$$

Taking this for granted, we may construct by induction a sequence (w_n) satisfying (9.10) and (9.11) for any n. Then $v(x) = \lim w_n(x)$ is the desired lifting. Indeed, $w_n(x)$ is Cauchy for any x in $\{x_1, x_2, \ldots\}$. Therefore $v(x) = \lim_n w_n(x)$ exists and satisfies $qv(x) = u(x)$. Since the norms $\|w_n\|$ are uniformly bounded (by 1), this still holds for any $x \in E$. We have $\|v\|_{cb} \leq \lim_n \|w_n\|_{cb} \leq 1$, and in the c.p. case v is c.p. as a limit of c.p. maps.

Thus it suffices to prove the claim. Let w_n be as in the claim. Going far enough in the net (v_γ), we can find $v : E \to A$ (that is also c.p. in the c.p. case) with $\|v\|_{cb} \leq 1$ such that

$$\|qvx_k - ux_k\| < 2^{-n-2} \qquad \forall k = 1, \ldots, n+1. \tag{9.12}$$

Let (σ_i) be an approximate unit in $\mathcal{I}$ as in §A.15.

For a suitable choice of i (to be specified later on) we will let

$$w_{n+1}(x) = \sigma_i^{1/2} w_n(x)\sigma_i^{1/2} + (1 - \sigma_i)^{1/2} v(x)(1 - \sigma_i)^{1/2}.$$

Note that since

$$w_{n+1}(x) = [\sigma_i^{1/2} \; (1 - \sigma_i)^{1/2}] \begin{bmatrix} w_n(x) & 0 \\ 0 & v(x) \end{bmatrix} \begin{bmatrix} \sigma_i^{1/2} \\ (1 - \sigma_i)^{1/2} \end{bmatrix}$$

we have

$$\|w_{n+1}\|_{cb} \leq \max\{\|w_n\|_{cb}, \|v\|_{cb}\} \leq 1.$$

Moreover, in the c.p. case, w_{n+1} is also c.p. By (A.23), for any given x and $\varepsilon > 0$, i can be chosen large enough so that

$$\|w_{n+1}(x) - [\sigma_i w_n(x) + (1 - \sigma_i)v(x)]\| < \varepsilon \tag{9.13}$$

and hence (since $\sigma_i w_n(x) - \sigma_i v(x) \in \mathcal{I}$) we have $\|q(w_{n+1}(x) - v(x))\| < \varepsilon$, which implies

$$\|qw_{n+1}(x) - u(x)\| < \varepsilon + \|qv(x) - u(x)\|.$$

So we can choose i large enough so that

$$\|qw_{n+1}(x_k) - u(x_k)\| < \varepsilon + 2^{-n-2} \qquad \forall k = 1, \ldots, n+1.$$

Moreover, using $w_{n+1}(x) - w_n(x) - w_{n+1}(x) - [\sigma_i w_n(x) + (1 - \sigma_i)w_n(x)]$ we find by (9.13)

$$\|w_{n+1}(x) - w_n(x)\| < \varepsilon + \|(1 - \sigma_i)[v(x) - w_n(x)]\|$$

hence for i large enough, by (A.20) we can ensure that

$$\|w_{n+1}(x) - w_n(x)\| < 2\varepsilon + \|q[v(x) - w_n(x)]\|$$
$$< 2\varepsilon + \|qv(x) - u(x)\| + \|qw_n(x) - u(x)\|.$$

Thus if we now make this last choice of i valid for any x in $\{x_1, \ldots, x_n\}$ and take $\varepsilon = 2^{-n-2}$ we obtain the announced estimates (9.10) and (9.11) for w_{n+1} (recalling (9.12) and (9.9)).

Lastly, the same proof yields the unital case. $\square$

We can now add a complement to Definition 7.46 concerning locally liftable unital c.p. maps.

Corollary 9.47 *Let C be a unital operator system and let $u : C \to A/\mathcal{I}$ be locally 1-liftable. If u is unital and c.p. then for any finite-dimensional operator system $E \subset C$ there is a unital c.p. map $v : E \to A$ such that $qv = u_{|E}$.*

Proof By (ii) in Proposition 9.42 applied to $u_{|E}$, for any $\varepsilon > 0$ there is a unital $v_\varepsilon \in CP(E, A)$ such that $\|qv_\varepsilon - u_{|E}\| \le \varepsilon$. By (ii) in Theorem 9.46, there is a unital $v \in CP(E, A)$ such that $qv = u_{|E}$. $\square$

Definition 9.48 Let $\lambda \ge 1$. We say that an operator space $E \subset B(H)$ has the λ-completely bounded approximation property (in short λ-CBAP) if there is a net of finite rank maps $u_i : E \to E$ with $\sup \|u_i\|_{cb} \le \lambda$ tending pointwise to the identity of E. We say that E has the CBAP if it satisfies this for some $1 \le \lambda < \infty$.

Corollary 9.49 *Let E be a separable operator space. Assume that E has the λ-CBAP for some $\lambda \ge 1$. Then any locally c-liftable $u \in CB(E, A/\mathcal{I})$ admits a (global) lifting $v \in CB(E, A)$ with $\|v\|_{cb} \le c\lambda$.*

Proof Let (u_i) be as in Definition 9.48. Let E_i be any finite-dimensional operator space such that $E_i \supset u_i(E)$. Since u locally c-lifts (see Definition 7.46), there is $w_i \in CB(E_i, A)$ with $\|w_i\|_{cb} \le c$ such that $qw_i = u$. Then $w_i u_i$ lifts $u u_i$ and $\|w_i u_i\|_{cb} \le c\lambda$. Clearly $qw_i u_i = u u_i$ tends to u pointwise. Then the net $v_i = (c\lambda)^{-1}w_i u_i$ is formed of complete contractions such that $qv_i \to (c\lambda)^{-1}u$ pointwise. Thus Theorem 9.46 gives us the conclusion. $\square$

Remark 9.50 There are purely Banach space analogues of Corollary 9.49. See [260] for a survey. For instance, for any *fixed* $n \geq 1$, there is an analogue for which the cb-norm of a map u is replaced everywhere by the norm of $u_n = Id_{M_n} \otimes u$. In the case $n = 1$ this reduces to the Banach space case. More explicitly, using $\|u_n\|$ in place of $\|u\|_{cb}$ in the definition of locally c-liftable maps, we get the definition of a locally (c, M_n)-liftable map $u : E \to A/\mathcal{I}$. Similarly, we define the M_n-BAP with constant $\lambda > 0$. Then if E is separable with the M_n-BAP with constant $\lambda > 0$, any locally (c, M_n)-liftable map $u : E \to A/\mathcal{I}$ admits a (global) lifting $v \in B(E, A)$ with $\|v_n\| \leq c\lambda$. This follows by a cosmetic adaptation of the proofs of Theorem 9.46 and Corollary 9.49.

Corollary 9.51 *Let E be a separable operator system. If E has the CPAP (in the sense of Definition 4.8), in particular if $\dim(E) < \infty$, then any unital and locally 1-liftable $u \in CP(E, A/\mathcal{I})$ admits a (global) lifting $v \in CP(E, A)$ with $\|v\| = \|u\|$.*

Proof Let (u_i) be a net of finite rank c.p. maps tending pointwise to Id_E. Recalling (1.27), we may assume $\|u\| = \|u\|_{cb} = 1$. Let E_i be any finite-dimensional operator system such that $E_i \supset u_i(E)$. Since u is locally 1-liftable (see Definition 7.46), by (ii) in Proposition 9.42 (or by Corollary 9.47) for any $\varepsilon > 0$ there is $w_i \in CP(E_i, A)$ with $\|w_i\| \leq 1$ such that $\|qw_i - u_{|E_i}\| < \varepsilon$. Then the composition $v_i = w_i u_i$ satisfies $\|qv_i - uu_i\| = \|(qw_i - u_{|E_i})u_i\| < \varepsilon$. Since $uu_i \to u$ pointwise (and ε is arbitrary), we can arrange so that the net (v_i) is such that $qv_i \to u$ pointwise on E. Then Theorem 9.46 allows us to conclude. □

Remark 9.52 Actually it suffices that u itself be approximable in the following sense: it is enough to assume that u is the pointwise limit of a net of finite rank maps $u_i \in CP(E, E)$ completely positively and completely contractively factorized through some $M_{n(i)}$. To check this, one uses the lifting property of the $M_{n(i)}$'s (see Remark 9.54).

We can now deduce the celebrated Choi–Effros lifting theorem.

Theorem 9.53 (Choi–Effros lifting theorem) *Let C be a unital separable nuclear C^*-algebra. Then any unital c.p. map $u : C \to A/\mathcal{I}$ (into an arbitrary unital quotient C^*-algebra) admits a c.p. lifting $v : C \to A$ with $\|v\| = \|u\|$.*

Proof Since C is nuclear, it has the LLP and (by Corollary 7.12) the CPAP, so by (i) ⇔ (iii) in Theorem 9.38 this can be viewed as a particular case of the preceding Corollary. □

Remark 9.54 (Lifting property of M_n) In particular, the preceding Theorem shows that any unital c.p. map $u : M_n \to A/\mathcal{I}$ admits a unital c.p. lifting $v : M_n \to A$.

We wish to emphasize again (see Remark 9.39) that the c.b. variant of the preceding lifting theorem is generally not valid. In general one cannot lift a complete contraction $u : M_n \to A/\mathcal{I}$ to a complete contration $v : M_n \to A$.

9.6 Linear maps with WEP or LLP

It might clarify certain features of the theory to consider the WEP and the LLP for linear maps, whence the following definitions, where we restrict for simplicity to the unit ball.

Definition 9.55 Let $E \subset B(H)$ be an operator space and B a C^*-algebra. Let us say that a linear mapping $u : E \to B$ is WEP if

$$\|Id_{\mathscr{C}} \otimes u : \mathscr{C} \otimes_{\min} E \to \mathscr{C} \otimes_{\max} B\| \leq 1.$$

The following is a corollary of Theorem 7.4.

Corollary 9.56 *In the situation of Definition 9.55, the following are equivalent:*

(i) *u is WEP.*
(ii) *For some H, there are $v \in CB(E, B(H))$ and $w \in D(B(H), B^{**})$ such that $wv = i_B u$ and $\|v\|_{cb}\|w\|_{dec} \leq 1$.*

Proof We have isometrically $\mathscr{C} \otimes_{\min} E \subset \mathscr{C} \otimes_{\min} B(H)$ and $\mathscr{C} \otimes_{\min} B(H) = \mathscr{C} \otimes_{\max} B(H)$ by Theorem 9.6. Thus we have isometrically $\mathscr{C} \otimes_{\min} E \subset \mathscr{C} \otimes_{\max} B(H)$, so this is indeed an immediate consequence of Theorem 7.4 and Remark 7.5. $\qquad\square$

This allows us to formulate the interesting variant of Theorem 9.22 that folllows.

Theorem 9.57 *Let $u : A \to B$ be a unital c.p. map between unital C^*-algebras. The following are equivalent.*

(i) *u is WEP.*
(ii) *The mapping $i_B u : A \to B^{**}$ factorizes completely positively and contractively through some $B(H)$, i.e. there are $v \in CP(A, B(H))$ and $w \in CP(B(H), B^{**})$ such that $wv = i_B u$ and $\|v\|\|w\| \leq 1$.*
(iii) *For some H, there are $v \in CB(A, B(H))$ and $w \in CP(B(H), B^{**})$ such that $wv = i_B u$ and $\|v\|_{cb}\|w\| \leq 1$.*

Proof Assume (i). Assume $A \subset B(H)$. By Corollary 9.56 u admits an extension $w : B(H) \to B^{**}$ with $\|w\|_{dec} \leq 1$. Since $w(1) = u(1) = 1$ and $\|w\|_{cb} \leq \|w\|_{dec} \leq 1$ (see (6.6)) w must be c.p. by Theorem 1.35. Thus we obtain (ii). (ii) $\Rightarrow$ (iii) is obvious. Assume (iii). We have by (4.30)

$$\|Id_{\mathscr{C}} \otimes i_B u : \mathscr{C} \otimes_{\min} A \to \mathscr{C} \otimes_{\max} B^{**}\|$$
$$\leq \|w\| \|Id_{\mathscr{C}} \otimes v : \mathscr{C} \otimes_{\min} A \to \mathscr{C} \otimes_{\max} B(H)\|.$$

By Theorem 9.6 this is

$$= \|w\| \|Id_{\mathscr{C}} \otimes v : \mathscr{C} \otimes_{\min} A \to \mathscr{C} \otimes_{\min} B(H)\|$$

and by (1.8) the latter is $\leq \|w\| \|v\|_{cb} \leq 1$. Lastly, since $B \subset B^{**}$ is max-injective (see Proposition 7.26), we have

$$\|Id_{\mathscr{C}} \otimes i_B u : \mathscr{C} \otimes_{\min} A \to \mathscr{C} \otimes_{\max} B^{**}\|$$
$$= \|Id_{\mathscr{C}} \otimes u : \mathscr{C} \otimes_{\min} A \to \mathscr{C} \otimes_{\max} B\|,$$

and we obtain (i). $\qquad\square$

To emphasize the parallelism WEP/LLP, we also introduce the LLP for linear maps.

Definition 9.58 Let $E \subset B(H)$ be an operator space and B a C^*-algebra. Let us say that a linear mapping $u : E \to B$ is LLP if

$$\|Id_{\mathscr{B}} \otimes u : \mathscr{B} \otimes_{\min} E \to \mathscr{B} \otimes_{\max} B\| \leq 1.$$

From this definition it is clear that $u : E \to B$ is LLP if and only if for any finite-dimensional subspace $E_1 \subset E$ the restriction $u_{|E_1} : E_1 \to B$ is LLP. Thus the assumption of finite dimensionality in the next result is not too restrictive.

Proposition 9.59 *Let $E \subset B(H)$ be finite dimensional and let $u : E \to B$ be a linear map into a C^*-algebra B. The following are equivalent:*

(i) *u is LLP.*
(ii) *There are $v \in CB(E, \mathscr{C})$ and $w \in D(\mathscr{C}, B)$ such that $wv = u$ and $\|v\|_{cb} \|w\|_{dec} \leq 1$.*

Proof Assume (i). Remark 7.49 and Theorem 7.48 show that (ii) holds with $\mathscr{C}$ replaced by $C^*(\mathbb{F})$ where $\mathbb{F}$ is a large enough free group so that $B = C^*(\mathbb{F})/\mathcal{I}$ for some ideal $\mathcal{I} \subset C^*(\mathbb{F})$. Then we can replace $C^*(\mathbb{F})$ by $\mathscr{C}$ using Remark 3.6.

Conversely, assume (ii). Since the fundamental pair $(\mathscr{B}, \mathscr{C})$ is nuclear, (i) is easily derived using (3) for v and (6.13) for w. $\qquad\square$

9.7 QWEP

We start with two consequences of Theorem 7.33 for QWEP C^*-algebras.

Corollary 9.60 *Let A, B be C^*-algebras, with B unital, for which there exists $\varphi \in CP(A, B)$ such that $\varphi(B_A) = B_B$. If a C^*-algebra C is such that (A, C) is a nuclear pair, the same is true for the pair (D_φ, C). In particular, if A is WEP, or merely QWEP, then B is QWEP.*

Proof The first assertion is obvious (by Remark 7.20). Thus, by our definition of WEP, if A is WEP, D_φ is WEP and hence B is QWEP. If A is a quotient of a WEP C^*-algebra, we may compose with the quotient mapping, then we are reduced to the case when A is WEP. $\qquad\square$

For further reference, we spell out a particular case:

Corollary 9.61 *Let $D \subset A$ be a C^*-subalgebra for which there is a c.p. projection $P : A \to D$ with $\|P\| = 1$. If A is QWEP, then D is QWEP.*

We now turn to the stability properties of the class of QWEP C^*-algebras. We first state an immediate consequence of Proposition 9.34.

Corollary 9.62 *For any family $\{A_i \mid i \in I\}$ of QWEP C^*-algebras the direct sum $\left(\oplus \sum_{i \in I} A_i\right)_\infty$ also is QWEP.*

Remark 9.63 Let $\{A_i \mid i \in I\}$ be a family, directed by inclusion, of C^*-subalgebras of $B(H)$ and let $A \subset B(H)$ be the norm closure of their union. If all the A_i's are WEP then A is WEP. Indeed, it suffices to show that $\|t\|_{\max} = \|t\|_{\min}$ for any $t \in (\cup A_i) \otimes \mathscr{C}$. But then $t \in A_i \otimes \mathscr{C}$ for some i and hence

$$\|t\|_{A \otimes_{\max} \mathscr{C}} \le \|t\|_{A_i \otimes_{\max} \mathscr{C}} = \|t\|_{A_i \otimes_{\min} \mathscr{C}} = \|t\|_{A \otimes_{\min} \mathscr{C}}.$$

Proposition 9.64 *Let $D \subset D_1$ be a C^*-subalgebra of a quotient one $D_1 = A_1/\mathcal{I}$. Let $q : A_1 \to D_1$ be the quotient map. Let $A = q^{-1}(D)$ so that $\mathcal{I} = q^{-1}(\{0\}) \subset A \subset A_1$ and $D = A/\mathcal{I}$.*

If the inclusion $D \subset D_1$ is max-injective, then the inclusion $A \subset A_1$ is also max-injective.

Proof Note $\mathcal{I} = \ker(q) = \ker(q_{|A})$. Let C be any C^*-algebra. We must show that the natural map $J : C \otimes_{\max} A \to C \otimes_{\max} A_1$ is injective. Let $x \in C \otimes_{\max} A$ be in its kernel. Then, if $D \subset D_1$ is max-injective, $(Id_C \otimes q_{|A})(x)$ must vanish in $C \otimes_{\max} D$, as the following commuting diagram shows.

$$
\begin{array}{ccc}
C \otimes_{\max} A & \xrightarrow{\ J\ } & C \otimes_{\max} A_1 \\
{\scriptstyle Id_C \otimes q_{|A}}\big\downarrow & & \big\downarrow {\scriptstyle Id_C \otimes q} \\
C \otimes_{\max} D & \longrightarrow & C \otimes_{\max} D_1
\end{array}
$$

Therefore by (7.6), $x \in C \otimes_{\max} \mathcal{I}$, but since $\mathcal{I}$ is an ideal in A_1, $\mathcal{I} \subset A_1$ is max-injective, and hence J restricted to $C \otimes_{\max} \mathcal{I}$ (which obviously coincides with the inclusion $C \otimes_{\max} \mathcal{I} \to C \otimes_{\max} A_1$) is injective. Therefore $x = 0$. This proves the proposition. $\square$

Corollary 9.65 *Let $D \subset D_1$ be a C^*-subalgebra of a QWEP C^*-algebra D_1. If the inclusion $D \subset D_1$ is max-injective, then D is QWEP.*

Proof Assume $D_1 = A_1/\mathcal{I}$ with A_1 WEP. Let $A = q^{-1}(D)$ where $q : A_1 \to D_1$ is again the quotient map. By Proposition 9.64 the inclusion $A \subset A_1$ is max-injective. By Remark 9.13 A has the WEP and hence $D = A/\mathcal{I}$ is QWEP. $\square$

Theorem 9.66 *A C^*-algebra B is QWEP if and only if its bidual B^{**} is also QWEP.*

Proof Assume B^{**} QWEP. By Corollary 7.27, the inclusion $B \to B^{**}$ is max-injective, and hence B is QWEP by Corollary 9.65.

To prove the converse, assume B QWEP and $B^{**} \subset B(H)$ (as a von Neumann subalgebra). Let I be the net of neighborhoods of 0 for the weak* topology of $B(H)$. Let $\mathcal{U}$ be an ultrafilter refining this net (see Remark A.6). By the weak* density of B_B, for any $x \in B_{B^{**}}$

$$\forall i \in I \ \exists x(i) \in (x + i) \cap B_B. \tag{9.14}$$

This implies that $x(i) \to x$ (weak*). For any $i \in I$ we set $A_i = B$. We define a linear map

$$\varphi : \left(\oplus \sum_{i \in I} A_i \right)_\infty \to B^{**}$$

by setting, for any $x = (x_i)$ in the unit ball of $\left(\oplus \sum_{i \in I} A_i \right)_\infty$

$$\varphi(x) = \lim_{\mathcal{U}} x_i,$$

where the limit is meant in the weak* sense. Clearly, φ is positive (meaning positivity preserving), and similarly (using $M_n \left(\left(\oplus \sum_{i \in I} A_i \right)_\infty \right) \cong \left(\oplus \sum_{i \in I} M_n(A_i) \right)_\infty$) we see that $\varphi_n = Id_{M_n} \otimes \varphi$ is positive for any n. Thus φ is c.p. Let $A = \left(\oplus \sum_{i \in I} A_i \right)_\infty$. By Corollary 9.62, A is QWEP. By (9.14), $\varphi(B_A) = B_{B^{**}}$. This shows that $\varphi \in CP(A, B_{B^{**}})$ satisfies the assumption of Corollary 9.60. The latter ensures that B^{**} is QWEP.

Actually, using the density of B_B in $B_{B^{**}}$ for the so-called strong* operator topology, one obtains directly a $*$-homomorphism in place of φ, and then B^{**} appears as a quotient of A. $\square$

Theorem 9.67 (Characterization of QWEP) *The following properties of a unital C*-algebra D are equivalent:*

(i) *D is QWEP.*
(ii) *For any LLP C*-algebras C and C_1, any decomposable map $u : C \to D$ with $\|u\|_{dec} \leq 1$, satisfies*

$$\|Id_{C_1} \otimes u : C_1 \otimes_{min} C \to C_1 \otimes_{max} D\| \leq 1.$$

(ii)' *Same as (ii) for $C = C_1 = \mathscr{C}$.*
(iii) *There is a free group $\mathbb{F}$ and a surjective unital *-homomorphism $\pi : C^*(\mathbb{F}) \to D$ such that*

$$\|Id_{\mathscr{C}} \otimes \pi : \mathscr{C} \otimes_{min} C^*(\mathbb{F}) \to \mathscr{C} \otimes_{max} D\| \leq 1,$$

in other words such that the quotient mapping $\pi : C^*(\mathbb{F}) \to D$ is WEP.

Proof Assume (i), so that $D = A/\mathcal{I}$ with A WEP. Let $q : A \to A/\mathcal{I}$ be the quotient map. Let u be as in (ii). Let $t \in C_1 \otimes C$. Let $E \subset C$ be a finite-dimensional subspace such that $t \in C_1 \otimes E$. By Theorem 9.38, if C is LLP u is locally 1-liftable, so there is $u^E \in CB(E, A)$ with $\|u^E\|_{cb} \leq 1$ such that $qu^E = u_{|E}$, and hence

$$\|(Id_{C_1} \otimes u^E)(t)\|_{C_1 \otimes_{min} A} \leq \|t\|_{C_1 \otimes_{min} C}.$$

By Corollary 9.40, the pair (C_1, A) is nuclear. Therefore we have

$$\|(Id_{C_1} \otimes u^E)(t)\|_{C_1 \otimes_{max} A} = \|(Id_{C_1} \otimes u^E)(t)\|_{C_1 \otimes_{min} A}$$

and hence, since (obviously) $(Id_{C_1} \otimes u)(t) = (Id_{C_1} \otimes u_{|E})(t)$, we find

$$\|(Id_{C_1} \otimes u)(t)\|_{C_1 \otimes_{max} A} = \|(Id_{C_1} \otimes qu^E)(t)\|_{C_1 \otimes_{max} A}$$
$$\leq \|(Id_{C_1} \otimes u^E)(t)\|_{C_1 \otimes_{min} A} \leq \|t\|_{C_1 \otimes_{min} C}.$$

This shows (i) $\Rightarrow$ (ii). (ii) $\Rightarrow$ (ii)' is trivial. (ii) $\Rightarrow$ (iii) is obvious since any D is a quotient of $C^*(\mathbb{F})$ for some $\mathbb{F}$ and the latter has the LLP by (9.5). In addition, (ii)' $\Rightarrow$ (iii) is easy to check using Lemma 3.8 since any $t \in \mathscr{C} \otimes C^*(\mathbb{F})$ lies in $\mathscr{C} \otimes E$ for some finite-dimensional $E \subset C^*(\mathbb{F})$.

Assume (iii). Let $C = C^*(\mathbb{F})$. By Theorem 9.57, we can write $\pi = wv$ with $v \in CP(C, B(H))$, $w \in CP(B(H), D^{**})$ of unit norm. By Theorem 8.1 w admits a weak* continuous extension $\ddot{w} \in CP(B(H)^{**}, D^{**})$ with $\|\ddot{w}\| = \|w\|$. Since π is onto, we know by Lemma A.33 that $\pi(B_C) = B_D$, and hence $w(B_{B(H)}) \supset B_D$. This implies by weak* density that $\ddot{w}(B_{B(H)^{**}}) = B_{D^{**}}$. Since $B(H)^{**}$ is QWEP (see Theorem 9.66), Corollary 9.60 shows that D^{**} is also QWEP. Since $D \subset D^{**}$ is max-injective (see Corollary 7.27), Corollary 9.65 shows that D is QWEP, and we obtain (i). □

Remark 9.68 Let G be a discrete group such that $C^*(G)$ has the LLP. If M_G is QWEP then the natural map $C^*(G) \otimes C^*(G) \to M_G \otimes M_G$ extends to $*$-homomorphism

$$C^*(G) \otimes_{\min} C^*(G) \to M_G \otimes_{\max} M_G.$$

A fortiori, G has the factorization property in the sense of Definition 7.36 (see also §11.8).

Indeed, this follows from (ii) in Theorem 9.67 with $C = C_1 = C^*(G)$ and $u = \dot{Q}_G : C^*(G) \to M_G$. Note that this remark typically applies when G is a free group (see Corollary 12.23).

Taking $C = D$ and $u = Id_D$ in (ii)' from Theorem 9.67, we immediately derive:

Corollary 9.69 *If a QWEP C^*-algebra D has the LLP then it has the WEP.*

Corollary 9.70 *Let A and D be C^*-algebras. Assume that there is a factorization of the identity of A of the form*

$$Id_A : A \xrightarrow{v} D \xrightarrow{w} A$$

where v, w are decomposable maps. If D is QWEP then A also is QWEP.

Proof Let $\lambda = \|v\|_{dec} \|w\|_{dec}$. Assume D is QWEP. Thus D satisfies (ii) in Theorem 9.67. Let $\pi : C^*(\mathbb{F}) \to A$ be a quotient map. By (6.13), $\|Id_{\mathscr{C}} \otimes \pi : \mathscr{C} \otimes_{\min} C^*(\mathbb{F}) \to \mathscr{C} \otimes_{\max} A\| \leq \lambda$. But since $Id_{\mathscr{C}} \otimes \pi$ is a $*$-homomorphism, the latter norm must be $= 1$. Thus A satisfies (iii) in Theorem 9.67 and hence is QWEP. $\square$

Remark 9.71 In Corollary 9.70 it suffices to assume that there are nets of decomposable maps $w_i : A \to D$ and $v_i : D \to A$ with $\sup_i \|v_i\|_{dec} \|w_i\|_{dec} < \infty$ such that $v_i w_i$ tends pointwise to $i_A : A \to A^{**}$. Indeed, the preceding proof leads to $\|Id_{\mathscr{C}} \otimes i_A \pi : \mathscr{C} \otimes_{\min} C^*(\mathbb{F}) \to \mathscr{C} \otimes_{\max} A^{**}\| \leq \lambda$, and Corollary 7.27 allows us to conclude.

Theorem 9.72 (Another characterization of QWEP) *The following properties of a unital C^*-algebra D are equivalent:*

(i) *D is QWEP.*

(ii) *For some H there is an embedding (as a von Neumann subalgebra) $D^{**} \subset B(H)^{**}$ admitting a contractive projection $P : B(H)^{**} \to D^{**}$.*

(iii) *There is a factorization of the identity of D^{**} of the form*

$$Id_{D^{**}} : D^{**} \xrightarrow{v} B(H)^{**} \xrightarrow{w} D^{**}$$

where v, w are decomposable maps.

Proof Assume $D = A/\mathcal{I}$ with A WEP. By Theorem 9.31 the algebra A^{**} certainly satisfies (ii). But by (A.37) we have $A^{**} \simeq \mathcal{I}^{**} \oplus D^{**}$. Therefore D^{**} also satisfies (ii). This shows (i) $\Rightarrow$ (ii). Then (ii) $\Rightarrow$ (iii) is obvious (recalling that P is automatically c.p. by Theorem 1.45) and (iii) $\Rightarrow$ (i) follows from Corollary 9.70 and Theorem 9.66. $\qquad\square$

Remark 9.73 (Warning on a trap) It is important to understand the difference between the last statement and Theorem 9.31. For that purpose, we urge the reader to look into Remark 9.33.

Theorem 9.74 *Let* $\{A_i \mid i \in I\}$ *a family, directed by inclusion, of* C^*-*subalgebras of* $B(H)$, *let* $A \subset B(H)$ *(resp.* $N \subset B(H)$) *be the norm (resp. weak*) closure of their union. If all the* A_i's *are QWEP then* A *and* N *are QWEP.*

Proof Assume A unital for simplicity. Using the unitary groups $U(A_i)$ it is easy to find a free group $\mathbb{F}$ and a surjective $*$-homomorphism $\pi : C^*(\mathbb{F}) \to A$ so that there is a family, directed by inclusion, of free subgroups G_i with union $= \mathbb{F}$, such that π restricted to $C^*(G_i)$ realizes A_i as a quotient of $C^*(G_i)$. Assume all the A_i's are QWEP. Then arguing as in Remark 9.63 one shows that π satifies (iii) in Theorem 9.67 and hence A is QWEP. We may view N as generated by A. Then by Theorem 9.66 A^{**} is QWEP, and since N is a quotient of A^{**} (see Theorem 8.1 (i)), it is also QWEP. One can also prove directly that N is QWEP by a modification of the proof of Theorem 9.66. $\quad\square$

Remark 9.75 (QWEP is separably determined) Let $A_1 \subset A$ and $A_2 \subset A$ be separable C^*-subalgebras of a C^*-algebra A. Then the C^*-algebra generated by $A_1 \cup A_2$ is still separable. Thus the family of separable C^*-subalgebras of A forms a directed net for inclusion and the union of all of them is equal to A. Thus, by Theorem 9.74 if all separable C^*-subalgebras are QWEP then A is QWEP. A fortiori, if all weak* separable subalgebras of a von Neumann algebra M are QWEP then M is QWEP.

9.8 Notes and remarks

As we already mentioned, Takesaki introduced the notion of nuclearity for C^*-algebras (initially under a different name), and of course implicitly also for pairs. The term was inspired by Grothendieck's work on nuclear locally convex spaces. Later on the subject was deepened by the works of Lance [165], Choi–Effros and Kirchberg [45, 154] (who independently proved that nuclearity is equivalent to the CPAP, see §10.2) and Effros–Lance [79]. A major step was taken thanks to Connes's work on injective factors [61] that allowed Choi and Effros to complete the proof that a C^*-algebra A is nuclear if and only if its bidual A^{**} is an injective von Neumann algebra (see §8.3). For most of this initial period, not much interest was devoted to what we call "nuclear pairs," except for Lance's question whether the nuclearity of (A, A^{op}) implies that A is nuclear, answered by Kirchberg in [155]. The latter constructed a counterexample, and at the same time started a deep study of nuclear pairs that led him to identify the WEP and the LLP for a C^*-algebra A as the nuclearity of the pairs respectively $(A, \mathscr{C})$ and $(A, \mathscr{B})$ (which in this volume we choose as defining WEP and LLP). Kirchberg's Theorem 9.6 (that tells us that $(\mathscr{C}, \mathscr{B})$ is a nuclear pair) was proved in [156], but we followed the simpler proof from [204]. We encourage the reader to compare the latter proof with the original one! Theorem 9.18 is due to Choi. The origin of the WEP is Lance's paper [165]. Our terminology is slightly different, as explained in Remark 9.23. Major advances were made in Kirchberg's [155]. Our presentation of the stability properties of the WEP (due to Lance and Kirchberg) in §9.3 is much influenced by Ozawa's [189]. Corollary 9.28 already appears in [77].

Lance raised the question whether there are examples of embeddings $A \subset B(H)$ admitting a weak expectation from $B(H)$ to $M = \overline{A}^\sigma$ but nevertheless not admitting a contractive projection from $B(H)$ onto M (in other words noninjective). Such examples were proposed by Blackadar in [23, 24]. More examples (where M is a free group factor and A a specially chosen weak*-dense Popa C^*-subalgebra) appear in later papers by Brown and Dykema [37, 38]. The definition of the LLP and the results in §9.4 are due to Kirchberg in [155]. The stability of the LLP under free products in Theorem 9.44 is due to the author [204]. A subsequent paper by Boca [31] contains variations on the same theme. The results of §9.5 are due to Kirchberg but are based on Arveson's ideas in [13]. Theorem 9.53 is due to Choi and Effros [49].

In §9.6 and 9.7 again the main ideas come from Kirchberg's work, but as usual we try to put forward properties of linear maps. Theorem 9.72 is easy to deduce from Ozawa's viewpoint in [189], but apparently was not formulated yet.

10

Exactness and nuclearity

We already gave a brief introduction to nuclear C^*-algebras in §4.2 and we already announced some of their main properties, like their characterization by the CPAP, which will be proved in the present chapter. Since, in the separable case, exact C^*-algebras are just C^*-subalgebras of nuclear ones, it is not surprising that many features of nuclear C^*-algebras and their approximation properties have analogues for exactness. We try to emphasize this "parallelism."

10.1 The importance of being exact

A C^*-algebra A is called exact if the phenomenon described in (7.16) cannot happen when we take the minimal tensor product with A. More precisely:

Definition 10.1 A C^*-algebra A is called exact if for any C^*-algebra B and any ideal $\mathcal{I} \subset B$ so that $B/\mathcal{I}$ is a C^*-algebra, the sequence

$$\{0\} \to \mathcal{I} \otimes_{\min} A \to B \otimes_{\min} A \to (B/\mathcal{I}) \otimes_{\min} A \to \{0\}$$

is exact. Equivalently, this holds if and only if the kernel of the mapping $B \otimes_{\min} A \to (B/\mathcal{I}) \otimes_{\min} A$ coincides with $\mathcal{I} \otimes_{\min} A$. In other words, this reduces to:

$$\text{the } *-\text{homomorphism } \frac{B \otimes_{\min} A}{\mathcal{I} \otimes_{\min} A} \to (B/\mathcal{I}) \otimes_{\min}$$

$$A \text{ is injective (or equivalently isometric).} \quad (10.1)$$

Even more explicitly this boils down (equivalently) to:

$$\forall x \in (B/\mathcal{I}) \otimes A \quad \|x\|_{(B \otimes_{\min} A)/(\mathcal{I} \otimes_{\min} A)} \leq \|x\|_{(B/\mathcal{I}) \otimes_{\min} A}. \quad (10.2)$$

Remark 10.2 Let $/B\mathcal{I}$ be a quotient C^*-algebra by an ideal $\mathcal{I}$. For brevity, let us say that A is $(B, \mathcal{I})$-exact if (10.1) or (10.2) holds. Let $B_0 \subset B$ be a

C^*-subalgebra, let $\mathcal{I}_0 = \mathcal{I} \cap B_0$ so that $\mathcal{I}_0 \subset \mathcal{I}$ and $B_0/\mathcal{I}_0 \subset B/\mathcal{I}$. Assume that there is a completely contractive projection $P : B \to \mathcal{I}$ such that $P(B_0) = \mathcal{I}_0$. It is easy to check that if A is $(B,\mathcal{I})$-exact then it is also $(B_0,\mathcal{I}_0)$-exact: indeed, we have isometric embeddings $(B_0/\mathcal{I}_0) \otimes_{\min} A \subset (B/\mathcal{I}) \otimes_{\min} A$ and $(B_0 \otimes_{\min} A)/(\mathcal{I}_0 \otimes_{\min} A) \subset (B \otimes_{\min} A)/(\mathcal{I} \otimes_{\min} A)$. For the latter we use the projection $P \otimes Id_A : B \otimes_{\min} A \to \mathcal{I} \otimes_{\min} A$.

The classical Calkin algebra $\mathscr{Q} = \mathscr{B}/\mathscr{K}$ is of special interest (here $\mathscr{K}$ is the ideal of compact operators in $\mathscr{B}$). Let (N_n) be any increasing sequence of integers, and let $B_0 = \left(\oplus \sum M_{N_n} \right)_\infty$ and let $\mathcal{I}_0 = \{b = (b_n) \in B_0 \mid \lim \|b_n\| = 0\}$. We have a classical block diagonal embedding $B_0 \subset \mathscr{B}$ such that $\mathcal{I}_0 = B_0 \cap \mathscr{K}$. Let $q_0 : B_0 \to B_0/\mathcal{I}_0$ denote the quotient map. It is easy to check that for any $b = (b_n) \in B_0$ we have $\|q_0(b)\| = \limsup \|b_n\|$. More generally, the same easy argument shows that for any finite-dimensional operator space E and any $b = (b_n) \in \left(\oplus \sum M_{N_n} \otimes_{\min} E \right)_\infty \simeq B_0 \otimes_{\min} E$ (for the last $\simeq$ see (1.14)), we have

$$\|(q_0 \otimes Id_E)(b)\|_{(B_0 \otimes_{\min} E)/(\mathcal{I}_0 \otimes_{\min} E)} = \limsup_{n \to \infty} \|b_n\|_{M_{N_n} \otimes_{\min} E}. \quad (10.3)$$

The main point of the next result is that exactness restricted to the Calkin algebra implies the general exactness.

Theorem 10.3 *The following properties of a C^*-algebra A are equivalent:*

(i) *The C^*-algebra A is $(\mathscr{B},\mathscr{K})$-exact.*

(ii) *For any finite-dimensional subspace $E \subset A$ and any bounded sequence of linear maps $u_n : E_n \to E$, where E_n are arbitrary operator spaces, we have*

$$\limsup_{n \to \infty} \|u_n\|_{cb} \le \sup_k \limsup_{n \to \infty} \|Id_{M_k} \otimes u_n : M_k(E_n) \to M_k(E)\|. \quad (10.4)$$

(iii) *For any finite-dimensional subspace $E \subset A$ and any $\varepsilon > 0$ there are an integer N, a subspace $\widetilde{E} \subset M_N$ and $u : \widetilde{E} \to E$ such that $\|u\|_{cb} < 1 + \varepsilon$ and $\|u^{-1}\|_{cb} \le 1$.*

(iv) *The C^*-algebra A is exact.*

Proof Assume (i). Choose $\varepsilon_n > 0$ such that $\varepsilon_n \to 0$. For any n there is N_n such that $\|u_n\|_{cb} \ge \|Id_{M_{N_n}} \otimes u_n\| > \|u_n\|_{cb} - \varepsilon_n$. We may clearly adjust so that (N_n) is increasing. For any n there is $x_n \in B_{M_{N_n}(E_n)}$ such that

$$\|u_n\|_{cb} \ge \|(Id_{M_{N_n}} \otimes u_n)(x_n)\|_{M_{N_n} \otimes_{\min} E} > \|u_n\|_{cb} - \varepsilon_n.$$

Let $t_n = (Id_{M_{N_n}} \otimes u_n)(x_n) \in M_{N_n} \otimes_{\min} E$. We use the notation in Remark 10.2 for B_0 and $\mathcal{I}_0$. Since the ranks of the u_n's are bounded we must have

$\sup \|u_n\|_{cb} < \infty$ (see Remark 1.56), and hence $(t_n) \in \left(\oplus \sum M_{N_n} \otimes_{\min} E\right)_\infty \simeq B_0 \otimes_{\min} E$. Let $\dot{t} = (q_0 \otimes Id_E)((t_n)) \in (B_0/\mathcal{I}_0) \otimes_{\min} E \subset (B_0/\mathcal{I}_0) \otimes_{\min} A$. By Remark 10.2, we may assume that A is $(B_0, \mathcal{I}_0)$-exact. Recalling Lemma 4.26, this gives us

$$\|\dot{t}\|_{(B_0 \otimes_{\min} E)/(\mathcal{I}_0 \otimes_{\min} E)} \leq \|\dot{t}\|_{(B_0/\mathcal{I}_0) \otimes_{\min} E}. \tag{10.5}$$

By (10.3) we have on one hand

$$\|\dot{t}\|_{(B_0 \otimes_{\min} E)/(\mathcal{I}_0 \otimes_{\min} E)} = \limsup \|t_n\|_{M_{N_n} \otimes_{\min} E} = \limsup \|u_n\|_{cb},$$

and on the other hand we claim that

$$\|\dot{t}\|_{(B_0/\mathcal{I}_0) \otimes_{\min} E} \leq \sup_k \limsup_{n \to \infty} \|Id_{M_k} \otimes u_n : M_k(E_n) \to M_k(E)\|.$$

The last two inequalities together with (10.5) imply (10.4), which proves (ii).

We now prove the claim. By (1.11) (recalling (1.6)) we have

$$\|\dot{t}\|_{(B_0/\mathcal{I}_0) \otimes_{\min} E} = \sup_k \sup\{\|(Id_{B_0/\mathcal{I}_0} \otimes v)(\dot{t})\|_{(B_0/\mathcal{I}_0) \otimes_{\min} M_k} \mid v \in B_{CB(E, M_k)}\}, \tag{10.6}$$

and since $(B_0/\mathcal{I}_0) \otimes_{\min} M_k = (B_0 \otimes_{\min} M_k)/(\mathcal{I}_0 \otimes_{\min} M_k)$ (indeed M_k is trivially exact), we find for any k and $v \in CB(E, M_k)$ by (10.3)

$$\|(Id_{B_0/\mathcal{I}_0} \otimes v)(\dot{t})\|_{(B_0/\mathcal{I}_0) \otimes_{\min} M_k} = \limsup_n \|(Id_{M_{N_n}} \otimes vu_n)(x_n)\|_{M_{N_n} \otimes_{\min} M_k}$$

$$\leq \limsup_n \|vu_n\|_{cb} \tag{10.7}$$

and since $\|vu_n\|_{cb} = \|Id_{M_k} \otimes vu_n\|$ by (1.18), we have

$$\limsup_n \|vu_n\|_{cb} = \limsup_n \|Id_{M_k} \otimes vu_n\| \leq \limsup_n \|Id_{M_k} \otimes u_n\|,$$

and the claim now follows from (10.6) and (10.7).

Assume (ii). Let $E \subset A$ with $\dim(E) < \infty$. Since E is separable, we may assume $E \subset B(H)$ (completely isometrically) with H separable. Let $H_n \subset H$ ($n \geq 1$) be an increasing sequence of subspaces with $\dim(H_n) < \infty$ whose union is dense in H. Then (as earlier for (1.9)) for any k and any $x \in M_k(E)$ we have $\|x\|_{M_k(E)} = \lim_n \uparrow \|Id_{M_k} \otimes v_n(x)\|$ where $v_n(e) = P_{H_n} e_{|H_n}$ ($e \in E$). Since the unit sphere S_E of E is compact (and $|\|v_n(x)\| - \|v_n(y)\|| \leq \|x - y\| \; \forall x, y \in E$), we have by Ascoli's theorem $\inf_{x \in S_E} \|v_n(x)\| \uparrow 1$ and hence $v_n : E \to v_n(E)$ is invertible for all n large enough. For such n, let $u_n = v_n^{-1} : v_n(E) \to E$ be its inverse. We have $\|u_n\| \downarrow 1$. Similarly, $\lim_n \downarrow \|Id_{M_k} \otimes u_n\| = 1$ for any k. By (ii) we have $\lim_n \|u_n\|_{cb} \leq 1$, and hence (iii) follows: choosing n large enough so that $\|u_n\|_{cb} < 1 + \varepsilon$ we may take $N = \dim(H_n)$ and $u = u_n$.

Assume (iii). Let $B/\mathcal{I}$ be a quotient C^*-algebra. To prove (10.2) it suffices to show that for any $E \subset A$ with $\dim(E) < \infty$ and any $\varepsilon > 0$ we have

$\|(B/\mathcal{I}) \otimes_{\min} E \to B \otimes_{\min} E/(\mathcal{I} \otimes_{\min} E)\| \leq 1 + \varepsilon$. Let $\widetilde{E} \subset M_N$ and u be as in (iii). Since M_N is exact we have isometrically $(B/\mathcal{I}) \otimes_{\min} M_N = (B \otimes_{\min} M_N)/(\mathcal{I} \otimes_{\min} M_N)$ and hence also isometrically $(B/\mathcal{I}) \otimes_{\min} \widetilde{E} = (B \otimes_{\min} \widetilde{E})/(\mathcal{I} \otimes_{\min} \widetilde{E})$ by Lemma 4.26. Therefore using $Id_E = u \circ u^{-1}$ we find $\|(B/\mathcal{I}) \otimes_{\min} E \to (B \otimes_{\min} E)/(\mathcal{I} \otimes_{\min} E)\| \leq \|u^{-1}\|_{cb} \|u\|_{cb} \leq 1 + \varepsilon$. Thus we obtain (iv) and (iv) $\Rightarrow$ (i) is trivial. □

Remark 10.4 By the results of the preceding §7.2 (see Proposition 7.15) it is clear that any nuclear C^*-algebra is exact.

Remark 10.5 (On the stability properties of exactness) By Lemma 4.26 (applied here with the roles of A and B exchanged) if A is exact then any C^*-subalgebra $D \subset A$ is also exact. This is also immediate from (iii) $\Leftrightarrow$ (iv) in Theorem 10.3. Incidentally, this shows that $\mathcal{B}$ is not exact.

Kirchberg proved that exactness also passes to quotient C^*-algebras but this lies much deeper (see [39, p. 297] for a detailed proof). However, as we will prove in §18.3, exactness is not stable under extensions.

By a simple iteration argument, one shows that the minimal tensor product of two exact C^*-algebras is exact. In sharp contrast, exactness is not preserved by the maximal tensor product. Indeed, by Theorem 4.11 there is an embedding $C^*(\mathbb{F}_n) \subset C^*_\lambda(\mathbb{F}_n) \otimes_{\max} C^*_\lambda(\mathbb{F}_n)$ and $C^*_\lambda(\mathbb{F}_n)$ is exact (see Remark 10.21) while $C^*(\mathbb{F}_n)$ is not exact when $n > 1$ (see Proposition 7.34).

Remark 10.6 (Embeddings in the Cuntz algebra) Kirchberg [158, 162] obtained a series of striking results on exact C^*-algebras culminating with his outstanding proof that any separable exact C^*-algebra A can be embedded (as a C^*-subalgebra) in a nuclear C^*-algebra, namely the Cuntz algebra O_2. Moreover, he showed that if A is nuclear the embedding $A \subset O_2$ can be made so that there is a contractive c.p. projection from O_2 onto A. See [3] for a nice presentation of this subject. Most of the tools for this topic, being related to C^*-algebraic K-theory, are quite far from the subject of the present volume, which explains why we do not discuss this fundamental embedding any further. Nevertheless, Theorem 10.3 as well as the next statement are important steps (which we fully prove) on the way to this achievement. In any case, with the applications related to the CBAP in Theorem 10.18, all this already shows that exactness is a certain form of "subnuclearity."

For (min $\to$ max)-tensorizing c.p. maps, see Theorem 7.11 (which is proved in the next section).

Theorem 10.7 *A C^*-algebra $A \subset B(H)$ is exact if and only if the inclusion mapping $j_A : A \to B(H)$ is* (min $\to$ max)-*tensorizing.*

Proof Assume A exact. Let B be any unital C^*-algebra. Then, by Proposition 3.39, for some free group $\mathbb{F}$, B is a quotient of $C = C^*(\mathbb{F})$, so that $B = C/\mathcal{I}$. Then the following $*$-homomorphism is contractive

$$A \otimes_{\min} B = \frac{A \otimes_{\min} C}{A \otimes_{\min} \mathcal{I}} \to \frac{B(H) \otimes_{\min} C}{B(H) \otimes_{\min} \mathcal{I}}$$

but since the pair $(C, B(H))$ is nuclear (see (9.4)) we have

$$\frac{B(H) \otimes_{\min} C}{B(H) \otimes_{\min} \mathcal{I}} = \frac{B(H) \otimes_{\max} C}{B(H) \otimes_{\max} \mathcal{I}}$$

and by (7.4) this last space coincides with $B(H) \otimes_{\max} (C/\mathcal{I}) = B(H) \otimes_{\max} B$. Therefore $A \to B(H)$ is (min $\to$ max)-tensorizing.

Conversely, if the latter inclusion is (min $\to$ max)-tensorizing, then for any C and any quotient $C/\mathcal{I}$ we have

$$A \otimes_{\min} (C/\mathcal{I}) \to B(H) \otimes_{\max} (C/\mathcal{I}) = \frac{B(H) \otimes_{\max} C}{B(H) \otimes_{\max} \mathcal{I}}$$

and a fortiori

$$A \otimes_{\min} (C/\mathcal{I}) \to \frac{B(H) \otimes_{\min} C}{B(H) \otimes_{\min} \mathcal{I}}.$$

Thus if we denote as before $Q[E] = \frac{E \otimes_{\min} C}{E \otimes_{\min} \mathcal{I}}$ we have a contractive map $A \otimes_{\min} (C/\mathcal{I}) \to Q[B(H)]$. But by Lemma 7.43 we know that the norm induced by $Q[B(H)]$ on $A \otimes (C/\mathcal{I})$ is the norm of $Q[A]$. Therefore we obtain $\|A \otimes_{\min} (C/\mathcal{I}) \to Q[A]\| \le 1$ which means that A is exact.

Alternative route: (iii) in Theorem 10.3 implies that j_A factors as in (iii) in Theorem 7.10.   □

Corollary 10.8 *If $A \subset B(H)$ is exact, any complete contraction $u : A \to W$ with values in a WEP C^*-algebra W is (min $\to$ max)-tensorizing.*

Proof If W is WEP, the inclusion $i_W : W \to W^{**}$ factors through some $B(K)$ via c.p. complete contractions. By the extension property of $B(K)$ (see Theorem 1.18) we have a factorization of $i_W u$ of the form $A \to B(H) \to W^{**}$, showing that $i_W u$ is (min $\to$ max)-tensorizing. Therefore (see Remark 7.28) u itself is (min $\to$ max)-tensorizing.   □

Applying this to $u = Id_A$ (and Recalling Remark 10.4) we immediately obtain

Corollary 10.9 *A C^*-algebra A is nuclear if and only if A is exact and has the WEP.*

Recall that we denote by $\mathscr{Q} = \mathscr{B}/\mathscr{K}$ (where $\mathscr{B} = B(\ell_2), \mathscr{K} = K(\ell_2)$) the classical "Calkin algebra." The analogue of the last corollary for the LLP involves $\mathscr{Q}$.

Corollary 10.10 *For a C^*-algebra A, the pair $(A, \mathscr{Q})$ is nuclear if and only if A is exact and has the LLP. In that case, (A, B) is nuclear for any QWEP C^*-algebra B.*

Proof If A is exact we have

$$A \otimes_{\min} \mathscr{Q} = (A \otimes_{\min} \mathscr{B})/(A \otimes_{\min} \mathscr{K}). \tag{10.8}$$

If A has the LLP, $(A, \mathscr{B})$ is nuclear by our definition of the LLP. Therefore

$$(A \otimes_{\min} \mathscr{B})/(A \otimes_{\min} \mathscr{K}) = (A \otimes_{\max} \mathscr{B})/(A \otimes_{\max} \mathscr{K}).$$

By (7.6) (exactness of the max-tensor product):

$$(A \otimes_{\max} \mathscr{B})/(A \otimes_{\max} \mathscr{K}) = A \otimes_{\max} \mathscr{Q}, \tag{10.9}$$

and hence $(A, \mathscr{Q})$ is nuclear. The same argument (with Corollary 9.40) shows that (A, B) is nuclear for any QWEP C^*-algebra B. Conversely, assume $(A, \mathscr{Q})$ nuclear. Then (10.8) holds by (10.9) (since the max-norm dominates the min-norm on $A \otimes \mathscr{B}$). By Theorem 10.3, (10.8) implies exactness. By (10.8) and (10.9) we have

$$(A \otimes_{\min} \mathscr{B})/(A \otimes_{\min} \mathscr{K}) = (A \otimes_{\max} \mathscr{B})/(A \otimes_{\max} \mathscr{K}),$$

and also $A \otimes_{\min} \mathscr{K} = A \otimes_{\max} \mathscr{K}$. One easily deduces from this that $A \otimes_{\min} \mathscr{B} = A \otimes_{\max} \mathscr{B}$, which means, by our definition of the LLP, that A has it. $\square$

Remark 10.11 Kirchberg conjectured (see Proposition 13.1) that LLP $\Rightarrow$ WEP, and also that all C^*-algebras are QWEP (see Proposition 13.1). By the preceding corollaries, this would show that $(A, \mathscr{Q})$ nuclear $\Rightarrow A$ nuclear.

We now return to the connection between exactness and local reflexivity (see Remark 8.34 for important additional information).

Proposition 10.12 *Property C' is equivalent to exactness.*

Proof Assume that A has property C'. Let $\mathcal{I} \subset B$ be an ideal. A preliminary observation is that by (A.37) the kernel of the natural mapping $B^{**} \otimes_{\min} A \to (B^{**}/\mathcal{I}^{**}) \otimes_{\min} A$ is equal to $\mathcal{I}^{**} \otimes_{\min} A$ (see also Corollary 7.52). Property C' tells us that we have an isometry $B^{**} \otimes_{\min} A \to (B \otimes_{\min} A)^{**}$. Let $Z = \ker[q \otimes Id_A : B \otimes_{\min} A \to (B/\mathcal{I}) \otimes_{\min} A]$. Viewing B as embedded in B^{**}, we view $B \otimes_{\min} A \subset B^{**} \otimes_{\min} A$. We also may view $\mathcal{I}^{**} \subset B^{**}$. Then our preliminary observation shows us that $Z \subset \mathcal{I}^{**} \otimes_{\min} A$. By property C' a fortiori $Z \subset \mathcal{I}^{**} \otimes_{\min} A \subset (\mathcal{I} \otimes_{\min} A)^{**}$. Therefore, any $z \in Z$ is in the

$\sigma((B \otimes_{\min} A)^{**}, (B \otimes_{\min} A)^*)$-closure of a bounded net in $\mathcal{I} \otimes_{\min} A$. But since both Z and $\mathcal{I} \otimes_{\min} A$ are included in $B \otimes_{\min} A$, this means that any element $z \in Z$ is the weak limit in $B \otimes_{\min} A$ of a bounded net in $\mathcal{I} \otimes_{\min} A$, and hence by (Mazur's) Theorem A.9 the norm limit of another such net, which implies that $z \in \mathcal{I} \otimes_{\min} A$. Thus $Z = \mathcal{I} \otimes_{\min} A$ and A is exact.

Conversely, assume $A \subset B(H)$ exact. Let B be a C^*-algebra. By Theorem 10.7 we have

$$\|A \otimes_{\min} B^{**} \to B(H) \otimes_{\max} B^{**}\| = 1.$$

By (8.12) (and (4.8)) we have $\|B(H) \otimes_{\max} B^{**} \to (B(H) \otimes_{\max} B)^{**}\| = 1$ and a fortiori $\|B(H) \otimes_{\max} B^{**} \to (B(H) \otimes_{\min} B)^{**}\| = 1$. Therefore $\|A \otimes_{\min} B^{**} \to (B(H) \otimes_{\min} B)^{**}\| = 1$. But since the inclusion $A \otimes_{\min} B \to B(H) \otimes_{\min} B$ is isometric, so is $(A \otimes_{\min} B)^{**} \to (B(H) \otimes_{\min} B)^{**}$. Therefore we conclude $\|A \otimes_{\min} B^{**} \to (A \otimes_{\min} B)^{**}\| = 1$, which means A has property C'. $\qquad \square$

Remark 10.13 The argument described in Remark 8.32 to show that local reflexivity (i.e. property C'') passes to quotients also shows the same for property C. However, it does not seem to work for property C'. Thus it does not lead to a proof of the same for exactness.

10.2 Nuclearity, exactness, approximation properties

The aim of this section is to give a reasonably direct proof that a C^*-algebra A is nuclear if and only if it has the approximation properties mentioned in Theorem 7.11 (but not proved yet). We expand on the same theme including several more general statements in the next section. The proofs will require the dual description of the maximal C^*-norm given by Theorem 6.15.

The next statement is the analogue for general linear maps of the characterization of (min $\to$ max)-tensorizing maps stated (but not proved yet) in Theorem 7.10.

Theorem 10.14 *Let λ be a positive constant. Consider two C^*-algebras A and B and an operator subspace $E \subset A$. Let $u : E \to B$ be a linear mapping. The following assertions are equivalent.*

(i) *For any C^*-algebra C, $Id_C \otimes u$ defines a bounded linear map from $C \otimes_{\min} E$ to $C \otimes_{\max} B$ with norm $\leq \lambda$. In other words, u is (min $\to$ max)-tensorizing with constant λ.*

(ii) *Same as (i) with $C = C^*\langle F \rangle$ for all finite-dimensional operator subspaces $F \subset E$.*

(iii) *For any finite-dimensional subspace $F \subset E$, the restriction $u_{|F}$ admits, for any $\varepsilon > 0$, a factorization of the form $F \xrightarrow{V} M_n \xrightarrow{w} B$ with $\|V\|_{cb}\|w\|_{dec} \leq \lambda + \varepsilon$.*

(iv) *There is a net of finite rank maps $u_i : E \to B$ admitting factorizations through matrix algebras of the form*

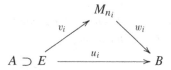

with $\|v_i\|_{cb}\|w_i\|_{dec} \leq \lambda$ such that $u_i = w_i v_i$ converges pointwise to u.

(v) *There is a net $u_i : A \to B$ of finite rank maps with $\sup \|u_i\|_{dec} \leq \lambda$ such that the restrictions $u_{i|E}$ tend pointwise to u.*

Proof (i) $\Rightarrow$ (ii) is trivial. Assume (ii). Let $F \subset E$ be an arbitrary finite-dimensional subspace and let $t_F \in F^* \otimes E$ be the tensor associated to the inclusion $j_F : F \to E$. Let $C = C^*\langle F^* \rangle$. By (ii) we have $\|(Id_C \otimes u)(t_F)\|_{C \otimes_{\max} B} \leq \lambda \|t_F\|_{C \otimes_{\min} E}$. But by the injectivity of the min-norm (see Remark 1.10) and by (2.15), we have

$$\|t_F\|_{C \otimes_{\min} E} = \|t_F\|_{F^* \otimes_{\min} E} = \|j_F\|_{CB(F,E)} = 1.$$

Hence we have $\|(Id_C \otimes u)(t_F)\|_{C \otimes_{\max} B} \leq \lambda$. By (6.18) and Remark 6.17, this implies that, for any $\varepsilon > 0$, there is a factorization of $u_{|F}$ of the following form

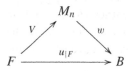

with $\|V\|_{cb} \leq 1$ and $\|w\|_{dec} \leq \lambda + \varepsilon$. This shows (ii) $\Rightarrow$ (iii). Assume (iii). By the extension property of M_n (see Theorem 1.18), we can extend V to a mapping $v : E \to M_n$ with $\|v\|_{cb} \leq \|V\|_{cb}$. Thus if we take for index set I the set of all finite-dimensional subspaces $F \subset E$ (directed by inclusion) we obtain nets $v_i : E \to M_{n_i}$ and $w_i : M_{n_i} \to B$ such that $\|w_i v_i(x) - u(x)\| \to 0$ for all x in E and such that (after a suitable renormalization) $\sup\{\|v_i\|_{cb}\|w_i\|_{dec}\} \leq 1$. This completes the proof that (iii) $\Rightarrow$ (iv). We may clearly assume (by the extension Theorem 1.18) that v_i is extended to A with the same cb norm, thus recalling (6.7) and (6.9), (iv) $\Rightarrow$ (v) is immediate.

Finally, the proof that (v) $\Rightarrow$ (i) is an immediate consequence of Proposition 6.13. $\qquad\square$

Recall that a C^*-algebra A is said to have the CPAP (for completely positive approximation property) if the identity on A is in the pointwise closure of the set of finite rank c.p. maps on A. To derive the known results on this approximation property, the following will be useful (here we follow closely [208, ch. 12]).

Lemma 10.15 *Let $E \subset B(H)$ be a finite-dimensional operator system and let A be a C^*-algebra. Consider a unital self-adjoint mapping $u : E \to A$ associated to a tensor $t \in E^* \otimes A$. Fix $\varepsilon > 0$. Then if $\delta(t) < 1 + \varepsilon$, we can decompose u as $u = \varphi - \psi$ with φ, ψ c.p. such that $\|\psi\| \le \varepsilon$ and φ admits for some n a factorization of the form*

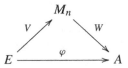

where V, W are c.p. maps with

$$\|V\| \le 1 + \varepsilon \text{ and } \|W\| \le 1.$$

Proof By the definition of the norm δ (see §6.2) and by Theorem 1.50, we can assume that $u = wv$ where $v : E \to M_n$ satisfies

$$\forall x \in E \quad v(x) = v_1^* \pi(x) v_2,$$

where $\pi : E \to B(\widehat{H})$ is the restriction of a $*$-homomorphism, v_1, v_2 are operators in $B(\ell_2^n, \widehat{H})$ with $\|v_1\| = \|v_2\| < (1 + \varepsilon)^{1/2}$, and $w : M_n \to A$ is defined by $w(e_{ij}) = a_i b_j$ with $\|\sum a_i a_i^*\| = \|\sum b_j^* b_j\| < 1$. Let b be a column matrix with entries $(b_1, \dots, b_n)$ and a^* a row matrix with entries $(a_1, \dots, a_n)$, so that, in matrix notation, we have $u(x) = a^* \cdot v_1^* \pi(x) v_2 \cdot b$.

Since u is self-adjoint we have $u(x) = u(x^*)^* = b^* v_2^* \pi(x) v_1 a$ and hence (by "polarization")

$$u(x) = \varphi(x) - \psi(x)$$

with

$$\varphi(x) = (1/4)[(v_1 a + v_2 b)^* \pi(x)(v_1 a + v_2 b)] \text{ and}$$
$$\psi(x) = (1/4)[(v_1 a - v_2 b)^* \pi(x)(v_1 a - v_2 b)].$$

Clearly φ and ψ are c.p. and $1 = u(1) = \varphi(1) - \psi(1)$ so that $\varphi(1) \ge 1$ and $\|\varphi(1)\| \le \|v_1 a + v_2 b\|^2 / 4 \le (1 + \varepsilon)$, which implies that $\|\varphi(1) - 1\| \le \varepsilon$. Thus we obtain (see (1.27)) $\|\psi\| = \|\psi(1)\| \le \varepsilon$. It remains to show that φ admits the announced factorization. Let (again in matrix notation)

$$V(x) = \frac{1}{2} \begin{pmatrix} v_1^* \pi(x) v_1 & v_1^* \pi(x) v_2 \\ v_2^* \pi(x) v_1 & v_2^* \pi(x) v_2 \end{pmatrix} = \frac{1}{2} \begin{pmatrix} v_1^* \\ v_2^* \end{pmatrix} \pi(x)(v_1, v_2).$$

Clearly V is c.p. from E to M_{2n} and $\|V\| \leq (1/2)(\|v_1\|^2 + \|v_2\|^2) \leq 1 + \varepsilon$. Moreover, if we define $W : M_{2n} \to A$ by $W(t) = \frac{1}{2}(a^* \ b^*)t\binom{a}{b}$ then W is c.p. with $\|W\| \leq 1$ and we have $\varphi(x) = W(V(x))$. $\qquad\square$

In the case of c.p. maps, Theorem 10.14 becomes:

Corollary 10.16 *Let $u : A \to B$ be a unital c.p. map between two unital C^*-algebras. The following assertions are equivalent.*

(i) *The mapping u is (min $\to$ max)-tensorizing.*
(ii) *For any C^*-algebra C, $Id_C \otimes u$ defines a completely positive linear map from $C \otimes_{\min} A$ to $C \otimes_{\max} B$ with norm $= 1$.*
(iii) *Same as (ii) with $C = C^*\langle E^* \rangle$ for all finite-dimensional operator systems $E \subset A$.*
(iv) *There is a net of finite rank maps (u_i) admitting factorizations through matrix algebras of the form*

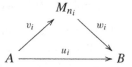

with c.p. maps v_i and w_i satisfying $\|v_i\| \|w_i\| \leq 1$ such that $u_i = w_i v_i$ converges pointwise to u.
(v) *There is a net $u_i : A \to B$ of finite rank c.p. maps that tends pointwise to u.*

Proof (i) $\Rightarrow$ (ii) was already observed in Remark 7.9 and (ii) $\Rightarrow$ (iii) is trivial. Assume (iii). Let E be a finite-dimensional operator system inside A. Let $B = C^*\langle E^* \rangle$. Let $t_E \in E^* \otimes A$ be the tensor associated to the inclusion map $j_E : E \to A$, so that $\|t_E\|_{\min} = \|j_E\|_{cb} = 1$. By (ii) we have $\|(Id_C \otimes u)(t_E)\|_{\max} \leq 1$. Note that the linear map from E to B associated to the tensor $(Id_C \otimes u)(t_E)$ is nothing but $u_{|E}$. Hence, by (6.18) (6.20) and by Lemma 10.15, for any $\varepsilon > 0$ we can write $u_{|E} = \varphi - \psi$ with $\varphi = WV$ as in Lemma 10.15. By the extension property of c.p. maps we may as well assume that V is a c.p. mapping of norm $\leq 1 + \varepsilon$ from A to M_n. Then, using the net (directed by inclusion) formed by the finite-dimensional operator systems $E \subset A$, and letting $\varepsilon \to 0$ we obtain (iv) (after a suitable renormalization of V). Then (iv) $\Rightarrow$ (v) is trivial.

Lastly, assume (v). Note that since $\|u_i\| = \|u_i(1)\|$ and $\|u_i(1)\| \to \|u(1)\| = 1$, we have "automatically" $\|u_i\| \to 1$. By (6.16) and (i) in Lemma 6.5, we have for any C^*-algebra C

$$\|Id_C \otimes u_i\|_{C \otimes_{\min} A \to C \otimes_{\max} B} \leq \|u_i\|,$$

and hence since $u_i \to u$ pointwise

$$\|Id_C \otimes u\|_{C \otimes_{\min} A \to C \otimes_{\max} B} \leq 1.$$

This shows that (v) $\Rightarrow$ (i).   □

We now recover the most classical characterization of nuclear C^*-algebras, due independently to Choi–Effros and Kirchberg [45, 154]:

Theorem 10.17 (Nuclear $\Leftrightarrow$ CPAP) *The following properties of a unital C^*-algebra A are equivalent:*

(i) *A is nuclear.*
(ii) *There is a net of finite rank maps of the form $A \xrightarrow{v_i} M_{n_i} \xrightarrow{w_i} A$ where v_i, w_i are c.p. maps with $\|v_i\| \leq 1$, $\|w_i\| \leq 1$, that tends pointwise to the identity.*
(iii) *A has the CPAP.*

Proof This follows from Corollary 10.16 with $A = B$ and $u = Id_A$.   □

Just like Theorem 10.7 the next statement is a characterization of exactness for C^*-algebras, that is parallel to that of nuclearity. We just replace the identity on A namely Id_A by the inclusion $j_A : A \to B(H)$. This already suggests to think of exactness as some sort of *"subnuclearity,"* at least in a somewhat local sense, but actually Kirchberg proved that separable exact C^*-algebras *globally embed* in nuclear ones (see Remark 10.6).

Note however, in sharp contrast, that the preceding paralellism does not extend to (iv) in Theorem 10.18, since the CBAP for Id_A does not imply nuclearity (see Remark 10.21).

Theorem 10.18 *The following properties of a unital C^*-algebra $A \subset B(H)$ are equivalent:*

(i) *A is exact.*
(ii) *There is a net of finite rank maps of the form $A \xrightarrow{v_i} M_{n_i} \xrightarrow{w_i} B(H)$ where v_i, w_i are c.p. maps with $\|v_i\| \leq 1$, $\|w_i\| \leq 1$, that tends pointwise to the inclusion mapping $j_A : A \subset B(H)$.*
(iii) *There is a net of finite rank c.p. maps $u_i : A \to B(H)$ tending pointwise to $j_A : A \subset B(H)$ (in other words, we might say that "j_A has the CPAP").*
(iv) *There is a net of finite rank c.b. maps $u_i : A \to B(H)$ with $\sup_i \|u_i\|_{cb} < \infty$ that tends pointwise to $j_A : A \subset B(H)$ (we might say that "j_A has the CBAP").*

Proof (i) $\Leftrightarrow$ (ii) $\Leftrightarrow$ (iii) follows from Corollary 10.16 with $B = B(H)$ and $u = j_A$.

(iii) $\Rightarrow$ (iv) is trivial. Assume (iv). By (6.9) $\|u_i\|_{dec} = \|u_i\|_{cb}$. By (i) $\Leftrightarrow$ (v) in Theorem 10.14, j_A is (min $\to$ max)-tensorizing with constant $\lambda = \sup_i \|u_i\|_{cb}$. But actually since j_A is a $*$-homomorphism, this automatically is true also with a constant $= 1$ (see Proposition A.24). Thus A is exact (meaning (i) holds) by Theorem 10.7. $\qquad\square$

Remark 10.19 [On unitizations] A C^*-algebra A is nuclear (resp. exact) if and only if its unitization is nuclear (resp. exact). Indeed, A is an ideal in its unitization $\widetilde{A}$ so that $\widetilde{A}/A = \mathbb{C}$. By Proposition 7.19, an ideal in a nuclear C^*-algebra is nuclear. Thus $\widetilde{A}$ nuclear implies A nuclear. The analogue for exactness is obvious (see Remark 10.5). The converses are easy and left as an exercise to the reader. Moreover, using an approximate unit of A, one easily shows that the CPAP of $\widetilde{A}$ implies that of A. This remark allows us to extend Theorem 10.17 to the nonunital case. A similar reasoning applies for exactness and Theorem 10.18.

Remark 10.20 [On the CBAP] Let $\Lambda(E)$ be the smallest λ for which E has the λ-CBAP in the sense of Definition 9.48. By Theorem 10.18 any C^*-algebra with the CBAP must be exact. We refer the reader to Cowling and Haagerup's [64] for an important example of a sequence of groups (G_n) (with property (T)) for which $\Lambda(C^*_\lambda(G_n)) = n + 1$ $(n = 1, 2, \dots)$.

Remark 10.21 [Exactness for free groups] By Haagerup's classical paper [103], for any free group $\mathbb{F}$, the reduced C^*-algebra $C^*_\lambda(\mathbb{F})$ has the 1-CBAP, and a fortiori is exact. Thus unlike the CPAP the CBAP does *not* imply nuclearity.

In sharp contrast, the full C^*-algebra $C^*(\mathbb{F}_n)$ is *not* exact whenever $n \geq 2$. The latter fact follows from (7.8) and (7.9).

Remark 10.22 (On the weak* CBAP) Let $\lambda > 0$ be a constant. We say that a von Neumann algebra M has the weak* λ-CBAP if there is a net of normal finite rank maps $T_i : M \to M$ with $\|T_i\|_{cb} \leq \lambda$ that tend weak* to the identity. Haagerup proved in 1986 (see [108]) that for a discrete group G, the λ-CBAP for $C^*_\lambda(G)$ is equivalent to the weak* λ-CBAP for M_G, and moreover when this holds we may always find the net (T_i) formed of finite rank multipliers (see Proposition 3.25 and Remark 3.29). In particular the latter net is then formed of maps such that $T_i(M_G) \subset C^*_\lambda(G)$. By Remark 10.21, M_G has the weak* 1-CBAP for any free group G. See [41, 110] for further developments on approximation properties for groups. Because of the equivalent reformulation with multipliers, the groups G for which $C^*_\lambda(G)$ has the CBAP are now called weakly amenable. Haagerup (1986, unpublished) proved that $SL_3(\mathbb{Z})$ is not weakly amenable (see also [190]), even though $C^*_\lambda(G)$ is exact when $G = SL_3(\mathbb{Z})$ as well as when G is any closed discrete subgroup of a connected Lie group. This last fact is attributed to A. Connes in [155, p. 453]. By [99] the same holds when G is any discrete linear group. See [109, 113, 114, 164] for more recent breakthroughs in this direction.

Remark 10.23 Haagerup also proved in [103] a different kind of multiplier approximation property for free groups, now called the Haagerup property, for which we refer to [42].

10.3 More on nuclearity and approximation properties

In this section we present a result (due to Junge and Le Merdy) that refines Theorem 6.15 using an original application of Kaplansky's density theorem (see Theorem A.47) discovered by Junge [137], as follows. We merely sketch the proof.

Lemma 10.24 *Let A, B be arbitrary C^*-algebras. Then any c.b. map $\theta : B^* \to A$ can be approximated in the point-norm topology by a net of weak*-continuous finite rank maps $\theta_i : B^* \to A$ with $\|\theta_i\|_{cb} \le \|\theta\|_{cb}$.*

Sketch Let $A^{**} \subset B(H)$, and $B^{**} \subset B(K)$ be embeddings (as von Neumann subalgebras). Let $M = A^{**}\overline{\otimes}B^{**}$ denote the von Neumann algebra generated by $A^{**} \otimes B^{**}$ in $B(H \otimes_2 K)$. The space $CB(B^*, A^{**})$ can be identified isometrically with M in a natural way (see e.g. [208, p. 49] for details). Let $t \in A^{**}\overline{\otimes}B^{**}$ be the tensor associated to $i_A\theta : B^* \to A^{**}$ (recall $i_A : A \to A^{**}$ is the canonical inclusion). Then, by Kaplansky's density Theorem A.47, there is a net (t_i) in $A \otimes B$ with $\|t_i\|_{\min} \le \|t\|$ such that $t_i \ \sigma(M, M_*)$-tends to t. Let $\theta_i : B^* \to A$ be the finite rank map associated to t_i. We have $\|\theta_i\|_{cb} = \|t_i\|_{\min} \le \|t\| = \|\theta\|_{cb}$. Moreover, for any ξ in B^*, $\theta_i(\xi)$ must $\sigma(A^{**}, A^*)$-tend to $\theta(\xi)$. But since $\theta_i(\xi)$ and $\theta(\xi)$ both lie in A, this means that $\theta_i(\xi)$ tends to $\theta(\xi)$ weakly in A. Passing to suitable convex hulls, we obtain (by Mazur's Theorem, A.9) a net (θ_i) such that, for any ξ, $\theta_i(\xi)$ tends to $\theta(\xi)$ in norm. $\qquad\square$

Remark 10.25 The same argument shows that, for any von Neumann algebra R, any c.b. map $\theta : R_* \to A$ can be approximated pointwise by a net of finite rank maps $\theta_i : R_* \to A$ with $\|\theta_i\|_{cb} \le \|\theta\|_{cb}$.

Remark 10.26 We will use the approximation property described in Lemma 10.24 via the following: Let A be a C^*-algebra. Assume that E is the dual B^* (resp. the predual R_*) of a C^*-algebra B (resp. von Neumann algebra R). Then for any $y \in E \otimes A$ the supremum defining the norm $\Delta(y)$ in (6.17) can be restricted to pairs (θ, π) where $\theta : E \to \pi(A)'$ is a complete contraction *of finite rank*. Moreover in case $E = B^*$, we can restrict to weak*-continuous *finite rank* maps $\theta_i : B^* \to A$. Indeed, this is an immediate application of Lemma 10.24 (resp. Remark 10.25).

The next statement shows that if E is a C^*-algebra and u a finite rank map, the "approximate factorizations" of u appearing in part (iv) of Theorem 10.14 become bona fide factorizations of u.

Theorem 10.27 ([137]) *Let* $u : B \to A$ *be a finite rank map between two C^*-algebras. Then, for any $\varepsilon > 0$, there is an integer n and a factorization $u = wv$ of the form*

$$B \xrightarrow{v} M_n \xrightarrow{w} A$$

with $(\|v\|_{cb}\|w\|_{cb} \leq) \|v\|_{cb}\|w\|_{dec} \leq \|u\|_{dec}(1 + \varepsilon)$.
  Therefore, if $y \in B^* \otimes A$ is the tensor associated to $u : B \to A$, we have

$$\|u\|_{dec} = \delta(y).$$

Proof If $u = wv$ as in Theorem 10.27, then, by (6.7) and (6.9), we have $\|u\|_{dec} \leq \|v\|_{dec}\|w\|_{dec} = \|v\|_{cb}\|w\|_{dec}$, hence by Remark 6.17,

$$\|u\|_{dec} \leq \delta(y).$$

We now turn to the converse. We may assume $u(x) = \sum_1^k \xi_j(x)a_j$ with $\xi_j \in B^*$ and $a_j \in A$, or equivalently $y = \sum \xi_j \otimes a_j$. By the equality $\delta = \Delta$ reinforced by the preceding Remark 10.26 it suffices to show the following claim: for any representation $\pi : A \to B(H)$ and weak* continuous finite rank map $\theta : B^* \to \pi(A)'$ with $\|\theta\|_{cb} \leq 1$, we have

$$\left\|\sum \theta(\xi_j)\pi(a_j)\right\| \leq 1.$$

Let θ be such a map and let $t \in \pi(A)' \otimes B$ be the tensor associated to it, so that $\|t\|_{\min} = \|\theta\|_{cb} \leq 1$. Let $C = \pi(A)'$. By (6.16), since $u : B \to A$ has finite rank, we have

$$\|(Id_C \otimes u)(t)\|_{C \otimes_{\max} A} \leq \|u\|_{dec}\|t\|_{C \otimes_{\min} B} = \|u\|_{dec}\|\theta\|_{cb} \leq \|u\|_{dec}.$$

Since $(Id_C \otimes u)(t) = \sum \theta(\xi_j) \otimes a_j$, this yields

$$\left\|\sum \theta(\xi_j)\pi(a_j)\right\| \leq \left\|\sum \theta(\xi_j) \otimes a_j\right\|_{\pi(A)' \otimes_{\max} A} \leq \|u\|_{dec}.$$

This proves the claim, and hence the equality $\delta(y) = \|u\|_{dec}$. Then, the first assertion follows from Remark 6.17 (applied with $E = B$). $\qquad\square$

Remark 10.28 The following question seems interesting: fix $\varepsilon > 0$, let k be the rank of u, can we obtain the preceding factorization with $n \leq f(k, \varepsilon)$ for some function f? In other words can we control n by a function f depending *only* on k and ε?

10.4 Notes and remarks

Section 10.1 is due to Kirchberg. In the separable case, exact C^*-algebras are just C^*-subalgebras of nuclear ones, but the latter fact was proved (by Kirchberg) as the crowning achievement of a long series of his own previous works [155, 159, 160]. We present in Theorems 10.3 and 10.18 only a simpler (but quite important) step that led him to that result. The fact that (iii) in Theorem 10.3 characterizes exactness is related to what Kirchberg called *locfin*(A) in [159]. These ideas make sense equally well for operator spaces, but in that generality one needs to introduce a constant of exactness, that replaces the constant 1 appearing in (10.2), see [208, ch. 17]. See [162] (or [3]) for a proof of the full result, namely that a separable exact C^*-algebra A embeds in O_2, and that if A is nuclear the embedding $A \subset O_2$ can be obtained together with a completely contractive projection onto A. In §10.2 the main points are due to Choi–Effros (Theorem 10.17) and Kirchberg (Theorem 10.18). The use of the δ-norm allows us to deduce them from results on general linear maps as in our previous book [208]. Theorem 10.27 is due to Junge and Le Merdy [137]. The latter statement explains transparently why the CPAP implies a reinforced approximation property by c.p. maps that uniformly factorize through M_n.

11

Traces and ultraproducts

11.1 Traces

We start with some preliminaries on noncommutative (or more appropriately not *necessarily* commutative) measure spaces.

Let M be a von Neumann algebra. Let M_+ denote the positive part of M. We recall that a *trace* on M is a map $\tau : M_+ \to [0, \infty]$ satisfying

- (i) $\tau(x + y) = \tau(x) + \tau(y), \forall\, x, y \in M_+$;
- (ii) $\tau(cx) = c\tau(x), \forall\, c \in [0, \infty), x \in M_+$;
- (iii) $\tau(a^*a) = \tau(aa^*), \forall\, a \in M$ (this is the "tracial" condition).

τ is said to be *faithful* if $\tau(x) = 0$ implies $x = 0$, *normal* if

$$\sup_i \tau(x_i) = \tau(\sup_i x_i) \tag{11.1}$$

for any bounded increasing net (x_i) in M_+. Note that since (x_i) is bounded there is x in M_+ such that, for any h in H, $\langle x_i h, h \rangle \uparrow \langle xh, h \rangle$, which implies that x_i tends to x weak* (see Remark A.11) and hence $x \in M_+$. The operator x being obviously the least upper bound of (x_i), it is natural to denote it by $\sup_i x_i$.

By considering the net of finite partial sums $\sum_{i \in \gamma} P_i$ ($\gamma \subset I$), we see that (11.1) implies that

$$\tau\left(\sum P_i\right) = \sum \tau(P_i)$$

for any mutually orthogonal family of (self-adjoint) projections $(P_i)_{i \in I}$ in M, which is analogous to the σ-additivity of measures.

When dealing with von Neumann algebras, it is customary to refer to *self-adjoint projections* simply as *"projections."* Since the confusion this abuse may create is very unlikely, we adopt this convention: *in a von Neumann algebra, the term projection will always mean a self-adjoint one.*

The trace τ is called *semifinite* if for any nonzero $x \in M_+$ there is $y \in M_+$, such that $0 \neq y \leq x$ and $\tau(y) < \infty$, and *finite* if $\tau(1) < \infty$ (1 denoting the identity of M). In the finite case, $\tau(x) < \infty$ for any $x \geq 0$.

Remark 11.1 If τ is semifinite there is a family of mutually orthogonal projections $(p_i)_{i \in I}$ in M with $\tau(p_i) < \infty$ for all $i \in I$ such that $\sum_{i \in I} p_i = 1$. Indeed, let $(p_i)_{i \in I}$ be a maximal such family (except for the latter condition). If $r = 1 - \sum_{i \in I} p_i \neq 0$ there is $y \in M$ nonzero such that $0 \leq y \leq r$ with $\tau(y) < \infty$. By spectral theory we have $y \geq \lambda q$ for some $\lambda > 0$ and some projection $0 \neq q \in M$. Then q contradicts the maximality of $(p_i)_{i \in I}$. Therefore $r = 0$. $\square$

A von Neumann algebra M is called *finite* if the family formed of the finite normal traces separates the points of M. Clearly this happens if M admits a single faithful normal finite trace, but a finite M may fail to have any faithful finite trace, for instance $M = \ell_\infty(\Gamma)$ with Γ uncountable. However, on a separable Hilbert space (i.e. if M is weak*-separable) the converse is also true; then M is finite if and only if it admits a faithful normal finite trace.

The definition of semifiniteness is simpler: A von Neumann algebra M is called *semifinite* if M admits a faithful normal semifinite trace.

For example, $B(H)$ is semifinite for any H. The trace $\tau : B(H)_+ \to [0, \infty]$ is the usual one defined for $T \in B(H)_+$ by $\tau(T) = \sum \langle e_i, T(e_i) \rangle$ where (e_i) is any orthonormal basis of H. The abundance of finite rank (and hence finite trace) operators guarantees semifiniteness. $B(H)$ is finite if and only if $\dim(H) < \infty$.

Remark 11.2 (Classification of von Neumann algebras in types I, II, III) It is traditional to classify von Neumann algebras in three main classes called type I, type II, and type III. A von Neumann algebra is called of type I if it is isomorphic to a direct sum of the form $(\oplus \sum_{i \in I} C_i \bar{\otimes} B(H_i))_\infty$ where $(H_i)_{i \in I}$ is a family of (mutually nonisomorphic) Hilbert spaces and $(C_i)_{i \in I}$ a family of commutative von Neumann algebras. It is easy to check that these are semifinite.

A von Neumann algebra M is called of type II if it is semifinite and if there is no nonzero projection p for which pMp is commutative (equivalently no $p \neq 0$ for which pMp is type I). Any semifinite M can be decomposed as $M = M_I \oplus M_{II}$ with M_I (resp. M_{II}) of type I (resp. II).

Then M is called of type III if there is no nonzero projection $p \in M$ for which pMp is semifinite. Any von Neumann algebra M can be decomposed as

$$M = M_I \oplus M_{II} \oplus M_{III}$$

with M_I (resp. M_{II}, rresp M_{III}) of type I (resp. II, rresp III).

See e.g. [240, ch. V] for a complete discussion.

In these notes we will be mainly concerned by finite or semifinite von Neumann algebras, (i.e. algebras of type I or II). However, in several instances we will have to consider the case of type III von Neumann algebras. Then an important structure theorem due to Takesaki will come to our rescue. We will use it as a "black box" and refer for the proof to Takesaki's book [242, th. XII.1.1 p. 364 and th. X.2.3] or to the original paper [239, th. 4.5 and §8]. See also [146, §13.3]. A very concise description can be found in [256]. See also [111] for useful information on the same theme.

Theorem 11.3 (Takesaki's duality theorem) *Let M be a von Neumann algebra. There is a semifinite von Neumann algebra $\mathcal{M}$ with the following two properties:*

(i) *There is an embedding $M \subset \mathcal{M}$ of M as a von Neumann subalgebra and a c.p. projection $P : \mathcal{M} \to M$ with $\|P\| = 1$.*

(ii) *There is an embedding $\mathcal{M} \subset M$ of $\mathcal{M}$ as a von Neumann subalgebra and a c.p. projection $Q : M \to \mathcal{M}$ with $\|Q\| = 1$.*

In its usual form, Takesaki's duality theorem asserts that Theorem 11.3 holds for any von Neumann algebra of type III. Schematically, in the latter case $\mathcal{M}$ is the crossed product of M with a one parameter automorphism group called the modular group. The algebra M appears as the fixed point algebra with respect to a certain one parameter (dual) automorphism group acting on $\mathcal{M}$. Then a well-known averaging argument based on the amenability of $\mathbb{R}$ (the parameter group) shows that there is a contractive projection P from $\mathcal{M}$ to M. To obtain (ii) a similar construction is applied to $\mathcal{M}$ (whence Q by the same averaging argument) but the resulting "double" crossed product is now isomorphic to M, so we obtain in this way the type III case. Since the semifinite case is trivial, one can use the decomposition $M = M_I \oplus M_{II} \oplus M_{III}$ and apply Takesaki's duality theorem for the type III case to produce a semifinite $\mathcal{M}_{III}$ associated to M_{III} as in Theorem 11.3. Then we obtain the general case by simply setting $\mathcal{M} = M_I \oplus M_{II} \oplus \mathcal{M}_{III}$.

Remark 11.4 In general the projection appearing in (i) or (ii) is *not* normal (see [248]), reflecting the fact that an invariant mean is not a measure. However, just like invariant means are pointwise limits of true measures there is a net of *normal* contractive c.p. maps $P_i : \mathcal{M} \to M$ that tend pointwise to P for the weak* topology on M, and similarly for Q (see [256, p. 45] for more details).

Remark 11.5 (L_p-spaces for semifinite traces) Let τ be a semifinite faithful normal trace on a von Neumann algebra M. Then $\mathcal{A} = \{x \in M \mid \tau(|x|) < \infty\}$ is a weak* dense *-subalgebra of M. When τ is finite, of course $\mathcal{A} = M$. For any $1 \le p < \infty$ one usually defines the space $L_p(\tau)$ as the completion of $\mathcal{A}$ equipped with the norm defined for any $x \in \mathcal{A}$ by $\|x\|_p = (\tau(|x|^p))^{1/p}$.

For example, when $M = B(H)$ with the usual trace the space $L_2(\tau)$ (resp. $L_1(\tau)$) can be identified with the Hilbert–Schmidt class (resp. the trace class). When $1 \le p < \infty$ we obtain the so-called Schatten p-class.

By Remark 11.1 in any semifinite M there is a directed increasing net (p_γ) of projections with finite traces tending weak* to 1 (just take $p_\gamma = \sum_{i \in \gamma} p_i$ for any finite subset $\gamma \subset I$). Let $M_\gamma = p_\gamma M p_\gamma$ and let τ_γ denote the restriction of τ to M_γ, which is clearly a *finite* faithful normal trace on M_γ. For any $\gamma \le \delta$ in the net we have $M_\gamma \subset M_\delta$ and there is a natural isometric embedding

$$L_p(M_\gamma, \tau_\gamma) \subset L_p(M_\delta, \tau_\delta).$$

Then it is not hard to check that the space $L_p(\tau)$ that we just defined can be identified with the completion of the union $\cup_\gamma L_p(M_\gamma, \tau_\gamma)$ with its natural norm. In this way the construction of the spaces $L_p(\tau)$ can be reduced to the finite trace case.

Since we will be using the semifinite case only much later in Chapter 22 we prefer to concentrate for now on the slightly less technical case of finite traces with $p = 2$ or $p = 1$.

11.2 Tracial probability spaces and the space $L_1(\tau)$

In general, a functional f on an algebra A is called *tracial* if $f(xy) = f(yx)$ for any $x, y \in A$.

Definition 11.6 By a "tracial probability space," we mean a von Neumann algebra M equipped with a trace $\tau : M_+ \to \mathbb{R}_+$ that is normalized (i.e. such that $\tau(1) = 1$) faithful and normal.

Then (see Remark 11.7) τ then extends to a normal (tracial) state on M, so that

$$\forall x, y \in M \quad \tau(xy) = \tau(yx).$$

Remark 11.7 Indeed, let M_{sa} denote the set of self-adjoint elements in M. Since any x in M_{sa} can be written $x = x_1 - x_2$ with $x_1, x_2 \in M_+$, τ uniquely extends to an $\mathbb{R}$-linear form on M_{sa} by setting $\tau(x) = \tau(x_1) - \tau(x_2)$. Since τ is additive on M_+, this definition is unambiguous. Note that since τ is nonnegative on M_+, this extension of τ preserves order, i.e. $x \le y$ implies $\tau(x) \le \tau(y)$. Then, by complexification, τ uniquely extends to a positive $\mathbb{C}$-linear form (and hence a state) on M. The latter state is normal by Theorem A.44. Moreover, by polarization, the tracial property of τ on M_+ implies $\tau(y^*x) = \tau(xy^*)$ for any $x, y \in M$, or equivalently (replace y by y^*) τ is tracial.

If M is commutative, M can be identified with $L_\infty(\Omega, \mathscr{A}, \mu)$ for some abstract probability space $(\Omega, \mathscr{A}, \mu)$, then projections are indicator functions of sets in $\mathscr{A}$ and τ corresponds to μ on $(\Omega, \mathscr{A})$, i.e. we have for all f in $L_\infty(\Omega, \mathscr{A}, \mu)$

$$\tau(f) = \int f(\omega) \, d\mu(\omega).$$

For this reason, (M, τ) is usually called a "noncommutative" probability space, but since we want to include the commutative case, we prefer the term tracial probability space.

The simplest example is of course the algebra M_n of all $n \times n$ matrices equipped with the usual normalized trace, $\tau(x) = n^{-1}\mathrm{tr}(x)$. As we will see in the next section, infinite tensor products and ultraproducts of matrix algebras of varying sizes give us examples of tracial probability spaces. Any discrete group G also provides us with another fundamental example namely (M_G, τ_G) (see §3.5).

Let (M, τ) be a tracial probability space, and let M_* be the predual of M. By definition (see §A.16), this is the subspace of M^* formed of all the normal (i.e. weak*-continuous) elements of M^*. It is well known that M_* can be identified with the ("noncommutative") L_1-space associated to (M, τ). Let us now explain how the latter is defined and how this identification works. For any x in M, we set

$$\|x\|_1 = \tau(|x|).$$

Proposition 11.8 *For all x in M we have $|\tau(x)| \le \tau(|x|)$. More generally*

$$\tau(|x|) = \sup\{|\tau(xy)| \mid y \in M \; \|y\| \le 1\} = \sup\{|\tau(yx)| \mid y \in M \; \|y\| \le 1\}. \tag{11.2}$$

In particular, this shows that $\| \;\; \|_1$ is a norm on M.

Proof Recall that as any positive functional, τ satisfies the Cauchy–Schwarz inequality, that is $|\tau(ba)| \le \tau(bb^*)\tau(a^*a)$ for all $a, b \in M$. Assume $\|y\| \le 1$. Let $x = u|x|$ be the polar decomposition of x. Let $a = u|x|^{1/2}$ and $b = |x|^{1/2}y$. Note $a^*a = |x|$ and $bb^* \le |x|$. By Cauchy–Schwarz we have

$$|\tau(xy)| = |\tau(ab)| = |\tau(ba)| \le (\tau(a^*a)\tau(bb^*))^{1/2} \le \tau(|x|).$$

In particular, taking $y = 1$, we obtain $|\tau(x)| \le \tau(|x|)$. Finally, $|x| = u^*x$ yields the first equality in (11.2). Since τ is tracial, the second one is clear. $\square$

The space $L_1(M, \tau)$ or simply $L_1(\tau)$ is defined as the completion of M with respect to the norm $\| \;\; \|_1$. Note that, by definition, we have an inclusion with dense range

$$M \subset L_1(M, \tau).$$

For any $x \in M$, let $f_x \in M^*$ be the linear form defined on M by

$$f_x(y) = \tau(xy). \tag{11.3}$$

Clearly, f_x is normal (indeed, since τ is normal, this follows from Remark A.37. Thus $f_x \in M_*$.

Remark 11.9 Let $\Lambda \subset M_*$ be the linear subspace defined by $\Lambda = \{f_x \mid x \in M\}$. We claim that Λ is dense in M_*. To see this it suffices to show that it is $\sigma(M_*, M)$-dense, or equivalently that any $\varphi \in (M_*)^* = M$ that vanishes on Λ must vanish on the whole of M_*. Equivalently, we must show that if $y \in M$ (corresponding to φ) is such that $f_x(y) = 0$ for any $x \in M$ then $y = 0$. Indeed this is clear: the choice of $x = y^*$ gives us $\tau(y^*y) = 0$, and hence $y = 0$.

Remark 11.10 More generally, let $V \subset M$ be any $\sigma(M, M_*)$-dense linear subspace. Then the same argument as for Remark 11.9 shows that V is dense in $L_1(\tau)$.

Theorem 11.11 *The map that takes $x \in M$ to f_x extends to an isometric isomorphism from $L_1(\tau)$ onto M_*. In short we have*

$$L_1(\tau) \simeq M_* \quad \text{(isometrically)}.$$

Proof By (11.2), $\|f_x\|_{M_*} = \tau(|x|) = \|x\|_1$. Therefore, since Λ is dense in M_*, the latter can be viewed as the completion of $(\Lambda, \|.\|_1)$. $\square$

Consequently, since $M = (M_*)^*$, we also have

$$L_1(\tau)^* \simeq M \quad \text{(isometrically)},$$

or more explicitly for the record:

Corollary 11.12 *The map that takes $y \in M$ to the functional $\varphi_y \in L_1(\tau)^*$ defined by $\varphi_y(x) = \tau(xy)$ is an isometric isomorphism from M onto $L_1(\tau)^*$. Thus we have*

$$\forall y \in M \quad \|y\| = \sup\{|\tau(xy)| \mid x \in M, \tau(|x|) \le 1\}. \tag{11.4}$$

11.3 The space $L_2(\tau)$

Let (M, τ) be a tracial probability space. We will denote by $L_2(\tau)$ the Hilbert space associated to (M, τ) by the GNS construction (see §A.13). More precisely, the space $L_2(\tau)$ is the completion of M with respect to the norm $x \mapsto \|x\|_2$ defined on M by

$$\|x\|_2 = (\tau(|x|^2))^{1/2},$$

or equivalently $\|x\|_2 = (\tau(x^*x))^{1/2}$. Note that the traciality of τ ensures that

$$\|x\|_2 = (\tau(x^*x))^{1/2} = (\tau(xx^*))^{1/2} = \|x^*\|_2.$$

The latter norm is derived from the scalar product

$$\langle y, x \rangle = \tau(y^*x) \qquad (x, y \in M), \tag{11.5}$$

which, as already mentioned, satisfies the Cauchy–Schwarz inequality

$$|\langle y, x \rangle| \le \tau(x^*x)^{1/2}\tau(y^*y)^{1/2} = \|x\|_2\|y\|_2. \tag{11.6}$$

Note that this shows in particular that $x \to \|x\|_2$ is subadditive, and since τ is assumed faithful (i.e. such that $\tau(x^*x) = 0 \Leftrightarrow x = 0$), we see that $\|\cdot\|_2$ is indeed a norm on M.

Remark 11.13 Let (M, τ) be a tracial probability space. If $x \in B_M$ is such that $\tau(x^*x) = 1$ then x is unitary (and conversely of course). Indeed, $\tau(1 - x^*x) = 0$ and $1 - x^*x \ge 0$ imply $x^*x = 1$ by the faithfulness of τ. Similarly $xx^* = 1$ so that $x \in U(M)$.

Remark 11.14 Recall that the "modulus" we constantly use is defined for all $x \in M$ by $|x| = (x^*x)^{1/2}$. It is worthwhile to emphasize that the triangle inequality $|x + y| \le |x| + |y|$ is *false* for this modulus. Here is an example: Let $x = \begin{pmatrix} 0 & 1 \\ 0 & 1 \end{pmatrix}$. Then $x - x^*$ is unitary so $|x - x^*| = I$. However, a simple calculation shows that $|x| = \sqrt{2}\begin{pmatrix} 0 & 0 \\ 0 & 1 \end{pmatrix}$ and $|x^*| = 2^{-1/2}\begin{pmatrix} 1 & 1 \\ 1 & 1 \end{pmatrix}$ so that $\langle(|x| + |x^*|)e_1, e_1\rangle = 2^{-1/2}$ while $\langle|x - x^*|e_1, e_1\rangle = 1$. Therefore $|x - x^*| \not\le |x| + |x^*|$.

Note however the following useful substitute from [1]: for any pair x, y in M there are isometries U, V in M such that $|x + y| \le U|x|U^* + V|y|V^*$.

From the GNS construction, we recall that for any a, x in M we have $a^*a \le \|a\|^2 1$ and hence $x^*a^*ax \le \|a\|^2 x^*x$ which implies $\|ax\|_2 \le \|a\|\|x\|_2$. Therefore, the left multiplication by a extends by density to a bounded operator

$$L(a) : L_2(\tau) \to L_2(\tau)$$

such that $\|L(a)\| \le \|a\|$ and $L : M \to B(L_2(\tau))$ is a $*$-homomorphism.

Let $j : M \to L_2(\tau)$ be the inclusion map. Note that j has dense range by definition of $L_2(\tau)$ and the unit vector $\xi = j(1)$ is cyclic for L i.e. $\overline{L(M)\xi} = L_2(\tau)$ (and $L(M)\xi = j(M)$). Note that

$$\forall a \in M \qquad \tau(a) = \langle 1, a \rangle = \langle \xi, L(a)\xi \rangle. \tag{11.7}$$

Up to now we did not make use of the trace property. We will now use it to show that right multiplications are also bounded on $L_2(\tau)$: For any a, x in $L_2(\tau)$ we have

$$\tau((xa)^*(xa)) = \tau((xa)(xa)^*) = \tau(xaa^*x^*) \leq \|a\|^2\tau(xx^*) = \|a\|^2\tau(x^*x),$$

and hence $\|xa\|_2 \leq \|a\|\|x\|_2$. Therefore the right multiplication by a extends to a bounded operator on $L_2(\tau)$ denoted by $R(a) : L_2(\tau) \to L_2(\tau)$. The mapping

$$R : M^{op} \to B(L_2(\tau))$$

is a $*$-homomorphism on the opposite algebra M^{op}, i.e. the same $*$-algebra but with the "opposite" product (defined by $a \cdot b = ba$).

In fact, what precedes holds more generally for the GNS construction associated to any tracial state on a C^*-algebra. But since τ is faithful and normal, L and R are isometric normal $*$-homomorphisms (see §A.13 and Remark A.42), which embed M (resp. M^{op}) as a von Neumann subalgebra of $B(L_2(\tau))$. The fact that $L(M)$ (resp. $R(M^{op})$) is a von Neumann subalgebra of $B(L_2(\tau))$ can be deduced e.g. from Kaplansky's density Theorem A.47 (because $B_{L(M)} = L(B_M)$ is weak* compact and similarly for R). The uniqueness of the predual then guarantees that L (resp. R) is a weak*-homeomorphism from M to $L(M)$ (resp. $R(M^{op})$).

Remark 11.15 Consider $x \in M$. Then $x \in B_M$ if and only if

$$\forall a,b \in M \quad |\tau(a^*xb)| \leq \|a\|_2\|b\|_2.$$

Indeed, this boils down to $\|x\| = \|L(x)\|$ (or $\|x\| = \|R(x)\|$).

Moreover, for any $x \in M$

$$x \geq 0 \Leftrightarrow \tau(a^*xa) \geq 0 \; \forall a \in M, \tag{11.8}$$

or equivalently

$$x \geq 0 \Leftrightarrow \tau(yx) \geq 0 \; \forall y \in M_+. \tag{11.9}$$

Indeed, (11.8) is the same as $\langle a, L(x)a \rangle \geq 0$ for any $a \in L_2(\tau)$.

Another way to see that $L(M)$ (resp. $R(M^{op})$) is a von Neumann subalgebra of $B(L_2(\tau))$ is via the next statement:

Proposition 11.16 *We have* $L(M) = R(M)'$ *and* $R(M) = L(M)'$, *and hence* $L(M) = L(M)''$ *and* $R(M^{op}) = R(M^{op})''$.

Proof Let $T \in R(M)'$. Assume $\|T\| = 1$ for simplicity. For any $a,b \in M$ we have $|\langle b, T(a)\rangle| \leq \|a\|_2\|b\|_2$ and also $T(a) = T(R(a)1) = R(a)T(1)$. Therefore $|\langle ba^*, T(1)\rangle| = |\langle R(a^*)b, T(1)\rangle| = |\langle b, R(a)T(1)\rangle| \leq \|a\|_2\|b\|_2$. Let $\varphi \in M^*$ be the linear form defined by $\varphi(x) = \langle x^*, T(1)\rangle$. Using the polar decomposition $x = U|x| = ab^*$ with $a = U|x|^{1/2}$ and $b = |x|^{1/2}$, we find $|\varphi(x)| \leq \tau(|x|)$ for any $x \in M$. By Corollary 11.12 (and the density of M in $L_1(\tau)$) there is $y \in B_M$ such that $\varphi(x) = \varphi_y(x)$ for any $x \in M$, and hence (since M is dense in $L_2(\tau)$) $T(1) = y$ in $L_2(\tau)$. Thus $T(1) \in M$ so that we may write $T(a) = R(a)T(1) = T(1)a = L(T(1))a$ for any $a \in M$, so that $T = L(T(1))$ and we conclude $R(M)' \subset L(M)$. The converse is obvious, and similarly for $L(M)'$. $\qquad\square$

Remark 11.17 We will use the following classical fact about the inclusion $M \subset L_2(\tau)$: the weak*-closure (i.e. $\sigma(M, M_*)$-closure) of a bounded convex subset of M coincides with its closure for the topology induced on M by the norm of $L_2(\tau)$. Indeed, since $L_2(\tau)$ is dense in M_*, the $\sigma(M, M_*)$-closure of a bounded subset of M coincides with its closure in the weak topology of $L_2(\tau)$, and by (Mazur's) Theorem A.9 for a convex set this is the same as the norm closure in $L_2(\tau)$.

Remark 11.18 In a tracial probability space (M, τ) certain inequalities are immediate consequences of the usual probabilistic ones. For instance for any $x \in M$ and any $0 < p < q < \infty$ we have for any $x \in M$

$$(\tau(|x|^p))^{1/p} \le (\tau(|x|^q))^{1/q} \text{ and } \lim_{p \to \infty} (\tau(|x|^p))^{1/p} = \|x\|_M. \qquad (11.10)$$

Indeed, let $N \subset M$ be the commutative von Neumann subalgebra generated by $|x|$. Since we may identify $(N, \tau_{|N})$ with $(L_\infty(\Omega, \mathscr{A}, \mu), \mu)$ for some abstract probability space $(\Omega, \mathscr{A}, \mu)$ (that of course depends on $|x|$), the preceding assertion reduces to the fact that for any $f \in L_\infty(\Omega, \mathscr{A}, \mu)$ we have $\|f\|_p \le \|f\|_q$ and $\lim_{p \to \infty} \|f\|_p = \|f\|_\infty$.

Proposition 11.19 *Let (M, τ) and (N, φ) be two tracial probability spaces. Let $T : L_2(\tau) \to L_2(\varphi)$ be a trace preserving isometry. Assume that there is a weak*-dense unital $*$-subalgebra $\mathcal{A} \subset M$ such that $T(\mathcal{A}) \subset N$ and $T_{|\mathcal{A}} : \mathcal{A} \to N$ is a unital $*$-homomorphism. Then, when restricted to M, T defines a normal (injective trace preserving) $*$-homomorphism embedding M into N (as a von Neumann subalgebra).*

Proof We claim that $\|T(a)\| = \|a\|$ for any a in $\mathcal{A}$. Indeed, since $(T(a)^*T(a))^m = T((a^*a)^m)$ and $\varphi(T(x)) = \tau(x)$ $(a, x \in \mathcal{A})$, we have by Remark 11.18

$$\|T(a)\| = \lim_{m \to \infty} \uparrow (\varphi((T(a)^*T(a))^m))^{\frac{1}{2m}} = \lim_{m \to \infty} \uparrow (\tau((a^*a)^m))^{\frac{1}{2m}} = \|a\|.$$

Let $\mathcal{A}_1 = T(\mathcal{A}) \subset N$. Then T defines an isometric isomorphism from $(\mathcal{A}, \|\ \|)$ to $(\mathcal{A}_1, \|\ \|)$. We may assume without loss of generality that $\mathcal{A}_1$ generates N. Thus, by Remark 11.17 and by Kaplansky's classical density theorem, we have $B_M = \overline{\mathcal{A} \cap B_M}^{L_2(\tau)}$ and similarly $B_N = \overline{\mathcal{A}_1 \cap B_N}^{L_2(\tau)}$. Since T is an isometry on $L_2(\tau)$ this clearly implies that $T(B_M) = B_N$. Moreover, since the product and the $*$-operation are continuous on $(B_M, \|\ \|_{L_2(\tau)})$, $T : M \to N$

is a $*$-homomorphism, which has kernel $= \{0\}$ and is necessarily surjective (since $T(B_M) = B_N$). Lastly, since $T^*(L_2(\tau)) \subset L_2(\tau)$, we have $T^*(N_*)$ $\subset M_*$, so T is normal. Actually $T : M \rightarrow N$, being a $*$-isomorphism between von Neuman algebras, is necessarily bicontinuous for the topologies $\sigma(M, M_*), \sigma(N, N_*)$ (see Remark A.38). □

Remark 11.20 Here is a typical application of Proposition 11.19. Let (M, τ) and (N, φ) be two tracial probability spaces. Fix $n \geq 1$. Let $(x_1, \ldots, x_n) \in M^n$ and $(y_1, \ldots, y_n) \in N^n$ be n-tuples that have "the same $*$-distribution" in the sense that for any polynomial (or equivalently for any monomial) $P(X_1, \ldots, X_n, X_1^*, \ldots, X_n^*)$ in noncommuting variables $(X_1, \ldots, X_n)$ and their adjoints we have

$$\tau(P(x_1, \ldots, x_n, x_1^*, \ldots, x_n^*)) = \varphi(P(y_1, \ldots, y_n, y_1^*, \ldots, y_n^*)).$$

Then the correspondence

$$T : P(x_1, \ldots, x_n, x_1^*, \ldots, x_n^*) \mapsto P(y_1, \ldots, y_n, y_1^*, \ldots, y_n^*)$$

extends to a trace preserving $*$-isomorphism between the (unital) von Neumann subalgebras M_x and N_y generated respectively by $(x_1, \ldots, x_n)$ and $(y_1, \ldots, y_n)$. Indeed, T clearly extends to an isometry from $L_2(M_x, \tau_{|M_x})$ onto $L_2(N_y, \varphi_{|N_y})$, and Proposition 11.19 can be applied to the unital $*$-algebra $\mathcal{A}_x$ generated by $(x_1, \ldots, x_n)$ in M_x.

More generally, the same can be applied for arbitrary families $(x_i)_{i \in I}$ and $(y_i)_{i \in I}$. In that case we say they have the same $*$-distribution if it is the case in the preceding sense for $(x_i)_{i \in I'}$ and $(y_i)_{i \in I'}$ for any finite subset $I' \subset I$. Then there is a trace preserving isomorphism between the von Neumann algebras generated by $(x_i)_{i \in I}$ and $(y_i)_{i \in I}$.

We will invoke several times the following well-known fact. It may be worthwhile to emphasize that this is special to the *finite trace* case. In general, for instance when $M = B(H)$ and $N \subset M$ there may not exist any bounded projection onto N.

Proposition 11.21 (Conditional expectations) *Let (M, τ) be a tracial probability space. Then for any von Neumann subalgebra $N \subset M$, there is a normal c.p. projection $P : M \rightarrow N$ with $\|P\| = \|P\|_{cb} = 1$, such that $P(axb) = aP(x)b$ for any $x \in M$ and $a, b \in N$.*

Proof By definition $L_2(N, \tau_{|N})$ can be naturally identified to a subspace of $L_2(M, \tau)$, namely the closure $\overline{N}$ of N in $L_2(M, \tau)$. Let P be the orthogonal projection from $L_2(M, \tau)$ to $\overline{N} \simeq L_2(N, \tau_{|N})$. Fix $x \in L_2(M, \tau)$. Just like in the classical commutative case, $P(x)$ is the unique $x' \in L_2(N, \tau_{|N})$ such that $\langle x', y \rangle = \langle x, y \rangle$ for any $y \in N$. Then since $ayb \in N$ for any $a, b \in N$, it follows that $P(axb) = aP(x)b$. By Remark 11.15, if $x \in M$ (resp. $x \in M_+$) then $P(x) \in N$ (resp. $P(x) \in N_+$) and $\|P(x)\| \le \|x\|$. Of course $P(x) = x$ if $x \in N$. Moreover, for any $a, b \in N$ the form $x \mapsto \langle a, P(x)b \rangle = \tau(a^*xb)$ is normal (since τ is normal) and hence P is normal by Remark A.42. Thus P is a normal positive projection from M onto N with $\|P\| = 1$. Its *complete* positivity is automatic by Theorem 1.45. But in any case, if we repeat the argument with $N \subset M$ replaced by $M_n(N) \subset M_n(M)$ with τ replaced by $\tau_n([x_{ij}]) = \sum \tau(x_{jj})$ we obtain that P is c.p. $\qquad\square$

Remark 11.22 (A recapitulation) We have natural inclusions $M \subset L_2(\tau)$ and $M \subset L_1(\tau)$, and since $\|x\|_1 \le \|x\|_2$ for any x in M, we also have a natural inclusion $L_2(\tau) \subset L_1(\tau)$ with norm $= 1$.

Let $j : M \to L_2(\tau)$ denote the natural inclusion. Then the (Banach space sense) adjoint $j^* : L_2(\tau)^* \to M^*$ actually takes values into M_*. Indeed, using the canonical identification $L_2(\tau)^* \simeq \overline{L_2(\tau)}$ we find that j^* takes $\bar{x} \in \overline{M} \subset \overline{L_2(\tau)} \simeq L_2(\tau)^*$ to the linear form f_{x^*} where f_x is as in (11.3), which is in M_*. Then it is easy to verify that the composition

$$T = \overline{j^*}j : M \to L_2(\tau) \simeq \overline{L_2(\tau)^*} \to \overline{M_*}$$

is the antilinear map that takes $x \in M$ to $f_{x^*} \in M_*$, or more rigorously the $\mathbb{C}$-linear map that takes $x \in M$ to $\overline{f_{x^*}} \in \overline{M_*}$.

Recall the canonical identification $\overline{M} \simeq M^{op}$ that takes $\bar{x} \in \overline{M}$ to x^*. Note $\overline{M_*} \simeq (M^{op})_*$. Using this we find that T can be identified to the mapping $T_0 : M \to (M^{op})_*$ defined by $T_0(x) = f_x$.

11.4 An example from free probability: semicircular and circular systems

In Voiculescu's free probability theory, stochastic independence of random variables is replaced by freeness of noncommutative analogues of random variables. We refer the reader to [254] for an account of this beautiful theory. Here we only want to introduce the basic tracial probability space that is the free analogue of $\mathbb{R}^n$ equipped with the standard Gaussian measure. The free analogue of a family of standard independent real (resp. complex) Gaussian variables is a free semicircular (resp. circular) family. They satisfy a similar distributional invariance under the orthogonal (resp. unitary) group.

Such families generate a tracial probability space that can be realized on the "full" Fock space, as follows. Let H be a (complex) Hilbert space. We denote by $\mathcal{F}(H)$(or simply by $\mathcal{F}$) the "full" Fock space associated to H, that is to say we set $\mathcal{H}_0 = \mathbb{C}$, $\mathcal{H}_n = H^{\otimes n}$ (Hilbertian tensor product) and finally

$$\mathcal{F} = \oplus_{n \geq 0} \mathcal{H}_n.$$

We consider from now on $\mathcal{H}_n$ as a subspace of $\mathcal{F}$. For every $h \in \mathcal{F}$, we denote by $\ell(h) : \mathcal{F} \to \mathcal{F}$ the operator defined by:

$$\ell(h)x = h \otimes x.$$

More precisely, if $x = \lambda 1 \in \mathcal{H}_0 = \mathbb{C}1$, we have $\ell(h)x = \lambda h$ and if $x = x_1 \otimes x_2 \cdots \otimes x_n \in \mathcal{H}_n$ we have $\ell(h)x = h \otimes x_1 \otimes x_2 \cdots \otimes x_n$. We will denote by Ω the unit element in $\mathcal{H}_0 = \mathbb{C}1$. The von Neumann algebra $B(\mathcal{F})$ is equipped with the vector state φ defined by

$$\varphi(T) = \langle \Omega, T\Omega \rangle,$$

called the vacuum state.

Let $(e_s)_{s \in S}$ be an orthonormal basis of H. Since $\varphi(\ell(h)\ell(h)^*) = 0$ and $\varphi(\ell(h)^*\ell(h)) = \|h\|^2$, we see that φ is *not* tracial on $B(\mathcal{F})$. However, it can be checked (a possible exercise for the reader) that *it is tracial when restricted to the von Neumann algebra M generated by the operators $\ell(e_s) + \ell(e_s)^*$ ($s \in S$)*, i.e. we have $\varphi(xy) = \varphi(yx)$ for all x, y in this $*$-subalgebra.

The pair (M, φ) is an example of a tracial probability space. Let

$$W_s = \ell(e_s) + \ell(e_s)^*. \tag{11.11}$$

Then the family $(W_s)_{s \in S}$ is the prototypical example of a free semicircular system, sometimes also called a "free-Gaussian" family. The term semicircular is used for any family with the same joint moments as $(W_s)_{s \in S}$. Such a family enjoys properties very much analogous to those of a standard independent Gaussian family. Indeed, let $[a_{ij}]$ $(i, j \in S)$ be an "orthogonal matrix," by which we mean that $a_{ij} \in \mathbb{R}$ $(i, j \in S)$ and that the associated $\mathbb{R}$-linear mapping is an isometric isomorphism on $\ell_2(S)$. Then if we let $W_i^a = \sum_j a_{ij} W_j$, we have

$$\varphi(P(W_i^a)) = \varphi(P(W_i)) \tag{11.12}$$

for any polynomial P in noncommuting variables $(X_i)_{i \in S}$. In other words the families $(W_s)_{s \in S}$ and $(W_s^a)_{s \in S}$ have the same joint moments. This is analogous to the rotational invariance of the usual Gaussian distributions on $\mathbb{R}^S$ when (say) S is finite.

In particular, for any fixed $s^0 \in S$ the variables W_{s^0} and $W_{s^0}^a$ have the same moments, so that $\varphi((W_{s^0}^a)^{2m}) = \varphi((W_{s^0})^{2m})$ for any $m \geq 0$, and letting $m \to \infty$ after taking the $2m$th root we find $\|W_{s^0}^a\| = \|W_{s^0}\|$. Since we can adjust a so that its s^0th row matches any element $(\alpha_s)_{s \in S}$ in the unit sphere of $\ell_2(S, \mathbb{R})$ we obtain

$$\forall (\alpha_s)_{s \in S} \in \ell_2(S, \mathbb{R}) \quad \left\| \sum_{s \in S} \alpha_s W_s \right\| = c_{\mathbb{R}} \left(\sum \alpha_s^2 \right)^{1/2},$$

where $c_{\mathbb{R}}$ is the common value of $\|W_{s^0}\|$. Thus, the real Banach space $\mathbb{R}$-linearly generated by $(W_s)_{s \in I}$ is isometric to a real Hilbert space. By (11.11) we have $c_{\mathbb{R}} \leq 2$, and with a little more work (see e.g. [254]) one shows that

$$c_{\mathbb{R}} = 2.$$

For any $a \in B_{B(H)}$, let $F(a) \in B_{B(\mathcal{F}(H))}$ denote the linear operator that fixes the vacuum vector Ω and acts like $a^{\otimes^n}$ on H^n (so called "second quantization" of a). To check (11.12) the simplest way is to observe that for all $h \in H$ and all $a \in B(H)$ we have

$$F(a)\ell(h) = \ell(ah)F(a).$$

If a is unitary then $F(a)$ is unitary and we have

$$F(a)\ell(h)F(a)^* = \ell(ah) \text{ and } F(a)\ell(h)^*F(a)^* = \ell(ah)^*.$$

Therefore, $F(a)(\ell(h) + \ell(h)^*)F(a)^* = \ell(ah) + \ell(ah)^*$, and hence when the coefficients a_{ij} are all real and $^t a$ denotes the transpose of a we find

$$F(^t a)W_s F(^t a)^* = W_s^a,$$

and hence for any polynomial P

$$F(^t a)P(W_s)F(^t a)^* = P(W_s^a).$$

Since all the $F(a)$'s and their adjoints preserve Ω, (11.12) follows.

Using the basic ideas of spectral theory and free probability one can show that if $|S| = n$ (resp. $S = \mathbb{N}$) then there is a trace preserving isomorphism from (M, φ) to $(M_{\mathbb{F}_n}, \tau_{\mathbb{F}_n})$ (resp. $(M_{\mathbb{F}_\infty}, \tau_{\mathbb{F}_\infty})$).

We now pass to the complex case. We need to assume that the index set is partitioned into two copies of the same set S, so we replace H by $\widetilde{H} = H \oplus H$ and assume given a (partitioned) orthonormal basis $\{e_s \mid s \in S\} \cup \{f_s \mid s \in S\}$ of $\mathcal{H}$. We then define $\widetilde{W}_s = \ell(e_s) + \ell(f_s)^* \in B(\mathcal{F}(\widetilde{H}))$. Let $\widetilde{M} \subset B(\mathcal{F}(\widetilde{H}))$ be the von Neumann algebra generated by $(\widetilde{W}_s)_{s \in S}$. Let $\widetilde{\varphi}$ be the vacuum state on $B(\mathcal{F}(\widetilde{H}))$. Then again $(\widetilde{M}, \widetilde{\varphi})$ is a tracial probability space and the family $(\widetilde{W}_s)_{s \in S}$ is the prototypical example of a so-called free circular system. It is the free analogue of an i.i.d. family of complex valued Gaussian variables with covariance equal to the 2×2 identity matrix. As the latter, the family $(\widetilde{W}_s)_{s \in S}$ is unitarily invariant. Indeed, let $[a_{ij}]$ $(i, j \in S)$ be the matrix of a unitary operator on $\ell_2(S)$. Then if we let $\widetilde{W}_i^a = \sum_j a_{ij} \widetilde{W}_j$, we have

$$\varphi \left(P \left(\widetilde{W}_i^a, \widetilde{W}_i^{a*} \right) \right) = \varphi \left(P \left(\widetilde{W}_i, \widetilde{W}_i^* \right) \right) \tag{11.13}$$

for any polynomial $P(X_i, X_i^*)$ in noncommuting variables $(X_i)_{i \in S}$ and $(X_i^*)_{i \in S}$. In other words the families $(\widetilde{W}_s, \widetilde{W}_s^*)$ and $(\widetilde{W}_s^a, \widetilde{W}_s^{a*})$ have the same joint moments. This is analogous to the unitary invariance of the usual standard Gaussian measure on $\mathbb{C}^S$ when S is finite. The proof of (11.13) is similar to that of (11.12) (hint: consider $F(\alpha)$ for $\alpha = a \oplus \bar{a}$ acting on $\widetilde{H}$).

As before, we have

$$\forall (\alpha_s)_{s \in S} \in \ell_2(S, \mathbb{C}) \quad \left\| \sum_{s \in S} \alpha_s \widetilde{W}_s \right\| = c_{\mathbb{C}} \left(\sum |\alpha_s|^2 \right)^{1/2},$$

where $c_{\mathbb{C}}$ is the common value of $\|\widetilde{W}_i\|$, and it can be shown (see [254]) that $c_{\mathbb{C}} = 2$.

11.5 Ultraproducts

Let $\{(M(i), \tau_i) \mid i \in I\}$ be a family of tracial probability spaces. Let $\mathcal{U}$ be an ultrafilter on the index set I (see Remark A.6). Then for any bounded family of numbers $(x_i)_{i \in I}$, the limit along $\mathcal{U}$ is well defined. We denote it by

$$\lim_{\mathcal{U}} x_i.$$

Let

$$B = \left\{ x \in \prod_{i \in I} M(i) \mid \sup_{i \in I} \|x(i)\|_{M(i)} < \infty \right\},$$

equipped with the norm

$$\|x\|_B = \sup_{i \in I} \|x(i)\|_{M(i)}.$$

As already mentioned, we adopt in these notes the notation

$$B = \left(\oplus \sum_{i \in I} M(i) \right)_{\infty}. \tag{11.14}$$

We define a functional $f_{\mathcal{U}} \in B^*$ by setting for all $t = (t_i)_{i \in I}$ in B

$$f_{\mathcal{U}}(t) = \lim_{\mathcal{U}} \tau_i(t_i).$$

Clearly $f_{\mathcal{U}}$ is a tracial state on B. Let $H_{\mathcal{U}}$ be the Hilbert space associated to the tracial state $f_{\mathcal{U}}$ in the GNS construction applied to B. We denote by

$$L : B \to B(H_{\mathcal{U}}) \quad \text{and} \quad R : B^{op} \to B(H_{\mathcal{U}})$$

the representations of B corresponding to left and right multiplication by an element of B. (We recall that B^{op} is the same C^*-algebra as B but with reverse multiplication.)

More precisely, let

$$p_{\mathcal{U}}(x) = \lim_{\mathcal{U}} \tau_i(x_i^* x_i)^{1/2}.$$

Clearly $p_\mathcal{U}$ is a Hilbertian seminorm on B. Let

$$\mathcal{I}_\mathcal{U} = \ker(p_\mathcal{U}) = \left\{x \in B \mid \lim_\mathcal{U} \tau_i(x_i^* x_i) = 0\right\}.$$

Then $\mathcal{I}_\mathcal{U}$ is a closed two-sided ideal, and $H_\mathcal{U}$ is defined as the completion of $B/\mathcal{I}_\mathcal{U}$ equipped with the Hilbertian norm associated to $p_\mathcal{U}$.

For any $t = (t_i)_i$ in B we denote by $\dot{t}$ the equivalence class of t in $B/\mathcal{I}_\mathcal{U}$.

Then L and R are defined by $L(x)\dot{t} = \hat{xt}$ and $R(x)\dot{t} = \hat{tx}$. Clearly since $f_\mathcal{U}$ is tracial, these are contractive representations of B and B^{op} on $H_\mathcal{U}$.

Let us record the following simple fact (see §A.4 for background on ultrafilter limits).

Lemma 11.23 *For any $t = (t_i)_i$ in B we have*

$$\|\dot{t}\|_{H_\mathcal{U}} = \lim_\mathcal{U} \|t_i\|_{L_2(\tau_i)}.$$

Moreover, for any $\varepsilon > 0$ there is $s = (s_i)_i$ in B such that $s - t \in \mathcal{I}_\mathcal{U}$ and

$$\sup_{i \in I} \|s_i\|_{L_2(\tau_i)} < \|\dot{t}\|_{H_\mathcal{U}} + \varepsilon = \|\dot{s}\|_{H_\mathcal{U}} + \varepsilon.$$

Proof It is immediate that for any $t, s \in B$ such that $t - s \in \mathcal{I}_\mathcal{U}$ we have $\lim_\mathcal{U} \|t_i\|_{L_2(\tau_i)} = \lim_\mathcal{U} \|s_i\|_{L_2(\tau_i)}$. By definition of the limit $\ell = \lim_\mathcal{U} \|t_i\|_{L_2(\tau_i)}$ for any $\varepsilon > 0$ the set

$$J_\varepsilon = \left\{i \in I \mid |\|t_i\|_{L_2(\tau_i)} - \ell| < \varepsilon\right\}$$

is such that $\lim_\mathcal{U} 1_{J_\varepsilon} = 1$ and $\lim_\mathcal{U} 1_{I \setminus J_\varepsilon} = 0$. See §A.4 for details if necessary. Thus we may simply set $s_i = t_i$ for any $i \in J_\varepsilon$ and $s_i = 0$ otherwise. Then we have $\lim_\mathcal{U} \|t_i - s_i\|_{L_2(\tau_i)} \le c \lim_\mathcal{U} 1_{I \setminus J_\varepsilon} = 0$, where $c = \sup_{i \in I} \|t_i - s_i\|_{L_2(\tau_i)}$. $\square$

Actually, it turns out that for many considerations the Hilbert space $H_\mathcal{U}$ is "too small." We need to embed it in the larger Hilbert space $\mathcal{H}_\mathcal{U}$ that is the ultraproduct of the family (H_i) defined by $H_i = L_2(\tau_i)$, as defined in §A.5.

We claim that we have a natural isometric inclusion

$$j_\mathcal{U} : H_\mathcal{U} \to \mathcal{H}_\mathcal{U}.$$

Indeed, to any $x \in B$ we can associate (x_i) in the space $X = \left(\oplus \sum_{i \in I} H_i\right)_\infty$ and we have by definition (see §A.5)

$$p_\mathcal{U}(\dot{x}) = \lim_\mathcal{U} \|x_i\|_{L_2(\tau_i)} = \|(x_i)_\mathcal{U}\|_{\mathcal{H}_\mathcal{U}}.$$

Thus the correspondence $\dot{x} \mapsto j_\mathcal{U}(\dot{x}) = (x_i)_\mathcal{U}$ extends to an isometric (linear of course) embedding $j_\mathcal{U} : H_\mathcal{U} \to \mathcal{H}_\mathcal{U}$.

Remark 11.24 For any $t = (t_i)_i$ in B, the mapping $(x_i) \mapsto (t_i x_i)$ is clearly in $B(X)$ and takes $\ker(\dot{p}_\mathcal{U})$ to itself. Thus it defines a mapping

$\pi_\ell(t) : \mathcal{H}_\mathcal{U} \rightarrow \mathcal{H}_\mathcal{U}$. The mapping $t \mapsto \pi_\ell(t)$ is clearly a $*$-homomorphism from $B/\mathcal{I}_\mathcal{U}$ to $B(\mathcal{H}_\mathcal{U})$. It is easy to check that if we view $H_\mathcal{U}$ as embedded in $\mathcal{H}_\mathcal{U}$ via $j_\mathcal{U}$, then $H_\mathcal{U} \subset \mathcal{H}_\mathcal{U}$ is an invariant subspace under $\pi_\ell(t)$ and we have

$$L(t) = \pi_\ell(t)_{|H_\mathcal{U}}.$$

See Remarks 11.28 and 11.29 for a description of $H_\mathcal{U}$ as the subset of $\mathcal{H}_\mathcal{U}$ formed of the elements admitting a "uniformly square integrable" representative. The reader who is already familiar with the latter can skip the next lemma, which is but a pedestrian reformulation of the same fact.

We will invoke the following simple observation.

Lemma 11.25 *For any $\beta \in H_\mathcal{U}$ and $\varepsilon > 0$, there is a family $(\beta_i) \in X$ such that*

$$j_\mathcal{U}(\beta) = (\beta_i)_\mathcal{U}, \tag{11.15}$$

$$\sup \|\beta_i\|_2 < \|\beta\|_{H_\mathcal{U}} + \varepsilon, \tag{11.16}$$

$$\forall t = (t_i) \in B \quad \|L(t)\beta\|_{H_\mathcal{U}} = \lim_\mathcal{U} \|t_i \beta_i\|_2 = \|\pi_\ell(t)((\beta_i)_\mathcal{U})\|_{\mathcal{H}_\mathcal{U}}, \tag{11.17}$$

and moreover, for any family of projections (p_i) $(p_i \in M(i))$ such that $\lim_\mathcal{U} \tau_i(p_i) = 0$, we have

$$\lim_\mathcal{U} \|p_i \beta_i\|_2 = \lim_\mathcal{U} \|\beta_i p_i\|_2 = 0. \tag{11.18}$$

Proof Let $\beta(m)$ be a sequence in $B/\mathcal{I}_\mathcal{U}$ such that $\beta = \sum \beta(m)$ and $\sum \|\beta(m)\|_{H_\mathcal{U}} < \|\beta\|_{H_\mathcal{U}} + \varepsilon/2$. Let $q : B \rightarrow B/\mathcal{I}_\mathcal{U}$ denote the quotient map. Then by Lemma 11.23 we may assume that, for each m, we have $\beta(m) = q(b(m))$ for some sequence $b(m) = (b_i(m))_{i \in I} \in B$ such that

$$\forall m \geq 0 \quad \sup_i \|b_i(m)\|_2 \leq \|\beta(m)\|_{H_\mathcal{U}} + 2^{-m}(\varepsilon/4). \tag{11.19}$$

Then, if we set

$$\beta_i = \sum_m b_i(m) \in L_2(\tau_i), \tag{11.20}$$

assuming $0 < \varepsilon < 1$, (11.16) clearly holds. From $\beta = \sum_m \beta(m)$ (absolutely convergent series in $H_\mathcal{U}$), since $j_\mathcal{U} : H_\mathcal{U} \rightarrow \mathcal{H}_\mathcal{U}$ is isometric we deduce

$$j_\mathcal{U}(\beta) = \sum_m j_\mathcal{U}(\beta(m)).$$

Note $j_\mathcal{U}(\beta(m)) = (b_i(m))_\mathcal{U}$. Therefore, we have (absolutely convergent series in $\mathcal{H}_\mathcal{U}$)

$$j_\mathcal{U}(\beta) = \sum_m (b_i(m))_\mathcal{U}.$$

But we have also by (11.20) (absolutely convergent series in $\mathcal{H}_\mathcal{U}$)

$$(\beta_i)_\mathcal{U} = \sum_m (b_i(m))_\mathcal{U}.$$

Therefore, we obtain (11.15). Also (11.17) is immediate by Remark 11.24.

To check (11.18), note that for each fixed k

$$\|p_i\beta_i\|_2 \leq \left\|p_i\left(\sum_{m\leq k} b_i(m)\right)\right\|_2 + \left\|p_i\left(\sum_{m>k} b_i(m)\right)\right\|_2$$

$$\leq \|p_i\|_2 \|\|\sum_{m\leq k} b(m)\|_{M(i)} + \sum_{m>k}\sup_i\|b_i(m)\|_2$$

and since $\sup_i\|\sum_{m\leq k} b_i(m)\|_{M(i)} \leq \|\sum_{m\leq k} b(m)\|_B < \infty$ and $\lim_\mathcal{U}\|p_i\|_2 = 0$ we have by (11.19)

$$\lim_\mathcal{U}\|p_i\beta_i\|_2 \leq \sum_{m>k}\|\beta(m)\|_{H_\mathcal{U}} + \sum_{m>k} 2^{-m}(\varepsilon/4)$$

and letting $k \to \infty$ we obtain one part of (11.18). The other part follows similarly. $\qquad\square$

The next result seems to go back to McDuff's 1969 early work. Note that the kernel $\mathcal{I}_\mathcal{U}$ is *not* weak* closed in B (and the quotient map is not normal), so the fact that the quotient $B/\mathcal{I}_\mathcal{U}$ is nevertheless (*-isomorphic to) a von Neumann algebra is a priori somewhat surprising.

Theorem 11.26 *The functional $f_\mathcal{U}: B \to \mathbb{C}$ vanishes on $\mathcal{I}_\mathcal{U}$. The associated functional*

$$\tau_\mathcal{U}: B/\mathcal{I}_\mathcal{U} \to \mathbb{C}$$

is a faithful tracial state on $B/\mathcal{I}_\mathcal{U}$ such that, if $q: B \to B/\mathcal{I}_\mathcal{U}$ denotes the quotient map, we have

$$\tau_\mathcal{U}(q(t)) = f_\mathcal{U}(t) \qquad \forall\, t \in B. \tag{11.21}$$

The kernels of L and R coincide with the set $\mathcal{I}_\mathcal{U}$. After passing to the quotient, L and R define isometric representations

$$L_\mathcal{U}: B/\mathcal{I}_\mathcal{U} \to B(H_\mathcal{U}) \quad and \quad R_\mathcal{U}: B^{op}/\mathcal{I}_\mathcal{U} \to B(H_\mathcal{U})$$

with commuting ranges. Lastly, the commutants satisfy

$$[L_\mathcal{U}(B/\mathcal{I}_\mathcal{U})]' = R_\mathcal{U}(B^{op}/\mathcal{I}_\mathcal{U}) \quad and \quad [R_\mathcal{U}(B^{op}/\mathcal{I}_\mathcal{U})]' = L_\mathcal{U}(B/\mathcal{I}_\mathcal{U}).$$

In particular, $L_\mathcal{U}(B/\mathcal{I}_\mathcal{U})$ and $R_\mathcal{U}(B^{op}/\mathcal{I}_\mathcal{U})$ are (mutually commuting) von Neumann subalgebras of $B(H_\mathcal{U})$.

Proof By Cauchy–Schwarz, we have $|f_\mathcal{U}(x)| \leq (f_\mathcal{U}(x^*x))^{1/2}(f_\mathcal{U}(1))^{1/2} = (f_\mathcal{U}(x^*x))^{1/2}$, and $\mathcal{I}_\mathcal{U} = \{x \in B \mid f_\mathcal{U}(x^*x) = 0\}$. Therefore, $f_\mathcal{U}$ vanishes on $\mathcal{I}_{\mathcal{U}}$. Passing to the quotient, we obtain a functional $\tau_{\mathcal{U}}$ unambiguously defined by (11.21), which is clearly a tracial state on $B/\mathcal{I}_{\mathcal{U}}$.

Let 1_i be the unit of $M(i)$ and let $\xi = (1_i)_{i \in I} \in B$. We have

$$f_{\mathcal{U}}(t) = \langle \dot{\xi}, L(t)\dot{\xi} \rangle = \langle \dot{\xi}, R(t)\dot{\xi} \rangle.$$

Let $t \in B$. If $L(t) = 0$, then $L(t^*t) = 0$ which by the preceding line implies $f_{\mathcal{U}}(t^*t) = 0$ hence $t \in \mathcal{I}_{\mathcal{U}}$. Conversely, if $t \in \mathcal{I}_{\mathcal{U}}$ then $x^*t^*tx \in \mathcal{I}_{\mathcal{U}}$ for any x in B, whence $f_{\mathcal{U}}(x^*t^*tx) = 0$ which means $\overset{\frown}{tx} = 0$ for all x in B, or equivalently $L(t) = 0$.

A similar argument applies for R, so we obtain that $\ker(L) = \ker(R) = \mathcal{I}_{\mathcal{U}}$. Then, after passing to the quotient by $\mathcal{I}_{\mathcal{U}}$, L and R define the isometric representations $L_{\mathcal{U}}$ and $R_{\mathcal{U}}$ with the same respective ranges as L and R. Therefore $L_{\mathcal{U}}$ and $R_{\mathcal{U}}$ still have commuting ranges.

Finally, let $T \in B(H_{\mathcal{U}})$ be an operator commuting with $L_{\mathcal{U}}(B/\mathcal{I}_{\mathcal{U}})$, i.e. $T \in L_{\mathcal{U}}(B/\mathcal{I}_{\mathcal{U}})'$. We will show that T must be in the range of $R_{\mathcal{U}}$. Let

$$\beta = T(\dot{\xi}) \in H_{\mathcal{U}}.$$

We will show that there is $b = (b_i)$ in B such that $\beta = \dot{b}$ and that

$$T = R(b) = R_{\mathcal{U}}(\dot{b}).$$

Indeed, we have for any $t = (t_i)$ in B

$$TL(t)\dot{\xi} = L(t)T\dot{\xi} = L(t)\beta \tag{11.22}$$

hence

$$\|L(t)\beta\|_{H_{\mathcal{U}}} \le \|T\| \, \|L(t)\dot{\xi}\|_{H_{\mathcal{U}}} = \|T\| \, \|\dot{t}\|_{H_{\mathcal{U}}}. \tag{11.23}$$

By Lemma 11.25 there is a family (β_i) with $\beta_i \in L_2(\tau_i)$ satisfying (11.18) and such that $\sup \|\beta_i\|_2 < \infty$, $(\beta_i)_{\mathcal{U}} = j_{\mathcal{U}}(\beta)$ and

$$\forall t = (t_i) \in B \quad \|L(t)\beta\|_{H_{\mathcal{U}}} = \lim_{\mathcal{U}} \|t_i\beta_i\|_2.$$

This, together with (11.23), implies that for any t in B

$$\lim_{\mathcal{U}} \tau_i(\beta_i\beta_i^*t_i^*t_i) \le \|T\|^2 \lim_{\mathcal{U}} \tau_i(t_i^*t_i). \tag{11.24}$$

Let $\beta_i = h_i v_i$ be the polar decomposition of β_i in $L_2(\tau_i)$ with $h_i \in L_2(\tau_i)$, $h_i \ge 0$, v_i partial isometry in $M(i)$ and $h_i = |\beta_i^*|$ (see Remark 11.29 for clarification). Fix $\varepsilon > 0$. Let p_i be the spectral projection of h_i associated to $]\|T\| + \varepsilon, \infty[$. Note that $\beta_i\beta_i^*p_i = h_i^2 p_i \ge (\|T\| + \varepsilon)^2 p_i$. Hence (11.24) implies (with $t_i = p_i$)

$$(\|T\| + \varepsilon)^2 \lim_{\mathcal{U}} \tau_i(p_i) \le \lim_{\mathcal{U}} \tau_i(\beta_i\beta_i^*p_i) \le \|T\|^2 \lim_{\mathcal{U}} \tau_i(p_i).$$

This forces $\lim_{\mathcal{U}} \tau_i(p_i) = 0$, and hence by (11.18) $\lim_{\mathcal{U}} \tau_i(\beta_i\beta_i^*p_i) = 0$.

Therefore, if we set finally $b_i = (1-p_i)h_i v_i$ we find $\|b_i\| \le \|(1-p_i)h_i\| \le \|T\| + \varepsilon$ and $\|\beta_i - b_i\|^2_{L_2(\tau_i)} \le \|p_i h_i v_i\|^2_{L_2(\tau_i)} \le \tau_i(\beta_i \beta_i^* p_i)$, hence $\lim_{\mathcal{U}} \|\beta_i - b_i\|_{L_2(\tau_i)} = 0$.

Let $b = (b_i)$. Then $(\beta_i)_{\mathcal{U}} = (b_i)_{\mathcal{U}}$, and $b \in B$ with $\|b\|_B \le \|T\| + \varepsilon$. Then going back to (11.22) we obtain finally

$$TL(t)\dot{\xi} = L(t)\beta = L(t)\dot{b} = \widehat{(t_i b_i)} = R_{\mathcal{U}}(b)L(t)\dot{\xi}.$$

This shows that $T = R_{\mathcal{U}}(b)$, which completes the proof that $L_{\mathcal{U}}(B/\mathcal{I}_{\mathcal{U}})' = R_{\mathcal{U}}(B^{op}/\mathcal{I}_{\mathcal{U}})$. The same argument clearly yields $R_{\mathcal{U}}(B^{op}/\mathcal{I}_{\mathcal{U}})' = L_{\mathcal{U}}(B/\mathcal{I}_{\mathcal{U}})$, and hence $L_{\mathcal{U}}(B/\mathcal{I}_{\mathcal{U}})'' = L_{\mathcal{U}}(B/\mathcal{I}_{\mathcal{U}})$, which proves that $L_{\mathcal{U}}(B/\mathcal{I}_{\mathcal{U}})$ is a von Neumann algebra. $\qquad\square$

Definition 11.27 Let $M_{\mathcal{U}} = L_{\mathcal{U}}(B/\mathcal{I}_{\mathcal{U}})$. The tracial probability space

$$(M_{\mathcal{U}}, \tau_{\mathcal{U}})$$

is called the ultraproduct of the family $(M(i), \tau_i)$ with respect to $\mathcal{U}$.

Remark 11.28 Let $\mathcal{H}_{\mathcal{U}}$ be the usual ultraproduct of the Hilbert spaces $H_i = L_2(\tau_i)$. Then, $\mathcal{H}_{\mathcal{U}}$ can be identified to the closure in $\mathcal{H}_{\mathcal{U}}$ of the subspace of all elements of the form $(b_i)_{\mathcal{U}}$ with $\sup_i \|b_i\|_{M(i)} < \infty$. Alternatively $H_{\mathcal{U}} \subset \mathcal{H}_{\mathcal{U}}$ can also be described as the subspace corresponding to the "uniformly square integrable" families. More precisely, let $\beta = (\beta_i)_{\mathcal{U}}$ be an element of $\mathcal{H}_{\mathcal{U}}$, with $\sup_i \|\beta_i\|_{H_i} < \infty$. Then β belongs to $H_{\mathcal{U}}$ (meaning rather $j_{\mathcal{U}}(H_{\mathcal{U}})$) if and only if

$$\lim_{c \to \infty} \lim_{\mathcal{U}} \tau_i(\beta_i \beta_i^* 1_{\{\beta_i \beta_i^* > c\}}) = 0$$

or if and only if

$$\lim_{c \to \infty} \lim_{\mathcal{U}} \tau_i(\beta_i^* \beta_i 1_{\{\beta_i^* \beta_i > c\}}) = 0,$$

where we have denoted (abusively) by $1_{\{h>c\}}$ the spectral projection of the Hermitian operator h for the interval (c, ∞).

Remark 11.29 In what precedes, we invoked the polar decomposition in $L_2(\tau_i)$, using its structure as a bimodule over $M(i)$. In general this involves dealing with unbounded operators. But actually, we will apply the preceding result only in the case when $M(i)$ is a (finite-dimensional) matrix algebra for which the polar decomposition is entirely elementary and classical, since $L_2(\tau_i)$ coincides with $M(i)$ itself.

In general, a unitary in a quotient C^*-algebra $A/\mathcal{I}$ does not lift to a unitary in A. However it is so when $A/\mathcal{I}$ is isomorphic to a von Neumann algebra, in particular for $B/\mathcal{I}_{\mathcal{U}}$.

Lemma 11.30 *Let* $q_{\mathcal{U}} : B \to M_{\mathcal{U}}$ *be the quotient map (given by* $q_{\mathcal{U}} = L_{\mathcal{U}}q$*).*
For any unitary u *in* $M_{\mathcal{U}}$*, there is a unitary* $\widehat{u} = (u_i)_{i \in I}$ *in* B *such that*
$q_{\mathcal{U}}(\widehat{u}) = u$.

Proof Since $M_{\mathcal{U}}$ is a von Neumann algebra (see Theorem 11.26), we can write
$u = \exp ix$ for some self-adjoint $x \in M_{\mathcal{U}}$ (see Remark A.39). Since $q_{\mathcal{U}}$ maps
B onto $M_{\mathcal{U}}$ and is a $*$-homomorphism, there is $\widehat{x} \in B$ with $\widehat{x}^* = \widehat{x}$ such that
$q_{\mathcal{U}}(\widehat{x}) = x$. Then $\widehat{u} = \exp i\widehat{x}$ is a unitary in B lifting u. $\square$

Remark 11.31 Any (self-adjoint) projection $Q \in M_{\mathcal{U}}$ admits a lifting $(Q_i) \in$
B such that Q_i is a (self-adjoint) projection for any i. Indeed, let $(x_i) \in B_B$
be a lifting of Q (see Lemma A.33). Replacing x_i by its real part, we may
assume all the x_i's self-adjoint. Observe that for any λ in the spectrum of x_i,
we have $|\lambda| \le 1$ and $d(\lambda, \{0, 1\}) \le 2|\lambda - \lambda^2|$. Using this it is easy to show
that there is a (self-adjoint) projection Q_i (in the commutative von Neumann
algebra generated by x_i) such that $|x_i - Q_i| \le 2|x_i - x_i^2|$. Since $Q = Q^2$
we have $(x_i - x_i^2) \in \mathcal{I}_{\mathcal{U}}$ and hence $(x_i - Q_i) \in \mathcal{I}_{\mathcal{U}}$. Thus we conclude as
announced $Q = q((Q_i))$.

Remark 11.32 We will be mostly interested with the case when the algebras
$M(i)$ are finite dimensional. The main and simplest case is when $M(i) =$
$M_{N(i)}$ (matrices of size $N(i) \times N(i)$) equipped with the normalized trace
$\tau_i(x) = N(i)^{-1}\text{tr}(x)$. In that case we refer to $M_{\mathcal{U}}$ as an ultraproduct of
matricial tracial probability spaces.

  If we merely assume that all the $M(i)$s are finite dimensional, the resulting
$M_{\mathcal{U}}$ can anyway be embedded in an ultraproduct of the preceding matricial
kind. Indeed, each finite-dimensional (M, τ) can be identified in a trace
preserving way with $M_{n(1)} \oplus \cdots \oplus M_{n(k)}$ equipped with the trace $\tau(x_1 \oplus \cdots \oplus$
$x_k) = w_1 n(1)^{-1}\text{tr}(x_1) + \cdots + w_k n(k)^{-1}\text{tr}(x_k)$ where the positive weights
satisfy $w_1 + \cdots + w_k = 1$. If these weights are all rational, say $w_j = p_j/N$
with $p_1 + \cdots + p_k = N$, then we can embed (M, τ) into (M_N, τ_N) (here τ_N
is the normalized trace on M_N) by a block diagonal embedding repeating each
factor $M_{n(j)}$ with multiplicity p_j. In the general case, when the weights are
arbitrary real numbers, we can approximate them by rationals, and form an
ultraproduct associated to these elementary numerical approximations. This
shows that any finite-dimensional (M, τ) (and hence any ultraproduct of such)
can be embedded in an ultraproduct of *matricial* tracial probability spaces. See
Corollary 12.7 for another way to justify this.

Remark 11.33 (On Dixmier's approximation theorem) A von Neumann alge-
bra M is called a factor if $M \cap M' = \mathbb{C}I$. For example, M_G is a factor if (and
only if) all the nontrivial conjugacy classes of G (i.e. the sets $\{gtg^{-1} \mid g \in G\}$

with $t \neq 1$) are infinite. Moreover, if all the $M(i)$s are factors, the ultraproduct $M_{\mathcal{U}}$ of the tracial probability spaces $(M(i), \tau_i)$ is also a factor. We leave the proofs as exercises for the reader.

Let (M, τ) be a tracial probability space. If M is a factor, a classical theorem asserts that τ is the *unique* tracial state on M. This is an immediate corollary of a more general result due to Dixmier (see e.g. [146, p. 523] for a proof): for any $x \in M$ the norm closure of the convex hull of $\{uxu^{-1} \mid u \in U(M)\}$ intersects the center $Z = M \cap M'$ (when M is a factor, the latter intersection is reduced to $\{\tau(x)1\}$). This implies that two tracial states on M that coincide on Z must be identical. In particular, given any other tracial probability space (N, φ), any unital embedding $\pi : M \to N$ automatically preserves the trace if M is a factor.

Remark 11.34 If (M, τ) is a tracial probability space then any $*$-homomorphism $\pi : A_1 \otimes A_2 \to M$ is continuous with respect to the minimal norm, and hence continuously extends to $A_1 \otimes_{\min} A_2$. We only sketch the proof. Replacing M by $\pi(A_1 \otimes A_2)''$, we may assume that $\pi(A_1 \otimes A_2)'' = M \subset B(L_2(\tau))$. By Remark A.27 it suffices to show that the tracial state $A_1 \otimes A_2 \ni x \mapsto \tau(\pi(x))$ continuously extends to $A_1 \otimes_{\min} A_2$. It is easy to show that the extreme points of the set of tracial states on a C^*-algebra A are all factorial states, i.e. states f for which $\pi_f(A)''$ is a factor. Indeed, if the center Z of $\pi_f(A)''$ is nontrivial we can find a nonzero projection $p \in \mathcal{P}_Z$ such that $0 \neq p \neq 1$ and then $\tau(\cdot) = \tau(p)[\tau(p)^{-1}\tau(p \cdot)] + \tau(1-p)[\tau(1-p)^{-1}\tau((1-p)\cdot)]$ shows that τ is not extreme. Using this for $A = A_1 \otimes_{\max} A_2$, we may assume that M is a factor. We may assume $\pi = \pi_1 \cdot \pi_2$ as in (4.4). Let $M_j = \pi_j(A_j)''$. Since π_1, π_2 have commuting ranges, the center of M_j is included in that of M. Therefore, if M is a factor, both M_1, M_2 are factors, and $\tau_j = \tau_{|M_j}$ is the tracial state of M_j. Applying Dixmier's approximation theorem (see Remark 11.33) to each of M_1 and M_2 we find that $\tau(x_1 x_2) = \tau_1(x_1)\tau_2(x_2)$ for all $(x_1, x_2) \in M_1 \times M_2$. Then we have $\tau \circ \pi = (\tau_1 \circ \pi_1) \otimes (\tau_2 \circ \pi_2)$, and hence $|\tau \circ \pi(x)| \leq \|x\|_{\min}$ for all $x \in A_1 \otimes A_2$, which completes the proof.

Remark 11.35 (On ascending unions of factors) Let (M, τ) be a tracial probability space. Let $M(i) \subset M$ be a family of von Neumann subalgebras directed by inclusion and let $N \subset M$ be the weak* closure (or equivalently the bicommutant by Theorem A.46) of their union. If each $M(i)$ is a factor then N is also a factor. To check this assertion, let $\tau_i = \tau_{|M(i)}$ and $E_i = L_2(M(i), \tau_i)$. We view E_i and $L_2(N, \tau_N)$ as subspaces of $L_2(\tau)$. Consider the orthogonal projection $P_i : L_2(\tau) \to E_i$. Clearly $P_i(x) \to x$ for all $x \in L_2(N, \tau)$. By Proposition 11.21, P_i is a conditional expectation, so that $P_i(axb) = aP_i(x)b$ for all $a, b \in M(i)$ and all $x \in M$. Since $N' \subset M(i)'$, this implies that

$P_i(M \cap N') \subset M(i)'$. Therefore $P_i(N \cap N') \subset M(i) \cap M(i)'$ and the assertion follows since $P_i(x) \to x$ in $L_2(N, \tau)$ for any $x \in N$ and we assume $M(i) \cap M(i)' = \mathbb{C}1$ for all $i \in I$.

Remark 11.36 (Reduction to factors) The notion of free product $M * N$ of two finite von Neumann algebras goes back to Ching [43] and Voiculescu (see [254]). Equivalently, they defined the free product $(M * N, \tau * \varphi)$ of two tracial probability spaces (M, τ), (N, φ). Moreover, the construction is done so that the canonical embeddings $M \to M * N$ and $N \to M * N$ are trace preserving. In [43] Ching proved that if M, N both admit an orthonormal basis formed of unitaries for their L_2 spaces (and $\dim(M) \geq 2$, $\dim(N) \geq 3$) then $M * N$ is automatically a factor. In [218, th. 4.1] Sorin Popa proves a very general result of this type from which it follows that if (M, τ) is *any* tracial probability space, and if for instance $(N, \varphi) = (M_{\mathbb{F}_2}, \tau_{\mathbb{F}_2})$ (namely the so-called free group factor) then $M * N$ is automatically a factor. More precisely, the relative commutant $N' \cap (M * N)$ is trivial. Actually, Popa proves this whenever (N, φ) is a nonatomic tracial probability space.

In particular, this shows that any tracial probability space (M, τ) embeds in a trace preserving way into one that is a factor and is separable if M is separable.

We note in passing that it is an interesting and fundamental open question whether for any tracial probability space (M, τ) (on a separable Hilbert space and "atomless," that is without nontrivial minimal projections) there is an orthonormal basis of $L_2(\tau)$ formed of unitaries. Of course this holds whenever $(M, \tau) = (M_G, \tau_G)$ with G a discrete group.

For more in-depth information on finite von Neumann algebras we strongly recommend [4] to the reader.

Remark 11.37 (GNS representations on $B(H)$ and ultraproducts) Let (P_n) be a sequence of mutually orthogonal (self-adjoint) projections in $B(H)$ with $\text{rk}(P_n) = n$ for all n. Let $\mathcal{U}$ be a nontrivial ultrafilter on $\mathbb{N}$ and let $f_{\mathcal{U}}$ be the state on $B(H)$ defined by $f_{\mathcal{U}}(x) = \lim_{\mathcal{U}} n^{-1} \text{tr}(P_n x P_n)$. Then, for some infinite-dimensional Hilbert space $K_{\mathcal{U}}$, we have $\pi_{f_{\mathcal{U}}}(B(H))'' \simeq B(K_{\mathcal{U}}) \bar{\otimes} M_{\mathcal{U}}$ where $M_{\mathcal{U}}$ is the ultraproduct of the tracial probability spaces (M_n, τ_n) (with $\tau_n(\cdot) = n^{-1} \text{tr}(\cdot)$). This is due to Anderson and Bunce, see [6, th. 5].

11.6 Factorization through $B(H)$ and ultraproducts

We will describe a simple criterion that guarantees that a $*$-homomorphism $\pi : A \to M$ from a C^*-algebra to a von Neumann algebra factorizes completely positively through $B(H)$. To adapt its use to various situations we consider more generally the restriction of π to a unital linear subspace $E \subset A$ spanned by unitaries.

Theorem 11.38 *Let A be a unital C^*-algebra, (M, τ) a tracial probability space. Let $\pi : A \to M$ be a unital $*$-homomorphism. Let $S \subset U(A)$ be a subset with $1 \in S$ and let $E = \mathrm{span}(S) \subset A$. Assume that*

$$\forall n \geq 1, \forall x_1, \ldots, x_n \in E \quad \sum \|\pi(x_j)\|_2^2 \leq \left\| \sum \overline{x_j} \otimes x_j \right\|_{\min}. \quad (11.25)$$

Then there is a state f on $\overline{M} \otimes_{\min} M$ such that

$$\forall x, y \in E \quad \tau(\pi(y)^* \pi(x)) = f(\overline{y} \otimes x). \quad (11.26)$$

More precisely, there is an embedding $A \subset B(H)$, a family of finite rank operators $(h_i)_{i \in I}$ on H with $\mathrm{tr}(h_i^* h_i) = 1$ and an ultrafilter $\mathcal{U}$ on I such that

$$\forall x, y \in E \quad \tau(\pi(y)^* \pi(x)) = \lim_{\mathcal{U}} \mathrm{tr}(h_i^* y^* h_i x). \quad (11.27)$$

Conversely (11.27) implies (11.25).

Remark 11.39 The proof will show that any family $(h_i)_{i \in I}$ of Hilbert–Schmidt operators on H satisfying (11.27) automatically also satisfies an approximate commutation condition as follows:

$$\forall x \in E \quad \lim_{\mathcal{U}} \mathrm{tr} |x h_i - h_i x|^2 = 0. \quad (11.28)$$

This is but a simple consequence of the equality case in Cauchy–Schwarz.

The rest of this section is devoted to the proof of Theorem 11.38 and the reinforced version in Theorem 11.42.

Proof We will assume (without loss of generality) that $A \subset B(H)$, with infinite multiplicity. More precisely, this means that, starting from an embedding $A \subset B(H_1)$, we replace H_1 by $H = \ell_2 \otimes_2 H_1$ and embed $B(H_1)$ into $B(H)$ (diagonally) by $T \mapsto Id_{\ell_2} \otimes T$. This gives us a new faithful $*$-homomorphism $\rho : A \to B(H)$. For simplicity we view ρ as an embedding i.e. we set $\rho(a) = a$ for all $a \in A$.

We identify $\bar{H} \otimes_2 H$ with the Hilbert–Schmidt class $S_2(H)$. Let $B_{\min}$ denote the unit ball of $(\bar{A} \otimes_{\min} A)^*$. Note that for any $T \in \bar{A} \otimes_{\min} A$ we have

$$\|T\|_{\min} = \sup \{ \Re(f(T)) \mid f \in B_{\min} \}.$$

By our assumption we have

$$\sum \|\pi(x_j)\|_{L_2(\tau)}^2 \leq \left\| \sum \overline{x_j} \otimes x_j \right\|_{\min} = \sup_{f \in B_{\min}} \sum \Re(f(\overline{x_j} \otimes x_j)).$$

We now apply Lemma A.16. Since $B_{\min}$ is convex and weak* compact, this gives us in the limit a functional f in $B_{\min}$ such that

$$\forall x \in E \quad \|\pi(x)\|^2_{L_2(\tau)} \leq \Re(f(\bar{x} \otimes x)). \tag{11.29}$$

Taking $x \in S$ we find $1 \leq \Re(f(\bar{x} \otimes x)) \leq 1$ and hence $\Re(f(\bar{x} \otimes x)) = 1$ for any $x \in S$. In particular, $1 = \Re(f(\bar{1} \otimes 1))$, and hence the real part of f is a state, which implies (see §A.23) that f itself is a state on $\bar{A} \otimes_{\min} A$. By Proposition 4.25 there is a net $(h_i)_{i \in I}$ in the unit sphere of $S_2(H)$ such that

$$\forall x \in E \quad f(\bar{x} \otimes x) = \lim \operatorname{tr}(x^* h_i x h_i^*).$$

Note (by the trace property) $\operatorname{tr}(x^* h_i x h_i^*) = \operatorname{tr}(h_i^* x^* h_i x)$ and hence $\operatorname{tr}(x^* h_i x h_i^*) = \langle x h_i, h_i x \rangle$ where the last inner product is relative to $S_2(H)$. Thus, for any (unitary) $x \in S$ we have

$$1 = \Re f(\bar{x} \otimes x) = \lim \Re \langle x h_i, h_i x \rangle,$$

and also $\|h_i x\|_{S_2(H)} = \|x h_i\|_{S_2(H)} = 1$. Therefore for any $x \in S$ and hence for any $x \in E$

$$\lim \|x h_i - h_i x\|_{S_2(H)} = 0. \tag{11.30}$$

This implies for any $x \in E$

$$f(\bar{x} \otimes x) = \lim \operatorname{tr}(h_i^* x^* h_i x) = \lim \operatorname{tr}(h_i^* x^* x h_i) \geq 0. \tag{11.31}$$

This allows us to rewrite (11.29) more simply as

$$\forall x \in E \quad \|\pi(x)\|^2_{L_2(\tau)} \leq f(\bar{x} \otimes x). \tag{11.32}$$

We claim that equality holds in (11.32), i.e.

$$\forall x \in E \quad \|\pi(x)\|^2_{L_2(\tau)} = f(\bar{x} \otimes x). \tag{11.33}$$

It clearly suffices to show equality for all $x \in \operatorname{span}(S_1)$ for any finite subset $S_1 \subset S$, so we assume $x \in \operatorname{span}(U_1, \ldots, U_r)$ with $\{U_1, \ldots, U_r\} \subset S$. Consider the matrices defined for $1 \leq i, j \leq r$ by $a_{ij} = \langle \pi(U_i), \pi(U_j) \rangle_{L_2(\tau)}$ and $b_{ij} = f(\bar{U}_i \otimes U_j)$. We then have the following situation: we have two (nonnegative) matrices $a, b \in M_r$ such that $a \leq b$ by (11.32) and also $a_{jj} = b_{jj}$ for all j, and this clearly implies that $a = b$ (because for $c = b - a \geq 0$, $\operatorname{tr}(c) = 0 \Rightarrow c = 0$). This proves our claim (11.33). By the polarization identity of sesquilinear forms, the latter and (11.31) yield (since $\operatorname{tr}(h_i^* y^* x h_i) = \operatorname{tr}(h_i h_i^* y^* x)$ by the trace property)

$$\forall x, y \in E \quad \tau(\pi(y^* x)) = f(\bar{y} \otimes x) = \lim \operatorname{tr}(h_i^* y^* h_i x) = \lim \operatorname{tr}(h_i h_i^* y^* x). \tag{11.34}$$

Since the finite rank operators are dense in $S_2(H)$ we may assume by perturbation that the h_i's are all of finite rank. Lastly, passing to an ultrafilter $\mathcal{U}$ refining the net (see §A.4) we may as well assume that the preceding limits are all with respect to $\mathcal{U}$. This completes the proof, since (recall Proposition 2.11) the converse direction is obvious. $\qquad\square$

Remark 11.40 (Complement to the proof) Note for further use that for any unitary $x \in B(H)$

$$\|xh_i - h_ix\|_{S_2(H)} = \|x^*(xh_i - h_ix)x^*\|_{S_2(H)}$$
$$= \|h_ix^* - x^*h_i\|_{S_2(H)}$$
$$= \|xh_i^* - h_i^*x\|_{S_2(H)},$$

and moreover (derivation rule) $xyh_i - h_ixy = x(yh_i - h_iy) + (xh_i - h_ix)y$. Therefore (11.30) still holds for any x in the group $G_S \subset U(A)$ generated by $S \cup S^{-1}$. Applying the derivation rule again we see that $\lim_{\mathcal{U}} \|xh_ih_i^* - h_ih_i^*x\|_{S_1(H)} = 0$. Thus we have

$$\forall x \in G_S \quad \lim_{\mathcal{U}} \|xh_ih_i^* - h_ih_i^*x\|_{S_1(H)} = 0. \tag{11.35}$$

By perturbation, we may clearly assume that the finite rank operators $h_ih_i^*$ have rational eigenvalues.

Remark 11.41 In [39], a state φ on $A \subset B(H)$ is called an "amenable trace" if it can be extended to a state $\widetilde{\varphi}$ on $B(H)$ such that $\widetilde{\varphi}(U^*bU) = \widetilde{\varphi}(b)$ for any $U \in U(A)$ and any $b \in B(H)$. In the situation of the preceding proof, assuming E dense in A, let $\varphi(x) = \tau(\pi(x))$ ($x \in A$), and $\widetilde{\varphi}(b) = \lim_{\mathcal{U}} \mathrm{tr}(h_i^*bh_i)$ ($b \in B(H)$). By (11.30) we have $\widetilde{\varphi}(b) = \widetilde{\varphi}(U^*bU)$ for any $U \in S$ and by (11.34) $\widetilde{\varphi}_{|E} = \varphi_{|E}$. Therefore, if E is dense in A, (11.25) implies that φ is an amenable trace on A. Conversely, if φ is an amenable trace then (11.25) holds for $E = A$. This is easy to check using the Powers–Størmer inequality (11.38) which comes next. The notion of "amenable trace" generalizes to C^*-algebras that of "hypertrace" that will be discussed for von Neumann algebras in §11.7.

Theorem 11.42 *The conclusion of Theorem 11.38 can be strengthened as follows:*

There are a Hilbert space $\mathcal{H}$, an embedding $\sigma : A \to B(\mathcal{H})$, a family of finite rank projections $(R_i)_{i\in I}$ on $\mathcal{H}$ and an ultrafilter $\mathcal{U}$ on I such that

$$\forall y,x \in E \quad \tau(\pi(y)^*\pi(x)) = \lim_{\mathcal{U}} (\mathrm{tr}(R_i))^{-1}\mathrm{tr}(R_i\sigma(y)^*R_i\sigma(x)), \tag{11.36}$$

and

$$\forall y,x \in E \quad \lim_{\mathcal{U}} \big|(\mathrm{tr}(R_i))^{-1}\mathrm{tr}(R_i\sigma(y)^*\sigma(x))$$
$$- (\mathrm{tr}(R_i))^{-1}\mathrm{tr}(R_i\sigma(y)^*R_i\sigma(x))\big| = 0. \tag{11.37}$$

Note that (11.36) is similar to (11.27) with h_i replaced by $h_i' = \mathrm{tr}(R_i)^{-1/2}R_i$.

To complete the proof we will need the Powers–Størmer inequality and Lemma 11.45.

Remark 11.43 Actually we will prove that (11.37) holds for all y in the linear span of the group generated by S in $U(A)$ and for all $x \in A$.

We need some technical preliminary to be able to complete the proof.

Lemma 11.44 (Powers–Størmer inequality) *Let* $s, t \geq 0$ *be trace class operators on* H. *Then*

$$\|s^{1/2} - t^{1/2}\|_2^2 \leq \|s - t\|_1.$$

In particular, for any unitary $U \in B(H)$

$$\|U^* t^{1/2} U - t^{1/2}\|_2^2 \leq \|U^* t U - t\|_1. \tag{11.38}$$

Proof Let e_k be an orthonormal basis of H consisting of eigenvectors of $s^{1/2} - t^{1/2}$ and let λ_k be the corresponding (real) eigenvalues. Note for later use that for any self-adjoint T we have $-|T| \leq T \leq |T|$ and hence for any $x \in H$

$$|\langle x, Tx \rangle| \leq \langle x, |T|x \rangle. \tag{11.39}$$

Note that since $\pm(s^{1/2} - t^{1/2}) \leq s^{1/2} + t^{1/2}$ it follows that $|\langle x, (s^{1/2} - t^{1/2})x \rangle| \leq \langle x, (s^{1/2} + t^{1/2})x \rangle$ for all $x \in H$, and hence

$$|\lambda_k| \leq \langle e_k, (s^{1/2} + t^{1/2})e_k \rangle.$$

Thus we have

$$\|s^{1/2} - t^{1/2}\|_2^2 = \mathrm{tr}(|s^{1/2} - t^{1/2}|^2) = \sum \langle e_k, |s^{1/2} - t^{1/2}|^2 e_k \rangle = \sum |\lambda_k|^2$$
$$\leq \sum |\lambda_k| \langle e_k, (s^{1/2} + t^{1/2})e_k \rangle.$$

Let ε_k be the sign of λ_k. Then $|\lambda_k| \langle e_k, (s^{1/2} + t^{1/2})e_k \rangle$ is the same as both

$$\varepsilon_k \langle (s^{1/2} - t^{1/2})e_k, (s^{1/2} + t^{1/2})e_k \rangle \quad \text{and} \quad \varepsilon_k \langle e_k, (s^{1/2} + t^{1/2})(s^{1/2} - t^{1/2})e_k \rangle.$$

Therefore

$$\sum |\lambda_k| \langle e_k, (s^{1/2} + t^{1/2})e_k \rangle = \sum 2^{-1} \varepsilon_k \big(\langle (s^{1/2} - t^{1/2})e_k, (s^{1/2} + t^{1/2})e_k \rangle$$
$$+ \langle e_k, (s^{1/2} + t^{1/2})(s^{1/2} - t^{1/2})e_k \rangle \big)$$
$$\leq 2^{-1} \sum |\langle e_k, (s^{1/2} - t^{1/2})(s^{1/2} + t^{1/2})$$
$$+ (s^{1/2} + t^{1/2})(s^{1/2} - t^{1/2})e_k \rangle|$$
$$= \sum |\langle e_k, (s - t)e_k \rangle|$$
$$\leq \sum \langle e_k, |s - t|e_k \rangle = \mathrm{tr}|s - t| = \|s - t\|_1,$$

where for the last $\leq$ we used (11.39). □

The next lemma is taken from [39, Lemma 6.2.5]. It is the most diffi-cult step.

Lemma 11.45 *For any* $x \in B(H)$ *we denote*

$$x' = x \otimes Id_{\ell_2} \in B(H \otimes_2 \ell_2).$$

Let $h \in B(H)$ *be a finite rank operator with* $\text{tr}(hh^*) = 1$. *Let* $t = hh^*$. *We assume that* $t = hh^*$ *has rational eigenvalues. Then there are an integer* r *and a projection* R *of rank* r *such that:*

$$\forall U \in U(H) \quad |\text{tr}(t) - r^{-1}\text{tr}(RU'^*RU')| \leq 2\|U^*tU - t\|_1^{1/2}, \qquad (11.40)$$

and more generally:

$$\forall U, V \in U(H) \quad \left|\text{tr}(U^*Vt) - r^{-1}\text{tr}\left(RU'^*RV'\right)\right|$$
$$\leq 2\left\|U^*tU - t\right\|_1^{1/4} \left\|V^*tV - t\right\|_1^{1/4}. \qquad (11.41)$$

Proof Let $t = \frac{p_1}{r}Q_1 + \frac{p_2}{r}Q_2 + \cdots + \frac{p_k}{r}Q_k$ be the spectral decomposition of t where the eigenvalues $\frac{p_j}{r}$ are in increasing order and the projections Q_j are mutually orthogonal. Note that $\sum_j p_j\text{tr}(Q_j) = \sum_j p_j\text{rk}(Q_j) = r\text{tr}(t) = r$. Let $K = \ell_2$. Let $P_1 \leq P_2 \leq \cdots \leq P_k$ be projections on K such that $\text{tr}(P_j) = \text{rk}(P_j) = p_j$ for any j. We then define a projection R on $H \otimes K$ by

$$R = \sum_{j=1}^k Q_j \otimes P_j.$$

Note $\text{tr}(R) = \text{rk}(R) = r$. Then we observe

$$\forall x \in B(H) \quad \text{tr}(xt) = \text{tr}(xhh^*) = r^{-1}\text{tr}(x'R). \qquad (11.42)$$

Indeed, $r^{-1}\text{tr}(x'R) = r^{-1}\sum \text{tr}(xQ_j)\text{tr}(P_j) = \text{tr}(xr^{-1}\sum Q_jp_j) = \text{tr}(xt) = \text{tr}(xhh^*)$.

We will first show that (11.40) implies (11.41). This uses an idea (an operator valued Cauchy–Schwarz inequality) similar to the one used to prove Theorem 5.1 about the multiplicative domain of a c.p. map (here the relevant map is $x \mapsto Rx'R$). Let

$$\forall x, y \in B(H) \quad F(x,y) = r^{-1}\text{tr}(x'^*y'R) - r^{-1}\text{tr}(Rx'^*Ry'). \qquad (11.43)$$

We claim that $F(x,x) \geq 0$. Indeed, $R \leq I$ implies

$$Rx'^*Rx'R = (x'R)^*R(x'R) \leq (x'R)^*I(x'R) = Rx'^*x'R,$$

and since $R^2 = R$ we have $\text{tr}(Rx'^*Rx') = \text{tr}(Rx'^*Rx'R)$ and $\text{tr}(x'^*x'R) = \text{tr}(Rx'^*x'R)$, from which the claim follows. Then by Cauchy–Schwarz we have

$$|F(x,y)| \leq F(x,x)^{1/2}F(y,y)^{1/2}.$$

In particular, $|F(U, V)| \le F(U, U)^{1/2} F(V, V)^{1/2}$. By (11.42) we have $\text{tr}(U^* V t) = r^{-1} \text{tr}(U'^* V' R)$. This shows that (11.40) $\Rightarrow$ (11.41).

We now turn to the more delicate verification of (11.40), for which we follow [39]. Since this elementary fact is implicitly used several times in the proof we remind the reader that $\text{tr}(xy) = \text{tr}(x^{1/2} y x^{1/2}) \ge 0$ whenever $x \ge 0$ and $y \ge 0$ are (say) Hilbert–Schmidt.

In the present situation since $\text{tr}(t) = 1$ the Powers–Størmer inequality (11.38) gives us

$$2(1 - \text{tr}(t^{1/2} U^* t^{1/2} U)) = \|t^{1/2} - U^* t^{1/2} U\|_2^2 \le \|U^* t U - t\|_1. \quad (11.44)$$

A simple verification shows that

$$r^{-1} \text{tr}(R U'^* R U') = \sum_{m,\ell} \frac{\min(p_m, p_\ell)}{r} \text{tr}(Q_m U^* Q_\ell U).$$

Plugging in the elementary inequality

$$\forall p, p' \ge 0 \quad \min(p, p') = \frac{1}{2} \left(p + p' - |p - p'| \right) \ge (pp')^{1/2} - \frac{1}{2} |p - p'|,$$

we find

$$r^{-1} \text{tr}(R U'^* R U') \ge \beta - \gamma, \quad (11.45)$$

where

$$\beta = \text{tr}(t^{1/2} U^* t^{1/2} U) \quad \text{and} \quad \gamma = (2r)^{-1} \sum_{m,\ell} |p_m - p_\ell| \, \text{tr}(Q_m U^* Q_\ell U).$$

Using $|p_m - p_\ell| = |p_m^{1/2} - p_\ell^{1/2}|(p_m^{1/2} + p_\ell^{1/2})$, we find by Cauchy–Schwarz

$$\gamma \le (2r)^{-1} \left(\sum_{m,\ell} (p_m^{1/2} - p_\ell^{1/2})^2 \text{tr}(Q_m U^* Q_\ell U) \right)^{1/2}$$
$$\times \left(\sum_{m,\ell} (p_m^{1/2} + p_\ell^{1/2})^2 \text{tr}(Q_m U^* Q_\ell U) \right)^{1/2}$$

and after expanding $(p_m^{1/2} - p_\ell^{1/2})^2$ and $(p_m^{1/2} + p_\ell^{1/2})^2$ we find

$$\gamma \le (2r)^{-1} (2r - 2r \text{tr}(t^{1/2} U^* t^{1/2} U))^{1/2} (2r + 2r \text{tr}(t^{1/2} U^* t^{1/2} U))^{1/2}.$$

Since the last term is $\le (4r)^{1/2}$ we obtain

$$\gamma \le (2 - 2 \text{tr}(t^{1/2} U^* t^{1/2} U))^{1/2},$$

and hence by (11.44)

$$\gamma \le \|U^* t U - t\|_1^{1/2}.$$

By (11.44) again $1 - \beta = 1 - \mathrm{tr}(t^{1/2}U^*t^{1/2}U) \leq 2^{-1}\|U^*tU - t\|_1 \leq 1$. A fortiori $1 - \beta \leq \|U^*tU - t\|_1^{1/2}$. Thus, recalling (11.43) and (11.45), we obtain

$$F(U,U) \leq 1 - \beta + \gamma \leq 2\|U^*tU - t\|_1^{1/2}.$$

This proves (11.40). $\qquad\qquad\qquad\square$

Proof of Theorem 11.42 Let $\mathcal{H} = H \otimes_2 \ell_2$ and $\sigma(a) = a'$ for $a \in A$. We will exploit Remark 11.40 and (11.35). Let (h_i) be as in Theorem 11.38. By perturbation we may assume that $h_ih_i^*$ has rational eigenvalues. By Lemma 11.45 we can find a net of projections R_i of rank $n(i) = \mathrm{tr}(R_i)$ on $\mathcal{H}$ so that (11.40) and (11.41) are satisfied for $t = h_ih_i^*$. Taking $U = 1$ in (11.41) (or invoking (11.42)) we find

$$\forall x \in U(B(H)) \quad n(i)^{-1}\mathrm{tr}(R_ix') = \mathrm{tr}(xh_ih_i^*). \qquad (11.46)$$

Then (11.41) and (11.35) imply for any $U \in G_S$ and $V \in U(B(H))$

$$\lim_{\mathcal{U}} |\mathrm{tr}(U^*Vh_ih_i^*) - n(i)^{-1}\mathrm{tr}(R_iU'^*R_iV')| = 0,$$

and hence by (11.46)

$$\lim_{\mathcal{U}} |n(i)^{-1}\mathrm{tr}(R_iU'^*V') - n(i)^{-1}\mathrm{tr}(R_iU'^*R_iV')| = 0,$$

which implies (11.37) and Remark 11.43. Recalling (11.34) we find

$$\forall U,V \in S \quad \tau(\pi(U^{-1}V)) = \lim_{\mathcal{U}} n(i)^{-1}\mathrm{tr}(R_iU'^*V')$$
$$= \lim_{\mathcal{U}} n(i)^{-1}\mathrm{tr}(R_iU'^*R_iV'),$$

which implies (11.36) since E is spanned by S. $\qquad\qquad\square$

Corollary 11.46 *In the situation of Theorem 11.38, assume that E generates A (as a C^*-algebra) and that $\pi(S) \subset U(M)$ is a subgroup generating M (i.e. such that $\pi(S)'' = M$). Then, for some H, π factors through $B(H)$ via unital c.p. maps. More precisely, there is a family $(M_{n(i)})_{i\in I}$ of matrix algebras and an ultrafilter $\mathcal{U}$ on I, so that π admits a factorization of the form*

$$A \xrightarrow{u} B \xrightarrow{q_{\mathcal{U}}} M_{\mathcal{U}} \xrightarrow{v} M \qquad (11.47)$$

where $B = (\oplus \sum_{i\in I} M_{n(i)})_\infty$, $q_{\mathcal{U}} : B \to M_{\mathcal{U}}$ is the quotient $*$-homomorphism and $u : A \to B$ as well as $v : M_{\mathcal{U}} \to M$ are unital c.p. maps (so that $\|u\|_{cb} = \|v\|_{cb} = 1$).

Moreover, $q_{\mathcal{U}}u$ is a $*$-homomorphism, v is normal and defines a contraction from $L_2(\tau_{\mathcal{U}})$ to $L_2(\tau)$.

Lastly, M embeds in $M_{\mathcal{U}}$.

Proof With the notation from the conclusion of Theorem 11.42, let $H_i \subset \mathcal{H}$ be the range of R_i and let $n(i) = \dim H_i$. For any $x \in E$, let $u_i(x) = R_i \sigma(x)_{|H_i} \in B(H_i)$. Choosing an orthonormal basis of the range of R_i, we may view $u_i(x)$ as an element of $M_{n(i)}$. Then let $u(x) = (u_i(x)) \in B$. Clearly, this defines a unital c.p. map $u : A \to B$. By (11.36) we have

$$\forall x, y \in S \quad \tau_{\mathcal{U}}(q_{\mathcal{U}}u(y)^* q_{\mathcal{U}}u(x)) = \tau(\pi(y)^* \pi(x)). \tag{11.48}$$

By sesquilinearity this remains valid for all $x, y \in E$ and hence

$$\forall x, y \in E \quad \|q_{\mathcal{U}}u(x) - q_{\mathcal{U}}u(y)\|_{L_2(\tau_{\mathcal{U}})} = \|\pi(x) - \pi(y)\|_{L_2(\tau)} \tag{11.49}$$

and also

$$\forall x \in E \quad \tau_{\mathcal{U}}(q_{\mathcal{U}}u(x)) = \tau(\pi(x)). \tag{11.50}$$

In particular, $\tau_{\mathcal{U}}(q_{\mathcal{U}}u(x)^* q_{\mathcal{U}}u(x)) = 1$ for any $x \in S$, which shows by Remark 11.13 that $q_{\mathcal{U}}u(x) \in U(M_{\mathcal{U}})$. By Proposition 9.7 (or because S is in its multiplicative domain), the linear map $q_{\mathcal{U}}u : A \to M_{\mathcal{U}}$ is a $*$-homomorphism on A. Let $\mathcal{A} \subset M$ be the linear span of $\pi(S)$ (i.e. $\mathcal{A} = \pi(E)$). By our assumption $\mathcal{A}$ is a weak*-dense subalgebra of M. By (11.49) the correspondence $\pi(x) \mapsto q_{\mathcal{U}}u(x)$ is a well-defined linear map from $\mathcal{A}$ to $M_{\mathcal{U}}$ and it is a $*$-homomorphism (since $q_{\mathcal{U}}u$ is one). By Remark 11.20, the "$*$-distribution equality" (11.50) implies that the von Neumann subalgebra $N_{\mathcal{U}}$ generated in $M_{\mathcal{U}}$ by $\{q_{\mathcal{U}}u(x) \mid x \in S\}$ is isomorphic to $M = \pi(S)''$, via the correspondence $T : M \to M_{\mathcal{U}}$ defined on $\mathcal{A}$ by $T(\pi(x)) = q_{\mathcal{U}}u(x)$. This gives us a $*$-isomorphism $T : M \to N_{\mathcal{U}} \subset M_{\mathcal{U}}$, and hence M embeds in $M_{\mathcal{U}}$. But, by Proposition 11.21, we also have a conditional expectation $P : M_{\mathcal{U}} \to N_{\mathcal{U}}$, thus setting $v = T^{-1}P$, we obtain the desired factorization. Note that (recall (11.49)) T also extends to an isometry from $L_2(\tau)$ to $L_2(\tau_{\mathcal{U}})$, so we could invoke Proposition 11.19 instead of Remark 11.20 in the preceding argument. We have clearly $\|v : L_2(\tau_{\mathcal{U}}) \to L_2(\tau)\| = 1$ (with the obvious abuse of notation), and $v : M_{\mathcal{U}} \to M$ is normal. Lastly, since B is injective, Id_B factors through $B(H)$ (for some H) via unital c.p. maps (see §1.5), which proves the first assertion. $\square$

Remark 11.47 To see that the factorization in (11.47) implies the embedding $M \subset M_{\mathcal{U}}$, let $\rho = q_{\mathcal{U}}u : A \to M_{\mathcal{U}}$, and let $\ddot{\rho} : A^{**} \to M_{\mathcal{U}}$ be the normal $*$-homomorphism extending ρ. Since v is normal, $\pi = v\rho$ implies $\ddot{\pi} = v\ddot{\rho}$. But (see (A.37)) we know $A^{**} \simeq M \oplus \ker(\ddot{\pi})$, so that with respect to the associated embedding $M \subset A^{**}$ we have $Id_M = \ddot{\pi}_{|M}$. Thus with $Id_M = v\ddot{\rho}_{|M}$ we obtain the embedding of M in $M_{\mathcal{U}}$.

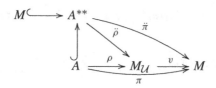

Corollary 11.48 *In the situation of Corollary 11.46, (or simply when $E = A$ in Theorem 11.38) assume given a unital embedding $A \subset A_1$ into another C^*-algebra A_1. Then any $*$-homomorphism $\pi : A \to M$ satisfying (11.25) extends to a unital c.p. mapping $\pi_1 : A_1 \to M$ still satisfying (11.25).*

Proof By the extension Theorem 1.39, $u : A \to B$ admits a u.c.p. extension $u_1 : A_1 \to B$. Then let $\pi_1 = vq_{\mathcal{U}}u_1$. Note that, by the converse direction in Theorem 11.38, $q_{\mathcal{U}}$ itself satisfies the property described by (11.25). Since $\|v : L_2(\tau_{\mathcal{U}}) \to L_2(\tau)\| = 1$ and $\|\overline{u_1} \otimes u_1 : \overline{A_1} \otimes_{\min} A_1 \to \overline{B} \otimes_{\min} B\| \le 1$, the map π_1 must also satisfy (11.25). $\qquad\square$

The extension property considered in Corollary 11.48 will be refined in a later chapter notably in Theorem 23.29.

We now relate injectivity to Theorem 11.38.

Corollary 11.49 *Let (M, τ) be a tracial probability space. The following are equivalent:*

(i) *M is injective.*

(ii) *For any finite subsets (x_j) and (y_j) in M we have*

$$\left| \sum \tau(y_j^* x_j) \right| \le \left\| \sum \overline{y_j} \otimes x_j \right\|_{\min}.$$

(iii) *For any finite subset (x_j) in M we have*

$$\sum \tau(x_j^* x_j) \le \left\| \sum \overline{x_j} \otimes x_j \right\|_{\min}.$$

Proof Assume (i). Let L, R be as in Proposition 11.16, so that $L(M) \simeq M$ is injective and $L(M)' = R(M^{op})$. By (i) $\Rightarrow$ (ii) in Theorem 8.14 applied to $L(M)$, for any n and any (y_j), (x_j) in M^n, since $\sum R(y_j^*)L(x_j)1 = \sum x_j y_j^*$, we have

$$\left| \tau \left(\sum y_j^* x_j \right) \right| \le \left\| \sum R(y_j^*)L(x_j) \right\|_{B(L_2(\tau))}$$
$$\le \left\| \sum R(y_j^*) \otimes L(x_j) \right\|_{R(M^{op}) \otimes_{\min} L(M)},$$

also $\| \sum R(y_j^*) \otimes L(x_j) \|_{R(M^{op}) \otimes_{\min} L(M)} = \| \sum y_j^* \otimes x_j \|_{M^{op} \otimes_{\min} M}$ and by (2.12)

$$\left\| \sum y_j^* \otimes x_j \right\|_{M^{op} \otimes_{\min} M} = \left\| \sum \overline{y_j} \otimes x_j \right\|_{\overline{M} \otimes_{\min} M}.$$

This shows (i) $\Rightarrow$ (ii). (ii) $\Rightarrow$ (iii) is trivial. Assume (iii). Equivalently (iii) means that (11.25) holds with $A = M$, $\pi = Id_M$ and $E = A(= M)$. The factorization in (11.47) implies that Id_M factors through B via u.c.p. maps. Therefore M is injective. $\square$

11.7 Hypertraces and injectivity

Let (M, τ) be a tracial probability space. The goal of this section is to present a particularly neat characterization of the injectivity of M, refining the last corollary when M is a factor, and its generalization involving the center of M when M is not. The proof uses the notion (due to Connes) of hypertrace.

Definition 11.50 A tracial state on $M \subset B(H)$ is called a hypertrace if it admits an extension to a state f on $B(H)$ such that

$$\forall U \in U(M), \forall x \in B(H), \quad f(UxU^*) = f(x). \qquad (11.51)$$

Note that this is the same as $\forall U \in U(M), \forall x \in B(H), f(Ux) = f(xU)$ or equivalently:

$$\forall y \in M, \forall x \in B(H), f(yx) = f(xy).$$

Hypertraces, or rather the states f satisfying (11.51), are analogous to invariant means for amenable groups. In this regard, the reader is invited to compare (11.52) to (iv) in Theorem 3.30.

For example, if $P : B(H) \to M$ is a contractive projection, then $f(x) = \tau(P(x))$ satisfies (11.51), because $P(UxU^*) = UP(x)U^*$ ($x \in B(H), U \in U(M)$) by Theorem 1.45. Thus τ is a hypertrace if M is injective. The converse is also true:

Proposition 11.51 *If τ is a hypertrace then M is injective.*

Proof Let $L_2(f)$ be the GNS Hilbert space associated to f and let $\pi_f : B(H) \to B(L_2(f))$ be the GNS representation with unit vector $\xi_f \in L_2(f)$ such that $f(x) = \langle \xi_f, \pi_f(x)\xi_f \rangle$. Since $\tau(x^*x) = f(x^*x)$ for any $x \in M$, we have an isometric embedding $L_2(\tau) \simeq \overline{\pi_f(M)\xi_f} \subset L_2(f)$. We view $M \subset L_2(\tau)$ and denote by $\psi : \overline{\pi_f(M)\xi_f} \to L_2(\tau)$ the isometric isomorphism that takes $\pi_f(a)\xi_f$ to a for any $a \in M$. Let $Q : L_2(f) \to \overline{\pi_f(M)\xi_f}$ be the orthogonal projection. We define $P : B(H) \to L_2(\tau)$ by

$P(x) = \psi Q(\pi_f(x)\xi_f)$ for any $x \in B(H)$. Then $P(x) = x$ for any $x \in M$. We claim that $P(x) \in M$ and $\|P(x)\| \le \|x\|$ for any $x \in B(H)$, which proves that M is injective. This will be checked by the same classical argument that was used to prove Proposition 11.21. To check the claim it suffices to show by Remark 11.15 that $|\tau(y_2^* P(x) y_1)| \le \|x\| \|y_1\|_{L_2(\tau)} \|y_2\|_{L_2(\tau)}$ for any $y_1, y_2 \in M$. Note $\tau(y_2^* P(x) y_1) = \tau(y_1 y_2^* P(x)) = \langle y_2 y_1^*, P(x) \rangle_{L_2(\tau)}$. Since ψ is isometric we have $\langle y_2 y_1^*, P(x) \rangle_{L_2(\tau)} = \langle \pi_f(y_2 y_1^*)\xi_f, Q(\pi_f(x)\xi_f) \rangle_{L_2(f)}$ and hence by the definition of Q and the hypertrace property of f

$$\langle y_2 y_1^*, P(x) \rangle_{L_2(\tau)} = \langle \pi_f(y_2 y_1^*)\xi_f, \pi_f(x)\xi_f \rangle_{L_2(f)}$$
$$= f(y_1 y_2^* x) = f(y_2^* x y_1)$$
$$= \langle \pi_f(y_2)\xi_f, \pi_f(x)\pi_f(y_1)\xi_f \rangle,$$

which yields $|\tau(y_2^* P(x) y_1)| \le \|x\| \|\pi_f(y_2)\xi_f\|_{L_2(f)} \|\pi_f(y_1)\xi_f\|_{L_2(f)} = \|x\| \|y_1\|_{L_2(\tau)} \|y_2\|_{L_2(\tau)}$. Thus P is a contractive projection onto M. □

Theorem 11.52 *Let (M, τ) be a tracial probability space. Let $Z = M \cap M'$ the center of M.*

(i) *If M is a factor (i.e. $Z = \mathbb{C}1$) then M is injective if and only if*

$$\forall n \ge 1, \forall U_j \in U(M) \quad n = \left\| \sum_1^n \overline{U_j} \otimes U_j \right\|_{\min}. \tag{11.52}$$

(ii) *In general, M is injective if and only if for any nonzero projection $q \in \mathcal{P}_Z$ we have*

$$\forall n \ge 1, \forall U_j \in U(M) \quad n = \left\| \sum_1^n \overline{qU_j} \otimes qU_j \right\|_{\min}. \tag{11.53}$$

Proof We first show that (11.53) holds if M is injective. Since we may replace M by qM which is still injective (with unit q and trace $x \mapsto \tau(q)^{-1}\tau(qx)$) it suffices to show this for $q = 1$. The latter case follows from (iii) in Corollary 11.49. This settles both "only if" parts.

(i) Assume (11.52). Then for any n and any $U_0, U_1, \ldots, U_{n-1}$ in $U(M)$ with $U_0 = 1$, for any $\varepsilon > 0$ by (2.9) there is a unit vector $h \in S_2(H)$ such that

$$n - \varepsilon^2/4n < \left\| \sum_1^n U_j h U_j^* \right\|_2,$$

and hence by (A.3) $\sup_j \|h - U_j h U_j^*\|_2 \le \varepsilon$. Thus there is a net (h_i) of unit vectors in $S_2(H)$ such that

$$\forall U \in U(M) \quad \|h_i - U h_i U^*\|_2 \to 0. \tag{11.54}$$

Let $\mathcal{U}$ be a nontrivial ultrafilter refining this net. We set $f(x) = \lim_{\mathcal{U}} \operatorname{tr}(x h_i^* h_i)$. Then (11.54) implies (11.51). A fortiori, $f_{|M}$ is a tracial state on M. Since (finite) factors have a *unique* tracial state (see Remark 11.33), we must have $f_{|M} = \tau$, so that τ is a hypertrace, and M is injective by Proposition 11.51.

(ii) Assume (11.53). We will use the classical fact that if two tracial states coincide on Z, then they are equal (see Remark 11.33). Consider the set I formed by all the disjoint partitions of 1_M as a finite sum $1_M = \sum_1^m q_k$ of mutually orthogonal projections in $\mathcal{P}_Z$. The set I is ordered by its natural order: $i \le i'$ means that each projection that is part of i is a sum of some of the projections in i'. Since Z is commutative, the set I is directed with respect to this order.

We claim that for any $i \in I$, $i = (q_1, \ldots, q_m)$, there is a state f_i on $B(H)$ satisfying (11.51) such that

$$\forall (\lambda_k) \in \mathbb{C}^m \quad f_i \left(\sum_1^m \lambda_k q_k \right) = \sum \lambda_k \tau(q_k). \tag{11.55}$$

We may view (f_i) as a net indexed by a directed ordered set. Let $\mathcal{U}$ be an ultrafilter refining this net and let $f = \lim_{\mathcal{U}} f_i$ (pointwise on $B(H)$). Taking the claim as granted for the moment, let us conclude. Clearly f still satisfies (11.51) and also for any $i = (q_1, \ldots, q_m)$ we have $f(\sum_1^m \lambda_k q_k) = \sum \lambda_k \tau(q_k)$ for any $(\lambda_k) \in \mathbb{C}^m$ (because for any $j \ge i$ (11.55) remains true if we replace f_i by f_j). This shows that $f_{|Z}$ and $\tau_{|Z}$ coincide on the linear span of $\mathcal{P}_Z$, and hence since the latter is norm-dense in Z (here Z is isomorphic to some L_∞-space) we conclude that $f_{|Z} = \tau|Z$. But since $f_{|M}$ is a tracial state, by the preceding classical fact $f_{|M} = \tau$. Thus τ is a hypertrace and again M is injective by Proposition 11.51.

To prove the claim, consider $i = (q_1, \ldots, q_m)$. Fix $1 \le k \le m$ and apply the preceding argument for (i) to the von Neumann algebra $q_k M \subset B(q_k H)$ (with unit q_k) instead of $M \subset B(H)$. This gives us a state f^k on $B(q_k H)$ satisfying (11.51) (with respect to $q_k H$ instead of H and $q_k M$ instead of M). Let $f_i(x) = \sum f^k(P_{q_k H} x_{|q_k H}) \tau(q_k)$ for any $x \in B(H)$. Then f_i is a state on $B(H)$ satisfying (11.51) and (11.55). This proves the claim. $\square$

Remark 11.53 Actually, any von Neumann algebra M satisfying (11.53) must be finite. Indeed, the preceding proof shows that if (11.53) holds then qM admits a tracial state for any $q \in \mathcal{P}_Z$. From this it is easy to deduce by structural arguments that M must be finite. If M is σ-finite this implies that M admits a faithful normal finite trace τ (see §23.1).

Remark 11.54 Let (M, τ) be a tracial probability space, with $M \subset B(H)$. Then τ is a hypertrace if and only if there is a net (h_i) of unit vectors in $S_2(H)$ such that $\|U h_i U^* - h_i\|_{S_2(H)} \to 0$ (or equivalently $\|U h_i - h_i U\|_{S_2(H)} \to 0$) for any $U \in U(M)$ and such that $\tau(x) = \lim \mathrm{tr}(x h_i^* h_i)$ for any $x \in M$.

Indeed, if τ is a hypertrace, let f be as in Definition 11.50. Let (t_i) be a net of unit vectors in $S_1(H) = B(H)_*$ with $t_i \ge 0$ such that $f(x) = \lim \mathrm{tr}(x t_i)$ for any $x \in B(H)$ (we may assume $t_i \ge 0$ because $\|t_i\|_{S_1(H)} \le 1$ and

$\mathrm{tr}(t_i) \to 1$ together imply that there is $t_i' \geq 0$ such that $\|t_i - t_i'\|_{S_1(H)} \to 0$). Then (11.51) implies that $Ut_iU^* - t_i \to 0$ for $\sigma(B(H)_*, B(H))$. By Mazur's Theorem A.9 after passing to a different net we may as well assume that $\|Ut_iU^* - t_i\|_{S_1(H)} \to 0$. Let $h_i = t_i^{1/2}$. By the Powers–Størmer inequality (11.38) we have $\|Uh_iU^* - h_i\|_{S_2(H)} \to 0$, while since $f_{|M} = \tau$ we have $\tau(x) = \lim \mathrm{tr}(xh_i^2)$ for any $x \in M$. No wonder if this argument rings a bell: we used an analogous one to prove (i) $\Rightarrow$ (ii) in Theorem 3.30. This proves the "only if" part. The converse is immediate (just set $f(x) = \lim_\mathcal{U} \mathrm{tr}(xh_i^*h_i)$ for any $x \in B(H)$).

Note that the existence of such a net (h_i) implies that $\pi = Id_M$ satisfies (11.27) and hence Proposition 11.51 can be deduced alternatively from Corollary 11.46 as in the proof of Corollary 11.49.

11.8 The factorization property for discrete groups

We already introduced the factorization property in Definition 7.36. We now give several equivalent properties.

Theorem 11.55 *The following properties of discrete group G are equivalent:*

(i) *The unitary representation $(s,t) \mapsto \lambda_G(s)\rho_G(t)$ on $G \times G$, extends to a (continuous) representation on $C^*(G) \otimes_{\min} C^*(G)$. In other words G has the factorization property.*

(ii) *The linear functional f defined on $\mathbb{C}[G] \otimes \mathbb{C}[G]$ by $f(x \otimes y) = \sum_{t \in G} x(t)y(t)$ extends to a linear form of norm 1 on $C^*(G) \otimes_{\min} C^*(G)$.*

(iii) *For any finite sequence $x_1, \ldots, x_n$ in $C^*(G)$ we have*

$$\sum \|\lambda_G(x_j)\|_{L_2(\tau_G)}^2 \leq \left\|\sum \overline{x_j} \otimes x_j\right\|_{\overline{C^*(G)} \otimes_{\min} C^*(G)}. \tag{11.56}$$

(iv) *The natural $*$-homomorphism $\dot{Q}_G : C^*(G) \to M_G$ (such that $\dot{Q}_G(U_G(t)) = \lambda_G(t)$ for all $t \in G$) factorizes via unital c.p. maps through $B(H)$ for some H.*

Proof Assume (i). Let f be as in (ii). Let $x, y \in \mathbb{C}[G]$. We set $\lambda_G(x) = \sum_{s \in G} x(s)\lambda_G(s)$ and $\rho_G(y) = \sum_{t \in G} y(t)\rho_G(t)$. Then $f(x \otimes y) = \langle \delta_e, \lambda_G(x)\rho_G(y)\delta_e \rangle$. This shows (i) $\Rightarrow$ (ii).

Assume (ii). The linear mapping taking $\sum \overline{x(t)}U_G(t)$ to $\overline{\sum x(t)U_G(t)}$ ($\forall x \in \mathbb{C}[G]$) extends to a $\mathbb{C}$-linear isomorphism $\Phi : C^*(G) \to \overline{C^*(G)}$. Therefore (ii) implies

$$\left|\sum_j \sum_t \overline{x_j(t)} y_j(t)\right| \le \left\|\sum \overline{x_j} \otimes y_j\right\|_{\overline{C^*(G)} \otimes_{\min} C^*(G)},$$

and hence taking $x_j = y_j$ we obtain (iii).

Assume (iii). We apply Corollary 11.46 to the case $A = C^*(G)$ with $S = \{U_G(t) \mid t \in G\}$ (i.e. essentially $S = G$) and $\pi = \dot{Q}_G$. This implies (iv).

Assume (iv). We first show that (iv) $\Rightarrow$ (iii). Unfortunately, we need to invoke an inequality satisfied by $B(H)$ that is proved only later on these notes, namely (22.15). The latter implies that if a $*$-homomorphism $\pi : C^*(G) \to M_G$ satisfies the factorization in (iv) then for any finite set (x_j) in $C^*(G)$

$$\left\|\sum \overline{\pi(x_j)} \otimes \pi(x_j)\right\|_{\overline{M_G} \otimes_{\max} M_G} \le \left\|\sum \overline{x_j} \otimes x_j\right\|_{\overline{C^*(G)} \otimes_{\min} C^*(G)}.$$

Since

$$\sum \left\|\pi(x_j)\right\|_{L_2(\tau_G)}^2 \le \left\|\sum \overline{\pi(x_j)} \otimes \pi(x_j)\right\|_{\overline{M_G} \otimes_{\max} M_G}$$

we obtain (iii).

Assume (iii). Let $A = C^*(G)$, $E = \text{span}[U_G(t) \mid t \in G]$ and $\pi = \dot{Q}_G$ as before. By Theorem 11.38 choosing a suitable embedding $A \subset B(H)$ there is a net (h_i) in the unit ball of $S_2(H)$ such that for any $x, y \in E$, say $x = \sum x(t) U_G(t)$, $y = \sum y(t) U_G(t)$ we have

$$\sum \overline{x(t)} y(t) = \tau_G(\pi(x^* y)) = \lim_{\mathcal{U}} \text{tr}(x^* h_i y h_i^*).$$

By Proposition 2.11 this implies

$$\left|\sum_j \sum_t \overline{x_j(t)} y_j(t)\right| = \lim_{\mathcal{U}} \left|\text{tr}(\sum x_j^* h_i y_j h_i^*)\right| \le \left\|\sum \overline{x_j} \otimes y_j\right\|_{\overline{A} \otimes_{\min} A}.$$

Thus, using again the $\mathbb{C}$-linear isomorphism $\Phi : C^*(G) \to \overline{C^*(G)}$, (ii) follows. To conclude we show (ii) $\Rightarrow$ (i). This is a routine argument based on the observation that the representation appearing in (i) is a GNS representation associated to the state appearing in (ii). Let $T = \sum x_j \otimes y_j \in A \otimes A$ with $\|T\|_{\min} \le 1$. Then $1 \otimes 1 - T^* T \in (A \otimes_{\min} A)_+$. Let $\kappa : A \otimes A \to B(\ell_2(G))$ be the $*$-homomorphism associated to $(s, t) \mapsto \lambda_G(s) \rho_G(t)$. Note for any $z \in A \otimes A$

$$f(z) = \langle \delta_e, \kappa(z) \delta_e \rangle.$$

For any $z \in A \otimes A$ we have $z^*(1 \otimes 1 - T^* T)z \in (A \otimes_{\min} A)_+$ and hence assuming (ii), we have

$$\|\kappa(T)\kappa(z)\delta_e\|_{\ell_2(G)}^2 = f(z^* T^* T z) \le f(z^* z) = \|\kappa(z)\delta_e\|_{\ell_2(G)}^2,$$

which shows $\|\kappa(T)\|_{B(\ell_2(G))} \le 1$. This completes the proof that (ii) $\Rightarrow$ (i). $\square$

Corollary 11.56 *If G has the factorization property, in particular if G is amenable (see Remark 7.37) then M_G embeds in a trace preserving way in an ultraproduct of matrix algebras.*

Proof This follows by applying Corollary 11.46 to the case when $A = C^*(G)$, $M = M_G = \lambda_G(G)''$, $S = G$ viewed as a subset of $U(A)$ and $\pi = \dot{Q}_G$. □

Remark 11.57 We will show in Corollary 12.23 that all free groups have the factorization property.

11.9 Notes and remarks

The construction of noncommutative measure theory outlined in §11.2 was motivated by quantum mechanics, and hence goes far back. The results of §11.2 are all classical facts. As for the origin of noncommutative L_p-spaces one usually attaches the names of Dixmier, Kunze, Segal, and more recently Nelson, see [215] for more information on this topic. The construction of ultraproducts goes back to McDuff [176]. See [4] for a much more complete treatment of the ramifications of this important topic. The reader interested in ultraproducts in the nontracial case is referred to [240, p. 115] and the more recent papers [8, 9]. Theorem 11.38 is a relatively easy fact reformulating ideas that can be traced back to Kirchberg's [155] in the style of a Pietsch factorization for 2-summing maps as in [205, §5]. The unital assumption (which guarantees that $\pi(S) \subset U(M)$) is the key to obtain an equality as in (11.27). The proof of Theorem 11.42 is more delicate. The main point comes from Brown and Ozawa's book [39, Lemma 6.2.5]. Its relevance to the situation considered in §11.6 was pointed out by Ozawa in [191]. The notion of hypertrace, together with Proposition 11.51 and Theorem 11.52 for factors are all due to Connes [61]. The generalization in Theorem 11.52 (ii) comes from Haagerup's [104] on which §11.7 is based.

The factorization property was introduced by Kirchberg in [155]. It is particularly interesting in connection with property (T) groups as in Theorem 17.5. The equivalence of the properties in Theorem 11.55 were surely known to Kirchberg [157]. For its proof we used some simplifications due to Ozawa [189]. See also [39, p. 219].

12

The Connes embedding problem

We now turn to the first of a series of problems that will turn out to be eventually all equivalent. Since it was formulated as a question (or, say, a problem) we do not refer to it as a conjecture.

12.1 Connes's question

In the classical von Neumann algebra terminology, a "II_1-factor" is an infinite-dimensional tracial probability space (M, τ) with trivial center, i.e. such that $M \cap M' = \mathbb{C}I$. In his famous paper [61], Connes observed that in addition to the case when G is amenable, in the somewhat "opposite" case when $G = \mathbb{F}_n$ ($2 \leq n \leq \infty$), the II_1-factor (M_G, τ_G) also embeds in a trace preserving way in $(M_{\mathcal{U}}, \tau_{\mathcal{U}})$ for some $\mathcal{U}$. Since the latter case was at the time the principal "bad apple" in the classification theory of factors, it was natural for Connes to wonder whether in fact the same embedding held for any discrete group G and any II_1-factor. But since it can be shown, using free products (see Remark 11.36) that any tracial probability space embeds (in a trace preserving way) in a II_1-factor, we can rephrase the problem more generally as follows.

Connes's question: *Is it true that any tracial probability space (M, τ) embeds in a trace preserving way in an ultraproduct of matricial tracial probability spaces $(M_{\mathcal{U}}, \tau_{\mathcal{U}})$ for some $\mathcal{U}$?*

A priori, the embedding in the preceding question must preserve much of the algebraic structure. But actually, much less is needed for the same conclusion: as we will show in Theorem 12.3, a mere isometric assumption (as opposed to a completely isometric one) implies a strong $*$-isomorphic conclusion. Not surprisingly, one of the key ingredients to prove this comes from the isometric theory (or the Jordan theory) of C^*-algebras, namely the following result, going back to Kadison [142] and Størmer [233]. See [123] for an excellent detailed account of this theory.

Theorem 12.1 *Let (M, τ) and (N, φ) be two tracial probability spaces, i.e. two von Neumann algebras equipped with faithful, normal, normalized traces. Let $T : L_2(\tau) \to L_2(\varphi)$ be an isometry such that $T(1) = 1$ and $T^*(1) = 1$. In other words, T is unital and trace preserving. Assume that $T(B_M) \subset B_N$ (i.e. T defines a mapping of norm 1 from M to N). Then $T : M \to N$ decomposes as a direct sum of a $*$-homomorphism and a $*$-antihomomorphism. More precisely, there is an orthogonal decomposition $I = P + Q$ in N ($P \perp Q$) with $P, Q \in T(M)' \cap N$ and an associated decomposition*

$$\forall x \in M \quad T(x) = PT(x)P + QT(x)Q$$

such that $x \mapsto PT(x)P$ is a $*$-homomorphism and $x \mapsto QT(x)Q$ is a $*$-antihomomorphism.

Proof By assumption $T : L_2(\tau) \to L_2(\varphi)$ is isometric but we assume $\|T : M \to N\| = 1$. From now on, we view T as a mapping from M to N. Note that the latter mapping is normal since its adjoint takes $L_2(\varphi)$ to $L_2(\tau)$, and hence by density (since $L_2(\varphi)$ is norm dense in $N = L_1(\varphi)$) it takes $L_1(\varphi)$ to $L_1(\tau)$, or equivalently N_* to M_*. Recapitulating, $T : M \to N$ is unital, normal, and preserves the trace. We will show that T takes unitaries to unitaries. Indeed, for any unitary $u \in M$, we have $\|T(u)\|_N \leq 1$ and $\|T(u)\|_2^2 = \|u\|_2 = 1$, so $T(u)$ is unitary by Remark 11.13.

We claim this implies that T is a Jordan $*$-morphism, i.e. that for any hermitian h, its image $T(h)$ is hermitian and we have $T(h^2) = T(h)^2$. Indeed, since $T(e^{ith}) = 1 + itT(h) + o(t)$ and $\|T(e^{ith})\| \leq 1$, $T(h)$ must be hermitian, and actually since $T(e^{ith})$ is unitary, if we develop further for small $t \in \mathbb{R}$

$$1 = (T(e^{ith}))^* T(e^{ith}) = 1 + t^2(T(h)^2 - T(h^2)) + o(t^2),$$

we find that necessarily $T(h)^2 = T(h^2)$. From the theory of Jordan representations as already used in Theorem 5.6 (see [123, p. 163] or also [147, pp. 588–589]), we know that there is an orthogonal decomposition $I = P + Q$ with $P, Q \in T(M)' \cap N$, $P \perp Q$ that gives us a decomposition

$$\forall x \in M \quad T(x) = PT(x)P + QT(x)Q$$

such that $x \mapsto PT(x)P$ is a $*$-homomorphism and $x \mapsto QT(x)Q$ is a $*$-antihomomorphism. (Of course these are a priori nonunital.) $\square$

Proposition 12.2 *Let (M, τ) and (N, φ) be two tracial probability spaces and let $\mathcal{W} \subset U(M)$ be a weak*-dense subset of the unitary group of M, i.e. such that $U(M)$ is the closure of $\mathcal{W}$ in $L_2(\tau)$. The following are equivalent:*

(i) *There is a unital trace preserving ($\mathbb{C}$-linear) isometry*
   $T : L_2(\tau) \to L_2(\varphi)$ *such that* $\|T : M \to N\| \le 1$.

(ii) *There is a function* $r : \{1\} \cup \mathcal{W} \to U(N)$ *such that* $r(1) = 1$ *and*

$$\forall u_1, u_2 \in \{1\} \cup \mathcal{W} \quad \tau(u_1^* u_2) = \varphi(r(u_1)^* r(u_2)). \tag{12.1}$$

(ii)' *There is a function* $r : \mathcal{W} \to B_N$ *such that*

$$\forall u_1, u_2 \in \mathcal{W} \quad \tau(u_1^* u_2) = \varphi(r(u_1)^* r(u_2)). \tag{12.2}$$

Proof Assume (i). The first step of the proof of Theorem 12.1 shows that $T(U(M)) \subset U(N)$. Then (i) $\Rightarrow$ (ii) is obvious: we just let $r = T_{|\{1\} \cup \mathcal{W}}$ and observe that

$$\forall u_1, u_2 \in U(M) \ \forall \mu \in \mathbb{C} \quad \|u_1 + \mu u_2\|_{L_2(\tau)}^2 = 1 + |\mu|^2 + 2\Re(\mu\tau(u_1^* u_2)).$$

Since $\|u_1 + \mu u_2\|_{L_2(\tau)}^2 = \|r(u_1) + \mu r(u_2)\|_{L_2(\varphi)}^2$, it follows that $r = T_{|\{1\} \cup \mathcal{W}}$ satisfies (ii).

   Assume (ii). Assuming for a moment that such a T exists, let $T : \mathrm{span}[\{1\} \cup \mathcal{W}] \to N$ be the linear mapping extending r. For any $x \in \mathrm{span}[\{1\} \cup \mathcal{W}]$, we have by (12.1)

$$\|T(x)\|_{L_2(\varphi)}^2 = \|x\|_{L_2(\tau)}^2,$$

and this shows that T is unambiguously well defined. Note that T preserves the trace (take $u_1 = 1$ in (12.1)), and $T(1) = r(1) = 1$. By our density assumption, the unitaries in $\mathcal{W}$ are dense for the $L_2(\tau)$-norm in the set $U(M)$ of unitaries of M (which linearly spans M), and hence $\mathrm{span}[\mathcal{W}]$ is dense in $L_2(\tau)$. Therefore T extends to a unital trace preserving isometry, still denoted by T, from $L_2(\tau)$ to $L_2(\varphi)$, such that $T(\mathcal{W}) \subset U(N)$. Moreover, since the set of unitary elements is closed in $L_2(\varphi)$ (see e.g. Remark 11.13), and $\mathcal{W}$ is assumed dense in $U(M)$, we have $T(U(M)) \subset U(N)$. By the Russo–Dye Theorem A.18, the convex hull of $U(M)$ is norm-dense (and a fortiori $L_2(\tau)$-dense) in the unit ball of M, and the unit ball of N is closed in $L_2(\varphi)$. Therefore $\|T : M \to N\| = 1$. This shows that (i) holds, and hence (i) $\Leftrightarrow$ (ii).

   (ii) $\Rightarrow$ (ii)' is trivial. Conversely, assume (ii)'. Then $\varphi(r(u)^* r(u)) = 1$ for any $u \in \mathcal{W}$. The fact that $r(\mathcal{W}) \subset U(N)$ is automatic by Remark 11.13. To take care of the condition $r(1) = 1$, we will change $\mathcal{W}$ and r. We pick a fixed $u_0 \in \mathcal{W}$ and we set $\mathcal{W}' = u_0^{-1}\mathcal{W} = \{u_0^{-1}u \mid u \in \mathcal{W}\}$ and $r'(u_0^{-1}u) = r(u_0)^{-1}r(u)$. Then $1 \in \mathcal{W}'$, $r'(1) = 1$, $(\mathcal{W}',r')$ satisfy (12.1) and $\mathcal{W}'$ is still dense in $U(M)$. Therefore, by the (already proved) implication (ii) $\Rightarrow$ (i) applied to $\mathcal{W}'$ and r', we see that (i) holds, and we already proved that (i) implies (ii). $\qquad\square$

The following criterion for M to embed in some $M_{\mathcal{U}}$ is due to Kirchberg.

Theorem 12.3 (Kirchberg's criterion) *Let (M, τ) be a tracial probability space. Let $\mathcal{W} \subset U(M)$ be a weak*-dense unital subset of the unitary group of M, i.e. we assume that $U(M)$ is the closure of $\mathcal{W}$ in $L_2(\tau)$. The following are equivalent:*

(i) *There is a (unital) trace preserving embedding of M in an ultraproduct of matrix algebras.*

(ii) *For any $\varepsilon > 0$, any n and any $u_1, \ldots, u_n \in \mathcal{W}$ there is an integer N and unitary $N \times N$ matrices $v_1, \ldots, v_n$ such that*

$$\forall i, j = 1, \ldots, n \quad |\tau(u_i^* u_j) - \tau_N(v_i^* v_j)| \le \varepsilon, \qquad (12.3)$$

and

$$\forall j = 1, \ldots, n \quad |\tau(u_j) - \tau_N(v_j)| \le \varepsilon. \qquad (12.4)$$

(iii) *For any $\varepsilon > 0$, any n and any $u_1, \ldots, u_n \in \mathcal{W}$ there is an integer N and $N \times N$ matrices $x_1, \ldots, x_n$ in the unit ball of M_N, equipped with its normalized trace τ_N, such that*

$$\forall i, j = 1, \ldots, n \quad |\tau(u_i^* u_j) - \tau_N(x_i^* x_j)| \le \varepsilon. \qquad (12.5)$$

Moreover, if $\mathcal{W} \subset U(M)$ is countable, (ii) $\Rightarrow$ (i) holds for an ultraproduct based on a sequence of matrix algebras.

Proof (i) $\Rightarrow$ (ii) is essentially obvious with $\mathcal{W} = U(M)$ (recall Lemma 11.30). (ii) $\Rightarrow$ (iii) is trivial. We give the rest of the proof assuming that $\mathcal{W}$ is countable. The general case can be treated similarly.

Assume (iii). To show that (i) holds we will use Proposition 12.2 and Theorem 12.1.

Let $\{u_1, u_2, \ldots\}$ be an enumeration of $\mathcal{W}$. Let I be the set of pairs (n, ε) with $n \in \mathbb{N}$ and $\varepsilon > 0$. We view it as a directed set for the order defined by $(n, \varepsilon) \le (n', \varepsilon')$ if $n \le n'$ and $\varepsilon' \le \varepsilon$. We may restrict ε to be in a countable sequence decreasing to 0 so that I be countable. Let $\mathcal{U}$ be an ultrafilter refining the resulting net (see Remark A.6). For each $i = (n, \varepsilon)$, we can find $N(i)$ and $(x_1(i), \ldots, x_n(i))$ in the unit ball of $M_{N(i)}$ such that (12.5) holds. We then set

$$\forall k \le n \quad v_k(i) = x_k(i),$$

$$\forall k > n \quad v_k(i) = 1 \text{ (say)}.$$

The values for $k > n$ will turn out to be irrelevant. Let $M(i) = M_{N(i)}$ equipped with its normalized trace. Let B and $M_{\mathcal{U}}$ be as before. We may define $V_k \in B_B$ by setting

$$V_k = (v_k(i))_{i \in I}.$$

Clearly, by (12.5), for any $k, \ell \in \mathbb{N}$

$$\lim_{\mathcal{U}} \tau_i (V_k(i)^* V_\ell(i)) = \tau(u_k^* u_\ell).$$

In other words, if we denote by $v_k^{\mathcal{U}} \in B_{M_{\mathcal{U}}}$ the equivalence class of V_k modulo $\mathcal{I}_{\mathcal{U}}$, we have

$$\tau_{\mathcal{U}}(v_k^{\mathcal{U}*} v_\ell^{\mathcal{U}}) = \tau(u_k^* u_\ell). \tag{12.6}$$

Thus if we define $r : \mathcal{W} \to B_{M_{\mathcal{U}}}$ by $r(u_k) = v_k^{\mathcal{U}}$, then (12.2) holds for $\varphi = \tau_{\mathcal{U}}$. By Proposition 12.2 and Theorem 12.1 there is a unital trace preserving isometry $T : M \to M_{\mathcal{U}}$ and an orthogonal decomposition $I = P + Q$ with $P, Q \in T(M)' \cap M_{\mathcal{U}}$ that induces a decomposition

$$\forall x \in M \quad T(x) = PT(x)P + QT(x)Q$$

such that $x \mapsto PT(x)P$ is a $*$-homomorphism and $x \mapsto QT(x)Q$ is a $*$-anti-homomorphism.

We claim that there is a unital trace preserving $*$-anti-isomorphism $\kappa : QM_{\mathcal{U}}Q \to QM_{\mathcal{U}}Q$. Using this claim we can obtain a bona fide unital and trace preserving $*$-homomorphism T' embedding M in $M_{\mathcal{U}}$ by setting $T'(x) = PT(x)P + \kappa(QT(x)Q)$, whence (i). To check the claim, first observe that $M_{\mathcal{U}}$ is clearly $*$-anti-isomorphic to itself by a trace preserving map $y \mapsto {}^t y$ (associated to matrix transposition). More precisely, for any $y = q_{\mathcal{U}}((y_i)) \in M_{\mathcal{U}}$ we set ${}^t y = q_{\mathcal{U}}(({}^t y_i))$. Note that the latter map gives us a $*$-anti-isomorphism from $QM_{\mathcal{U}}Q$ (with unit Q) to ${}^t Q M_{\mathcal{U}} {}^t Q$ (with unit ${}^t Q$). Let $(Q_i) \in B$ be a representative of Q such that Q_i is a projection in $M(i) = M_{N(i)}$ for all $i \in I$ (see Remark 11.31). Since Q_i and ${}^t Q_i$ have the same trace, their ranges have the same dimension, so there is a unitary matrix Υ_i such that $Q_i = \Upsilon_i {}^t Q_i \Upsilon_i^*$. Then if $\Upsilon \in M_{\mathcal{U}}$ is the unitary associated to (Υ_i) the mapping κ defined by $\kappa(y) = \Upsilon {}^t y \Upsilon^*$ is the desired $*$-anti-isomorphism. $\qquad \square$

Remark 12.4 By (i) $\Leftrightarrow$ (iii) in Theorem 12.3 to show that a tracial probability space (M, τ) embeds in a trace preserving way in an ultraproduct of matrix algebras, it suffices to check that this holds for any finitely generated (and a fortiori) weak* separable von Neumann subalgebra of M.

Remark 12.5 (Separable factors suffice) Recall that, by definition, a von Neumann algebra M is a factor if its center is trivial. By Remark 11.36, to answer positively Connes's question for *all* finite von Neumann algebras, it actually suffices to answer it for finite "factors" on a separable Hilbert space (which incidentally is the original question raised in [61]). Indeed, by Remark 11.36 any weak* separable tracial probability space embeds in a trace preserving way into one that is a factor. Moreover, by Remark 12.4 the weak* separable case of the Connes embedding problem implies the general one. Actually, any embedding of a factor M as a C^*-subalgebra of an ultraproduct $M_{\mathcal{U}}$ of matrix algebras is automatically trace preserving (and hence normal) since M has a unique tracial state (see Remark 11.33). Thus for the Connes embedding problem it suffices to show that any weak* separable finite factor embeds as a C^*-subalgebra of a matricial $M_{\mathcal{U}}$.

Using Lemma 11.30, one easily deduces the following fact (which can also be proved by observing that an ultraproduct of ultraproducts is again an ultraproduct).

Corollary 12.6 *Let $(M(i), \tau_i)_{i \in I}$ be a family of tracial probability spaces. Assume that each one of them embeds in a trace preserving way into an ultraproduct of matrix algebras. Then the same is true for their ultraproduct $(M_{\mathcal{U}}, \tau_{\mathcal{U}})$ relative to any ultrafilter $\mathcal{U}$ on I.*

For future reference it may be worthwhile to formulate the following obvious consequence:

Corollary 12.7 *Let (M, τ) and (N, φ) be two tracial probability spaces. Assume that there is a trace preserving embedding of M in an ultraproduct of matrix algebras. For the same to hold for (N, φ) the following condition is sufficient: For any $\varepsilon > 0$, any n and any $u_0, \ldots u_n \in U(N)$ there are $v_0, v_1, \ldots v_n \in B_M$ such that*

$$\forall i, j = 0, \ldots, n \quad |\varphi(u_i^* u_j) - \tau(v_i^* v_j)| \leq \varepsilon. \tag{12.7}$$

Proof We may replace M by $M_{\mathcal{U}}$. Then N satisfies (iii) in Theorem 12.3 with $\mathcal{W} = U(N)$. $\qquad \square$

The next variant is more involved. Here we use Lemma 11.45 to further refine the preceding criterion.

Theorem 12.8 *The conditions* (i)–(iii) *in Theorem 12.3 are equivalent to the following ones:*

(iv) *For any $\varepsilon > 0$, any n and any $u_1, \ldots u_n \in \mathcal{W}$, there are an integer N, matrices $x_1, \ldots x_n$ in the unit ball of M_N and $\eta \in M_N$ with $\tau_N(\eta^* \eta) = 1$ such that*

$$\forall i, j = 1, \ldots, n \quad |\tau(u_i^* u_j) - \tau_N(x_i^* \eta x_j \eta^*)| < \varepsilon. \tag{12.8}$$

(iv)' *For any $\varepsilon > 0$, any n and any $u_1, \ldots u_n \in \mathcal{W}$, there are $x_1, \ldots x_n$ in the unit ball of $B(\ell_2)$ and a Hilbert–Schmidt operator $h \in B(\ell_2)$ with $\mathrm{tr}(h^* h) = 1$ such that*

$$\forall i, j = 1, \ldots, n \quad |\tau(u_i^* u_j) - \mathrm{tr}(x_i^* h x_j h^*)| < \varepsilon. \tag{12.9}$$

(v) *For any $\varepsilon > 0$, any n and any $u_1, \ldots u_n \in \mathcal{W}$ there are an integer N, $N \times N$ unitary matrices $v_1, \ldots v_n$ and $\eta \in M_N$ with $\tau_N(\eta^*\eta) = 1$ such that*

$$\forall i, j = 1, \ldots, n \quad |\tau(u_i^* u_j) - \tau_N(v_i^* \eta v_j \eta^*)| < \varepsilon. \qquad (12.10)$$

Proof (iii) $\Rightarrow$ (iv) is obvious (with $\eta = 1$) and (iv) $\Rightarrow$ (iv)' is trivial (with $h = N^{-1/2}\eta$). Conversely, if (iv)' holds we may assume by density of the finite rank operators in $S_2(\ell_2)$ that there is a projection P of finite rank N such that $h = PhP$. We can replace x_j by $Px_j P$ and setting $\eta = N^{1/2}h$ we see that (12.9) becomes (12.8). This shows (iv) $\Leftrightarrow$ (iv)'.

Assume (iv). By polar decomposition we can write $x_j = v_j|x_j|$ with v_j unitary and also $x_j^* = v_j^*(v_j|x_j|v_j^*) = v_j^*|x_j^*|$. Then

$$\tau_N(x_i^* \eta x_j \eta^*) = \tau_N(v_i^*|x_i^*|\eta v_j|x_j|\eta^*).$$

Thus to show (v) it suffices to prove that

$$|\tau_N(v_i^*|x_i^*|\eta v_j|x_j|\eta^*) - \tau_N(v_i^* \eta v_j \eta^*)| \le f_1(\varepsilon) \qquad (12.11)$$

where f_1 depends only on ε and $f_1(\varepsilon) = o(\varepsilon)$. We will denote $f_2, f_3, \ldots$ functions of the same kind. To prove (12.11) it suffices to show that

$$\||x_j|\eta^* - \eta^*\|_{L_2(\tau_N)} \le f_1(\varepsilon)/2 \text{ and } \||x_i^*|\eta - \eta\|_{L_2(\tau_N)} \le f_1(\varepsilon)/2. \qquad (12.12)$$

Taking $i = j$ in (12.8) we find

$$|1 - \langle x_j\eta^*, \eta^* x_j \rangle_{L_2(\tau_N)}| \le \varepsilon, \qquad (12.13)$$

and hence $\|\eta^* x_j - x_j\eta^*\|^2_{L_2(\tau_N)} \le 2\varepsilon$. Therefore $|1 - \langle x_j\eta^*, x_j\eta^* \rangle_{L_2(\tau_N)}| \le f_2(\varepsilon) = \varepsilon + \sqrt{2\varepsilon}$, which means $|1 - \tau_N(\eta x_j^* x_j \eta^*)| \le f_2(\varepsilon)$. Since $|x_j|^2 \le |x_j|$ we have $0 \le \tau_N(\eta x_j^* x_j \eta^*) \le \tau_N(\eta|x_j|\eta^*) \le 1$ and hence $|1 - \tau_N(\eta|x_j|\eta^*)| \le f_2(\varepsilon)$. The latter means $|1 - \langle \eta^*, |x_j|\eta^* \rangle_{L_2(\tau_N)}| \le f_2(\varepsilon)$, and hence $\|\eta^* - |x_j|\eta^*\|^2_{L_2(\tau_N)} \le 2f_2(\varepsilon)$, which proves the first part of (12.12). We now reapply the same argument starting from $|1 - \langle x_j^*\eta, \eta x_j^* \rangle_{L_2(\tau_N)}| \le \varepsilon$ instead of (12.13) and this gives us the second part of (12.12), so that (v) follows.

It remains to show (v) $\Rightarrow$ (iii). Assume (v). We have for any j

$$|1 - \langle v_j\eta, \eta v_j \rangle_{L_2(\tau_N)}| = |1 - \tau_N(v_j^* \eta v_j \eta^*)| \le \varepsilon,$$

and hence $\|\eta v_j - v_j\eta\|^2_{L_2(\tau_N)} \le 2\varepsilon$. Using $\eta\eta^* - v_j\eta\eta^* v_j^* = (\eta v_j - v_j\eta)v_j^*\eta^* + v_j\eta(v_j^*\eta^* - \eta^* v_j^*)$ we find

$$\|\eta\eta^* - v_j\eta\eta^* v_j^*\|_{L_1(\tau_N)} \le 2(2\varepsilon)^{1/2},$$

and also

$$|\tau_N(v_i^*(\eta v_j)\eta^*) - \tau_N(v_i^*(v_j\eta)\eta^*)| \le (2\varepsilon)^{1/2}. \qquad (12.14)$$

We may assume that $\eta\eta^*$ has rational eigenvalues. Let $H = \ell_2^N$ and let $v' = v \otimes Id_{\ell_2} \in B(H \otimes_2 \ell_2)$ as in Lemma 11.45. The latter associates to $t = N^{-1}\eta\eta^*$ a projection $R \in B(H \otimes_2 \ell_2)$ of finite rank r, such that

$$|\tau_N(v_i^* v_j \eta\eta^*) - r^{-1}\mathrm{tr}(Rv_i' R^* v_j')| \le f_3(\varepsilon).$$

By (12.14)

$$|\tau_N(v_i^* \eta v_j \eta^*) - r^{-1}\mathrm{tr}(Rv_i'^* Rv_j')| \le (2\varepsilon)^{1/2} + f_3(\varepsilon),$$

and hence lastly

$$|\tau(u_i^* u_j) - r^{-1}\mathrm{tr}(Rv_i'^* Rv_j')| \le \varepsilon + (2\varepsilon)^{1/2} + f_3(\varepsilon).$$

Let $\mathcal{H} \subset H \otimes_2 \ell_2$ be the range of R. We conclude that (iii) holds with r now playing the role of N and $x_j = Rv_{j|\mathcal{H}}'$ viewed as an operator on $\mathcal{H}$, or equivalently as an element of M_r.   □

12.2 The approximately finite-dimensional (i.e. "hyperfinite") II_1-factor

Although we prefer not to use this in the sequel, we wish to briefly mention here another way to formulate the Connes question in terms of the so-called approximately finite-dimensional (i.e. "hyperfinite") II_1-factor (R, τ_0), which can be defined as the unique countably generated, *approximately finite-dimensional*, tracial probability space with trivial center and no atom (i.e. no nonzero minimal projection).

One can legitimately think of (R, τ_0) as a noncommutative analogue of the Lebesgue interval $([0, 1], dt)$; indeed, the latter is also the unique countably generated atomless probability space. The uniqueness of (R, τ_0) (up to isomorphism) goes back to Murray and von Neumann who proved that any two countably generated finite factors with no atom (the so-called II_1-factors) are isomorphic if each is approximately finite dimensional. For a detailed proof see [146, p. 896]. Many years later, solving a longstanding problem, Connes [61] proved that for such algebras injective implies (and hence is equivalent to) approximately finite dimensional. Thus (R, τ_0) can be described as the unique (up to isomorphism) *injective*, tracial probability space with trivial center and no atom on a separable Hilbert space. We will not prove any of these uniqueness theorems. We briefly describe one classical construction by which (R, τ_0) can be produced, thus showing its "existence" but we take its uniqueness for granted. The quick construction we outline highlights that there is a copy of (R, τ_0) inside an ultraproduct of matricial tracial probability spaces. The Connes embedding problem can then be reformulated using the "ultrapowers" of (R, τ_0). Let $(R(i))_{i \in I}$ be a family of copies of R. Then the ultraproduct of $(R(i), \tau_0)$ with respect to an ultrafilter $\mathcal{U}$ is called an ultrapower of (R, τ_0); we denote it by $(R^{\mathcal{U}}, \tau_0^{\mathcal{U}})$. We will show that a tracial probability space (M, τ) embeds (trace preservingly) in $R^{\mathcal{U}}$ for some $\mathcal{U}$ if and only if it similarly embeds in an ultraproduct of matricial tracial probability spaces.

Let $(m_i)_{i \in \mathbb{N}}$ be a sequence of integers ≥ 2. The traditional way is to define (R, τ_0) as an infinite tensor product (often called "ITPFI" in the literature) of the form

$$\bigotimes_{i \in \mathbb{N}} (M_{m_i}, \tau_{m_i}). \tag{12.15}$$

By the uniqueness results just mentioned, up to isomorphism, the resulting algebra does not depend on the choice of the sequence (m_i), and the simplest choice is clearly to take $m_i = 2$ for all i, but we will not prove this either. Let $N(i) = \prod_{k \leq i} m_k$. Let

$$M(i) = M_{N(i)} \simeq \otimes_{k \leq i} M_{m_k}$$

and let τ_i be the normalized trace on $M(i)$. Note that $M(k+1) \simeq M(k) \otimes M_{m_{k+1}}$ and more generally, for any $i > k$

$$M(i) \simeq M(k) \otimes M_{m_{k+1}} \otimes \cdots \otimes M_{m_i}$$

so that we have a natural trace preserving embedding $M(k) \to M(i)$ taking $x \in M(k)$ to $x \otimes 1 \otimes \cdots \otimes 1 \in M(i)$. Using these embeddings we may think of $M(k)$ as a $*$-subalgebra of $M(i)$ and form the unital $*$-algebra that is the union (i.e. formally the inductive limit) $\mathcal{A} = \cup M(i)$. It is convenient to think of a typical element x of $\mathcal{A}$ as $x = a \otimes 1 \otimes 1 \otimes \cdots$ with $a \in M(i)$ for some i, followed by an infinite sequence of $\otimes 1$'s. Then $\tau(x) = \tau_i(a)$ defines a linear functional on $\mathcal{A}$, such that $\tau(x^*x) = \tau(xx^*) \geq 0$ and $\tau(1) = 1$. The GNS construction applied to $\mathcal{A}$ produces a Hilbert space H and an injective $*$-homomorphism $\pi : \mathcal{A} \to B(H)$, such that $\tau(x) = \langle 1, \pi(x)1 \rangle$ $(x \in \mathcal{A})$. We then define $\otimes_{i \in \mathbb{N}}(M_{m_i}, \tau_{m_i})$ as $R = \pi(\mathcal{A})''$ equipped with the natural extension of τ, that is a faithful normal trace τ_0 on $\pi(\mathcal{A})''$. We have a natural identification $H \simeq L_2(\tau_0)$ with which π becomes the representation L of left multiplication on $L_2(\tau_0)$.

We now relate this construction to ultraproducts. Let $\mathcal{U}$ be a non trivial ultrafilter on $\mathbb{N}$ and let $(M_{\mathcal{U}}, \tau_{\mathcal{U}})$ be the ultraproduct of $(M(i), \tau_i)$. We will show that $\otimes_{i \in \mathbb{N}}(M_{m_i}, \tau_{m_i})$ embeds in a trace preserving way into $M_{\mathcal{U}}$.

As before, let $B = \left(\oplus \sum_{i \in I} M(i) \right)_{\infty}$, with quotient map $q_{\mathcal{U}} : B \to M_{\mathcal{U}}$, let $v_k : M(k) \to B$ be the map taking $a \in M(k)$ to $b = (b_i) \in B$ defined by $b_i = a \otimes 1 \otimes \cdots \otimes 1$ for all $i \geq k$ and (this is actually somewhat irrelevant)

$b_i = 0$ for all $i < k$. Then let $u_k : M(k) \to M_{\mathcal{U}}$ be the map taking $a \in M(k)$ to $q_{\mathcal{U}}(v_k(a))$. Note $\tau_{\mathcal{U}}(u_k(a)) = \tau_k(a)$ and $u_k : M(k) \to M_{\mathcal{U}}$ is an embedding. Let $A_k = u_k(M(k)) \subset M_{\mathcal{U}}$. With the obvious identification (corresponding to $a \mapsto a \otimes 1$) we have $A_k \subset A_{k+1}$. This gives us an embedding $\psi : \mathcal{A} \subset M_{\mathcal{U}}$ such that $\tau_{\mathcal{U}}(\psi(x)) = \tau_0(x)$ for any $x \in \mathcal{A}$. Clearly, this extends to an isometry $T : L_2(R, \tau_0) \to L_2(M_{\mathcal{U}}, \tau_{\mathcal{U}})$. By Proposition 11.19 that same map defines a trace preserving embedding of (R, τ_0) into $(M_{\mathcal{U}}, \tau_{\mathcal{U}})$.

Proposition 12.9 *Let (M, τ) be a tracial probability space on a separable Hilbert space. Then there is a trace preserving embedding of M into an ultraproduct of matrix algebras if and only if there is one of M into $R^{\mathcal{U}}$ for some ultrafilter $\mathcal{U}$.*

Proof By Corollary 12.6 there is a trace preserving embedding of $R^{\mathcal{U}}$ into an ultraproduct of matrix algebras. This settles the "if part." For the converse, recall that (12.15) gives us a copy of R no matter what (m_i) is. Thus $M_{N(i)}$ embeds in R for each i, and hence any ultraproduct of $(M_{N(i)})$ relative to $\mathcal{U}$ embeds in $R^{\mathcal{U}}$. □

Remark 12.10 As most ultraproducts, the von Neumann algebra $R^{\mathcal{U}}$ is defined on a nonseparable Hilbert space and this is unavoidable. The appearance of large cardinals suggests that issues from logic should play a role. Consider for instance the very natural question whether the ultrapowers $R^{\mathcal{U}}$ (for varying nontrivial ultrafilters on $\mathbb{N}$) are all isomorphic: a positive answer turns out to be equivalent to the continuum hypothesis [86]. See [86–89] where the analogous question for ultraproducts of $(M_n)_{n \geq 1}$ is discussed as well as other similar issues.

12.3 Hyperlinear groups

Definition 12.11 A discrete group G will be called approximately linear (also called "hyperlinear") if (M_G, τ_G) embeds in a trace preserving way in an ultraproduct of matricial tracial probability spaces, such as $(M_{\mathcal{U}}, \tau_{\mathcal{U}})$ with all $M(i)$'s matricial.

In other words, G is approximately linear (or "hyperlinear") if for (M_G, τ_G) the answer to the Connes question is positive. We already know (see Remark 11.56) that this holds if G is amenable or has the factorization property.

Just like "hyperfinite," the term "hyperlinear" is hardly a good choice since it does not imply linear in any reasonable sense, so we prefer to use "approximately linear" which seems more appropriate.

Theorem 12.12 *The following properties of a discrete group G are equivalent:*

(i) *The group G is approximately linear (so-called hyperlinear).*
(ii) *There is a group representation $\pi : G \to U(M_{\mathcal{U}})$ embedding G into the unitary group $U(M_{\mathcal{U}})$ of an ultraproduct of matricial tracial probability spaces and satisfying*

$$\forall t \in G \quad \tau_{\mathcal{U}}(\pi(t)) = \tau_G(\lambda_G(t)). \tag{12.16}$$

(iii) *For any finite subset $S \subset G$ containing the unit e and any $\varepsilon > 0$ there is an integer $N < \infty$ and a function $\psi : S \to \mathbb{U}_N$ with values in the group $\mathbb{U}_N = U(M_N)$ of $N \times N$-unitary matrices such that*

$$\forall s, t \in S \quad \|\psi(s)\psi(t) - \psi(st)\|_{L_2(\tau_N)} < \varepsilon,$$

and

$$|\tau_N(\psi(e)) - 1| < \varepsilon \quad \text{and} \quad \forall t \in S, t \neq e \quad |\tau_N(\psi(t))| < \varepsilon.$$

Proof Assume (i). Let $\Phi : M_G \to M_{\mathcal{U}}$ the embedding. Let $\pi(t) = \Phi(\lambda_G(t))$ $(t \in G)$. Then (ii) is immediate.

Assume (ii). By Lemma 11.30 each $\pi(t)$ has a unitary representative modulo $\mathcal{I}_{\mathcal{U}}$. Thus, for each i there is $\pi_i(t) \in U(M(i))$ such that $q_{\mathcal{U}}((\pi_i(t))) = \pi(t)$. Since $\pi(st) = \pi(s)\pi(t)$ $(s, t \in G)$, we have $\lim_{\mathcal{U}} \|\pi_i(s)\pi_i(t) - \pi_i(st)\|_{L_2(\tau_i)} = 0$ and since $\tau(\pi(e)) = 1$ and $\tau(\pi(t)) = 0$ if $t \neq e$, we have $\lim_{\mathcal{U}} |\tau_i(\pi_i(e)) - 1| = 0$ and $\lim_{\mathcal{U}} |\tau_i(\pi_i(t))| = 0$ if $t \neq e$. Recall that $(M(i), \tau_i) = (M_{N(i)}, \tau_{N(i)})$ for some $N(i) < \infty$ Thus, it suffices to take $\psi = \pi_i$ and to choose i far enough relative to $\mathcal{U}$ to obtain (iii).

Assume (iii). Let I be the set of pairs (S, ε) with $S \subset G$ finite subset and $\varepsilon > 0$, with the usual ordering $(S, \varepsilon) \leq (S', \varepsilon')$ if $S \subset S'$ and $\varepsilon' \leq \varepsilon$. Let $\mathcal{U}$ be an ultrafilter refining the net associated to this directed set (see Remark A.6). For any $i = (S, \varepsilon)$, when $t \in S$ we set $\pi_i(t) = \psi(t)$ where ψ is the function given by (iii); and when $t \notin S$ (this is actually irrelevant) we set $\pi_i(t) = 1$. Let $\pi(t) = q_{\mathcal{U}}((\pi_i(t)))$. Then it is easy to check that (ii) holds, but our goal is (i).

Let $\mathcal{A} = \text{span}[\lambda_G(t) \mid t \in G] \subset L_2(\tau_G)$. Let $T : \mathcal{A} \to M_{\mathcal{U}}$ be the $*$-homomorphism taking $\lambda_G(t)$ to $\pi(t)$. Note that $\|T(a)\|_{L_2(\tau_{\mathcal{U}})}^2 = \tau_{\mathcal{U}}(\pi(a^*a))$ and $\|a\|_{L_2(\tau_G)}^2 = \tau_G(a^*a)$. By (12.16), for any $a \in \mathcal{A}$, we have $\|T(a)\|_{L_2(\tau_{\mathcal{U}})} = \|a\|_{L_2(\tau_G)}$, and hence T extends to an isometric embedding from $L_2(\tau_G)$ to $L_2(\tau_{\mathcal{U}})$. Therefore, by Proposition 11.19, T also extends to a normal (trace preserving) embedding on M_G into $M_{\mathcal{U}}$, showing that (i) holds. Alternatively, for (ii) $\Rightarrow$ (i) we could invoke Remark 11.20, observing that (ii) simply means that $(\lambda_G(t))_{t \in G}$ and $(\pi(t))_{t \in G}$ have the same $*$-distribution. $\qquad \square$

Remark 12.13 If we take $s = t = e$ in $\|\psi(s)\psi(t) - \psi(st)\|_{L_2(\tau_N)} < \varepsilon < 1$ it follows that the unitary matrix $\psi(e)$ satisfies $|\tau_N(\psi(e)) - 1| < \sqrt{2}\varepsilon$, so the condition $|\tau_N(\psi(e)) - 1| < \varepsilon$ could be omitted.

12.4 Residually finite groups and Sofic groups

We denote by $\mathbb{S}_N$ the group of permutations of the set with N elements, i.e. the "symmetric group." Note the embedding $\mathbb{S}_N \subset \mathbb{U}_N$ obtained by identifying a permutation σ with the $N \times N$ unitary matrix u defined by $u(e_j) = e_{\sigma(j)}$. Note that

$$\tau_N(u) = N^{-1}|\{j \in \{1, \ldots, N\} \mid \sigma(j) = j\}|$$

is the proportion of the number of fixed points of σ.

Definition 12.14 A discrete group G is called "sofic" if it satisfies the condition (iii) in Theorem 12.12 with a function ψ taking values in permutation matrices.

Equivalently, and more explicitly, G is sofic if for any finite subset $S \subset G$ with $e_G \in S$ and any $\varepsilon > 0$ there is an integer $N < \infty$ and a function $\psi : S \to \mathbb{S}_N$ such that

$$\forall s, t \in S \quad |\{j \in \{1, \ldots, N\} \mid (\psi(s)\psi(t))(j) \neq \psi(st)(j)\}| \leq \varepsilon N,$$

and

$$|\{j \in \{1, \ldots, N\} \mid \psi(e)(j) \neq j\}| \leq \varepsilon N \text{ and } \forall t \in S \setminus \{e\},$$

$$|\{j \in \{1, \ldots, N\} \mid \psi(t)(j) = j\}| \leq \varepsilon N.$$

Remark 12.15 (About examples of sofic groups) Using Følner sequences (see Remark 3.33), we show next that amenable groups are sofic, but also as we will soon show (see Lemma 12.19) all free groups are sofic, which leads one to the important open question whether *every* group is sofic. This seems to be the group theoretic analogue of the Connes problem. The notion was introduced by Gromov and the term "sofic" was coined by B. Weiss (sofi means finite in hebrew). We refer to a series of papers by Elek and Szabó (for instance [81]) for more information on sofic groups.

Lemma 12.16 *Amenable groups are sofic.*

Proof By Remark 3.33 there is a net (B_i) formed of finite subsets of our amenable G such that

$$\forall t \in G \quad \lim |B_i \setminus t^{-1}B_i||B_i|^{-1} = 0.$$

Let $\psi_i(t)$ be a permutation of B_i that is equal to $x \mapsto tx$ for any $x \in B_i \cap t^{-1}B_i$ and that is extended to $B_i \backslash t^{-1}B_i$ in such a way that $\psi(t) : B_i \to B_i$ is bijective. Then for any unital finite set $S \subset G$ and $\varepsilon > 0$ when i is far enough in the net we will have $|B_i \setminus (st)^{-1}B_i| < (\varepsilon/3)|B_i|$ for any $(s,t) \in S \times S$ (and hence also for (s, e_G) and (e_G, t)). It is then easy to check that $\psi = \psi_i$ satisfies the conditions required in Definition 12.14 for G to be sofic. $\qquad \square$

Definition 12.17 A group G is called residually finite if there exists a collection of finite groups (Γ_i) and homomorphisms $\varphi_i : G \to \Gamma_i$ separating the points of G, i.e. for any finite subset $S \subset G$ there is an i for which the restriction of φ_i to S is injective. Without loss of generality, we may assume that $\Gamma_i = G/N_i$ where each $N_i \subset G$ is a normal subgroup with finite index and φ_i is the canonical quotient map. Thus, G is residually finite if and only if it admits a family of normal subgroups with finite index (N_i), directed by (downward) inclusion and such that $\bigcap_{i \in I} N_i = \{e_G\}$.

Proposition 12.18 *Any residually finite group is sofic and any sofic group is approximately linear (so-called hyperlinear).*

Proof If G is residually finite, let $S \subset G$ be a finite subset. There is a finite group Γ and a group homomorphism $\psi : G \to \Gamma$ that is injective on S. Let $N = |\Gamma|$. We may view Γ as acting on itself by translation (and hence any $t \neq e$ acts without fixed points), so that $\Gamma \subset \mathbb{S}_N$. Then, viewing ψ as acting into $\mathbb{S}_N$, we obtain the properties in Definition 12.14 with $\varepsilon = 0$. Therefore G is clearly sofic. The implication sofic $\Rightarrow$ approximately linear (so-called hyperlinear) is obvious given the definition of sofic and (iii) $\Rightarrow$ (i) in Theorem 12.12. $\qquad \square$

The following fact is classical.

Lemma 12.19 *Free groups are residually finite (and a fortiori sofic).*

Proof Let $G = F_I$. Let $\{g_i \mid i \in I\}$ be the (free) generators. Let $C \subset G$ be a finite subset. It suffices to produce a (group) homomorphism $h : G \to \Gamma$ into a finite group Γ such that, for any c in C, we have $h(c) \neq e_\Gamma$ if $c \neq e$, where e_Γ denotes the unit in Γ and e the unit in G (i.e. the "empty word").

We may assume that $C \subset G'$ where G' is the subgroup generated by a finite subset $\{g_i \mid i \in J\}$ of the generators. Let $k = \max\{|c| \mid c \in C\}$ (here $|c|$ denotes the length of the reduced word associated to c, i.e. the number of elements in $\{g_i, g_i^{-1} \mid i \in I\}$ used to express c in reduced form). We then set

$$S = \{t \in G' \mid |t| \leq k\}.$$

We will take for Γ the (finite) group of all permutations of the (finite) set S. For any i in J, we introduce

$$S_i = \{t \in S \mid g_i t \in S\}.$$

Then clearly $S_i \subset S$ and $g_i S_i \subset S$. Hence (since $|S_i| = |g_i S_i|$ and S is finite) there is a permutation $\sigma_i : S \to S$ such that $\sigma_i(s) = g_i s$ for any s in S_i. Then if $s, t \in S$ and if $g_i t = s$ (or equivalently $t = g_i^{-1} s$) we have $\sigma_i(t) = s$ (or equivalently $t = \sigma_i^{-1}(s)$). Thus it is easy to check that if a reduced word $t = g_{i_1}^{\varepsilon_1} g_{i_2}^{\varepsilon_2} \ldots g_{i_m}^{\varepsilon_m}$ $(m \le k \; \varepsilon_i = \pm 1)$ lies in S (note that, by definition of S, e and all the subwords of t also lie in S) we have

$$\sigma_{i_1}^{\varepsilon_1} \sigma_{i_2}^{\varepsilon_2} \ldots \sigma_{i_m}^{\varepsilon_m}(e) = t.$$

Therefore, if we define $h : G \to \Gamma$ as the unique homomorphism such that $h(g_i) = \sigma_i \; \forall i \in J$ and $\sigma(g_i) = e_\Gamma \; \forall i \notin J$, we find, for t as before, $h(t) = \sigma_{i_1}^{\varepsilon_1} \sigma_{i_2}^{\varepsilon_2} \ldots \sigma_{i_m}^{\varepsilon_m}$ and $h(t)(e) = t$, in particular we have $h(t) \ne e_\Gamma$ whenever $t \in S$ and $t \ne e$. Since $C \subset S$, we obtain the announced result. $\square$

Remark 12.20 By a famous result due to Malcev [177], finitely generated *linear groups* are residually finite. Using this, Lemma 12.19 could be deduced from Choi's Theorem 9.18.

Consequently, we obtain the following important fact, which was mentioned in passing by Connes in [61, p. 105] as motivation for his question discussed in §12.

Theorem 12.21 ([255]) *There is a trace preserving embedding of the von Neumann algebra of the free groups $\mathbb{F}_n$ or $\mathbb{F}_\infty$ into an ultraproduct of matrix algebras (in other words free groups are approximately linear).*

Corollary 12.22 *When G is a free group, there is a factorization of the canonical $*$-homomorphism $\dot{Q}_G : C^*(G) \to M_G$ of the form $\dot{Q}_G : C^*(G) \overset{w}{\to} B \overset{v}{\to} M_G$ where w is a $*$-homomorphism, v is c.p. with $\|v\|_{cb} \le 1$, and B is a von Neumann algebra with the WEP (actually B is injective).*

Proof By Lemma 12.19 and Proposition 12.18, G is approximately linear (so-called hyperlinear). We have $M_G \subset M_\mathcal{U}$ with a c.p. projection (the conditional expectation) $P : M_\mathcal{U} \to M_G$. The unitary representation $\pi : G \to U(M_\mathcal{U})$ appearing in property (ii) in Theorem 12.12 admits a lifting to a unitary representation $\hat{\pi} : G \to U(B)$ where $B = \left(\oplus \sum_{i \in I} M(i) \right)_\infty$ as in (11.14). Indeed, by the freeness of G, it suffices for this to be able to lift the images under π of each free generator, and this is guaranteed by Lemma 11.30. Then $\hat{\pi}$ extends to a $*$-homomorphism $C^*(G) \overset{w}{\to} B$ and, denoting as before by $q : B \to M_\mathcal{U}$ the quotient map, we can take $v = Pq$. Note that B is injective and hence has the WEP (see Proposition 1.48 and Corollary 9.26). $\square$

By Theorem 11.55, we deduce from Corollary 12.22:

Corollary 12.23 *Free groups have the factorization property described in §11.8.*

12.5 Random matrix models

The term "matrix model" is frequently used with respect to a tracial probability space (M, τ). This is an alternative way to discuss the embedding in the Connes question. More precisely, assume that M is generated by a family of elements $(x_s)_{s \in S}$ (indexed by some set S), consider a sequence of matricial sizes $(n_N)_{N \geq 1}$ and families $(x_s^{(N)})_{s \in S}$ in M_{n_N} such that for any polynomial $P(X_s, X_s^*)$ in the noncommuting formal variables $(X_s)_{s \in S}$ we have

$$\lim_{N \to \infty} \tau_{n_N}(P(x_s^{(N)}, x_s^{(N)*})) = \tau(P(x_s, x_s^*)).$$

We then say that $(x_s^{(N)})_{s \in S}$ is a matrix model for $(x_s)_{s \in S}$, or (somewhat abusively) for (M, τ). Thus to say that (M, τ) admits a matrix model is but another way of saying that it embeds in a trace preserving way in an ultraproduct of matricial tracial probability spaces. However, the matrix model terminology is better adapted to the theory of random matrices. The latter provides very interesting and fruitful examples of matrix models, in connection with Voiculescu's free probability theory (see [254]). For instance, the convergence of both the unitary and Gaussian models in the following statements, in which $n_N = N$ and $S = \{1, 2, \dots\}$, is due to Voiculescu:

Theorem 12.24 (Random unitary matrix model) *Let $(U_s^{(N)})_{s \geq 1}$ be an independent, identically distributed (i.i.d. in short) sequence of random matrices uniformly distributed over the unitary group $\mathbb{U}_N$. We assume (for convenience) all random elements defined on a probability space $(\Omega, \mathbb{P})$. Let $(M, \tau) = (M_{\mathbb{F}_\infty}, \tau_{\mathbb{F}_\infty})$. Consider the sequence $x_s = \lambda_{\mathbb{F}_\infty}(g_s)$ $(s \geq 1)$ in M. Then for almost all $\omega \in \Omega$, the sequence $(U_s^{(N)}(\omega))_{s \geq 1}$ is a matrix model for $(x_s)_{s \geq 1}$ (with respect to $N \to \infty$).*

Remark 12.25 (Random permutation matrix model) The same result is valid if we assume that $(U_s^{(N)})_{s \geq 1}$ is an i.i.d. family uniformly distributed over the subgroup of $\mathbb{U}_N$ formed of matrices of permutation of size $N \times N$. This is due to Nica (see [179]).

Fix a nontrivial ultrafilter $\mathcal{U}$ on $\mathbb{N}$. Let $M_{\mathcal{U}}$ be the ultraproduct of the family (M_N, τ_N) with respect to $\mathcal{U}$. Let $\mathbb{B} = (\oplus \sum_{N \geq 1} M_N)_\infty$. Let $q_{\mathcal{U}} : \mathbb{B} \to M_{\mathcal{U}}$ be the quotient map. Fix ω such that $(U_s^{(N)}(\omega))_{N \geq 1}$ is a matrix model for $(\lambda_{\mathbb{F}_\infty}(g_s))$. We know by Remark 11.20 that the correspondence $\lambda_{\mathbb{F}_\infty}(g_s) \mapsto q_{\mathcal{U}}((U_s^{(N)}(\omega))_{N \geq 1})$ extends to a trace preserving $*$-homomorphism $J_\omega^{\mathcal{U}} : M_{\mathbb{F}_\infty} \to M_{\mathcal{U}}$ embedding $M_{\mathbb{F}_\infty}$ as a von Neumann subalgebra of $M_{\mathcal{U}}$. Moreover, by Proposition 11.21 there is a c.p. contractive projection $P_\omega^{\mathcal{U}}$ from $M_{\mathcal{U}}$ onto $J_\omega^{\mathcal{U}}(M_{\mathbb{F}_\infty})$, whence the following statement.

Corollary 12.26 *For almost all* ω, *the preceding map* $J_\omega^{\mathcal{U}} : M_{\mathbb{F}_\infty} \to M_{\mathcal{U}}$ *is a trace preserving (von Neumann sense) embedding and there is a c.p. contractive projection* $P_\omega^{\mathcal{U}}$ *from* $M_{\mathcal{U}}$ *onto* $J_\omega^{\mathcal{U}}(M_{\mathbb{F}_\infty})$.

The case of Gaussian random matrices is central. To state the result in that case one needs the notion of a free semicircular sequence (x_s), and that of a circular one (y_s), for which we refer to §11.4. It is known that the von Neumann algebra generated by either (x_s) or (y_s) is isomorphic to the von Neumann algebra $M = M_{\mathbb{F}_\infty}$.

Let $(X_s^{(N)})_{s \geq 1}$ (resp. $(Y_s^{(N)})_{s \geq 1}$) be an independent, identically distributed (i.i.d. in short) sequence of random matrices each with the same distribution as a Gaussian model $X^{(N)}$ (resp. $Y^{(N)}$). By definition its entries $X^{(N)}(i, j)$ (for $1 \leq i \leq j \leq N$) are all independent mean zero (real valued) Gaussian variables with $\mathbb{E}|X^{(N)}(i, j)|^2 = 1/N$ for all $i < j$ and $\mathbb{E}|X^{(N)}(j, j)|^2 = 2/N$ for all j; the other entries are determined by $X^{(N)}(i, j) = X^{(N)}(j, i)$ so that $X^{(N)}$ is a symmetric random matrix. This is known as the GOE random matrix model.

By definition the entries $Y^{(N)}(i, j)$ (for $1 \leq i, j \leq N$) are i.i.d. complex valued Gaussian variables such that $\Re(Y^{(N)}(i, j))$ and $\Im(Y^{(N)}(i, j))$ are independent (real valued) Gaussian with mean zero and such that

$$\mathbb{E}|\Re(Y^{(N)}(i, j))|^2 = \mathbb{E}|\Im(Y^{(N)}(i, j))|^2 = (2N)^{-1} = (1/2)\mathbb{E}|Y^{(N)}(i, j)|^2.$$

Theorem 12.27 (Gaussian random matrix model) *Let* $N \geq 1$ *be any matrix size. Let* $(x_s)_{s \geq 1}$ *(resp.* $(y_s)_{s \geq 1}$*) be a free semicircular (resp. circular) sequence in* M. *Then for almost all* $\omega \in \Omega$, *the sequence* $(X_s^{(N)}(\omega))_{s \geq 1}$ *(resp.* $(Y_s^{(N)}(\omega))_{s \geq 1}$*) is a matrix model for* $(x_s)_{s \geq 1}$ *(resp.* $(y_s)_{s \geq 1}$*).*

We refer the reader to [7, 254] for the proofs.

12.6 Characterization of nuclear von Neumann algebras

It is easy to see from the definition that for any family $(H_i)_{i \in I}$ of Hilbert spaces the von Neumann algebra

$$M = \left(\oplus \sum\nolimits_{i \in I} B(H_i) \right)_\infty$$

is injective. In the particular case $I = \mathbb{N}$ with H_n n-dimensional, the von Neumann algebra

$$\mathbb{B} = \left(\oplus \sum\nolimits_{n \geq 1} M_n \right)_\infty$$

is injective (and a fortiori WEP). However, it is *not* nuclear. This was proved by S. Wassermann:

Lemma 12.28 ([255]) *The von Neumann algebra $\mathbb{B}$ is not nuclear.*

Proof Let $G = \mathbb{F}_n$ with $1 < n \leq \infty$. By Theorem 12.21, M_G embeds in an ultraproduct $M_{\mathcal{U}}$ of matrix algebras and by Proposition 11.21 there is a contractive c.p. projection (conditional expectation) $P : M_{\mathcal{U}} \to M_G$. A priori $M_{\mathcal{U}}$ is a quotient of $(\oplus_{i \in I} \sum B(H_i))_\infty$ for some family of *finite-dimensional* Hilbert spaces. Since G is countable and residually finite, from the proof of Theorem 12.21, we may as well assume that $M_{\mathcal{U}}$ is a quotient of $\mathbb{B}$. Now the nuclearity of $\mathbb{B}$ would imply by Theorem 8.15 the injectivity of $M_{\mathcal{U}}$, and hence also of M_G (recall Corollary 1.47). Since we know by Theorem 3.30 (and the remarks after it) that M_G is not injective (because G is not amenable) we conclude that $\mathbb{B}$ is not nuclear. $\square$

More precisely, we have:

Theorem 12.29 ([255]) *Let M be a von Neumann algebra. The following are equivalent:*

  (i) *M is nuclear.*
 (ii) *M does not contain a copy of $\mathbb{B}$ as a von Neumann subalgebra.*
(iii) *There is a finite set I, integers $n(i) \geq 1$ ($i \in I$) and commutative von Neumann algebras C_i such that M is isomorphic to $(\oplus \sum_{i \in I} C_i \otimes_{\min} M_{n(i)})_\infty$.*

Sketch of proof Since $\mathbb{B}$ is injective, (i) $\Rightarrow$ (ii) follows from the preceding Lemma (recall Remark 9.3). Assume (iii). By Remarks 4.9 and 4.10, M is nuclear, so (iii) $\Rightarrow$ (i).

The remaining implication (ii) $\Rightarrow$ (iii) lies deeper. Its fully detailed proof requires classical results from the structural theory of von Neumann algebras that would take us too far off to cover in these notes. We merely outline the argument for the convenience of the reader. We have a decomposition $M \simeq M_I \oplus M_{II} \oplus M_{III}$ into three (possibly vanishing) parts called respectively of type I, II, III. By general results, it can be shown that if either $M_{II} \neq \{0\}$ (resp. $M_{III} \neq \{0\}$) then $\mathbb{B}$ embeds in M_{II} (resp. M_{III}). Thus we may assume $M = M_I$, i.e. that M is of type I. Moreover, assuming (ii) we know that $B(\ell_2)$ does not embed in $M = M_I$. By the classification of type I von Neumann algebras, M is isomorphic to a direct sum $(\oplus_{n \geq 1} \sum C_n \otimes_{\min} M_n)_\infty$ where each C_n is a commutative von Neumann algebra (i.e. $C_n = L_\infty(\Omega_n, \mu_n)$ for some measure space (Ω_n, μ_n) and $C_n \otimes_{\min} M_n = L_\infty(\Omega_n, \mu_n; M_n)$). Let $I = \{n \geq 1 \mid C_n \neq 0\}$. The assumption (ii) implies that I is a finite set. Thus we obtain (iii). $\square$

12.7 Notes and remarks

As already mentioned in the text, the ideas for Theorem 12.1 go back to Kadison [142] and Størmer [233] (see [123]). Proposition 12.2 is an elementary fact formulated for the convenience of our presentation. Kirchberg's criterion in Theorem 12.3 is much more substantial and Theorem 12.8 is even more so. The latter refined criterion is essentially due to Ozawa [191, th. 29]. It will be crucially used to prove Theorem 14.7. Concerning injective factors and in particular the uniqueness of the injective factor R, the fundamental reference is Connes's paper [61]. The notion of sofic group has recently become quite popular. The names of Gromov and Weiss are associated with it. Initial work by Elek and Szabo [81] has been influential. Theorem 12.21 and its corollary are due to S. Wassermann [255] and independently to Connes [61]. The results in §12.5 are due to Voiculescu (see [254]), except for the random permutation model due to Nica [179]. The results of §12.6 are due to S. Wassermann [255].

13

Kirchberg's conjecture

We now turn to the second problem of our series, which actually was explicitly formulated as a conjecture by Kirchberg.

13.1 LLP $\Rightarrow$ WEP?

At the end of his landmark paper [155] Eberhard Kirchberg formulated several conjectures about the properties WEP and LLP. Essentially, he asked whether they are equivalent. However, the implication WEP $\Rightarrow$ LLP was soon disproved by Marius Junge and the author in [141], where it was proved that the prototypical WEP C^*-algebra, namely $\mathscr{B} = B(\ell_2)$ fails the LLP. We return to this in more detail in §18.1.

This left open the remaining conjecture, namely the implication LLP $\Rightarrow$ WEP. Given that $\mathscr{C}$ is the prototypical example of a C^*-algebra with LLP, one can reformulate the conjecture like this:

Kirchberg's Conjecture: *The C^*-algebra $\mathscr{C}$ has the Weak Expectation Property (WEP).*

Because of its equivalence with Connes's problem, this is now widely considered as one of the most important open problems in Operator Algebra theory (if not the most important one).

At this point, it is worthwhile to list several equivalent forms of Kirchberg's conjecture.

Proposition 13.1 *The following conjectures are all equivalent:*

- (i) $\mathscr{C}$ *is WEP.*
- (i)' *The pair* $(\mathscr{C}, \mathscr{C})$ *is a nuclear pair.*
- (ii) $\mathscr{C} \otimes_{\max} \mathscr{C}$ *is residually finite dimensional.*
- (iii) $\mathscr{C} \otimes_{\max} \mathscr{C}$ *has a faithful tracial state.*

(iv) *For any free group* $\mathbb{F}$, $C^*(\mathbb{F})$ *has the WEP.*

(v) *Any unital C^*-algebra is isomorphic to a quotient of a WEP*
     C^*-*algebra. (We call these QWEP.)*

(vi) *Any von Neumann algebra is QWEP.*

(vii) *LLP $\Rightarrow$ WEP.*

Proof (i) $\Leftrightarrow$ (i)' is tautological in view of our definition of WEP. We first consider (i)–(iii).

(i) $\Rightarrow$ (ii) follows from Theorem 9.18 and Remark 9.19.

(ii) $\Rightarrow$ (iii) follows from Remark 9.17 since, of course, each M_n has a faithful tracial state.

(iii) $\Rightarrow$ (i)' follows from Remark 11.34 applied to the GNS representation of the faithful tracial state, which is isometric on $\mathscr{C} \otimes_{\max} \mathscr{C}$. Whence (i)–(iii) are equivalent.

(i)' $\Rightarrow$ (iv) follows from (9.4) (applied with $B = \mathscr{C}$). (iv)$\Rightarrow$(v) is clear since, by Proposition 3.39, any unital C^*-algebra is a quotient of $C^*(\mathbb{F})$ for some $\mathbb{F}$, and (v)$\Rightarrow$ (vi) is trivial.

Let us show (vi)$\Rightarrow$ (vii). Assume (vi). Let C be a C^*-algebra with the LLP. By (vi), C^{**} is QWEP. By (i) $\Rightarrow$ (ii) in Theorem 9.67, the linear map $i_C : C \to C^{**}$ is WEP. Since the latter is max-injective, it follows that Id_C and hence C itself is WEP, so that (vii) holds.

Lastly we have (vii)$\Rightarrow$(i) since $\mathscr{C}$ has the LLP (by Theorem 9.6). $\square$

Remark 13.2 (One-for-all ...) By Corollary 9.69 if $\mathscr{C}$ is QWEP, then it has the WEP and by Proposition 13.1 *every* C^*-algebra is QWEP. Motivated by this, we will say that a unital C^*-algebra A is "one-for-all" if the property:

(viii) A is QWEP

implies that all C^*-algebras are QWEP (in other words implies the Kirchberg conjecture).

If A has a one-for-all quotient, then A itself is one-for-all. Moreover, by Corollary 9.70 if the identity of a one-for-all C^*-algebra B factors with unital c.p. maps through A, then A is also one-for-all (more generally see Remark 13.2).

The obvious example of one-for-all is $C^*(\mathbb{F})$ when $\mathbb{F}$ is any non-Abelian free group. More generally any C^*-algebra that admits $C^*(\mathbb{F}_2)$ as a quotient is also an example, but it turns out there are more noteworthy examples. For instance the universal unital C^*-algebra of a contraction described in Remark 2.27 is one-for-all. Clearly $C_u^* \langle \mathbb{C} \rangle$ is generated (as a unital C^*-algebra) by the single polynomial $P(X) = X$, and any singly generated unital C^*-algebra is a quotient of $C_u^* \langle \mathbb{C} \rangle$. It follows that any C^*-algebra that is generated by a pair of (a priori noncommuting) hermitian contractions x_1, x_2 is a quotient of $C_u^*\langle\mathbb{C}\rangle$ since it is generated by $(x_1 + ix_2)/2$. In particular, if C_j is generated by x_j then the full (unital) free product $C_1 * C_2$ is a quotient of $C_u^*\langle\mathbb{C}\rangle$. It is a simple exercise to check that the C^*-algebra $C^*(G)$ of a finite Abelian group G can be generated by a single hermitian element. Thus (see (4.13)) if G_1, G_2 are finite Abelian groups then $C^*(G_1 * G_2)$ is a quotient of $C_u^*\langle\mathbb{C}\rangle$. This shows e.g. that $C^*(\mathbb{Z}_2 * \mathbb{Z}_3)$ is a quotient of $C_u^*\langle\mathbb{C}\rangle$, but it is well known (see Remark 3.40) that $\mathbb{F}_\infty$ is a subgroup of $\mathbb{Z}_2 * \mathbb{Z}_3$, and hence by Proposition 3.5 there is a unital c.p. factorization of the identity of $\mathscr{C}$ through $C^*(\mathbb{Z}_2 * \mathbb{Z}_3)$. Therefore, we can now conclude: if $C_u^*\langle\mathbb{C}\rangle$ is QWEP then so is $C^*(\mathbb{Z}_2 * \mathbb{Z}_3)$, and by Corollary 9.70 so is $\mathscr{C}$. This shows that $C_u^*\langle\mathbb{C}\rangle$ is one-for-all. We remind the reader that $\mathbb{Z}_2 * \mathbb{Z}_2$ being amenable, we do need $\mathbb{Z}_3$ here, see Remark 3.40.

The same argument shows that the unital free product $A_1 * A_2$ of two unital C^*-algebras is one-for-all if (say) A_1 (resp. A_2) admits $C^*(\mathbb{Z}_n)$ (resp. $C^*(\mathbb{Z}_m)$) as a quotient with $n \geq 2$ and $m \geq 3$. More generally, by Boca's Theorem 2.24, the same holds if we have a unital c.p. factorization of the identity of $C^*(\mathbb{Z}_n)$ (resp. $C^*(\mathbb{Z}_m)$) through A_1 (resp. A_2). For instance, $C^*(\mathbb{Z}_n)$ admits such a factorization through M_n (because $C^*(\mathbb{Z}_n) \simeq \ell_\infty^n$ and the latter can be identified with diagonal matrices in M_n, see (3.17)). This shows that $M_n *$ $C^*(\mathbb{Z})$ is one-for-all if $n \geq 2$. It is easy to show that $M_n * C^*(\mathbb{Z}) = M_n(B_n)$ for some unital C^*-algebra B_n called the Brown algebra (see Remark 6.31). Clearly B is QWEP if and only if $M_n(B)$ is QWEP. Therefore B_n is one-for-all. It can be shown that B_n is the unital C^*-algebra generated by the entries of a universal unitary block matrix in $M_n(B(H))$. In passing we observe that $M_n * C^*(\mathbb{Z})$ and hence B_n has the LLP since, by Theorem 9.44, the LLP is stable by free products.

The free product of Cuntz algebras $O_n * O_m$ is one-for-all if $n, m \geq 1$.

Another interesting example is $\mathscr{B} \otimes_{\max} \mathscr{B}$. Indeed, let $G = \mathbb{F}_\infty$. By Theorem 4.11 the identity of $\mathscr{C}$ factorizes with unital c.p. maps through $M_G \otimes_{\max} M_G$. More precisely, the latter factorization is of the form

$$\mathscr{C} \to \mathscr{C} \otimes_{\max} \mathscr{C} \to M_G \otimes_{\max} M_G \to \mathscr{C}. \tag{13.1}$$

But since by Corollary 12.22 the natural map $\mathscr{C} \to M_G$ factorizes with unital c.p. maps through some $B(H)$, it follows that in (13.1) the mapping $\mathscr{C} \otimes_{\max}$ $\mathscr{C} \to M_G \otimes_{\max} M_G$ factorizes with unital c.p. maps through $B(H) \otimes_{\max} B(H)$. It follows that the identity of $\mathscr{C}$ factorizes similarly through $B(H) \otimes_{\max} B(H)$. In fact (see the proof of Corollary 12.22) we can take here $H = \ell_2$ so we obtain that $\mathscr{C}$ factorizes with unital c.p. maps through $\mathscr{B} \otimes_{\max} \mathscr{B}$. Thus by Corollary 9.70 if $\mathscr{B} \otimes_{\max} \mathscr{B}$ is QWEP so is $\mathscr{C}$. In other words $\mathscr{B} \otimes_{\max} \mathscr{B}$ is one-for-all.

13.2 Connection with Grothendieck's theorem

Curiously, there seems to be a connection between Grothendieck's theorem (in short GT) (or Grothendieck's inequality) and the Kirchberg conjecture. Indeed, the latter problem can be phrased as the identity of two norms on $\ell_1 \otimes \ell_1$, while GT implies that these two norms are equivalent, and their ratio is bounded by the Grothendieck constant K_G.

Here is the simplest formulation of the classical GT.

Theorem 13.3 (GT) *Let $[a_{ij}]$ be an $n \times n$ scalar matrix ($n \geq 1$). Assume that for any n-tuples of scalars (α_i), (β_j) we have*

$$\left| \sum a_{ij} \alpha_i \beta_j \right| \leq \sup_i |\alpha_i| \sup_j |\beta_j|. \tag{13.2}$$

Then for any Hilbert space H and any n-tuples (x_i), (y_j) in H, we have

$$\left| \sum a_{ij} \langle x_i, y_j \rangle \right| \leq K \sup \|x_i\| \sup \|y_j\|, \tag{13.3}$$

where K is a numerical constant. The best K (valid for all H and all n) is denoted by K_G.

In this statement the scalars can be either real or complex, but that affects the constant K_G, so we must distinguish its value in the real case $K_G^{\mathbb{R}}$ and in the complex case $K_G^{\mathbb{C}}$. To this day, its exact value is still unknown although it is known that $1 < K_G^{\mathbb{C}} < K_G^{\mathbb{R}} \leq 1.782$, see [210] for more information. In our context (in connection with spectral theory) it is natural to restrict to the case of complex scalars.

Remark 13.4 If we restrict to positive semidefinite matrices $[a_{ij}]$, then the best constant in (13.3) is known to be exactly $4/\pi$ in the complex case (and $\pi/2$ in the real case). This is called the "little" GT, see [210, §5] for details.

Let (e_i) denote the canonical basis of ℓ_1. Let $t = \sum a_{ij} e_i \otimes e_j \in \ell_1 \otimes \ell_1$ be a tensor in the linear span of $\{e_i \otimes e_j\}$. We denote

$$\|t\|_{H'} = \sup \left\{ \left| \sum a_{ij} \langle x_i, y_j \rangle \right| \right\} \tag{13.4}$$

where the supremum is over all Hilbert spaces H and all x_i, y_j in the unit ball of H.

We identify $\ell_1^n \subset \ell_1$ with the linear span of $\{e_i \mid 1 \leq i \leq n\}$ in ℓ_1, so we may consider that $t \mapsto \|t\|_{H'}$ is a norm on $\ell_1^n \otimes \ell_1^n$ for each integer $n \geq 1$.

The classical "injective" Banach space tensor norm for an element $t = \sum a_{ij} e_i \otimes e_j \in \ell_1^n \otimes \ell_1^n \subset \ell_1 \otimes \ell_1$ is given by the following formula.

$$\|t\|_\vee = \sup\left\{\left|\sum a_{ij}\alpha_i\beta_j\right| \;\middle|\; \alpha_i,\beta_j \in \mathbb{C},\ \sup_i |\alpha_i| \le 1,\ \sup_j |\beta_j| \le 1\right\}.$$
(13.5)

With this notation, GT in the form (13.3) can be restated as follows: there is a constant K such that for any n and any t in $\ell_1^n \otimes \ell_1^n$ we have

$$\|t\|_{H'} \le K\|t\|_\vee.$$
(13.6)

We will also need another norm introduced by Grothendieck as follows. We abusively denote again by $\{e_i \mid 1 \le i \le n\}$ the canonical basis of ℓ_∞^n. Note that we have isometrically both $\ell_\infty^n = (\ell_1^n)^*$ and $\ell_1^n = (\ell_\infty^n)^*$. Let $t' = \sum a_{ij}' e_i \otimes e_j \in \ell_\infty^n \otimes \ell_\infty^n$ ($a_{ij}' \in \mathbb{C}$). We define

$$\|t'\|_H = \inf\{\sup_i \|x_i\| \sup_j \|y_j\|\}$$
(13.7)

where the infimum runs over all Hilbert spaces H and all x_i, y_j in H such that $a_{ij}' = \langle x_i, y_j \rangle$ for all $i,j = 1,\dots,n$. It is an easy exercise to check directly that this is a norm but actually this follows from Proposition 1.57.

We denote by $\ell_1^n \otimes_{H'} \ell_1^n$ (resp. $\ell_\infty^n \otimes_H \ell_\infty^n$) the space $\ell_1^n \otimes \ell_1^n$ (resp. $\ell_\infty^n \otimes \ell_\infty^n$) equipped with the H'-norm (resp. H-norm). By definition of $\|t\|_{H'}$, we have clearly

$$\|t\|_{H'} = \sup\left\{\left|\sum a_{ij}a_{ij}'\right| \;\middle|\; \|t'\|_H \le 1\right\} = \sup\{|\langle t,t'\rangle| \mid \|t'\|_H \le 1\}, \quad (13.8)$$

so we have isometrically

$$\ell_1^n \otimes_{H'} \ell_1^n = (\ell_\infty^n \otimes_H \ell_\infty^n)^* \quad \text{and} \quad \ell_\infty^n \otimes_H \ell_\infty^n = (\ell_1^n \otimes_{H'} \ell_1^n)^*.$$

In operator theory terms, the norm $\|a\|_{H'}$ can be rewritten as follows. For any n and any $a_{ij} \in \mathbb{C}$ ($1 \le i,j \le n$) we have

$$\left\|\sum a_{ij}e_i \otimes e_j\right\|_{H'} = \sup\left\{\left\|\sum a_{ij}u_i v_j\right\|_{B(H)}\right\}$$
(13.9)

where the sup runs over all Hilbert spaces H and all n-tuples (u_i), (v_j) in the unit ball of $B(H)$. Indeed, it is an easy exercise (left to the reader) to check that (13.4) and (13.9) are equal.

Equivalently (by the Russo–Dye Theorem A.18)

$$\left\|\sum a_{ij}e_i \otimes e_j\right\|_{H'} = \sup\left\{\left\|\sum a_{ij}U_i V_j\right\|_{B(H)}\right\}$$
(13.10)

where the sup runs over all Hilbert spaces H and all n-tuples (U_i), (V_j) of unitaries in $B(H)$. In both cases it suffices to consider $H = \ell_2$.

Remark 13.5 Consider the free product $\mathbb{F}_\infty * \mathbb{F}_\infty$ and its associated (full) C^*-algebra $C^*(\mathbb{F}_\infty * \mathbb{F}_\infty) = \mathscr{C} * \mathscr{C}$. Let $(U_j^{(1)})$ (resp. $(U_j^{(2)})$) denote the free unitary generators of the first (second) of the two "factors" of the free product. Then it is easy to check that (13.10) is equivalent to:

$$\left\| \sum a_{ij} e_i \otimes e_j \right\|_{H'} = \left\| \sum a_{ij} U_i^{(1)} U_j^{(2)} \right\|_{\mathscr{C} * \mathscr{C}}. \tag{13.11}$$

To explain the connection with GT, we first give several equivalent reformulations of the Kirchberg conjecture. As before, we set $\mathscr{C} = C^*(\mathbb{F}_\infty)$. Let $(U_j)_{j \geq 1}$ denote the unitaries in $\mathscr{C}$ that correspond to the free generators of $\mathbb{F}_\infty$. For convenience of notation we set $U_0 = 1$ (i.e. the unit in $\mathscr{C}$). We recall that the closed linear span $E \subset \mathscr{C}$ of $\{U_j \mid j \geq 0\}$ is isometric to ℓ_1 (see Remark 3.14) and that E generates $\mathscr{C}$ as a C^*-algebra.

Fix $n \geq 1$. Consider a family of matrices $\{a_{ij} \mid i, j \geq 0\}$ with $a_{ij} \in M_N$ for all $i, j \geq 0$ such that $|\{(i, j) \mid a_{ij} \neq 0\}| < \infty$. We denote

$$a = \sum a_{ij} \otimes U_i \otimes U_j \in M_N \otimes E \otimes E.$$

We denote, again for $\mathscr{C} = C^*(\mathbb{F}_\infty)$:

$$\|a\|_{\min} = \left\| \sum a_{ij} \otimes U_i \otimes U_j \right\|_{M_N(\mathscr{C} \otimes_{\min} \mathscr{C})} \quad \text{and}$$

$$\|a\|_{\max} = \left\| \sum a_{ij} \otimes U_i \otimes U_j \right\|_{M_N(\mathscr{C} \otimes_{\max} \mathscr{C})}.$$

Then, on one hand, going back to the definitions, it is easy to check that

$$\|a\|_{\max} = \sup \left\{ \left\| \sum a_{ij} \otimes u_i v_j \right\|_{M_N(B(H))} \right\} \tag{13.12}$$

where the supremum runs over all H and all possible unitaries u_i, v_j on the same Hilbert space H such that $u_i v_j = v_j u_i$ for all i, j. On the other hand, using the known fact that $\mathscr{C}$ embeds into a direct sum of matrix algebras (see Theorem 9.18), one can check that

$$\|a\|_{\min} = \sup \left\{ \left\| \sum a_{ij} \otimes u_i v_j \right\|_{M_N(B(H))} \,\middle|\, \dim(H) < \infty \right\} \tag{13.13}$$

where the sup is as in (13.12) except that we restrict it to all *finite-dimensional* Hilbert spaces H. Indeed, since $\dim(H) < \infty \Rightarrow B(H) \otimes_{\min} B(H) = B(H) \otimes_{\max} B(H)$, the sup in (13.13) is the same as the supremum over all finite-dimensional H's and all unitaries u_i, v_j on H of

$$\left\| \sum a_{ij} \otimes (u_i \otimes v_j) \right\|_{M_N(B(H) \otimes_{\min} B(H))}$$

and by Theorem 9.18, the latter is the same as $\|a\|_{\min}$.

We may ignore the restriction $u_0 = v_0 = 1$ because we can always replace (u_i, v_i) by $(u_0^{-1} u_i, v_i v_0^{-1})$ without changing either (13.12) or (13.13).

The following was observed in [204].

Proposition 13.6 *Let $\mathscr{C} = C^*(\mathbb{F}_\infty)$. The following assertions are equivalent:*

(i) $\mathscr{C} \otimes_{\min} \mathscr{C} = \mathscr{C} \otimes_{\max} \mathscr{C}$ *(i.e. Kirchberg's conjecture is correct).*

(ii) *For any $N \geq 1$ and any $\{a_{ij} \mid i, j \geq 0\} \subset M_N$ as previously the norms (13.12) and (13.13) coincide i.e. $\|a\|_{\min} = \|a\|_{\max}$.*

(iii) *The identity $\|a\|_{\min} = \|a\|_{\max}$ holds for all $N \geq 1$ but merely for all families $\{a_{ij}\}$ in M_N supported in the union of $\{0\} \times \{0, 1, 2\}$ and $\{0, 1, 2\} \times \{0\}$.*

Note that (iii) reduces the Kirchberg conjecture to a statement about an operator space of dimension 5! But it requires to control the whole operator space structure of this five-dimensional space, so the size N of the five matrix coefficients is unbounded in (iii).

It was conceivable that the equality $\|a\|_{\max} = \|a\|_{\min}$ might hold when $a_{ij} \in \mathbb{C}$ i.e. in the case $N = 1$ in the previous (ii). But recently in [191, th. 29] Ozawa proved that this is actually equivalent to the Kirchberg conjecture. More precisely, he showed that the conditions in Proposition 13.6 are equivalent to:

(iv) *For any $\{a_{ij} \mid i, j \geq 0\} \subset \mathbb{C}$ as previously the norms (13.12) and (13.13) coincide, i.e. we have*

$$\|a\|_{\min} = \|a\|_{\max}, \text{ for all } a \text{ in the } \mathbb{C}\text{-linear span of } U_i \otimes U_j.$$

Note that Ozawa's argument requires infinitely many generators in (iv).

This result will be restated and proved later on as Theorem 14.7. It should be compared with the next two statements.

Theorem 13.7 (Tsirelson [251]) *We consider the case $N = 1$ and $a_{ij} \in \mathbb{R}$ for all i, j. Then*

$$\|a\|_{\max} = \|a\|_{\min} = \|a\|_{H'},$$

where we have denoted by $\|a\|_{H'}$ the norm appearing either in (13.9), (13.10) or (13.11). Moreover, these norms are all equal to

$$\sup \left\| \sum a_{ij} u_i v_j \right\| \tag{13.14}$$

where the sup runs over all $d \geq 1$ and all self-adjoint unitary $d \times d$ matrices u_i, v_j such that $u_i v_j = v_j u_i$ for all i, j.

Proof Recall (see (13.4)) that $\|a\|_{H'} \leq 1$ if and only if for any unit vectors x_i, y_j in a Hilbert space we have

$$\left| \sum a_{ij} \langle x_i, y_j \rangle \right| \leq 1.$$

Note that, since $a_{ij} \in \mathbb{R}$, whether we work with real or complex Hilbert spaces does not affect this condition. The resulting H'-norm is the same. We have trivially

$$\|a\|_{\min} \leq \|a\|_{\max} \leq \|a\|_{H'},$$

so it suffices to check $\|a\|_{H'} \leq \|a\|_{\min}$. Consider unit vectors x_i, y_j in a *real* Hilbert space H. We may assume that $\{a_{ij}\}$ is supported in $[1, \ldots, n] \times [1, \ldots, n]$ and that $\dim(H) = n$. From classical facts on "spin systems" (Pauli matrices, Clifford algebras and so on), we claim that there are self-adjoint unitary matrices u_i, v_j (of size 2^n) such that $u_i v_j = v_j u_i$ for all i, j and a (vector) state f such that $f(u_i v_j) = \langle x_i, y_j \rangle \in \mathbb{R}$. Indeed, let $H = \mathbb{R}^n$, $\widehat{H} = \mathbb{C}^n$ and let $\widehat{H}^{\wedge k}$ denote the k-fold antisymmetric tensor product, which admits $\{e_{i_1} \wedge \cdots \wedge e_{i_k} \mid i_1 < \cdots < i_k\}$ as orthonormal basis. Let $\mathcal{F} = \mathbb{C} \oplus \widehat{H} \oplus \widehat{H}^{\wedge 2} \oplus \cdots$ denote the (2^n-dimensional) antisymmetric Fock space associated to $\widehat{H}$ with vacuum vector Ω ($\Omega \in \mathcal{F}$ is the unit in $\mathbb{C} \subset \mathcal{F}$). For any $x, y \in H = \mathbb{R}^n$, let $c(x), c(y) \in B(\mathcal{F})$ (resp. $d(x), d(y) \in B(\mathcal{F})$) be the left (resp. right) creation operators defined by $c(x)\Omega = x$ (resp. $d(x)\Omega = x$) and $c(x)t = x \wedge t$ (resp. $d(x)t = t \wedge x$) for $x \in \widehat{H}^{\wedge n}$ with $n > 0$. Let $v_y = d(y) + d(y)^*$ and $u_x = c(x) + c(x)^*$. Then u_x, v_y are commuting self-adjoint unitaries and setting $f(T) = \langle \Omega, T\Omega \rangle$, we find $f(u_x v_y) = \langle x, y \rangle$. So applying this to x_i, y_j yields the claim. Thus we obtain by (13.13)

$$\left| \sum a_{ij} \langle x_i, y_j \rangle \right| = \left| f\left(\sum a_{ij} u_{x_i} v_{y_j} \right) \right| \leq \|a\|_{\min},$$

and hence $\|a\|_{H'} \leq \|a\|_{\min}$. This proves $\|a\|_{H'} = \|a\|_{\min}$ but also $\|a\|_{H'} \leq$ (13.14). Since, by (13.13), we have (13.14) $\leq \|a\|_{\min}$, (13.14) must be also equal to the number $\|a\|_{H'} = \|a\|_{\min} = \|a\|_{\max}$. $\qquad\square$

Remark 13.8 When $a_{ij} \in \mathbb{C}$, the preceding argument shows that

$$\sup\left\{ \left| \sum a_{ij} \Re\langle x_i, y_j \rangle \right| \;\middle|\; x_i, y_j \in B_{\ell_2} \right\} \leq \left\| \sum a_{ij} U_i \otimes U_j \right\|_{\min}.$$

Indeed, the supremum remains the same when reduced to unit vectors $x_j, y_i \in \ell_2^n(\mathbb{C})$, and then $\Re\langle x_i, y_j \rangle$ can be thought of as the scalar product in the "real" Hilbert space $\ell_2^{2n}(\mathbb{R}) \simeq \ell_2^n(\mathbb{C})$.

This inequality fails (for $a_{ij} \in \mathbb{C}$) if one replaces $\Re\langle x_i, y_j \rangle$ by $\langle x_i, y_j \rangle$ (see Proposition 13.10). However, we have

Proposition 13.9 *Assume $N = 1$ and $a_{ij} \in \mathbb{C}$ for all $i, j \geq 0$ then $\|a\|_{\max} \leq \|a\|_{H'}$ and $\|a\|_{\vee} \leq \|a\|_{\min}$. Therefore $\|a\|_{\max} \leq K_G^{\mathbb{C}} \|a\|_{\min}$.*

Moreover, for any positive semidefinite matrix $[a_{ij}]$ we have $\|a\|_{\max} \leq (4/\pi)\|a\|_{\min}$.

Proof By (13.13) we have

$$\sup\left\{\left|\sum a_{ij}s_it_j\right| \;\middle|\; s_i, t_j \in \mathbb{C} \quad |s_i| = |t_j| = 1\right\} = \|a\|_\vee \leq \|a\|_{\min}.$$

For any unit vectors x, y in H, we have

$$\left|\sum a_{ij}\langle y, u_iv_jx\rangle\right| = \left|\sum a_{ij}\langle u_i^*y, v_jx\rangle\right| \leq \|a\|_{H'},$$

so that

$$\left\|\sum a_{ij}u_iv_j\right\| \leq \|a\|_{H'}$$

and actually this holds without any extra assumption (such as mutual commutation) on the $\{u_i\}$'s and the $\{v_j\}$'s. A fortiori, we have $\|a\|_{\max} \leq \|a\|_{H'}$, and by Theorem 13.3 we conclude that $\|a\|_{\max} \leq K_G^{\mathbb{C}}\|a\|_{\min}$. The last assertion follows from Remark 13.4. $\qquad\square$

Remark The equalities $\|a\|_{H'} = \|a\|_{\min}$ and $\|a\|_{H'} = \|a\|_{\max}$ in Theorem 13.7 do not extend to the case of matrices with complex entries. This was proved by Éric Ricard. See the subsequent Proposition 13.10. In sharp contrast, if $[a_{ij}]$ is a 2×2 matrix, they do extend because, by [69, 249] for 2×2 matrices in the complex case (13.3) happens to be valid with $K = 1$ (while in the real case, for 2×2 matrices, the best constant is $\sqrt{2}$).

See [57, 75, 95, 138, 222, 223] for related contributions.

Recall that we view $\ell_1 \subset C^*(\mathbb{F}_\infty)$ completely isometrically. In this embedding the canonical basis of ℓ_1 is identified with the canonical free unitaries $\{U_j \mid j \geq 0\}$ generating $C^*(\mathbb{F}_\infty)$, with the convention $U_0 = 1$ (see Remark 3.14). Similarly, for any $n = 1, 2, \ldots$, we may restrict to $\{U_j \mid 0 \leq j \leq n - 1\}$ and identify $\ell_1^n \subset C^*(\mathbb{F}_\infty)$ as the span of $\{U_j \mid 0 \leq j \leq n - 1\}$. Equivalently, we could replace $\mathbb{F}_\infty$ by $\mathbb{F}_{n-1}$ and view $\ell_1^n \subset C^*(\mathbb{F}_{n-1})$.

The following proposition and its proof were kindly communicated by Éric Ricard.

Proposition 13.10 (Ricard) *For any $n \geq 4$, the H'-norm on $\ell_1^n \otimes \ell_1^n$ does not coincide with the norm induced by the* max*-norm on $C^*(\mathbb{F}_\infty) \otimes C^*(\mathbb{F}_\infty)$ (or equivalently on $C^*(\mathbb{F}_{n-1}) \otimes C^*(\mathbb{F}_{n-1})$).*

Proof The proof is inspired by some of the ideas in [115]. Let us consider a family $x_0, x_1, \ldots$ in the unit sphere of $\mathbb{C}^2$ and let $a_{jk} = \langle x_j, x_k\rangle$. We identify ℓ_1^n with the span of the canonical free unitaries $1 = U_0, U_1, \ldots, U_{n-1}$ generating $C^*(\mathbb{F}_{n-1})$. Then consider the element $a = \sum_{j,k=0}^{n-1} a_{jk}e_j \otimes e_k$. Note that a is clearly in the unit ball of $\ell_\infty^n \otimes_H \ell_\infty^n$. For simplicity let $A = C^*(\mathbb{F}_{n-1})$. Assume that the max-norm coincides with the H'-norm on $\ell_1^n \otimes \ell_1^n$. Then any such a defines a functional of norm at most 1 on $\ell_1^n \otimes \ell_1^n \subset A \otimes_{\max} A$ (with induced norm). Therefore (Hahn–Banach) there is f in the unit ball of $(A \otimes_{\max} A)^*$ such that $f(U_j \otimes U_k) = a_{jk}$, and since $f(1) = f(U_0 \otimes U_0) = a_{00} = \langle x_0, x_0 \rangle = 1$, f is a state (see Remark 1.33). By the GNS representation of states (A.16) there is a Hilbert space H, a unit vector $h \in H$ and commuting unitaries u_j, v_k^* on H (associated to a unital representation of $A \otimes_{\max} A$) such that $u_0 = v_0 = 1$ and

$$\langle x_j, x_k \rangle = f(U_j \otimes U_k) = \langle h, v_k^* u_j h \rangle = \langle v_k h, u_j h \rangle \quad (0 \le j, k \le n-1).$$
(13.15)

Moreover the vector h can be chosen cyclic for the $*$-algebra generated by u_j, v_k^* (or equivalently of course u_j, v_k). Since $a_{jj} = \langle x_j, x_j \rangle = 1$, we have $u_j h = v_j h$ for all j, and hence h must also be cyclic for *each* of the $*$-algebras A_u and A_v generated by (u_j) and (v_j) respectively (indeed note that $A_u h = A_v h$ implies $A_u A_v h = A_u h = A_v h$). But by a classical elementary fact (see Lemma A.62) this implies that h is separating for each of the commutants A_u' and A_v', and since we have $A_u \subset A_v'$ and $A_v \subset A_u'$, it follows that h is separating for *each* of the $*$-algebras A_u and A_v.

We claim that there is an isomorphism $T : \mathrm{span}[u_j] \to \mathrm{span}[x_j]$ such that $T(u_j) = x_j$ and in particular the linear span of (u_j) is at most two-dimensional. Indeed, first observe that by (13.15) (since $u_j h = v_j h$) any linear combination $\sum b_j x_j$ ($b_j \in \mathbb{C}$) satisfies $\| \sum b_j x_j \|^2 = \| \sum b_j u_j h \|^2 \le \| \sum b_j u_j \|^2$, and hence $T : \mathrm{span}[u_j] \to \mathrm{span}[x_j]$ such that $T(u_j) = x_j$ is well defined. But now any linear relation $\sum b_j x_j = 0$ ($b_j \in \mathbb{C}$) implies $\| \sum b_j u_j h \|^2 = 0$ and (since h is separating on A_u) $\sum b_j u_j = 0$. This shows that $T : \mathrm{span}[u_j] \to \mathrm{span}[x_j]$ is injective, and hence an isomorphism so the linear span of (u_j) is at most two-dimensional.

Suppose that x_0, x_1 are linearly independent. Then so is $1, u_1$ (recall $u_0 = 1$), and the other u_j's are linear combinations of $1, u_1$ (so that A_u is the unital commutative C^*-algebra generated by u_1). So let us choose simply for x_0, x_1 the canonical basis of $\mathbb{C}^2$, and recall $T^{-1} x_j = u_j$. Choosing (here $i = \sqrt{-1}$) $x_2 = (1, i)2^{-1/2} = (x_0 + ix_1)2^{-1/2}$ and $x_3 = (1, 1)2^{-1/2} = (x_0 + x_1)2^{-1/2}$, by the linearity of T^{-1} we must have $T^{-1}(x_2) = (1 + iu_1)2^{-1/2}$. The latter must be unitary, equal to u_2, as well as $T^{-1}(x_3) = (1 + u_1)2^{-1/2}$ equal to u_3. But this is impossible because $(1 + iu_1)2^{-1/2}$ unitary requires the spectrum of u_1 included in $\{\pm 1\}$, while $(1 + u_1)2^{-1/2}$ unitary requires that it is included in $\{\pm i\}$. Thus we obtain a contradiction for any $n \ge 4$. $\square$

This should be compared with Dykema and Juschenko's results in [75].

13.3 Notes and remarks

§13.1 is due to Kirchberg. Remark 13.2 on "one-for-all" C^*-algebras combines various observations, some of them already in [208]. The remark that the Brown algebra is such an example appears in [128]. For more remarks on a similar theme, see [91], [128], and [63].

For §13.2 most references are credited in the text. The papers [91, 152, 153] contain several equivalent formulations of the WEP as a Riesz interpolation property and in terms of matrix completion problems. The formulation of Grothendieck's inequality in Theorem 13.3 was put forward by Lindenstrauss and Pełczyński in 1968, see [210] for more (and more precise) references.

14

Equivalence of the two main questions

We will now prove the equivalence of both problems.

14.1 From Connes's question to Kirchberg's conjecture

Theorem 14.1 *If the answer to Connes's question is positive, then any von Neumann algebra is QWEP and consequently Kirchberg's conjecture follows.*

The proof of Theorem 14.1 is based on the three steps described in the following lemmas.

Lemma 14.2 *If the answer to Connes's question is positive then any finite von Neumann algebra M is QWEP.*

Proof Assume $M \subset M_{\mathcal{U}}$ in a trace preserving way. Since we are dealing with finite traces, there is a c.p. conditional expectation $P : M_{\mathcal{U}} \to M$ with $\|P\| = 1$ (see Proposition 11.21). Clearly $M_{\mathcal{U}}$ is QWEP by definition. Thus the result follows from Corollary 9.61. □

Lemma 14.3 *If all finite von Neumann algebras are QWEP, then all semifinite von Neumann algebras M are QWEP.*

Proof If M is semifinite we can write

$$M = \overline{\cup M_\gamma}^{\text{weak}*}$$

where the von Neumann (unital) subalgebras M_γ are all finite and form a net directed by inclusion. Indeed, let P_γ be an increasing net of projections with finite trace tending weak* to the identity on M (see Remark 11.1). Then the unital subalgebras $M_\gamma = P_\gamma M P_\gamma + \mathbb{C}(1 - P_\gamma)$ are finite and their increasing union is weak*-dense in M. Since all the M_γ's are QWEP, we conclude that M is QWEP by Theorem 9.74. □

Lemma 14.4 *If all semifinite von Neumann algebras are QWEP, then all von Neumann algebras are QWEP.*

Proof By Takesaki's Theorem 11.3 any von Neumann algebra M can be embedded in a semifinite algebra N, so we have $M \subset N$, in such a way that there is a c.p. projection $P : N \to M$ with $\|P\| = 1$. Using this, the lemma follows immediately from Corollary 9.61.   □

Proof of Theorem 14.1 By the three preceding Lemmas, if the answer to Connes's question is positive then any von Neumann algebra is *QWEP*. By Proposition 13.1 this implies Kirchberg's conjecture.   □

Remark 14.5 Let $(M(i), \tau_i)$ be a family of tracial probability spaces, with ultraproduct $M_{\mathcal{U}}$. If all the $M(i)$'s are QWEP, then $M_{\mathcal{U}}$ is also QWEP by Corollary 9.62. Moreover, any von Neumann subalgebra $M \subset M_{\mathcal{U}}$ is also QWEP. Indeed, since the conditional expectation is a c.p. projection $P : M_{\mathcal{U}} \to M$, this follows from Corollary 9.61.

Remark 14.6 (On the Effros–Maréchal topology) In a series of remarkable papers, Haagerup with Winsløw and Ando ([9, 121, 122]) make a deep study of the Effros–Maréchal topology on the set $vN(H)$ of all von Neumann algebras $M \subset B(H)$ (on a given Hilbert space H). The latter topology is defined as the weakest topology for which the map $M \mapsto \|\varphi_{|M}\|$ is continuous for every $\varphi \in B(H)_*$. In particular, they show that a von Neumann algebra is QWEP if and only if it is in the closure (for that topology) of the set of injective factors. Thus their work shows that the Kirchberg conjecture is equivalent to the density of the latter set in $vN(H)$, say for $H = \ell_2$. Actually, it is also equivalent to the density of the set of type I factors and, as expected, to the density of the set of finite-dimensional factors (i.e. matricial factors).

14.2 From Kirchberg's conjecture to Connes's question

Let

$$E = \operatorname{span}\{U_j \mid j \geq 0\} \subset \mathscr{C}$$

be the linear span of the generators $\{U_j \mid j \geq 1\}$ and the unit (recall the convention $U_0 = 1$) in $\mathscr{C} = C^*(\mathbb{F}_\infty)$.

We will work with $\bar{E} \otimes E \subset \bar{\mathscr{C}} \otimes \mathscr{C}$.

In $\bar{E} \otimes E$ we distinguish the cone $(\bar{E} \otimes E)^+$ formed by all the "positive definite" tensors, more precisely

$$(\bar{E} \otimes E)^+ = \left\{ \sum_1^r \overline{x_k} \otimes x_k \in \bar{E} \otimes E \mid x_k \in E, r \geq 1 \right\}.$$

Alternatively:

$$(\bar{E} \otimes E)^+ = \left\{ \sum a_{ij}\overline{U_i} \otimes U_j \mid n \geq 1, a \in (M_n)_+ \right\}. \tag{14.1}$$

We call the elements of $(M_n)_+$ "positive definite" matrices (they are often called positive semidefinite). To check (14.1) observe that (by classical linear algebra) any such matrix can be written as a finite sum $a_{ij} = \sum_{k=1}^{r} \overline{x_k(i)}x_k(j)$ $(x_k(j) \in \mathbb{C})$, and conversely any matrix of the latter form is positive definite. Then we have $\sum a_{ij}\overline{U_i} \otimes U_j = \sum_1^r \overline{x_k} \otimes x_k$ where $x_k = \sum_j x_k(j)U_j$.

We define analogously

$$(E \otimes E)^+ = \left\{ \sum a_{ij}U_i \otimes U_j \mid n \geq 1, a \in (M_n)_+ \right\}. \tag{14.2}$$

We also define

$$(\bar{E} \otimes E)^+_{\min} = \{t \in (\bar{E} \otimes E)^+ \mid \|t\|_{\min} \leq 1\} \quad \text{and}$$
$$(\bar{E} \otimes E)^+_{\max} = \{t \in (\bar{E} \otimes E)^+ \mid \|t\|_{\max} \leq 1\}.$$

We have then

Theorem 14.7 (Ozawa [191]) *If the* min *and* max *norms coincide on* $E \otimes E \subset \mathscr{C} \otimes \mathscr{C}$ *(or equivalently if they coincide on* $\bar{E} \otimes E \subset \bar{\mathscr{C}} \otimes \mathscr{C}$*), then the Connes problem has a positive solution.*

Actually with the same idea as Ozawa's we will prove the following refinement:

Theorem 14.8 *If* $(\bar{E} \otimes E)^+_{\min} = (\bar{E} \otimes E)^+_{\max}$*, then the Connes problem has a positive solution.*

Since a positive answer to Connes's question implies Kirchberg's conjecture (see §14.1), we have:

Corollary 14.9 *If the* min *and* max *norms coincide on* $E \otimes E \subset \mathscr{C} \otimes \mathscr{C}$ *(resp. on* $(\bar{E} \otimes E)^+ \subset \bar{\mathscr{C}} \otimes \mathscr{C}$*) then they coincide on* $\mathscr{C} \otimes \mathscr{C}$ *(resp. on* $\bar{\mathscr{C}} \otimes \mathscr{C}$*), meaning the Kirchberg conjecture holds.*

Remark 14.10 Curiously there is currently no direct proof available for the preceding corollary.

Remark 14.11 For any C^*-algebra A we have $\bar{A} \simeq A^{op}$. The mapping being $\bar{x} \mapsto x^*$ (see §2.3). So the assumption $(\bar{E} \otimes E)^+_{\min} = (\bar{E} \otimes E)^+_{\max}$ can be rewritten as

$$\left\| \sum x_k^* \otimes x_k \right\|_{A^{op} \otimes_{\min} A} = \left\| \sum x_k^* \otimes x_k \right\|_{A^{op} \otimes_{\max} A}.$$

Remark 14.12 Recall (see Remark 3.7) that when $A = C^*(\mathbb{F}_\infty)$ we have $\bar{A} \simeq A$ and we have a $\mathbb{C}$-linear isomorphism $\Phi : A \to \bar{A}$ that takes $U_{\mathbb{F}_\infty}(t)$ to $\overline{U_{\mathbb{F}_\infty}(t)}$. In particular, this isomorphism takes U_i to $\bar{U}_i$ and $\mathscr{C}$ onto $\bar{\mathscr{C}}$. Therefore, the assumption $(\bar{E} \otimes E)^+_{\min} = (\bar{E} \otimes E)^+_{\max}$ is equivalent to the following one:

For any n and any $a \in (M_n)_+$ we have

$$\left\| \sum a_{ij} U_i \otimes U_j \right\|_{\mathscr{C} \otimes_{\min} \mathscr{C}} = \left\| \sum a_{ij} U_i \otimes U_j \right\|_{\mathscr{C} \otimes_{\max} \mathscr{C}}.$$

The main point is as follows:

Theorem 14.13 *Let (M, τ) be a tracial probability space on a separable Hilbert space. Let $\pi : \mathscr{C} \to M$ be a $*$-homomorphism taking the free generators $\{U_j\}$ to an $L_2(\tau)$-dense subset of $U(M)$. Assume that for any finite set (x_j) in $\mathrm{span}[U_j] \subset \mathscr{C}$ we have*

$$\sum \|\pi(x_j)\|_2^2 \leq \left\| \sum \overline{x_j} \otimes x_j \right\|_{\overline{\mathscr{C}} \otimes_{\min} \mathscr{C}}. \tag{14.3}$$

Then M embeds in a trace preserving way in a matricial $M_{\mathcal{U}}$ for some $\mathcal{U}$ on $\mathbb{N}$.

Proof Let $U_0 = 1$. We apply Theorem 11.38 with $S = \{U_j \mid j \geq 0\}$ and $W = \pi(S)$. This shows that W satisfies the condition (iv)' in Kirchberg's criterion (see Theorem 12.8). By the $L_2(\tau)$-density of W in $U(M)$ it follows that there is a trace preserving embedding of M in an ultraproduct of matrix algebras, which completes the proof. $\square$

Remark 14.14 (A variant) If we assume that $\pi(S)$ is an $L_2(\tau)$-dense *subgroup* of $U(M)$ then we can conclude using Corollary 11.46 instead of Kirchberg's criterion (Theorem 12.3).

Corollary 14.15 *A finite von Neumann algebra (on a separable Hilbert space) is QWEP if and only if it embeds in a matricial $M_{\mathcal{U}}$ for some $\mathcal{U}$ on $\mathbb{N}$.*

Proof If M is QWEP, let π be as in Theorem 14.13 (or Remark 14.14). By Theorem 9.67,

$$\|Id_{\overline{\mathscr{C}}} \otimes \pi : \overline{\mathscr{C}} \otimes_{\min} \mathscr{C} \to \overline{\mathscr{C}} \otimes_{\max} M\| \leq 1,$$

and hence a fortiori

$$\|\overline{\pi} \otimes \pi : \overline{\mathscr{C}} \otimes_{\min} \mathscr{C} \to \overline{M} \otimes_{\max} M\| \leq 1.$$

From this, since $\sum \tau(\pi(x_j)^* \pi(x_j)) = \sum \langle 1, L(\pi(x_j)^*) R(\pi(x_j)) 1 \rangle$, we deduce

$$\sum \tau(\pi(x_j)^* \pi(x_j)) \leq \left\| \sum \overline{\pi(x_j)} \otimes \pi(x_j) \right\|_{\overline{M} \otimes_{\max} M} \leq \left\| \sum \overline{x_j} \otimes x_j \right\|_{\overline{\mathscr{C}} \otimes_{\min} \mathscr{C}}.$$

By Theorem 14.13 (or Remark 14.14) M embeds in some matricial $M_{\mathcal{U}}$. Conversely if M embeds in some matricial $M_{\mathcal{U}}$, the latter is obviously QWEP, and there is a unital c.p. projection (conditional expectation) from $M_{\mathcal{U}}$ onto M. By Corollary 9.61 M inherits QWEP from $M_{\mathcal{U}}$. $\qquad\square$

We can now easily complete the proof of Theorems 14.8 and 14.7. All the difficult ingredients have been already proved mainly in §11.6.

Proof of Theorems 14.8 and 14.7 It clearly suffices to prove Theorem 14.8. Let (M, τ) be a tracial probability space on a separable Hilbert space. Then there is a countable subgroup $\mathcal{W} \subset U(M)$ that is $L_2(\tau)$-dense in $U(M)$. We may assume that there is a unital bijection from $S = \{U_j \mid j \geq 0\}$ onto $\mathcal{W}$. Let $\pi : \mathscr{C} \to M$ be the $*$-homomorphism extending the latter bijection (this exists by the freeness of S). Since left-hand and right-hand multiplications on $L_2(\tau)$ have commuting ranges we have for any $x_j \in \mathscr{C}$

$$\sum \tau(\pi(x_j)^* \pi(x_j)) \leq \left\| \sum \overline{\pi(x_j)} \otimes \pi(x_j) \right\|_{\overline{M} \otimes_{\max} M} \leq \left\| \sum \overline{x_j} \otimes x_j \right\|_{\overline{\mathscr{C}} \otimes_{\max} \mathscr{C}}.$$
$$(14.4)$$

Thus, if the min and max norms coincide on $(\bar{E} \otimes E)^+$, (14.3) holds, and by Theorem 14.13, M trace preservingly embeds in some matricial $M_{\mathcal{U}}$, so that the Connes question has a positive answer. $\qquad\square$

Remark 14.16 (A slight refinement) If the min and max norms coincide on

$$\{T + T^* \mid T \in (\bar{E} \otimes E)^+\} = \left\{ \sum \overline{x_j} \otimes x_j + \sum \overline{x_j^*} \otimes x_j^* \mid x_j \in E \right\}$$

then they coincide on $\overline{\mathscr{C}} \otimes \mathscr{C}$.

Indeed, with the notation of the preceding proof, we have

$$\sum \tau(\pi(x_j)^* \pi(x_j)) + \tau(\pi(x_j)\pi(x_j)^*)$$
$$\leq \left\| \sum \overline{\pi(x_j)} \otimes \pi(x_j) + \sum \overline{\pi(x_j^*)} \otimes \pi(x_j^*) \right\|_{\overline{M} \otimes_{\max} M}$$
$$\leq \left\| \sum \overline{x_j} \otimes x_j + \sum \overline{x_j^*} \otimes x_j^* \right\|_{\max}$$
$$= \left\| \sum \overline{x_j} \otimes x_j + \sum \overline{x_j^*} \otimes x_j^* \right\|_{\min} \leq 2 \left\| \sum \overline{x_j} \otimes x_j \right\|_{\overline{\mathscr{C}} \otimes_{\min} \mathscr{C}}.$$

But since $\tau(\pi(x_j)^* \pi(x_j)) = \tau(\pi(x_j)\pi(x_j)^*)$ (trace property) we again obtain (14.3), and we conclude as in the preceding proof.

Remark 14.17 (A slight refinement (bis)) If the min and max norms coincide on

$$\{T + T^* \mid T \in (E \otimes E)^+\}$$

then they coincide on $\mathscr{C} \otimes \mathscr{C}$.

Indeed, by Remark 14.12 we have $\Phi((E \otimes E)^+) = (\bar{E} \otimes E)^+$.

14.3 Notes and remarks

Most of the results are due to Kirchberg. One major exception is Ozawa's Theorem 14.7 derived from [191, th. 29] which is a quite significant improvement over Proposition 13.6. The refinement from Theorem 14.7 to Theorem 14.8 or Theorem 14.13 is obtained by an easy adaptation of Ozawa's ideas. Corollary 14.15 is due to Kirchberg.

15

Equivalence with finite representability conjecture

Definition 15.1 A Banach space X is finitely representable in another Banach space Y if for any $\varepsilon > 0$ and any finite-dimensional subspace $E \subset X$ there is a subspace $\widetilde{E} \subset Y$ that is $(1+\varepsilon)$-isomorphic to E. In that case we write X f.r. Y.

We discuss equivalent forms of finite representability in §A.6.

We denote by $S_1(H)$ the Banach space formed of all the trace class operators on H (see §A.10). The space $S_1(H)$ is isometric to the predual of $B(H)$. Recall the latter space is also isometric to the dual of $K(H)$. When $H = \ell_2$ we set $S_1 = S_1(\ell_2)$.

15.1 Finite representability conjecture

In the late 1970s, the following conjecture began to circulate:

Finite representability conjecture: A^* f.r. S_1 for any C^*-algebra A.

Equivalently, M_* f.r. S_1 for any von Neumann algebra M.

This conjecture was popularized in the Banach space community by talks and papers notably by Pełczyński and Garling, but it probably can also be traced back to Haagerup.

Remark 15.2 Let $S_1(H)$ be the space of trace class operators on a Hilbert space H. Let $x \in S_1(H)$ with H infinite dimensional. Then there is clearly a subspace $K \simeq \ell_2$ such that $x = P_K x P_K$. More generally, given a finite set $x_1, \ldots, x_n \in S_1(H)$ there is a K such that this holds for any $x \in \{x_1, \ldots, x_n\}$. Therefore any finite-dimensional subspace $E \subset S_1(H)$ embeds isometrically in $S_1 = S_1(\ell_2)$. A fortiori, $S_1(H)$ f.r. S_1.

Theorem 15.3 (Kirchberg) *The finite representability conjecture is equivalent to the Kirchberg conjecture. More precisely, a C^*-algebra A is QWEP if and only if A^* f.r. S_1.*

Proof We first show that if A is QWEP then A^* f.r. S_1. The converse is much more delicate. Assume that A is QWEP. By Theorem 9.72, A^{**} is a 1-complemented subspace of $B(H)^{**}$, and hence A^{***}, and a fortiori $A^* \subset A^{***}$, embeds isometrically in $B(H)^{***}$. By Proposition A.15 (i.e. the local reflexivity principle) we know that $B(H)^{***}$ f.r. $B(H)^*$, and also (applying it again) that $B(H)^*$ f.r. $B(H)_*$, while Remark 15.2 tells us $B(H)_*$ f.r. S_1. Thus we obtain that A^* f.r. S_1.

Assume A^* f.r. S_1. By Lemma A.8 there is a set I and a (metric) surjection $Q : \ell_\infty(I; B(\ell_2)) \to A^{**}$ taking the closed unit ball of $\ell_\infty(I; B(\ell_2))$ onto that of A^{**}. Let $W = \ell_\infty(I; B(\ell_2))$ (which is WEP by, say, Proposition 9.34) and $N = W^{**}$. Note that N is QWEP by Theorem 9.66. Replacing Q by $\varphi = \ddot{Q} : W^{**} \to A^{**}$ (as defined in (A.32)) we find a normal (metric) surjection $\varphi : N \to A^{**}$ taking the closed unit ball of N onto that of A^{**}. To conclude we call *Jordan algebras* to the rescue via Theorem 5.7 applied to $M = A^{**}$, as follows. For any projection p in N, the C^*-algebras pNp and $(pNp)^{op}$ are QWEP by Corollary 9.70. If q is another such projection, with $q \perp p$, then $pNp \oplus (qNq)^{op}$ is QWEP (recall Remark 9.16). By Theorem 5.7, the identity of $M = A^{**}$ factors completely positively through $pNp \oplus (qNq)^{op}$ for some p, q. Therefore, by Corollary 9.70 A^{**} is QWEP, and a fortiori by Theorem 9.66, A itself is QWEP. $\square$

Remark 15.4 An operator space X is finitely representable in the operator sense in another one Y if for any finite-dimensional subspace $E \subset X$ and any $\varepsilon > 0$ there is $\widetilde{E} \subset Y$ that is *completely* $(1 + \varepsilon)$-isomorphic to E. We say that X is strictly locally reflexive if X^{**} is finitely representable in the operator sense in X. By [78] this holds whenever X^* is completely isometric to a C^*-algebra (i.e. for any so-called noncommutative L_1-space), in particular for $X = B(H)^*$ and $X = S_1$. Combined with the first part of the preceding argument, this shows that A^* is finitely representable in the operator sense in S_1 whenever A is QWEP.

In the same spirit as the finite representability conjecture, we note one more characterization of QWEP von Neumann algebras involving only their Banach space structure:

Theorem 15.5 (Kirchberg) *A von Neumann algebra N is QWEP if and only if it is isometric as a Banach space to a (Banach space sense) quotient of $B(H)$ for some Hilbert space H.*

Proof Consider first the special case $N = B(H)^{**}$. In that case Proposition A.12 tells us that N is isometric to a quotient of some $(\oplus \sum_{i \in I} X_i)_\infty$ with $X_i = B(H)$ for all $i \in I$. Let $\mathcal{H} = \oplus \sum_{i \in I} H_i$ with $H_i = H$ for all $i \in I$.

The projection onto the diagonal defines a metric surjection $B(\mathcal{H}) \rightarrow (\oplus \sum_{i \in I} X_i)_\infty$. Therefore $(\oplus \sum_{i \in I} X_i)_\infty$, and a fortiori N, is isometric to a quotient of $B(\mathcal{H})$.

If N is QWEP, by Theorem 9.72 its bidual N^{**} is isometric to a quotient of some $B(H)^{**}$ and hence by what precedes to a quotient of some $B(\mathcal{H})$. By Remark A.52 the same is true for N.

Conversely, assume N is isometric to a quotient of $B(H)$. Then N^* embeds isometrically in $B(H)^*$. A fortiori N^* f.r. $B(H)^*$. By the local reflexivity principle (Proposition A.15) and Remark 15.2 we know that $B(H)^*$ f.r. $B(H)_*$ f.r. S_1. A fortiori N^* f.r. S_1, and N is QWEP by Theorem 15.3. $\qquad \square$

15.2 Notes and remarks

The results here come from Kirchberg's [155], but Ozawa's expository paper [189] greatly clarified the picture.

16

Equivalence with Tsirelson's problem

It is difficult to introduce Tsirelson's problem without mentioning the unusual saga that led to it. In a remarkable paper [251] in 1980, Tsirelson observed that the famous Bell inequality, widely celebrated in quantum mechanics, could be viewed as a particular instance of Grothendieck's inequality equally celebrated by Banach space theorists. Both inequalities involve matrices of the form $[\langle x_i, y_j \rangle]$ where x_i, y_j are in the unit ball of a Hilbert space H. Tsirelson was particularly interested in the case when $H = L_2(M, \tau)$ on a tracial probability space (M, τ). In his discussion he asserted without proof that for his specific purpose (to be described) the case of a general (M, τ) could be reduced to the matricial one. Years later, in connection with the development of Quantum Information Theory it was pointed out to him that this reduction was not clear and he himself advertised the problem widely. The goal of this section is to show that the latter Tsirelson problem is actually equivalent to the Connes embedding problem. The proof of this equivalence was completed in several steps, in the papers [95] and [138], the final step being taken by Ozawa in [191].

16.1 Unitary correlation matrices

As a preliminary, it seems worthwhile to describe several reformulations of the Connes and Kirchberg problem in terms of unitary correlation matrices.

Let $R_c(d)$ be the set of $d \times d$-matrices $x = [x_{ij}]$ of the form

$$x_{ij} = \langle \xi, u_i^* v_j \xi \rangle \qquad (16.1)$$

where (u_i) and (v_i) are d-tuples of unitaries in $B(H)$ such that $u_i v_j = v_j u_i$ (or equivalently $u_i^* v_j = v_j u_i^*$) for all i, j, where H is arbitrary and $\xi \in H$ is a unit vector. We then define the analogous set $R_s(d)$ but restricted to finite-dimensional H's.

Explicitly: let $R_s(d)$ be the set of $d \times d$-matrices $x = [x_{ij}]$ of the form $x_{ij} = \langle \xi, u_i^* v_j \xi \rangle$ where u_i, v_j are unitaries on H with $\dim(H) < \infty$ such that $u_i v_j = v_j u_i$ for all i, j and $\xi \in H$ is a unit vector.

Let $E(d) \subset \mathscr{C}$ be the linear span of $\{U_1, \ldots, U_d\}$. For any state f on $\mathscr{C} \otimes_{\max} \mathscr{C}$ (resp. $\mathscr{C} \otimes_{\min} \mathscr{C}$) let $x^f \in M_d$ be the matrix defined by:

$$x_{ij}^f = f(U_i \otimes U_j).$$

Clearly $x^f \mapsto f_{|E(d) \otimes E(d)}$ is a linear isomorphism. Let $S^{\max}$ (resp. $S^{\min}$) be the set of states on $\mathscr{C} \otimes_{\max} \mathscr{C}$ (resp. $\mathscr{C} \otimes_{\min} \mathscr{C}$). Then

$$R_c(d) = \{x^f \mid f \in S^{\max}\}. \tag{16.2}$$

Indeed, if x is as in (16.1), let $\pi : \mathscr{C} \otimes \mathscr{C} \to B(H)$ be a unital $*$-homomorphism taking $U_i \otimes U_j$ to $u_i^* v_j$. We have $x = x^f$ with $f(\cdot) = \langle \xi, \pi(\cdot)\xi \rangle$ in $S^{\max}$. Conversely, the GNS factorization of f (see §A.13) shows that $x^f \in R_c(d)$ for any $f \in S^{\max}$. Since $S^{\max}$ is weak* closed, (16.2) implies that $R_c(d)$ is closed (this can also be shown using ultraproducts as in (16.9)). Analogously, we claim that

$$\overline{R_s(d)} = \{x^f \mid f \in S^{\min}\}. \tag{16.3}$$

Indeed, this is easy to deduce from the same reasoning but using Proposition 9.20.

Obviously, $R_s(d) \subset R_c(d)$, and since $R_c(d)$ is closed we have $\overline{R_s(d)} \subset R_c(d)$.

Proposition 16.1 *The Kirchberg conjecture is equivalent to the assertion that $R_s(d)$ is dense in $R_c(d)$ for any d.*

Proof The Kirchberg conjecture implies $S^{\max} = S^{\min}$ and hence $\overline{R_s(d)} = R_c(d)$ by (16.2) and (16.3). We turn to the converse. Let $[a_{ij}]$ be a positive definite matrix in M_d. In particular $a = a^*$ so that $a_{ij} = \overline{a_{ji}}$. Let $T = \sum a_{ij} U_i \otimes U_j \in (E \otimes E)^+ \subset \mathscr{C} \otimes \mathscr{C}$. We claim that

$$\sup \left\{ \left| \sum a_{ij}(x_{ij} + \overline{x_{ji}}) \right| \,\middle|\, x \in R_c(d) \right\} = \|T + T^*\|_{\max} \tag{16.4}$$

while

$$\sup \left\{ \left| \sum a_{ij}(x_{ij} + \overline{x_{ji}}) \right| \,\middle|\, x \in R_s(d) \right\} \le \|T + T^*\|_{\min}. \tag{16.5}$$

In fact we will see that (16.5) is an equality. Taking these claims for granted, the density of $R_s(d)$ in $R_c(d)$ implies $\|T + T^*\|_{\max} = \|T + T^*\|_{\min}$. Then the Kirchberg conjecture follows by Corollary 14.9 and Remark 14.17.

We now verify the claim. By definition of the max-norm we have

$$\|T + T^*\|_{\max} = \sup \left\{ \left\| \sum a_{ij} u_i^* v_j + \overline{a_{ij}} u_i v_j^* \right\| \right\},$$

where the sup runs over all $(u_i), (v_j)$ as in the definition of $R_c(d)$. Thus

$$\|T + T^*\|_{\max} = \sup\left\{\left|\left\langle\xi, \sum a_{ij}u_i^* v_j \xi\right\rangle + \left\langle\xi, \sum \overline{a_{ij}}u_i v_j^* \xi\right\rangle\right|\right\}$$

where the sup runs over all unit vectors ξ. Equivalently

$$\|T + T^*\|_{\max} = \sup\left\{\left|\sum x_{ij}a_{ij} + \sum \overline{x_{ij}a_{ij}}\right| \,\middle|\, x \in R_c(d)\right\}$$

which (since $a = a^*$) is the same as (16.4).

Let A_u (resp. A_v) denote the C^*-algebra generated by (u_i) (resp. (v_i)). We have by a similar argument

$$\sup\left\{\left|\sum a_{ij}(x_{ij} + \overline{x_{ji}})\right| \,\middle|\, x \in R_s(d)\right\} = \sup\left\{\left\|\sum a_{ij}u_i^* v_j + \overline{a_{ij}}u_i v_j^*\right\|\right\}$$

where the sup runs over all $(u_i), (v_j)$ on a finite-dimensional H as in the definition of $R_s(d)$; moreover, since A_u is nuclear if $\dim(H) < \infty$, the latter $(u_i), (v_j)$ satisfy

$$\left\|\sum a_{ij}u_i^* v_j + \overline{a_{ij}}u_i v_j^*\right\| \le \left\|\sum a_{ij}\overline{u_i} \otimes v_j + \overline{a_{ij}}\overline{u_i^*} \otimes v_j^*\right\|_{\overline{A_u} \otimes_{\max} A_v}$$

$$= \left\|\sum a_{ij}\overline{u_i} \otimes v_j + \overline{a_{ij}}\overline{u_i^*} \otimes v_j^*\right\|_{\overline{A_u} \otimes_{\min} A_v},$$

and we have (using the $*$-homomorphisms $U_i \mapsto \overline{u_i}, U_j \mapsto v_j$)

$$\left\|\sum a_{ij}\overline{u_i} \otimes v_j + \overline{a_{ij}}\overline{u_i^*} \otimes v_j^*\right\|_{\overline{A_u} \otimes_{\min} A_v} \le \left\|\sum a_{ij}U_i \otimes U_j + \overline{a_{ij}}U_i^* \otimes U_j^*\right\|_{\mathscr{C} \otimes_{\min} \mathscr{C}}.$$

Thus we obtain $\sup\{|\sum a_{ij}(x_{ij} + \overline{x_{ji}})| \mid x \in R_s(d)\} \le \|T + T^*\|_{\min}$, which is (16.5). Actually (we do not need this in the sequel) equality holds in (16.5). This is easy to check using Proposition 9.20.   □

Remark 16.2 Let $R_\otimes(d)$ be the set of $d \times d$-matrices of the form

$$x_{ij} = \langle\xi, (u_i \otimes v_j)\xi\rangle$$

where (u_i) and (v_i) are d-tuples of unitaries in $B(H)$ where H is any finite-dimensional Hilbert space and $\xi \in H \otimes_2 H$ is a unit vector.

We claim $R_\otimes(d) = R_s(d)$. Obviously $R_\otimes(d) \subset R_s(d)$ (since $u_i \otimes 1$ commutes with $1 \otimes v_j$). To check the converse consider $x \in R_s(d)$, say $x_{ij} = \langle\xi, u_i^* v_j \xi\rangle$ with $u_i, v_j \in U(B(H))$ such that $u_i v_j = v_j u_i$ for all i, j and $\dim(H) < \infty$. Let A_u, A_v be again the C^*-algebras generated in $B(H)$ by $\{u_i^*\}$ and $\{v_j\}$. Let f be the state defined on $A_u \otimes_{\max} A_v$ by $f(z \otimes y) = \langle\xi, zy\xi\rangle$. Since the algebras are finite dimensional this is a state on the min-tensor product, and hence extends to a state on $B(H) \otimes_{\min} B(H) = B(H \otimes_2 H)$. Since $\dim(H) < \infty$ this implies that there is a finite family of unit vectors ξ_k in

$H \otimes_2 H$ and $w_k > 0$ $(1 \leq k \leq m)$ with $\sum w_k = 1$ such that $f(z \otimes y) = \sum_k w_k \langle \xi_k, (z \otimes y) \xi_k \rangle$ for all $z, y \in B(H)$. Let $\mathcal{H} = \ell_2^m \otimes_2 [H \otimes_2 H]$. We now use the multiplicity trick to write:

$$f(z \otimes y) = \left\langle \left(\sum_k \sqrt{w_k} e_k \otimes \xi_k \right), \left(\sum_k e_{kk} \otimes [z \otimes y] \right) \left(\sum_k \sqrt{w_k} e_k \otimes \xi_k \right) \right\rangle.$$
(16.6)

In particular, we find

$$x_{ij} = f(u_i^* \otimes v_j) = \langle \xi', (u_i' \otimes v_j') \xi' \rangle$$

with a unit vector $\xi' = \sum_k \sqrt{w_k} e_k \otimes \xi_k \in \mathcal{H}$, $u_i' = \sum_k e_{kk} \otimes u_i \otimes 1 \in U(B(\mathcal{H}))$ and $v_j' = \sum_k e_{kk} \otimes 1 \otimes v_j \in U(B(\mathcal{H}))$. This shows that $x \in R_\otimes(d)$, which completes the proof.

Remark 16.3 It is easy to see that both $R_s(d)$ and $R_c(d)$ are convex sets. Indeed, this can be checked by the same multiplicity trick as in (16.6).

16.2 Correlation matrices with projection valued measures

The Tsirelson problem asks whether the analogue of Proposition 16.1 holds for projection valued measures, in short PVMs. In quantum theory, a PVM with m outputs is an m-tuple of mutually orthogonal projections $(P_j)_{1 \leq j \leq m}$ on a Hilbert space H with $\sum P_j = I$. The terminology reflects the fact that if we set $\mu_P(A) = \sum_{j \in A} P_j$, then μ_P is indeed a projection valued measure on $\{1, \ldots, m\}$. Seen from another viewpoint, we have a $*$-homomorphism $\pi_P : \ell_\infty^m \to B(H)$ defined by

$$\pi_P(x) = \sum_{j=1}^m x_j P_j.$$
(16.7)

Let $(Q_j)_{1 \leq j \leq m}$ be another PVM on H, such that Q_j commutes with P_i for all i, j, or equivalently the ranges of π_P and π_Q commute. In quantum mechanics, the commutation corresponds to the independence of the corresponding experimental measurements. When we are given a state of the system in the form of a unit vector $\xi \in H$ the probability of the event (i, j) is given by $\langle \xi, P_i Q_j \xi \rangle$ and the matrix $[\langle \xi, P_i Q_j \xi \rangle]$ represents their correlation. It is then natural to wonder whether the same numbers can be obtained by modeling the system on a finite-dimensional Hilbert space $\mathcal{H}$. The first and simplest question is whether the $m \times m$ "covariance matrix" $[\langle \xi, P_i Q_j \xi \rangle]$ can be approximated by $m \times m$-matrices of the same form but realized on a finite-dimensional $\mathcal{H}$. In the present simplest situation, the answer is yes. To justify this we choose a pedantic but hopefully instructive way. Let $f : \ell_\infty^m \otimes_{\max} \ell_\infty^m \to \mathbb{C}$ be the linear form defined by $f(x \otimes y) = \langle \xi, \pi_P(x) \pi_Q(y) \xi \rangle$. Clearly f is a state and of course $\ell_\infty^m \otimes_{\max} \ell_\infty^m = \ell_\infty^m \otimes_{\min} \ell_\infty^m$ can be identified with $\ell_\infty^{m^2}$. A state of $\ell_\infty^{m^2}$ is simply an element $z = (z_{ij})$ with $z_{ij} \geq 0$ such that $\sum_i \sum_j z_{ij} = 1$. Thus we may describe f as $f = (f_{ij})$ with $f_{ij} \geq 0$ such that $\sum_i \sum_j f_{ij} = 1$ given by $f_{ij} = \langle \xi, P_i Q_j \xi \rangle$. Now let $\mathcal{H} = \ell_2^{m^2}$ with o.n. basis e_{ij}, and let p_i (resp. q_j) denote the orthogonal projection onto the span of $\{e_{ij} \mid 1 \leq j \leq m\}$ (resp. $\{e_{ij} \mid 1 \leq i \leq m\}$). Let $\xi' = \sum_i \sum_j f_{ij}^{1/2} e_{ij}$. Then ξ' is a unit vector, (p_i), (q_j) are commuting PVMs on $\mathcal{H}$ with $\dim(\mathcal{H}) < \infty$ and

$$\forall i, j \quad \langle \xi, P_i Q_j \xi \rangle = \langle \xi', p_i q_j \xi' \rangle. \tag{16.8}$$

A fortiori our approximation problem is solved: we even obtain an equality.

A more serious difficulty appears when we consider the same problem for several PVMs. We give ourselves two d-tuples of PVMs $(P^1, \ldots, P^d)$, $(Q^1, \ldots, Q^d)$ (each with m outputs) on the same H and we assume that P_i^k commutes with Q_j^l for any $1 \leq i, j \leq m$, $1 \leq k, l \leq d$.

Let $Q_c(m, d)$ denote the set of all "covariance matrices" $x = [x(k, i; l, j)]$ of the form

$$x(k, i; l, j) = \langle \xi, P_i^k Q_j^l \xi \rangle$$

(on an arbitrary H), and let $Q_s(m, d)$ be the same set of matrices but restricted to finite-dimensional Hilbert spaces H. We first claim that $Q_c(m, d)$ is closed. Let us briefly sketch the easy ultraproduct argument for this: Assume that $a = [a_{ij}^{kl}]$ is the limit when $\gamma \to \infty$ of $\langle \xi(\gamma), P_i^k(\gamma) Q_j^l(\gamma) \xi(\gamma) \rangle$ that we assume relative to $H(\gamma)$. Let $H(\mathcal{U})$ be the Hilbert space that is the ultraproduct of the $H(\gamma)$'s for a nontrivial $\mathcal{U}$, let $\xi(\mathcal{U}), P_i^k(\mathcal{U}), Q_j^l(\mathcal{U})$ be the associated objects on $H(\mathcal{U})$, then we find

$$a_{ij}^{kl} = \lim_{\mathcal{U}} \langle \xi(\gamma), P_i^k(\gamma) Q_j^l(\gamma) \xi(\gamma) \rangle = \langle \xi(\mathcal{U}), P_i^k(\mathcal{U}) Q_j^l(\mathcal{U}) \xi(\mathcal{U}) \rangle \tag{16.9}$$

and we conclude $a \in Q_c$.

Since $Q_s(m, d) \subset Q_c(m, d)$ is obvious, it follows that

$$\overline{Q_s(m, d)} \subset Q_c(m, d).$$

We can now state Tsirelson's question.

Tsirelson's problem: *Is it true that $\overline{Q_s(m, d)} = Q_c(m, d)$ for all m, d ?*

As we already mentioned in Remark 3.40, the free product $\mathbb{Z}_2 * \mathbb{Z}_2$ is amenable and hence $\ell_\infty^2 * \ell_\infty^2$ is nuclear. It is not difficult to deduce from this (as should be clear from the sequel) that $Q_s(m, d) = Q_c(m, d)$ when $m = d = 2$, and of course we already checked in (16.8) that this also holds for any m if $d = 1$. For the other values the problem is open, and in fact:

Theorem 16.4 ([95, 138, 191]) *The Tsirelson problem is equivalent to the Kirchberg conjecture whether the min and max norms coincide on $\mathcal{C} \otimes \mathcal{C}$.*

To clarify the connection we need more notation. Let

$$A(m,d) = \ell_\infty^m * \cdots * \ell_\infty^m \quad (d \text{ times}).$$

Let $A_j \subset A(m,d)$ denote the jth copy of ℓ_∞^m in the (unital) free product $A(m,d)$.

Let $E(m,d) \subset A(m,d)$ be the linear subspace formed of all the sums $x_1 + \cdots + x_d$ where $x_j \in A_j$ for all $1 \leq j \leq d$. Note that being unital and self-adjoint, $E(m,d)$ is an operator system.

We start with an elementary observation.

Remark 16.5 (From PVMs to $A(m,d) \otimes_{\max} A(m,d)$) Consider two d-tuples of PVMs $(P^1, \ldots, P^d)$, $(Q^1, \ldots, Q^d)$ in $B(H)$ such that all P_i^k commute with all Q_j^l as before ($1 \leq k,l \leq d$, $1 \leq i,j \leq m$). Let $\pi_{P^k} : \ell_\infty^m \to B(H)$ (resp. $\pi_{Q^l} : \ell_\infty^m \to B(H)$) be the associated $*$-homomorphism as in (16.7)). Let $\pi_p : A(m,d) \to B(H)$ (resp. $\pi_q : A(m,d) \to B(H)$) be the $*$-homomorphism on the free product canonically defined by $(\pi_{P^k})_{1 \leq k \leq d}$ (resp. $(\pi_{Q^l})_{1 \leq l \leq d}$). Our commutation assumption implies that π_p and π_q have commuting ranges. Therefore $\pi_p.\pi_q$ is a $*$-homomorphism on $A(m,d) \otimes_{\max} A(m,d)$, and hence the linear map $F : A(m,d) \otimes A(m,d) \to \mathbb{C}$ defined by $F(x \otimes y) = \langle \xi, \pi_p(x)\pi_q(y)\xi \rangle$ extends to a state on $A(m,d) \otimes_{\max} A(m,d)$.

This gives us:

Proposition 16.6 *Let F be a state on $A(m,d) \otimes_{\max} A(m,d)$. Let e_j^k denote the jth basis vector of ℓ_∞^m in the kth copy of ℓ_∞^m inside $A(m,d) = \ell_\infty^m * \cdots * \ell_\infty^m$. Let $x_F(k,i;l,j) = F(e_i^k \otimes e_j^l)$. Then*

$$Q_c(m,d) = \{x_F \mid F \text{ state on } A(m,d) \otimes_{\max} A(m,d)\}. \tag{16.10}$$

Proof The preceding remark shows $Q_c \subset \{x_F\}$. To show the converse, just consider the GNS construction applied to F on $A(m,d) \otimes_{\max} A(m,d)$. $\qquad\square$

The next lemma records an elementary (linear algebraic) observation.

Lemma 16.7 *Let $(P_j)_{1 \leq j \leq m}$ be a PVM on a Hilbert space $\widehat{H}$. Let $H \subset \widehat{H}$ be finite dimensional. There is a finite-dimensional K_0 with $H \subset K_0 \subset \widehat{H}$ such that for any finite-dimensional K with $K_0 \subset K \subset \widehat{H}$ there is a PVM $(P_j')_{1 \leq j \leq m}$ on K such that $P_H P_{j|H}' = P_H P_{j|H}$ for all $1 \leq j \leq m$.*

Proof Let $E_j = P_j(H)$. Of course P_j and P_{E_j} commute and they coincide on H. Let $K_0 = E_1 + \cdots + E_m$. Note $H \subset K_0$. Assume $K_0 \subset K \subset \widehat{H}$. Let

$$p_j = P_{E_j|K} : K \to E_j \subset K.$$

In order to replace (p_j) by an m-tuple (P_j') with sum $= Id_K$, we set $P_j' = p_j$ for $j < m$ and $P_m' = p_m + [Id_K - (p_1 + \cdots + p_m)]$. Then for any $h \in H$ we have $P'_j h = p_j h = P_{E_j} h = P_j h$ for $j < m$ and $P'_m h \in P_m h + K \ominus K_0 \subset P_m h + H^\perp$. This yields the desired property. $\qquad\square$

Lemma 16.8 *Let* $(P_i)_{1 \le i \le m}$ *and* $(Q_j)_{1 \le j \le m}$ *be PVM's respectively on* $\widehat{H}_1$ *and* $\widehat{H}_2$. *Let* $H_1 \subset \widehat{H}_1$ *and* $H_2 \subset \widehat{H}_2$ *be finite-dimensional subspaces. There are finite-dimensional subspaces* K_1, K_2 *with* $H_1 \subset K_1 \subset \widehat{H}_1$ *and* $H_2 \subset K_2 \subset \widehat{H}_2$ *and PVM's* P'^k *and* Q'^l $(1 \le k, l \le d)$ *respectively on* K_1 *and* K_2 *such that for all* $1 \le i, j \le m$ *and all* $1 \le k, l \le d$ *we have*

$$P_{H_1 \otimes_2 H_2}[P'^k_i \otimes Q'^l_j]_{|H_1 \otimes_2 H_2} = P_{H_1 \otimes_2 H_2}[P^k_i \otimes Q^l_j]_{|H_1 \otimes_2 H_2}.$$

Proof We apply Lemma 16.7. This tells us that if K_1 and K_2 are chosen large enough there are P'^k_i, Q'^l_j as required. $\qquad\square$

Lemma 16.9 *Let* H_1, H_2 *be finite dimensional. Let* $u : A(m, d) \to B(H_1)$ *and* $v : A(m, d) \to B(H_2)$ *be unital c.p. maps. There are finite-dimensional Hilbert spaces* K_1, K_2 *with* $K_1 \supset H_1$, $K_2 \supset H_2$ *and* $*$-*homomorphisms* $\pi_1 : A(m, d) \to B(K_1)$, $\pi_2 : A(m, d) \to B(K_2)$ *such that*

$$\forall x \in E(m, d) \otimes E(m, d) \quad (u \otimes v)(x) = P_{H_1 \otimes_2 H_2}(\pi_1 \otimes \pi_2)(x)_{|H_1 \otimes_2 H_2}.$$

Proof By Stinespring's theorem we know that this holds (on the whole of $A(m, d) \otimes A(m, d)$) with arbitrary Hilbert spaces, say $\widehat{H}_1, \widehat{H}_2$ and $*$-homomorphisms $\pi'_1 : A(m, d) \to B(\widehat{H}_1)$, $\pi'_2 : A(m, d) \to B(\widehat{H}_2)$ in place of H_1, H_2. Let then P^k (resp. Q^l) be the PVM's associated respectively to the restrictions of π'_1 (resp. π'_2) to the kth (resp. lth) copy of ℓ^m_∞ in $A(m, d)$. Let K_1, K_2 and P'^k_i, Q'^l_j be as in Lemma 16.8. Then the $*$-homomorphisms π_1, π_2 with values in $B(K_1)$ and $B(K_2)$, associated respectively to (P'^k_i) and (Q'^l_j) satisfy the required equality on $E(m, d) \otimes E(m, d)$. $\qquad\square$

Remark 16.10 Let $Q_\otimes(m, d)$ be the set of matrices $x = [x(k, i; l, j)]$ of the form $x_f(k, i; l, j) = \langle \xi, (P^k_i \otimes Q^l_j)\xi \rangle$ where P^k and Q^l are PVM's on a finite-dimensional space H and ξ is a unit vector in $H \otimes_2 H$. The same reasoning as for Remark 16.6 shows that $Q_\otimes(m, d) = Q_s(m, d)$.

Proposition 16.11 *Let* f *be a state on* $A(m, d) \otimes_{\min} A(m, d)$. *Let* $x_f(k, i; l, j) = f(e^k_i \otimes e^l_j)$. *Then*

$$\overline{Q_s(m, d)} = \overline{Q_\otimes(m, d)} = \{x_f \mid f \text{ state on } A(m, d) \otimes_{\min} A(m, d)\}. \quad (16.11)$$

Proof By (iii) in Theorem 4.21 f is the pointwise limit of a net (F_γ) of states of the form

$$F^\gamma(x \otimes y) = \psi^\gamma(u_\gamma(x) \otimes v_\gamma(y)),$$

where ψ^γ is a state on $B(H_1^\gamma \otimes_2 H_2^\gamma)$ for finite-dimensional Hilbert spaces H_1^γ, H_2^γ and where u_1^γ, u_2^γ are unital c.p. maps from $A(m,d)$ to $B(H_1^\gamma)$ and $B(H_2^\gamma)$. By the same reasoning as for (16.6) (see also Remark 4.24), we may assume that ψ^γ is a vector state. By Lemma 16.9 there are $*$-homomorphisms $\pi_1^\gamma : A(m,d) \to B(K_1^\gamma)$ and $\pi_2^\gamma : A(m,d) \to B(K_2^\gamma)$ with K_1^γ, K_2^γ finite dimensional such that for all $x, y \in E(m,d)$

$$u_1^\gamma(x) \otimes u_2^\gamma(y) = P_{H_1^\gamma \otimes_2 H_2^\gamma}(\pi_1^\gamma(x) \otimes \pi_2^\gamma(y))_{|H_1^\gamma \otimes_2 H_2^\gamma}.$$

Applying that to $x = e_i^k$ and $y = e_j^l$ we obtain

$$x_f(k,i;l,j) = \lim_\gamma x^\gamma(k,i;l,j) \text{ where}$$
$$x^\gamma(k,i;l,j) = F^\gamma(e_i^k \otimes e_j^l)$$
$$= \psi^\gamma(u_1^\gamma(e_i^k) \otimes u_2^\gamma(e_j^l))$$

and we claim that $x^\gamma \in Q_\otimes(m,d)$. Indeed, if $\psi^\gamma(t) = \langle \xi, t\xi \rangle$ for some unit vector $\xi \in H_1^\gamma \otimes_2 H_2^\gamma$ we find $x^\gamma(k,i;l,j) = \langle \xi, (P_i^k \otimes Q_j^l)\xi \rangle$ with $P_i^k = \pi_1^\gamma(e_i^k)$ and $Q_j^l = \pi_2^\gamma(e_j^l)$. This shows that $\{x_f\} \subset \overline{Q_s}$. To show the converse, by the preceding Remark 16.10 it suffices to show $Q_\otimes \subset \{x_f\}$ (since the latter set is obviously closed). This inclusion is immediate: let $x \in Q_\otimes$ of the form $x(k,i;l,j) = \langle \xi, (P_i^k \otimes Q_j^l)\xi \rangle$. Then $x = x_f$ where f is the state on $A(m,d) \otimes_{\min} A(m,d)$ defined by $f(x \otimes y) = \langle \xi, (\pi_{P^k}(x) \otimes \pi_{Q^l}(y))\xi \rangle$. $\qquad \square$

Putting together Proposition 16.6 and the last one we obtain:

Proposition 16.12 *The following assertions are equivalent for any fixed* (m,d).

(i) $\overline{Q_s(m,d)} = Q_c(m,d)$.
(ii) *The restrictions to $E(m,d) \otimes E(m,d)$ of the states of* $A(m,d) \otimes_\alpha A(m,d)$ *for $\alpha = \min$ and $\alpha = \max$ form identical subsets of* $(E(m,d) \otimes E(m,d))^*$.

Moreover, these assertions imply

(iii) *The min and max norms of $A(m,d) \otimes A(m,d)$ coincide on the subspace* $\{T + T^* \mid T \in E(m,d) \otimes E(m,d)\}$.

Proof Observe that $(e_i^k \otimes e_j^l)$ is a basis of the linear space $E(m,d) \otimes E(m,d)$. Therefore the restriction of a state f of $A(m,d) \otimes_\alpha A(m,d)$ to $E(m,d) \otimes E(m,d)$ is entirely determined by x_f. The equivalence of (i) and (ii) is then clear by (16.10) and (16.11).

Let $T \in E(m,d) \otimes E(m,d)$. Clearly $\|T+T^*\|_\alpha = \sup\{|f(T+T^*)|\}$ where the sup runs over all states f on $A(m,d) \otimes_\alpha A(m,d)$. Thus (ii) $\Rightarrow$ (iii). $\quad\square$

We will now relate the last statement to $C^*(\mathbb{F}_d) \otimes C^*(\mathbb{F}_d)$.

Lemma 16.13 *The C^*-algebra $A(m,d)$ has the LLP. Moreover, the identity of $A(m,d)$ factors via unital c.p. maps through $C^*(\mathbb{F}_d)$.*

Proof The first assertion is clear since the LLP is stable by free products (see Theorem 9.44). For the second one (which actually also implies the first one) we use Boca's Theorem 2.24. It is easy to see that the identity of $\ell_\infty^m = C^*(\mathbb{Z}/m\mathbb{Z})$ factors via unital c.p. maps through $C^*(\mathbb{Z})$ (this follows by applying Remark 9.54 to the natural quotient $*$-homomorphism $C^*(\mathbb{Z}) \to C^*(\mathbb{Z}/m\mathbb{Z})$). By Theorem 2.24 the identity of $A(m,d)$ factors through $C^*(\mathbb{Z}) * \cdots * C^*(\mathbb{Z}) = C^*(\mathbb{Z} * \cdots * \mathbb{Z})$ (d times) and $\mathbb{Z} * \cdots * \mathbb{Z} = \mathbb{F}_d$. $\quad\square$

Lemma 16.14 *Let $E(d) \subset C^*(\mathbb{F}_d)$ be the linear span of the d unitary generators and the unit. There are unital c.p. maps*

$$v_m : C^*(\mathbb{F}_d) \to A(m,d) \text{ and } w_m : A(m,d) \to C^*(\mathbb{F}_d)$$

such that $v_m(E(d)) \subset E(m,d)$ for any m and such that $w_m v_m \to Id_{C^*(\mathbb{F}_d)}$ pointwise when $m \to \infty$.

Proof It is easy to check this for $d = 1$ (see Remark A.20 and recall Remark 1.32). Then by Boca's Theorem 2.24 the general case follows. $\quad\square$

We can now complete the proof of the equivalence of the Tsirelson problem with the Connes–Kirchberg one.

Proof of Theorem 16.4 Assume $\overline{Q_s(m,d)} = Q_c(m,d)$ for all m. Recall that any pair of unital c.p. maps gives rise to a contraction on the maximal tensor product (see Corollary 4.18). Then by Proposition 16.12 and Lemma 16.14 the min and max norms coincide on $\{T + T^* \mid T \in E(d) \otimes E(d)\} \subset C^*(\mathbb{F}_d) \otimes C^*(\mathbb{F}_d)$. Since this holds for all d it holds for $d = \infty$, i.e. with $E \subset \mathscr{C}$ in place of $E(d) \subset C^*(\mathbb{F}_d)$. By Corollary 14.9 and Remark 14.17 we conclude that the min and max norms coincide on $\mathscr{C} \otimes \mathscr{C}$, which means the Kirchberg conjecture holds. Conversely, if the Kirchberg conjecture holds, by Proposition 13.1 we know from *LLP $\Rightarrow$ WEP* that the min and max norms coincide on $A \otimes A$ for any A with the LLP. Therefore, by Lemma 16.13 and Proposition 16.12 again we conclude that $\overline{Q_s(m,d)} = Q_c(m,d)$ for all m,d. $\quad\square$

Remark 16.15 (About POVMs) A positive operator valued measure (in short POVM) with m outputs is an m-tuple of self-adjoint operators $(a_j)_{1 \le j \le m}$ on a Hilbert space H with $\sum a_j = I$. Although this notion is more general than

PVMs, it is easy to reduce the study of correlation matrices of POVMs to that of PVMs. Indeed, consider two d-tuples of POVMs $(a^1, \ldots, a^d)$, $(b^1, \ldots, b^d)$ (each with m outputs) on the same H and assume that a_i^k commutes with b_j^l for any $1 \leq i, j \leq m$, $1 \leq k, l \leq d$. We wish to address the same problem as before but for

$$x(k, i; l, j) = \langle \xi, a_i^k b_j^l \xi \rangle.$$

There is a 1–1 correspondence between POVMs and unital positive (and hence c.p. by Remark 1.32) linear maps from ℓ_∞^m to $B(H)$. Just consider for any $a = (a_i)_{1 \leq i \leq m}$ the map $u_a : \ell_\infty^m \to B(H)$ defined by $u_a(e_i) = a_i$. If two such POVMs $(a_i)_{1 \leq i \leq m}$ and $(b_j)_{1 \leq j \leq m}$ mutually commute, the ranges of u_a and u_b also commute. By Boca's Theorem 2.24 and by Corollary 4.19 we have a unital c.p. map $u : A(m, d) \otimes_{\max} A(m, d) \to B(H)$ such that $u(e_i^k \otimes e_j^l) = a_i^k b_j^l$ or any $1 \leq i, j \leq m$, $1 \leq k, l \leq d$. By Stinespring's dilation Theorem 1.22 we have $\widehat{H} \supset H$ and a $*$-homomorphism $\pi : A(m, d) \otimes_{\max} A(m, d) \to B(\widehat{H})$ such that $u(\cdot) = P_H \pi(\cdot)_H$. It follows that for any $\xi \in H$ we have

$$\langle \xi, a_i^k b_j^l \xi \rangle = \langle \xi, P_i^k Q_j^l \xi \rangle$$

where P^k and Q^l are the PVMs associated to π by setting $\pi(e_i^k \otimes 1) = P_i^k$ and $\pi(1 \otimes e_j^l) = Q_j^l$. This shows that the Tsirelson problem for POVMs reduces to the case of PVMs.

16.3 Strong Kirchberg conjecture

In [192] Ozawa introduced a strong version of Kirchberg's conjecture that he proved to be equivalent to the assertion that

$$\mathscr{C} \otimes_{\min} \mathscr{C} \otimes_{\min} \mathscr{B} = \mathscr{C} \otimes_{\max} \mathscr{C} \otimes_{\max} \mathscr{B}.$$

It is an easy exercise to see that this is the same as the two assertions that $\mathscr{C} \otimes_{\min} \mathscr{C} = \mathscr{C} \otimes_{\max} \mathscr{C}$ together with $(\mathscr{C} \otimes_{\max} \mathscr{C}) \otimes_{\min} \mathscr{B} = (\mathscr{C} \otimes_{\max} \mathscr{C}) \otimes_{\max} \mathscr{B}$. In other words, this is the same as the Kirchberg conjecture together with the assertion that $\mathscr{C} \otimes_{\max} \mathscr{C} = C^*(\mathbb{F}_\infty \times \mathbb{F}_\infty)$ has the LLP.

The strong Kirchberg conjecture from [192] asserts the following for any $d \geq 2$:

For any $\delta > 0 \; \exists \varepsilon > 0$ such that given unitaries $U_1, \ldots, U_d$, $V_1, \ldots, V_d$ in $B(H)$ with $\dim(H) < \infty$, satisfying

$$\forall i, j \quad \|[U_i, V_j]\| \leq \varepsilon$$

there is $\widehat{H} \supset H$ with $\dim(\widehat{H}) < \infty$ and unitaries $\widehat{U}_1, \ldots, \widehat{U}_d$, $\widehat{V}_1, \ldots, \widehat{V}_d$ in $B(\widehat{H})$ satisfying

$$\forall i, j \quad [\widehat{U}_i, \widehat{V}_j] = 0 \text{ and } \|U_i - P_H \widehat{U}_{i|H}\| \le \delta, \quad \|V_j - P_H \widehat{V}_{j|H}\| \le \delta.$$

Actually, its validity for *some* $d \ge 2$ implies the same for *all* $d \ge 2$.
The next lemma somewhat explains the role of the LLP.

Lemma 16.16 *Consider the assertion of the strong Kirchberg conjecture without the requirement that $\widehat{H}$ be finite dimensional. Then this is equivalent to the LLP for $C^*(\mathbb{F}_d \times \mathbb{F}_d)$.*

In [192], Ozawa also considers an analogous equivalent conjecture involving positive operator valued measures (POVMs) in the style of Tsirelson's conjecture. At this stage we refer to [192] for detailed proofs.

16.4 Notes and remarks

The elementary results of §16.1 are only formulated to clarify (hopefully) those of the subsequent §16.2. See [139, 140] for more recent information on Bell's inequalities and their connections with operator spaces. See Dykema and Juschenko's [75] for more results on unitary correlation matrices. In [129] Harris and Paulsen study a different kind of unitary correlation matrices, derived from consideration of the Brown algebra (see Remark 13.2) and prove an analogue of Theorem 16.4 in their setting. See Slofstra's [230, 231] (and Dykema and Paulsen's [76]) for proofs that the set $Q_s(m, d)$ is not closed in general.

In a different direction, the papers [115, 116] by Haagerup and Musat relate what they call the asymptotic quantum Birkhoff property for factorizable Markov maps to the Connes embedding problem.

17

Property (T) and residually finite groups

Thom's example

Definition 17.1 A finitely generated discrete group G, with generators $g_1, g_2, \ldots, g_n$, is said to have property (T) if the trivial representation is isolated in the set of all unitary representations of G. More precisely, this means that there is a number $\delta > 0$ such that, for any unitary representation π the condition

$$(\exists \xi \in H_\pi, \ \|\xi\| = 1, \ \sup_{j \le n} \|\pi(g_j)\xi - \xi\| < \delta)$$

suffices to conclude that π admits a nonzero invariant vector, or in other words that π contains the trivial representation (as a subrepresentation).

It is easy to see that this property actually does not depend on the choice of the set of generators, but of course the corresponding δ does depend on that choice.

The classical example of discrete group with property (T) is $G = SL_3(\mathbb{Z})$. This goes back to D. Kazhdan (1967) and property (T) groups are often also called "Kazhdan groups."

We will use the following basic fact.

Lemma 17.2 *If G with generators $S = \{g_1, g_2, \ldots, g_n\}$ has property (T) then there is a function $f : (0, \infty) \to (0, \infty)$ with $\lim_{\varepsilon \to 0} f(\varepsilon) = 0$ such that, for any unitary representation $\pi : G \to B(H_\pi)$, if $\xi \in H_\pi$ is a unit vector satisfying*

$$\sup_{j \le n} \|\pi(g_j)\xi - \xi\| < \varepsilon$$

there is a unit vector $\xi' \in H_\pi$ such that

$$\pi(g)\xi' = \xi' \ \forall g \in G \text{ and } \|\xi - \xi'\| < f(\varepsilon).$$

Proof Let $H_\pi^{\text{inv}} \subset H_\pi$ be the subspace of π-invariant vectors, i.e. vectors $x \in H_\pi$ such that $\pi(g)x = x$ for all $g \in G$. Let P be the orthogonal projection from H_π onto H_π^{inv}. Clearly, $H_\pi^{\mathrm{inv}\perp}$ is invariant under π and the restriction of π to it has no nonzero invariant vector. Therefore for any $\xi \in H_\pi^{\mathrm{inv}\perp}$ with $\|\xi\| = 1$ we must have $\sup_{j \le n} \|\pi(g_j)\xi - \xi\| \ge \delta$, where δ is as in Definition 17.1. By homogeneity this implies $\sup_{j \le n} \|\pi(g_j)\xi - \xi\| \ge \delta\|\xi\|$ for all ξ and since $\pi(g_j)P\xi = P\xi$ we have $\pi(g_j)\xi - \xi = \pi(g_j)(\xi - P\xi) - (\xi - P\xi)$ for all $1 \le j \le n$, and hence

$$\forall \xi \in H_\pi \quad \sup_{j \le n} \|\pi(g_j)\xi - \xi\| \ge \delta\|\xi - P\xi\|.$$

Therefore

$$\forall \xi \in H_\pi \quad d(\xi, H_\pi^{\mathrm{inv}}) \le \delta^{-1} \sup_{j \le n} \|\pi(g_j)\xi - \xi\|.$$

Let $\xi' = P\xi\|P\xi\|^{-1}$. If $\sup_{j \le n} \|\pi(g_j)\xi - \xi\| < \varepsilon$ and $\|\xi\| = 1$ we have $\|\xi - P\xi\| \le \varepsilon/\delta$ and hence $\|P\xi\| \ge 1 - \varepsilon/\delta$. Assuming $0 < \varepsilon < \delta$, we obtain $\|\xi' - \xi\| \le \|\xi' - \xi\|P\xi\|^{-1}\| + \|\xi\|P\xi\|^{-1} - \xi\| \le (\varepsilon/\delta)(1 - \varepsilon/\delta)^{-1}$. $\qquad\square$

Property (T) can also be reformulated in terms of spectral gap, as follows. We will return to that theme in §19.3.

Proposition 17.3 *A discrete group G generated by a finite subset $S \subset G$ containing the unit has property (T) if and only if there is $\varepsilon \in (0, 1)$ such that for any unitary representation $\pi : G \to B(H_\pi)$ we have*

$$\left\| |S|^{-1} \sum_{s \in S} \pi(s) - P_{H_\pi^{\mathrm{inv}}} \right\| \le 1 - \varepsilon. \tag{17.1}$$

Proof Let $n = |S|$. By (A.6) if

$$n^{-1} \left\| \sum_{s \in S} \pi(s) P_{|H_\pi^{\mathrm{inv}\perp}} \right\| > 1 - \varepsilon \tag{17.2}$$

there is a unit vector ξ in $H_\pi^{\mathrm{inv}\perp}$ such that $\|\pi(s)\xi - \pi(t)\xi\| < 2\sqrt{2\varepsilon n}$ for any $s, t \in S$. In particular since the unit of G is in S we have $\sup_{s \in S} \|\pi(s)\xi - \xi\| < 2\sqrt{2\varepsilon n}$.

Now assume that G has property (T). Let δ be as in Definition 17.1. Then if $2\sqrt{2\varepsilon n} = \delta$, i.e. if $\varepsilon = \delta^2/8n$, we conclude that (17.2) is impossible: otherwise the restriction of π to $H_\pi^{\mathrm{inv}\perp}$ would have an invariant unit vector and this is absurd by the very definition of H_π^{inv}. This shows that (17.1) holds for $\varepsilon = \delta^2/8n$.

Conversely, if (17.1) holds for some $\varepsilon > 0$ and if $\sup_{s \in S} \|\pi(s)\xi - \xi\| < \delta$ for some unit vector ξ then

$$\|\xi - P_{H_\pi^{\mathrm{inv}}}\xi\| \le 1 - \varepsilon + \delta.$$

Thus if we choose $\delta < \varepsilon$, we find $P_{H_\pi^{\mathrm{inv}}}\xi \ne 0$ so that the vector $P_{H_\pi^{\mathrm{inv}}}\xi$ is a nonzero invariant vector for π, and G has property (T). $\qquad\square$

When the group G has property (T) the structure of $C^*(G)$ is very special, in particular it splits as a direct sum $pC^*(G) \oplus (1 - p)C^*(G)$, as shown by the next statement.

Proposition 17.4 *Let G be a finitely generated group with Property (T). Then there is a self-adjoint projection p in the center of $C^*(G)$ such that for any $*$-homomorphism $\pi : C^*(G) \to B(H_\pi)$ associated to a unitary representation $\underline{\pi} : G \to B(H_\pi)$ we have*

$$\pi(p) = P_{H_{\underline{\pi}}^{inv}},$$

in particular, $\pi(p) = 1$ if $\underline{\pi}$ is the trivial representation and $\pi(p) = 0$ if $\underline{\pi}$ is any nontrivial irreducible unitary representation.

Proof Note that $H_{\underline{\pi}}^{inv}$ is also the set of invariant vectors of the range of π so we will denote it simply by H_π^{inv}. Let S be a finite symmetric set of generators. Let $t = |S|^{-1} \sum_{s \in S} U_G(s)$, so that $\pi(t) = |S|^{-1} \sum_{s \in S} \underline{\pi}(s)$. The projection p is simply the limit in $C^*(G)$ of t^m when $m \to \infty$. Let us show that this limit exists. Note $\pi(t)^m - P_{H_\pi^{inv}} = \pi(t)^m(1 - P_{H_\pi^{inv}})$ for any $m \geq 1$ and $\pi(t)$ commutes with $(1 - P_{H_\pi^{inv}})$ so that $\pi(t)^m - P_{H_\pi^{inv}} = (\pi(t)(1 - P_{H_\pi^{inv}}))^m = (\pi(t) - P_{H_\pi^{inv}})^m$. By Property (T) as in (17.1) there is $\varepsilon \in (0,1)$ such that $\|\pi(t) - P_{H_\pi^{inv}}\| \leq 1 - \varepsilon$ for any π, and hence

$$\|\pi(t)^m - P_{H_\pi^{inv}}\| \leq (1 - \varepsilon)^m \tag{17.3}$$

for any $m \geq 1$. A fortiori, $\|\pi(t)^m - \pi(t)^{m+1}\| \leq 2(1 - \varepsilon)^m$, and since this holds for any π we have $\|t^m - t^{m+1}\|_{C^*(G)} \leq 2(1 - \varepsilon)^m$. Therefore by the Cauchy criterion t^m converges to a limit $p \in C^*(G)$. Since $t^* = t$ we have $p^* = p$, and also $p^2 = \lim t^{2m} = p$. Moreover, by (17.3) for any π we have $\pi(p) = \lim \pi(t^m) = \lim \pi(t)^m = P_{H_\pi^{inv}}$, and since $P_{H_\pi^{inv}}$ obviously commutes with the range of π, we have $\pi(p)\pi(x) = \pi(x)\pi(p)$ for any $x \in C^*(G)$. In particular, taking $\pi = Id_{C^*(G)}$ we see that p is in the center of $C^*(G)$. $\square$

We now connect the factorization property, introduced in Definition 7.36 and further studied in Theorem 11.55, with property (T). For simplicity, we still denote by $\lambda_G : C^*(G) \to C_\lambda^*(G)$ the $*$-homomorphism that extends the unitary representation λ_G, so that $\lambda_G(U_G(t)) = \lambda_G(t)$ for $t \in G$.

Theorem 17.5 *Let G be a discrete group with property (T). Let $S \subset G$ be an arbitrary generating set with $1 \in S$. Let $E \subset C^*(G)$ be the linear span of S. Assume that*

$$\forall n \, \forall x_1, \ldots, x_n \in E \quad \sum \tau_G(|\lambda_G(x_j)|^2)) \leq \left\| \sum \overline{x_j} \otimes x_j \right\|_{min}. \tag{17.4}$$

Then S is residually finite dimensional (RFD in short), i.e. S is separated by the set of finite-dimensional unitary representations. In particular, if (17.4) holds for S = G, then G is RFD.

Before giving the proof we start with comments and consequences to motivate Theorem 17.5.

Remark 17.6 By a classical Theorem of Malcev [177], any finitely generated linear group is RF. Thus for finitely generated groups RFD implies RF. Actually, the groups G that are separated by their finite-dimensional unitary representations are characterized as subgroups of compact groups. They are usually called "maximally almost periodic."

By Theorem 11.55, the factorization property implies (17.4) with $S = G$, therefore:

Corollary 17.7 *If G has property (T) and the factorization property, then G is residually finite.*

Corollary 17.8 *If $C^*(G)$ has the WEP and G has property (T), then G is residually finite.*

Proof If $C^*(G)$ has the WEP, then the min and max norms coincide on the set $\{\sum \overline{x_j} \otimes x_j\} \subset \overline{C^*(G)} \otimes C^*(G)$. Therefore we again have (17.4) for $S = G$. Property T implies that G is finitely generated, thus by Malcev's theorem (see Remark 17.6) G is RF. $\qquad\square$

Proof of Theorem 17.5 Let $C = C^*(G)$. By Theorem 11.38 and Remark 11.39 for a suitable embeddding $C \subset B(H)$ there is a net (h_i) of Hilbert–Schmidt operators on H with $\mathrm{tr}(h_i^* h_i) = 1$ such that

$$\forall y, x \in E \quad \tau_G(\lambda_G(y)^* \lambda_G(x)) = \lim_i \mathrm{tr}(y^* x h_i^* h_i)$$

and such that, denoting by $g \to U(g) \in B(H)$ the unitary representation of G obtained by restricting to G the embedding $C \to B(H)$, we have

$$\forall s \in S \quad \|U(s)h_i - h_i U(s)\|_2 = \|U(s)h_i U(s)^* - h_i\|_2 \to 0.$$

It follows that (h_i) is an approximately invariant unit vector for the representation $\overline{U} \otimes U$ of G on $S_2(H) = \overline{H} \otimes_2 H$. Since G has property (T) there is, by Lemma 17.2, a net of *invariant* unit vectors (h_i') in $S_2(H)$ such that $\|h_i - h_i'\|_2 \to 0$. Let $T_i = h_i'^* h_i'$. We have then

$$\forall y, x \in E \quad \tau_G(\lambda_G(y)^* \lambda_G(x)) = \lim_i \mathrm{tr}(y^* x T_i),$$

$\mathrm{tr}(T_i) = 1$ and $U(s)T_i = T_i U(s)$ for all $s \in S$, and hence for all $g \in G$.

The latter condition implies that $U(g)$ commutes with all the spectral projections of $T_i \geq 0$, and since T_i is compact these are finite dimensional. Thus if we write the spectral decomposition of T_i as

$$T_i = \sum_k \lambda_k^i P_k^i$$

with

$$\lambda_k^i > 0 \qquad \sum_k \lambda_k^i \mathrm{tr}(P_k^i) = 1$$

we find that $\pi_k^i : g \to U(g)_{|P_k^i(H)}$ is a finite-dimensional unitary representation of G on $P_k^i(H)$, such that for any $y, x \in E$

$$\tau(\lambda_G(y)^* \lambda_G(x)) = \lim_i \sum_k \lambda_k^i \mathrm{tr}(\pi_k^i(y)^* \pi_k^i(x)).$$

In particular for any $s, t \in S$

$$\delta_{s,t} = \lim_i \sum_k \lambda_k^i \mathrm{tr}(\pi_k^i(s^{-1}t)).$$

From which it is clear that $\{\pi_k^i\}$ separates the points of S. $\qquad\square$

In [244], Andreas Thom exhibited a group G with property (T), that is approximately linear (i.e. "hyperlinear"), and even sofic, but *not* residually finite. This is a remarkable example (or counterexample) because:

Proposition 17.9 *Let G be a group with property (T), that is approximately linear (i.e. "hyperlinear") but not residually finite. Then $C^*(G)$ has neither the WEP nor the LLP.*

Proof That $C^*(G)$ fails the WEP follows from Corollary 17.8. Since G is approximately linear, we know (see Corollary 9.60 or Remark 14.5) that we can write M_G as a quotient $M_G = A/\mathcal{I}$ of a WEP C^*-algebra A. Let $E \subset C^*(G)$ be any finite-dimensional subspace. If $C^*(G)$ had the LLP, the natural map $u : C^*(G) \to A/\mathcal{I}$ (which is the same as $\dot{Q}_G$) would be locally 1-liftable. We claim that this would imply (17.4). Indeed, there would be a map $u^E : E \to A$ with $\|u^E\|_{cb} \leq 1$ such that $qu^E = u_{|E}$. Let $x_j \in E$ and $y_j = u^E(x_j) \in A$ ($1 \leq j \leq n$). Then we would have $u(x_j) = q(y_j)$ and hence

$$\sum \|u(x_j)\|_{L_2(\tau_G)}^2 \leq \left\| \sum \overline{u(x_j)} \otimes u(x_j) \right\|_{\overline{M_G} \otimes_{\max} M_G}$$

$$\leq \left\| \sum \overline{x_j} \otimes y_j \right\|_{\overline{C^*(G)} \otimes_{\max} A}.$$

By Corollary 9.40 (general form of Kirchberg's Theorem) and since $y_j = u^E(x_j)$ and $\|u^E\|_{cb} \leq 1$ we would have

$$\left\| \sum \overline{x_j} \otimes y_j \right\|_{\overline{C^*(G) \otimes_{\max} A}} = \left\| \sum \overline{x_j} \otimes y_j \right\|_{\overline{C^*(G) \otimes_{\min} A}}$$

$$\leq \left\| \sum \overline{x_j} \otimes x_j \right\|_{\overline{C^*(G) \otimes_{\min} E}},$$

and since $\left\| \sum \overline{x_j} \otimes x_j \right\|_{\overline{C^*(G) \otimes_{\min} E}} = \left\| \sum \overline{x_j} \otimes x_j \right\|_{\overline{C^*(G) \otimes_{\min} C^*(G)}}$, our claim (17.4) would follow. By Theorem 17.5 this would contradict the fact that G is *not* residually finite, thus showing that $C^*(G)$ fails the LLP. $\quad\square$

Remark 17.10 Let us say that an operator $u : C \to B$ between C^*-algebras locally factors (completely boundedly) through a C^*-algebra A if there is a constant c such that for any finite-dimensional subspace $E \subset C$ the restriction $u_{|E}$ admits a factorization $u_{|E} : E \xrightarrow{v} A \xrightarrow{w} B$ with $\|v\|_{cb}\|w\|_{cb} \leq c$. Using the subsequent inequality (22.16) the preceding argument can be modified to show more generally that the inclusion $u : C^*(G) \to M_G$ does *not* locally factor (completely boundedly) through a WEP C^*-algebra. This negates at the same time both WEP and LLP for $C^*(G)$, when we know that M_G is QWEP and G has (T).

17.1 Notes and remarks

Theorem 17.5 originally comes from [157]. A simpler proof appears in [17] which we recommend to the reader for (much) more information on Property (T) (see also [127]). Proposition 17.4 is due to Valette [252]. A. Thom's construction of a group G as in Proposition 17.9 appears in [244] to which we refer the reader for full details. This is the first example for which $C^*(G)$ fails the LLP.

18

The WEP does not imply the LLP

Although S. Wassermann had proved in his 1976 paper [255] that $B(H)$ is not nuclear (assuming $\dim(H) = \infty$), the problem whether $A \otimes_{\min} B = A \otimes_{\max} B$ when $A = B = B(H)$ remained open until [141]. In the latter paper, several different proofs were given. Taking into account the most recent information from [120], we now know that

$$\sup \left\{ \frac{\|t\|_{\max}}{\|t\|_{\min}} \;\middle|\; t \in B(H) \otimes B(H), \quad \mathrm{rk}(t) \leq n \right\} \geq \frac{n}{2\sqrt{n-1}}. \tag{18.1}$$

This estimate is rather sharp asymptotically, since it can be shown that the supremum appearing in (18.1) is $\leq \sqrt{n}$ (see [141] or [208, p. 353]).

Remark 18.1 We need to clarify a few points regarding complex conjugation, which we already discussed in §2.3. In general, we will need to consider the conjugate $\bar{A}$ of a C^*-algebra A. This is the same object but with the complex multiplication changed to $(\lambda, a) \rightarrow \bar{\lambda}a$, so that $\bar{A}$ is anti-isomorphic to A. Recall that for any $a \in A$, we denote by $\bar{a}$ the same element viewed as an element of $\bar{A}$. Recall (see Remark 2.13) that $\bar{A} \simeq A^{op}$ via the mapping $\bar{a} \rightarrow a^*$.

The distinction between A and $\bar{A}$ is necessary in general, but not for $A = B(H)$ since in that case, using $H \simeq \bar{H}$, we have $\overline{B(H)} \simeq B(\bar{H}) \simeq B(H)$, and in particular $\overline{M_N} \simeq M_N$. In the case of M_N, the mapping $\bar{a} \mapsto [\bar{a}_{ij}]$ is an embedding of $\overline{M_N}$ into M_N, taking $\overline{e_{ij}}$ to e_{ij}. As a consequence, for any matrix a in $M_N(A)$ we have

$$\left\| \sum \overline{e_{ij}} \otimes a_{ij} \right\|_{\overline{M_N} \otimes_{\min} A} = \left\| \sum e_{ij} \otimes a_{ij} \right\|_{M_N \otimes_{\min} A} = \left\| [a_{ij}] \right\|_{M_N(A)}. \tag{18.2}$$

By (2.13) this also implies

$$\| [\bar{a}_{ij}] \|_{M_N(\bar{A})} = \| [a_{ij}] \|_{M_N(A)}.$$

Note however that $H \simeq \overline{H}$ depends on the choice of a basis so the isomorphism $\overline{B(H)} \simeq B(H)$ is not canonical. Nevertheless, this shows that the problem whether the min and max norms are the same is identical for $B(H) \otimes B(H)$ and for $\overline{B(H)} \otimes B(H)$.

Remark 18.2 Consider $a_1, \ldots, a_n$ in A and $b_1, \ldots, b_n$ in B. Using the preceding remark we have

$$\left\| \sum \overline{a_j} \otimes b_j \right\|_{\overline{A} \otimes_\alpha B} = \left\| \sum a_j^* \otimes b_j \right\|_{A^{op} \otimes_\alpha B}$$

for any "reasonably" well behaved C^*-norm, in particular for $\alpha = \min$ or $\max$. Moreover, we have

$$\left\| \sum \overline{a_j} \otimes b_j \right\|_{\overline{A} \otimes_{\max} B} = \left\| \sum a_j^* \otimes b_j \right\|_{A^{op} \otimes_{\max} B} = \sup \left\{ \left\| \sum \pi(a_j^*) \sigma(b_j) \right\| \right\}$$

where the supremum runs over all commuting range pairs $\pi : A \to B(H)$, $\sigma : B \to B(H)$ with σ a representation and π an anti-representation on the same (arbitrary) Hilbert space H.

Remark 18.3 Let M be a C^*-algebra equipped with a *tracial* state τ. Then the GNS construction (see §A.13) associated to (M, τ) produces a Hilbert space $H = L_2(\tau)$, a cyclic unit vector ξ in H associated to 1_M and commuting left-hand and right-hand actions of M induced by the corresponding multiplications on M. As earlier, we denote them by $L(a)h = a \cdot h$ (L is what we denoted π_f in §A.13) and $R(a)h = h \cdot a$. Then L (resp. R) is a representation of M (resp. M^{op}) on $B(H)$ and the ranges of L and R commute (see the beginning of §11).

We have then for any n-tuple $(u_1, \ldots, u_n)$ of unitaries in M

$$\left\| \sum_1^n \overline{u_j} \otimes u_j \right\|_{\overline{M} \otimes_{\max} M} = \left\| \sum_1^n u_j^* \otimes u_j \right\|_{M^{op} \otimes_{\max} M} = n. \qquad (18.3)$$

Indeed, this is

$$\geq \left\| \sum_1^n L(u_j^*) R(u_j) \right\| \geq \left\| \sum_1^n u_j^* \cdot \xi \cdot u_j \right\| \qquad (18.4)$$

but since $\xi \in L_2(\tau)$ is the element associated to 1_M we have $u_j^* \cdot \xi \cdot u_j = \xi$ hence (18.4) is $\geq n$ and $\leq n$ is trivial by the triangle inequality.

In particular, for any unitary matrices $u_1, \ldots, u_n$ in M_N we have

$$\left\| \sum_1^n \overline{u_j} \otimes u_j \right\|_{\min} = \left\| \sum_1^n \overline{u_j} \otimes u_j \right\|_{\max} = n. \qquad (18.5)$$

18.1 The constant $C(n)$: WEP $\not\Rightarrow$ LLP

In [141], a crucial role is played by a certain constant $C(n)$, defined as follows: $C(n)$ is the infimum of the constants C such that for each $m \geq 1$, there is $N_m \geq 1$ and an n-tuple $[u_1(m), \ldots, u_n(m)]$ of unitary $N_m \times N_m$ matrices such that

$$\sup_{m \neq m'} \left\| \sum_{j=1}^{n} \overline{u_j(m)} \otimes u_j(m') \right\|_{\min} \leq C. \tag{18.6}$$

By (18.2) the preceding min-norm can be understood either in $M_{N_m} \otimes_{\min} M_{N_{m'}} \simeq M_{N_m N_{m'}}$ (with $\overline{u_j(m)}$ denoting the usual matrix with conjugate entries) or in $\overline{M_{N_m}} \otimes_{\min} M_{N_{m'}}$ with $\overline{u_j(m)} \in \overline{M_{N_m}}$.

The connection of $C(n)$ to $B(H) \otimes B(H)$ goes through the following statement.

Theorem 18.4 ([141]) *For any $n \geq 1$ and $\varepsilon > 0$, there is a tensor t of rank n in $\mathscr{B} \otimes \mathscr{B}$ such that*

$$\|t\|_{\max} / \|t\|_{\min} \geq n / C(n) - \varepsilon.$$

Remark 18.5 By Corollaries 22.13 and 22.16 we have $\|t\|_{\max} = \|t\|_{\min}$ for any tensor $t \in \overline{B(H)} \otimes B(H)$ of the form $t = \sum \overline{x_j} \otimes x_j$. This perhaps explains why Theorem 18.4 is not so easy.

We have trivially $C(n) \leq n$ for all n. The crucial fact to show that $B(H) \otimes_{\min} B(H) \neq B(H) \otimes_{\max} B(H)$ is that $C(n) < n$ for at least one $n > 1$. The final word on this is now:

Theorem 18.6 ([120])

$$\forall n \geq 2 \quad C(n) = 2\sqrt{n-1}. \tag{18.7}$$

The (much easier) lower bound $2\sqrt{n-1} \leq C(n)$ was proved in [206]. The complete proof of the upper bound uses a delicate random matrix ingredient (namely Theorem 18.16) the proof of which is beyond the scope of these notes, but we will give the proof of (18.7) modulo this ingredient in the next section. See the next chapter for simpler proofs that $C(n) < n$.

Theorem 18.4 is an immediate consequence of the next Theorem 18.9. To prove the latter we will use a compactness argument for "convergence in distribution" (or rather in moments) of n-tuples of operators, that is described in the next two lemmas. By "distribution" we mean the collection of all "moments" in the operators (viewed as noncommutative random variables) and their adjoints. This is the same notion as that of $*$-distribution used by Voiculescu in free probability (see [254]), but our notation is slightly different.

Let $\mathscr{S}$ be the set consisting of the disjoint union of the sets

$$\mathscr{S}_k = [1, \ldots, n]^k \times \{1, *\}^k.$$

For any $w = ((i_1, \ldots, i_k), (\varepsilon_1, \ldots, \varepsilon_k))$ in $\mathscr{S}_k$ and any n-tuple $x = (x_1, \ldots, x_n)$ in $B(H)$ we denote

$$w(x) = x_{i_1}^{\varepsilon_1} x_{i_2}^{\varepsilon_2} \cdots x_{i_k}^{\varepsilon_k}$$

(where $x^\varepsilon = x$ if $\varepsilon = 1$ and $x^\varepsilon = x^*$ if $\varepsilon = *$). Let $x = (x_1, \ldots, x_n)$ be an n-tuple in a von Neumann algebra M equipped with a tracial state τ. By "the distribution of x," we mean the function

$$\mu_x : \mathscr{S} \to \mathbb{C}$$

defined by

$$\mu_x(w) = \tau(w(x)).$$

When $x = (x_1, \ldots, x_n)$ is an n-tuple of unitary operators, we may as well consider that μ_x is a function defined on $\mathbb{F}_n$ (free group with generators $g_1, \ldots, g_n$) by setting $\mu_x(w) = \tau(\pi_x(w))$ for any "word" $w \in \mathbb{F}_n$, where $\pi_x : \mathbb{F}_n \to M$ is the unitary representation defined by $\pi(g_j) = u_j$.

The following is elementary and well known.

Lemma 18.7 *Fix $n \geq 1$. Let $(M(m), \tau_m)$ be a sequence of von Neumann algebras equipped with (tracial) states. Let $x(m) = (x_1(m), \ldots, x_n(m))$ be a bounded sequence of n-tuples with $x(m) \in M(m)^n$. Then there is a subsequence $\{m_k\}$ such that the distributions of $x(m_k)$ converge pointwise on $\mathscr{S}$ when $k \to \infty$.*

To identify the limit of a sequence of distribution, it will be convenient to use ultraproducts. Let $(M(m), \tau_m)$ be as before ($m \in \mathbb{N}$). Let $\mathcal{U}$ be a nontrivial ultrafilter on $\mathbb{N}$. Let $\mathbb{B} = (\oplus \sum_m M(m))_\infty$. Let $H_\mathcal{U}$ be the GNS Hilbert space for the state $\tau_\mathcal{U}$ defined on $\mathbb{B}$ by $\forall y = (y_m) \in \mathbb{B}$ $\tau_\mathcal{U}(y) = \lim_\mathcal{U} \tau_m(y_m)$. Let $\mathcal{I}_\mathcal{U} = \{y = (y_m) \mid \lim_\mathcal{U} \tau_m(y_m^* y_m) = 0\}$. As explained earlier (see §11.2), since $\tau_\mathcal{U}$ vanishes on $\mathcal{I}_\mathcal{U}$ it defines a tracial state on $\mathbb{B}/\mathcal{I}_\mathcal{U}$, that we still denote (albeit abusively) by $\tau_\mathcal{U}$. We have an isometric representation: $a \to L(a) \in B(L_2(\tau_\mathcal{U}))$ of $\mathbb{B}/\mathcal{I}_\mathcal{U}$ on $L_2(\tau_\mathcal{U})$ (associated to left-hand multiplication) and an isometric representation $a \to R(a) \in B(L_2(\tau_\mathcal{U}))$ of $(\mathbb{B}/\mathcal{I}_\mathcal{U})^{op}$ (associated to right-hand multiplication). We already know (cf. Theorem 11.26) that $M_\mathcal{U} = L(\mathbb{B}/\mathcal{I}_\mathcal{U})$ is a von Neumann subalgebra of $B(H_\mathcal{U})$ and that we have $M'_\mathcal{U} = R(\mathbb{B}/\mathcal{I}_\mathcal{U})$ and $R(\mathbb{B}/\mathcal{I}_\mathcal{U})' = M_\mathcal{U}$. We will view (abusively) $\tau_\mathcal{U}$ as a functional on $M_\mathcal{U}$ by setting for any $y \in M_\mathcal{U}$

$$\tau_\mathcal{U}(y) = \lim_\mathcal{U} \tau_m(y_m),$$

where $(y_m) \in \mathbb{B}$ is any element of the equivalence class of $L^{-1}(y) \in \mathbb{B}/\mathcal{I}_\mathcal{U}$.

Let $x = \{x(m) \mid m \in \mathbb{N}\}$ be a bounded sequence of n-tuples with $x(m) \in M(m)^n$ as before. Equivalently, x is a sequence in $\mathbb{B}^n$. Let $\widehat{x} = (\widehat{x}_1, \ldots, \widehat{x}_n)$ be the associated n-tuple in $M_{\mathcal{U}}^n$. Then, for any "word" w in $\mathscr{S}$, we clearly have

$$\tau_{\mathcal{U}}(w(\widehat{x})) = \lim_{\mathcal{U}} \tau_m(w(x(m))).$$

Hence the distribution of $x(m)$ tends pointwise to that of $\widehat{x}$ along $\mathcal{U}$, so we can write $\lim_{\mathcal{U}} \mu_{x(m)} = \mu_{\widehat{x}}$. The next (again elementary and well known) lemma connects limits in distribution with ultrafilters.

Lemma 18.8 *Let $\{x(m) \mid m \in \mathbb{N}\}$ be a sequence of n-tuples as in the preceding Lemma. The following are equivalent.*

(i) *The distributions of $x(m)$ converge pointwise when $m \to \infty$.*

(ii) *For any nontrivial ultrafilter $\mathcal{U}$ on $\mathbb{N}$, the associated n-tuple $\widehat{x} = (\widehat{x}_1, \ldots, \widehat{x}_n)$ in $(M_{\mathcal{U}}, \tau_{\mathcal{U}})$ has the same distribution (i.e. its distribution does not depend on $\mathcal{U}$).*

(iii) *There is a tracial probability space (M, τ) and $y = (y_1, \ldots, y_n)$ in M^n such that $x(m) \to y$ in distribution.*

Proof (i) $\Rightarrow$ (ii) is essentially obvious. (ii) $\Rightarrow$ (iii) is proved by picking any fixed nontrivial ultrafilter $\mathcal{U}$ and taking $(M, \tau) = (M_{\mathcal{U}}, \tau_{\mathcal{U}})$ (see Remark A.4 for clarification). Lastly (iii) $\Rightarrow$ (i) is again obvious. $\square$

With the notation in (11.14), let $\{[u_1(m), \ldots, u_n(m)], m \in \mathbb{N}\}$ be a sequence of n-tuples of unitary matrices as in (18.6) (recall $u_1(m), \ldots, u_n(m)$ are of size $N_m \times N_m$).

For any subset $\omega \subset \mathbb{N}$, let

$$\mathbb{B}_\omega = \left(\oplus \sum_{m \in \omega} M_{N_m} \right)_\infty. \tag{18.8}$$

Let $\mathbb{N} = \omega(1) \cup \omega(2)$ be any disjoint partition of $\mathbb{N}$ into two infinite subsets, and let

$$u_j^1 = \bigoplus_{m \in \omega(1)} u_j(m) \in \mathbb{B}_{\omega(1)} \qquad u_j^2 = \bigoplus_{m' \in \omega(2)} u_j(m') \in \mathbb{B}_{\omega(2)}. \tag{18.9}$$

Theorem 18.9 *Suppose that $[u_1(m), \ldots, u_n(m)]$ converges in distribution when $m \to \infty$ and satisfies (18.6). Let*

$$t = \sum_{j=1}^n \overline{u_j^1} \otimes u_j^2.$$

We have then

$$\|t\|_{\min} \le C \quad and \quad \|t\|_{\max} = n,$$

and hence $\|t\|_{\max}/\|t\|_{\min} \geq n/C$, *where the min and max norms are relative to* $\overline{\mathbb{B}_{\omega(1)}} \otimes \mathbb{B}_{\omega(2)}$.

Proof We have obviously

$$\|t\|_{\min} = \sup_{m \in \omega(1), m' \in \omega(2)} \left\| \sum \overline{u_j(m)} \otimes u_j(m') \right\|$$

hence $\|t\|_{\min} \leq C$. We now turn to $\|t\|_{\max}$. Let $\mathcal{U}$ be a nontrivial ultrafilter on $\omega(1)$ and let $\mathcal{V}$ be one on $\omega(2)$. We construct the ultraproducts $M_{\mathcal{U}}$ and $M_{\mathcal{V}}$ as previously. Since the quotient mappings $\mathbb{B}_{\omega(1)} \to M_{\mathcal{U}}$ and $\mathbb{B}_{\omega(2)} \to M_{\mathcal{V}}$ are $*$-homomorphisms, we have

$$\|t\|_{\max} \geq \left\| \sum \overline{u_j} \otimes v_j \right\|_{\overline{M_{\mathcal{U}}} \otimes_{\max} M_{\mathcal{V}}}$$

where u_j (resp. v_j) is the equivalence class modulo $\mathcal{U}$ (resp. $\mathcal{V}$) of $\bigoplus_{m \in \omega(1)} u_j(m)$ (resp. $\bigoplus_{m \in \omega(2)} u_j(m)$).

Now, since we assume that $[u_1(m), \ldots, u_n(m)]$ converges in distribution when $m \to \infty$, the two n-tuples $(u_1, \ldots, u_n)$ and $(v_1, \ldots, v_n)$ must have the same distribution relative respectively to $\tau_{\mathcal{U}}$ and $\tau_{\mathcal{V}}$. But this implies (see Remark 11.20) that there is a $*$-isomorphism π from the von Neumann algebra $N_u \subset M_{\mathcal{U}}$ generated by $(u_1, \ldots, u_n)$ to the one $N_v \subset M_{\mathcal{V}}$ generated by $(v_1, \ldots, v_n)$, defined simply by $\pi(u_j) = v_j$. Moreover, since we are dealing here with *finite* traces, there is a conditional expectation P from $M_{\mathcal{U}}$ onto N_u (see Proposition 11.21). Therefore the composition $T = \pi P$ is a unital completely positive map from $M_{\mathcal{U}}$ to $N_v \subset M_{\mathcal{V}}$ such that $T(u_j) = v_j$. Thus we find by (4.30)

$$\left\| \sum \overline{u_j} \otimes v_j \right\|_{\overline{M_{\mathcal{U}}} \otimes_{\max} M_{\mathcal{V}}} \geq \left\| \sum \overline{T(u_j)} \otimes v_j \right\|_{\overline{M_{\mathcal{V}}} \otimes_{\max} M_{\mathcal{V}}}$$

$$= \left\| \sum \overline{v_j} \otimes v_j \right\|_{\overline{M_{\mathcal{V}}} \otimes_{\max} M_{\mathcal{V}}}. \qquad (18.10)$$

But then by (18.3) we conclude that $\|t\|_{\max} \geq n$. $\qquad \square$

Remark 18.10 The same reasoning shows that

$$\|t\|_{\min} \geq \left\| \sum \overline{v_j} \otimes v_j \right\|_{\overline{M_{\mathcal{V}}} \otimes_{\min} M_{\mathcal{V}}}.$$

Proof of Theorem 18.4 Since there exist max-injective inclusions $\mathbb{B}_{\omega(1)} \subset \mathcal{B}$ and $\mathbb{B}_{\omega(2)} \subset \mathcal{B}$, Theorem 18.9 gives us a tensor $t' \in \overline{\mathcal{B}} \otimes \mathcal{B}$ of rank n with $\|t'\|_{\max}/\|t'\|_{\min} \geq n/C$. Using $\overline{\mathcal{B}} \simeq \mathcal{B}$, we find a similar tensor in $\mathcal{B} \otimes \mathcal{B}$. $\qquad \square$

We now exploit the mere fact that $C(n) < n$ for some n, which we know by Theorem 18.6:

Corollary 18.11 *Recall* $\mathbb{B} = \left(\oplus \sum_{n\geq 1} M_n\right)_\infty$. *Then the pair* $(\mathbb{B}, \mathbb{B})$ *(or the pair* $(\overline{\mathbb{B}}, \mathbb{B})$*) is not nuclear and* $\mathbb{B}$ *(although it has the WEP) fails the LLP.*

Proof Since $C(n) < n$ for some n, Theorem 18.9 tells us that $(\overline{\mathbb{B}_{\omega(1)}}, \mathbb{B}_{\omega(2)})$ is not a nuclear pair. We have inclusions $\mathbb{B}_{\omega(1)} \subset \mathbb{B}$ and $\mathbb{B}_{\omega(2)} \subset \mathbb{B}$ each admitting a c.p. contractive projection. By Proposition 7.19 (i), these inclusions are max-injective, therefore we have an isometric embedding

$$\overline{\mathbb{B}_{\omega(1)}} \otimes_{\max} \mathbb{B}_{\omega(2)} \subset \overline{\mathbb{B}} \otimes_{\max} \mathbb{B}$$

so the min and max norms cannot coincide on $\overline{\mathbb{B}} \otimes \mathbb{B}$ or equivalently (recall $\mathbb{B} \cong \overline{\mathbb{B}}$) on $\mathbb{B} \otimes \mathbb{B}$. Since for the inclusion $\mathbb{B} \subset B(\ell_2)$ there is also a c.p. contractive projection from $B(\ell_2)$ onto $\mathbb{B}$, the same argument shows they do not coincide on $B(\ell_2) \otimes \mathbb{B}$, which means that $\mathbb{B}$ fails the LLP. $\square$

Similarly:

Corollary 18.12 *For* $\mathscr{B} = B(\ell_2)$, *we have*

$$\mathscr{B} \otimes_{\min} \mathscr{B} \neq \mathscr{B} \otimes_{\max} \mathscr{B}$$

or equivalently $\mathscr{B}$ *(although it has the WEP) fails the LLP.*

Proof Since we have max-injective inclusions $\mathbb{B} \subset \mathscr{B}$ this follows from Corollary 18.11. $\square$

More generally, we have

Corollary 18.13 *A von Neumann algebra M has the LLP if and only if M is nuclear.*

A pair (M, N) of von Neumann algebras is nuclear if and only if one of them is nuclear.

Proof Let $\mathbb{B} = \left(\oplus \sum_{n\geq 1} M_n\right)_\infty$. If M is not nuclear, by Theorem 12.29 there is an embedding $\mathbb{B} \subset M$, and since $\mathbb{B}$ is injective, the embedding is max-injective. Since $\mathbb{B}$ fails the LLP (by Corollary 18.11), so does M (see Remark 9.3 or Remark 9.13).

If, in addition, N is not nuclear, we have $\mathbb{B} \cong \overline{\mathbb{B}} \subset N$. Thus again, since $(\mathbb{B}, \mathbb{B})$ is not nuclear, we find that (M, N) is not nuclear if none of M, N is nuclear, and the converse is trivial. $\square$

18.2 Proof that $C(n) = 2\sqrt{n-1}$ using random unitary matrices

We start with an easy result from [206] estimating $C(n)$ from below. The alternate proof we give here of (18.11) is due to Szarek.

Proposition 18.14 *Let* $u_1, \ldots, u_n$ *be arbitrary unitary operators in* $B(H)$ *(H any Hilbert space), then*

$$2\sqrt{n-1} \leq \left\| \sum_{j=1}^{n} \overline{u_j} \otimes u_j \right\|_{\min}. \tag{18.11}$$

Proof Let $\mathcal{S}^+ = \{t \in S_2(H) \mid t \geq 0 \quad \|t\|_2 = 1\}$. We will use the self-adjoint case of (2.11). Note:

$$\forall a \in B(H), \forall t \in \mathcal{S}^+ \quad \operatorname{tr}(tata^*) = \operatorname{tr}([t^{1/2}at^{1/2}][t^{1/2}at^{1/2}]^*) \geq 0. \tag{18.12}$$

Let $T = \sum_{1}^{n} \overline{u_j} \otimes u_j$ and let $S = \sum_{j=1}^{n} \lambda_{\mathbb{F}_n}(g_j)$. In accordance with the identification of T with the operator $t \mapsto \sum u_j t u_j^*$ acting on $S_2(H)$, we denote $\langle t, Tt \rangle = \operatorname{tr}(\sum t^* u_j t u_j^*)$ for any $t \in S_2(H)$. The idea of the proof is to show that for any integer $m \geq 1$ and any t in $\mathcal{S}^+$ we have

$$\langle t, (T^*T)^m t \rangle \geq \langle \delta_e, (S^*S)^m \delta_e \rangle = \tau_{\mathbb{F}_n}((S^*S)^m) \tag{18.13}$$

where δ_e denotes the basis vector in $\ell_2(\mathbb{F}_n)$ indexed by the unit element of $\mathbb{F}_n$. We can expand $(T^*T)^m$ as a sum of the form $\sum_{w \in I} \overline{u^w} \otimes u^w$ where the u^w's are unitaries of the form $u_{i_1}^* u_{j_1} u_{i_2}^* u_{j_2} \ldots$.

Now for certain ws, we have $u^w = I$ (and hence $\operatorname{tr}(u^w t u^{w*} t) = 1$) by formal cancellation (no matter what the u_js are). Let us denote by $I' \subset I$ the set of all such w's. Then by (18.12) we have for all t in $\mathcal{S}^+$

$$\langle t, (T^*T)^m t \rangle = \sum_{w \in I} \operatorname{tr}(u^w t u^{w*} t) \geq \sum_{w \in I'} 1 = \operatorname{card}(I').$$

An elementary counting argument shows that $\operatorname{card}(I') = \langle (S^*S)^m \delta_e, \delta_e \rangle = \tau_{\mathbb{F}_n}((S^*S)^m)$. Thus we obtain (18.13). Therefore (recalling (11.10))

$$\|T^*T\| \geq \limsup_{m \to \infty} \langle t, (T^*T)^m t \rangle^{1/m} \geq \limsup_{m \to \infty} (\tau_{\mathbb{F}_n}((S^*S)^m))^{1/m} = \|S^*S\|,$$

so that we obtain $\|T\| \geq \|S\|$, whence (18.11) by (3.20). $\square$

Corollary 18.15

$$2\sqrt{n-1} \leq C(n). \tag{18.14}$$

Proof Let $(u_j(m))$ be as in (18.6). Let $B = (\oplus \sum M_{N_m})_\infty$. Let $\mathcal{U}$ be a non trivial ultrafilter on $\mathbb{N}$. Let $\mathcal{I} \subset B$ be the (closed self-adjoint two-sided) ideal formed of all $x \in B$ such that $\lim_{\mathcal{U}} \|x_i\| = 0$. Let $q : B \to B/\mathcal{I}$ be the quotient morphism. Let $u_j = q((u_j(m))$. We claim that

$$\left\| \sum_{j=1}^{n} \overline{u_j} \otimes u_j \right\|_{\min} \leq \sup_{m \neq m'} \left\| \sum_{1}^{n} \overline{u_j(m)} \otimes u_j(m') \right\|.$$

Using the claim we deduce (18.14) from (18.11). To check the claim note that for any C^*-algebra A we have $A \otimes_{\min} B = (\oplus \sum_m A \otimes_{\min} M_{N_m})_\infty$ and hence for any (a_j) in A^n

$$\left\|\sum_1^n a_j \otimes u_j\right\|_{A \otimes_{\min}(B/\mathcal{I})} \leq \lim_{\mathcal{U},m'} \left\|\sum_1^n a_j \otimes u_j(m')\right\|_{A \otimes_{\min} M_{N_{m'}}}.$$

Indeed, since $Id_A \otimes q : A \otimes_{\min} B \to A \otimes_{\min}(B/\mathcal{I})$ is a contraction we observe that the left-hand side is $\leq \sup_{m'} \|\sum_1^n a_j \otimes u_j(m')\|_{\min}$. Moreover, let $\gamma \subset \mathbb{N}$ be a subset that belongs to $\mathcal{U}$. If we multiply $m' \mapsto \sum_1^n a_j \otimes u_j(m')$ by the indicator of γ and apply the same observation we find

$$\left\|\sum_1^n a_j \otimes u_j\right\|_{A \otimes_{\min}(B/\mathcal{I})} \leq \sup_{m' \in \gamma} \left\|\sum_1^n a_j \otimes u_j(m')\right\|_{A \otimes_{\min} M_{N_{m'}}},$$

and since this holds for any γ the claim follows. (Note: the claim merely spells out the fact that the natural morphism $(A \otimes_{\min} B)/(A \otimes_{\min} \mathcal{I}) \to A \otimes_{\min}(B/\mathcal{I})$ is always contractive, see the discussion in §10.1.) Applying the claim twice gives us

$$\left\|\sum_1^n \overline{u_j} \otimes u_j\right\|_{\min} \leq \lim_{\mathcal{U},m} \left\|\sum_1^n \overline{u_j(m)} \otimes u_j\right\|_{\min}$$

$$\leq \lim_{\mathcal{U},m} \lim_{\mathcal{U},m'} \left\|\sum_1^n \overline{u_j(m)} \otimes u_j(m')\right\|_{\min},$$

and obviously the last term is $\leq \sup_{m \neq m'} \|\sum_1^n \overline{u_j(m)} \otimes u_j(m')\|$, which proves the claim.

Alternate proof: Passing to a subsequence we may assume that $(u_j(m))$ converges in distribution when $m \to \infty$. Then with the notation from Remark 18.10 we have

$$\sup_{m \neq m'} \left\|\sum_1^n \overline{u_j(m)} \otimes u_j(m')\right\| \geq \|t\|_{\min} \geq \left\|\sum \overline{v_j} \otimes v_j\right\|_{M_\mathcal{V} \otimes_{\min} M_\mathcal{V}}$$

and we again deduce (18.14) from (18.11). $\qquad\square$

The proof that $C(n) \leq 2\sqrt{n-1}$ is much more delicate. The first proof by Haagerup and Thorbjørnsen in [119] was based on a fundamental limit theorem for Gaussian random matrices, which was a considerable strengthening of Theorem 12.24. We will use the following refinement due to Collins and Male [56], valid for unitary matrices, which gives a more direct approach.

Theorem 18.16 ([56]) *Let $\mathbb{U}_N$ denote the group of all $N \times N$ unitary matrices ($N \geq 1$). Let $U_1^{(N)}, \dots, U_n^{(N)}$ be a sequence of independent matrix valued random variables, each having as its distribution the normalized Haar measure on $\mathbb{U}_N$. Let $g_1, \dots, g_n$ be the free generators of the free group $\mathbb{F}_n$. For convenience we set $U_0^{(N)} = I$ (unit in $\mathbb{U}_N$) and we denote by g_0 the unit in $\mathbb{F}_n$. Then, for all k and for all $a_0, \dots, a_n$ in M_k, we have, for almost all ω*

$$\lim_{N \to \infty} \left\|\sum_0^n a_j \otimes U_j^{(N)}(\omega)\right\|_{\min} = \left\|\sum_0^n a_j \otimes \lambda_{\mathbb{F}_n}(g_j)\right\|_{\min}. \tag{18.15}$$

In particular, if $a_1, \ldots, a_n$ are all unitary, for almost all ω

$$\lim_{N \to \infty} \left\| \sum_1^n a_j \otimes U_j^{(N)}(\omega) \right\|_{\min} = 2\sqrt{n-1}. \qquad (18.16)$$

The implication $(18.15) \Rightarrow (18.16)$ follows from Remark 3.17 and (3.20).

Proof of Theorem 18.6 By (18.14) it suffices to show that $C(n) \le 2\sqrt{n-1}$. Fix $\varepsilon > 0$. Obviously it suffices to construct a sequence of n-tuples $\{(u_j(m))_{1 \le j \le n} \mid m \ge 1\}$ of unitary matrices (we emphasize that $(u_j(m))_{1 \le j \le n}$ is assumed to be an n-tuple of matrices of size $N_m \times N_m$) such that, for any integer $p \ge 1$, we have

$$\sup_{1 \le m \ne m' \le p} \left\| \sum \overline{u_j(m)} \otimes u_j(m') \right\|_{\min} < 2\sqrt{n-1} + \varepsilon. \qquad (18.17)$$

We will construct this sequence and the sizes N_m by induction on p. Assume that we already know the result up to p. That is, we already know a family $\{(u_j(m))_{1 \le j \le n} \mid 1 \le m \le p\}$ formed of p n-tuples satisfying (18.17). We need to produce an additional n-tuple $(u_j(p+1))_{1 \le j \le n}$ of unitary matrices (possibly of some larger size $N_{p+1} \times N_{p+1}$) such that (18.17) still holds for the enlarged family $\{(u_j(m))_{1 \le j \le m} \mid 1 \le m \le p+1\}$ formed of one more n-tuple. By (18.16), for any $1 \le m \le p$, we have for almost all ω

$$\lim_{N \to \infty} \left\| \sum_1^n \overline{u_j(m)} \otimes U_j^{(N)}(\omega) \right\|_{\min} = 2\sqrt{n-1}.$$

Hence, if N is chosen large enough, we can ensure that, for all $1 \le m \le p$ simultaneously, we can find ω such that

$$\left\| \sum_1^n \overline{u_j(m)} \otimes U_j^{(N)}(\omega) \right\|_{\min} < 2\sqrt{n-1} + \varepsilon.$$

But then, if we set $N_{p+1} = N$ and $u_j(p+1) = U_j^{(N)}(\omega)$, the extended family $\{(u_j(m))_{1 \le j \le n} \mid 1 \le m \le p+1\}$ clearly still satisfies (18.17). This proves $C(n) \le 2\sqrt{n-1}$. $\qquad \square$

Fix $\varepsilon > 0$. Actually, by concentration of measure arguments (see e.g. [7, §4.4]), there is a sequence of sizes $N_1 < N_2 < \cdots$, for our random unitary matrices, such that

$$\mathbb{P} \left\{ \omega \in \Omega \mid \sup_{m \ne m'} \left\| \sum_{j=1}^n \overline{U_j^{(N_m)}(\omega)} \otimes U_j^{(N_{m'})}(\omega) \right\|_{\min} \le 2\sqrt{n-1} + \varepsilon \right\} > 1 - \varepsilon.$$

Thus, provided the sizes grow sufficiently fast, with close to full probability a random choice yields (18.6) with almost the best C. We refer to [211, 212] for more details and for related estimates.

18.3 Exactness is not preserved by extensions

In this section we indicate a quick way to produce an example of a separable nonexact C^*-algebra A and a closed ideal $\mathcal{I} \subset A$ such that both $\mathcal{I}$ and the quotient $C = A/\mathcal{I}$ are exact. In this situation one usually says that A is an extension of C by $\mathcal{I}$. Thus exactness is *not stable* by extension. When $\mathcal{I}$ and C are both nuclear, then A is nuclear (see Corollary 8.18). Thus in sharp contrast, nuclearity is preserved by extensions.

Recall $\mathbb{B} = (\oplus \sum_{N \geq 1} M_N)_\infty$. For any $V \in \mathbb{B}$, for notational convenience, we will denote in this section by $V^{(N)} \in M_N$ the N'th coordinate of V.

Remark 18.17 Let $\mathcal{I}_0 \subset \mathbb{B}$ be the subset of all $b = (b^{(N)}) \in \mathbb{B}$ such that $\lim_N \|b^{(N)}\| = 0$. Let $A \subset \mathbb{B}$ be a unital C^*-subalgebra. Let $q : A \to A/A \cap \mathcal{I}_0$ be the quotient map. Then for any $b_0, \ldots, b_n \in A$ and $a_0, \ldots, a_n$ in M_k $(n, k \geq 1)$ we have

$$\limsup_{N \to \infty} \left\| \sum_0^n a_j \otimes b_j^{(N)} \right\|_{M_k(M_N)} = \left\| \sum_0^n a_j \otimes q(b_j) \right\|_{M_k(A/A \cap \mathcal{I}_0)} . \quad (18.18)$$

To justify this, let $Q : \mathbb{B} \to \mathbb{B}/\mathcal{I}_0$ be the quotient morphism. Since

$$A/A \cap \mathcal{I}_0 = A/\ker(Q_{|A}) \simeq Q(A) \subset Q(\mathbb{B}) \simeq \mathbb{B}/\mathcal{I}_0, \quad (18.19)$$

it suffices to check (18.18) in the case $A = \mathbb{B}$, for which the easy verification is left to the reader.

Kirchberg gave in [155] the first example of a nonexact extension of an exact C^*-algebra by the algebra of compact operators on ℓ_2. Our example, based on the random unitary matrix model, will be deduced from Theorem 18.16, but first we analyze the underlying deterministic matrix model.

Remark 18.18 Let $V_0 = 1_\mathbb{B}$ and let g_0 be as before the unit in $\mathbb{F}_n$. Let $V_1, \ldots, V_n$ be unitaries in $\mathbb{B}$ that form a matrix model for $M_{\mathbb{F}_n}$ in the sense of §12.5. This implies that for any nontrivial ultrafilter $\mathcal{U}$ on $\mathbb{N}$, denoting as before by $q_\mathcal{U} : \mathbb{B} \to M_\mathcal{U}$ the quotient map, the correspondence $\lambda_{\mathbb{F}_n}(g_j) \mapsto q_\mathcal{U}(V_j)$ $(0 \leq j \leq n)$ extends to an isometric normal $*$-homomorphism embedding $M_{\mathbb{F}_n}$ into $M_\mathcal{U}$. In particular this implies that the following holds:

$$\forall k \geq 1 \; \forall a_j \in M_k \quad \left\| \sum_0^n a_j \otimes \lambda_{\mathbb{F}_n}(g_j) \right\|_{M_k(M_{\mathbb{F}_n})} \quad (18.20)$$

$$= \left\| \sum_0^n a_j \otimes q_\mathcal{U}(V_j) \right\|_{M_k(M_\mathcal{U})} .$$

By Proposition 9.7 the latter property (18.20) implies conversely that the previous correspondence extends to an isometric $*$-homomorphism from $C_\lambda^*(\mathbb{F}_n)$ into $M_\mathcal{U}$.

Theorem 18.19 *Let $V_0 = 1$ and let $V_1, \ldots, V_n$ be unitaries in $\mathbb{B}$. Assume that they satisfy*

$$\forall k \geq 1 \, \forall a_j \in M_k \quad \left\| \sum_0^n a_j \otimes \lambda_{\mathbb{F}_n}(g_j) \right\|_{M_k(C_\lambda^*(\mathbb{F}_n))}$$

$$= \limsup_{N \to \infty} \left\| \sum_0^n a_j \otimes V_j(N) \right\|_{M_k(M_N)}. \tag{18.21}$$

Let $A \subset \mathbb{B}$ be the (unital) C^*-algebra generated by $V_0, V_1, \ldots, V_n$, and let $\mathcal{I} = A \cap \mathcal{I}_0$. Then $A/\mathcal{I} \simeq C_\lambda^*(\mathbb{F}_n)$, so that $A/\mathcal{I}$ and $\mathcal{I}$ are exact but A is not exact.

Proof First observe that since $\mathcal{I} \subset \mathcal{I}_0$ and $\mathcal{I}_0$ is obviously nuclear (e.g. because it has the CPAP, see Corollary 7.12) it is immediate that $\mathcal{I}$ is exact (see Remarks 10.4 and 10.5). Let $q : A \to A/\mathcal{I}$ be the quotient morphism. Let $E = \mathrm{span}[\lambda_{\mathbb{F}_n}(g_j) \mid 0 \leq j \leq n] \subset C_\lambda^*(\mathbb{F}_n)$. Similarly let $E_q = \mathrm{span}[q(V_j) \mid 0 \leq j \leq n] \subset A/\mathcal{I}$, and let $u : E \to E_q$ be the linear map defined by $u(\lambda_{\mathbb{F}_n}(g_j)) = q(V_j)$ $(0 \leq j \leq n)$. By (18.21) and (18.18) u is a unital completely isometric isomorphism. By Proposition 9.7 it extends to a $*$-isomorphism $\pi : C_\lambda^*(\mathbb{F}_n) \to A/\mathcal{I}$. Thus $A/\mathcal{I}$ is exact by Remark 10.21. As observed in (18.19) $A/\mathcal{I} \subset \mathbb{B}/\mathcal{I}_0$ and since $\mathcal{I}_0 \subset \mathcal{I}_\mathcal{U}$ we have obviously a canonical map $r_\mathcal{U} : \mathbb{B}/\mathcal{I}_0 \to \mathbb{B}/\mathcal{I}_\mathcal{U} = M_\mathcal{U}$, such that $q_\mathcal{U} = r_\mathcal{U} Q$. Let $\sigma : C^*(\mathbb{F}_n) \to A$ be the $*$-homomorphism defined by $\sigma(U_{\mathbb{F}_n}(g_j)) = V_j$ $(0 \leq j \leq n)$. Let $J : A \to \mathbb{B}$ denote the canonical inclusion. Let $\psi = q_\mathcal{U} J : A \to M_\mathcal{U}$. Note the factorization through $\mathbb{B}$:

$$\psi : A \xrightarrow{J} \mathbb{B} \xrightarrow{q_\mathcal{U}} M_\mathcal{U}.$$

Assume for contradiction that A was exact. Then J would be (min $\to$ max)-tensorizing by Corollary 10.8, and hence so would be $\psi = q_\mathcal{U} J : A \to M_\mathcal{U}$. It would follow (see the diagram at the end of the proof) that the $*$-homomorphism $\Psi : C^*(\mathbb{F}_n) \to M_\mathcal{U}$ defined by $\Psi(U_{\mathbb{F}_n}(g_j)) = q_\mathcal{U}(V_j)$ (that can be factorized as $C^*(\mathbb{F}_n) \xrightarrow{\sigma} A \xrightarrow{\psi} M_\mathcal{U}$) would also be (min $\to$ max)-tensorizing. But we claim that this is not true when $n \geq 2$, a contradiction that shows that A is not exact. To check the claim, observe that

$$\left\| \sum_0^n \overline{q_\mathcal{U}(V_j)} \otimes \Psi(U_{\mathbb{F}_n}(g_j)) \right\|_{\overline{M_\mathcal{U}} \otimes_{\max} M_\mathcal{U}} = \left\| \sum_0^n \overline{q_\mathcal{U}(V_j)} \otimes q_\mathcal{U}(V_j) \right\|_{\overline{M_\mathcal{U}} \otimes_{\max} M_\mathcal{U}}$$

$$= n + 1$$

where the last equality follows from (18.3), and also since $q_\mathcal{U}(V_j) = r_{\mathcal{U}|A/\mathcal{I}}(q(V_j))$

$$\left\| \sum_0^n \overline{q\mathcal{U}(V_j)} \otimes U_{\mathbb{F}_n}(g_j) \right\|_{\overline{M_{\mathcal{U}} \otimes_{\min} C^*(\mathbb{F}_n)}}$$

$$\leq \left\| \sum_0^n \overline{q(V_j)} \otimes U_{\mathbb{F}_n}(g_j) \right\|_{\overline{A/\mathcal{I} \otimes_{\min} C^*(\mathbb{F}_n)}}$$

$$= \left\| \sum_0^n \overline{u(\lambda_{\mathbb{F}_n}(g_j))} \otimes U_{\mathbb{F}_n}(g_j) \right\|_{\overline{A/\mathcal{I} \otimes_{\min} C^*(\mathbb{F}_n)}}$$

$$= \left\| \sum_0^n \overline{\lambda_{\mathbb{F}_n}(g_j)} \otimes U_{\mathbb{F}_n}(g_j) \right\|_{\overline{C^*_\lambda(\mathbb{F}_n) \otimes_{\min} C^*(\mathbb{F}_n)}} = \sqrt{n},$$

where the last inequality follows from (3.21) and (3.13). Since $2\sqrt{n} < n+1$ this proves the claim. $\qquad\square$

The preceding proof is summarized by the following diagram:

$$
\begin{array}{ccccccc}
C^*(\mathbb{F}_n) & \xrightarrow{\ \sigma\ } & A & \xrightarrow{\ J\ } & \mathbb{B} & \xrightarrow{\ q\mathcal{U}\ } & M_{\mathcal{U}} \\
& & \downarrow{\scriptstyle q} & & \downarrow{\scriptstyle Q} & \nearrow{\scriptstyle r\mathcal{U}} & \\
& & C^*_\lambda(\mathbb{F}_n) \simeq A/\mathcal{I} & \longrightarrow & \mathbb{B}/\mathcal{I}_0 & &
\end{array}
$$

Corollary 18.20 *Let* $1 = U_0^{(N)}, U_1^{(N)}, \dots, U_n^{(N)}$ *be the random unitaries in Theorem 18.16, assumed defined on a probability space* $(\Omega, \mathbb{P})$. *Let* $\mathbb{B} = (\oplus_{N \geq 1} M_N)_\infty$. *Let*

$$U_j(\omega) = (U_j^{(N)}(\omega))_{N \geq 1} \in \mathbb{B}.$$

Let $A_\omega \subset \mathbb{B}$ *be the (unital) C^*-algebra generated by* $U_0(\omega), \dots, U_n(\omega)$, *and let* $\mathcal{I}_\omega = A_\omega \cap \mathcal{I}_0$. *Then for almost all* ω *we have*

$$A_\omega/\mathcal{I}_\omega \simeq C^*_\lambda(\mathbb{F}_n), \tag{18.22}$$

and $A_\omega/\mathcal{I}_\omega$ *and* $\mathcal{I}_\omega$ *are exact but* A_ω *is not.*

Proof By the preceding statement, it suffices to show that $(U_j^{(N)}(\omega))$ satisfies (18.21) for almost all ω. To check the latter assertion just observe that there is $\Omega' \subset \Omega$ with $\mathbb{P}(\Omega') = 1$ such that for any $\omega \in \Omega'$ we have (18.15) for all k and all $a_0, \dots, a_n \in M_k$. Indeed, we may restrict by density to $a_0, \dots, a_n$ with rational entries, and then Ω' appears as an intersection of a *countable* family of events each having full probability by (18.15). $\qquad\square$

18.4 A continuum of C^*-norms on $\mathbb{B} \otimes \mathbb{B}$

The preceding results from [141] only showed that there is more than one C^*-norm on $\mathscr{B} \otimes \mathscr{B}$ or $\mathbb{B} \otimes \mathbb{B}$. In [193] N. Ozawa and the author proved that there is actually a continuum of distinct such norms. The proof for $\mathbb{B} \otimes \mathbb{B}$ is very simple and it yields a family of (maximal) cardinality $2^{2^{\aleph_0}}$ of distinct C*-norms. We include it in this section. Curiously, there does not seem to be a simple argument to transplant the result to $\mathscr{B} \otimes \mathscr{B}$ (as we did for min $\neq$ max in Corollary 18.12). The latter case is more delicate, and we refer the reader to [193] for full details.

Let (N_m) be any sequence of positive integers tending to ∞ and let

$$B = \left(\oplus \sum_m M_{N_m} \right)_\infty .$$

Theorem 18.21 *There is a family of cardinality $2^{2^{\aleph_0}}$ of mutually distinct (and hence inequivalent) C*-norms on $B \otimes M$ for any von Neumann algebra M that is not nuclear.*

Remark 18.22 Assuming $M \subset B(\ell_2)$ nonnuclear, we note that the cardinality of $B(\ell_2)$ and hence of $B(\ell_2) \otimes M$ is $c = 2^{\aleph_0}$, so the set of *all* real valued functions of $M \otimes B(\ell_2)$ into $\mathbb{R}$ has the same cardinal $2^{2^{\aleph_0}}$ as the set of C*-norms.

Fix $n > 2$, and let $[u_1(m), \ldots, u_n(m)]$ be a sequence of n-tuples of unitary $N_m \times N_m$ matrices satisfying (18.6).

By compactness (see Lemmas 18.7 and 18.8) we may and do assume (after passing to a subsequence) that the n-tuples $[u_1(m), \ldots, u_n(m)]$ converge in distribution (i.e. in moments) to an n-tuple $[u_1, \ldots, u_n]$ of unitaries in a von Neumann algebra M equipped with a faithful normal trace τ. Then, if $\mathcal{U}$ is any nontrivial ultrafilter, we can take for (M, τ) the ultraproduct $(M_\mathcal{U}, \tau_\mathcal{U})$, and the resulting limit distribution along $\mathcal{U}$ does not depend on $\mathcal{U}$.

For any subset $s \subset \mathbb{N}$ and any $u \in B$ we denote by $u[s] = \oplus_m u[s](m) \in B$ the element of B defined by $u[s](m) = u(m)$ if $m \in s$ and $u[s](m) = 0$ otherwise.

We denote by

$$\pi_\mathcal{U} : B \to M_\mathcal{U} \quad (\text{or } \overline{\pi_\mathcal{U}} : \overline{B} \to \overline{M_\mathcal{U}})$$

the natural quotient map.

Recall that if $\mathcal{U}, \mathcal{V}$ are ultrafilters on $\mathbb{N}$, then $\mathcal{U} \neq \mathcal{V}$ if and only if there are disjoint subsets $s \subset \mathbb{N}$ and $s' \subset \mathbb{N}$ with $s \in \mathcal{U}$ and $s' \in \mathcal{V}$ (see Remark A.5). In that case we have

$$\forall u \in B \quad \pi_\mathcal{U}(u[s']) = \pi_\mathcal{V}(u[s]) = 0. \tag{18.23}$$

Lemma 18.23 *Let $\mathcal{U} \neq \mathcal{V}$ be ultrafilters on $\mathbb{N}$. Consider disjoint subsets $s \subset \mathbb{N}$ and $s' \subset \mathbb{N}$ with $s \in \mathcal{U}$ and $s' \in \mathcal{V}$, and let*

$$t(s, s') = \sum_{k=1}^n \overline{u_k[s]} \otimes u_k[s'] \in \overline{B} \otimes B.$$

Then

$$\|t(s,s')\|_{\overline{B} \otimes_{\min} B} \leq C \quad \text{and} \quad \|[\overline{\pi_{\mathcal{U}}} \otimes \pi_{\mathcal{V}}](t(s,s'))\|_{\overline{M_{\mathcal{U}}} \otimes_{\max} M_{\mathcal{V}}} = n.$$

Proof We have obviously

$$\|t(s,s')\|_{\min} = \sup_{(m,m') \in s \times s'} \left\| \sum \overline{u_k(m)} \otimes u_k(m') \right\|$$

hence $\|t(s,s')\|_{\min} \leq C$. We now turn to the max tensor product.

Let $u_k = \pi_{\mathcal{U}}(u_k[s])$ and $v_k = \pi_{\mathcal{V}}(u_k[s'])$ so that we have

$$\|[\overline{\pi_{\mathcal{U}}} \otimes \pi_{\mathcal{V}}](t(s,s'))\|_{\overline{M_{\mathcal{U}}} \otimes_{\max} M_{\mathcal{V}}} = \left\| \sum \overline{u}_k \otimes v_k \right\|_{\overline{M_{\mathcal{U}}} \otimes_{\max} M_{\mathcal{V}}}.$$

Since we assume that $[u_1(m), \ldots, u_n(m)]$ converges in distribution, $(u_1, \ldots, u_n)$ and $(v_1, \ldots, v_n)$ must have the same distribution relative respectively to $\tau_{\mathcal{U}}$ and $\tau_{\mathcal{V}}$. Arguing as for (18.10) we obtain $\|[\overline{\pi_{\mathcal{U}}} \otimes \pi_{\mathcal{V}}](t(s,s'))\|_{\max} = n$. $\qquad \square$

For any nontrivial ultrafilter $\mathcal{U}$ on $\mathbb{N}$ we denote by $\alpha_{\mathcal{U}}$ the norm defined on $\overline{B} \otimes B$ by

$$\forall t \in \overline{B} \otimes B \qquad \alpha_{\mathcal{U}}(t) = \max\{\|t\|_{\overline{B} \otimes_{\min} B}, \|[\overline{\pi_{\mathcal{U}}} \otimes Id](t)\|_{\overline{M_{\mathcal{U}}} \otimes_{\max} B}\}.$$

Theorem 18.24 *There is a family of cardinality $2^{2^{\aleph_0}}$ of mutually distinct (and hence inequivalent) C*-norms on $\overline{B} \otimes B$. More precisely, the family $\{\alpha_{\mathcal{U}}\}$ indexed by nontrivial ultrafilters on $\mathbb{N}$ is such a family on $\overline{B} \otimes B$.*

Proof Let $(\mathcal{U}, \mathcal{V})$ be two distinct nontrivial ultrafilters on $\mathbb{N}$. Let $s \subset \mathbb{N}$ and $s' \subset \mathbb{N}$ be disjoint subsets such that $s \in \mathcal{U}$ and $s' \in \mathcal{V}$. By Lemma 18.23 we have

$$\alpha_{\mathcal{U}}(t(s,s')) \geq \|[\overline{\pi_{\mathcal{U}}} \otimes \pi_{\mathcal{V}}](t(s,s'))\|_{\overline{M_{\mathcal{U}}} \otimes_{\max} M_{\mathcal{V}}} = n$$

but since $(\pi_{\mathcal{V}} \otimes Id)(t(s,s')) = 0$ by (18.23) we have $\alpha_{\mathcal{V}}(t(s,s')) \leq C < n$. This shows $\alpha_{\mathcal{U}}$ and $\alpha_{\mathcal{V}}$ are different, and hence (automatically for C*-norms) inequivalent. Lastly, it is well known (see e.g. [58, p. 146]) that the cardinality of the set of nontrivial ultrafilters on $\mathbb{N}$ is $2^{2^{\aleph_0}}$. $\qquad \square$

Proof of Theorem 18.21 If M is not nuclear, by Theorem 12.29 there is an embedding $B \subset M$. Moreover, since B is injective, there is a conditional expectation from M to B, which guarantees that, for any A, the max norm on $A \otimes B$ coincides with the restriction of the max norm on $A \otimes M$ (see Corollary 4.18 or Proposition 7.19). Thus we can extend $\alpha_{\mathcal{U}}$ to a C*-norm $\tilde{\alpha}_{\mathcal{U}}$ on $\overline{B} \otimes M$ by setting

$$\forall t \in \overline{B} \otimes M \qquad \tilde{\alpha}_{\mathcal{U}}(t) = \max\{\|t\|_{\overline{B} \otimes_{\min} M}, \|[\overline{\pi_{\mathcal{U}}} \otimes Id](t)\|_{\overline{M_{\mathcal{U}}} \otimes_{\max} M}\}.$$

Since $\tilde{\alpha}_\mathcal{U} = \alpha_\mathcal{U}$ on $\overline{B} \otimes B$, this gives us a family of distinct C^*-norms on $\overline{B} \otimes M$. Since $\overline{B} \simeq B$, we can replace $\overline{B}$ by B if we wish. $\qquad\square$

Remark 18.25 It is easy to see that Theorem 18.21 remains valid for any choice of the sequence (N_m) and in particular it holds if $N_m = m$ for all m, i.e. for $B = \mathbb{B}$.

18.5 Notes and remarks

The main sources for §18.1 are [141] with the simplifications brought by [209]. The first proof in [141] that $(\mathscr{B}, \mathscr{B})$ is not nuclear was more indirect. It went through first proving that if $\mathscr{B}$ had the LLP then the set OS_n of n-dimensional ($n \geq 3$) operator spaces would be separable for the metric d_{cb} (see §20), and (after some more topological considerations) that would contradict a certain operator space version of Grothendieck's theorem. A second proof was proposed in a revision of the same paper [141]. The latter proof used property (T) groups to show that $C(n) < n$ for $n \geq 3$, and deduced from that the nonseparability of (OS_n, d_{cb}) (see §20 for more on this). Later on, A. Valette observed that the Lubotzky–Philips–Sarnak results were exactly what was needed to prove that $C(n) \leq 2\sqrt{n-1}$ for any $n = p + 1$ with p prime. Finally, the paper [120] proved using random matrices that this bound remains valid for any $n \geq 2$, and hence (by the easy lower bound in [206]) that $C(n) = 2\sqrt{n-1}$ for all $n \geq 2$. §18.4 comes from [193].

19

Other proofs that $C(n) < n$: quantum expanders

In what precedes, we used random matrices to show that $C(n) < n$. In this section, we will describe a different way to prove the latter fact using more explicit examples based on the theory of "expanders" or "expanding graphs." Actually, we only use graphs that are Cayley graphs of finite groups. We will see that the more recent notion of "quantum expander" is particularly well adapted for our purposes. We should warn the reader that there is also a notion of quantum graph that seems to have little to do with quantum expanders.

19.1 Quantum coding sequences. Expanders. Spectral gap

To prove that $C(n) < n$ we must produce a sequence of n-tuples $(u(m))_{m \geq 1}$ of unitary matrices of the same size N_m such that (18.6) holds for some $C < n$. Following (and modifying) the terminology from [253] *we call such a sequence a "quantum coding sequence"* of degree n. See §19.4 for an explanation of our terminology.

To introduce expanders we first recall some notation.

Let $\pi : G \to B(H_\pi)$ be a unitary representation. Let

$$H_\pi^{\mathrm{inv}} = \{\xi \in H_\pi \mid \pi(t)\xi = \xi \quad \forall t \in G\}$$

be the set of π-invariant vectors. Let $S \subset G$ be a finite subset generating G (i.e. G is the smallest subgroup containing S). We denote

$$\varepsilon(\pi, S) = 1 - |S|^{-1} \left\| \sum\nolimits_{s \in S} \pi(s) P_{H_\pi^{\mathrm{inv}\perp}} \right\|. \tag{19.1}$$

Definition 19.1 Let (G_m, S_m) be a sequence of finite groups with generating sets S_m such that $|S_m| = n$ for all $m \geq 1$. The sequence of associated Cayley graphs is called "an expander" or an "expanding family" (the terminology is not so well established) if $|G_m| \to \infty$ and

$$\inf_{m \geq 1} \varepsilon(G_m, S_m) > 0.$$

The notion of "expanding graph" has had a major impact far beyond graph theory. For our purposes, we will only discuss Cayley graphs of groups. Given a group G with a finite symmetric set of generators $S \subset G$, the associated Cayley graph is defined as having G as its vertex set and having as edges the pairs (x, y) in G^2 such that $y^{-1}x \in S$.

For a general group G the spectral gap $\varepsilon(G, S)$ can be defined by setting

$$\varepsilon(G, S) = \inf \varepsilon(\pi, S) \tag{19.2}$$

with the infimum running over all unitary representations π of G. Equivalently, this is the spectral gap of the universal representation. As anticipated in Proposition 17.3, $\varepsilon(G, S) > 0$ characterizes property (T) (see §19.3), and we will see in §19.3 that certain property (T) groups lead to expanding families of graphs.

For the moment, let us assume that G is finite and that $|S| = n$. In that case we will show that we may restrict consideration to $\pi = \lambda_G$ and we have

$$1 - \varepsilon(G, S) = \left\| n^{-1} \sum_{s \in S} \lambda_G(s)_{|1^\perp} \right\| \tag{19.3}$$

where $\mathbf{1}$ denotes the constant function on G (i.e. the element $\xi \in \ell_2(G)$ such that $\xi(t) = 1$ for all $t \in G$). The latter is an eigenvector for the eigenvalue 1 for the so-called Markov operator $n^{-1} \sum_{s \in S} \lambda_G(s)$. The number $\varepsilon(G, S)$ measures the gap between that extreme eigenvalue and the rest of the spectrum.

Since the only vectors invariant under λ_G are those in $\mathbb{C}\mathbf{1}$, we have (using the notation (19.1))

$$\varepsilon(G, S) = \varepsilon(\lambda_G, S).$$

Indeed, as is well known, when $|G| < \infty$, λ_G decomposes as a direct sum of a family formed of all the irreducible representations (each with the same multiplicity as its dimension). Let $\widehat{G}$ denote the set of irreducible representations on G and let T be the trivial representation on G. As usual we identify two representations if they are unitarily equivalent. With this notation we may write

$$\lambda_G \simeq \bigoplus_{\pi \in \widehat{G}} \pi$$

and also

$$\lambda_{G|1^\perp} \simeq \bigoplus_{\pi \in \widehat{G} \setminus \{T\}} \pi. \tag{19.4}$$

Therefore

$$1 - \varepsilon(G, S) = \sup_{\pi \in \widehat{G} \backslash \{T\}} |S|^{-1} \left\| \sum_{s \in S} \pi(s) \right\|. \tag{19.5}$$

Remark 19.2 For any finite group G we have

$$\varepsilon(G, S) = \inf\{\varepsilon(\pi, S)\}, \tag{19.6}$$

where the infimum runs over all unitary representations π of G without invariant vectors.

Indeed, by decomposing π into irreducibles the infimum remains unchanged if we restrict it to $\pi \in \widehat{G} \setminus \{T\}$. Thus (19.6) follows from (19.5).

Remark 19.3 By the Schur Lemma A.69 and (19.6), for any pair $\pi \not\simeq \sigma \in \widehat{G}$ we have

$$|S|^{-1} \left\| \sum_{s \in S} \overline{\pi(s)} \otimes \sigma(s) \right\| \le 1 - \varepsilon(G, S). \tag{19.7}$$

Proposition 19.4 *Let* $\{\pi \mid \pi \in \mathcal{T}\}$ *be a finite set of distinct irreducible representations of G on a common N-dimensional Hilbert space. Let* $\varepsilon = \varepsilon(G, S)$ *and* $n = |S|$. *Then* $|\mathcal{T}| \le (1 + 2/\sqrt{\varepsilon})^{2nN^2}$.

Proof By (19.7), for any $\pi \ne \sigma \in \mathcal{T}$ we have $|\sum_{s \in S} \text{tr}(\pi(s)^* \sigma(s))| \le n(1 - \varepsilon)$, and hence

$$\forall \pi \ne \sigma \in \mathcal{T} \quad \left(n^{-1} \sum_{s \in S} \text{tr} \, |\pi(s) - \sigma(s)|^2 \right)^{1/2} \ge \sqrt{2\varepsilon}.$$

Thus we have a set of $|\mathcal{T}|$ unit vectors in a Hilbert space of dimension nN^2 (and hence Euclidean dimension $2nN^2$) that are mutually at distance $\ge \sqrt{2\varepsilon}$. The proposition follows by a well-known volume argument: the open balls centered at these points with radius $r = \sqrt{2\varepsilon}/2$ being disjoint, the volume of their union is $= |\mathcal{T}| r^{2nN^2} \text{vol}(B)$ where B is the Euclidean ball of dimension $2nN^2$, and the latter union being included in a ball of radius $1 + r$ has volume at most $(1 + r)^{2nN^2} \text{vol}(B)$. This implies $|\mathcal{T}| \le (1 + 1/r)^{2nN^2}$. $\quad\square$

Corollary 19.5 *Let* $(G_m, S_m)_{m \ge 1}$ *be an expanding family with* $|S_m| = n$ *for all* $m \ge 1$. *Let* N_m *be the largest dimension d_π of a representation $\pi \in \widehat{G_m}$. Then* $N_m \to \infty$.

Proof Fix a number N. By the preceding proposition, $\sum_{\pi \in \widehat{G_m}, d_\pi \le N} d_\pi^2$ remains bounded. Since $\sum_{\pi \in \widehat{G_m}} d_\pi^2 = |G_m| \to \infty$, when m is large enough we must have $d_\pi > N$ for some $\pi \in \widehat{G_m}$. In other words $N_m > N$ when m is large enough. $\quad\square$

19.2 Quantum expanders

Let H, K be Hilbert spaces. Let $S_2(K, H)$ denote the Hilbert space of Hilbert–Schmidt operators $x: K \to H$ equipped with the norm $\|x\|_{S_2} = (\operatorname{tr}(x^*x))^{1/2} = (\operatorname{tr}(xx^*))^{1/2}$. Let $u = (u_j)_{1 \le j \le n} \in B(H)^n$ and $v = (v_j)_{1 \le j \le n} \in B(K)^n$. We denote by $T_{u,v}: S_2(K, H) \to S_2(K, H)$ the mapping defined by

$$T_{u,v}(x) = n^{-1} \sum\nolimits_1^n u_j^* x v_j.$$

Since $\overline{K} \simeq K^*$ (canonically), we may identify $\overline{K} \otimes_2 H$ with $S_2(K, H)$. With this identification, $T_{u,v}$ corresponds to $n^{-1} \sum_1^n \overline{v}_j^* \otimes u_j^*$ which has the same norm as $n^{-1} \sum_1^n \overline{u}_j \otimes v_j$ (see Proposition 2.11) so that

$$\|T_{u,v}\| = n^{-1} \left\| \sum\nolimits_1^n \overline{u}_j \otimes v_j \right\|$$

where the last norm is in $\overline{B(K)} \otimes_{\min} B(H)$ or equivalently in $B(\overline{K} \otimes_2 H)$.

Let us denote by $L_2(\tau_N)$ the space M_N equipped with the norm

$$\|x\|_{L_2(\tau_N)} = (N^{-1} \operatorname{tr}(x^*x))^{1/2}$$

associated to the normalized trace τ_N on M_N (namely $\tau_N(x) = N^{-1} \operatorname{tr}(x)$). Except for the normalization of the trace this is the same as $S_2(K, H)$ when $K = H = \ell_2^N$. Therefore for any $v = (v_j)_{1 \le j \le n} \in M_N^n$ we have

$$\|T_{v,v}\| = n^{-1} \left\| \sum \overline{v}_j \otimes v_j \right\|_{\overline{M_N} \otimes_{\min} M_N}$$

$$= \sup \left\| n^{-1} \sum v_j^* \xi v_j \right\|_{L_2(\tau_N)} \,\Big|\, \|\xi\|_{L_2(\tau_N)} \le 1 \}.$$

Note that for any $\xi, \eta \in B_{L_2(\tau_N)}$ we have

$$\left| n^{-1} \sum \tau_N(v_j^* \xi^* v_j \eta) \right| = \left| \left\langle n^{-1} \sum v_j^* \xi v_j \,,\, \eta \right\rangle_{L_2(\tau_N)} \right| \le \|T_{v,v}\|. \qquad (19.8)$$

Let $T: L_2(\tau_N) \to L_2(\tau_N)$ be a linear map with $\|T\| = 1$ such that $T(I) = I$ and $T^*(I) = I$ where $I \in M_N$ denotes the identity matrix, so that 1 is an eigenvalue (with eigenvector I) of T and $I^\perp$ is invariant under T. The "spectral gap" of T is defined as

$$e(T) = 1 - \|T_{|I^\perp}\|. \qquad (19.9)$$

Let u be an n-tuple in $\mathbb{U}_N$. We then set

$$\varepsilon(u) = e(T_{u,u}). \qquad (19.10)$$

Equivalently $\varepsilon(u)$ is the largest number $\varepsilon \ge 0$ such that for any ξ, η in $B_{L_2(\tau_N)}$ with $\tau_N(\xi) = \tau_N(\eta) = 0$ we have

$$\left| n^{-1} \tau_N \left(\sum\nolimits_1^n u_j^* \xi^* u_j \eta \right) \right| \le 1 - \varepsilon. \qquad (19.11)$$

Lemma 19.6 *Let* $u = (u_j)_{1 \leq j \leq n} \in \mathbb{U}_N^n$. *Then for any* $k \leq N$ *and any* $v = (v_j)_{1 \leq j \leq n} \in M_k^n$ *we have*

$$\|T_{u,v}\| \leq (k/N + 1 - \varepsilon(u))^{1/2} \|T_{v,v}\|^{1/2}. \tag{19.12}$$

Proof For simplicity we replace v_j by $v_j \oplus 0 \in M_N$. Thus we assume that the v_js are $N \times N$ matrices for which there is an orthogonal projection $P \in M_N$ of rank k such that

$$v_j P = P v_j = v_j \text{ for all } j.$$

To prove (19.12) it suffices to show that for any ξ, η in $B_{L_2(\tau_N)}$ we have

$$\left| n^{-1} \sum_1^n \tau_N(u_j^* \xi^* v_j \eta) \right| \leq (k/N + 1 - \varepsilon(u))^{1/2} \|T_{v,v}\|^{1/2}.$$

Since we may replace ξ, η by $P\xi, P\eta$, we may assume ξ, η of rank $\leq k$. Let $\xi = U|\xi|$ and $\eta = V|\eta|$ be the polar decomposition. Using the identity

$$\tau_N(u_j^* \xi^* v_j \eta) = \tau_N \left((|\eta|^{1/2} u_j^* |\xi|^{1/2})(|\xi|^{1/2} U^* v_j V |\eta|^{1/2}) \right)$$

and Cauchy–Schwarz we find by (19.8)

$$\left| n^{-1} \sum_1^n \tau_N(u_j^* \xi^* v_j \eta) \right| \leq \left| n^{-1} \sum_1^n \tau_N(|\eta| u_j^* |\xi| u_j) \right|^{1/2} \|T_{v,v}\|^{1/2}. \tag{19.13}$$

Now let $\xi' = |\xi| - \tau_N(|\xi|)$ and $\eta' = |\eta| - \tau_N(|\eta|)$ so that $\xi', \eta' \in B_{L_2(\tau_N)}$ but now $\tau_N(\xi') = \tau_N(\eta') = 0$. We have

$$n^{-1} \sum_1^n \tau_N(|\eta| u_j^* |\xi| u_j) \leq I + II \tag{19.14}$$

where $I = \tau_N(|\xi|) \tau_N(|\eta|)$ and $II = n^{-1} \sum_1^n \tau_N(\eta' u_j^* \xi' u_j)$. Since $|\xi|$ and $|\eta|$ have rank $\leq k$ (and $\tau(P) = k/N$) we have $\tau_N(|\xi|) \leq \tau_N(|\xi|^2)^{1/2}(k/N)^{1/2}$ and similarly for $|\eta|$. It follows that $|I| \leq k/N$. Now using the definition of the spectral gap $\varepsilon(u)$ in (19.11) we find $|II| \leq 1 - \varepsilon(u)$. Putting these bounds together with (19.13) and (19.14) we obtain (19.12). $\qquad\square$

In analogy with the theory of expanding graphs (or expanders) the following definition was recently introduced:

Definition 19.7 Let $(N_m)_{m \geq 1}$ be a nondecreasing sequence of integers. For each m, consider

$$u(m) = (u_1(m), \ldots, u_n(m)) \in \mathbb{U}_{N_m}^n.$$

The sequence $(u(m))_{m \geq 1}$ is called a quantum expander if $N_m \to \infty$ and $\inf_{m \geq 1} \varepsilon(u(m)) > 0$. We call n the degree of $(u(m))_{m \geq 1}$.

The link between quantum expanders and the constant $C(n)$ goes through the following.

Proposition 19.8 *Let $(u(m))_{m \geq 1}$ be a quantum expander. There is a subsequence of $(u(m))_{m \geq 1}$ that is a quantum coding sequence.*

Proof Assume $\varepsilon(u(m)) \geq \varepsilon > 0$ for all m. Just choose the subsequence $m_1 < m_2 < \cdots$ so that

$$[N_{m_k}/N_{m_{k+1}}] + 1 - \varepsilon \leq 1 - \varepsilon/2$$

for all $k \geq 1$. Then by Lemma 19.6 we obtain (18.6) with $C = (1 - \varepsilon/2)n$, which means $(u(m))_{m \geq 1}$ is a quantum coding sequence. $\square$

Corollary 19.9 *To show that $C(n) < n$, it suffices to know that there is a quantum expander of degree n.*

Remark 19.10 (From expanders to quantum expanders) Let $(G_m, S_m)_{m \geq 1}$ be an expander in the sense of Definition 19.1. Recall $|S_m| = n$ for any $m \geq 1$. Let $\pi_m : G_m \to B(H_m)$ be irreducible representations such that $\dim(H_m) \to \infty$ (see Corollary 19.5). Then, by Schur's lemma A.69, the representation $[\overline{\pi_m} \otimes \pi_m]_{|1^\perp}$ has no nonzero invariant vector. This implies that its decomposition into irreducible components has only nontrivial irreducible representations. By (19.4), the latter are all contained in the restriction of λ_{G_m} to $1^\perp$, and hence

$$n^{-1} \left\| \sum\nolimits_{s \in S_m} [\overline{\pi_m(s)} \otimes \pi_m(s)]_{|1^\perp} \right\| \leq 1 - \varepsilon(G_m, S_m).$$

In other words

$$\varepsilon((\pi_m(s))_{s \in S_m}) \geq \varepsilon(G_m, S_m),$$

so that the sequence of n-tuples $(\pi_m(s))_{s \in S_m}$ $(m \geq 1)$ forms a quantum expander.

19.3 Property (T)

We already introduced groups with property (T) in Definition 17.1. We will now show that the existence of such groups leads to that of expanders, from which that of quantum expanders and quantum coding sequences follows. Compared with the random approach, the advantage of this method is that it produces explicit examples.

Proposition 19.11 *A discrete group G generated by a finite subset $S \subset G$ containing the unit has property (T) if and only if it has a nonzero spectral gap $\varepsilon(G, S)$ (as defined in (19.2)) or equivalently if there is $\varepsilon > 0$ such that for any unitary representation $\pi : G \to B(H_\pi)$ we have*

$$\varepsilon(\pi, S) \geq \varepsilon. \tag{19.15}$$

Proof This is but a restatement of Proposition 17.3. □

Lemma 19.12 *Let G, S be as in Proposition 19.11, with property (T). Let $\varepsilon > 0$ be as in (19.15).*

(i) *For any irreducible unitary representation $\pi : G \to B(H)$ with $\dim(H) < \infty$ we have*

$$\varepsilon((\pi(s))_{s \in S}) \geq \varepsilon. \tag{19.16}$$

(ii) *Let σ be another finite-dimensional irreducible unitary representation that is not unitarily equivalent to π, then*

$$|S|^{-1} \left\| \sum_{s \in S} \overline{\pi(s)} \otimes \sigma(s) \right\| \leq 1 - \varepsilon. \tag{19.17}$$

Proof (i) Let $\rho(t) = \overline{\pi(t)} \otimes \pi(t)$. Observe that $H_\rho^{\mathrm{inv}} = \mathbb{C}I$ by the irreducibility of π (see §A.21). Therefore by (19.15)

$$|S|^{-1} \left\| \sum_{s \in S} \overline{\pi(s)} \otimes \pi(s)_{|I^\perp} \right\| \leq 1 - \varepsilon. \tag{19.18}$$

Equivalently we have (i).

(ii) By Schur's lemma A.69, the representation $\overline{\pi} \otimes \sigma$ does not have any invariant vector $\xi \neq 0$. Thus (19.17) follows from (19.15) applied with $\overline{\pi} \otimes \sigma$ in place of π. □

Proposition 19.13 *Let G be a property (T) group with S as in Proposition 19.11. Let $n = |S|$ and let $\varepsilon_0 = \varepsilon(G, S)$. Assume that G admits a sequence $(\pi_m)_{m \geq 1}$ of finite-dimensional distinct irreducible unitary representations with dimension tending to ∞. Then the sequence*

$$\{(\pi_m(s))_{s \in S} \mid m = 1, 2, \ldots\}$$

is both a quantum expander and a quantum coding sequence and $C(n) \leq (1 - \varepsilon_0)n$.

Proof Let $S = \{t_0, \ldots, t_{n-1}\}$ with $t_0 = 1$. Let $N_m = \dim(H_{\pi_m})$. Let

$$u_j(m) = \pi_m(t_j).$$

For any $m \neq m'$ we have $\pi_m \neq \pi'_m$. By (19.17) this implies

$$n^{-1} \left\| \sum_0^{n-1} \overline{u_j(m)} \otimes u_j(m') \right\| \leq 1 - \varepsilon_0,$$

and hence we have a quantum coding sequence. By (19.16) or (19.18), $(\pi_m(s))_{s \in S}$ is a quantum expander. □

Remark 19.14 (From property (T) to expanders) In the preceding situation, assume that the group G admits a sequence of finite quotient groups G_m with quotient maps denoted by $q_m : G \to G_m$. Let $S_m = q_m(S)$. We assume that $|S_m| = n$. Let $\sigma_m : G_m \to B(H_m)$ be a unitary representation without invariant unit vector (e.g. a nontrivial irreducible one) ($m \geq 1$). Then $\pi_m = \sigma_m q_m$ is a unitary representation without invariant unit vector on G. Moreover, $\|n^{-1} \sum_S \pi_m\| = \|n^{-1} \sum_{S_m} \sigma_m\|$, and hence $\varepsilon(\sigma_m, S_m) \geq \varepsilon(G, S)$. Since this holds for any such σ_m on G_m this implies $\varepsilon(G_m, S_m) \geq \varepsilon(G, S)$. Therefore if $|G_m| \to \infty$ the sequence (G_m, S_m) is an expanding family (i.e. "an expander"). This shows that we can deduce the existence of expanding families from that of a property (T) group with the required properties. We give an example in the next remark. Note that for the present remark we could content ourselves with property (τ) for which we refer the reader to [173].

Remark 19.15 (($SL_3(\mathbb{Z}_p)$) is an expander) By Remarks 19.14 and 19.10, to prove that $C(n) < n$ (or to produce quantum expanders) it suffices to produce a group G with property (T) admitting a sequence of distinct finite-dimensional irreducible unitary representations with unbounded dimensions. The classical example for this phenomenon is $G = SL_3(\mathbb{Z})$ (or $SL_d(\mathbb{Z})$ for $d \geq 3$). For any prime number p let $\mathbb{Z}_p = \mathbb{Z}/p\mathbb{Z}$. Recall this is a field with p elements. The group $SL_3(\mathbb{Z}_p)$ is a finite quotient of G. Indeed, we have a natural homomorphism $q_p : SL_3(\mathbb{Z}) \to SL_3(\mathbb{Z}_p)$ that takes a matrix $[a_{ij}]$ to the matrix $[\dot{a}_{ij}]$ where $\dot{a} \in \mathbb{Z}/p\mathbb{Z}$ denotes the congruence equivalence class of $a \in \mathbb{Z}$ modulo p. It is known that this maps $SL_3(\mathbb{Z})$ onto $SL_3(\mathbb{Z}_p)$. This follows for instance from the well-known fact that, for any field $\mathbf{k}$, $SL_3(\mathbf{k})$ is generated by the set $S_{\mathbf{k}}$ formed of the unit and the matrices with 1 on the diagonal and only one nonzero entry elsewhere equal to 1. When $\mathbf{k} = \mathbb{Z}/p\mathbb{Z}$, it is obvious that such matrices are in the range of the preceding homomorphism. Therefore the latter is onto $SL_3(\mathbb{Z}_p)$. We will use $S = S_{\mathbb{Z}}$ so that $n = |S| = 6$. Clearly $|q_p(S)| = |S| = 6$ if $p > 1$. Thus Remark 19.14 shows that the sequence $(SL_3(\mathbb{Z}_p), q_p(S))$ indexed by prime numbers $p > 1$ is an expanding family (i.e. "an expander").

Actually, by Remark 19.14 for any finite generating set S in $SL_3(\mathbb{Z})$ the sequence $(SL_3(\mathbb{Z}_p), q_p(S))$ indexed by large enough prime numbers $p > 1$ is an expanding family (i.e. "an expander"). Indeed, by taking p large enough we can clearly ensure that q_p is injective on S and hence $|q_p(S)| = |S|$.

The irreducible unitary representations on $SL_3(\mathbb{Z}_p)$ have unbounded dimensions when $p \to \infty$. This can be deduced from the property (T) of $SL_3(\mathbb{Z})$ using the same idea as for Corollary 19.5. More explicitly, consider for instance the action of $SL_3(\mathbb{Z}_p)$ on the set of "lines" $\mathcal{L}(p)$ in $\mathbb{Z}_p^3$, or equivalently the set of 1-dimensional subspaces in the vector space $\mathbb{Z}_p^3$ (over the field $\mathbb{Z}_p$). By standard linear algebra, $SL_3(\mathbb{Z}_p)$ acts transitively on the latter set and this action defines a unitary representation π_p of $SL_3(\mathbb{Z}_p)$ on $\ell_2(\mathcal{L}(p))$ that permutes the canonical basis vectors. Since the action on $\mathcal{L}(p)$ is transitive, the constant functions on $\mathcal{L}(p)$ are the only invariant vectors. Furthermore, again by linear algebra, the action is actually bitransitive, and hence (see Lemma A.68) the restriction $\pi_p^0 = \pi_p|\mathbf{1}^\perp$ to the orthogonal of constant functions is irreducible, and of course its dimension, equal to $|\mathcal{L}(p)| - 1 = p^2 + p$ (see Remark 24.29), tends to ∞ when $p \to \infty$. Thus we conclude by Remark 19.10 that for any finite unital generating set S in $SL_3(\mathbb{Z})$ with $|S| = n$ the sequence of n-tuples $(\pi_p^0(s))_{s \in S}$ indexed by large enough primes $p > 1$ is a quantum expander.

Remark 19.16 Since it is known (see [250]) that $SL_3(\mathbb{Z})$ is generated by a pair of elements (together with the unit), we may take $|S| = 3$. The preceding remark then gives us $C(3) < 3$. Since it can be shown (exercise) that $C(2) = 2$ this is optimal.

19.4 Quantum spherical codes

We would like to motivate the terminology that we adopted to emphasize the analogy with certain questions in coding theory. In [62, ch. 9], Conway and Sloane define a spherical code $\mathcal{SC}$ of dimension n, size k and minimal angle $0 < \theta < \pi/2$ as a set of k points of the unit sphere in $\mathbb{R}^n$ with the property that

$$\forall x, y \in \mathcal{SC}, \ x \neq y \quad x \cdot y \leq \cos \theta. \tag{19.19}$$

They discuss the reasons why it is of interest to find the maximal size $A(n, \theta)$ of such a code, and give estimates for it.

A similar variant of that problem is the search for a maximal set of vectors $\{\xi(m)\}$ in the sphere of radius $n^{1/2}$ in ℓ_2^n with coordinates all unimodular (e.g. equal to ± 1) that are such that $\sup_{m \neq m'} |\langle \xi(m), \xi(m') \rangle| < C$ for some $C < n$.

Such (finite) sequences are useful in coding theory: the family $\{\xi(m)\}$ itself can be thought of as a code. If we know that the message (i.e. a length n sequence of ± 1s) consists of one of the $\xi(m)$s then even if there are erroneous digits (=coordinates) but fewer than $(n-C)/8$ of them, we can recognize which $\xi(m)$ was sent in the message. (We leave the easy verification as an exercise). It is thus quite useful to have a number as large as possible of vectors $\xi(m)$ given n and $C < n$. This becomes all the more useful when C/n is small, typically when $C = \delta n$ with $0 < \delta < n$. For this (classical) problem, of course there can only be *finitely* many such $\xi(m)$s since the sphere is compact, but in the quantum version, the effect of compactness diminishes when the matrix size $N \to \infty$, and the significant problem becomes to produce an infinite sequence such as the ones we call quantum coding sequences. One can also fix the matrix size N and try to estimate (as a function of N, n, C) the maximal number of n-tuples of $N \times N$-unitaries that are separated as in (18.6) for some fixed $C < n$. Now the main point is the asymptotic behavior of the latter number when $N \to \infty$ as in the forthcoming Theorem 19.17.

To illustrate better the analogy that we wish to emphasize, we revise our notation as follows.

For any $x = (x_1, \dots, x_n)$ and $y = (y_1, \dots, y_n)$ in M_N^n we set $\bar{x} = (\overline{x_1}, \dots, \overline{x_n}) \in \overline{M_N}^n$ and denote

$$x \cdot y = \sum_1^n x_j \otimes y_j \in M_N \otimes M_N \quad \bar{x} \cdot y = \sum_1^n \overline{x_j} \otimes y_j \in \overline{M_N} \otimes M_N.$$

We view the set of x's such that $\|\bar{x} \cdot x\|_{\min} \le 1$ as a quantum analogue of the Euclidean unit sphere.

Note that if $x = (x_j) \in n^{-1/2} \mathbb{U}_N^n$ then $\|\bar{x} \cdot x\|_{\min} = 1$.

When $\|\bar{x} \cdot x\|_{\min} \le 1$ and $\|\bar{y} \cdot y\|_{\min} \le 1$ we say that x, y are δ-separated if $\|\sum_1^n \overline{x_j} \otimes y_j\|_{\min} \le 1 - \delta$. This is analogous to the preceding separation condition (19.19) for spherical codes with $\cos\theta = 1 - \delta$.

The next result is a matricial analogue of some of the known estimates for spherical codes. We interpret it as an upper estimate of the size of quantum spherical codes of dimension n, angle $\theta = \arccos(1 - \delta)$ and (this is the novel parameter) matrix size N. In addition, it gives us a rather large number of δ-separated quantum expanders.

Theorem 19.17 ([212]) *There are absolute constants $\beta > 0$ and $\delta > 0$ such that for each $0 < \varepsilon < 1$ and for all sufficiently large integers n and N, more precisely such that $n \ge n_0$ and $N \ge N_0$ with n_0 depending on ε, and N_0 depending on n and ε, there is a subset $\mathcal{T} \subset M_N^n$ with cardinal*

$$|\mathcal{T}| \ge \exp \beta n N^2,$$

such that

$$\forall x \in \mathcal{T} \quad (n^{1/2} x_j) \in \mathbb{U}_N^n \quad and \quad \varepsilon(x) \ge 1 - (2\sqrt{n-1}/n + \varepsilon)$$

$$\forall x \ne y \in \mathcal{T} \quad \|\bar{x} \cdot y\|_{\min} \le (1 - \delta).$$

Note that $2\sqrt{n-1} < n$ for all $n \ge 3$ and hence for $0 < \varepsilon < 1$ small enough $1 - (2\sqrt{n-1}/n + \varepsilon) > 0$.

Remark 19.18 As for lower estimates, by the same volume argument as for Proposition 19.4, for any set $\mathcal{T}$ with the property in Theorem 19.17 we have $|\mathcal{T}| \le (1 + 2/\sqrt{\delta})^{2nN^2} \le \exp(4nN^2/\sqrt{\delta})$.

We refer to [212] for the proof, which makes crucial use of the following result of Hastings [132]:

Theorem 19.19 (Hastings) *If we equip $\mathbb{U}_N^n$ with its normalized Haar measure $\mathbb{P}_{N,n}$, then for each n and $\varepsilon > 0$ we have*

$$\lim_{N\to\infty} \mathbb{P}_{N,n}(\{u \in \mathbb{U}_N^n \mid 1 - \varepsilon(u) \leq 2\sqrt{n-1}/n + \varepsilon\}) = 1.$$

This is best possible in the sense that Theorem 19.19 fails if $2\sqrt{n-1}$ is replaced by any smaller number. The proof of Theorem 19.19 in [132] is rather delicate; however a simpler proof based on a comparison with Gaussian random matrices is given in [212] with $2\sqrt{n-1}$ replaced by $c\sqrt{n-1}$ where c is a numerical constant.

19.5 Notes and remarks

The first explicit expanding graphs were discovered by Margulis around 1973. See [227] for a very concise introduction to the subject. The main reference for expanders and Kazhdan's property (T) is Lubotzky's book [172]. See Lubotzky's survey [174] for a more recent update. The main interest for us is the notion of spectral gap. We return to that theme with some more references in §24.4. Proposition 19.4 and Corollary 19.5 originate in S. Wassermann's [259]. The term "coding sequences" was coined by Voiculescu in [253] for the sequences that we choose to call quantum coding sequences to emphasize their noncommutative nature. While they suffice for our main goal to prove that $C(n) < n$, the notion of quantum expanders turns out to be more convenient because of the analogy with the usual expanders. Quantum expanders were introduced independently by Hastings [132] – a mathematical physicist – and by two computer scientists Ben-Aroya and Ta-Shma [18]. See [19] for more on the connection with computer science. Hastings [132] proved the crucial bound appearing in Theorem 19.19 to exhibit quantum expanders using random unitaries. See also [212]. Quantum expander theory was further developed by Aram Harrow (see [130, 131]) who in particular observed the content of Remark 19.10.

20

Local embeddability into $\mathscr{C}$ and nonseparability of (OS_n, d_{cb})

In this section, we tackle various issues concerning finite-dimensional operator spaces. While all the spaces of the same dimension are obviously completely isomorphic there is a natural "distance" that measures to what degree they are really close.

When dealing with just Banach spaces E, F one defines classically the Banach–Mazur "distance" $d(E, F)$ as equal to ∞ if E, F are not isomorphic and otherwise as

$$d(E, F) = \inf\{\|u\|\|u^{-1}\| \mid u : E \to F \text{ isomorphism}\}.$$

Given two operator spaces $E \subset B(H)$ and $F \subset B(K)$ the analogous "distance" (called the cb-distance) is defined as

$$d_{cb}(E, F) = \inf\{\|u\|_{cb}\|u^{-1}\|_{cb} \mid u : E \to F \text{ complete isomorphism}\}.$$

If E, F are not completely isomorphic we set $d_{cb}(E, F) = \infty$.

In contrast, we have clearly $d_{cb}(E, F) < \infty$ if $\dim(E) = \dim(F) < \infty$.

This is a "multiplicative distance" meaning that the triangle inequality takes the following form: for any operator spaces E, F, G we have

$$d_{cb}(E, G) \leq d_{cb}(E, F)d_{cb}(F, G).$$

Thus if we wanted to insist to have a bona fide distance we could replace d_{cb} by $\delta_{cb} = \log d_{cb}$ and then we would have the usual triangle inequality for δ_{cb}. Moreover, since the axioms of a distance include Hausdorff separation, it is natural to identify the spaces E and F if $\delta_{cb}(E, F) = 0$ or equivalently if $d_{cb}(E, F) = 1$. If E, F are finite dimensional $d_{cb}(E, F) = 1$ (or $\delta_{cb}(E, F) = 0$) if and only if E, F are *completely isometric*. The last assertion is an easy exercise based on the compactness of the unit ball of $CB(E, F)$.

Let us denote by OS_n the set of all operator spaces, with the convention to identify two spaces when they are completely isometric, and let us equip it with the distance δ_{cb}. Then we obtain a bona fide metric space. Again a simple exercise shows that it is complete. The Banach space analogue (that we could denote by (B_n, δ)) is called the "Banach–Mazur compactum" and as the name indicates it is a compact metric space for each dimension n (for a proof see [208, p. 334]). In sharp contrast (OS_n, δ_{cb}) is *not compact* and actually *not even separable!* The main goal of this chapter is to establish this by exhibiting for each n large enough a continuous family of elements of OS_n that are uniformly separated.

But in practice we will usually not bother to replace d_{cb} by $\delta_{cb} = \log d_{cb}$ and we will state the "distance estimates" in terms of d_{cb} alone. The reader should remember that $d_{cb}(E, F) = 1$ is the "shortest distance" so that $d_{cb}(E, F) > 1 + \varepsilon$ with $\varepsilon > 0$ means E, F are separated.

In analogy with the Banach space case, it is known (see [208, p. 133]) that

$$\forall E, F \in OS_n \quad d_{cb}(E, F) \le n. \tag{20.1}$$

20.1 Perturbations of operator spaces

We include here several simple facts from the Banach space folklore which have been easily transferred to the operator space setting.

We start with a well-known fact (the proof is the same as for ordinary norms of operators).

Lemma 20.1 *Let* $v : X \to Y$ *be a complete isomorphism between operator spaces. Then clearly any map* $w : X \to Y$ *with* $\|v - w\|_{cb} < \|v^{-1}\|_{cb}^{-1}$ *is again a complete isomorphism and if we let* $\Delta = \|v - w\|_{cb} \|v^{-1}\|_{cb}$ *we have*

$$\|w^{-1}\|_{cb} \le \|v^{-1}\|_{cb}(1 - \Delta)^{-1} \quad \text{and} \quad \|w^{-1} - v^{-1}\|_{cb} \le \|v^{-1}\|_{cb}^2 (1 - \Delta)^{-1}.$$

Lemma 20.2 (Perturbation Lemma) *Fix* $0 < \varepsilon < 1$. *Let* X *be an operator space. Consider a biorthogonal system* (x_j, x_j^*) $(j = 1, 2, \ldots, n)$ *with* $x_j \in X$, $x_j^* \in X^*$ *and let* $y_1, \ldots, y_n \in X$ *be such that*

$$\sum \|x_j^*\| \, \|x_j - y_j\| < \varepsilon.$$

Then there is a complete isomorphism $w : X \to X$ *such that* $w(x_j) = y_j$,

$$\|w\|_{cb} \le 1 + \varepsilon \quad \text{and} \quad \|w^{-1}\|_{cb} \le (1 - \varepsilon)^{-1}.$$

In particular, if $E_1 = \text{span}(x_1, \ldots, x_n)$ *and* $E_2 = \text{span}(y_1, \ldots, y_n)$, *we have*

$$d_{cb}(E_1, E_2) \le (1 + \varepsilon)(1 - \varepsilon)^{-1}.$$

Proof Recall (1.3). Let $\xi : X \to X$ be the map defined by setting $\xi(x) = \sum x_j^*(x)(y_j - x_j)$ for all x in X. Then $\|\xi\|_{cb} \le \sum \|x_j^*\| \, \|y_j - x_j\| < \varepsilon$. Let

$w = I + \xi$. Note that $w(x_j) = y_j$ for all $j = 1, 2, \ldots, n$, $\|w\|_{cb} \le 1 + \|\xi\|_{cb} \le 1 + \varepsilon$ and by the preceding lemma we have $\|w^{-1}\|_{cb} \le (1 - \varepsilon)^{-1}$.   □

Corollary 20.3 *Let X be any separable operator space. Then, for any n, the set denoted by $OS_n(X)$ of all the n-dimensional subspaces of X is separable for the "distance" associated to d_{cb}.*

Proof Let $(x_1(m), \ldots, x_n(m))$ be a dense sequence in the set of all linearly independent n-tuples of elements of X. Let $E_m = \mathrm{span}(x_1(m), \ldots, x_n(m))$. Then, by the preceding lemma, for any $\varepsilon > 0$ and any n-dimensional subspace $E \subset X$, there is an m such that $d_{cb}(E, E_m) \le 1 + \varepsilon$.   □

Lemma 20.4 *Consider an operator space E and a family of subspaces $E_i \subset E$ directed by inclusion and such that $\overline{\cup E_i} = E$. Then for any $\varepsilon > 0$ and any finite-dimensional subspace $S \subset E$, there exists i and $\widetilde{S} \subset E_i$ such that $d_{cb}(S, \widetilde{S}) < 1 + \varepsilon$. Let $u : F_1 \to F_2$ be a linear map between two operator spaces. Assume that u admits the following factorization $F_1 \xrightarrow{a} E \xrightarrow{b} F_2$ with c.b. maps a, b such that a is of finite rank. Then for each $\varepsilon > 0$ there exists i and a factorization $F_1 \xrightarrow{\widetilde{a}} E_i \xrightarrow{\widetilde{b}} F_2$ of u with $\|\widetilde{a}\|_{cb} \|\widetilde{b}\|_{cb} < (1 + \varepsilon) \|a\|_{cb} \|b\|_{cb}$, and $\widetilde{a}$ of finite rank.*

Proof For the first part let $x_1, \ldots, x_n$ be a linear basis of S and let x_j^* be the dual basis extended (by Hahn–Banach) to elements of E^*. Fix $\varepsilon' > 0$. Choose i large enough and $y_1, \ldots, y_n \in E_i$ such that $\sum \|x_j^*\| \|x_j - y_j\| < \varepsilon'$. Let $\widetilde{S} = \mathrm{span}(y_1, \ldots, y_n)$. Then, by the preceding lemma, there is a complete isomorphism $w : E \to E$ with $\|w\|_{cb} \|w^{-1}\|_{cb} < (1 + \varepsilon')(1 - \varepsilon')^{-1}$ such that $w(S) = \widetilde{S} \subset E_i$. In particular, $d_{cb}(S, \widetilde{S}) \le (1 + \varepsilon')(1 - \varepsilon')^{-1}$ so it suffices to adjust ε' to obtain the first assertion.

Now consider a factorization $F_1 \xrightarrow{a} E \xrightarrow{b} F_2$ and let $S = a(F_1)$. Note that S is finite dimensional by assumption. Applying the preceding to this S, we find i and a complete isomorphism $w : E \to E$ with $\|w\|_{cb} \|w^{-1}\|_{cb} < 1 + \varepsilon$ such that $w(S) \subset E_i$. Thus, if we take $\widetilde{a} = wa : F_1 \to E_i$ and $\widetilde{b} = bw^{-1}_{|E_i}$, we obtain the announced factorization.   □

20.2 Finite-dimensional subspaces of $\mathscr{C}$

The results of this section are derived from [141].

For any operator space $\mathcal{X}$ (actually $\mathcal{X}$ will often be a C^*-algebra), and any finite-dimensional operator space E, we introduce

$$d_{S\mathcal{X}}(E) = \inf\{d_{cb}(E, F) \mid F \subset \mathcal{X}\}. \tag{20.2}$$

Of course if $\mathcal{X} = B(H)$ with $\dim(H) = \infty$, $d_{S\mathcal{X}}(E) = 1$ for all E.

We will concentrate on the special case when $\mathcal{X} = \mathscr{C} = C^*(\mathbb{F}_\infty)$ and to simplify the notation we set

$$d_f(E) = d_{SC^*(\mathbb{F}_\infty)}(E). \tag{20.3}$$

Theorem 20.5 *Let $c \geq 0$ be a constant and let $X \subset B(\mathcal{H})$ be an operator space. The following are equivalent.*

(i) *$d_f(E) \leq c$ for all finite-dimensional subspaces $E \subset X$.*

(ii) *For any finite-dimensional subspace $E \subset X$ and any $\varepsilon > 0$ the inclusion $E \subset B(\mathcal{H})$ admits a factorization through $\mathscr{C}$ of the form*
   $E \xrightarrow{v^E} \mathscr{C} \xrightarrow{w^E} B(\mathcal{H})$ *with $\|v^E\|_{cb}\|w^E\|_{cb} \leq c + \varepsilon$.*

(iii) *For any H and any operator space $F \subset B(H)$, we have*

$$\|t\|_{B(\mathcal{H}) \otimes_{\max} B(H)} \leq c\|t\|_{\min}. \qquad \forall t \in X \otimes F$$

(iv) *Same as (iii) with $H = \ell_2$ and $F = B(\ell_2)$.*

(v) *Any mapping $u : X \to A/I$ into a quotient C^*-algebra that factorizes through $B(K)$ (for some K) as $X \xrightarrow{v} B(K) \xrightarrow{w} A/I$ with $\|v\|_{cb} \leq 1$ and $\|w\|_{dec} \leq 1$ is locally c-liftable.*

Proof Assume (i). Let $u : X \to B(\mathcal{H})$ be the inclusion map. Let $E \subset X$ be a finite-dimensional subspace. Let $\varepsilon > 0$. Since $d_f(E) \leq c$ there is $\widehat{E} \subset \mathscr{C}$ such that $d_{cb}(E, \widehat{E}) < c + \varepsilon$. By the extension property of $B(\mathcal{H})$ there is a factorization of $u_{|E} : E \to B(\mathcal{H})$ of the form $u_{|E} : E \xrightarrow{v^E} \mathscr{C} \xrightarrow{w^E} B(\mathcal{H})$ with $\|v^E\|_{cb}\|w^E\|_{cb} \leq c + \varepsilon$. Thus (ii) holds. (ii) $\Rightarrow$ (iii) follows from Kirchberg's Theorem 9.6, (6.13) and (6.9). (iii) $\Leftrightarrow$ (iv) is essentially trivial.

Assume (iv). Let u be as in (v) of the form $u : X \xrightarrow{v} B(K) \xrightarrow{w} A/I$, i.e. $u = wv$ with $\|v\|_{cb} \leq 1$ and $\|w\|_{dec} \leq 1$. Let $\widetilde{v} : B(\mathcal{H}) \to B(K)$ be such that $\widetilde{v}_{|X} = v$ and $\|\widetilde{v}\|_{cb} = \|v\|_{cb} \leq 1$. Let $T = w\widetilde{v} : B(\mathcal{H}) \to A/I$. Note $T_{|X} = u$. Moreover $\|\widetilde{v}\|_{cb} = \|\widetilde{v}\|_{dec}$ (see (6.9)) and hence $\|T\|_{dec} \leq 1$ (see (6.7)). By (6.13) we have for any $t \in X \otimes \mathscr{B}$

$$\|(u \otimes Id)(t)\|_{(A/I) \otimes_{\max} \mathscr{B}} = \|(T \otimes Id)(t)\|_{(A/I) \otimes_{\max} \mathscr{B}} \leq \|t\|_{B(\mathcal{H}) \otimes_{\max} \mathscr{B}},$$

and by (7.6)

$$\|(u \otimes Id)(t)\|_{(A \otimes_{\min} \mathscr{B})/(I \otimes_{\min} \mathscr{B})} \leq \|(u \otimes Id)(t)\|_{(A \otimes_{\max} \mathscr{B})/(I \otimes_{\max} \mathscr{B})}$$
$$= \|(u \otimes Id)(t)\|_{(A/I) \otimes_{\max} \mathscr{B}},$$

therefore using (iv) we obtain

$$\|(u \otimes Id)(t)\|_{(A \otimes_{\min} \mathscr{B})/(I \otimes_{\min} \mathscr{B})} \leq c\|t\|_{\min}.$$

By (ii) $\Leftrightarrow$ (iii) from Proposition 7.48 u is locally c-liftable. In other words (v) holds.

Assume (v). Assume $B(\mathcal{H}) = C^*(\mathbb{F})/\mathcal{I}$. Then by (v) the inclusion $X \to C^*(\mathbb{F})/\mathcal{I}$ is locally c-liftable. Clearly this implies there is $\widehat{E} \subset C^*(\mathbb{F})$ such that $d_{cb}(E, \widehat{E}) \le c$. We may as well assume $\widehat{E} \subset C^*(\mathbb{F}_\infty)$ (see Lemma 3.8). Thus we obtain (i). $\qquad\qquad\qquad\qquad\qquad\qquad\qquad\qquad\qquad\qquad\quad\square$

Applying the preceding theorem with $X = E$, we obtain:

Corollary 20.6 *Let* $E \subset B(\mathcal{H})$ *be an n-dimensional operator space. Then*

$$d_f(E) = \sup \left\{ \left. \frac{\|u\|_{B(\mathcal{H}) \otimes_{\max} B(\ell_2)}}{\|u\|_{\min}} \right| u \in E \otimes B(\ell_2) \right\}.$$

By Theorem 10.7 this implies:

Corollary 20.7 *For any exact (in particular any nuclear)* C^**-algebra* A *we have*

$$d_f(E) \le d_{SA}(E) \qquad\qquad\qquad (20.4)$$

for any finite-dimensional operator space E.

Remark 20.8 (Completely isometric embeddings in $\mathscr{C}$) If a separable operator space X is such that $d_f(E) = 1$ for every finite-dimensional subspace $E \subset X$ and if in addition X admits a net of completely contractive finite rank maps tending pointwise to the identity (in particular if X itself is finite dimensional), then X embeds completely isometrically into $\mathscr{C}$. Indeed, assume $X \subset B(\mathcal{H})$, and also $B(\mathcal{H}) = C^*(\mathbb{F})/\mathcal{I}$. Then by (v) in Theorem 20.5, the inclusion $X \to C^*(\mathbb{F})/\mathcal{I}$ is locally 1-liftable. By Corollary 9.49 it is 1-liftable, and hence X embeds completely isometrically into $C^*(\mathbb{F})$, or equivalently (since X is separable, see Remark 9.14) into $\mathscr{C}$. In particular, the operator Hilbert space OH from [205] embeds completely isometrically into $\mathscr{C}$. Unfortunately however, we cannot describe the embedding more explicitly.

Remark 20.9 Let A be a nuclear C^*-algebra. By the preceding remark the condition $d_f(A) = 1$ implies that A is completely isometric to a subspace of $\mathscr{C}$. Note however that A need not embed as a C^*-algebra into $\mathscr{C}$. For instance, let A be the Cuntz algebra, being nuclear it is a fortiori exact, so that $d_f(A) = d_{S\mathscr{K}}(A) = 1$, however A does not embed into $\mathscr{C}$, because $\mathscr{C}$ embeds into a direct sum of matrix algebras (see Theorem 9.18), hence left invertible elements in it are right invertible, and the latter property obviously fails in the Cuntz algebra. Nevertheless, by the Choi–Effros lifting Theorem 9.53 there is a unital completely positive (and completely contractive) factorization of the identity of the Cuntz algebra (or any separable nuclear C^*-algebra) through $\mathscr{C}$.

Let us record here an obvious consequence of Corollary 20.6:

Corollary 20.10 *Let $H = \ell_2$. For any $n \geq 1$ we have*

$$\sup \left\{ \left. \frac{\|u\|_{\max}}{\|u\|_{\min}} \right| u \in B(H) \otimes B(H), \ \mathrm{rk}(u) \leq n \right\} = \sup\{d_f(E) \mid \dim(E) \leq n\}.$$

(20.5)

Theorem 20.5 naturally leads us to introduce a new tensor product $E_1 \otimes_M E_2$, both for operator spaces and C^*-algebras, as follows.

Definition 20.11 Let $E_1 \subset B(H_1), E_2 \subset B(H_2)$ be arbitrary operator spaces. We will denote by $\| \ \|_M$ the norm induced on $E_1 \otimes E_2$ by $B(H_1) \otimes_{\max} B(H_2)$, and by $E_1 \otimes_M E_2$ its completion with respect to this norm. Clearly $E_1 \otimes_M E_2$ can be viewed as an operator space embedded into $B(H_1) \otimes_{\max} B(H_2)$.

It can be checked easily, using the extension property of c.b. maps into $B(H)$ (see Theorem 1.18), that $\| \ \|_M$ and $E_1 \otimes_M E_2$ do not depend on the particular choices of complete embeddings $E_1 \subset B(H_1), E_2 \subset B(H_2)$. Indeed, this is an immediate consequence of the following Lemma.

Lemma 20.12 *Let $E_1 \subset B(H_1), E_2 \subset B(H_2), F_1 \subset B(K_1), F_2 \subset B(K_2)$ be operator spaces.*

(i) *Consider c.b. maps $u_1 : E_1 \to F_1$ and $u_2 : E_2 \to F_2$. Then $u_1 \otimes u_2$ defines a c.b. map from $E_1 \otimes_M E_2$ to $F_1 \otimes_M F_2$ with*
$$\|u_1 \otimes u_2\|_{CB(E_1 \otimes_M E_2, F_1 \otimes_M F_2)} \leq \|u_1\|_{cb}\|u_2\|_{cb}.$$
(ii) *If u_1 and u_2 are complete isometries, then*
$$u_1 \otimes u_2 : E_1 \otimes_M E_2 \to F_1 \otimes_M F_2 \text{ also is a complete isometry.}$$

Proof By Theorem 1.18 we may assume that each u_j admits an extension $\tilde{u}_j \in CB(B(H_j), B(K_j))$ with the same cb-norm. By Proposition 6.7 $\|\tilde{u}_j\|_{cb} = \|\tilde{u}_j\|_{dec}$, therefore (i) follows from (6.15) and (6.6). Applying (i) to the inverse mappings, we obtain (ii). $\qquad\square$

Remark When E_1, E_2 are C^*-algebras, then $E_1 \otimes_M E_2$ can be identified with a C^*-subalgebra of $B(H_1) \otimes_{\max} B(H_2)$, so that this tensor product $\otimes_M$ makes sense in both categories, operator spaces and C^*-algebras.

The next result analyzes more closely the significance of $\|u\|_M = 1$ for $u \in E \otimes F$. It turns out to be closely connected to the factorizations of the associated linear operator $U : F^* \to E$ through a subspace of $C^*(\mathbb{F}_\infty)$.

Proposition 20.13 *Let $E \subset B(H)$ and $F \subset B(K)$ be operator spaces, let $t \in E \otimes F$ and let $T : F^* \to E$ be the associated finite rank linear operator.*

Consider a finite-dimensional subspace $S \subset C^*(\mathbb{F}_\infty)$ and a factorization of T of the form $T = ba$ with bounded linear maps $a : F^* \to S$ and $b : S \to E$, where $a : F^* \to S$ is weak* continuous. Then

$$\|t\|_M = \inf\{\|a\|_{cb}\|b\|_{cb}\} \tag{20.6}$$

where the infimum, which is actually attained, runs over all such factorizations of T.

Proof It clearly suffices (recalling Lemma 2.16) to prove (20.6) in the case when E and F are both finite dimensional, so we do assume that. Then since both sides of (20.6) are finite we may assume by homogeneity that $\|t\|_M = 1$. Assume first T factorized for some S as previously. We claim that $\|a\|_{cb}\|b\|_{cb} \geq 1$. Indeed, by Kirchberg's Theorem 9.6, the min and max norms are equal on $C^*(\mathbb{F}_\infty) \otimes B(K)$. Hence, by (ii) in Lemma 20.12, we have isometrically $S \otimes_{\min} F = S \otimes_M F$, so that if $\widehat{a}$ is the element of $S \otimes_{\min} F$ associated to a, we have $\|\widehat{a}\|_M = \|a\|_{cb}$ and $t = (b \otimes Id_F)(\widehat{a})$. Therefore, by (i) in Lemma 20.12 we have $1 = \|t\|_M \leq \|b\|_{cb}\|\widehat{a}\|_M \leq \|a\|_{cb}\|b\|_{cb}$ which proves the claim.

We will now show that equality holds. Let $\mathbb{F}$ be a large enough free group so that $B(H)$ is a quotient of $C^*(\mathbb{F})$ (see Proposition 3.39) and let $q : C^*(\mathbb{F}) \to B(H)$ be the quotient $*$-homomorphism with kernel $\mathcal{I}$. By the exactness of the maximal tensor product (see Proposition 7.15), if we view t as sitting in $B(H) \otimes B(K)$, we have

$$1 = \|t\|_{B(H) \otimes_{\max} B(K)} = \|t\|_{(C^*(\mathbb{F})/\mathcal{I}) \otimes_{\max} B(K)} = \|t\|_{(C^*(\mathbb{F}) \otimes_{\max} B(K))/(\mathcal{I} \otimes_{\max} B(K))}$$

$$\geq \|t\|_{(C^*(\mathbb{F}) \otimes_{\min} B(K))/(\mathcal{I} \otimes_{\min} B(K))}.$$

By Lemma 7.43, since $t \in B(H) \otimes F$, we find

$$1 \geq \|t\|_{(C^*(\mathbb{F}) \otimes_{\min} F)/(\mathcal{I} \otimes_{\min} F)}$$

and by Lemma 7.44 the tensor $t \in B(H) \otimes F$ admits a lifting $\widehat{t}$ in $C^*(\mathbb{F}) \otimes F$ with $\|\widehat{t}\|_{\min} \leq 1$. Let $S \subset C^*(\mathbb{F})$ be a finite-dimensional subspace such that $\widehat{t} \in S \otimes F$. Note that since S is separable, by Remark 3.6 there is a subgroup $G_1 \subset \mathbb{F}$ isomorphic to $\mathbb{F}_\infty$ such that $S \subset C^*(G_1) \simeq C^*(\mathbb{F}_\infty)$. Let $a : F^* \to S$ be the linear map associated to $\widehat{t}$, and let b be the restriction of q to S. Since $\widehat{t}$ lifts t, we have $T = ba$, $\|b\|_{cb} \leq \|q\|_{cb} \leq 1$ and $\|a\|_{cb} = \|\widehat{t}\|_{\min} \leq 1$. Thus we obtain $\|a\|_{cb}\|b\|_{cb} = 1$, which proves (20.6) and at the same time that the infimum is attained. $\qquad\square$

Remark 20.14 The preceding proof can be shortened using Proposition 9.59 applied to the operator $u : F^* \to B(H)$ that is the same as T viewed as acting into $B(H)$. We just need to observe that

$$\|Id_{\mathscr{B}} \otimes u : \mathscr{B} \otimes_{\min} F^* \to \mathscr{B} \otimes_{\max} B(H)\| = \|t\|_M. \quad (20.7)$$

Indeed, for any $t' \in B_{F^* \otimes_{\min} \mathscr{B}}$ with associated linear operator $T' \in B_{CB(F, \mathscr{B})}$ we have

$$(Id_{B(H)} \otimes T')(t) = (u \otimes Id_{\mathscr{B}})(t')$$

and hence

$$\sup_{t' \in B_{F^* \otimes_{\min} \mathscr{B}}} \|(u \otimes Id_{\mathscr{B}})(t')\|_{B(H) \otimes_{\max} \mathscr{B}}$$

$$= \sup_{T' \in B_{CB(F, \mathscr{B})}} \|(Id_{B(H)} \otimes T')(t)\|_{B(H) \otimes_{\max} \mathscr{B}} = \|t\|_M,$$

and since the left-hand side is equal to the one in (20.7) up to a transposition we obtain (20.7).

Applying the last result to $T = Id_E$, we obtain

Corollary 20.15 *Let E be a finite-dimensional operator space. Let $I_E \in E \otimes E^*$ be the tensor associated to the identity on E. Then $\|I_E\|_M = d_f(E)$. In particular we have*

$$d_f(E) = d_f(E^*). \quad (20.8)$$

Corollary 20.16 *For any finite-dimensional operator space E, there is a subspace $\widehat{E} \subset \mathscr{C} = C^*(\mathbb{F}_\infty)$ and an isomorphism $u : E \to \widehat{E}$ such that*

$$\|u\|_{cb} \|u^{-1}\|_{cb} = d_f(E).$$

In particular, E satisfies $d_f(E) = 1$ if and only if E is completely isometric to a subspace of $\mathscr{C}$.

20.3 Nonseparability of the metric space OS_n of n-dimensional operator spaces

Using the number $C(n)$ it was proved in [141] that (OS_n, δ_{cb}) is nonseparable for any $n \geq 3$ (the case $n = 2$ remains open). See [208, ch. 21] for a detailed proof. More precisely, there is a continuous family $(E_t)_{t \in [0,1]}$ in OS_n and a constant $c_n > 1$ such that $d_{cb}(E_s, E_t) \geq c_n$ for any $s \neq t \in [0,1]$. When n is large the method in [141] based on $C(n)$ gives this with $c_n \approx \sqrt{n}$. The variant $C_u(n)$ introduced next will lead us to the order of growth $c_n \approx n$ when $n \to \infty$, which is the optimal one by (20.1).

We denote by $C_u(n)$ the infimum of the numbers $C > 0$ for which there is a sequence of sizes $(N_m)_{m \geq 1}$ and a sequence $u(m) = (u_j(m))_{1 \leq j \leq n}$ of n-tuples in $\mathbb{U}_{N_m}^n$ such that for any unitary matrix $a \in \mathbb{U}_n$ we have

$$\sup_{m \neq m'} \left\| \sum_{i,j=1}^{n} a_{ij} \overline{u_i(m')} \otimes u_j(m) \right\| \leq C. \tag{20.9}$$

Clearly $C(n) \leq C_u(n)$ and $C_u(n) \leq n$ by (2.3).

Theorem 20.17 *For any $n \geq 5$ there is a continuous family $(E_t)_{t \in [0,1]}$ in OS_n such that*

$$\forall \varepsilon > 0 \; \forall s \neq t \quad d_{cb}(E_s, E_t) \geq (n/C_u(n))^2 - \varepsilon \geq n/4 - \varepsilon. \tag{20.10}$$

Corollary 20.18 *The metric space (OS_n, δ_{cb}) is not separable for any $n \geq 5$.*

To prove Theorem 20.17 we will need several lemmas.

Lemma 20.19 *Let $U_1^{(N)}, \ldots, U_n^{(N)}$ be a sequence of independent random unitary matrices uniformly distributed over $\mathbb{U}_N$ as in Theorem 18.16. Then for any k and any $x_1, \ldots, x_n \in \mathbb{U}_k$ we have almost surely*

$$\lim_{N \to \infty} \sup_{a \in \mathbb{U}_n} \left\| \sum_{i,j=1}^{n} a_{ij} \overline{x_i} \otimes U_j^{(N)} \right\|_{\min} \leq 2\sqrt{n}. \tag{20.11}$$

Proof Let us first observe that for any $a, b \in \mathbb{U}_n$ we have for any k, k' and any $x_i \in B_{M_k}, \; y_j \in B_{M_{k'}}$

$$\left\| \sum_{i,j=1}^{n} (a_{ij} - b_{ij}) \overline{x_i} \otimes y_j \right\|_{\min} \leq \sum_{i,j=1}^{n} |a_{ij} - b_{ij}|. \tag{20.12}$$

By compactness there is a finite ε-net $\mathcal{N}_\varepsilon \subset \mathbb{U}_n$ for the distance appearing on the right-hand side of (20.12). By the triangle inequality it follows that for any k, k' and any $x_i \in B_{M_k}, \; y_j \in B_{M_{k'}}$

$$\sup_{a \in \mathbb{U}_n} \left\| \sum_{i,j=1}^{n} a_{ij} \overline{x_i} \otimes y_j \right\|_{\min} \leq \sup_{a \in \mathcal{N}_\varepsilon} \left\| \sum_{i,j=1}^{n} a_{ij} \overline{x_i} \otimes y_j \right\|_{\min} + \varepsilon. \tag{20.13}$$

By (18.15) and Corollary 3.38 for each fixed $a \in \mathcal{N}_\varepsilon$ we have a.s.

$$\lim_{N \to \infty} \left\| \sum_{i,j=1}^{n} a_{ij} \overline{x_i} \otimes U_j^{(N)} \right\|_{\min} \leq 2\sqrt{n}$$

and hence also (since $\mathcal{N}_\varepsilon$ is finite)

$$\lim_{N \to \infty} \sup_{a \in \mathcal{N}_\varepsilon} \left\| \sum_{i,j=1}^{n} a_{ij} \overline{x_i} \otimes U_j^{(N)} \right\|_{\min} \leq 2\sqrt{n}.$$

By (20.13) we obtain (20.11) since $\varepsilon > 0$ is arbitrary. $\square$

Lemma 20.20 *For any $n \geq 5$ the obvious bound $C_u(n) \leq n$ can be improved to*

$$C_u(n) \leq 2\sqrt{n}.$$

Sketch Since the argument is very similar to the one we gave in §18.2 we will be brief. Fix $\varepsilon > 0$. Obviously it suffices to construct a sequence of n-tuples $\{(u_j(m))_{1 \leq j \leq n} \mid m \geq 1\}$ of unitary matrices (the mth one being of size $N_m \times N_m$) such that, for any integer $p \geq 1$, we have

$$\sup_{a \in \mathbb{U}_n} \sup_{1 \leq m \neq m' \leq p} \left\| \sum a_{ij} \overline{u_i(m)} \otimes u_j(m') \right\|_{\min} < 2\sqrt{n} + \varepsilon. \qquad (20.14)$$

We will construct this sequence and the sizes N_m by induction on p. Assume that we already know the result up to p. That is, we already know a family $\{(u_j(m))_{1 \leq j \leq n} \mid 1 \leq m \leq p\}$ formed of p n-tuples satisfying (20.14). We need to produce an additional n-tuple $(u_j(p+1))_{1 \leq j \leq n}$ of unitary matrices (a priori of some larger size $N_{p+1} \times N_{p+1}$) such that (20.14) still holds for the enlarged family $\{(u_j(m))_{1 \leq j \leq m} \mid 1 \leq m \leq p+1\}$ formed of one more n-tuple. By (20.11) such a choice is possible by simply choosing N_{p+1} large enough. $\qquad \square$

The next lemma is very useful to compute $d_{cb}(E, F)$. See [208, ch. 10] for illustrations of this. It is proved by a rather simple averaging argument.

Lemma 20.21 *Let E, F be n-dimensional operator spaces. Let (x_j) (resp. (y_j)) be a linear basis of E (resp. F). For any $n \times n$ matrix a we denote by $\pi_e(a): E \to E$ (resp. $\pi_f(a): F \to F$) the linear map associated as usual to a, that is defined by $\pi_e(a)(x_j) = \sum_i a_{ij} x_i$ (resp. $\pi_f(a)(y_j) = \sum_i a_{ij} y_i$). Let $u_0: E \to F$ be the linear map defined by $u_0(x_j) = y_j$ for all $1 \leq j \leq n$. We assume that there is a constant $c > 0$ such that for any $a, b \in \mathbb{U}_n$ and any $u \in CB(E, F)$ we have*

$$\|\pi_f(b) u \pi_e(a)\|_{cb} \leq c \|u\|_{cb}.$$

Then

$$c^{-2} \|u_0\|_{cb} \|u_0^{-1}\|_{cb} \leq d_{cb}(E, F) \leq \|u_0\|_{cb} \|u_0^{-1}\|_{cb}. \qquad (20.15)$$

Proof It suffices to prove the first inequality. Let $u: E \to F$ be an isomorphism and let $x \in M_n$ be the matrix representing u with respect to the given bases. By the polar decomposition and classical linear algebra we can write $x = a_0 b_0 D b_0^{-1}$ where D is a diagonal matrix with positive coefficients and $a_0, b_0 \in \mathbb{U}_n$. By abuse of notation we view D as a linear mapping from E to F. Then $u = \pi_f(a_0 b_0) D \pi_e(b_0^{-1})$. Let

$$v = \int \pi_f(a a_0^{-1}) u \pi_e(a^{-1}) dm(a) \qquad (20.16)$$

where m is normalized Haar measure on $\mathbb{U}_n$. By translation invariance of the integral we have $v = \int \pi_f(a) D \pi_e(a)^{-1} dm(a)$. In other words $v: E \to F$ is the linear map associated to the matrix $\int a D a^{-1} dm(a) = n^{-1}\mathrm{tr}(D)I$, which means that $v = n^{-1}\mathrm{tr}(D)u_0$. From (20.16) and Jensen's inequality we get $\|v\|_{cb} \leq c\|u\|_{cb}$ and hence since $\mathrm{tr}D = \mathrm{tr}|x|$ we find

$$n^{-1}(\mathrm{tr}|x|)\|u_0\|_{cb} \leq c\|u\|_{cb}.$$

But obviously x^{-1} is the representing matrix for u^{-1} and if we apply the same argument to u^{-1} we obtain

$$n^{-1}(\mathrm{tr}|x^{-1}|)\|u_0^{-1}\|_{cb} \leq c\|u^{-1}\|_{cb},$$

and taking the product of the last two inequalities we find

$$n^{-2}(\mathrm{tr}|x|\mathrm{tr}|x^{-1}|)\|u_0\|_{cb}\|u_0^{-1}\|_{cb} \leq c^2\|u\|_{cb}\|u^{-1}\|_{cb}. \qquad (20.17)$$

Now a simple verification shows that since $x^{-1} = b_0 D^{-1} b_0^{-1} a_0^{-1}$ we have $|x^{-1}| = (x^{-1*}x^{-1})^{1/2} = a_0 b_0 D^{-1}(a_0 b_0)^{-1}$ and hence $\mathrm{tr}|x^{-1}| = \mathrm{tr}(D^{-1})$. Then if we denote by $D_1, \ldots, D_n$ the diagonal coefficients of D we have by Cauchy–Schwarz

$$n = \sum_k D_k^{1/2} D_k^{-1/2} \leq (\mathrm{tr}D\mathrm{tr}(D^{-1}))^{1/2}$$

and hence

$$n^2 \leq \mathrm{tr}D\mathrm{tr}(D^{-1}).$$

Thus (20.17) implies (20.15).   □

Remark 20.22 Given a C^*-algebra $B \subset B(H)$ and an n-dimensional operator space $E \subset B$ with a specific basis $(x_j)_{1\leq j\leq n}$ we wish to define a unitarily invariant operator space denoted by $\widehat{E}$, still n-dimensional, with a basis $(\widehat{x}_j)_{1\leq j\leq n}$ such that for any k and any $y_j \in M_k$ we have

$$\left\|\sum_i y_i \otimes \widehat{x}_i\right\|_{\min} = \sup_{a\in\mathbb{U}_n}\left\|\sum_{ij} a_{ji}y_i \otimes x_j\right\|_{\min}. \qquad (20.18)$$

To do this we consider the C^*-algebra $C = C(\mathbb{U}_n; B)$ of all continuous functions on $\mathbb{U}_n$ with values in B and we define $\widehat{x}_i \in C$ by

$$\forall 1 \leq i \leq n \quad \widehat{x}_i(a) = \sum_j a_{ji}x_j.$$

Then (20.18) clearly holds. The space $\widehat{E}$ is unitarily invariant in the following sense: let $\pi_{\widehat{e}}(a) : \widehat{E} \to \widehat{E}$ be the linear map defined for any $a \in \mathbb{U}_n$ by $\pi_{\widehat{e}}(a)(\widehat{x}_j) = \sum_i a_{ij}\widehat{x}_i$. Then by translation invariance on $\mathbb{U}_n$ one checks using (20.18) that $\pi_{\widehat{e}}(a)$ is a complete isometry.

Now let F be another n-dimensional operator space with a specific basis, $\widehat{F}$ be the similar associated space. Then the linear map $\pi_{\widehat{f}}(a) : \widehat{F} \to \widehat{F}$ analogous to $\pi_{\widehat{e}}(a)$ is a complete isometry for any $a \in \mathbb{U}_n$. Let $u_0 : \widehat{E} \to \widehat{F}$ be the linear map associated to the identity matrix. Then by Lemma 20.21 (applied with $c = 1$) we have

$$d_{cb}(\widehat{E}, \widehat{F}) = \|u_0\|_{cb}\|u_0^{-1}\|_{cb}. \tag{20.19}$$

Proof of Theorem 20.17 Let $\{[u_1(m), \ldots, u_n(m)], m \in \mathbb{N}\}$ be a sequence of n-tuples of unitary matrices satisfying (20.9) for some constant C (recall $u_1(m), \ldots, u_n(m)$ are of size $N_m \times N_m$).

For any subset $\omega \subset \mathbb{N}$, let again

$$\mathbb{B}_\omega = \left(\oplus \sum\nolimits_{m \in \omega} M_{N_m} \right)_\infty.$$

Let $\omega(1) \subset \mathbb{N}$ and $\omega(2) \subset \mathbb{N}$ be any pair of infinite subsets, and let

$$u_i^1 = \bigoplus\nolimits_{m \in \omega(1)} u_i(m) \in \mathbb{B}_{\omega(1)} \qquad u_i^2 = \bigoplus\nolimits_{m' \in \omega(2)} u_i(m') \in \mathbb{B}_{\omega(2)}.$$

Let $E_k = \mathrm{span}[u_j^k \mid 1 \leq j \leq n]$ where $k = 1$ or $k = 2$. Let $\widehat{E}_1$ and $\widehat{E}_2$ be the unitarily invariant spaces as defined in Remark 20.22. Let $u_0 : \widehat{E}_1 \to \widehat{E}_2$ be the linear map (associated to the identity matrix) defined by $u_0(u_i^1) = u_i^2$ $(1 \leq i \leq n)$. By Remark 20.22 we have $d_{cb}(\widehat{E}_1, \widehat{E}_2) = \|u_0\|_{cb}\|u_0^{-1}\|_{cb}$. We claim that if $\omega(2) \not\subset \omega(1)$ then

$$\|u_0\|_{cb} \geq n/C. \tag{20.20}$$

Indeed, let $m \in \omega(2) \setminus \omega(1)$. On one hand by (20.9) (applied with the transposed of a) we have

$$\left\| \sum \overline{u_i^2(m)} \otimes \widehat{u_i^1} \right\|_{\min} \leq C.$$

And on the other hand by (18.3)

$$\left\| \sum \overline{u_i^2(m)} \otimes \widehat{u_i^2} \right\|_{\min} \geq \sup_{a \in \mathbb{U}_n} \left\| \sum a_{ji} \overline{u_i^2(m)} \otimes u_j^2(m) \right\|_{\min}$$

$$\geq \left\| \sum\nolimits_j \overline{u_j^2(m)} \otimes u_j^2(m) \right\|_{\min} = n.$$

By the definition of the cb-norm (here we view $\sum \overline{u_j^2(m)} \otimes \widehat{u_j^1}$ as an element of $M_{N_m}(\widehat{E}_1)$), this implies our claim (20.20). Now if we assume moreover that $\omega(1) \not\subset \omega(2)$ then we find $\|u_0^{-1}\|_{cb} \geq n/C$, and hence by (20.19) we have $d_{cb}(\widehat{E}_1, \widehat{E}_2) \geq (n/C)^2$ and since we can take for C any value $> C_u(n)$ for any $\varepsilon > 0$ fixed in advance we obtain

$$d_{cb}(\widehat{E}_1, \widehat{E}_2) \geq (n/C_u(n))^2 - \varepsilon.$$

By Lemma 20.20 we have $C_u(n) \leq 2\sqrt{n}$ and hence $(n/C_u(n))^2 \geq n/4$. It remains to observe (exercise) that there is a set T of subsets of $\mathbb{N}$ with the cardinality of the continuum such that for any $s \neq t \in T$ we have both $s \not\subset t$

and $t \not\subset s$. Hint: consider the set V of vertices of an infinite binary tree, the set T of subsets of V forming an infinite branch from the root has the required properties, and V has the same cardinal as $\mathbb{N}$. Actually the set of numbers in $[0, 1]$ that do not have a finite dyadic expansion is in $1 - 1$ correspondence with a subset of T. Thus we may as well take $[0, 1]$ as our index set in (20.10).   □

By Corollary 20.3 the following is an immediate consequence of Theorem 20.17.

Corollary 20.23 *There does not exist a separable C^*-algebra (or operator space) X such that $d_{SX}(E) = 1$ for any finite-dimensional operator space. More precisely for any $n \geq 5$ we have*

$$\sup\{d_{SX}(E) \mid \dim(E) \leq n\} \geq \sqrt{n}/2.$$

Proof Assume $\sup\{d_{SX}(E) \mid \dim(E) \leq n\} < c$. By definition for any $t \in [0, 1]$ and $\varepsilon > 0$ there is a subspace $F_t \subset X$ such that $d_{cb}(E_t, F_t) < c$. Since $OS_n(X)$ is d_{cb}-separable we can restrict F_t to belong to a countable dense subset of $OS_n(X)$. But then (pidgeon hole principle) the assigment $t \mapsto F_t$ cannot be injective so there must be $s \neq t$ such that $F_s = F_t$. The latter implies $d_{cb}(E_s, E_t) \leq d_{cb}(E_s, F_s)d_{cb}(F_t, E_t)$ and hence $n/4 - \varepsilon < c^2$. This gives the announced bound.   □

In particular, taking $X = \mathscr{C}$:

Corollary 20.24 *For any $n \geq 5$ we have*

$$\sup\{d_f(E) \mid \dim(E) \leq n\} \geq \sqrt{n}/2.$$

Remark 20.25 By the identity in (20.5) the last corollary can be restated as: for any $n \geq 5$ we have

$$\sup\left\{\frac{\|u\|_{\max}}{\|u\|_{\min}} \,\middle|\, u \in B(H) \otimes B(H), \text{ rk}(u) \leq n\right\} \geq \sqrt{n}/2,$$

which is slightly less precise than Theorem 18.4.

Remark 20.26 These estimates are asymptotically best possible. Indeed, if $A = K(\ell_2)$ (the algebra of compact operators on ℓ_2) for any $E \in OS_n$ we have $d_{SA}(E) \leq \sqrt{n}$ (see [208, p. 133]). By Corollary 20.7 this implies $d_f(E) \leq \sqrt{n}$, which shows that the numbers appearing in either Corollary 20.24 or Remark 20.25 are $\leq \sqrt{n}$.

Remark 20.27 In sharp contrast with the nonseparability of OS_n, Ozawa proved in [184] that the subset formed of all the n-dimensional subspaces of a so-called noncommutative L_1-space (i.e. a von Neumann algebra predual) is separable and even *compact* for the metric δ_{cb}.

20.4 Notes and remarks

§20.1 collects basic facts on operator spaces easily adapted from the corresponding well-known statements for Banach spaces. The main source for §20.2 is [141] but the results were somewhat anticipated by Kirchberg in his questions at the end of [155] (see conjecture (A7) in [155, p. 483]). Remark 20.8 which is based on Arveson's ideas on operator systems appears in [208, p. 352] (see also [208, ex. 2.4.2]) but is due to Ozawa. In [124] Harcharras studies the stability properties of the class of spaces E such that $d_f(E) = 1$.

§20.3 is based on [141] and [180]. In the former we obtained a continuous family separated by $c\sqrt{n}$ for some $c > 0$ independent of n. The optimal result with $c\sqrt{n}$ replaced by $n/4$ stated in Theorem 20.17 is due to Oikhberg and Ricard in the latter paper [180]. For Corollaries 20.23, 20.24, and Remark 20.25, the lower bounds previously obtained in [141] (by $n(2\sqrt{n-1})^{-1}$) are slightly better than the ones we gave in the text (by $\sqrt{n}/2$). See [208, ch. 21] for more details. Results such as Lemma 20.21 are due to Zhang [264]. They are presented in detail in [208, p. 217]. It is natural to try to estimate the metric entropy of the metric space OS_n when equipped with natural metrics for which it becomes compact. This question is considered in [213] both for B_n (the Banach space case) and OS_n.

21

WEP as an extension property

We will first show (see Theorem 21.3) that the WEP of a C^*-algebra is characterized by a certain form of extension property for maps into it but defined on a special class of *finite-dimensional* subspaces of C^*-algebras. It is interesting that such a weak form of extension property already implies the WEP, because as the next result shows (see Theorem 21.4) the WEP implies a much stronger extension property.

Remark 21.1 Fix an integer $N \geq 1$. Let us recall here the content of Remark 3.14 when $|I| = N$. We will view the Banach space ℓ_1^N as an operator space, as follows: denoting $U_j = U_{\mathbb{F}_N}(g_j)$, we let $E_1^N = \mathrm{span}[U_1, \ldots, U_N] \subset C^*(\mathbb{F}_N)$. Then E_1^N is isometric to ℓ_1^N as a Banach space. We use the notation E_1^N to distinguish the Banach space ℓ_1^N from the operator space we just defined.

When $A = M_n$, the identity (3.7) describes the norm in $M_n(E_1^N)$ for arbitrary $n \geq 1$. From (3.6) it is immediate that the space $CB(E_1^N, M_n)$ can be identified with $\ell_\infty^N(M_n)$, and $\|u\|_{cb} = \|u\| = \sup_j \|u(U_j)\|$ for any u from E_1^N to any operator space. This means (see (2.14)) that $E_1^{N*} \simeq \ell_\infty^N$ completely isometrically. Moreover, the last assertion in Lemma 3.10 shows that E_1^N is also completely isometric to $\mathrm{span}\{1, U_1, \ldots, U_{N-1}\} \subset C^*(\mathbb{F}_{N-1})$. The latter is thus an alternative way to define the same operator space structure.

21.1 WEP as a local extension property

We first consider extension properties for maps defined on finite-dimensional subspaces.

Theorem 21.2 *Let $W \subset B$ be an inclusion of C^*-algebras. The following are equivalent:*

(i) *There is a generalized weak expectation $T : B \to W^{**}$ with $\|T\|_{cb} \leq 1$.*

(ii) *For any $N \geq 1$, $\varepsilon > 0$ and any subspace $E \subset E_1^N$, every $u : E \to W$ that is the restriction of a complete contraction $v : E_1^N \to B$ admits an extension $\widetilde{u} : E_1^N \to W$ with $\|\widetilde{u}\|_{cb} \leq 1 + \varepsilon$.*

Proof Assume (i). Let $E \subset E_1^N$, $v : E_1^N \to B$ with $\|v\|_{cb} \leq 1$ such that $v(E) \subset W$ and let $u = v_{|E} \in CB(E, W)$. Let $\widehat{u} = Tv : E_1^N \to W^{**}$. Note $B(E_1^N, A) = \ell_\infty^N(A)$ for any Banach space A, and the unit ball of $\ell_\infty^N(A)$ is clearly weak*-dense in that of $\ell_\infty^N(A^{**})$. Therefore there is a net of maps $v_i : E_1^N \to W$ with $\|v_i\| \leq 1$ tending weak* to $\widehat{u}$. In particular for any $x \in E$ the net $v_i(x)$ converges weak* to $Tv(x) = u(x)$, but since they both belong to W, weak convergence holds. By Mazur's Theorem A.9, we can form convex combinations of the $v_{i|E}$'s that converge in norm to u (see Remark A.10). The corresponding convex combinations of the v_i's give us a modified net v_i' in the unit ball of $B(E_1^N, W)$ such that $v_{i|E}'$ tends in norm to u. Thus for any $\delta > 0$ there is v' in the unit ball of $B(E_1^N, W)$ such that $\|u - v_{|E}'\| < \delta$. By the perturbation Lemma 20.2 if we choose $\delta > 0$ small enough there is an isomorphism $w_\delta : W \to W$ such that $u = w_\delta v_{|E}'$ and $\|w_\delta\|_{cb}\|w_\delta^{-1}\|_{cb} < 1 + \varepsilon$. Then $\widetilde{u} = w_\delta v'$ satisfies (ii).

Assume (ii). Let $j : W \to B$ denote the inclusion. We will apply the dual extension criterion formulated in Proposition A.2. With the latter we will extend the mapping $i_W : W \to W^{**}$ (a priori only defined on $W \subset B$) to a contraction on the whole of B. Let $v : W^{**} \to W$ be a weak* continuous finite rank map such that $N_\wedge(jv) < 1$. By (A.3) jv can be rewritten as $jv = T_2 D T_1$ with $T_1 : W^{**} \to \ell_\infty^N$, $D : \ell_\infty^N \to \ell_1^N$ and $T_2 : \ell_1^N \to B$ as in (A.2) satisfying (by homogeneity) $\|T_1\| = \|T_2\| = 1$ and $\|D\| < 1$. Note that, using $\ell_1^N \simeq E_1^N$, we have $\|T_2\| = \|T_2\|_{cb}$ by (3.6) (and similarly for T_1 and D). Let then $E = DT_1(W^{**}) \subset E_1^N$. Let $u = T_{2|E} : E \to W$. Note $\|u\|_{cb} \leq \|T_2\|_{cb} = \|T_2\|$. By the assumed extension property, there is $\widetilde{u} : E_1^N \to W$ extending u with $\|\widetilde{u}\|_{cb} \leq (1 + \varepsilon)\|T_2\|_{cb} = 1 + \varepsilon$. It follows that the operator $\widetilde{u}DT_1 : W^{**} \to W$ satisfies $N_\wedge(\widetilde{u}DT_1) \leq (1 + \varepsilon)\|T_1\|\|D\|\|T_2\| < 1 + \varepsilon$ and since $\widetilde{u}DT_1 = v$ we obtain

$$|\mathrm{tr}(i_W v)| \leq N_\wedge(i_W v) \leq N_\wedge(v) < 1 + \varepsilon.$$

Since $\varepsilon > 0$ was arbitrary, this shows by homogeneity that $|\mathrm{tr}(i_W v)| \leq N_\wedge(v) \leq N_\wedge(jv)$ for any weak* continuous finite rank $v : W^{**} \to W$. Then Proposition A.2 implies the existence of a contraction T as in (i). By Remark 7.31 (essentially Tomiyama's theorem) T is automatically a complete contraction. $\qquad\square$

In the case of the inclusion $W \subset B(H)$, since $B(H)$ is injective this gives us as an immediate consequence:

Theorem 21.3 *A C^*-algebra W has the WEP if and only if for any $N \geq 1$, $\varepsilon > 0$ and any subspace $E \subset E_1^N$, every $u : E \to W$ admits an extension $\widetilde{u} : E_1^N \to W$ with $\|\widetilde{u}\|_{cb} \leq (1 + \varepsilon)\|u\|_{cb}$.*

We now come to the local extension property satisfied by C^*-algebras with the WEP.

Theorem 21.4 *Let C be a separable C^*-algebra with the LLP and let W be another one with the WEP. Then for any finite-dimensional subspace $E \subset C$ and any $\varepsilon > 0$, any $u \in CB(E, W)$ admits an extension $\widetilde{u} \in CB(C, W)$ such that $\|\widetilde{u}\|_{cb} \leq (1 + \varepsilon)\|u\|_{cb}$.*

$$
\begin{array}{ccc}
C & & \\
\uparrow & \searrow^{\widetilde{u}} & \\
\downarrow & & \\
E & \xrightarrow{\ u\ } & W
\end{array}
$$

Moreover, if $E \subset C$ is a finite-dimensional operator system (assuming C unital) and u is a unital c.p. map then we can find a unital $v \in CP(C, W)$ such that $\|v_{|E} - u\| < \varepsilon$.

Remark 21.5 The proof will show the following:
   Assume that C is an operator space such that

$$d_f(C) = \sup\{d_f(E) | E \subset C, \dim(E) < \infty\} < \infty.$$

Then (assuming W has the WEP) for any $E \subset C$ with $\dim(E) < \infty$ and any $\varepsilon > 0$, any $u \in CB(E, W)$ admits an extension $\widetilde{u} \in CB(C, W)$ such that $\|\widetilde{u}\|_{cb} \leq (1 + \varepsilon)d_f(C)\|u\|_{cb}$.

Proof We first claim that it suffices to show the following "local" extension property:
   For any finite-dimensional $F \subset C$ such that $E \subset F \subset C$ and any $u \in CB(E, W)$ there is $\widetilde{u} : F \to W$ extending u with $\|\widetilde{u}\|_{cb} \leq (1 + \varepsilon)\|u\|_{cb}$. Indeed, assuming this, consider a sequence $F_0 \subset \cdots F_n \subset F_{n+1} \subset \cdots$ with $F_0 = E$ and $\overline{\cup F_n} = C$ and $\varepsilon_n > 0$ such that $\prod(1 + \varepsilon_n) < 1 + \varepsilon$. Then, by repeated application of the preceding "local" extension property, we obtain the announced result. Thus it suffices to establish the latter property.
   Let $E \subset F \subset C$ with $\dim F < \infty$. Since W has the WEP, by Theorem 9.22 we have a factorization

$$i_W : W \xrightarrow{\pi} B(H) \xrightarrow{T} W^{**}$$

where π is an isometric $*$-homomorphism and T is c.p. with $\|T\| = 1$. Let $u \in CB(E, W)$. By Theorem 1.18 there is a mapping $v \in CB(F, B(H))$ extending πu, so that $v_{|E} = \pi u$ with $\|v\|_{cb} = \|u\|_{cb}$. We may identify v with

$t \in F^* \otimes B(H)$ such that $\|t\|_{\min} = \|v\|_{cb}$. Since by (20.8) $d_f(F^*) = d_f(F) = 1$ there is a completely isometric embedding $F^* \subset \mathscr{C}$. With respect to the latter embedding, by Kirchberg's Theorem 9.6 we have $\|t\|_{\max} = \|t\|_{\min} = \|v\|_{cb} = \|u\|_{cb}$, and hence a fortiori by (4.30)

$$\|(Id \otimes T)(t)\|_{\mathscr{C} \otimes_{\max} W^{**}} \leq \|u\|_{cb}.$$

We now invoke §8.4: by (8.12) (applied to $\mathscr{C} \otimes_{\max} W^{**}$ instead of $A^{**} \otimes_{\max} B$), this implies

$$\|(Id \otimes T)(t)\|_{(\mathscr{C} \otimes_{\max} W)^{**}} \leq \|u\|_{cb}$$

and a fortiori (since $F^* \otimes_{\min} W \subset \mathscr{C} \otimes_{\min} W$ isometrically)

$$\|(Id \otimes T)(t)\|_{(F^* \otimes_{\min} W)^{**}} = \|(Id \otimes T)(t)\|_{(\mathscr{C} \otimes_{\min} W)^{**}} \leq \|u\|_{cb}. \quad (21.1)$$

By homogeneity we may assume $\|u\|_{cb} = 1$. Let $Z = F^* \otimes_{\min} W$. Then (21.1) means that there is a net (t_i) in B_Z tending to $(Id \otimes T)(t)$ with respect to $\sigma(Z^{**}, Z^*)$. Let $v_i : F \to W$ be the linear maps in the unit ball of $CB(F, W)$ associated to t_i. Since $x \otimes \xi \in Z^*$ for any $(x, \xi) \in F \times W^*$, we have $v_i(x) \to Tv(x)$ with respect to $\sigma(W^{**}, W^*)$. Restricting to E we find

$$\forall e \in E \quad v_i(e) \to Tv(e) = u(e).$$

And hence, since $v_i(e) - u(e) \in W$, we have

$$\forall e \in E \quad v_i(e) - u(e) \to 0 \quad \sigma(W, W^*).$$

Since $\dim(E) < \infty$, this means $v_i \to u$ in the weak topology of $CB(E, W)$. Therefore by the classical Mazur Theorem A.9, u is in the norm closure of the convex hull of $\{v_i | i \in I\}$. Note that $\|v_i\|_{cb} = \|t_i\|_{\min} \leq 1$. Thus

$$\forall \varepsilon > 0 \quad \exists w \in \text{conv}\{v_i | i \in I\} \subset B_{CB(F, W)} \quad \text{such that} \quad \|w_{|E} - u\|_{cb} < \varepsilon.$$

By the perturbation Lemma 20.2 there is $\delta(\varepsilon) > 0$ and a complete isomorphism $\Phi : W \to W$ with $\|\Phi\|_{cb} \|\Phi^{-1}\|_{cb} \leq 1 + \delta(\varepsilon)$ such that $\Phi w_{|E} = u$ and $\lim_{\varepsilon \to 0} \delta(\varepsilon) = 0$. Then, the mapping $\tilde{u} = \Phi w : F \to W$ satisfies $\|\tilde{u}\|_{cb} \leq 1 + \delta(\varepsilon)$ and $\tilde{u}_{|E} = u$.

If E is a finite-dimensional operator system and u is unital and c.p., by a suitably modified iteration argument, we can again reduce to finding $v \in CP(F, W)$ such that $\|v_{|E} - u\| < \varepsilon$ when $F \supset E$ is an arbitrary finite-dimensional operator system. Let us assume u c.p. and unital. Then (by (1.27)) $\|u\|_{cb} = 1$. The first part of the proof gives us an extension $\tilde{u} : F \to W$ with $\|\tilde{u}\| \leq 1 + \varepsilon$. Of course $\tilde{u}$ is still unital. By the perturbation Theorem 2.28, when F remains fixed and $\varepsilon \to 0$, $\tilde{u}$ becomes "close" to a c.p. map, so the conclusion follows. $\qquad \square$

21.2 WEP versus approximate injectivity

We already know that c.b. maps into a WEP C^*-algebra A can be extended to c.b. maps into the larger algebra A^{**} (see Corollary 9.24). Our goal in this section is to describe situations where the passage to A^{**} can be avoided.

Definition 21.6 A unital C^*-algebra A is said to be approximately injective if for any pair $\mathcal{S}_0 \subset \mathcal{S}_1$ of finite-dimensional operator systems, any unital completely positive map $u_0 : \mathcal{S}_0 \to A$ can be nearly extended to a completely positive map on $\mathcal{S}_1$, meaning that for any $\varepsilon > 0$ there is $u_1 \in CP(\mathcal{S}_1, A)$ such that $\|u_{1|\mathcal{S}_0} - u_0\| < \varepsilon$.

   In [77] where this property was introduced, it is proved that a unital C^*-algebra A is approximately injective if and only if the preceding property holds for not necessarily unital c.p. maps $u_0 : \mathcal{S}_0 \to A$.
   A general C^*-algebra will be called approximately injective if its unitization is approximately injective.
   We will see (following [77] and [155, Lemma 2.5]), that any approximately injective C^*-algebra has the WEP. But the converse is apparently not true if one believes the equivalence of the conjectures $(A2) \Leftrightarrow (A4)$ of [155]. Indeed, the latter claims[1] that if WEP $\Rightarrow$ approximately injective, then WEP $\Rightarrow$ LLP, but by the results of [141] (presented in the present volume in §18.1), there exists a WEP unital C^*-algebra (namely $\mathscr{B}$!) which is *not* LLP.
   Nevertheless, it turns out that if we restrict to operator spaces $\mathcal{S}_1$ that are subspaces of $\mathscr{C}$, i.e. such that $\mathcal{S}_1 \subset \mathscr{C}$, then the resulting property that we call approximate $\mathscr{C}$-injectivity is equivalent to the WEP.

Definition 21.7 Let C be a unital C^*-algebra. A C^*-algebra A will be called approximately C-injective if for any pair $E_0 \subset E_1$ of finite-dimensional operator spaces with $E_1 \subset C$, for any $u_0 \in CB(E_0, A)$ and any $\varepsilon > 0$ there is $u_1 \in CB(E_1, A)$ extending u_0, i.e. satisfying $u_{1|E_0} = u_0$, and such that $\|u_1\|_{cb} \le \|u_0\|_{cb}(1 + \varepsilon)$.

Proposition 21.8 *If A is approximately C-injective in the sense of the preceding Definition 21.7, then for any pair $\mathcal{S}_0 \subset \mathcal{S}_1 \subset C$ of finite-dimensional operator systems, any unital completely positive map $u_0 : \mathcal{S}_0 \to A$ can be nearly extended to a completely positive map on $\mathcal{S}_1$ in the sense of Definition 21.6, that is for any $\varepsilon > 0$ there is $u_1 \in CP(\mathcal{S}_1, A)$ such that $\|u_{1|\mathcal{S}_0} - u_0\| < \varepsilon$.*

Proof Indeed, by Definition 21.7, for any $\delta > 0$ there is $\varphi \in CB(\mathcal{S}_1, A)$ extending u_0 and such that $\|\varphi\|_{cb} \le 1 + \delta$. Then by part (ii) in Theorem 2.28

---

[1] The author does not understand the proof that $(A6) \Rightarrow (A2)$ or that $(A4) \Rightarrow (A2)$ in [155, p. 485], and the related final remark 8.2 in [155, p. 485] does not seem correct.

the announced result holds with $\varepsilon = 8\delta \dim(\mathcal{S}_1)$. Since $\delta > 0$ is arbitrary, this completes the argument. □

We use a slightly different terminology from [155]: there it is said that A is approximately injective in $\mathscr{C}$ when A is approximately $\mathscr{C}$-injective in the previous sense. Our exposition includes several remarks from [124].

Lemma 21.9 *If $M_2(A)$ is approximately injective, then A is approximately C-injective for any C.*

Proof Let $E_0 \subset E_1 \subset C$ be finite-dimensional operator spaces. We restrict ourselves, without loss of generality, to the case when $C = B(H)$ and A is unital. Let $u_0 \in CB(E_0, A)$. We may assume by homogeneity that $\|u_0\|_{cb} = 1$. Let $\mathcal{S}_0 \subset \mathcal{S}_1 \subset M_2(C)$ be the operator systems associated to $E_0 \subset E_1$ as in Lemma 1.38. Then the mapping $V_0 : \mathcal{S}_0 \to M_2(A)$ defined by

$$V_0\left(\begin{pmatrix} \lambda & x \\ y^* & \mu \end{pmatrix}\right) = \begin{pmatrix} \lambda & u_0(x) \\ u_0(y^*) & \mu \end{pmatrix}$$

is unital and c.p. If $M_2(A)$ is approximately injective, then for any $0 < \varepsilon < 1$ there is $V_1 \in CP(\mathcal{S}_1, M_2(A))$ such that $\|V_{1|\mathcal{S}_0} - V_0\| < \varepsilon$. By (1.20) we have $\|V_1\|_{cb} = \|V_1(1)\|$ and $V_0(1) = 1$, therefore $\|V_1\|_{cb} < 1 + \varepsilon$. Let $v : E_1 \to A$ be the $(1, 2)$ entry of V_1. Clearly $\|v\|_{cb} \leq \|V_1\|_{cb} < 1 + \varepsilon$. Since $\|v_{|E_0} - u_0\| \leq \|V_{1|\mathcal{S}_0} - V_0\|$ we have $\|v_{|E_0} - u_0\| < \varepsilon$. Since $\dim(E_0) < \infty$, the norm and the cb-norm on $CB(E_0, A)$ are equivalent (see Remark 1.56), so (using a different ε) we can find a v satisfying as well $\|v_{|E_0} - u_0\|_{cb} < \varepsilon$. Note that $\|v - v/(1 + \varepsilon)\|_{cb} \leq \varepsilon$. Thus, replacing v by $v/(1 + \varepsilon)$ we can find $v \in CB(E_1, A)$ with $\|v\|_{cb} \leq 1$ and $\|v_{|E_0} - u_0\| < 2\varepsilon$. Thus the linear mapping $v \mapsto v_{|E_0}$ takes the unit ball of $CB(E_1, A)$ to a dense subset of the unit ball of $CB(E_0, A)$. Then a standard iteration argument gives the announced result: the latter mapping is onto and for any $\varepsilon > 0$ there is $u_1 \in CB(E_1, A)$ such that $u_{1|E_0} = u_0$ and $\|u_1\|_{cb} \leq 1 + \varepsilon$. □

We will now concentrate on the case when $C = \mathscr{C}$. In that case we note that the condition $\mathcal{S}_1 \subset \mathscr{C}$ can be rephrased in terms of the quantity d_f defined by (20.3) (see Corollary 20.16). Using the latter, we may rewrite $\mathcal{S}_1 \subset \mathscr{C}$ as $d_f(\mathcal{S}_1) = 1$.

Recapitulating, we may state:

Theorem 21.10 *A C^*-algebra is approximately $\mathscr{C}$-injective if and only if it has the WEP.*

Proof By Theorem 21.3 approximate $\mathscr{C}$-injectivity implies the WEP. The converse direction was already proved in Theorem 21.4. □

21.3 The (global) lifting property LP

Combining several of our earlier statements, we obtain a global lifting theorem for the pair (C, W) as in Theorem 21.4 when C is separable.

Theorem 21.11 (Global lifting from LLP to QWEP) *Let C, W be unital C^*-algebras and let $q : W \to W/I$ be a surjective $*$-homomorphism. If C is separable with LLP and if W has the WEP, then any unital $u \in CP(C, W/I)$ admits a unital lifting $v \in CP(C, W)$ (i.e. we have $qv = u$) with $\|v\| = \|u\|$.*

Proof By Theorem 9.38, the c.p. map u is locally 1-liftable. By (ii) in Proposition 9.42, for any $\varepsilon > 0$ and any finite-dimensional operator system $E \subset C$ there is $v_E : E \to W$ unital c.p. such that $\|qv_E - u_{|E}\| \le \varepsilon$. By Theorem 21.4 we may find a unital c.p. map $w^E : C \to W$ such that $\|w_{|E}^E - v_E\| \le \varepsilon$. A fortiori $\|(qw^E - u)_{|E}\| \le 2\varepsilon$. Let us denote $i = (E, \varepsilon)$ and let $v_i = w^E : C \to W$. We view the set of i's as directed in the usual way so that E tends to C and ε to zero when $i \to \infty$. Then $\|qv_i(x) - u(x)\| \to 0$ for any given $x \in C$ when $i \to \infty$. By part (ii) of Theorem 9.46 (applied to $v_i : C \to W$) there is a unital c.p. map lifting u. $\qquad\square$

We will now briefly discuss the (global) lifting property.

Definition 21.12 A unital C^*-algebra C is said to have the lifting property (LP) if any unital c.p. $u : C \to A/I$, into an arbitrary quotient C^*-algebra, admits a unital c.p. lifting $\widehat{u} : C \to A$ with the same norm as u.

Remark 21.13 The property makes sense in the nonunital case as well. Just omit "unital" in the preceding definition. We restrict to the unital case for simplicity.

The main example is $\mathscr{C}$ for which Kirchberg proved the LP in [155]. See [189] or [39, p.376] for a detailed proof.

Corollary 21.14 *If Kirchberg's conjecture holds, then LLP $\Rightarrow$ LP for separable unital C^*-algebras.*

Proof Assume C and A unital. Let C be separable with LLP. Consider a unital c.p. map $u : C \to A/I$. Let $q : A \to A/I$ denote the quotient map. Let $\mathbb{F}$ be a free group such that there is a surjective $*$-homomorphism $Q : C^*(\mathbb{F}) \to A$ (see Proposition 3.39). Let $W = C^*(\mathbb{F})$ and $J = \ker(qQ)$. Then $A/I = W/J$ and W has the LLP. Thus if the Kirchberg conjecture is valid, W has the WEP and hence by Theorem 21.11 the map $u : C \to A/I = W/J$ admits a unital c.p. lifting $v : C \to W$, but then $Qv : C \to A$ is a unital c.p. lifting of u, which proves that C has the LP. $\qquad\square$

21.4 Notes and remarks

Approximate injectivity was introduced by Effros and Haagerup in [77].
Among many results, they show there that it implies the WEP. §21.2 comes
from [77] and [155] although we changed slightly the terminology. Theorem
21.4 is essentially [155, Lemma 2.5]. §21.3 comes from [155].

22

Complex interpolation and maximal
tensor product

22.1 Complex interpolation

The complex interpolation method is a very useful way to produce intermediate Banach spaces, or intermediate norms, between two given ones (see e.g. [22]). Starting with a pair of norms $\| \cdot \|_0$, $\| \cdot \|_1$ one defines a norm $\| \cdot \|_\theta$ ($0 < \theta < 1$) that can be thought of as a deformation of a special kind, providing a privileged path from $\| \cdot \|_0$ to $\| \cdot \|_1$, somewhat analogous to a geodesic. Geometrically, in many examples $\| \cdot \|_\theta$ appears as a sort of "geometric mean" of $\| \cdot \|_0$ and $\| \cdot \|_1$. The classical example is the pair $L_{p(0)}, L_{p(1)}$ with their usual norms (on a fixed measure space) for which the norm $\| \cdot \|_\theta$ is the norm in the space $L_{p(\theta)}$ for $p(\theta)$ determined by $1/p(\theta) = (1-\theta)/p(0) + \theta/p(1)$. It is known that this extends to noncommutative L_p-spaces (see e.g. [215]), but our interest here will be in a different (although not totally unrelated) direction.

To state the precise definitions, we will need some specific notation. Let

$$S = \{z \in \mathbb{C} \mid 0 < \mathrm{Re}(z) < 1\},$$

$$\partial_0 = \{z \in \mathbb{C} \mid \mathrm{Re}(z) = 0\} \quad \text{and} \quad \partial_1 = \{z \in \mathbb{C} \mid \mathrm{Re}(z) = 1\}. \tag{22.1}$$

Note that $\partial S = \partial_0 \cup \partial_1$. Given a subset $\Omega \subset \mathbb{C}$ and a Banach space B, we denote by $C_b(\Omega; B)$ the space of bounded continuous functions $f : Omega \to B$ equipped with the norm

$$\|f\|_{C_b(\Omega; B)} = \sup_{z \in \Omega} \|f(z)\|_B.$$

The following celebrated and classical three line lemma is the crucial tool to develop complex interpolation (the extension from the $\mathbb{C}$-valued to the B-valued case is straightforward).

Lemma 22.1 (Three line lemma) *Let $f : ovlS \to B$ be a bounded continuous function with values in a Banach space B that is analytic on S. Then for any $0 < \theta < 1$*

$$\|f(\theta)\| \leq (\sup_{\partial_0} \|f\|)^{1-\theta} (\sup_{\partial_1} \|f\|)^{\theta}. \tag{22.2}$$

To formally define interpolation methods between two Banach spaces, we always need to assume that the initial pair (B_0, B_1) is "compatible." This means that we are given a (Hausdorff) topological vector space V and continuous injections

$$j_0 : B_0 \rightarrow V \quad \text{and} \quad j_1 : B_1 \rightarrow V.$$

This rudimentary structure is just what is needed to define the intersection $B_0 \cap B_1$ and the sum $B_0 + B_1$.

The space $B_0 \cap B_1$ is defined as $j_0(B_0) \cap j_1(B_1)$ equipped with the norm

$$\|x\| = \max\{\|j_0^{-1}(x)\|_{B_0}, \|j_1^{-1}(x)\|_{B_1}\}.$$

The space $B_0 + B_1$ is defined as the setwise sum $j_0(B_0) + j_1(B_1)$ equipped with the norm

$$\|x\|_{B_0+B_1} = \inf\{\|x_0\|_{B_0} + \|x_1\|_{B_1} \mid x = j_0(x_0) + j_1(x_1)\}.$$

It is an easy exercise to check that $B_0 \cap B_1$ and $B_0 + B_1$ are Banach spaces. Following a well-established tradition, we will identify B_0 and B_1 with $j_0(B_0)$ and $j_1(B_1)$, so that j_0 and j_1 become the inclusion mappings $B_0 \subset V$ and $B_1 \subset V$. We then have $\forall i = 0, 1$

$$B_0 \cap B_1 \subset B_i \subset B_0 + B_1$$

and these inclusions have norm ≤ 1. Note that if we wish we may now replace V by $B_0 + B_1$, so that we may as well assume that V is a Banach space.

Actually, it seems worthwhile to warn the reader that, in the present notes, the main case of interest is the technically much easier case of a pair (B_0, B_1) consisting of the same Banach space B but equipped with two distinct norms $\| \cdot \|_0$ and $\| \cdot \|_1$ (see Theorem 22.10). In that case, of course we take $V = B$, and complex interpolation is used here only to define a family of norms $\| \cdot \|_\theta$ on B indexed by $0 < \theta < 1$.

In this section all Banach spaces will be over the complex field of scalars $\mathbb{C}$.

We will denote by $\mathcal{F}(B_0, B_1)$ (or often simply by $\mathcal{F}$) the space of all functions f in $C_b(\overline{S}; B_0 + B_1)$ such that $f_{|S} : S \rightarrow B_0 + B_1$ is analytic and $f_{|\partial_0} \in C_b(\partial_0, B_0)$, $f_{|\partial_1} \in C_b(\partial_1, B_1)$. We set

$$\|f\|_{\mathcal{F}} = \max_{j=0,1} \{\|f_{|\partial_j}\|_{C_b(\partial_j; B_j)}\}.$$

Let $0 < \theta < 1$. The complex interpolation space $(B_0, B_1)_\theta$ is defined as follows

$$(B_0, B_1)_\theta = \{x \in B_0 + B_1 \mid \exists f \in \mathcal{F}, \; f(\theta) = x\}.$$

It is equipped with the norm

$$\|x\|_{(B_0, B_1)_\theta} = \inf\{\|f\|_{\mathcal{F}} \mid f \in \mathcal{F}, \ f(\theta) = x\}. \tag{22.3}$$

It is easy to check that it can be identified isometrically with the quotient of $\mathcal{F}$ by the closed subspace $\{f \in \mathcal{F} \mid f(0) = 0\}$, and hence it is a Banach space.

In analogy with (22.2) it can be shown that

$$\|x\|_{(B_0, B_1)_\theta} = \inf\{(\sup_{\partial_0} \|f\|_{B_0})^{1-\theta} (\sup_{\partial_1} \|f\|_{B_1})^\theta \mid f \in \mathcal{F}, \ f(\theta) = x\}.$$

The latter formula leads us to describe the unit ball of $(B_0, B_1)_\theta$ roughly as some sort of geometric mean of those of B_0 and B_1 (in analogy with their "arithmetic mean" $(1 - \theta)B_{B_0} + \theta B_{B_1}$).

Remark 22.2 We have

$$B_0 \cap B_1 \subset (B_0, B_1)_\theta \subset B_0 + B_1,$$

and $B_0 \cap B_1$ is dense in $(B_0, B_1)_\theta$. Moreover, for any x in $B_0 \cap B_1$

$$\|x\|_{B_0+B_1} \leq \|x\|_{(B_0, B_1)_\theta} \leq \|x\|_{B_0}^{1-\theta} \|x\|_{B_1}^\theta \leq \|x\|_{B_0 \cap B_1}. \tag{22.4}$$

Remark 22.3 When $B_0 \cap B_1$ is dense both in B_0 and B_1 we may embed both B_0^* and B_1^* in $(B_0 \cap B_1)^*$, in order to view them as a compatible couple, to which we can apply the complex method. A classical interpolation theorem (see [21]) then says that the space $(B_0^*, B_1^*)_\theta$ can be identified isometrically with the closure of $B_0^* \cap B_1^*$ in the dual of $(B_0, B_1)_\theta$. In the situation of interest to us in the sequel, B_0 and B_1 are the same space equipped with two different norms. Then of course this duality theorem simply says that we can identify isometrically $(B_0^*, B_1^*)_\theta$ and $(B_0, B_1)_\theta^*$.

Iterating this fact, we can also identify isometrically $(B_0^{**}, B_1^{**})_\theta$ and $(B_0, B_1)_\theta^{**}$.

The basic application of complex interpolation is the following result (see [22]).

Theorem 22.4 (Classical interpolation theorem) *Let* (B_0, B_1) *and* (A_0, A_1) *be two compatible pairs in the sense just defined. Let* $A_\theta = (A_0, A_1)_\theta$ *and* $B_\theta = (B_0, B_1)_\theta$. *Let* $T : A_0 + A_1 \to B_0 + B_1$ *be a bounded linear operator such that* $T(A_j) \subset B_j$ *for* $j = 0, 1$. *Let* $C_j = \|T : A_j \to B_j\|$. *Then for any* $0 < \theta < 1$ *we have* $T(A_\theta) \subset B_\theta$ *and*

$$\|T : A_\theta \to B_\theta\| \leq C_0^{1-\theta} C_1^\theta.$$

Remark 22.5 (Stein's interpolation principle) This is a variant of the preceding Theorem 22.4 where the operator T is replaced by an operator valued analytic function $z \mapsto T(z)$. While much more general versions are valid (see e.g. [66])

we will use only the simplest (and straightforward) case when $z \mapsto T(z)$ is a function with values in $B(A_0 + A_1, B_0 + B_1)$ that is bounded and analytic in a strip that contains $\bar{S}$ in its interior, and also is such that its restrictions to ∂_j ($j = 0, 1$) and to A_j are in $C_b(\partial_j; B(A_j, B_j))$ and satisfy the following bounds for $j = 0, 1$

$$\forall z \in \partial_j \quad \|T(z) : A_j \to B_j\| \leq C_j.$$

Then Stein's interpolation principle says that

$$\|T(\theta) : (A_0, A_1)_\theta \to (B_0, B_1)_\theta\| \leq C_0^{1-\theta} C_1^\theta.$$

The underlying idea is rather obvious: given $f \in \mathcal{F}(A_0, A_1)$ such that $f(\theta) = x$ just observe that the function $g : z \mapsto T(z) f(z)$ is in $\mathcal{F}(B_0, B_1)$ and satisfies $g(\theta) = T(\theta)x$.

Remark 22.6 (Interpolation of multilinear maps) It is well known that Theorem 22.4 can be generalized easily to multilinear maps T, even if they are antilinear in some variables. We only need in the sequel the case of sesquilinear ones, as follows. Let (X_0, X_1) be another compatible couple. Let $T : (X_0 + X_1) \times (A_0 + A_1) \to B_0 + B_1$ be a sesquilinear mapping (antilinear in the first variable and linear in the second one). Assume that T restricts to a bounded sesquilinear map from $X_j \times A_j$ to B_j with norm C_j. Then T defines a bounded sesquilinear map from $X_\theta \times A_\theta$ to B_θ with norm $\leq C_0^{1-\theta} C_1^\theta$.

This follows from the observation that if $f \in \mathcal{F}(X_0, X_1)$ and $g \in \mathcal{F}(A_0, A_1)$ with $f(\theta) = x$ and $g(\theta) = a$ then the function $h : z \mapsto T(f(\bar{z}), g(z))$ is in $\mathcal{F}(B_0, B_1)$ and $h(\theta) = F(x, a)$.

Alternatively, this can be reduced to the bilinear case modulo the isometric identity

$$(\overline{X_0}, \overline{X_1})_\theta = \overline{(X_0, X_1)_\theta}.$$

The most classical application of complex interpolation is to pairs of L_p-spaces. For brevity we denote here simply by L_p the space $L_p(\Omega, \mathcal{A}, m)$ relative to a given arbitrary measure space.

Theorem 22.7 *Let* $1 \leq p_0, p_1 \leq \infty$. *Then for any* $0 < \theta < 1$ *we have*

$$(L_{p_0}, L_{p_1})_\theta = L_{p_\theta} \tag{22.5}$$

with identical norms, where p_θ *is defined by the equality*

$$p_\theta^{-1} = (1 - \theta) p_0^{-1} + \theta p_1^{-1}.$$

More generally, the same procedure can be applied to construct the so-called noncommutative L_p-spaces associated to a von Neumann algebra M equipped with a semifinite faithful normal trace τ. By the semifiniteness of τ, the

∗-subalgebra $\mathcal{A} = \{x \in M \mid \tau(|x|) < \infty\}$ is weak*-dense in M (see Remark 11.1). When $1 \le p < \infty$ the space $L_p(\tau)$ can be defined as the completion of $\mathcal{A}$ with respect to the norm

$$\forall x \in \mathcal{A} \quad \|x\|_p = \tau(|x|^p)^{1/p}.$$

The space $L_1(\tau)$ can be identified isometrically with the predual M_* of M, for the duality defined by $\langle x, y \rangle = \tau(xy)$ for $x \in L_1(\tau), y \in M$. More precisely, we define this for $x \in M$ and then extend it to $x \in L_1(\tau)$ by continuity and density. We have a natural inclusion $M \subset L_1(\tau)$ if τ is finite.

The noncommutative version of Theorem 22.7 identifies the space $L_p(\tau)$ with the complex interpolation space $(M, M_*)_{1/p}$, but to make sense of this we must first say how we turn the pair (M, M_*) into a compatible couple. When τ is finite this is easy: since we have a canonical inclusion $M \subset L_1(\tau)$ we can take $V = L_1(\tau)$ to define compatibility. Equivalently we can take $V = M + L_1(\tau)$ (that is the same space with a different norm) to obtain natural norm 1 inclusions $M \subset V$, $L_1(\tau) \subset V$. When τ is infinite, there are several known equivalent ways to do this. A quick one, perhaps less conventional, is like this: let (p_i) be an increasing directed net of projections with finite trace in M. Note $(p_i M p_i)_* = p_i M_* p_i$. Let $M_i = p_i M p_i$ equipped with the restriction of the trace $\tau_i = \tau_{|p_i M p_i}$. Then since τ_i is finite, $(M_i, L_1(M_i, \tau_i))$ can be viewed as compatible.

Let $V_i = M_i + L_1(M_i, \tau_i)$. We have natural norm 1 inclusions

$$M_i \subset V_i \quad L_1(M_i, \tau_i) \subset V_i.$$

Let

$$V = \left(\oplus \sum\nolimits_{i \in I} V_i\right)_\infty.$$

Then we have natural norm 1 inclusions $M \subset V$ and $M_* \subset V$, that allow us to define compatibility. To emphasize the analogy with the classical L_p-spaces we denote M by $L_\infty(\tau)$ in the sequel.

Theorem 22.8 *In the preceding situation, we have for $\theta = 1/p$ an isometric identity*

$$(L_\infty(\tau), L_1(\tau))_\theta = L_p(\tau). \tag{22.6}$$

In particular,

$$L_2(\tau) = (L_\infty(\tau), L_1(\tau))_{1/2}.$$

There is also a generalization of (22.5) to the noncommutative case. We refer the reader to [215] for more information in that direction.

22.2 Complex interpolation, WEP and maximal tensor product

Let A be a C^*-algebra. Fix an integer $n \geq 1$. Let X_0 (resp. X_1) be the space A^n equipped with the norm

$$\|(x_1, \ldots, x_n)\|_{X_0} = \left\| \sum_1^n x_j x_j^* \right\|^{1/2} \left(\text{resp. } \|(x_1, \ldots, x_n)\|_{X_1} = \left\| \sum_1^n x_j^* x_j \right\|^{1/2} \right). \tag{22.7}$$

Let $0 < \theta < 1$. We denote for brevity

$$\forall (x_j) \in A^n \quad \|(x_1, \ldots, x_n)\|_{\theta, A} = \|(x_1, \ldots, x_n)\|_{(X_0, X_1)_\theta}. \tag{22.8}$$

Let us first observe that the embedding $A \subset A^{**}$ preserves the norms just defined:

Lemma 22.9 *For any $(x_j) \in A^n$ and any $0 < \theta < 1$ we have*

$$\forall (x_j) \in A^n \quad \|(x_1, \ldots, x_n)\|_{\theta, A} = \|(x_1, \ldots, x_n)\|_{\theta, A^{**}}. \tag{22.9}$$

More generally if we denote by $H_{n,\theta}(A)$ the Banach space A^n equipped with the norm $(x_j) \mapsto \|(x_1, \ldots, x_n)\|_{\theta, A}$, we have the following isometric identification:

$$H_{n,\theta}(A)^{**} = H_{n,\theta}(A^{**}). \tag{22.10}$$

Proof This is an immediate consequence of Remark 22.3. □

The following statement comes from [203, 205]. As usual we denote by $L(x)$ (resp. $R(x)$) the operator of left (resp. right) multiplication by $x \in M$.

Theorem 22.10 *Fix an integer $n \geq 1$. Let (M, τ) be a von Neumann algebra equipped with a faithful normal semifinite trace τ. Let $0 < \theta < 1$. Then we have*

$$\|(x_1, \ldots, x_n)\|_{\theta, M} = \left\| \sum R(x_j^*) L(x_j) \right\|_{B(L_p(\tau))}^{1/2} \tag{22.11}$$

where $\theta = 1/p$.

We will prove this theorem later on in this section. We first discuss its consequences and some preliminary results. The next statement follows by a direct application of complex interpolation (independent of Theorem 22.10).

Lemma 22.11 *Let (M, τ) and (N, φ) be von Neumann algebras equipped with faithful normal semifinite traces. Then for any n, any $(x_j) \in M^n$ and any $(y_j) \in N^n$ we have*

$$\left\| \sum \overline{x_j} \otimes y_j \right\|_{\overline{M} \otimes_{\max} N} \leq \|(x_j)\|_{1/2, M} \|(y_j)\|_{1/2, N}. \tag{22.12}$$

Proof Let $a_j = \overline{x_j} \otimes 1, b_j = 1 \otimes y_j$. Note $\sum \overline{x_j} \otimes y_j = \sum a_j b_j = \sum b_j a_j$. By the operator variant of Cauchy–Schwarz (see (2.2)) we have

$$\left\| \sum \overline{x_j} \otimes y_j \right\|_{\max} = \left\| \sum a_j b_j \right\|_{\max} \leq \left\| \left(\sum a_j a_j^* \right)^{1/2} \right\|_{\max} \left\| \left(\sum b_j^* b_j \right)^{1/2} \right\|_{\max}$$

$$\left\| \sum \overline{x_j} \otimes y_j \right\|_{\max} = \left\| \sum b_j a_j \right\|_{\max} \leq \left\| \left(\sum a_j^* a_j \right)^{1/2} \right\|_{\max} \left\| \left(\sum b_j b_j^* \right)^{1/2} \right\|_{\max}$$

and hence we have $\| \sum \overline{x_j} \otimes y_j \|_{\max} \leq \|(x_j)\|_{X_0} \|(y_j)\|_{Y_1}$ and also $\| \sum \overline{x_j} \otimes y_j \|_{\max} \leq \|(x_j)\|_{X_1} \|(y_j)\|_{Y_0}$ where (X_0, X_1) is as before (but here with $A = M$) and (Y_0, Y_1) is the analogous pair of norms on N^n. Observe that $(Y_0, Y_1)_{1/2} = (Y_1, Y_0)_{1/2}$ isometrically. We will apply the interpolation theorem described in Remark 22.6 to the sesquilinear mapping

$$(x, y) \mapsto \sum \overline{x_j} \otimes y_j.$$

Since the latter is of norm 1 both from $X_0 \times Y_1$ to $\overline{M} \otimes_{\max} N$ and from $X_1 \times Y_0$ to $\overline{M} \otimes_{\max} N$, we have $\left\| \sum \overline{x_j} \otimes y_j \right\|_{\max} \leq \|(x_j)\|_{1/2, M} \|(y_j)\|_{1/2, N}$, which establishes (22.12). $\square$

We then obtain the following important consequence of the preceding theorem:

Corollary 22.12 *In the situation of Theorem 22.10, for any $(x_j) \in M^n$ we have*

$$\|(x_j)\|_{1/2, M} = \left\| \sum \overline{x_j} \otimes x_j \right\|_{\overline{M} \otimes_{\max} M}^{1/2} \tag{22.13}$$

and hence

$$\left\| \sum \overline{x_j} \otimes x_j \right\|_{\overline{M} \otimes_{\max} M}^{1/2} = \left\| \sum R(x_j^*) L(x_j) \right\|_{B(L_2(\tau))}^{1/2}. \tag{22.14}$$

Proof Taking $N = M$ and $y_j = x_j$ in (22.12) we find

$$\left\| \sum \overline{x_j} \otimes x_j \right\|_{\overline{M} \otimes_{\max} M} \leq \|(x_j)\|_{1/2, M}^2.$$

By Theorem 22.10 we have $\| \sum \overline{x_j} \otimes x_j \|_{\overline{M} \otimes_{\max} M} \leq \| \sum R(x_j^*) L(x_j) \|_{B(L_2(\tau))}$, and by definition of the max-norm, since L, R are representations with commuting ranges

$$\left\| \sum R(x_j^*) L(x_j) \right\|_{B(L_2(\tau))} \leq \left\| \sum \overline{x_j} \otimes x_j \right\|_{\overline{M} \otimes_{\max} M}.$$

Thus we conclude $\| \sum \overline{x_j} \otimes x_j \|_{\overline{M} \otimes_{\max} M} = \left\| \sum R(x_j^*) L(x_j) \right\| = \|(x_j)\|_{1/2, M}^2$. $\square$

In the particular case when $M = B(H)$ we have:

Corollary 22.13 *Let H, K be Hilbert spaces. For any $(x_j) \in B(H)^n$ we have*

$$\|(x_j)\|_{1/2,\, B(H)} = \left\| \sum \overline{x_j} \otimes x_j \right\|_{\overline{B(H)} \otimes_{\min} B(H)}^{1/2} = \left\| \sum \overline{x_j} \otimes x_j \right\|_{\overline{B(H)} \otimes_{\max} B(H)}^{1/2},$$

and for any $(y_j) \in B(K)^n$

$$\left\| \sum \overline{x_j} \otimes y_j \right\|_{\overline{B(H)} \otimes_{\max} B(K)}$$

$$\leq \left\| \sum \overline{x_j} \otimes x_j \right\|_{\overline{B(H)} \otimes_{\min} B(H)}^{1/2} \left\| \sum \overline{y_j} \otimes y_j \right\|_{\overline{B(K)} \otimes_{\min} B(K)}^{1/2}.$$

Proof Indeed, when τ is the ordinary trace on $M = B(H)$, $\bar{H} \otimes_2 H$ can be identified with $S_2(H) = L_2(\tau)$, and the norm of $\sum R(x_j^*)L(x_j)$ coincides with the norm of $\sum \overline{x_j} \otimes x_j$ on $\bar{H} \otimes_2 H$, i.e. with the min norm as in (2.11). This yields the first inequality. Then the second one follows from Lemma 22.11 with $N = B(K)$. □

Corollary 22.14 *For any von Neumann algebra and any finite set (x_j) in M we have*

$$\left\| \sum \overline{x_j} \otimes x_j \right\|_{\overline{M} \otimes_{\max} M}^{1/2} = \left\| \sum \overline{x_j} \otimes x_j \right\|_{\overline{M} \otimes_{\mathrm{bin}} M}^{1/2}.$$

Proof The semifinite case follows from (22.14): indeed, since the representations L and R are normal the right-hand side of (22.14) is clearly $\leq \left\| \sum \overline{x_j} \otimes x_j \right\|_{\overline{M} \otimes_{\mathrm{bin}} M}^{1/2}$. In the general case we may assume that $M \subset \mathcal{M}$ with $\mathcal{M}$ semifinite such that there is a contractive c.p. projection from $\mathcal{M}$ to M (see Theorem 11.3). Then for (x_j) in M we have $\left\| \sum \overline{x_j} \otimes x_j \right\|_{\overline{M} \otimes_{\max} M}^{1/2} = \left\| \sum \overline{x_j} \otimes x_j \right\|_{\overline{M} \otimes_{\max} \mathcal{M}}^{1/2}$, and hence (since the inclusion $M \subset \mathcal{M}$ is normal)

$$\left\| \sum \overline{x_j} \otimes x_j \right\|_{\overline{M} \otimes_{\max} M}^{1/2} = \left\| \sum \overline{x_j} \otimes x_j \right\|_{\overline{\mathcal{M}} \otimes_{\mathrm{bin}} \mathcal{M}}^{1/2} \leq \left\| \sum \overline{x_j} \otimes x_j \right\|_{\overline{M} \otimes_{\mathrm{bin}} M}^{1/2}.$$

By the maximality of the max-norm, the last inequality must be an equality. □

The next corollary was observed by Haagerup in [107] (unpublished).

Corollary 22.15 *The identity (22.13) remains valid when M is an arbitrary C^*-algebra.*

Proof We first treat the case when M is an arbitrary von Neumann algebra. The semifinite case was already settled in (22.13). By Takesaki's Theorem 11.3 M can be embedded in a semifinite von Neumann algebra $\mathcal{M}$ admitting a c.p. projection $P : \mathcal{M} \to M$ with $\|P\|_{cb} = 1$. The latter implies that for any $(x_j) \in M^n$ we have $\|(x_j)\|_{1/2,M} = \|(x_j)\|_{1/2,\mathcal{M}}$. Indeed, by Theorem 22.4, since P induces a norm 1 map both from Y_0 to X_0 and from Y_1 to X_1, it also does from $(Y_0, Y_1)_{1/2}$ to $(X_0, X_1)_{1/2}$. Similarly, since P is c.p. and c.p. maps are (max $\to$ max)-tensorizing we have $\left\| \sum \overline{x_j} \otimes x_j \right\|_{\overline{M} \otimes_{\max} M} = \left\| \sum \overline{x_j} \otimes x_j \right\|_{\overline{\mathcal{M}} \otimes_{\max} \mathcal{M}}$ for any $(x_j) \in M^n$. Thus, since (22.13) is valid for $\mathcal{M}$ (semifinite case) it must hold also for M. Let A be an arbitrary C^*-algebra. By what precedes (22.13) is valid in particular for the von Neumann algebra A^{**}. By Proposition 7.26, for any $(x_j) \in A^n$ we have $\left\| \sum \overline{x_j} \otimes x_j \right\|_{\overline{A} \otimes_{\max} A} = \left\| \sum \overline{x_j} \otimes x_j \right\|_{\overline{A^{**}} \otimes_{\max} A^{**}}$. Similarly, by Lemma 22.9 for any $(x_j) \in A^n$ we have $\|(x_j)\|_{1/2,A} = \|(x_j)\|_{1/2,A^{**}}$, and we conclude that $\|(x_j)\|_{1/2,A} = \left\| \sum \overline{x_j} \otimes x_j \right\|_{\overline{A} \otimes_{\max} A}^{1/2}$. $\qquad \square$

Corollary 22.16 *For any C^*-algebra A with the WEP (in particular if $A = B(H)$ or if A is injective), for any $(x_1, \ldots, x_n)$ in A we have*

$$\left\| \sum \overline{x_j} \otimes x_j \right\|_{\overline{A} \otimes_{\min} A}^{1/2} = \left\| \sum \overline{x_j} \otimes x_j \right\|_{\overline{A} \otimes_{\max} A}^{1/2}. \qquad (22.15)$$

Proof We already know this for $A = B(H)$ by Corollary 22.13. Using the c.p. factorization $i_A : A \to B(H) \to A^{**}$ (see Theorem 9.22) we find $\left\| \sum \overline{x_j} \otimes x_j \right\|_{\overline{A^{**}} \otimes_{\max} A^{**}}^{1/2} \leq \left\| \sum \overline{x_j} \otimes x_j \right\|_{\overline{A} \otimes_{\min} A}^{1/2}$ and we conclude by Proposition 7.26. $\qquad \square$

Remark 22.17 We will prove later on in Theorem 23.7 that (22.15) characterizes the WEP.

Applying the second parts of Corollary 22.13 and Proposition 7.26, we obtain by a similar reasoning:

Corollary 22.18 *If A, B are WEP C^*-algebras, then for any $(x_j) \in A^n$, $(y_j) \in B^n$ we have*

$$\left\| \sum \overline{x_j} \otimes y_j \right\|_{\overline{A} \otimes_{\max} B} \leq \left\| \sum \overline{x_j} \otimes x_j \right\|_{\overline{A} \otimes_{\min} A}^{1/2} \left\| \sum \overline{y_j} \otimes y_j \right\|_{\overline{B} \otimes_{\min} B}^{1/2}.$$

The next corollary is a bit surprising in view of the fact that a priori only decomposable mappings are (max $\to$ max)-tensorizing (see Theorem 7.6).

Corollary 22.19 *Let $u : A \to B$ be a c.b. map between C^*-algebras with $\|u\|_{cb} \leq 1$. Then for any n and any $(x_j) \in A^n$*

$$\left\| \sum \overline{u(x_j)} \otimes u(x_j) \right\|_{\overline{B} \otimes_{\max} B} \leq \left\| \sum \overline{x_j} \otimes x_j \right\|_{\overline{A} \otimes_{\max} A}. \qquad (22.16)$$

Proof We will use the previously defined notation in (22.7) for (X_0, X_1) with underlying space A^n, and denote by (Y_0, Y_1) the analogous pair of Banach spaces with underlying space B^n. We have by Proposition 2.1

$$\|(u(x_j))\|_{Y_0} \le \|(x_j)\|_{X_0} \text{ and } \|(u(x_j))\|_{Y_1} \le \|(x_j)\|_{X_1}. \tag{22.17}$$

By Theorem 22.4, this implies $\|(u(x_j))\|_{1/2, B} \le \|u\|_{cb}\|(x_j)\|_{1/2, A}$, and (22.16) now follows from Corollary 22.15. $\qquad\square$

Remark 22.20 For (22.16) to hold it suffices to assume a priori much less than $\|u\|_{cb} \le 1$. Firstly it suffices, as expressed in (22.17), that u tensorizes with constant 1 with both the row and the column operator spaces, namely R and C. Secondly, (22.16) also holds if the c.b. norm of u is computed as a map from A to B^{op}. The reason being that in that case, we have $\|(u(x_j))\|_{Y_1} \le \|u\|_{cb}\|(x_j)\|_{X_0}$ and $\|(u(x_j))\|_{Y_0} \le \|u\|_{cb}\|(x_j)\|_{X_1}$ but since $(Y_0, Y_1)_{1/2} = (Y_1, Y_0)_{1/2}$, we conclude just the same that (22.16) holds.

The next result was conjectured by Wigner, Yanase, and Dyson and proved by Lieb [169]. We sketch a quick proof based on the Stein interpolation theorem.

Lemma 22.21 (WYDL concavity inequality) *Let (M, τ) be a semifinite von Neumann algebra. Let $x \in M$. Let $F : L_1(\tau)_+ \times L_1(\tau)_+ \to \mathbb{R}_+$ be the function defined by*

$$F(f, g) = \tau(x^* f^{1/2} x g^{1/2}).$$

Then F is jointly concave. Explicitly for any $0 < \theta < 1$ and any $(f_0, g_0), (f_1, g_1) \in L_1(\tau)_+ \times L_1(\tau)_+$ we have

$$(1 - \theta)F(f_0, g_0) + \theta F(f_1, g_1) \le F(f_\theta, g_\theta) \tag{22.18}$$

where $(f_\theta, g_\theta) = (1 - \theta)(f_0, g_0) + \theta(f_1, g_1)$.

Proof By a density argument (based on the definition of semifiniteness) we may assume that τ is finite. Then we may view 1 as the element of $L_1(\tau)_+$ corresponding to τ. Since M is dense in $L_1(\tau)$, we may assume again by density (or actually by a simple truncation) argument that f_0, g_0, f_1, g_1 are all in M_+. Noting that

$$F(f, g) = \lim_{\varepsilon \to 0^+} \tau(x^*(f^{1/2} + \varepsilon 1)x(g^{1/2} + \varepsilon 1)),$$

we may reduce even further to the case when f_0, g_0, f_1, g_1 are all invertible elements in M_+. Then we may substitute x for $f_\theta^{-1/4} x g_\theta^{-1/4}$ and (22.18) reduces to

$$(1 - \theta)\tau(g_\theta^{-1/4} x^* f_\theta^{-1/4} f_0^{1/2} f_\theta^{-1/4} x g_\theta^{-1/4} g_0^{1/2})$$
$$+ \theta\tau(g_\theta^{-1/4} x^* f_\theta^{-1/4} f_1^{1/2} f_\theta^{-1/4} x g_\theta^{-1/4} g_1^{1/2}) \le \|x\|_{L_2(\tau)}^2,$$

which is identical to

$$(1 - \theta)\|f_0^{1/4} f_\theta^{-1/4} x g_\theta^{-1/4} g_0^{1/4}\|_{L_2(\tau)}^2$$
$$+ \theta\|f_1^{1/4} f_\theta^{-1/4} x g_\theta^{-1/4} g_1^{1/4}\|_{L_2(\tau)}^2 \le \|x\|_{L_2(\tau)}^2. \tag{22.19}$$

By convention we set $L_p(\tau) = M$ when $p = \infty$. For any $1 \le p \le \infty$ we introduce the space $X_{p,\theta}$ that is $L_p(\tau) \oplus L_p(\tau)$ equipped with the norm defined for $1 \le p < \infty$ by

$$\|(x_0, x_1)\|_{X_{p,\theta}} = ((1 - \theta)\|x_0\|_p^p + \theta\|x_1\|_p^p)^{1/p},$$

and for $p = \infty$ by

$$\|(x_0, x_1)\|_{X_{p,\theta}} = \max\{\|x_0\|_M, \|x_1\|_M\}.$$

Since the parameter θ will remain fixed throughout the argument we will denote $X_{p,\theta}$ simply by X_p. For any $z \in S$ let $T(z): M \to X_p$ be defined for all $x \in M$ by

$$T(z)(x) = (T_0(z)(x), T_1(z)(x))$$

where

$$T_0(z)(x) = f_0^{z/2} f_\theta^{-z/2} x g_\theta^{-z/2} g_0^{z/2} \quad \text{and} \quad T_1(z)(x) = f_1^{z/2} f_\theta^{-z/2} x g_\theta^{-z/2} g_1^{z/2}.$$

When $z = it \in \partial_0$ ($t \in \mathbb{R}$) then $f_0^{z/2}, f_1^{z/2}, g_0^{z/2}, g_1^{z/2}, f_\theta^{-z/2}, g_\theta^{-z/2}$ are all unitary so that

$$\|T(z)(x)\|_{X_\infty} \le \|x\|_{L_\infty(\tau)}$$

is immediate.

When $z = 1 + it \in \partial_1$ ($t \in \mathbb{R}$) we claim that

$$\|T(z)(x)\|_{X_1} \le \|x\|_{L_1(\tau)}.$$

Taking the claim for granted, the proof can be completed using the Stein interpolation theorem (see Remark 22.5): we have

$$\|T(1/2)(x)\|_{X_2} \le \|x\|_{L_2(\tau)}$$

for any $x \in M$, which is the same as (22.19). This yields (22.19).

Let $u_{0,t} = f_0^{it/2}$, $u_{1,t} = f_1^{it/2}$ and $y_{0,t} = g_0^{it/2}$, $y_{1,t} = g_1^{it/2}$. Again these are all unitaries. To check the claim observe that $\|T(1 + it)(x)\|_{X_1}$ can be rewritten as

$$(1 - \theta)\|u_{0,t} f_0^{1/2} f_\theta^{-1/2} (v_t x w_t) g_\theta^{-1/2} g_0^{1/2} y_{0,t}\|_{L_1(\tau)}$$
$$+ \theta \|u_{1,t} f_1^{1/2} f_\theta^{-1/2} (v_t x w_t) g_\theta^{-1/2} g_1^{1/2} y_{1,t}\|_{L_1(\tau)}$$

where v_t, w_t are unitary. By unitary invariance

$$\|T(1 + it)(x)\|_{X_1} = (1 - \theta)\|f_0^{1/2} f_\theta^{-1/2} (v_t x w_t) g_\theta^{-1/2} g_0^{1/2}\|_{L_1(\tau)}$$
$$+ \theta \|f_1^{1/2} f_\theta^{-1/2} (v_t x w_t) g_\theta^{-1/2} g_1^{1/2}\|_{L_1(\tau)}.$$

Let

$$a_0 = (1 - \theta)^{1/2} f_0^{1/2} f_\theta^{-1/2}, \quad b_0 = (1 - \theta)^{1/2} g_\theta^{-1/2} g_0^{1/2},$$
$$a_1 = \theta^{1/2} f_1^{1/2} f_\theta^{-1/2}, \quad b_1 = \theta^{1/2} g_\theta^{-1/2} g_1^{1/2}.$$

Then

$$\|T(1 + it)(x)\|_{X_1} = \|a_0 (v_t x w_t) b_0\|_{L_1(\tau)} + \|a_1 (v_t x w_t) b_1\|_{L_1(\tau)},$$

with

$$\tau(a_0^* a_0 + a_1^* a_1) = \tau(f_\theta^{-1/2}((1 - \theta) f_0 + \theta f_1) f_\theta^{-1/2}) = 1$$

and similarly $\tau(b_0 b_0^* + b_1 b_1^*) \leq 1$. By (2.4) we have

$$\|a_0 (v_t x w_t) b_0\|_{L_1(\tau)} + \|a_1 (v_t x w_t) b_1\|_{L_1(\tau)} \leq \|v_t x w_t\|_{L_1(\tau)} = \|x\|_{L_1(\tau)}.$$

This establishes the claim. $\qquad\square$

The next lemma gives us a simple formula to compute the norm of a *self-adjoint* $t \in (\overline{M} \otimes M)^+$. As explained in Remark 22.23, it can be viewed as a noncommutative variant of the classical Perron–Frobenius theorem. See e.g. [97] for more on that theme.

Lemma 22.22 *Let M be a semifinite von Neumann algebra equipped with a (normal faithful semifinite) trace τ. Then for any $t = \sum \overline{x_j} \otimes x_j \in \overline{M} \otimes M$ such that $t = t^*$*

$$\left\|\sum \overline{x_j} \otimes x_j\right\|_{\max} = \sup \sum \tau(x_j^* f^{1/2} x_j f^{1/2}), \qquad (22.20)$$

where the sup runs over all normal states $f \in L_1(\tau)$. Moreover, if M is assumed σ-finite, the sup does not change if we restrict it to faithful normal states $f \in L_1(\tau)$.

Proof By (22.14) we have $\left\|\sum \overline{x_j} \otimes x_j\right\|_{\max} = \sup \left|\sum \tau(x_j^* \eta x_j \xi)\right|$, where the sup runs over all η, ξ in the unit ball of $L_2(\tau)$. Since any such ξ admits a decomposition $\xi = \xi_1 - \xi_2 + i(\xi_3 - \xi_4)$ with $\xi_k \in L_2(\tau)_+$ ($1 \leq k \leq 4$) such that $\sum_k \|\xi_k\|_2^2 \leq 1$ and similarly for η, it is easy to check that we may without change restrict the last sup to all $\eta, \xi \geq 0$ in the unit sphere of $L_2(\tau)$. This gives us

$$\left\| \sum \overline{x_j} \otimes x_j \right\|_{\max} = \sup \sum \tau(x_j^* f^{1/2} x_j g^{1/2}),$$

where the sup runs over all normal states $f, g \in L_1(\tau)$. Let $F(f, g) = \sum \tau(x_j^* f^{1/2} x_j g^{1/2})$. Since $t = t^*$ we have $F(f, g) = F(g, f)$, and hence $F(f, g) = (F(f, g) + F(g, f))/2$. By the WYDL concavity inequality (22.18) we have $F(f, g) = (F(f, g) + F(g, f))/2 \leq F(\varphi, \varphi)$ with $\varphi = (f + g)/2$. From this (22.20) becomes clear.

The last assertion is immediate by a perturbation argument since, if M is σ-finite, the set of $f^{1/2}$s with f a faithful state is dense in the set of $f \geq 0$ in the unit sphere of $L_2(\tau)$. □

Remark 22.23 In the situation of the preceding lemma, let $T : L_2(\tau) \to L_2(\tau)$ denote the operator defined by $T(f) = \sum x_j f x_j^*$ $(x_j \in M)$. Note that T is self-adjoint when $t^* = t$. Moreover T is positivity preserving. Equivalently we have $\langle f, T(g) \rangle \geq 0$ for any $f, g \geq 0$ and in particular $\langle f, T(f) \rangle \geq 0$ for any $f \geq 0$. Indeed, this holds by the following simple identity

$$\forall a, b \in L_2(\tau)_+ \quad \tau(ax^*bx) = \|b^{1/2} x a^{1/2}\|_{L_2(\tau)}^2. \tag{22.21}$$

Assume that $\|T\| = 1$ and the supremum in (22.20) is attained on some $f \geq 0$ in the unit sphere of $L_2(\tau)$ (which is automatic e.g. in the finite-dimensional case). Then $\langle f, T(f) \rangle = 1$ forces $T(f) = f$, which means that f is an eigenvector for the eigenvalue 1.

To prove Theorem 22.10 we will need some results from the theory of operator valued Hardy spaces. Let H be a Hilbert space. We denote by $H^2(H)$ the space of H-valued functions f on the unit disc $D \subset \mathbb{C}$ that are analytic on D with Taylor series $f = \sum_0^\infty x_n z^n$ such that $\sum \|x_n\|^2 < \infty$. This is a Hilbert space with the norm $\|f\| = \left(\sum_0^\infty \|x_n\|^2 \right)^{1/2}$. For a.a. points $z \in \mathbb{T} = \partial D$ the radial limit $F(z) = \lim_{r \to 1} f(rz)$ exists and defines a function in $L_2(\mathbb{T}; H)$ with Fourier transform vanishing on the negative integers. Conversely, any function $F \in L_2(\mathbb{T}; H)$ with Fourier transform vanishing on the negative integers extends to an analytic function $f \in H^2(H)$ inside D. Moreover the correspondence $f \mapsto F$ is isometric. When $H = \mathbb{C}$, we denote simply $H^2 = H^2(\mathbb{C})$.

We will use Szegö's classical factorization theorem which says that under a nonvanishing condition, a positive function W in $L_1(\mathbb{T})$ can always be written as $W = |F|^2$ ($W = \overline{F} F$ is more suggestive in view of the noncommutative case) for some F in H^2. Moreover, this can be done with F "outer," so that $z \to 1/F(z)$ is analytic inside the disc, and if we additionally require $F(0) > 0$

then F is unique. Actually, we will need an extension of this theorem (due to Devinatz) valid for $B(H)$-valued functions. The following consequence of Devinatz's theorem will be enough for our purposes (cf. [70, 133]).

Theorem 22.24 *Let H be a separable Hilbert space and let $W : \mathbb{T} \to B(H)$ be a function such that, for all x, y in H, the function $t \to \langle W(t)x, y \rangle$ is in $L_1(\mathbb{T})$. Assume that there is $\delta > 0$ such that $W(t) \geq \delta I$ for all t. Then there is a unique analytic function $F : D \to B(H)$ such that*

(i) *for all x in H, $z \to F(z)x$ is in $H^2(H)$ and its boundary values satisfy almost everywhere on $\mathbb{T}$*

$$\langle W(t)x, y \rangle = \langle F(t)x, F(t)y \rangle,$$

(ii) $F(0) \geq 0$,

(iii) $z \to F(z)^{-1}$ *exists and is bounded analytic on D.*

Corollary 22.25 *Consider a von Neumann subalgebra $M \subset B(H)$. Then in the situation of Theorem 22.24, if W is M-valued, F necessarily also is M-valued.*

Proof Indeed, for any unitary u in the commutant M', the function $z \to u^* F(z)u$ still satisfies the conclusions of Theorem 22.24, and hence (by uniqueness) we must have $F = u^* Fu$, which implies by the bicommutant Theorem A.46 that $F(z) \in M'' = M$. $\qquad\square$

Proof of Theorem 22.10 Let $\theta = 1/p$. Recall that we denote $L_\infty(\tau) = M$. By Theorem 22.8 when $\theta = 1/p$ we have isometrically $(L_\infty(\tau), L_1(\tau))_\theta = L_p(\tau)$. We will use the spaces (X_0, X_1) defined in (22.7) but for $A = M$. Clearly for all $(x_j) \in M^n$ we have by (2.1)

$$\left\| \sum R(x_j^*)L(x_j) \right\|_{B(L_\infty(\tau))} = \|(x_1, \dots, x_n)\|_{X_0}.$$

Similarly, it is easy to check by transposition that

$$\left\| \sum R(x_j^*)L(x_j) \right\|_{B(L_1(\tau))} = \|(x_1, \dots, x_n)\|_{X_1}.$$

Hence, if (x_j) is in the unit ball of $(X_0, X_1)_\theta$ and $\theta = 1/p$, we have by Theorem 22.4 and (22.6)

$$\left\| \sum R(x_j^*)L(x_j) \right\|_{B(L_p(\tau))} \leq 1.$$

This is the easy direction. To prove the converse, we assume that

$$\left\| \sum R(x_j^*)L(x_j) \right\|_{B(L_p(\tau))} \leq 1. \tag{22.22}$$

We will proceed by duality. Let B denote the open unit ball in the space $(X_{0*}, X_{1*})_\theta$. Note that X_{0*} (resp. X_{1*}) coincides with M_*^n equipped with the norm

$$\|(\xi_1, \ldots, \xi_n)\|_{0*} = \tau\left[\left(\sum \xi_j^* \xi_j\right)^{1/2}\right],$$

$$\left(\text{resp.} \quad \|(\xi_1, \ldots, \xi_n)\|_{1*} = \tau\left[\left(\sum \xi_j \xi_j^*\right)^{1/2}\right]\right).$$

Let B^o be the polar of B in the duality between M^n and M_*^n. By the duality property of interpolation spaces (cf. Remark 22.3) B^o coincides with the unit ball of $(X_0, X_1)_\theta$. Hence to conclude it suffices to show that (22.22) implies $(x_1, \ldots, x_n) \in B^o$. Equivalently, to complete the proof it suffices to show that, if (22.22) holds, then for any $(\xi_1, \ldots, \xi_n)$ in B we have $\left|\sum \xi_j(x_j)\right| \le 1$. The rest of the proof is devoted to the verification of this.

By the definition of semifiniteness there is an increasing net of (finite trace) projections q_i tending weak* to the identity for which $q_i M q_i$ is a finite von Neumann algebra with unit q_i. If we identify again M_* with $L_1(\tau)$ in the usual way, we have $\|\xi - q_i \xi q_i\|_{M_*} \to 0$ for any $\xi \in M_*$. Thus, by perturbation we may assume ξ_j of the form $\xi_j(x) = \tau(b_j x)$ for some b_j in qMq where q is a projection in M, such that qMq is finite. In that case we have $\xi_j(x_j) = \xi_j(q x_j q)$. Note that (22.22) remains true if we replace (x_j) by $(q x_j q)$. Therefore, at this point we may as well replace M by the finite von Neumann algebra qMq (with unit q) on $L_2(qMq, \tau_{|qMq})$ so that we are reduced to the finite case. For simplicity, we assume in the rest of the proof that M is finite with unit I and that ξ_j lies in M viewed as a subspace of $L_1(\tau)$ (i.e. that the elements b_j appearing in the preceding step are in M and $q = I$). By definition of $(X_{0*}, X_{1*})_\theta$, since (ξ_j) is in B there are functions $f_j : \bar{S} \to L_1(\tau)$ that are bounded, continuous on $\bar{S}$ and analytic on S such that $\xi_j = f_j(\theta)$ for $j = 1, \ldots, n$ and moreover such that

$$\sup_{z \in \partial_0} \tau\left[\left(\sum f_j(z)^* f_j(z)\right)^{1/2}\right] < 1 \quad \text{and} \quad \sup_{z \in \partial_1} \tau\left[\left(\sum f_j(z) f_j(z)^*\right)^{1/2}\right] < 1.$$

$$(22.23)$$

Since ξ_j is in $M \subset L_1(\tau)$ and M^n is dense in M_*^n, we may as well assume, by a well-known fact due to Stafney (see [214, Lemma 8.11] for full details), that the functions $f_1, \ldots, f_n$ take their values in a fixed finite-dimensional subspace of $M \subset L_1(\tau)$. In that case the ranges $\{f_j(z) \mid z \in \bar{S}\}$ of the f_j's all lie in a weak* separable von Neumann subalgebra of M (that can be embedded in $B(\ell_2)$ if we wish). We are then in a position to use Theorem 22.24 and its corollary.

Let $\delta > 0$ (to be specified later). We define functions W_1 and W_2 on $\partial S = \partial_0 \cup \partial_1$ by setting

$$\forall\, z \in \partial_1 \qquad W_1(z) = \left(\left(\sum f_j(z) f_j(z)^* \right)^{1/2} + \delta I \right)^{1/2}, \qquad (22.24)$$

$$\forall\, z \in \partial_0 \qquad W_1(z) = I, \qquad (22.25)$$

$$\forall\, z \in \partial_1 \qquad W_2(z) = I, \qquad (22.26)$$

$$\forall\, z \in \partial_0 \qquad W_2(z) = \left(\left(\sum f_j(z)^* f_j(z) \right)^{1/2} + \delta I \right)^{1/2}. \qquad (22.27)$$

By (22.23) we can choose δ small enough so that

$$\sup_{z \in \partial_1} \tau(W_1^2) < 1 \quad \text{and} \quad \sup_{z \in \partial_0} \tau(W_2^2) < 1. \qquad (22.28)$$

By Theorem 22.24 and Corollary 22.25 , using a conformal mapping from S onto D, we find bounded M-valued analytic functions F and G on S with (nontangential) boundary values satisfying

$$FF^* = W_1^2 \quad \text{and} \quad G^*G = W_2^2. \qquad (22.29)$$

Moreover, F^{-1} and G^{-1} are analytic and bounded on S. Therefore we can write

$$f_j(z) = F(z) g_j(z) G(z)$$

where

$$g_j(z) = F(z)^{-1} f_j(z) G(z)^{-1}. \qquad (22.30)$$

We claim that

$$\forall\, z \in S \qquad \sum \|g_j(z)\|_{L_2(\tau)}^2 \le 1. \qquad (22.31)$$

By the three lines lemma, to verify this it suffices to check it on the boundary of S. (Note that we know a priori that $\sup_{z \in S} \|g_j(z)\|_{L_2(\tau)} < \infty$ since $\|F^{-1}\| < \delta^{-1/2}$ and $\|G^{-1}\| \le \delta^{-1/2}$, hence g_j is an H^∞ function with values in $L_2(\tau)$, and its nontangential boundary values still satisfy (22.30) a.e. on the boundary of S.) We have

$$\forall\, z \in \partial_1 \quad \sum \|g_j(z)\|_{L_2(\tau)}^2 = \tau \left(\sum g_j(z) g_j(z)^* \right)$$

$$= \tau \left(F(z)^{-1} \sum f_j(z) f_j(z)^* F(z)^{-1*} \right)$$

$$= \tau((FF^*)^{-1}(W_1^2 - \delta I)^2)$$

hence by (22.29) and (22.28)

$$\le \tau(W_1^2) < 1.$$

Similarly, we find

$$\forall z \in \partial_0 \qquad \sum \|g_j(z)\|^2_{L_2(\tau)} \leq \tau(W_2^2) < 1.$$

This proves our claim (22.31). Finally, if $\theta = 1/p$ we have

$$L_{2p}(\tau) = (M, L_2(\tau))_\theta \quad \text{and} \quad L_{2p'}(\tau) = (L_2(\tau), M)_\theta.$$

Hence by definition of the latter complex interpolation spaces, since $\|F(z)\|_M = \|W_1\|_M \leq 1$ on ∂_0 and (by (22.28)) $\|F(z)\|_{L_2(\tau)} < 1$ on ∂_1, we have

$$\|F(\theta)\|_{L_{2p}(\tau)} \leq 1$$

and similarly $\|G(\theta)\|_{L_{2p'}(\tau)} \leq 1$. To conclude we have $\xi_j = f_j(\theta) = F(\theta)g_j$ $(\theta)G(\theta)$, and hence if (x_j) satisfies (22.22) we have by (22.31) (and Cauchy–Schwarz)

$$\left|\sum \xi_j(x_j)\right| = \left|\sum \tau(F(\theta)g_j(\theta)G(\theta)x_j)\right|$$

$$\leq \left(\sum \|G(\theta)x_j F(\theta)\|^2_{L_2(\tau)}\right)^{1/2}$$

$$\leq \left\|\sum x_j F(\theta)F(\theta)^* x_j^*\right\|^{1/2}_{L_p(\tau)}$$

$$\leq \left\|\sum R(x_j^*)L(x_j)\right\|^{1/2}_{B(L_p(\tau))} \leq 1. \qquad (22.32)$$

Thus we have verified that (22.22) implies $(x_j) \in B^o$. This concludes the proof of Theorem 22.10. $\qquad\qquad\square$

22.3 Notes and remarks

The standard reference on complex interpolation in §22.1 is the book [22]. We also need some less standard facts notably by Bergh [21], Stafney and others, which the reader can find treated in some detail in [214]. To see how non-commutative L_p-spaces can be constructed using complex interpolation, see e.g. [215].

In §22.2 the main results come from [203, 205], with several improvements due to Haagerup, through personal communications at the time [203] was being published. This motivated Haagerup for the results described in the next chapter. The inequality in Corollary 22.18 has been communicated (in the main case $A = B = B(H)$) by Haagerup to Junge and the author for inclusion in the earlier paper [141].

The equalities in Theorem 22.10 and Corollaries 22.12 and 22.13 were greatly influenced by the discovery (due to O. Kouba, see [205]) that a certain kind of tensor product, in particular the Haagerup tensor product, behaves very nicely under complex interpolation. The main technical ingredient is an operator valued version of a classical theorem of Szegö that tells us that under a suitable nonvanishing assumption any integrable nonnegative matrix valued function $W : \mathbb{T} \to M_n$ can be written in the form $W = F^*F$ where $F : \mathbb{T} \to M_n$ is the boundary value of an *analytic* function in the Hardy space H^2. In Theorem 22.24 and Corollary 22.25 we use a generalization of Szegö's theorem with $B(\ell_2)$ in place of M_n due to Devinatz [70]. Such results have a long history, starting with Masani–Wiener and Helson–Lowdenslager, see e.g. [133]. The subject was later on investigated for operator algebras by Arveson [11] (see also the discussion of noncommutative Hardy spaces in our survey with Q. Xu [215]). Related results appear in a paper of Haagerup and the author [117]. We refer to Blecher and Labuschagne's work [26] for a more recent update on the state of the art on generalizations of Szegö's theorem in operator algebra theory.

23

Haagerup's characterizations of the WEP

In this chapter we give two new characterizations of the WEP that are significantly more involved than the preceding ones.

23.1 Reduction to the σ-finite case

In both cases, the main point consists in proving that if an inclusion of von Neumann algebras $M \subset \mathcal{M}$ satisfies a certain property, say property $\mathscr{P}$, then there is a completely contractive projection $P : \mathcal{M} \to M$. In this section we will show that modulo a simple assumption we may restrict to the case when M is σ-finite. The proof will use the structural Theorem A.65.

The assumptions on $\mathscr{P}$ are as follows: if $M \subset \mathcal{M}$ has property $\mathscr{P}$ then for any projection $q \in M$ the inclusion $qMq \subset q\mathcal{M}q$ (unital with unit q) also has property $\mathscr{P}$. Moreover, if $\pi : \mathcal{M} \to \mathcal{M}^1$ is an isomorphism of von Neumann algebras taking M onto a subalgebra $M^1 \subset \mathcal{M}^1$, then we assume that the "isomorphic inclusion" $M^1 \subset \mathcal{M}^1$ also has property $\mathscr{P}$.

Proposition 23.1 *Under the preceding assumptions, to show that every inclusion $M \subset \mathcal{M}$ with property $\mathscr{P}$ admits a completely contractive projection $P : \mathcal{M} \to M$, it suffices to settle the case when M is σ-finite.*

Proof Consider a general inclusion $M \subset \mathcal{M}$. By the structural Theorem A.65 we may assume

$$M = \left(\oplus \sum\nolimits_{i \in I} B(\mathcal{H}_i) \bar{\otimes} N_i \right)_\infty, \tag{23.1}$$

with the N_i's σ-finite. Let $M_i = B(\mathcal{H}_i) \bar{\otimes} N_i$. Let q_i be the (central) projection in M corresponding to M_i in (23.1) so that $M_i = q_i M = q_i M q_i$. Let $\mathcal{M}_i = q_i \mathcal{M} q_i$. By our first assumption on $\mathscr{P}$ the inclusion $M_i \subset \mathcal{M}_i$ satisfies $\mathscr{P}$. We claim that we have a von Neumann algebra $\mathcal{N}_i$, with a subalgebra $N_i^1 \subset \mathcal{N}_i$ and an isomorphism $\pi_i : \mathcal{M}_i \to B(\mathcal{H}_i) \bar{\otimes} \mathcal{N}_i$ such that $\pi_i(M_i) = B(\mathcal{H}_i) \bar{\otimes} N_i^1$. In other words, the inclusion $M_i \subset \mathcal{M}_i$ is "isomorphic" in the preceding sense to the inclusion $B(\mathcal{H}_i) \bar{\otimes} N_i^1 \subset B(\mathcal{H}_i) \bar{\otimes} \mathcal{N}_i$. Indeed, since $B(\mathcal{H}_i) \simeq B(\mathcal{H}_i) \otimes 1 \subset \mathcal{M}_i$, by Proposition 2.20, for some $\mathcal{N}_i$ we have an isomorphism $\pi_i : \mathcal{M}_i \to B(\mathcal{H}_i) \bar{\otimes} \mathcal{N}_i$ so that $\pi_i(x \otimes 1) = x \otimes 1$ for any $x \in B(\mathcal{H}_i)$. Then the subalgebra $1 \otimes N_i \subset M_i \subset \mathcal{M}_i$ is mapped by $\pi_i : \mathcal{M}_i \to B(\mathcal{H}_i) \bar{\otimes} \mathcal{N}_i$ to a subalgebra that commutes with $B(\mathcal{H}_i) \otimes 1_{\mathcal{N}_i}$, and hence is included in $1 \otimes \mathcal{N}_i$. Thus we find N_i^1 such that $\pi_i(1 \otimes N_i) = 1 \otimes N_i^1$, and an isomorphism $\psi_i : N_i \to N_i^1$ such that $\pi_i(1 \otimes y) = 1 \otimes \psi_i(y)$ for all $y \in N_i$. It follows that $\pi_i(B(\mathcal{H}_i) \otimes N_i) = B(\mathcal{H}_i) \otimes N_i^1$, and since π_i is bicontinuous for the weak* topology, we have $\pi_i(B(\mathcal{H}_i) \bar{\otimes} N_i) = B(\mathcal{H}_i) \bar{\otimes} N_i^1$. This proves the claim.

By our second assumption on $\mathscr{P}$, the inclusion $B(\mathcal{H}_i) \bar{\otimes} N_i^1 \subset B(\mathcal{H}_i) \bar{\otimes} \mathcal{N}_i$ satisfies $\mathscr{P}$. Let r_i be a rank 1 projection in $B(\mathcal{H}_i)$. Let $q_i' = r_i \otimes 1$. By our first assumption again, the inclusion $q_i'[B(\mathcal{H}_i) \bar{\otimes} N_i^1]q_i' \subset q_i'[B(\mathcal{H}_i) \bar{\otimes} \mathcal{N}_i]q_i'$ (with unit q_i') also satisfies $\mathscr{P}$. The latter being clearly "isomorphic" to the inclusion $N_i^1 \subset \mathcal{N}_i$ we conclude that $N_i^1 \subset \mathcal{N}_i$ satisfies $\mathscr{P}$. But now, at last, since $N_i^1 \simeq N_i$ is σ-finite, if we accept the σ-finite case, we find that there is a completely contractive projection $P_i : \mathcal{N}_i \to N_i^1$. By (2.19) and (2.22), $Id_{B(\mathcal{H}_i)} \otimes P_i$ defines a completely contractive projection from $B(\mathcal{H}_i) \bar{\otimes} \mathcal{N}_i$ to $B(\mathcal{H}_i) \bar{\otimes} N_i^1$. Then $Q_i = \pi_i^{-1}[Id_{B(\mathcal{H}_i)} \otimes P_i]\pi_i$ is a completely contractive projection from $\mathcal{M}_i$ to M_i, and hence the mapping $x \mapsto (Q_i(q_i x q_i))_{i \in I} \in \left(\oplus \sum_{i \in I} M_i \right)_\infty$ gives us a completely contractive projection from $\mathcal{M}$ onto M. $\qquad \square$

23.2 A new characterization of generalized weak expectations and the WEP

The main result is the following characterization of generalized weak expectations (see Definition 9.21), in terms of decomposable maps. It may be viewed as a refinement of Kirchberg's characterization of the latter in Theorem 7.6.

Theorem 23.2 *Let B be a C^*-algebra. Let $i : A \to B$ be the inclusion mapping from a C^*-subalgebra $A \subset B$. The following are equivalent:*

(i) *For any $n \geq 1$ and any $u : \ell_\infty^n \to A$ we have*

$$\|u\|_{D(\ell_\infty^n, A)} = \|iu\|_{D(\ell_\infty^n, B)}.$$

(ii) *For any $n \geq 1$ and any $v : \ell_\infty^n \to A^{**}$ we have*

$$\|v\|_{D(\ell_\infty^n, A^{**})} = \|i^{**}v\|_{D(\ell_\infty^n, B^{**})}.$$

(iii) *There is a completely contractive c.p. projection* $P : B^{**} \to A^{**}$ *(in other words by Remark 9.32 there is a generalized weak expectation from B to A^{**}).*

Proof We first claim (i) $\Leftrightarrow$ (ii). This is an immediate consequence of Theorem 8.25. Indeed, let $X_n = D(\ell_\infty^n, A)$ and $Y_n = D(\ell_\infty^n, B)$, viewed as Banach spaces. Then, the assertion that $X_n \subset Y_n$ isometrically, which is a reformulation of (i), is equivalent to $X_n^{**} \subset Y_n^{**}$ isometrically. This follows from the classical fact (see Remark A.13) that a mapping between Banach spaces is isometric if and only if its bitranspose is isometric. By Theorem 8.25 we have $X_n^{**} = D(\ell_\infty^n, A^{**})$ and $Y_n^{**} = D(\ell_\infty^n, B^{**})$. Thus, (i) $\Leftrightarrow$ (ii) follows. Then the equivalence (ii) $\Leftrightarrow$ (iii) will follow from the next statement about von Neumann algebras applied to the inclusion $A^{**} \subset B^{**}$. $\qquad\square$

Theorem 23.3 *Let $\mathcal{M}$ be a von Neumann algebra. Let $i : M \to \mathcal{M}$ be the inclusion mapping from a von Neumann subalgebra $M \subset \mathcal{M}$. The following are equivalent:*

(i) *For any $n \geq 1$ and any $u : \ell_\infty^n \to M$ we have*

$$\|u\|_{D(\ell_\infty^n, M)} = \|iu\|_{D(\ell_\infty^n, \mathcal{M})}.$$

(ii) *There is a completely contractive c.p. projection $P : \mathcal{M} \to M$.*

Proof We first show (i) $\Rightarrow$ (ii). We will use the reduction to the σ-finite case. Let $\mathscr{P}$ be the property appearing in (i). By the results of §6.1 it is easy to check that $\mathscr{P}$ satisfies the assumptions of Theorem 23.1. Therefore, to show (i) $\Rightarrow$ (ii) we may assume M σ-finite. Then (see Theorem A.63) there is a realization of M in some $B(H)$ such that M has a cyclic vector. Let $M' \subset B(H)$ be the commutant of M in $B(H)$. Let $I \subset U(M') \setminus \{1\}$ be a set of unitaries in M' that jointly generate M' as a von Neumann algebra, and let $\dot{I} = I \cup \{1\}$. Let $\mathbb{F}$ be a free group with (free) generators $(g_x)_{x \in I}$. Let $U_x = U_{\mathbb{F}}(g_x) \in C^*(\mathbb{F})$ $(x \in I)$, set also $U_1 = 1_{C^*(\mathbb{F})}$, and let $\sigma : C^*(\mathbb{F}) \to M'$ be the unital $*$-homomorphism defined by $\sigma(U_x) = x$ for all $x \in I$. Let $E = \mathrm{span}[U_x \mid x \in \dot{I}]$. Consider then the linear mapping $\widehat{u} : E \otimes M \to B(H)$ defined for any $e \in E, m \in M$ by $\widehat{u}(e \otimes m) = \sigma(e)m$ (and extended by linearity to $E \otimes M$). Then for any $t \in E \otimes M$ we have clearly by (4.6)

$$\|\widehat{u}(t)\| \leq \|t\|_{C^*(\mathbb{F}) \otimes_{\max} M}.$$

By (6.37) (i) implies that for any $t \in E \otimes M$ we have

$$\|t\|_{C^*(\mathbb{F}) \otimes_{\max} M} = \|t\|_{C^*(\mathbb{F}) \otimes_{\max} \mathcal{M}}.$$

This shows that (i) implies

$$\|\widehat{u} : E \otimes_{\overline{\max}} M \to B(H)\| \leq 1$$

where $E \otimes_{\overline{\max}} M$ is viewed as a subspace of $C^*(\mathbb{F}) \otimes_{\max} \mathcal{M}$ equipped with the induced norm.

By Theorem 2.6 since M has a cyclic vector we have

$$\|\widehat{u} : E \otimes_{\overline{\max}} M \to B(H)\| = \|\widehat{u} : E \otimes_{\overline{\max}} M \to B(H)\|_{cb}.$$

By the extension Theorem 1.18 there is $\widetilde{u} : C^*(\mathbb{F}) \otimes_{\max} \mathcal{M} \to B(H)$ extending $\widehat{u}$ with $\|\widetilde{u}\|_{cb} \le 1$.

$$
\begin{array}{ccc}
C^*(\mathbb{F}) \otimes_{\max} \mathcal{M} & & \\
\uparrow & \diagdown \; \widetilde{u} & \\
\downarrow & & \diagdown \\
E \otimes_{\overline{\max}} M & \xrightarrow{\ \widehat{u}\ } & B(H)
\end{array}
$$

Since $\widehat{u}$ is unital so is $\widetilde{u}$ and hence $\widetilde{u}$ is c.p. by Theorem 1.35. We claim that $E \otimes 1$ (and hence actually $C^*(\mathbb{F}) \otimes 1$) is included in the multiplicative domain $D_{\widetilde{u}}$. Indeed, since $\widetilde{u}(U_x \otimes 1) = \widehat{u}(U_x \otimes 1) = x \in U(M')$ for any $x \in \dot{I}$, we have $U_x \otimes 1 \in D_{\widetilde{u}}$ for any $x \in \dot{I}$ and the claim follows. Let $P : \mathcal{M} \to B(H)$ be defined by $P(b) = \widetilde{u}(1 \otimes b)$. Then P is completely contractive and c.p. Since $U_x \otimes 1 \in D_{\widetilde{u}}$ for any $x \in \dot{I}$ and since, by Theorem 5.1, $\widetilde{u}$ is bimodular with respect to $D_{\widetilde{u}}$ we have (by the trick we used previously many times) for any $x \subset \dot{I} = U(M')$

$$xP(b) = x\widetilde{u}(1 \otimes b) = \widetilde{u}((U_x \otimes 1)(1 \otimes b)) = \widetilde{u}(U_x \otimes b)$$
$$= \widetilde{u}((1 \otimes b)(U_x \otimes 1)) = \widetilde{u}(1 \otimes b)x = P(b)x.$$

Since the unitaries in I generate M', this shows that $P(b) \in (M')' = M$ and completes the proof that (i) $\Rightarrow$ (ii).

The converse (ii) $\Rightarrow$ (i) is an immediate consequence of (6.7) (recalling (i) in Lemma 6.5). $\qquad\square$

Remark 23.4 (The case $n = 3$) In the situation of the preceding Theorem 23.3 let us merely assume that for any $u : \ell_\infty^3 \to M$ we have $\|u\|_{D(\ell_\infty^3, M)} = \|iu\|_{D(\ell_\infty^3, \mathcal{M})}$. If we assume in addition that $M \subset B(H)$ is cyclic and that M' is generated by a pair of unitaries, then the same proof (now with $\mathbb{F} = \mathbb{F}_2$ and $|\dot{I}| = 3$) shows that there is a completely contractive c.p. projection $P : \mathcal{M} \to M$. Thus when $\mathcal{M} = B(H)$ we conclude that M is injective.

We recall in passing that it is a longstanding open problem whether any von Neumann algebra on a separable Hilbert space is generated by a single element or equivalently by two unitaries. Important partial results are known, notably by Carl Pearcy, see [74] for details and references. See Sherman's paper [229] for the current status of that problem. Actually this single generation problem is open for $M_{\mathbb{F}_\infty}$, which is a natural candidate for a counterexample. Note that since $M_{\mathbb{F}_2}$ is clearly singly generated a negative answer would have the spectacular consequence that $M_{\mathbb{F}_2}$ and $M_{\mathbb{F}_\infty}$ are not isomorphic, which is another famous open question.

We come to the characterization of the WEP:

Corollary 23.5 *Let $A \subset B(H)$ be a C^*-algebra. The following are equivalent:*

(i) *For any $n \geq 1$ and any $u : \ell_\infty^n \to A$ we have*

$$\|u\|_{D(\ell_\infty^n, A)} = \|u\|_{cb}.$$

(ii) *The C^*-algebra A has the WEP.*

Proof We apply Theorem 23.2 with $B = B(H)$. Note that in that case $\|u\|_{D(\ell_\infty^n, B(H))} = \|u\|_{cb}$ by (6.9) so that (i) in Corollary 23.5 is the same as (i) in Theorem 23.2. By Theorem 9.31 when $B = B(H)$ (iii) in Theorem 23.2 means that A has the WEP.   □

Since WEP and injectivity are equivalent for von Neumann algebras (see Corollary 9.26), we can now recover Haagerup's original result (see [104]) on this section's theme.

Corollary 23.6 *When A is a von Neumann algebra, the assertion (i) in Corollary 23.5 holds if and only if A is injective.*

23.3 A second characterization of the WEP and its consequences

Haagerup's unpublished characterization of the WEP follows as Theorem 23.7. This important result is closely related to the complex interpolation results presented in §22. It will be fully proved later on in §23.5. For the moment we only give indications on the easy parts of the proof.

Theorem 23.7 *The following properties of a C^*-algebra A are equivalent:*

(i) *A has the WEP.*

(ii) *For any n and any $a_1, \ldots, a_n$ in A we have*

$$\left\| \sum \overline{a_j} \otimes a_j \right\|_{\overline{A} \otimes_{\max} A} = \left\| \sum \overline{a_j} \otimes a_j \right\|_{\overline{A} \otimes_{\min} A}.$$

(iii) *There is a constant C such that for all n and all $a_1, \ldots, a_n$ in A we have*

$$\left\| \sum \overline{a_j} \otimes a_j \right\|_{\overline{A} \otimes_{\max} A} \leq C \left\| \sum \overline{a_j} \otimes a_j \right\|_{\overline{A} \otimes_{\min} A}.$$

(iv) *The inclusion $i_A : A \to A^{**}$ factors completely boundedly through $B(\mathcal{H})$ for some $\mathcal{H}$.*

First part of the proof of Theorem 23.7 Note that (iii) $\Rightarrow$ (ii) is elementary: let $x \in \bar{A} \otimes A$ be of the form appearing in (iii) or (ii). Assuming (iii) we have $\|x\|_{\max}^{2m} = \|(x^*x)^m\|_{\max} \leq C\|(x^*x)^m\|_{\min} = C\|x\|_{\min}^{2m}$, and hence $\|x\|_{\max} \leq C^{1/2m}\|x\|_{\min}$. Letting $m \to \infty$ we obtain (ii). The converse is trivial. For the proof of (i) $\Rightarrow$ (ii) see Corollary 22.16.

The proof of (ii) $\Rightarrow$ (i) is more delicate. It is analogous to that of Theorem 14.8. We prove this remaining part of the proof at the end of §23.5 after Theorems 23.34 and 23.35.

Taking this for granted this shows that (i), (ii), and (iii) are equivalent. We now turn to (iv).

(i) $\Rightarrow$ (iv) is immediate from the condition (iv) in Theorem 9.22.

Assume (iv). By Corollary 22.16, $B(\mathcal{H})$ satisfies the property (ii) in Theorem 23.7. Then, by (22.16) applied to the c.b. map from $B(\mathcal{H})$ to A^{**} in the factorization in (iv), there is a constant C such that for all n and all $a_1, \ldots, a_n$ in A we have

$$\left\| \sum \bar{a}_j \otimes a_j \right\|_{\overline{A^{**}} \otimes_{\max} A^{**}} \leq C \left\| \sum \bar{a}_j \otimes a_j \right\|_{\overline{A} \otimes_{\min} A}.$$

By part (ii) in Proposition 7.26, we deduce (iii), and hence (iv) is equivalent to (i), (ii), and (iii). $\qquad\square$

Remark 23.8 The remarkable advantage of the characterizations of the WEP in Theorem 23.7 is that they use only the operator space structure of A, because by Corollary 22.19 for any C^*-algebra B, any $u \in CB(A, B)$ and any $a_1, \ldots, a_n$ in A we have

$$\left\| \sum \overline{u(a_j)} \otimes u(a_j) \right\|_{\overline{B} \otimes_{\max} B} \leq \|u\|_{cb}^2 \left\| \sum \bar{a}_j \otimes a_j \right\|_{\overline{A} \otimes_{\max} A}. \tag{23.2}$$

Moreover the analogous inequality for the min-norm is obvious. Thus if B is completely isomorphic to a C^*-algebra A with the WEP, it must satisfy (iii) in Theorem 23.7, and hence have the WEP.

Corollary 23.9 *The properties in Theorem 9.22 (and in Theorem 23.7) are equivalent to:*

(v) *Any c.b. map $u : A \to M$ into a von Neumann algebra admits for some $\mathcal{H}$ a factorization of the form*

$$u : A \xrightarrow{v} B(\mathcal{H}) \xrightarrow{w} M$$

with $\|v\|_{cb}\|w\|_{cb} = \|u\|_{cb}$.

Proof Indeed, Theorem 8.1 (iii) shows that u admits an extension $\ddot{u} : A^{**} \to M$ with $\|\ddot{u}\|_{cb} = \|u\|_{cb}$, so we have a factorization of u of the form $A \xrightarrow{i_A} A^{**} \xrightarrow{\ddot{u}} M$. This shows that the property (iv) in Theorem 9.22 implies (v).

Conversely, assume (v). Then (v) holds for $u = i_A : A \to A^{**}$. Theorem 23.7 shows that A has the WEP. $\qquad\square$

A first consequence of the implication (iii) $\Rightarrow$ (i) in Theorem 23.7 and of (23.2) is a different approach to a result due independently to Christensen-Sinclair and the author (see [208, p. 273]):

Corollary 23.10 *Let $A \subset B(H)$ be a C^*-subalgebra such that there is a c.b. projection $P : B(H) \to A$ then A has the WEP. Thus if A is a von Neumann algebra, it must be injective.*

Proof Since it has the WEP, $B(H)$ satisfies (ii) in Theorem 23.7. Then by (23.2) A satisfies (iii) in Theorem 23.7 with $C = \|P\|_{cb}^2$. $\qquad\square$

We give more results in the same vein, on c.b. projections from a general von Neumann algebra onto a subalgebra, at the end of §23.5.

23.4 Preliminaries on self-polar forms

It will be crucial to consider von Neumann algebras $M \subset B(H)$ for which the commutant M' appears as a mirror copy of M. The following simple (Radon–Nikodym type) lemma will be useful.

Lemma 23.11 *Let $M \subset B(H)$ be a von Neumann algebra. Fix a unit vector $\xi \in H$. Let $\varphi(x) = \langle \xi, x\xi \rangle$ for all $x \in M$. For any $\psi \in M^*$ such that $0 \le \psi \le \varphi$ there is $a \in M'$ with $0 \le a \le 1$ such that*

$$\forall x \in M \qquad \psi(x) = \langle \xi, ax\xi \rangle.$$

Proof Let $H_\xi = \overline{M\xi} \subset H$. We have

$$\forall x, y \in M \quad |\psi(y^*x)| \le (\psi(x^*x)\psi(y^*y))^{1/2} \le \|x\xi\| \, \|y\xi\|.$$

Therefore there is a (unique) linear map $a : H_\xi \to H_\xi$ with $\|a\| \le 1$ such that

$$\forall x, y \in M \qquad \psi(y^*x) = \langle y\xi, ax\xi \rangle. \qquad (23.3)$$

We extend a to H by setting $a = 0$ on $H_\xi^\perp$. Clearly (23.3) still holds. Since

$$\forall z \in M \quad \psi((z^*y)^*x) = \psi(y^*zx) = \psi(y^*(zx))$$

and $\langle z^*y\xi, ax\xi \rangle = \langle y\xi, zax\xi \rangle$, we have

$$\langle y\xi, zax\xi \rangle = \langle y\xi, azx\xi \rangle,$$

and hence $(za - az)_{|H_\xi} = 0$. Also $za - az = 0$ on $H_\xi^\perp$ (because $z^* H_\xi \subset H_\xi$ implies $zH_\xi^\perp \subset H_\xi^\perp$). Therefore $a \in M'$ and we have $\psi(x) = \langle \xi, ax\xi \rangle$ for any

x in M. Lastly $\psi(x^*x) = \langle x\xi, ax\xi \rangle$ shows $a \geq 0$ on H_ξ. Since a vanishes on $H_\xi^\perp$ we have $0 \leq a \leq 1$ as announced. $\qquad\square$

We will work with sesquilinear forms $s : M \times M \to \mathbb{C}$ on a complex vector space M. This means that s is antilinear (resp. linear) in the first (resp. second) variable. Clearly such forms are in $1 - 1$ correspondence with *linear* forms on $\overline{M} \otimes M$. A sesquilinear form $s : M \times M \to \mathbb{C}$ will be called positive definite if $s(x,x) \geq 0$ for any $x \in M$. If moreover $s(x,x) = 0 \Rightarrow x = 0$ we say that s is strictly positive definite (or nondegenerate).

Remark 23.12 Let $u : A \to H$ be a bounded linear map. Then $s(y,x) = \langle uy, ux \rangle$ is a bounded positive definite sesquilinear form on $A \times A$. Conversely, any bounded positive sesquilinear form $s : A \times A \to \mathbb{C}$ is of this form (and we may replace H by $\overline{u(A)}$ to ensure that u has dense range if we wish). Indeed by a well-known construction (after passing to the quotient by the kernel of s and completing) we find a Hilbert space H associated to the inner product $(y,x) \to s(y,x)$ and a linear map $u : A \to H$ with dense range such that $s(y,x) = \langle uy, ux \rangle$.

Remark 23.13 Let s be a positive definite bounded sesquilinear form on a von Neumann algebra M. If (x_i) is a bounded net in M converging weak* to $x \in M$ and if s is separately weak* continuous then $s(x,x) \leq \liminf s(x_i, x_i)$.

Indeed, by Cauchy–Schwarz $s(x,x) = \lim s(x,x_i) \leq \liminf s(x,x)^{1/2} s(x_i, x_i)^{1/2}$.

Definition 23.14 Let A be a unital C^*-algebra. A (sesquilinear) form $s : A \times A \to \mathbb{C}$ will be called "bipositive" if it is both positive definite and such that

$$s(a,b) \geq 0 \qquad \forall a,b \in A_+. \tag{23.4}$$

We call it normalized if $s(1,1) = 1$.

For example, let (M,τ) be a tracial probability space. Assume $A \subset M$. Recall that for all $a,b \in M_+$ we have by the trace property $\tau(ab) = \tau(a^{1/2}ba^{1/2}) \geq 0$. This shows that the form $(y,x) \mapsto \tau(y^*x)$ is bipositive.

More generally, given $\xi \geq 0$ in $L_2(\tau)$ the sesquilinear form s defined on $A \times A$ by

$$s(y,x) = \tau(y^*\xi x\xi) = \langle \xi^{1/2} y\xi^{1/2}, \xi^{1/2} x\xi^{1/2} \rangle_{L_2(\tau)} \tag{23.5}$$

is bipositive.

The condition (23.4) means that the linear map $u : A \to \overline{A}^*$ associated to s is positivity preserving. For example this holds whenever s is associated to a state on $\overline{A} \otimes_{\max} A$ since that means u is c.p. (see Theorem 4.16). However, not every state on $\overline{A} \otimes_{\max} A$ gives rise to a *positive definite* form s. We call those which satisfy this "positive definite states" on $\overline{A} \otimes_{\max} A$. In other words a state for $\overline{A} \otimes_{\max} A$ is called positive definite if $f(\overline{x} \otimes x) \geq 0$ for any $x \in A$.

Remark 23.15 If a bipositive form s is nondegenerate then the state $\varphi : x \mapsto s(1,x)$ is faithful. Indeed, if $x \geq 0$ and $\varphi(x) = 0$ then $0 \leq s(x,x) \leq s(\|x\|1,x) = 0$, which shows that s nondegenerate implies φ faithful. In the converse direction, if φ is faithful and $s(x,x) = 0$ for some $x \geq 0$ then by Cauchy–Schwarz $s(y,x) = 0$ for any $y \in M$ and hence $\varphi(x) = 0$, so that $x = 0$, which is a weaker form of nondegeneracy.

Remark 23.16 Let $s : M \times M$ be a bipositive form. Let $\varphi \in M^*$ be the functional defined by $\varphi(x) = s(1,x)$. If φ is weak* continuous (i.e. normal) on M, then s is separately weak* continuous. Indeed, for any $a \in M_+$ we have $0 \leq s(a,x) \leq s(\|a\|1,x) = \|a\|\varphi(x)$ for all $x \in M_+$, and hence (see Remark A.45) $x \mapsto s(a,x)$ is normal for any $a \in M_+$. Since M is linearly generated by M_+ this remains true for any $a \in M$.

We will be mainly interested in the following notion.

Definition 23.17 Let M be a von Neumann algebra. A (sesquilinear) form s on $M \times M$ such that $x \mapsto s(1,x)$ is a normal state on M will be called self-polar if it is bipositive and such that for any $\psi \in M_+^*$ such that $0 \leq \psi(x) \leq s(1,x)$ for all x in M_+, there is $a \in M$ with $0 \leq a \leq 1$ such that $\psi(x) = s(a,x)$.

Remark 23.18 For example, if $\xi \geq 0$ and $\tau(\xi^2) = 1$ the form s in (23.5) is self-polar. Indeed, by Lemma 23.11 the condition $0 \leq \psi(x) \leq \tau(\xi^2 x) = \langle \xi, L(x)\xi \rangle$ $(x \in M)$ implies the existence of $0 \leq a \leq 1$ in M such that $\psi(x) = \langle \xi, R(a)L(x)\xi \rangle = s(a,x)$.

The following theorem plays a key role in the sequel. It is due to Woronowicz and Connes (see [59, 263]).

Theorem 23.19 *Let M be a von Neumann algebra equipped with a faithful normal state φ. Let s, s_1 be normalized bipositive forms on $M \times M$ such that*

$$\forall x \in M_+ \qquad s_1(1,x) \leq s(1,x) = \varphi(x).$$

If s is self-polar and strictly positive definite then $s_1(x,x) \leq s(x,x)$ for any $x \in M$.

Proof For any $a \in M$ with $0 \leq a \leq 1$, note

$$\forall x \in M_+ \qquad 0 \leq s_1(a,x) \leq s_1(1,x) \leq \varphi(x).$$

Let $\psi_a(x) = s_1(a,x)$. By the self-polar property of s, there is $b \in M$ with $0 \leq b \leq 1$ such that

$$\forall x \in M \quad \psi_a(x) = s(b,x).$$

Morever, such a b is unique. Indeed, since s is strictly positive definite we know $s(b,x) = 0\ \forall x \Rightarrow b = 0$. Thus we may set $b = T(a)$. In particular $0 \leq T(1) \leq 1$. The correspondence $a \to Ta$ is clearly additive and hence extends by scaling to M_+ and by linearity to the whole of M (see (A.11)): we set first $T(a^+ - a^-) = T(a^+) - T(a^-)$ if $a^* = a$, and $T(a) = T(\Re(a)) + iT(\Im(a))$ for an arbitrary $a \in M$.

Then for any $a \in M$ the element $Ta \in M$ is characterized by the identity

$$\forall x \in M \qquad s(Ta,x) = s_1(a,x).$$

It follows that $T : M \to M$ is a positive linear map with $T(1) \leq 1$. Therefore $\|T\| \leq 1$ by Corollary A.19. Furthermore, since s_1 is positive definite it satisfies $s_1(a,x) = \overline{s_1(x,a)}$ for all $a,x \in M$. From this we deduce $s(Ta,x) = s(a,Tx)$. We claim that

$$\forall x \in M \qquad |s(Tx,x)| \leq s(x,x).$$

Indeed assuming $x \neq 0$ let

$$\lambda(k) = \left| \frac{s(T^k x, x)}{s(x,x)} \right|.$$

Then by Cauchy–Schwarz we have

$$\lambda(k) \leq \lambda(2k)^{1/2}$$

and hence $\lambda(1) \leq \lambda(2)^{1/2} \leq \cdots \leq \lambda(2^n)^{1/2^n}$ for any $n \geq 1$. Since for any $0 \leq a \leq 1$ we have $0 \leq T(a) \leq 1$, by iteration we have $0 \leq T^k(a) \leq 1$ for any k and hence $s(T^k a, a) \leq 1$. Note that $a \mapsto s(T^k(a), T^k(a))^{1/2} = s(T^{2k}(a), a)^{1/2}$ is a subadditive functional on M. It follows using $a = a^+ - a^-$ (as in (A.11)) that $|s(T^{2k} a, a)| \leq 4\|a\|^2$ and $|s(T^{2k} x, x)| \leq 16\|x\|^2$ for any x in M. This gives us $\lambda(1) \leq (16\|x\|^2 s(x,x)^{-1})^{1/2^n}$ and letting $n \to \infty$ we find $\lambda(1) \leq 1$. Since the case $x = 0$ is trivial this proves our claim and a fortiori that $s_1(x,x) \leq s(x,x)$ for any $x \in M$. $\qquad\square$

Corollary 23.20 *Let M be a von Neumann algebra equipped with a self-polar form s such that the state $x \mapsto s(1,x)$ is faithful and normal. Then any normalized bipositive form s_1 such that*

$$\forall x \in M \quad s(x,x) \leq s_1(x,x)$$

must be identical to s.

Proof Since the forms are normalized, $x \mapsto s_1(1,x)$ is a state and $s(1 + tx, 1 + tx) \leq s_1(1+tx, 1+tx)$ implies $2t\Re s(1,x) + t^2 s(x,x) \leq 2t\Re s_1(1,x) + t^2 s_1(x,x)$. Letting $t \to 0$ we obtain $\Re s(1,x) = \Re s_1(1,x)$ for all $x \in M$ and hence $s(1,x) = s_1(1,x)$ for all $x \in M_+$. Then Theorem 23.19 tells us that $s_1(x,x) \leq$

$s(x,x)$ and hence $s_1(x,x) = s(x,x)$ for any $x \in M$. By polarization $s_1(y,x) = s(y,x)$ for any $y, x \in M$. $\square$

The preceding theorem involves positivity with respect to two distinct cones in $M \times M$, namely $M_+ \times M_+$ and $\{(x,x) \mid x \in M\}$. For the latter case the associated order relation for two positive definite forms s_1, s_2 on $M \times M$ means that $s_1 \leq s_2$ if $s_1(x,x) \leq s_2(x,x)$ for all $x \in M$. We will refer to this order as the pointwise order. In other words this is the pointwise ordering of the associated quadratic forms. The following statement is then an immediate consequence of Theorem 23.19:

Corollary 23.21 *Let φ and s be as in Theorem 23.19. Then s is the largest element (for the pointwise order) in the set of bipositive forms s' on $M \times M$ such that $s'(1,x) = \varphi(x)$ for all $x \in M$.*

Proposition 23.22 *Let A, B be unital C^*-algebras with $A \subset B$. Let s be a positive definite form on $A \times A$ such that $x \to s(1,x)$ is a state on A. The following are equivalent:*

(i) *We have*

$$\forall n, \ \forall x_1, \ldots, x_n \in A, \quad \sum s(x_j, x_j) \leq \left\| \sum \overline{x_j} \otimes x_j \right\|_{\overline{B} \otimes_{\max} B}.$$

(ii) *There is a state f on $\overline{B} \otimes_{\max} B$ such that*

$$\forall x \in A \qquad s(x,x) \leq \Re(f(\overline{x} \otimes x)) \tag{23.6}$$

and moreover for any self-adjoint $x \in A$

$$s(1,x) = (f(\overline{1} \otimes x) + f(\overline{x} \otimes 1))/2. \tag{23.7}$$

Proof Assume (i). We may assume that $\overline{B} \otimes_{\max} B \subset B(H)$ and that the embedding is of the form $\overline{y} \otimes x \to \sigma(y)^* \pi(x)$ where $\pi : B \to B(H)$ (resp. $\sigma : B \to B(H)$) is a $*$-homomorphism (resp. antihomomorphism) and π, σ have commuting ranges. Then we have

$$\sum s(x_j, x_j) \leq \sup_{f \in C} \sum \Re(f(\overline{x_j} \otimes x_j))$$

where C is the (weak* compact) unit ball of $(\overline{B} \otimes_{\max} B)^*$. By Hahn–Banach (see Lemma A.16) there is a net of finitely supported probabilities (λ_i) on C such that $s(x,x) \leq \lim_i \int \Re(f(\overline{x} \otimes x)) d\lambda_i(f)$. Let $f_i = \int f d\lambda_i \in C$. Passing to a subnet we may ensure by the weak* compactness of C that $f_i \to f$ pointwise and hence we have

$$\forall x \in A \qquad s(x,x) \leq \Re(f(\overline{x} \otimes x)).$$

But $1 = s(1, 1) \leq \Re(f(\bar{1} \otimes 1)) = (\Re f)(\bar{1} \otimes 1)$ implies that $\Re f$ is a state, and since $f \in C$ we must have $\Im(f) = 0$, so that f is a state (see Remark A.23). Replacing x by $1 + tx$ in (23.6) and letting $t \to 0$ we obtain (23.7). This proves (i) $\Rightarrow$ (ii). The converse is obvious. □

23.5 max$^+$-injective inclusions and the WEP

Our goal here is Haagerup's theorem [107] that asserts that if an inclusion $A \to B$ of C^*-algebras is such that the map $\bar{A} \otimes_{\max} A \to \bar{B} \otimes_{\max} B$ is injective when restricted to the "positive definite" tensors then that map is injective and actually $A \to B$ is max-injective. Note the analogy with the previous Corollary 14.9. When applied to the inclusion $A \subset B(H)$ this gives a new characterization of the WEP. This result may seem at first sight of limited scope, but it turns out to lie rather deep. In particular, despite our efforts we could not avoid the use of ingredients from the Tomita–Takesaki theory in the proof, which makes it less self-contained than we hoped for. In the case of the inclusion $A \subset B(H)$, one must show that any $*$-homomorphism $\pi : A \to M$ into a von Neumann algebra factors completely contractively through $B(H)$. When M is finite or semifinite, we do give a self-contained proof (see the proof of Theorem 23.30), but the general case eludes us, we need to use the so-called standard form of M or some fact of similar nature, for which including a complete proof would take us too far out.

Let A, B be C^*-algebras with $A \subset B$. We say that the inclusion $A \subset B$ is max$^+$-injective if

$$\forall n \; \forall x_1, \ldots, x_n \in A \quad \left\| \sum \overline{x_j} \otimes x_j \right\|_{\bar{A} \otimes_{\max} A} = \left\| \sum \overline{x_j} \otimes x_j \right\|_{\bar{B} \otimes_{\max} B}. \tag{23.8}$$

Let us denote again by $(\bar{A} \otimes A)^+$ the subset of the "positive definite" elements in $\bar{A} \otimes A$, i.e. the subset consisting of all the finite sums of the form $\sum \overline{x_j} \otimes x_j \; (x_j \in A)$.

Warning: this should not be confused with the set $(\bar{A} \otimes A)_+ = \{t^* t \mid t \in \bar{A} \otimes A\}$.

Remark 23.23 Let $t \in \bar{A} \otimes A$. It is an easy exercise in linear algebra to check that $t \in (\bar{A} \otimes A)^+$ if and only if $(\bar{\varphi} \otimes \varphi)(t) \geq 0$ for any $\varphi \in A^*$. In particular, whenever $A \subset B$, by the Hahn–Banach theorem, we have

$$(\bar{A} \otimes A)^+ = (\bar{A} \otimes A) \cap (\bar{B} \otimes B)^+. \tag{23.9}$$

With this notation the inclusion $A \subset B$ is $\max^+$-injective if the inclusion $\overline{A} \otimes_{\max} A \to \overline{B} \otimes_{\max} B$ is isometric when restricted to $(\overline{A} \otimes A)^+$.

Remark 23.24 Actually for this to hold it suffices that there exists a constant C such that

$$\forall n \ \forall x_1, \ldots, x_n \in A \quad \left\| \sum \overline{x_j} \otimes x_j \right\|_{\overline{A} \otimes_{\max} A} \le C \left\| \sum \overline{x_j} \otimes x_j \right\|_{\overline{B} \otimes_{\max} B}.$$
$$(23.10)$$

Indeed, let $t = \sum \overline{x_j} \otimes x_j$. Observe that $(t^*t)^m \in (\overline{A} \otimes A)^+$ for any $m \ge 1$. Therefore, (23.10) implies

$$\|t\|^{2m}_{\overline{A} \otimes_{\max} A} = \|(t^*t)^m\|_{\overline{A} \otimes_{\max} A} \le C \|(t^*t)^m\|_{\overline{B} \otimes_{\max} B} = C \|t\|^{2m}_{\overline{B} \otimes_{\max} B}$$

and hence

$$\|t\|_{\overline{A} \otimes_{\max} A} \le C^{1/2m} \|t\|_{\overline{B} \otimes_{\max} B}.$$

Letting $m \to \infty$ we obtain (23.8).

Remark 23.25 If $B \subset C$ is another inclusion between C^*-algebras, and if both inclusions $A \subset B$ and $B \subset C$ are $\max^+$-injective, then the same is true for $A \subset C$.

The following fact is an immediate consequence of (22.10) and Corollary 22.15.

Lemma 23.26 *If $A \subset B$ is $\max^+$-injective then $A^{**} \subset B^{**}$ is also $\max^+$-injective.*

Remark 23.27 If $A \subset B$ is max-injective then it is $\max^+$-injective. Indeed, since $\overline{A} \subset \overline{B}$ is clearly max-injective as well, the $*$-homomorphisms $\overline{A} \otimes_{\max} A \to \overline{A} \otimes_{\max} B$ and $\overline{A} \otimes_{\max} B \to \overline{B} \otimes_{\max} B$ are each isometric, and hence their composition $\overline{A} \otimes_{\max} A \to \overline{B} \otimes_{\max} B$ is isometric too. A fortiori the inclusion $A \subset B$ is $\max^+$-injective.

Our main goal is to show that conversely

$$\max^+\text{-injective} \Rightarrow \max\text{-injective},$$

but the proof will be quite indirect. In fact we will prove that $\max^+$-injective implies that there is a contractive projection $P : B^{**} \to A^{**}$. Since P is automatically c.p. this holds if and only if the inclusion $A \subset B$ is max-injective (see Theorem 7.29). By arguments that have been already discussed it suffices to show the following:

For any von Neumann algebra M any $*$-homomorphism $\pi : A \to M$ admits a c.p. extension $\upsilon : B \to M$.

The next lemma will allow us to reduce to the case when π is injective.

Lemma 23.28 *Let $A \subset B$ be a von Neumann subalgebra of a von Neumann algebra B. Let $\pi : A \to M$ be a normal *-homomorphism onto a von Neumann algebra M, let $\mathcal{I} = \ker(\pi)$ and let $[\pi] : A/\mathcal{I} \to M$ be the *-homomorphism canonically associated to π. We view $A/\mathcal{I} \subset B$ via the chain of embeddings*

$$A/\mathcal{I} \subset A/\mathcal{I} \oplus \mathcal{I} \simeq A \subset B.$$

(i) *If $A \subset B$ is max+-injective then the inclusion $A/\mathcal{I} \to B$ is max+-injective.*

(ii) *If $[\pi] : A/\mathcal{I} \to M$ admits a contractive positive (resp. c.p.) extension to B, then π admits a contractive positive (resp. c.p.) extension to B.*

Proof By (A.24) we have a decomposition

$$A = (A/\mathcal{I}) \oplus \mathcal{I}.$$

Let p, q be the central projections in A corresponding to this decomposition, so that

$$p + q = 1 \quad A/\mathcal{I} = qA, \qquad \mathcal{I} = pA.$$

Let $Q : A \to A/\mathcal{I}$ be the quotient map. For any $x \in A$ the inclusion $A/\mathcal{I} \subset A$ takes $Q(x)$ to qx (see Remark A.34). To show (i) we simply observe that the inclusion $A/\mathcal{I} \subset A$ is max-injective by Proposition 7.19. Then (i) becomes clear by Remark 23.25.

We turn to (ii). Let $w : B \to M$ be a contractive positive (resp. c.p.) extension of $[\pi] : A/\mathcal{I} \to M$, i.e. we have $w(qa) = \pi(a)$ for any a in A. Let $v : B \to M$ be defined by

$$\forall x \in B \qquad v(x) = w(qxq).$$

Clearly v is contractive positive (resp. c.p.) and for any $a \in A$ since $qa = aq$ we have

$$v(a) = w(qaq) = w(qa) = \pi(a),$$

which proves (ii). $\qquad\qquad\qquad\qquad\qquad\qquad\qquad\qquad\qquad\qquad$ □

The key step is the following. See §A.20 for background on σ-finiteness.

Theorem 23.29 *Let B be a C^*-algebra and $A \subset B$ a C^*-subalgebra. Let $\pi : A \to M$ be a *-homomorphism into a (σ-finite) von Neumann algebra M with a faithful normal state φ. If $A \subset B$ is max+-injective, then π admits a contractive c.p. extension $\tilde{\pi} : B \to M$.*

The key ingredient to prove this is the next statement, which guarantees the existence of self-polar forms associated to any faithful normal state φ. This fact (due to Connes) for a fully general M requires knowledge of the Tomita–Takesaki Theory, and unfortunately we will have to accept it without proof. We merely give indications on its proof.

Theorem 23.30 *Let φ be a faithful normal state on a (σ-finite) von Neumann algebra M. There is a unique strictly positive definite self-polar form s_φ on $M \times M$ such that*

$$s_\varphi(1,x) = \varphi(x) \qquad \forall x \in M.$$

*In addition, φ "extends" to a state Φ on $\overline{M} \otimes_{\max} M$ in the sense that $\Phi(\overline{x} \otimes x) = s_\varphi(x,x) \ \forall x \in M$. More precisely, there is a Hilbert space H, an embedding $\sigma : M \to B(H)$ (as a von Neumann subalgebra), a unit vector $\xi_\varphi \in H$ and a *-homomorphism $\rho_\varphi : \overline{M} \otimes_{\max} M \to B(H)$ such that*

$$s_\varphi(y,x) = \langle \xi_\varphi, \rho_\varphi(\overline{y} \otimes x)\xi_\varphi \rangle \tag{23.11}$$

$$\forall x \in M \quad \rho_\varphi(\overline{1} \otimes x) = \sigma(x) \tag{23.12}$$

and moreover $\overline{y} \mapsto \rho_\varphi(\overline{y} \otimes 1)$ is an isomorphism from $\overline{M}$ to $\sigma(M)'$.

Indications on the proof of Theorem 23.30 The unicity follows from Theorem 23.19 (or Corollary 23.21), so we will concentrate on the existence of such a form.

The case when M admits a faithful tracial state τ is easy. In that case, we may identify φ with an element of $L_1(\tau)_+$ so that $\varphi(x) = \tau(\varphi x)$ (see §11.2). Then $\varphi^{1/2} \in L_2(\tau)_+$ and, as we already mentioned (see Remark 23.18), it is not hard to show that the form s_φ defined by

$$s_\varphi(y,x) = \tau(y^* \varphi^{1/2} x \varphi^{1/2})$$

which is bipositive by (22.21) is self-polar. Letting $\xi_\varphi = \varphi^{1/2}$, $H = L_2(\tau)$ and $\rho_\varphi(\overline{y} \otimes x) = R_{y^*} L_x$ where L_x (resp. R_x) denotes left- (resp. right-)hand multiplication by x on $L_2(\tau)$, we obtain all the other properties.

Although this is technically more involved, the same idea works if τ is merely semifinite. However, in the general case, despite our efforts we could not find a shortcut to avoid the use (without proof) of the Tomita–Takesaki modular theory. We will use it via the following basic fact (see [241, p. 151], and Haagerup's landmark paper [102]):

Any von Neumann algebra admits a "standard form," which means that there is a triple $(H, J, P^\natural)$ consisting of a Hilbert space H such that $M \subset B(H)$ (as a von Neumann algebra), an antilinear isometric involution $J : H \to H$ and a cone $P^\natural \subset H$ such that

(i) $JMJ = M'$ and $JxJ = x^*$ $\quad \forall x \in M \cap M'$.

(ii) $P^\natural \subset \{\xi \in H \mid J\xi = \xi\}$ and $P^\natural$ is self-dual, i.e.

$$P^\natural = \{\xi \in H \mid \langle \xi, \eta \rangle \geq 0 \; \forall \eta \in P^\natural\}.$$

(iii) $\forall x \in M \quad JxJx(P^\natural) \subset P^\natural$.

(iv) For any φ in M_*^+ there is a unique ξ_φ in $P^\natural$ such that $\varphi(x) = \langle \xi_\varphi, x\xi_\varphi \rangle$ for any x in M.

Let φ be a normal faithful state so that $\|\xi_\varphi\| = 1$. We set $\sigma(x) = x$ and we define

$$s_\varphi(y, x) = \langle \xi_\varphi, \, JyJx\xi_\varphi \rangle.$$

By (iii) and (ii) we have $s_\varphi(x, x) \geq 0$ for any x in M. Moreover, since φ is faithful, the vector ξ_φ is separating for both M and M' (since $JMJ = M'$). By a well-known reasoning (see Lemma A.62) this implies that ξ_φ is cyclic for both M and M'. In particular $\overline{M'\xi_\varphi} = H$. Now assume $s_\varphi(x, x) = 0$. By Cauchy–Schwarz we have $s_\varphi(y, x) = 0$ for any y in M and hence $x\xi_\varphi \perp M'\xi_\varphi$, which means $x\xi_\varphi = 0$ and hence $x = 0$. Thus s_φ is nondegenerate.

To check the self-polarity, we will use Lemma 23.11. Assume $0 \leq \psi \leq \varphi$. By Lemma 23.11 there is b' in M' with $0 \leq b' \leq 1$ such that

$$\forall x \in M \qquad \psi(x) = \langle \xi_\varphi, b'x\xi_\varphi \rangle.$$

Since $JMJ = M'$, we may write $b' = JbJ$ with $0 \leq b \leq 1$ in M and we obtain $\psi(x) = s_\varphi(b, x)$. Lastly, since $y \to JyJ$ and $x \to x$ are commuting $*$-homomorphisms on $\overline{M}$ and M respectively, the map ρ_φ defined for $y, x \in M$ by $\rho_\varphi(\overline{y} \otimes x) = JyJx$ extends to a $*$-homomorphism on $\overline{M} \otimes_{\max} M$, and s_φ extends (so to speak) to a state Φ on $\overline{M} \otimes_{\max} M$ defined by $\Phi(\overline{y} \otimes x) = \langle \xi_\varphi, \rho_\varphi(\overline{y} \otimes x)\xi_\varphi \rangle$. In particular, this shows that s_φ is bipositive. $\qquad\square$

Remark 23.31 The preceding proof shows that for any unit vector $\xi \in P^\natural$ the form s defined on $M \times M$ by $s(y, x) = \langle \xi, JyJx\xi \rangle$ is a (normalized and separately normal) self-polar form.

Remark 23.32 The standard form of a von Neumann algebra $M \subset B(H)$ relative to $(J, P^\natural)$ is unique in the following very strong sense. Let $\widetilde{M} \subset B(\widetilde{H})$ be in standard form on $\widetilde{H}$, relative to $(\widetilde{J}, \widetilde{P}^\natural)$. If M is isomorphic to $\widetilde{M}$ via an isomorphism $\pi : M \to \widetilde{M}$, there is a *unique* unitary $U : H \to \widetilde{H}$ such that $\pi(x) = UxU^{-1}$ for all $x \in M$, $\widetilde{J} = UJU^{-1}$ and $\widetilde{P}^\natural = U(P^\natural)$.

Proof of Theorem 23.29 Let $A \subset B$ be a max$^+$-injective inclusion. By Lemma 23.26 $A^{**} \subset B^{**}$ is also max$^+$-injective. Assume A, B and π unital. Since π extends to a normal $*$-homomorphism $\ddot\pi : A^{**} \to M$, it suffices to show that $\ddot\pi$ admits a contractive c.p. extension to B^{**}. In other words, it suffices to prove Theorem 23.29 when A is a von Neumann subalgebra of a von Neumann algebra B and π is normal. By Lemma 23.28, we may assume that $\pi : A \to M$ is an isomorphism.

To simplify the notation, we observe that it suffices to prove the following.

Claim: Let $M \to B$ be an injective $*$-homomorphism such that the corresponding inclusion is $\max^+$-injective then there is a contractive c.p. projection $P : B \to M$.

Indeed, going back to the preceding situation, $M \ni x \mapsto \pi^{-1}(x) \in A \subset B$ defines clearly a $\max^+$-injective inclusion of M into B, and if there is P as in the claim then πP is a contractive c.p. extension of π as required in Theorem 23.29.

We now turn to the proof of the claim. For simplicity we assume $M \subset B$. Let s_φ be the self-polar form on M given by Theorem 23.30. By Proposition 23.22 there is a state f on $\overline{B} \otimes_{\max} B$ such that

$$\forall x \in M \qquad s_\varphi(x,x) \leq \Re f(\overline{x} \otimes x), \qquad (23.13)$$

and for any self-adjoint $x \in M$

$$\varphi(x) = s_\varphi(1,x) = (f(\overline{1} \otimes x) + f(\overline{x} \otimes 1))/2.$$

Therefore, decomposing in real and imaginary parts, we find for any $x \in M$

$$\varphi(x) = s_\varphi(1,x) = (f(\overline{1} \otimes x) + \overline{f(\overline{x} \otimes 1)})/2.$$

Let $s : B \times B \to \mathbb{C}$ be the sesquilinear form defined by

$$s(y,x) = (f(\overline{y} \otimes x) + \overline{f(\overline{x} \otimes y)})/2.$$

Note that $s(x,x) = \Re(f(\overline{x} \otimes x))$ for any x in B and $s(1,x) = \varphi(x)$ for any x in M. Moreover, s satisfies the positivity condition (23.4) on B (we even have complete positivity, see Theorem 4.16). Thus the restriction of s to $M \times M$ is bipositive. Since s_φ is self-polar (by Theorem 23.30), Corollary 23.20 implies $s(y,x) = s_\varphi(y,x)$ for any $y,x \in M$, and hence for any $t = \sum \overline{y_j} \otimes x_j \in \overline{M} \otimes M$

$$\left| \sum s_\varphi(y_j, x_j) \right| = \left| \left(f\left(\sum \overline{y_j} \otimes x_j\right) + \overline{f\left(\sum \overline{x_j} \otimes y_j\right)} \right)/2 \right|$$

whence since f is a state on $\overline{B} \otimes_{\max} B$

$$\leq (1/2) \left(\left\| \sum \overline{y_j} \otimes x_j \right\|_{\max} + \left\| \sum \overline{x_j} \otimes y_j \right\|_{\max} \right) = \left\| \sum \overline{y_j} \otimes x_j \right\|_{\overline{B} \otimes_{\max} B}.$$

Let $\rho_\varphi : \overline{M} \otimes M \to B(H)$ be the $*$-homomorphism described in Theorem 23.30. By (23.11) we have $\langle \xi_\varphi, \rho_\varphi(t)\xi_\varphi \rangle = \sum s_\varphi(y_j, x_j)$. By what we just proved $t \mapsto \langle \xi_\varphi, \rho_\varphi(t)\xi_\varphi \rangle$ has unit norm as a functional on $\overline{M} \otimes M$ equipped with the norm induced by $\overline{B} \otimes_{\max} B$. Obviously, since ξ_φ is cyclic for M it is a cyclic vector for ρ_φ, and hence by Remark A.27 we have $\|\rho_\varphi(t)\| \leq \|t\|_{\overline{B} \otimes_{\max} B}$ for any $t \in \overline{M} \otimes M$. Let $\mathcal{E}$ be the closure of $\overline{M} \otimes M$ in $\overline{B} \otimes_{\max} B$. Since ρ_φ is a $*$-homomorphism it automatically defines a (completely contractive) c.p. map v from $\mathcal{E}$ to $B(H)$, itself admitting a c.p. contractive extension $\tilde{v} : \overline{B} \otimes_{\max} B \to B(H)$, containing $\overline{M} \otimes M$ in its multiplicative domain. We now conclude as in Theorem 6.20. Let $\sigma : M \to B(H)$ be the inclusion map so that $\sigma(x) = \rho_\varphi(\bar{1} \otimes x)$ for any $x \in M$. Let $P_1 : B \to B(H)$ be defined by $P_1(b) = \tilde{v}(\bar{1} \otimes b)$ for $b \in B$. Then P_1 extends σ and by the usual multiplicative domain argument $P_1(b)$ commutes with $\tilde{v}(\bar{y} \otimes 1) = \rho_\varphi(\bar{y} \otimes 1)$ for any $y \in M$. Since $\{\rho_\varphi(\bar{y} \otimes 1) \mid y \in M\} = \sigma(M)'$, we conclude that $P_1(b) \in \sigma(M)'' = \sigma(M)$. Thus $P : B \to M$ defined by $P(b) = \sigma^{-1}P_1(b)$ is the desired projection, proving the claim. This proves the unital case. The argument is easily modified to cover the nonunital case. We leave this to the reader. $\square$

Corollary 23.33 *Let $\mathcal{M}, M$ be von Neumann algebras with $M \subset \mathcal{M}$. Assume that M is σ-finite or equivalently admits a faithful normal state φ. If the inclusion $M \subset \mathcal{M}$ is max$^+$-injective, then there is a contractive c.p. projection $P : \mathcal{M} \to M$.*

Proof By Theorem 23.29 the identity of M admits a contractive c.p. extension $P : \mathcal{M} \to M$. $\square$

We can now reach our goal:

Theorem 23.34 *Let A be a C^*-subalgebra of a C^*-algebra B. The following are equivalent:*

  (i) *The inclusion $A \subset B$ is max$^+$-injective.*
 (i)' *The (bitransposed) inclusion $A^{**} \subset B^{**}$ is max$^+$-injective.*
 (ii) *There is a contractive c.p. projection $P : B^{**} \to A^{**}$.*
(iii) *The inclusion $A \subset B$ is max-injective.*

Proof By Lemma 23.26 we know (i) $\Rightarrow$ (i)'. We already know that (ii) and (iii) are equivalent by Theorem 7.29, and (iii) $\Rightarrow$ (i) is clear by Remark 23.27. Thus it only remains to show the implication (i)' $\Rightarrow$ (ii). This is settled by the next statement, in which we remove the σ-finiteness assumption from Corollary 23.33. $\square$

Theorem 23.35 *Let $\mathcal{M}$ be a von Neumann algebra. Let $M \subset \mathcal{M}$ be a von Neumann subalgebra of $\mathcal{M}$. The following are equivalent:*

  (i) *The inclusion $M \subset \mathcal{M}$ is max$^+$-injective.*
 (ii) *There is a completely contractive c.p. projection $P : \mathcal{M} \to M$.*

Proof Assume (i). If M is σ-finite, (ii) follows by Corollary 23.33. Let $\mathscr{P}$ be the property of max$^+$-injectivity for inclusions such as $M \subset \mathcal{M}$. By the reduction in §23.1 to prove (i) $\Rightarrow$ (ii) in general it suffices to show that $\mathscr{P}$ satisfies the assumptions of Proposition 23.1. This can be done by a routine diagram chasing verification as follows.

Let q be a projection in M. We claim that the inclusion $q M q \subset q \mathcal{M} q$ is max$^+$-injective. This can be checked on the following commuting diagram:

$$
\begin{array}{ccc}
q\mathcal{M}q & \subset & \mathcal{M} \\
\cup & & \cup \\
qMq & \subset & M
\end{array}
$$

Indeed, by (i) in Proposition 7.19 the horizontal arrows are max-injective (and a fortiori max$^+$-injective by Remark 23.27) and the second vertical one is max$^+$-injective by our assumption. The latter means that the inclusion $(\overline{M} \otimes_{\max} M)^+ \subset (\overline{\mathcal{M}} \otimes_{\max} \mathcal{M})^+$ is isometric. Therefore the commuting diagram

$$
\begin{array}{ccc}
(\overline{q\mathcal{M}q} \otimes_{\max} q\mathcal{M}q)^+ & \subset (\overline{\mathcal{M}} \otimes_{\max} \mathcal{M})^+ \\
\uparrow & \cup \\
(\overline{qMq} \otimes_{\max} qMq)^+ & \subset (\overline{M} \otimes_{\max} M)^+
\end{array}
$$

must have its first vertical arrow isometric. This proves the claim.

Let $\pi : \mathcal{M} \to \mathcal{M}^1$ be an isomorphism of von Neumann algebras, and let $M^1 = \pi(M)$. Then $\bar{\pi} \otimes \pi : \overline{\mathcal{M}} \otimes_{\max} \mathcal{M} \to \overline{\mathcal{M}^1} \otimes_{\max} \mathcal{M}^1$ is an isometric isomorphism (indeed its inverse is $\bar{\pi}^{-1} \otimes \pi^{-1}$). Let $i : M \to \mathcal{M}$ and $i_1 : M^1 \to \mathcal{M}^1$ be the inclusions. We have

$$\pi i = i_1 \text{ and hence also } (\bar{\pi} \otimes \pi)(\bar{i} \otimes i) = \overline{i_1} \otimes i_1.$$

This shows that if $\bar{i} \otimes i : (\overline{M} \otimes_{\max} M)^+ \to (\overline{\mathcal{M}} \otimes_{\max} \mathcal{M})^+$ is isometric, then $\overline{i_1} \otimes i_1 : (\overline{M^1} \otimes_{\max} M^1)^+ \to (\overline{\mathcal{M}^1} \otimes_{\max} \mathcal{M}^1)^+$ is also isometric. This shows that $\mathscr{P}$ satisfies the required assumptions. $\qquad\square$

End of the proof of Theorem 23.7 It remains only to prove (ii) $\Rightarrow$ (i). Assume (ii). Let $A \subset B(H)$ be any embedding of A. Then (ii) implies (and by (22.15) is the same as saying) that the inclusion $A \subset B(H)$ is max$^+$-injective. By Theorem 23.34 it is max-injective and hence A has the WEP by Theorem 9.22. $\qquad\square$

The next Corollary was obtained in various steps by Uffe Haagerup and the author and independently by Christensen and Sinclair. See [51, 52, 107, 202, 203].

Corollary 23.36 *Let* $M, \mathcal{M}$ *be von Neumann algebras with* $M \subset \mathcal{M}$. *If there is a c.b. projection from* $\mathcal{M}$ *onto* M, *then there is a completely contractive (and c.p.) one.*

Proof Let $P : \mathcal{M} \to M$ be a c.b. projection. Let $C = \|P\|_{cb}^2$. By Corollary 22.19 the inclusion $M \subset \mathcal{M}$ satisfies the property in (23.10). By Remark 23.24, this means it is $\max^+$-injective, and the corollary follows. $\qquad \square$

Remark 23.37 By Remark 22.20 we obtain the same conclusion if we merely assume that P tensorizes boundedly with both the row and the column operator spaces. More precisely, if we have $\|Id_X \otimes P : X \otimes_{\min} \mathcal{M} \to X \otimes_{\min} M\| \le C^{1/2}$ for both $X = R$ and $X = C$ (row and column operator spaces), then the inclusion $M \subset \mathcal{M}$ satisfies (23.10).

Remark 23.38 However, it remains an open problem whether the mere existence of a *bounded* linear projection $P : \mathcal{M} \to M$ is enough for the conclusion of Corollary 23.36. An affirmative answer is given in [201] in case $M \bar{\otimes} M \simeq M$. It is proved in [118] that if G is any group containing $\mathbb{F}_2$ there is no bounded projection from $B(\ell_2(G))$ to M_G.

23.6 Complement

Our goal in this section is to prove the following extension of Theorem 22.10 beyond the semifinite case.

Theorem 23.39 *Let* $M \subset B(H)$ *be a von Neumann algebra assumed in standard form on* H *with respect to* $(J, P^\natural)$ *as described in the proof of Theorem 23.30. Then for any finite set* $(x_1, \ldots x_n)$ *in* M *we have*

$$\left\| \sum \overline{x_j} \otimes x_j \right\|_{\max} = \left\| \sum J x_j J x_j \right\|$$
$$= \sup \left\{ \left\langle \xi, \sum J x_j J x_j \xi \right\rangle \mid \xi \in P^\natural, \|\xi\|_H = 1 \right\}, \tag{23.14}$$

where for the second equality we assume in addition that $\sum \overline{x_j} \otimes x_j$ *is self-adjoint in* $\overline{M} \otimes M$.

Moreover, we have

$$\left\| \sum \overline{x_j} \otimes x_j \right\|_{\max} \ge \sup \sum s(x_j, x_j) \tag{23.15}$$

where the sup runs either over all separately normal, normalized bipositive forms s *on* $M \times M$, *or over all self-polar forms, and equality holds if* $\sum \overline{x_j} \otimes x_j$ *is self-adjoint.*

Proof We will first assume M σ-finite and show that both parts of Theorem 23.39 hold. By Takesaki's Theorem 11.3 we have an embedding $M \subset \mathcal{M}$ with $\mathcal{M}$ semifinite and σ-finite such that there is a contractive c.p. projection $P : \mathcal{M} \to M$. It follows (by (i) in Proposition 7.19) that for all (x_j) in M we have $\left\| \sum \overline{x_j} \otimes x_j \right\|_{\overline{M} \otimes_{\max} M} = \left\| \sum \overline{x_j} \otimes x_j \right\|_{\overline{M} \otimes_{\max} \mathcal{M}}$. Assume that $t = \sum \overline{x_j} \otimes x_j$ is self-adjoint in $\overline{M} \otimes M$. Let τ be a faithful normal semifinite trace on $\mathcal{M}$. By (22.20), for any $\varepsilon > 0$ there is a faithful normal state f on $\mathcal{M}$ such that $\left\| \sum \overline{x_j} \otimes x_j \right\|_{\overline{M} \otimes_{\max} M} - \varepsilon < \sum \tau(x_j^* f^{1/2} x_j f^{1/2})$. Let $s : M \times M \to \mathbb{C}$ be the form defined by $s(x,x) = \tau(x^* f^{1/2} x f^{1/2})$, which is bipositive by (22.21). Let $\varphi(x) = \tau(fx)$ for any $x \in M$. Then φ is a normal faithful state on M. Let s_φ be the corresponding self-polar form as in Theorem 23.30 and let $\rho = \rho_\varphi$ as in (23.11). By Theorem 23.19 we have $s(x,x) \leq s_\varphi(x,x)$ for any $x \in M$. This gives us

$$\left\| \sum \overline{x_j} \otimes x_j \right\|_{\overline{M} \otimes_{\max} M} - \varepsilon < \sum s_\varphi(x_j, x_j) = \langle \xi_\varphi, \rho(t) \xi_\varphi \rangle$$

and hence we obtain

$$\|t\|_{\max} \leq \sup\{\langle \xi, \rho(t)\xi \rangle \mid \xi \in P^\natural, \|\xi\|_H = 1\} \leq \|\rho(t)\|.$$

Since $\|\rho(t)\| \leq \|t\|_{\max}$ is obvious we obtain (23.14). The latter is proved assuming $t = t^*$, but since we may replace a general t by t^*t, we obtain $\|\rho(t)\| = \|t\|_{\max}$ for all $t \in \overline{M} \otimes M$. This also proves the second equality in (23.14) when $t = t^*$.

Now let s be a normalized bipositive separately normal form and let s_1 be any nondegenerate self-polar form such that $x \mapsto s_1(1,x)$ is normal. For any $0 < \varepsilon < 1$ consider the normalized bipositive form $s_\varepsilon = (1 - \varepsilon)s + \varepsilon s_1$. Let $\varphi_\varepsilon = s_\varepsilon(1,x)$ $(x \in M)$. Then φ_ε is normal and faithful. By Theorem 23.30 we have $s_\varepsilon \leq s_{\varphi_\varepsilon}$ in the pointwise order, and hence

$$(1 - \varepsilon)\sum s(x_j, x_j) \leq \sum s_\varepsilon(x_j, x_j)$$
$$\leq \sum s_{\varphi_\varepsilon}(x_j, x_j) = \langle \xi_{\varphi_\varepsilon}, \rho(t) \xi_{\varphi_\varepsilon} \rangle \leq \|\rho(t)\|.$$

Letting $\varepsilon \to 0$ we find $\sum s(x_j, x_j) \leq \|\rho(t)\|$. This yields the second part of Theorem 23.39 (for self-polar forms recall Remark 23.31).

We now turn to the general case. We start by observing that

$$\left\| \sum \overline{x_j} \otimes x_j \right\|_{\max} = \sup \left\| \sum \overline{px_j p} \otimes px_j p \right\|_{\max} \tag{23.16}$$

where the sup runs over all projections $p \in M$ such that pMp is σ-finite. Indeed, this can be deduced fairly easily from the structure Theorem A.65. Let $p \in M$ be any projection with pMp σ-finite. We will show that $\left\| \sum \overline{px_j p} \otimes px_j p \right\|_{\max} \leq \left\| \sum Jx_j Jx_j \right\|$.

Let $q \in B(H)$ be the projection defined by

$$q = JpJp = \rho(\bar{p} \otimes p).$$

Let $p' = JpJ$. Note p' is a projection in M' so that $q = pp' = p'p$. Note also $p = Jp'J$ (since $J^2 = 1$). A simple verification shows that $qJ = Jq = pJp$. By known results on the standard form (see [102, Lemma 2.6] for a detailed argument) pMp is isomorphic to qMq via the correspondence $pxp \mapsto qxq$ and the embedding $qMq \subset B(q(H))$ is a realization of qMq (with unit q) in standard form on the Hilbert space $q(H)$ with respect to qJq (restricted to $q(H)$) and the cone $q(P^\natural)$. By the first part of the proof, we have

$$\left\| \sum \overline{px_j p} \otimes px_j p \right\|_{\max} = \left\| \sum qJqx_j q J qx_j q \right\|. \qquad (23.17)$$

But since $p' = JpJ \in M'$ commutes with p we have $qxq = (pxp)p' = p'(pxp)$ for any $x \in M$ so that $qJqxqJqxq = q(Jp'(pxp)J)(pxp)p'$ and hence since $qJp' = qJp'JJ = qpJ = qJ$ we have

$$qJqxqJqxq = q(J(pxp)J)(pxp)q. \qquad (23.18)$$

Recalling that ρ is a $*$-homomorphism on $\overline{M} \otimes M$, we may write

$$J(pxp)J(pxp) = \rho(\overline{pxp} \otimes pxp) = \rho(\bar{p} \otimes p)\rho(\bar{x} \otimes x)\rho(\bar{p} \otimes p)$$

$$= q\rho(\bar{x} \otimes x)q$$

so that by (23.18)

$$\sum qJqx_j q J qx_j q = q\rho(t)q. \qquad (23.19)$$

Thus, if we denote $t' = \sum \overline{px_j p} \otimes px_j p$, the identity (23.17) gives us

$$\|t'\|_{\max} \le \|\rho(t)\|.$$

Using (23.16) we conclude $\|t\|_{\max} \le \|\rho(t)\|$ and since the converse is trivial, the first part of (23.14) follows. By the third property in the definition of a standard form we have $q(P^\natural) \subset P^\natural$. Using this with (23.19) the preceding argument allows us to extend the second equality in (23.14) when $t^* = t$.

Now let s be a separately normal, normalized bipositive form on $M \times M$. Let (x_j) be a finite set in M and let $t = \sum \bar{x}_j \otimes x_j$. Again invoking Theorem A.65 via Corollary A.66, one can find a net p_i of projections in M such that $p_i M p_i$ is σ-finite for any i, and such that $p_i x p_i \to x$ weak* for any $x \in M$. Then (see Remark 23.13) since s is separately normal $\sum s(x_j, x_j) \le \liminf_i \sum s(p_i x_j p_i, p_i x_j p_i)$. By the first part of the proof, we have $\sum s(p_i x_j p_i, p_i x_j p_i) \le \|(\overline{p_i} \otimes p_i)t(\overline{p_i} \otimes p_i)\|_{\max} \le \|t\|_{\max}$ and hence we obtain $\sum s(x_j, x_j) \le \|t\|_{\max}$. This shows

$$\sup \sum s(x_j, x_j) \le \|t\|_{\max},$$

where the sup runs over all separately normal, normalized bipositive forms on $M \times M$. But by (23.14) and Remark 23.31 if we restrict the sup to the subset of those s of the form $s(y,x) = \langle \xi, JyJx\xi \rangle$ with ξ unit vector in $P^\natural$, we obtain equality in (23.15) when $t = t^*$. $\qquad\square$

The proof of Theorem 23.39 actually proves the following fact.

Theorem 23.40 *Let* $\| \ \|_\alpha$ *be a* C^**-norm on* $\overline{M} \otimes M$. *Recall the notation*

$$(\overline{M} \otimes M)^+ = \left\{ \sum_1^n \overline{x_j} \otimes x_j \ \middle| \ n \geq 1, x_j \in M \right\}.$$

If $\|t\|_{\max} \leq \|t\|_\alpha$ *for all* $t \in (\overline{M} \otimes M)^+$ *then* $\|\rho(t)\| \leq \max\{\|t\|_\alpha, \|^t t\|_\alpha\}$ *for all* $t \in \overline{M} \otimes M$, *where* $t \mapsto {}^t t$ *denotes the linear map on* $\overline{M} \otimes M$ *taking* $\overline{y} \otimes x$ *to* $\overline{x^*} \otimes y^*$.

Sketch Since the argument is the same as earlier, we only outline it. Assume M in standard form. By (23.14) we have for any unit vector $\xi \in P^\natural$

$$\sum \langle \xi, Jx_j Jx_j\xi \rangle \leq \sup \left\{ \Re f\left(\sum \overline{x_j} \otimes x_j \right) \ \middle| \ f \in B_{(\overline{M} \otimes_\alpha M)^*} \right\}.$$

Therefore (as in Prop. 23.22) there is a state f on $(\overline{M} \otimes_\alpha M)^*$ (depending on ξ of course) such that for all $x \in M$

$$\langle \xi, JxJx\xi \rangle \leq \Re f(\overline{x} \otimes x) = (1/2)f(\overline{x} \otimes x + \overline{x^*} \otimes x^*).$$

Observing that the form $s(y,x) = (1/2)f(\overline{y} \otimes x + \overline{x^*} \otimes y^*)$ is normalized, bipositive and such that $\langle \xi, JxJx\xi \rangle \leq s(x,x)$ it follows from Corollary 23.20 that $\langle \xi, JyJx\xi \rangle = s(y,x)$ for any $y,x \in M$. Therefore for any $t = \sum_1^n \overline{y_j} \otimes x_j$ we have

$$|\langle \xi, \rho(t)\xi \rangle| = \left| \sum \langle \xi, Jy_j Jx_j\xi \rangle \right| = (1/2)f(t + {}^t t) \leq \max\{\|t\|_\alpha, \|^t t\|_\alpha\}.$$

In particular, applying this to t^*t gives us $\|\rho(t)\xi\|^2 \leq \max\{\|t\|_\alpha^2, \|^t t\|_\alpha^2\}$, and we conclude if ξ is cyclic for M. This settles the σ-finite case. The general case is proved by the same reduction as in the preceding proof. We skip the details. $\qquad\square$

The preceding naturally leads us to introduce a new C^*-norm as follows. For any von Neumann algebra $M \subset B(H)$ in standard form on $B(H)$ with the preceding notation we set (temporarily)

$$\forall t = \sum \overline{y_j} \otimes x_j \in \overline{M} \otimes M \quad \|t\|_{\mathrm{vns}} = \left\| \sum Jy_j Jx_j \right\|$$

or equivalently $\|t\|_{\mathrm{vns}} = \|\rho(t)\|$. Note this is only a seminorm. If (M, τ) is a tracial probability space, viewing $M \subset B(L_2(\tau))$ via left multiplications

$x \mapsto L(x)$ as usual, and denoting by $y \mapsto R(y)$ the right-hand side multiplications, this means that

$$\forall t = \sum \overline{y_j} \otimes x_j \in \overline{M} \otimes M \quad \|t\|_s = \left\| \sum R(y_j^*) L(x_j) \right\|.$$

However, we need the generality of the standard form to make sense of the following.

Definition 23.41 For any C^*-algebra A and any $t \in \overline{A} \otimes A$ we define

$$\|t\|_s = \max\{\|t\|_{\overline{A^{**}} \otimes_{\text{vns}} A^{**}}, \|t\|_{\min}\}. \tag{23.20}$$

If $A = M$ is a von Neumann algebra, then for some central projection $p \in M^{**}$ we have $M \simeq pM^{**}$ and $M^{**} = pM^{**} \oplus (1 - p)M^{**}$. The analysis of the standard form of $pM^{**}p$ done previously (and here p being central this is much simpler) shows that the restriction to $\overline{M} \otimes M$ of the vns-seminorm of $\overline{M^{**}} \otimes M^{**}$ coincides with the vns-seminorm of $\overline{M} \otimes M$. Thus when $A = M$ we can replace M^{**} by M in (23.20). Of course s stands here for standard (or self-polar), and it would be natural to call $\| \ \|_s$ the standard C^*-norm on $\overline{A} \otimes A$.

Remark 23.42 With this notation, Theorem 23.40 can be reformulated as saying this:

For any symmetric C^*-norm $\| \ \|_\alpha$ on $\overline{M} \otimes M$, if $\|t\|_s \leq \|t\|_\alpha$ for any $t \in (\overline{M} \otimes M)^+$ then $\|t\|_s \leq \|t\|_\alpha$ for any $t \in \overline{M} \otimes M$.

All this brings us to the following nice sounding refinement of Haagerup's characterization.

Theorem 23.43 *A C^*-algebra A has the WEP if and only if*

$$\overline{A} \otimes_s A = \overline{A} \otimes_{\min} A.$$

Proof If $\|t\|_s = \|t\|_{\min}$ for any $t \in \overline{A} \otimes_s A$, then a fortiori it holds for any $t \in (\overline{A} \otimes_s A)^+$ and hence the min and max norms coincide on $(\overline{A} \otimes A)^+$ by (23.14). The WEP follows by Theorem 23.7. Conversely, if A has the WEP the min and max norms coincide on $(\overline{A} \otimes A)^+$, and by Lemma 23.26 assuming $A \subset B(H)$ the inclusion $A^{**} \subset B(H)^{**}$ is max$^+$-injective. By (23.14), this implies that the s norm on $(\overline{A^{**}} \otimes A^{**})^+$ is less than (actually equal to) the norm induced on it by $\overline{B(H)^{**}} \otimes_s B(H)^{**}$. The latter being a (symmetric) C^*-norm by Theorem 23.40 the same domination must be true on the whole of $\overline{A^{**}} \otimes_s A^{**}$. Therefore since $\overline{A} \otimes_s A \subset \overline{A^{**}} \otimes_s A^{**}$ isometrically (by definition) and similarly for $A = B(H)$ we have isometrically $\overline{A} \otimes_s A \subset \overline{B(H)} \otimes_s B(H)$. Lastly we observe that obviously (since the natural representation $B(H) \mapsto B(\overline{H} \otimes_2 H)$ by left multiplication is in standard form) $\overline{B(H)} \otimes_s B(H) = \overline{B(H)} \otimes_{\min} B(H)$ and we conclude by injectivity of the min-norm that $\overline{A} \otimes_s A = \overline{A} \otimes_{\min} A$. $\qquad \square$

23.7 Notes and remarks

This chapter contains "new" results in the sense that they have not been published yet. The characterization of the WEP in Corollary 23.5 was claimed by Haagerup in personal communication to Junge and Le Merdy while they completed their paper [137]. They do not have a written trace of the proof. Similarly the author, who had just written [203] and was at that time in close contact with Haagerup in connection with the latter's related unpublished manuscript [107] does not remember being informed about the content of Corollary 23.5. Thus we are left guessing what his argument was, but the results of §23.2 seem very likely to be close to what Haagerup had in mind. Note that the question whether Corollary 23.5 holds is implicit in Haagerup's previous fundamental (published) paper [104], where he proves Corollary 23.6 and then asks explicitly whether for a von Neumann algebra M the isometric identity $D(\ell_\infty^3, M) = CB(\ell_\infty^3, M)$ implies its injectivity (we discussed this briefly in Remark 6.34). In other words he asks whether (i) in Corollary 23.5 with $n = 3$ suffices to imply the same for all n. This is still open, but it holds if $M \subset B(H)$ is cyclic and M' generated by a pair of unitaries (see Remark 23.4). As observed by Junge and Le Merdy in [137] it also holds if the equality $D(\ell_\infty^3, M) = CB(\ell_\infty^3, M)$ is meant in the completely isometric sense (the proof uses the main idea of [204], or equivalently Theorem 9.8 in the present volume). The reduction in §23.1 is directly inspired from the reasoning from [107] but it involves only fairly standard ideas.

The results in §23.5 are all included in Haagerup's unpublished paper [107], but he does not use the terms max-injective and $\max^+$-injective, which we introduce for convenience. Our presentation deliberately emphasizes the parallel between the two characterizations of the WEP in Theorems 23.2 and 23.34. We draw the reader's attention to the analogy between the norm of $D(\ell_\infty^n, A)$ as described in (6.32) and the norm in (22.13) (with Corollary 22.15). Both are derived very directly from the "column" norm $A^n \ni (x_j) \mapsto \left\| \left(\sum x_j^* x_j \right)^{1/2} \right\|$.

In §23.5 we use a new ingredient: self-polar forms. We prove some of the basic facts we need about them in §23.4. The main references in that direction are Connes's [59] (his "thèse de 3ème cycle") and Woronowicz's work, in part joint with Pusz [219, 220, 263]. Their work generalizes Araki's previous work in that direction. Theorem 23.19 and its corollary are due to them. Proposition 23.22 is an easy consequence of the Hahn–Banach theorem in the form described in §A.9. Corollary 23.36 was proved first for $\mathcal{M} = B(H)$ by the author in [202] and independently by Christensen–Sinclair in [51]. The case when M was semifinite was obtained in [203] and one of Haagerup's motivation for [107] was to prove the general case, which was also obtained independently by Christensen and Sinclair [52]. The papers [118, 201] discuss the situation when there is merely a *bounded* projection $P : B(H) \to M$ when $M \subset B(H)$.

In the appendix of [16], Haagerup's results from [107] (specifically Theorem 23.35) are used to prove the equivalence of several notions of co-amenability for inclusions of von Neumann algebras.

24

Full crossed products and failure of WEP for $\mathscr{B} \otimes_{\min} \mathscr{B}$

Our goal for this section is to present Ozawa's result that $\mathscr{B} \otimes_{\min} \mathscr{B}$ (or $M \otimes_{\min} N$ for any pair (M, N) of nonnuclear von Neumann algebras) fails the WEP. We follow Ozawa's main idea in [188] to study full crossed products and exploit Selberg's spectral bound (in place of (18.7)) but we broaden his viewpoint and we take advantage of the shortcut indicated in [209].

24.1 Full crossed products

The definition of the maximal tensor product of two C^*-algebras $B \otimes_{\max} C$ involves all pairs $r = (\sigma, \pi)$ of $*$-homomorphisms from B, C respectively into $B(H)$ with commuting ranges. So the fundamental relation imposed on $r = (\sigma, \pi)$ is $\sigma(x)\pi(y) = \sigma(x)\pi(y)$ for all x, y. It is natural to wonder whether analogous properties hold if we impose different relations to the pair $r = (\sigma, \pi)$. When $C = C^*(G)$ for some group G, and G acts on B, the crossed products that we define next are an illustration of this quest, for the set of relations (24.1). The latter are inspired from the ones that appear for semi-direct products of groups.

Let B be a C^*-algebra and G a group. Any homomorphism $g \to \alpha_g$ of G into the group of $*$-automorphisms of B will be called simply an action of G on B.

The triple (B, G, α) is usually called a C^*-dynamical system.

By definition, a covariant representation of (B, G, α) on a Hilbert space H is a pair $r = (\sigma, \pi)$ where $\sigma : B \to B(H)$ is a $*$-homomorphism and $\pi : G \to B(H)$ a unitary representation such that

$$\forall g \in G, \forall b \in B, \qquad \sigma(\alpha_g b) = \pi(g)\sigma(b)\pi(g)^{-1}. \tag{24.1}$$

Let $\mathcal{R}_\alpha$ be the set of all such pairs. For $r \in \mathcal{R}_\alpha$, we denote by H_r the corresponding Hilbert space. Let $B[G]$ denote the set of all finitely supported functions $f : G \to B$. We define a linear mapping

$$\Phi_\alpha : B[G] \to \left(\oplus \sum\nolimits_{r \in \mathcal{R}_\alpha} B(H_r) \right)_\infty \subset B(\oplus_{r \in \mathcal{R}_\alpha} H_r)$$

by setting:

$$\Phi_\alpha(f) = \left(\sum\nolimits_{g \in G} \sigma(f(g))\pi(g) \right)_{r=(\sigma,\pi) \in \mathcal{R}_\alpha}.$$

Then the image of Φ_α is a $*$-subalgebra of $B(\mathcal{H})$ where $\mathcal{H} = \oplus_{r \in \mathcal{R}_\alpha} H_r$.

Indeed, this follows from the relations imposed to the covariant representation $r = (\sigma, \pi)$.

The closure of $\Phi_\alpha(B[G])$ in $B(\mathcal{H})$ is a C^*-algebra called the full (or maximal) crossed product of B by G with respect to α. Equivalently, this is the completion of $B[G]$ for the norm defined by

$$\forall f \in B[G], \quad \|f\|_\rtimes = \|\Phi_\alpha(f)\|.$$

We denote it by $B \rtimes_\alpha G$ or simply by $B \rtimes G$ when there is no ambiguity.

Remark 24.1 (Full crossed products generalize the max-tensor product) If α acts trivially on B, i.e. $\alpha_g(b) = b$ for all $g \in G$, $b \in B$, then it is easy to check that $B \rtimes_\alpha G$ can be identified with $B \otimes_{\max} C^*(G)$. Indeed, covariant pairs boil down to pairs with commuting ranges. See Proposition 24.8 for a more general statement.

It will be convenient to use the following

Notation: $f \in B[G]$ will be denoted as a formal sum $\sum_{g \in G} f(g)g$. Then we have

$$\|f\|_\rtimes = \left\| \sum\nolimits_{g \in G} f(g)g \right\|_\rtimes = \sup_{r=(\sigma,\pi) \in \mathcal{R}_\alpha} \left\| \sum \sigma(f(g))\pi(g) \right\|_{B(H_r)}.$$

Moreover, using Φ_α we define a $*$-algebra structure on $B[G]$ by transplanting that of $\Phi_\alpha(B[G]) \subset B(\mathcal{H})$ where $\mathcal{H} = \left(\oplus \sum_r H_r \right)_2$. This means that we define f^* and $f_1 f_2$ for $f, f_1, f_2 \in B[G]$ simply by the identities

$$\Phi_\alpha(f^*) = \Phi_\alpha(f)^* \quad \text{and} \quad \Phi_\alpha(f_1 f_2) = \Phi_\alpha(f_1)\Phi_\alpha(f_2).$$

With this convention we have

$$\forall g \in G, \forall b \in B, \quad (bg)^* = \alpha_{g^{-1}}(b^*)g^{-1}.$$

$$\forall g_1, g_2 \in G, \forall b_1, b_2 \in B, \quad (b_1 g_1)(b_2 g_2) = (b_1 \alpha_{g_1}(b_2))(g_1 g_2),$$

and in particular

$$\forall b \in B, \quad gbg^{-1} = \alpha_g(b).$$

Remark 24.2 We have a canonical $*$-homomorphism $B \to B \rtimes_\beta G$ defined by $b \mapsto b e_G$ and, assuming B unital, a group homomorphism from G to $U(B \rtimes_\beta G)$ defined by $g \mapsto 1_B g$.

We saw previously that c.p. maps preserve the maximal tensor product (see Corollary 4.18). The next statement and its corollaries generalize this property to full crossed products. Indeed, by Remark 24.1 if β and α act trivially on B and L respectively then $B \rtimes_\beta G = B \otimes_{\max} C^*(G)$ and $L \rtimes_\alpha G = L \otimes_{\max} C^*(G)$.

Theorem 24.3 (Stinespring's factorization for full crossed products) *Let (B, G, β) be a C^*-dynamical system. Consider a pair (φ, π) where $\pi : G \to B(H)$ is a unitary representation and $\varphi : B \to B(H)$ a c.p. map that is "covariant" in the sense that*

$$\pi(g)\varphi(b)\pi(g)^{-1} = \varphi(\beta_g b) \qquad (24.2)$$

for any $b \in B, g \in G$.

(i) *There are a Hilbert space $\widehat{H}$, an isometry $V : H \to \widehat{H}$, a homomorphism $\widehat{\pi} : G \to B(\widehat{H})$ and a $*$-homomorphism $\widehat{\sigma} : B \to B(\widehat{H})$ such that the pair $(\widehat{\sigma}, \widehat{\pi})$ is covariant and we have*

$$\forall b \in B, \forall g \in G, \quad \varphi(b)\pi(g) = V^*\widehat{\sigma}(b)\widehat{\pi}(g)V. \qquad (24.3)$$

(ii) *The mapping defined on $B[G]$ by*

$$\widehat{\varphi}(bg) = \varphi(b)\pi(g)$$

extends to a c.p. map $\widehat{\varphi} : B \rtimes G \to B(H)$ with $\|\widehat{\varphi}\| = \|\varphi\|$.

Moreover, if B and φ are unital, $\widehat{\varphi}$ is also unital.

Proof (i) Assume B unital for simplicity. The proof is exactly the same as that of Stinespring's Theorem 1.22: We equip $B \otimes H$ with the scalar product defined for $t \in B \otimes H, t = \sum b_j \otimes h_j$ by $\langle t, t \rangle = \sum_{i,j} \langle h_i, \varphi(b_i^* b_j)h_j \rangle$. This leads to a Hilbert space $\widehat{H}$, a unital $*$-homomorphism $\widehat{\sigma} : B \to B(\widehat{H})$ and an isometry $V : H \to \widehat{H}$ defined by $\widehat{\sigma}(b)t = \sum bb_j \otimes h_j$ and (recall we assume B unital) $Vh = 1 \otimes h$. Note $\|V\|^2 = \|V^*V\| \le \|\varphi\|$.

Let $\widehat{\pi} : G \to B(\widehat{H})$ be the mapping defined by

$$\widehat{\pi}(g)(t) = \sum \beta_g b_j \otimes \pi(g)h_j.$$

By our assumption (24.2), $g \to \widehat{\pi}(g)$ defines a *unitary* representation on $\widehat{H}$. Moreover, we have

$$\widehat{\pi}(g)\widehat{\sigma}(b)\widehat{\pi}(g)^{-1}(t) = \widehat{\sigma}(\beta_g(b))(t).$$

Therefore, the pair $(\widehat{\sigma}, \widehat{\pi})$ is a covariant representation for (B, G, β). Moreover, we have (24.3). Indeed, for any $h \in H$,

$$\langle h,\, V^*\widehat{\sigma}(b)\widehat{\pi}(g)Vh \rangle = \langle Vh,\, \widehat{\sigma}(b)\widehat{\pi}(g)Vh \rangle$$
$$= \langle 1 \otimes h,\, b \otimes \pi(g)h \rangle = \langle h,\, \varphi(b)\pi(g)h \rangle.$$

(ii) Let $\widehat{\Phi} : B \rtimes G \to B(H)$ be the $*$-homomorphism associated to the covariant pair $(\widehat{\sigma}, \widehat{\pi})$. We have $\widehat{\varphi}(\cdot) = V^*\widehat{\Phi}(\cdot)V$. Therefore $\widehat{\varphi} : B \rtimes G \to B(H)$ is (unital and) c.p. with $\|\widehat{\varphi}\| \le \|V\|^2 = \|\varphi\|$.

The nonunital case can be proved as we did for Corollary 4.18. $\qquad \square$

Corollary 24.4 (Equivariant c.p. maps preserve full crossed products) *Let B, L be C^*-algebras (resp. unital C^*-algebras) with actions of G denoted by $g \to \beta_g$ and $g \to \alpha_g$ respectively on B and L. Let $\varphi \in CP(B, L)$ (resp. such that $\varphi(1) = 1$). Assume that φ is equivariant in the sense that*

$$\forall g \in G, \forall b \in B, \quad \varphi(\beta_g(b)) = \alpha_g(\varphi(b)). \tag{24.4}$$

Then the linear mapping $\widehat{\varphi}$ defined on $B[G]$ by

$$\widehat{\varphi}(bg) = \varphi(b)g$$

extends to a c.p. map (resp. unital) from $B \rtimes G$ to $L \rtimes G$ with norm $\|\widehat{\varphi}\| = \|\varphi\|$ (resp. $\|\widehat{\varphi}\| = 1$).

Proof Let (σ, π) be a covariant representation of (L, G, α) on H, such that the linear map $\Phi : L \rtimes G \to B(H)$ defined by $\Phi(lg) = \sigma(l)\pi(g)$ is an embedding. It suffices to show that $\Phi \circ \widehat{\varphi}$ is c.p. with norm $\le \|\widehat{\varphi}\|$. In other words, if we replace φ by $\varphi' = \sigma \circ \varphi$, we are reduced to the case when $L = B(H)$ with $\varphi' : B \to B(H)$ c.p. such that $\varphi(\beta_g b) = \pi(g)\varphi'(b)\pi(g)^{-1}$, which is covered by Theorem 24.3. $\qquad \square$

Corollary 24.5 (Equivariant conditional expectations) *Let $L \subset B$ be a unital C^*-subalgebra. Let $g \to \beta_g$ be an action of G on B and $g \to \alpha_g$ an action of G on L. Assume that*

$$\forall g \in G, \quad \beta_{g|L} = \alpha_g.$$

Assume moreover that there is a c.p. projection $P : B \to L$ such that $P(\beta_g b) = \alpha_g P(b), \forall b \in B$. Then the natural inclusion $L \subset B$ defines an isometric $*$-homomorphism

$$L \rtimes_\alpha G \subset B \rtimes_\beta G.$$

Proof We have obviously a (contractive) unital $*$-homomorphism $L \rtimes_\alpha G \to B \rtimes_\beta G$. To show that it is isometric, let $\varphi = P$. Then Corollary 24.4 shows that $\|f\|_{L \rtimes G} = \|\widehat{\varphi}(f)\|_{L \rtimes G} \le \|f\|_{B \rtimes G}$ for any $f \in L[G]$. $\qquad \square$

Proposition 24.6 ($\rtimes$ and $\otimes_{\max}$ commute) *Assume B unital. Let C be another C*-algebra. We use* $\check{\beta}_g = \beta_g \otimes Id_C$ *as action of G on* $B \otimes_{\max} C$. *Then the natural map*

$$\Psi : (B \otimes_{\max} C) \rtimes_{\check{\beta}} G \to (B \rtimes_{\beta} G) \otimes_{\max} C$$

taking $(b \otimes c)g$ *to* $(bg) \otimes c$ *is a* $*$*-isomorphism.*

Proof Consider an isometric embedding $\rho : (B \rtimes_{\beta} G) \otimes_{\max} C \to B(H)$. Let $\rho_1 : B \rtimes_{\beta} G \to B(H)$ and $\rho_2 : C \to B(H)$ be the associated commuting pair as in (4.4) so that $\|t\|_{\max} = \|(\rho_1 \cdot \rho_2)(t)\|_{B(H)}$ for any t in $(B \rtimes_{\beta} G) \otimes_{\max} C$. Let $\rho_1' = \rho_{1|B}$, $\sigma = \rho_1' \cdot \rho_2 : B \otimes_{\max} C \to B(H)$ and $\pi(g) = \rho_1(g)$. By Remark 24.2, σ is a $*$-homomorphism on $B \otimes_{\max} C$, and π a unitary representation of G on H with range commuting with that of ρ_2. For all $b \otimes c$ in $B \otimes C$, we have

$$\pi(g)\sigma(b \otimes c)\pi(g)^{-1} = \pi(g)\rho_1(b)\rho_2(c)\pi(g)^{-1} = \pi(g)\rho_1(b)\pi(g)^{-1}\rho_2(c)$$

$$= \rho_1(gbg^{-1})\rho_2(c) = \rho_1'(\beta_g(b))\rho_2(c) = \sigma(\check{\beta}_g(b \otimes c)).$$

Thus (σ, π) is a covariant representation for $(B \otimes_{\max} C, \check{\beta}_g, G)$. This implies that the linear mapping defined by

$$(b \otimes c)g \longmapsto \sigma(b \otimes c)\pi(g) = (\rho_1 \cdot \rho_2)(\Psi((b \otimes c)g))$$

extends to a contractive $*$-homomorphism on $(B \otimes_{\max} C) \rtimes_{\check{\beta}} G$. Therefore $\|\Psi(\chi)\|_{\max} \le \|\chi\|$ for any $\chi \in (B \otimes_{\max} C) \rtimes_{\check{\beta}} G$. By the maximality of the max-norm, Ψ must be a $*$-isomorphism. $\qquad\square$

24.2 Full crossed products with inner actions

Here we show that when the action is inner the full crossed product can be identified to a maximal tensor product.

Definition 24.7 Let B be a unital C^*-algebra. Let $\beta : G \to \text{Aut}(B)$ be an action as before. We say that β is inner if there is a unitary representation $\rho : G \to U(B)$ (into the unitary group of B) such that

$$\forall b \in B, \forall g \in G, \quad \beta_g b = \rho(g)b\rho(g)^{-1}. \tag{24.5}$$

Proposition 24.8 *Let* $g \to U(g)$ *be the universal representation of G. If* β *is inner then the mapping*

$$\Psi : bg \to b\rho(g) \otimes U(g),$$

where ρ *is the representation in (24.5), extends to a* $*$*-isomorphism from* $B \rtimes_{\beta} G$ *to* $B \otimes_{\max} C^*(G)$.

Proof The pair $r = (\sigma, \pi)$ defined by

$$\sigma(b) = b \otimes 1, \quad \pi(g) = \rho(g) \otimes U(g)$$

is clearly a covariant representation for (B, G, β) on $H_\rho \otimes H_U$. Therefore

$$\forall f \in B[G] \quad \|\Psi(f)\|_{\max} \le \|f\|_\rtimes.$$

By density (on both sides of the isomorphism) it clearly suffices to prove that

$$\|\Psi(f)\|_{\max} = \|f\|_\rtimes$$

for any $f \in B[G]$. To check this it suffices to observe that the norm defined on $B \otimes \mathbb{C}[G]$ by setting

$$\forall x \in B \otimes \mathbb{C}[G] \quad \|x\| = \|\Psi^{-1}(x)\|_\rtimes$$

or more explicitly setting for any $x = \sum_{g \in G} b_g \otimes U(g)$ with $g \mapsto b_g \in B$ finitely supported,

$$\left\| \sum b_g \otimes U(g) \right\| \overset{\text{def}}{=} \left\| \Psi^{-1}\left(\sum b_g \otimes U(g) \right) \right\| = \left\| \sum (b_g \rho(g)^{-1}) g \right\|_\rtimes$$

defines a C^*-norm on $B \otimes C^*(G)$. Indeed, one can check that if (σ, π) is a covariant pair then the pair π_1, π_2 defined by $\pi_1(b) = \sigma(b)$ and $\pi_2(g) = \sigma(\rho(g^{-1}))\pi(g)$ extends to a commuting pair of $*$-homomorphisms on B and $C^*(G)$ respectively. This gives us $\|\Psi^{-1}(x)\|_\rtimes \le \|x\|_{\max}$ and hence $\|f\|_\rtimes = \|\Psi(f)\|_{\max}$ for any $f \in B[G]$. $\qquad\square$

We now state an application of the preceding results to a certain "tensorizing" property, in the spirit of §7.1.

Corollary 24.9 *Let (B_1, G, β_1) and (B_2, G, β_2) be two unital C^*-dynamical systems, and let $\varphi \in CP(B_1, B_2)$ be equivariant in the sense that $\beta_2 \circ \varphi = \varphi \circ \beta_1$. Assume β_1, β_2 both inner with respect to homomorphisms $\rho_1 : G \to B_1$ and $\rho_2 : G \to B_2$. Then:*

(i) *The mapping $T_\varphi : B_1 \otimes C^*(G) \to B_2 \otimes C^*(G)$ defined by*

$$T_\varphi(b_1 \otimes U(g)) = \varphi(b_1 \rho_1(g)^{-1})\rho_2(g) \otimes U(g)$$

extends to a c.p. map

$$T_\varphi : B_1 \otimes_{\max} C^*(G) \to B_2 \otimes_{\max} C^*(G)$$

with norm $\|T_\varphi\| \le \|\varphi\|$.

(ii) *Let C be another unital C^*-algebra. Assume that $Id_C \otimes \varphi$ extends to a (necessarily c.p.) mapping $\Phi : C \otimes_{\min} B_1 \to C \otimes_{\max} B_2$ with $\|\Phi\| \le 1$. Then $Id_C \otimes T_\varphi$ extends to a bounded map*

$$\widetilde{Id_C \otimes T_\varphi} : (C \otimes_{\min} B_1) \otimes_{\max} C^*(G) \to (C \otimes_{\max} B_2) \otimes_{\max} C^*(G)$$

with norm ≤ 1.

Proof (i) By Corollary 24.4 we have a c.p. map $\widehat{\varphi} : B_1 \rtimes G \to B_2 \rtimes G$ with norm $= \|\varphi\|$. Composing with the $*$-isomorphisms $\psi_j : B_j \rtimes G \to B_j \otimes_{\max} C^*(G)$ ($j = 1, 2$) given by Proposition 24.8, we find that $T_\varphi = \psi_2 \widehat{\varphi} \psi_1^{-1}$ is a c.p. map with norm $= \|\varphi\|$.

(ii) Given an action β on B, let us denote by $\dot{\beta}$ (resp. $\check{\beta}$) the action of G on $C \otimes_{\min} B$ (resp. $C \otimes_{\max} B$) defined on the algebraic tensor product as $Id_C \otimes \beta$ and then extended by density to $C \otimes_{\min} B$ (resp. $C \otimes_{\max} B$). Note that $\dot{\beta}$ and $\check{\beta}$ are both inner if β is inner. Thus we may apply the first part to the mapping Φ. Since $T_\Phi = Id_C \otimes T_\varphi$ we obtain the announced result. □

Corollary 24.10 *Let (B, G, β) be a unital C^*-dynamical system that is inner with respect to a homomorphism $\rho : G \to B$ and let C be a C^*-algebra. Let $P \in CP(B, B)$ be equivariant (in the sense that $\beta \circ P = P \circ \beta$), and such that $\|Id_C \otimes P : C \otimes_{\min} B \to C \otimes_{\max} B\| \le 1$. Then the mapping θ defined on $C \otimes B \otimes C^*(G)$ by*

$$\theta(c \otimes b \otimes U(g)) = c \otimes P(b\rho(g)^{-1})\rho(g) \otimes U(g)$$

extends to a map $\widetilde{\theta}$ satisfying

$$\|\widetilde{\theta} : (C \otimes_{\min} B) \otimes_{\max} C^*(G) \to (C \otimes_{\max} B) \otimes_{\max} C^*(G)\| \le 1.$$

Proof Apply the preceding Corollary with $B_1 = B_2 = B$, $\beta_1 = \beta_2 = \beta$ and $\varphi = P$. □

Corollary 24.11 *In the situation of the preceding Corollary let $E = \{x \in B \mid Px = x\}$. Let*

$$F = \mathrm{span}[x\rho(g) \otimes U(g) \mid x \in E, g \in G] \subset B \otimes C^*(G).$$

Then the norms of $(C \otimes_{\min} B) \otimes_{\max} C^*(G)$ and $(C \otimes_{\max} B) \otimes_{\max} C^*(G)$ coincide on $C \otimes F$.

Proof The mapping $\widetilde{\theta}$ is the identity on $C \otimes F$. □

Theorem 24.12 *Let C be a unital C^*-algebra. In the situation of Corollary 24.5, assume:*

$$\beta \text{ is inner,} \tag{24.6}$$

$$(L, C) \text{ is a nuclear pair.} \tag{24.7}$$

Then

$$(B \otimes_{\min} C, C^*(G)) \text{ nuclear} \Rightarrow (L \rtimes G, C) \text{ nuclear.} \tag{24.8}$$

Proof We start by observing that the assumption in (24.8) implies a fortiori

$$(B, C^*(G)) \text{ is a nuclear pair.} \tag{24.9}$$

Indeed, this follows from Remark 7.21 (where we replace D by B and B by $C^*(G)$).

Recall $\dot{\beta}$ denotes the inner action of G on $B \otimes_{\min} C$, defined by $\dot{\beta}_g = \beta_g \otimes Id_C$, and similarly for $\dot{\alpha}$ on $L \otimes_{\min} C = L \otimes_{\max} C$. We first claim that we have an isometric embedding

$$(L \rtimes_\alpha G) \otimes_{\max} C \subset (B \otimes_{\min} C) \otimes_{\max} C^*(G). \qquad (24.10)$$

Indeed, by Proposition 24.6 (applied to $L \otimes_{\max} C$) we have

$$(L \rtimes_\alpha G) \otimes_{\max} C = (L \otimes_{\max} C) \rtimes_{\tilde{\alpha}} G,$$

and by (24.7) and Corollary 24.5 (applied to $P \otimes Id_C : B \otimes_{\min} C \to L \otimes_{\min} C$) we have an isometric embedding

$$(L \otimes_{\max} C) \rtimes_{\tilde{\alpha}} G = (L \otimes_{\min} C) \rtimes_{\dot{\alpha}} G \subset (B \otimes_{\min} C) \rtimes_{\dot{\beta}} G. \qquad (24.11)$$

Furthermore, by Proposition 24.8, since $\dot{\beta}$ is inner on $B \otimes_{\min} C$, we have

$$(B \otimes_{\min} C) \rtimes_{\dot{\beta}} G = (B \otimes_{\min} C) \otimes_{\max} C^*(G).$$

This proves the claim (24.10). We now turn to $(L \rtimes_\alpha G) \otimes_{\min} C$.

By Corollary 24.5 applied this time to $P : B \to L$ we have an isometric embedding

$$L \rtimes G \to B \rtimes G$$

and hence the following map is also isometric

$$(L \rtimes_\alpha G) \otimes_{\min} C \to (B \rtimes_\beta G) \otimes_{\min} C.$$

By Proposition 24.8, since β is inner this produces an isometric embedding

$$(L \rtimes_\alpha G) \otimes_{\min} C \to (B \otimes_{\max} C^*(G)) \otimes_{\min} C$$

and using (24.9) (and permuting factors using (1.6) and (1.5)) we find an isometric embedding

$$(L \rtimes_\alpha G) \otimes_{\min} C \subset (B \otimes_{\min} C) \otimes_{\min} C^*(G). \qquad (24.12)$$

Moreover, it is easy to trace back the identifications we made to check that (24.10) and (24.12) coincide on $(L \rtimes_\alpha G) \otimes C$. Thus the implication in (24.8) follows from (24.10) and (24.12). □

Corollary 24.13 *In the situation of Corollary 24.5 with β inner, assume that $C^*(G)$ and L have the LLP. If $B \otimes_{\min} \mathscr{B}$ has the WEP then $L \rtimes G$ has the LLP.*

Proof We apply Theorem 24.12 with $C = \mathscr{B}$. In that case (24.7) and (by our assumption on $B \otimes_{\min} \mathscr{B}$) the left-hand side of (24.8) hold by the generalized Kirchberg Theorem 9.40. □

24.3 $\mathcal{B} \otimes_{\min} \mathcal{B}$ fails WEP

Consider again the constant $C(n)$ defined previously as the infimum of the C's in (18.6).

We need a modified version, as follows: $C_0(n)$ is the infimum of the constants $C \leq n$ such that for each $m \geq 1$, there is $N_m \geq 1$ and an n-tuple $[u_1(m), \ldots, u_n(m)]$ of unitary $N_m \times N_m$ matrices of *permutation* such that

$$\sup_{m \neq m'} \left\| \left[\sum_{j=1}^n \overline{u_j(m)} \otimes u_j(m') \right]_{|[\overline{\chi} \otimes \chi']^{\perp}} \right\|_{\min} \leq C, \qquad (24.13)$$

where χ and χ' are the constant unit vectors in $\mathbb{C}^{N_m}$ and $\mathbb{C}^{N_{m'}}$ respectively, i.e.

$$\chi = N_m^{-1/2} \sum_1^{N_m} e_k \text{ and } \chi' = N_{m'}^{-1/2} \sum_1^{N_{m'}} e_k.$$

Equivalently, since

$$[\overline{\chi} \otimes \chi']^{\perp} = [\overline{\chi}^{\perp} \otimes \chi'^{\perp}] \oplus [\overline{\chi}^{\perp} \otimes \chi'] \oplus [\overline{\chi} \otimes \chi'^{\perp}]$$

and $u_j(m)(\chi) = \chi$ for all j, m, (24.13) means that we have

$$\sup_{m \neq m'} \left\| \left[\sum_{j=1}^n \overline{u_j(m)}_{|\chi^{\perp}} \otimes u_j(m')_{|\chi'^{\perp}} \right] \right\|_{\min} \leq C, \qquad (24.14)$$

together with

$$\sup_m \left\| \sum_{j=1}^n u_j(m)_{|\chi^{\perp}} \right\| \leq C. \qquad (24.15)$$

Note that the matrix associated to $u_j(m)_{|\chi^{\perp}}$ is a unitary matrix of size $N_m - 1$ (with respect to any orthonormal basis of $\chi^{\perp}$). Thus if we neglect (24.15), which in our framework will be easy to verify, the definition of $C_0(n)$ is the same as the previous one of $C(n)$ in (18.6) but with the additional requirement that the unitary matrices must be obtained by restricting permutation matrices to the orthogonal of the constant vector. Moreover, since the latter have real entries, we could drop the complex conjugation sign, but to preserve the analogy with $C(n)$, we prefer not to do that. In any case, we have

$$C(n) \leq C_0(n).$$

Remark 24.14 In the case $m = m'$ the operators $\{\overline{u_j(m)} \otimes u_j(m) \mid 1 \leq j \leq n\}$, which obviously admit $J = \overline{\chi} \otimes \chi$ as a common invariant vector, also admit the "identity" namely $I = \sum_1^{N_m} \overline{e_k} \otimes e_k$, as another one. Actually, when the permutations giving rise to the unitaries $(u_j(m))_{1 \leq j \leq n}$ generate $\mathbb{S}_{N_m}$, all the common invariant vectors lie in span$[J, I]$. This will be the case in most of the examples that follow. Thus in the quantum expander context it will be natural to consider the restriction of $\sum_1^{N_m} \overline{u_j(m)} \otimes u_j(m)$ to $\mathrm{span}[J, I]^\perp$, or equivalently $\{J, I\}^\perp$. Let us observe for later use the orthogonal decomposition

$$\{J, I\}^\perp = [(\overline{\chi}^\perp \otimes \chi^\perp) \cap I^\perp] \oplus [\overline{\chi}^\perp \otimes \chi] \oplus [\overline{\chi} \otimes \chi^\perp],$$

where the sum of the dimensions, which are respectively $(N_m - 1)^2 - 1$, $N_m - 1$ and $N_m - 1$, is equal to $\dim(\{J, I\}^\perp) = N_m^2 - 2$.

Using the latter decomposition, we find

$$\left\| \left[\sum_1^{N_m} \overline{u_j(m)} \otimes u_j(m) \right]_{|\{J, I\}^\perp} \right\|$$
$$= \max \left\{ \left\| \left[\sum_1^{N_m} \overline{u_j(m)_{\chi^\perp}} \otimes u_j(m)_{\chi^\perp} \right]_{|I^\perp} \right\|, \left\| \sum_1^{N_m} u_j(m)_{\chi^\perp} \right\| \right\}. \tag{24.16}$$

The crucial ingredient to show that $B(H) \otimes_{\min} B(H)$ fails the WEP is that $C_0(n) < n$ for at least one $n > 1$.

We will prove this later on with $n = 3$ using a fundamental result due to Selberg on the group $SL_2(\mathbb{Z})$. This way seems to produce the most explicit example. However, here again one can use probabilistic methods (such as those of [93, 94]), or the fact that the family of all permutation groups forms an expander ([148]). We describe these alternative ways in §24.5.

The general approach is similar to the one we used to prove Theorem 18.7. The main new ingredient is a very special property of permutation matrices with spectral gap (the following Lemma 24.18) that is somewhat implicit in Ozawa's [188].

We need some specific notation. We denote by $\Pi(n, N)$ the set of n-tuples of $N \times N$ unitary permutation matrices $u = (u_1, \ldots, u_n)$, such that $u_1 = I$. For any such u the operators u_j all admit $\chi_N = N^{-1/2} \sum_1^N e_k$ as a common invariant vector, and hence their sum $\sum_1^n u_j$ admits χ_N as eigenvector for the eigenvalue n. We denote by $\varepsilon_0(u)$ the "spectral gap" of the latter operator beyond n. More precisely, the number $\varepsilon_0(u) \in [0, n]$ is defined by

$$\left\| \left[\sum_1^n u_j \right]_{|\chi_N^\perp} \right\| = n - \varepsilon_0(u).$$

Remark 24.15 Let $[u_1(m), \ldots, u_n(m)]$ be a sequence of n-tuples of permutation matrices satisfying (24.13). Let $v_j(m) = u_1(m)^{-1} u_j(m)$ for all $1 \le j \le n$. Then $[v_1(m), \ldots, v_n(m)]$ belongs to $\Pi(n, N_m)$ and the sequence $[v_1(m), \ldots, v_n(m)]$ still satisfies (24.13).

Remark 24.16 Consider $u \in \Pi(n, N)$, $u' \in \Pi(n, N')$. We denote by $\overline{u} \otimes u'$ the n-tuple $(\overline{u_j} \otimes u_j')$. If we identify ℓ_2^N and $\overline{\ell_2^N}$ (using the canonical basis) we may view $(\overline{u_j})$ as an n-tuple of permutation matrices, and hence view $\overline{u} \otimes$

$u' = (\overline{u_j} \otimes u'_j)$ as an n-tuple of matrices of size $N \times N'$. It is easy to check that the latter are still permutation matrices, so that with our identification $\overline{u} \otimes u' \in \Pi(n, NN')$. Note also that if u_0 and u'_0 are diagonal matrices with size respectively N and N', then $\overline{u_0} \otimes u'_0$ is identified with a diagonal matrix of size NN'. Moreover $\chi_{NN'}$ can be identified with $\chi_N \otimes \chi_{N'}$ (or $\overline{\chi_N} \otimes \chi_{N'}$). Therefore, the assumption appearing in (24.13) can be equivalently rewritten as

$$\inf_{m \neq m'} \varepsilon_0(\overline{u(m)} \otimes u(m')) \geq n - C.$$

Remark 24.17 We will be interested in the consequences of $\varepsilon_0(u) > 0$ for an n-tuple $u \in \Pi(n, N)$. More precisely let us assume $\varepsilon_0(u) \geq \varepsilon_0$ for some fixed $\varepsilon_0 > 0$, or equivalently

$$\left\| \left[\sum_1^n u_j \right]_{|\chi_N^\perp} \right\| \leq n - \varepsilon_0.$$

Note that $\sum_1^n u_j$ admits both $\mathbb{C}\chi_N$ and $[\mathbb{C}\chi_N]^\perp$ as invariant subspaces. From this it is easy to check that there is a function $f_1 : (0, 1) \to \mathbb{R}_+$ (depending only on ε_0 and n) with $\lim_{\varepsilon \to 0} f_1(\varepsilon) = 0$ such that for any unit vector $x \in \ell_2^N$

$$\left\| \sum_1^n u_j(x) \right\| \geq n - \varepsilon \Rightarrow \inf_{z \in \mathbb{C}} \|x - z\chi_N\| \leq f_1(\varepsilon). \tag{24.17}$$

Indeed, let $\theta = \|P_{\chi_N^\perp}x\|$. Then $(1 - \varepsilon/n)^2 \leq \|n^{-1}\sum_1^n u_j(x)\|^2 \leq (1 - \theta^2) + (1 - \varepsilon_0/n)^2\theta^2$ which implies

$$\theta^2 \leq 2\varepsilon(2\varepsilon_0 - \varepsilon_0^2/n)^{-1} \leq 2\varepsilon/\varepsilon_0,$$

so that we can take $f_1(\varepsilon) = (2\varepsilon/\varepsilon_0)^{1/2}$.

Let us now assume that the unit vector $x \in \ell_2^N$ has its coordinates in $\mathbb{R}_+$. There is then a function $f_2 : (0, 1) \to \mathbb{R}_+$ (depending only on ε_0 and n) with $\lim_{\varepsilon \to 0} f_2(\varepsilon) = 0$ such that

$$\left\| \sum_1^n u_j(x) \right\| \geq n - \varepsilon \Rightarrow \|x - \chi_N\| \leq f_1(\varepsilon) + f_2(\varepsilon). \tag{24.18}$$

Indeed, since x and χ_N are unit vectors and x has nonnegative coordinates, an elementary argument shows that the optimal z in (24.17) necessarily satisfies both $|1 - |z|| \leq f_1(\varepsilon)$ and $d(z, \mathbb{R}_+) \leq |z - \langle \chi_N, x \rangle| \leq f_1(\varepsilon)$, whence an estimate $|1 - z| \leq f_2(\varepsilon)$ for some function f_2 with $\lim_{\varepsilon \to 0} f_2(\varepsilon) = 0$.

As already mentioned, the next key lemma is implicit in [188].

Lemma 24.18 *Fix $n \geq 1$ and $\varepsilon_0 > 0$.*

(i) *For each $\delta > 0$ there is $\varepsilon > 0$ (depending only on δ, ε_0 and n) such that the following holds:*

for any $N \geq 1$, any $u \in \Pi(n, N)$ such that $\varepsilon_0(u) \geq \varepsilon_0$ satisfies the following property:

for any Hilbert space H and for any n-tuple $(V_j)_{1 \leq j \leq n}$ of unitaries on H, if a unit vector $\xi \in \ell_2^N(H)$ is such that

$$\left\| \left(\sum_{j=1}^{n} u_j \otimes V_j \right) (\xi) \right\| > n - \varepsilon \tag{24.19}$$

then

$$\| |\xi| - \chi_N \|_{\ell_2^N} < \delta, \tag{24.20}$$

where χ_N is the (constant) unit vector of ℓ_2^N defined by $\chi_N = N^{-1/2} 1_{[1,...,N]}$, and where $|\xi| \in \ell_2^N$ is the unit vector defined by

$$\forall i \in [1, \dots, N] \quad |\xi|(i) = \|\xi(i)\|_H.$$

(ii) *There is $\varepsilon_0' > 0$ (depending only on ε_0 and n) such that for any $N \geq 1$, any $u \in \Pi(n, N)$ satisfying $\varepsilon_0(u) \geq \varepsilon_0$ satisfies the following property:*

for any H, for any $V_j \in U(B(H))$ $(1 \leq j \leq n)$ with $V_1 = 1$, and for any diagonal unitary operator $D \in M_N$ with zero trace, we have

$$\left\| D \otimes I + \sum_{j=1}^{n} u_j \otimes V_j \right\| \leq n + 1 - \varepsilon_0'. \tag{24.21}$$

Proof (i) By (A.6), the assumption (24.19) implies that there is unit vector $\xi' \in \ell_2^N(H)$ such that $\sup_j \|(u_j \otimes V_j)(\xi) - \xi'\| < f(\varepsilon)$ for some function $f(\varepsilon)$ such that $\lim_{\varepsilon \to 0} f(\varepsilon) = 0$. Let θ_j be the permutation represented by u_j so that $u_j e_i = e_{\theta_j(i)}$. Note for any $i \in [1, \dots, N]$

$$\|[(u_j \otimes V_j)(\xi) - \xi'](i)\|_H \geq | \|V_j \xi(\theta_j(i))\|_H - \|\xi'(i)\|_H |$$
$$= | \|\xi(\theta_j(i))\|_H - \|\xi'(i)\|_H |$$
$$= |[u_j(|\xi|) - |\xi'|](i)|$$

and hence

$$f(\varepsilon) > \|(u_j \otimes V_j)(\xi) - \xi'\| \geq \|u_j(|\xi|) - |\xi'|\|, \tag{24.22}$$

and furthermore

$$\left\| \sum_{1}^{n} u_j(|\xi|) \right\| \geq \|n\xi'\| - \left\| \sum_{1}^{n} u_j(|\xi|) - |\xi'| \right\|$$
$$\geq n\|\xi'\| - nf(\varepsilon) = n - nf(\varepsilon).$$

By (24.18) we have $\| |\xi| - \chi_N \| \leq f_3(\varepsilon)$ for some function f_3 (independent of N) with $\lim_{\varepsilon \to 0} f_3(\varepsilon) = 0$. Thus if we adjust ε so that $f_3(\varepsilon) < \delta$ we obtain (24.20).

(ii) Assume $\left\| D \otimes I + \sum_{j=1}^{n} u_j \otimes V_j \right\| > n + 1 - \varepsilon$. We will reach a contradiction when ε is small enough. By our assumption there is ξ as in (i) satisfying

$$\left\| \left[D \otimes I + \sum_{j=1}^{n} u_j \otimes V_j \right](\xi) \right\| > n + 1 - \varepsilon. \qquad (24.23)$$

Then (by the triangle inequality) (24.19) and hence (24.20) holds. By the uniform convexity of Hilbert space again, (24.23) implies that $\max_j \| [D \otimes I](\xi) - [u_j \otimes V_j](\xi) \| \leq f_2(\varepsilon)$ for some function f_2 such that $\lim_{\varepsilon \to 0} f_2(\varepsilon) = 0$, and hence (using $j = 1$) that $\| \xi - [D \otimes I](\xi) \| \leq f_2(\varepsilon)$. But $\| \xi - [D \otimes I](\xi) \| = \| |\xi - [D \otimes I](\xi)| \|$ and a moment of thought shows that $|\xi - [D \otimes I](\xi)| = |I - D|(|\xi|)$. Therefore $\| \xi - [D \otimes I](\xi) \| = \| (I - D)(|\xi|) \|$. By (24.20) $\| \xi - [D \otimes I](\xi) \| \geq \| (I - D)(\chi_N) \| - 2\delta$, and since $\chi_N \perp D(\chi_N)$, this implies $\| \xi - [D \otimes I](\xi) \| \geq \sqrt{2} - 2\delta$. Thus we obtain $f_2(\varepsilon) \geq \sqrt{2} - 2\delta$. This is the desired contradiction: when ε is small enough the number δ becomes arbitrarily small, so $\lim_{\varepsilon \to 0} f_2(\varepsilon) = 0$ is impossible.   □

We now state the crucial fact on which the proof rests, but postpone its proof to §24.4.

Theorem 24.19 *We have $C_0(3) < 3$.*

We recall the notation introduced in (18.8) and (18.9).

Let $\{ [u_1(m), \ldots, u_n(m)], m \in \mathbb{N} \}$ be a sequence of n-tuples of unitary matrices of size $N_m \times N_m$. By Remark 24.15 we may always assume as we do in the sequel that $u_1(m) = 1$ for all m and we will work with n-tuples in $\Pi(n, N_m)$. For any subset $\omega \subset \mathbb{N}$, let

$$\mathbb{B}_\omega = \left(\oplus \sum_{m \in \omega} M_{N_m} \right)_\infty.$$

In addition we denote by

$$L_\omega \subset \mathbb{B}_\omega$$

the commutative (and hence nuclear) C^*-subalgebra formed of those $x = (x_m)_{m \in \omega}$ such that x_m is a diagonal matrix for all m.

Let $\mathbb{N} = \omega(1) \cup \omega(2)$ be any disjoint partition of $\mathbb{N}$ into two infinite subsets, and let

$$u_j^1 = \bigoplus_{m \in \omega(1)} u_j(m) \in \mathbb{B}_{\omega(1)} \qquad u_j^2 = \bigoplus_{m' \in \omega(2)} u_j(m') \in \mathbb{B}_{\omega(2)}.$$
$$(24.24)$$

We now deduce Ozawa's result by a shorter route (although based on similar ingredients as his).

Theorem 24.20 *Fix $\varepsilon_0 > 0$. Let $\{u(m), m \in \mathbb{N}\}$ be a sequence of n-tuples of matrices with $u(m) \in \Pi(n, N_m)$ for each m satisfying*

$$\inf_{m \neq m'} \varepsilon_0(\overline{u(m)} \otimes u(m')) \geq \varepsilon_0 > 0. \tag{24.25}$$

For each m, we choose a diagonal unitary matrix $u_0(m)$ with zero trace and size $N_m \times N_m$. Suppose that $\{(u_j(m))_{0 \leq j \leq n}, m \in \mathbb{N}\}$ converges in distribution when $m \to \infty$. Let $(U_j)_{2 \leq j \leq n}$ be the unitary generators of $\mathbb{F}_{n-1}$ with the convention $U_0 = U_1 = 1$. Let $\mathbb{N} = \omega(1) \cup \omega(2)$ be any disjoint partition of $\mathbb{N}$ into two infinite subsets. With the preceding notation (24.24), we set

$$t = \sum_{j=0}^{n} \overline{u_j^1} \otimes u_j^2 \otimes U_j \in [\overline{\mathbb{B}_{\omega(1)}} \otimes_{\min} \mathbb{B}_{\omega(2)}] \otimes C^*(\mathbb{F}_{n-1}). \tag{24.26}$$

We have then

$$\|t\|_{\min} \leq n + 1 - \varepsilon_0' \quad and \quad \|t\|_{\max} = n + 1, \tag{24.27}$$

where $\varepsilon_0' > 0$ is given by (24.21), and hence

$$[\overline{\mathbb{B}_{\omega(1)}} \otimes_{\min} \mathbb{B}_{\omega(2)}] \otimes_{\min} C^*(\mathbb{F}_{n-1}) \neq [\overline{\mathbb{B}_{\omega(1)}} \otimes_{\min} \mathbb{B}_{\omega(2)}] \otimes_{\max} C^*(\mathbb{F}_{n-1}).$$

Proof The estimate $\|t\|_{\min} \leq n + 1 - \varepsilon_0'$ follows from Remark 24.16 and Lemma 24.18.

We now turn to $\|t\|_{\max}$. As explained in the proof of Theorem 18.9 we have that

$$\left\| \sum_0^n \overline{u_j^1} \otimes u_j^2 \right\|_{\overline{\mathbb{B}_{\omega(1)}} \otimes_{\max} \mathbb{B}_{\omega(2)}} = n + 1. \tag{24.28}$$

Let $G = \mathbb{F}_{n-1}$, and for convenience let $g_2, g_3, \ldots, g_n$ denote the $n - 1$ free generators, with the notational convention $g_0 = g_1 = 1$. We consider the unitary representation $\rho_2 : G \to \mathbb{B}_{\omega(2)}$ defined by $\rho_2(g_j) = u_j^2$ for $j = 2, 3, \ldots, n$ (and of course $\rho_2(g_j) = 1$ for $j = 0, 1$). Let β be the action of G on $\mathbb{B}_{\omega(2)}$ defined by

$$\beta_g(b) = \rho_2(g) b \rho_2(g)^{-1}.$$

Recall $L_{\omega(2)} \subset \mathbb{B}_{\omega(2)}$ is the diagonal subalgebra. Since the $\rho_2(g)$'s are permutation matrices, this action preserves diagonal matrices and hence its restriction to $L_{\omega(2)}$ gives us an action (a priori *not* inner) on $L_{\omega(2)}$, but the diagonal projection $P : \mathbb{B}_{\omega(2)} \to L_{\omega(2)}$ is clearly equivariant. Moreover, $L_{\omega(2)}$ being commutative is nuclear (see Remark 4.10), and of course included in $\mathbb{B}_{\omega(2)}$. Therefore, for any C^*-algebra C, $Id_C \otimes P$ defines a c.p. contraction $\Phi : C \otimes_{\min} \mathbb{B}_{\omega(2)} \to C \otimes_{\max} \mathbb{B}_{\omega(2)}$, such that $\Phi(c \otimes b) = c \otimes P(b)$. Thus we are in the situation to apply Corollary 24.11. Using $C = \overline{\mathbb{B}_{\omega(1)}}$ and $B = \mathbb{B}_{\omega(2)}$, we observe that, with the notation in Corollary 24.11, since $P(1) = 1$ and $P(u_0^2) = u_0^2$, we have

$$t \in C \otimes F.$$

Therefore Corollary 24.11 tells us that

$$\|t\|_{[\overline{\mathbb{B}_{\omega(1)}} \otimes_{\max} \mathbb{B}_{\omega(2)}] \otimes_{\max} C^*(\mathbb{F}_{n-1})} = \|t\|_{[\overline{\mathbb{B}_{\omega(1)}} \otimes_{\min} \mathbb{B}_{\omega(2)}] \otimes_{\max} C^*(\mathbb{F}_{n-1})}.$$

Applying to t the $*$-homomorphism $Id_{\overline{\mathbb{B}_{\omega(1)}} \otimes_{\max} \mathbb{B}_{\omega(2)}} \otimes T$ where T is the linear map associated to the trivial representation on G that takes all $U(g)$'s to 1, we find

$$n + 1 = \left\| \sum \overline{u_j^1} \otimes u_j^2 \right\|_{\overline{\mathbb{B}_{\omega(1)}} \otimes_{\max} \mathbb{B}_{\omega(2)}} \leq \|t\|_{\max},$$

and the proof is complete.   □

Remark 24.21 It is probably worthwhile for the reader to emphasize that we cannot replace t by $t_1 = \sum_{j=1}^{n} \overline{u_j^1} \otimes u_j^2 \otimes U_j$, because we have $\|t_1\|_{[\overline{\mathbb{B}_{\omega(1)}} \otimes_{\min} \mathbb{B}_{\omega(2)}] \otimes_{\min} C^*(\mathbb{F}_{n-1})} = n$. Indeed, pick $m \in \omega(1), m' \in \omega(2)$ and let $v_j = \overline{u_j^1}(m) \otimes u_j^2(m')$. Using the map taking U_j to $\overline{v_j}$, we have

$$\|t_1\|_{[\overline{\mathbb{B}_{\omega(1)}} \otimes_{\min} \mathbb{B}_{\omega(2)}] \otimes_{\min} C^*(\mathbb{F}_{n-1})} \geq \left\| \sum_{j=1}^{n} v_j \otimes \overline{v_j} \right\|_{\min}$$

and, since the v_j's are unitary matrices, we have by (18.5) $\left\| \sum_{j=1}^{n} v_j \otimes \overline{v_j} \right\|_{\min} = n$.

Note that if we try to apply that same argument to t, recalling $U_0 = U_1 = 1$, we are led to write

$$\|t\|_{\min} \geq \left\| (v_0 + v_1) \otimes \overline{1} + \sum_{j=2}^{n} v_j \otimes \overline{v_j} \right\|_{\min}$$

and since $\text{tr}(v_0 + v_1) = 0$, this leads to $\|t\|_{\min} \geq n - 1$, but what really matters is that, as Theorem 24.20 shows us, $\|t\|_{\min} < n + 1$.

Corollary 24.22 $\mathscr{B} \otimes_{\min} \mathscr{B}$ *fails the WEP.*

Proof We base the proof directly on the information that $C_0(n) < n$. By Selberg's bound this holds at least for $n = 3$. We use similar ingredients as for the proof that $\mathscr{B} \otimes_{\min} \mathscr{B} \neq \mathscr{B} \otimes_{\max} \mathscr{B}$ given previously in Theorem 18.9. By Remarks 24.15 and 24.16 we may assume that we have a sequence $u(m)$ with $u(m) \in \Pi(n, N_m)$ such that $\varepsilon_0(\overline{u(m)} \otimes u(m')) \geq \varepsilon_0 > 0$. For each m, we choose a *diagonal* unitary matrix $u_0(m)$ with zero trace. The preceding theorem, then shows that the pair $(\overline{\mathbb{B}_{\omega(1)}} \otimes_{\min} \mathbb{B}_{\omega(2)}, C^*(\mathbb{F}_2))$ is not nuclear, so $\overline{\mathbb{B}_{\omega(1)}} \otimes_{\min} \mathbb{B}_{\omega(2)}$ is not WEP. As in Corollary 18.12, it follows that $\mathscr{B} \otimes_{\min} \mathscr{B}$ fails the WEP.   □

More generally, using Theorem 12.29 we have

Corollary 24.23 *If* (M, N) *are nonnuclear von Neumann algebras,* $M \otimes_{\min} N$ *fails the WEP.*

Proof By Theorem 12.29 and the injectivity of $\mathbb{B}$ we have a completely positive factorization of the identity of $\mathbb{B} \otimes_{\min} \mathbb{B}$ through $M \otimes_{\min} N$. □

Although we opted for tackling directly the failure of WEP for $\overline{\mathbb{B}_{\omega(1)}} \otimes_{\min} \mathbb{B}_{\omega(2)}$, we could have first proved the following (in accordance with Theorem 24.12):

Corollary 24.24 *In the situation of Theorem 24.20, the space $L_{\omega(2)} \rtimes \mathbb{F}_{n-1}$ fails the LLP.*

Proof We continue to use the same notation. In particular, β is the action of G on $\mathbb{B}_{\omega(2)}$, the crossed products are with respect to β and with respect to the natural extensions of β to $C \otimes \mathbb{B}_{\omega(2)}$ with $C = \overline{\mathbb{B}_{\omega(1)}}$. Recall the convention $g_0 = g_1 = 1$. Note in passing that $g_2, \ldots, g_n$ and $L_{\omega(2)}$ together generate $L_{\omega(2)} \rtimes \mathbb{F}_{n-1}$. We define $s_j \in L_{\omega(2)}$ and $s'_j \in L_{\omega(2)} \rtimes \mathbb{F}_{n-1}$ by setting:

$$s_0 = u_0^2 \in L_{\omega(2)}, \quad s_j = 1 \in L_{\omega(2)} \text{ for } 1 \le j \le n,$$

$$\text{and} \quad s'_j = s_j.g_j \in L_{\omega(2)} \rtimes \mathbb{F}_{n-1} \text{ for } 0 \le j \le n.$$

Consider the tensors

$$s' = \sum\nolimits_{j=0}^{n} \overline{u_j^1} \otimes s'_j \in \overline{\mathbb{B}_{\omega(1)}} \otimes [L_{\omega(2)} \rtimes \mathbb{F}_{n-1}],$$

and

$$s = \sum\nolimits_{j=0}^{n} [\overline{u_j^1} \otimes s_j].g_j \in [\overline{\mathbb{B}_{\omega(1)}} \otimes_{\max} L_{\omega(2)}] \rtimes \mathbb{F}_{n-1}.$$

We claim that $\|s'\|_{\max} = \|t\|_{\max} = n + 1$ and $\|s'\|_{\min} = \|t\|_{\min} \le n + 1 - \varepsilon'_0$.

By Proposition 24.6 we know that

$$\|s'\|_{\max} = \|s\|_{\overline{[\overline{\mathbb{B}_{\omega(1)}} \otimes_{\max} L_{\omega(2)}] \rtimes \mathbb{F}_{n-1}}}.$$

By the nuclearity of $L_{\omega(2)}$

$$\|s'\|_{\max} = \|s\|_{\overline{[\overline{\mathbb{B}_{\omega(1)}} \otimes_{\min} L_{\omega(2)}] \rtimes \mathbb{F}_{n-1}}},$$

since $L_{\omega(2)} \subset B_{\omega(2)}$ (equivariantly)

$$\|s'\|_{\max} \ge \|s\|_{\overline{[\overline{\mathbb{B}_{\omega(1)}} \otimes_{\min} B_{\omega(2)}] \rtimes \mathbb{F}_{n-1}}}.$$

Recall that presently $C = \overline{\mathbb{B}_{\omega(1)}}$ and $G = \mathbb{F}_{n-1}$. Since the action of G on $C \otimes_{\min} B_{\omega(2)}$ is inner, the map Ψ associated to $(C \otimes_{\min} B_{\omega(2)}) \rtimes G$ as in Proposition 24.8 is such that $\Psi(s) = t$ (because here $\rho = \rho_2$ and $s_j \rho_2(g_j) = u_j^2$ for all $0 \le j \le n$), where t is as in (24.26). Therefore

$$\|s'\|_{\max} \ge \|t\|_{\max} = n + 1.$$

We now turn to $\|s'\|_{\min}$. By Corollary 24.5 the conditional expectation implies

$$[L_{\omega(2)} \rtimes \mathbb{F}_{n-1}] \subset [B_{\omega(2)} \rtimes \mathbb{F}_{n-1}],$$

by Proposition 24.8 (i.e. by innerness)

$$[B_{\omega(2)} \rtimes \mathbb{F}_{n-1}] \simeq B_{\omega(2)} \otimes_{\max} C^*(\mathbb{F}_{n-1})$$

and by Kirchberg's Theorem (see Corollary 9.40)

$$B_{\omega(2)} \otimes_{\max} C^*(\mathbb{F}_{n-1}) = B_{\omega(2)} \otimes_{\min} C^*(\mathbb{F}_{n-1}).$$

This shows we have

$$\|s'\|_{\overline{\overline{B_{\omega(1)}} \otimes_{\min} [L_{\omega(2)} \rtimes \mathbb{F}_{n-1}]}} = \|t\|_{\overline{[\overline{B_{\omega(1)}} \otimes_{\min} \mathbb{B}_{\omega(2)}] \otimes_{\min} C^*(\mathbb{F}_{n-1})}}.$$

By (24.27) this proves our claim and hence that $L_{\omega(2)} \rtimes \mathbb{F}_{n-1}$ fails the LLP. $\square$

We now show

Corollary 24.25 *In the situation of Theorem 24.20, the pair* $(\overline{L_{\omega(1)}} \rtimes \mathbb{F}_{n-1}, L_{\omega(2)} \rtimes \mathbb{F}_{n-1})$ *is not nuclear.*

Proof By the conditional expectation argument (see Corollary 24.5) we know that

$$L_{\omega(2)} \rtimes \mathbb{F}_{n-1} \subset B_{\omega(2)} \rtimes \mathbb{F}_{n-1} \text{ and } \overline{L_{\omega(1)}} \rtimes \mathbb{F}_{n-1} \subset \overline{B_{\omega(1)}} \rtimes \mathbb{F}_{n-1},$$

with conditional expectations onto each of them. Therefore both the min and max norms on

$$[\overline{L_{\omega(1)}} \rtimes \mathbb{F}_{n-1}] \otimes [L_{\omega(2)} \rtimes \mathbb{F}_{n-1}]$$

are equal to the norms induced respectively by the min and max norms on

$$[\overline{B_{\omega(1)}} \rtimes \mathbb{F}_{n-1}] \otimes [B_{\omega(2)} \rtimes \mathbb{F}_{n-1}].$$

By Proposition 24.8 (applied twice) these can be identified with the min and max norms on

$$[\overline{B_{\omega(1)}} \otimes_{\max} C^*(\mathbb{F}_{n-1})] \otimes [B_{\omega(2)} \otimes_{\max} C^*(\mathbb{F}_{n-1})].$$

By Kirchberg's theorem (see Corollary 9.40) these are the same as the min and max norms on

$$[\overline{B_{\omega(1)}} \otimes_{\min} C^*(\mathbb{F}_{n-1})] \otimes [B_{\omega(2)} \otimes_{\min} C^*(\mathbb{F}_{n-1})].$$

Let us denote by $s_j(2) \in L_{\omega(2)}$ the element previously denoted by s_j, and by $s_j(1) \in L_{\omega(1)}$ the analogous element. Consider the element

$$z = \sum_{j=0}^{n} \overline{(s_j(1).g_j)} \otimes (s_j(2).g_j) \in [\overline{L_{\omega(1)}} \rtimes \mathbb{F}_{n-1}] \otimes [L_{\omega(2)} \rtimes \mathbb{F}_{n-1}].$$

The element z' corresponding to z in $[\overline{B_{\omega(1)}} \otimes_{\min} C^*(\mathbb{F}_{n-1})] \otimes [B_{\omega(2)} \otimes_{\min} C^*(\mathbb{F}_{n-1})]$ is

$$z' = \sum_{j=0}^{n} \overline{[s_j(1)\rho_1(g_j) \otimes U(g_j)]} \otimes [s_j(2)\rho_2(g_j) \otimes U(g_j)],$$

where ρ_1 is like ρ_2 (defined in the proof of Theorem 24.20) but relative to $\omega(1)$. We have after permuting factors

$$\|z'\|_{\min} = \left\| \sum_{j=0}^{n} \overline{s_j(1)\rho_1(g_j)} \otimes s_j(2)\rho_2(g_j) \right.$$
$$\left. \otimes U(g_j) \otimes U(g_j) \right\|_{\overline{B_{\omega(1)}} \otimes_{\min} B_{\omega(2)} \otimes_{\min} C^*(\mathbb{F}_{n-1}) \otimes_{\min} C^*(\mathbb{F}_{n-1})}$$

and since $U \otimes U \simeq U$, the latter is the same as

$$\left\| \sum_{j=0}^{n} \overline{s_j(1)\rho_1(g_j)} \otimes s_j(2)\rho_2(g_j) \otimes U(g_j) \right\|_{\overline{B_{\omega(1)}} \otimes_{\min} B_{\omega(2)} \otimes_{\min} C^*(\mathbb{F}_{n-1})}.$$

Note that for all $0 \le j \le n$

$$s_j(1)\rho_1(g_j) = u_j^1 \text{ and } s_j(2)\rho_2(g_j) = u_j^2.$$

Thus we obtain by (24.27)

$$\|z'\|_{\min} = \|t\|_{\min} \le n + 1 - \varepsilon_0'.$$

We now turn to

$$\|z'\|_{\max} = \|z'\|_{\overline{B_{\omega(1)}} \otimes_{\max} C^*(\mathbb{F}_{n-1}) \otimes_{\max} B_{\omega(2)} \otimes_{\max} C^*(\mathbb{F}_{n-1})}.$$

Composing with the trivial representation on $\mathbb{F}_{n-1}$ we find

$$\|z'\|_{\max} \ge \left\| \sum_{j=0}^{n} \overline{s_j(1)\rho_1(g_j)} \otimes s_j(2)\rho_2(g_j) \right\|_{\overline{B_{\omega(1)}} \otimes_{\max} B_{\omega(2)}},$$

and the latter is $= n + 1$ by (24.28). This proves that $\|z'\|_{\max} \ne \|z'\|_{\min}$ and hence that the pair $(\overline{L_{\omega(1)}} \rtimes \mathbb{F}_{n-1}, L_{\omega(2)} \rtimes \mathbb{F}_{n-1})$ is not nuclear. $\qquad\square$

24.4 Proof that $C_0(3) < 3$ (Selberg's spectral bound)

Our goal in this section is to describe how Theorem 24.19 follows from a famous result due to Selberg [228] that we will now describe without proof. Lubotzky, Philip, and Sarnak [175] first observed that Selberg's results on the spectrum of the Laplacian on the hyperbolic space imply that the finite groups $\{SL_2(\mathbb{Z}/p\mathbb{Z})\}$ form an expander family. We refer the reader to [172, pp. 51–54] and [243] for more information. Tao's book [243] contains a detailed proof on the deduction of the spectral gaps for $\{SL_2(\mathbb{Z}/p\mathbb{Z})\}$ from the ones for the Laplacian on a certain family of "arithmetic Riemann surfaces" traditionally denoted by $\{X(p)\}$(see [243, pp. 75–76]).

Selberg's Theorem [228] says that the trivial representation of $SL_2(\mathbb{Z})$ is isolated in the set of representations that factor through $SL_2(\mathbb{Z}/p\mathbb{Z})$ for some integer $p \geq 2$. Note that the kernel of the natural quotient map $SL_2(\mathbb{Z}) \to SL_2(\mathbb{Z}/p\mathbb{Z})$ is the set

$$\left\{ \begin{pmatrix} 1+a & b \\ c & 1+d \end{pmatrix} \middle| a,b,c,d \in p\mathbb{Z} \right\}.$$

Thus a representation factors through $SL_2(\mathbb{Z}/p\mathbb{Z})$ if and only if it is trivial (i.e. $=1$) on the latter subset.

It is well known that $SL_2(\mathbb{Z})$ (the "modular group") is generated by $\{t_2, t_3\}$ where

$$t_2 = \begin{pmatrix} 1 & 1 \\ 0 & 1 \end{pmatrix} \text{ and } t_3 = \begin{pmatrix} 0 & 1 \\ -1 & 0 \end{pmatrix} \text{ (see [224, p. 9] or [65, p. 94]).}$$

For convenience we set $t_1 = 1$. Thus the (unital) subset $S = \{t_1, t_2, t_3\} \subset SL_2(\mathbb{Z})$ generates $SL_2(\mathbb{Z})$.

Equivalently the Selberg property for $SL_2(\mathbb{Z})$ means that for some $\varepsilon_0 > 0$ we have

$$\sup_\rho \left\| \sum_{j=1}^3 \rho(t_j) \right\| \leq 3 - \varepsilon_0, \tag{24.29}$$

where the sup runs over all the unitary representations ρ that factor through $SL_2(\mathbb{Z}/p\mathbb{Z})$ for some integer $p \geq 2$ and do not admit any invariant vector.

We will content ourselves with representations of a special form. We assume that ρ is associated to an action of $SL_2(\mathbb{Z})$ by permutation on a finite index set Λ_ρ, so that $H_\rho = \ell_2(\Lambda_\rho)$, we assume moreover that ρ factors through $SL_2(\mathbb{Z}/p\mathbb{Z})$ for some p and that the constant function 1 on Λ_ρ is its only invariant vector (up to scaling) in $H_\rho = \ell_2(\Lambda_\rho)$.

We denote by Π the set of such ρ's.

Let $p > 1$ be a prime number. Consider the action of $SL_2(\mathbb{Z})$ on $\mathbb{Z}_p^2$, viewed as a vector space over the field $\mathbb{Z}_p = \mathbb{Z}/p\mathbb{Z}$. Let Λ_p denote the set of lines in $\mathbb{Z}_p^2$. In other words Λ_p is the set of one-dimensional subspaces (i.e. the projective space) in the vector space $\mathbb{Z}_p^2$. The group $SL_2(\mathbb{Z})$ acts (via $SL_2(\mathbb{Z}_p)$) by permutation on Λ_p. This induces a representation ρ_p of $SL_2(\mathbb{Z})$ on $H_p = \ell_2(\Lambda_p)$. Thus we have in this case $\Lambda_{\rho_p} = \Lambda_p$. For simplicity we also denote by $\chi^p \in \ell_2(\Lambda_p)$ the constant vector with all coordinates equal to $|\Lambda_p|^{-1/2}$.

Lemma 24.26 *For any pair p, q of distinct prime numbers $\overline{\rho_p} \otimes \rho_q \in \Pi$ (up to unitary equivalence).*

Proof Recall $H_p = \ell_2(\Lambda_p)$. Using the canonical basis of H_p we may identify H_p and $\overline{H_p}$. Then $\overline{H_p} \otimes H_q \simeq \ell_2(\Lambda_p \times \Lambda_q)$ and $\overline{\rho_p} \otimes \rho_q$ acts by permutation on $\Lambda_p \times \Lambda_q$. Note that any two distinct elements of Λ_p produce a linear basis of $\mathbb{Z}_p^2$. Therefore, by classical linear algebra over the field $\mathbb{Z}_p$, the action of $SL_2(\mathbb{Z}_p)$ (by permutation) on Λ_p is bitransitive, and hence by Lemma A.68 the representation $\rho_p^0 = \rho_{p|\chi^{p\perp}}$ is irreducible. Since $|\Lambda_p| = p + 1$ (see Remark 24.29) $|\Lambda_p| \neq |\Lambda_q|$ whenever $p \neq q$. This implies that ρ_p^0 and ρ_q^0 are *distinct* irreducible representations. Let T denote the trivial representation. By Schur's classical Lemma A.69, $\overline{\rho_p^0} \otimes \rho_q^0$, as well as $\overline{\rho_p^0} \otimes T$ and $\overline{T} \otimes \rho_q^0$, have no invariant vector. Therefore, the only invariant vector of $\overline{\rho_p} \otimes \rho_q$ is $\overline{\chi^p} \otimes \chi^q$. Since $\overline{\rho_p} \otimes \rho_q$ is trivial on matrices with entries divisible by both p and q, it factors through $SL_2(\mathbb{Z}/\mathbb{Z}_{pq})$ and hence belongs to Π. $\qquad\square$

Proof of Theorem 24.19 By (24.29) and the preceding lemma we have

$$\sup\nolimits_{p \neq q} \left\| \sum\nolimits_{j=1}^{3} (\overline{\rho_p} \otimes \rho_q)(t_j)_{|[\overline{\chi^p} \otimes \chi^q]^\perp} \right\| \leq 3 - \varepsilon_0.$$

Thus if we set

$$\forall j = 1, 2, 3 \quad u_j(p) = \rho_p(t_j)$$

we have

$$\varepsilon_0(\overline{u(p)} \otimes u(q)) \geq \varepsilon_0,$$

and hence $C_0(3) \leq 3 - \varepsilon_0$. $\qquad\square$

Remark 24.27 By (i) in Lemma A.69, (24.29) also implies:

$$\sup\nolimits_p \left\| \sum\nolimits_{j=1}^{3} (\overline{\rho_p^0} \otimes \rho_p^0)(t_j)_{|I^\perp} \right\| \leq 3 - \varepsilon_0.$$

Thus the family $\{(\rho_p^0(t_j))_{1 \leq j \leq 3} \mid p \text{ prime}\}$ is a quantum expander.

24.5 Other proofs that $C_0(n) < n$

In the next remarks we give several alternative ways to prove that $C_0(n) < n$.

Remark 24.28 (Using property (T) for $SL_d(\mathbb{Z})$ for $d \geq 3$) This is similar to what we just did with $SL_2(\mathbb{Z}_p)$. We use the fact that $SL_d(\mathbb{Z})$ has property (T) for $d \geq 3$ (and only for $d \geq 3$). Let $S \subset SL_d(\mathbb{Z})$ be a finite generating set containing the unit. By Proposition 17.3 there is $\varepsilon > 0$ such that any unitary representation π on $SL_d(\mathbb{Z})$ without nonzero invariant vector satisfies

$$|S|^{-1} \left\| \sum\nolimits_{s \in S} \pi(s) \right\| \leq 1 - \varepsilon. \qquad (24.30)$$

For any prime p we let $SL_d(\mathbb{Z}_p)$ act on the set of lines (i.e. one dimensional linear subspaces) in $\mathbb{Z}_p^d$. By classical linear algebra over the field $\mathbb{Z}_p$, this action is bitransitive, and hence yields by Lemma A.68 (after composition with the surjection $SL_d(\mathbb{Z}) \to SL_d(\mathbb{Z}_p)$) an irreducible representation π_p^0 defined on $SL_d(\mathbb{Z})$.

Remark 24.29 (How many lines in $\mathbb{Z}_p^d$?) The set of lines in $\mathbb{Z}_p^d$ has cardinality $(p^d - 1)/(p - 1)$. Indeed each point in $\mathbb{Z}_p^d \setminus \{0\}$ determines a line, each line contains $p - 1$ nonzero points, and two distinct lines intersect only at 0.

Arguing as for Lemma 24.26 (comparing the respective dimensions) we find that $\pi_p^0 \neq \pi_q^0$ for any pair of distinct primes $p \neq q$. By Schur's lemma (see (ii) in Lemma A.69) $\overline{\pi_p^0} \otimes \pi_q^0$ has no invariant vector if $p \neq q$ and hence by (24.30)

$$\sup_{p \neq q} |S|^{-1} \left\| \sum_{s \in S} \overline{\pi_p^0(s)} \otimes \pi_q^0(s) \right\| \leq 1 - \varepsilon.$$

Let $n = |S|$. Arguing as in the preceding proof of Theorem 24.19 we obtain $C_0(n) \leq n - \varepsilon n$.

Moreover, we also have (see (i) in Lemma A.69)

$$\sup_p |S|^{-1} \left\| \left[\sum_{s \in S} \overline{\pi_p^0(s)} \otimes \pi_p^0(s) \right]_{|I^\perp} \right\| \leq 1 - \varepsilon,$$

which shows that the family $\{(\pi_p^0(s))_{s \in S} \mid p \text{ prime}\}$ forms a quantum expander.

Since it is known (see [250] or [262, prop. 5]) that $SL_d(\mathbb{Z})$ can be generated just by two elements and the unit, we may take $n = 3$ in what precedes, so we obtain again $C_0(3) < 3$.

Remark 24.30 (Using Kassabov's expander) In answer to a longstanding question, Kassabov proved in [148] that with a suitable choice of generators of a fixed size n the sequence of the permutation groups forms an expander. From this one can deduce easily that there are quantum expanders formed of permutation matrices restricted to the orthogonal of the constant vector as in (24.14) and (24.15), and hence that $C_0(n) < n$ for that same n. I am grateful to Aram Harrow for pointing this out to me. More precisely, let $\mathbb{S}_m$ denote the symmetric group of all permutations of an m element set. Kassabov [148] proved that the family $\{\mathbb{S}_m \mid m \geq 1\}$ forms an expanding family with respect to subsets $S_m \subset \mathbb{S}_m$ of a fixed size n and a fixed spectral gap $\delta > 0$. The construction detailed in [148] leads to a rather large value of n, but in [148, Rem. 5.1] it is asserted that, assuming m large enough, one can obtain $n = 20$ at the expense of a smaller gap $\delta > 0$, and also that one can obtain generating sets formed of involutions.

To produce quantum expanders coming from permutation matrices, we invoke Lemma A.68: the natural representation $\pi_m : \mathbb{S}_m \to B(\ell_2^m)$ that acts on ℓ_2^m by permuting the basis vectors (i.e. $\pi_m(\sigma)(e_j) = e_{\sigma(j)}$) decomposes as the sum of the trivial representation on $\mathbb{C}\chi_m$ and an irreducible representation π_m^0 that is the restriction of π_m to $\chi_m^{\perp}$. By Remark 19.10 the sequence $\{(\pi_m^0(s))_{s \in S_m} \mid m \geq 1\}$ forms a quantum expander, relative to the dimensions $N_m = m - 1$. By Proposition 19.8, we can extract from it a quantum coding sequence. This implies that $C_0(n) < n$.

24.6 Random permutations

Let us assume that $n \geq 4$ is an even integer. We choose an n-tuple of random permutation matrices in the following way: we simply select $u_1^{(N)}, \ldots, u_{n/2}^{(N)}$ independently and uniformly over the group of permutation matrices of size $N \times N$. We then define $u_{j+n/2}^{(N)} = (u_j^{(N)})^{-1}$ for any $1 \leq j \leq n/2$. A priori this allows some repetitions, but when N is much larger than n we obtain with (very) high probability an n-tuple of distinct permutation matrices. Indeed, the probability that $u_i^{(N)} = u_j^{(N)}$ for some $i \neq j$ in $[1, n/2]$ is less than $(n/2)^2/N!$.

Let

$$V_j^{(N)} = u_j^{(N)}{}_{|\chi^{\perp}},$$

where again $\chi = N^{-1/2} \sum_1^N e_k$. We view $(V_j^{(N)})$ as random $(N-1) \times (N-1)$ unitary matrices (with respect to any orthonormal basis of $\chi_N^{\perp}$).

In [93], Joel Friedman proved that for any fixed $\varepsilon > 0$ when $N \to \infty$ we have

$$\mathbb{P}\left(\left\{\left\|\sum_1^n V_j^{(N)}\right\| > 2\sqrt{n-1} + \varepsilon\right\}\right) \to 0, \tag{24.31}$$

which had been conjectured by Noga Alon.

Using a quite different approach, Bordenave and Collins recently proved an analogue of Theorem 18.16 for the same model of independent random permutation matrices (restricted to $\chi^{\perp}$), which leads to an optimal estimate of $C_0(n)$. They prove the following result:

Theorem 24.31 ([32]) *Fix an even integer $n \geq 4$. Let $g_1, \ldots, g_{n/2}$ be the free generators of the free group $\mathbb{F}_{n/2}$. We set $\lambda_j = \lambda_{\mathbb{F}_n}(g_j)$ and $\lambda_{j+n/2} = \lambda_{\mathbb{F}_n}(g_j^{-1})$ for all $1 \leq j \leq n/2$. Then, for all k and for all $a_0, \ldots, a_n$ in M_k such that $a_0 = a_0^*$ and $a_{j+n/2} = a_j^*$ for all $1 \leq j \leq n/2$ we have*

$$\forall \varepsilon > 0 \quad \lim_{N \to \infty} \mathbb{P}\left(\left\{\left\|a_0 \otimes I + \sum_1^n a_j \otimes V_j^{(N)}\right\|\right.\right.$$
$$\left.\left. > \left\|a_0 \otimes I + \sum_1^n a_j \otimes \lambda_j\right\| + \varepsilon\right\}\right) = 0.$$

Corollary 24.32 $C_0(n) = 2\sqrt{n-1}$ *for all even $n \geq 4$.*

Proof We will use the preceding result assuming that the a_j's are all unitary matrices with $a_0 = 0$. In that case by the absorption principle (3.13) and by (3.18) we have $\| \sum_1^n a_j \otimes \lambda_j \| = 2\sqrt{n-1}$. Then, we may proceed exactly as we did to prove Theorem 18.6 in §18.2 to prove that $C_0(n) \le 2\sqrt{n-1}$ for all even $n \ge 4$. Equality holds by (18.14).   □

Bordenave and Collins [32] also prove a result that yields quantum expanders derived from permutation matrices. They prove that Theorem 24.31 still holds if we replace $V_j^{(N)} = u_j^{(N)}{}_{|\chi^\perp}$ by $\overline{u_j^{(N)}} \otimes u_j^{(N)}{}_{|\{J,I\}^\perp}$ where J, I are as in (24.16). By (24.16), we may deduce from the latter result the following consequences which refine the analogous result of Hastings [132] for random unitaries.

Theorem 24.33 ([32]) *In the situation of Theorem 24.31, we have*

$$\forall \varepsilon > 0 \quad \lim_{N \to \infty} \mathbb{P}\left(\left\{ \left\| \left[\sum_1^n \overline{V_j^{(N)}} \otimes V_j^{(N)} \right]_{|I^\perp} \right\| > 2\sqrt{n-1} + \varepsilon \right\}\right) = 0,$$

where I stands here for the tensor associated to the identity on the $N-1$-dimensional space $\chi^\perp$.

Corollary 24.34 *For any $\delta > 0$ and $\varepsilon > 0$ such that $2\sqrt{n-1} + \varepsilon < n$, there is a sequence $N_m \to \infty$ such that, with probability $> 1 - \delta$, the family $(V_j^{(N_m)})_{1 \le j \le n}$ forms a quantum expander such that*

$$\sup_m \left\| \left[\sum_1^n \overline{V_j^{(N_m)}} \otimes V_j^{(N_m)} \right]_{|I^\perp} \right\| \le 2\sqrt{n-1} + \varepsilon.$$

Proof By Theorem 24.33 we can choose a sequence $N_m \to \infty$ such that

$$\sum_m \mathbb{P}\left(\left\{ \left\| \left[\sum_1^n \overline{V_j^{(N_m)}} \otimes V_j^{(N_m)} \right]_{|I^\perp} \right\| > 2\sqrt{n-1} + \varepsilon \right\}\right) < \delta.$$

Then $\mathbb{P}(\{ \sup_m \| [\sum_1^n \overline{V_j^{(N_m)}} \otimes V_j^{(N_m)}]_{|I^\perp} \| \le 2\sqrt{n-1} + \varepsilon \}) > 1 - \delta.$ □

Remark 24.35 The preceding theorem improves an earlier result from [94] (see also [131]). The same paper [94] also contains bounds for the norm of sums of the form $\sum V_j^{(N)} \otimes \cdots \otimes V_j^{(N)}$ of degree r with $r > 2$.

24.7 Notes and remarks

This chapter is based on [188]. Crossed products (like tensor products) are more often considered in the literature in the reduced case than in the "full" one as we do here. The results stated in §24.1 are rather standard facts. Those of

§24.2 are easy variants of the approach Ozawa uses in [188] to relate the WEP of $\mathscr{B} \otimes_{\min} \mathscr{B}$ with the LLP of a certain (full) crossed product. The presentation we adopt in §24.3 emphasizes the parallel between Ozawa's proof in [188] that $\mathscr{B} \otimes_{\min} \mathscr{B}$ fails WEP and that in [141] that $\mathscr{B} \otimes_{\min} \mathscr{B} \neq \mathscr{B} \otimes_{\max} \mathscr{B}$. In [209] we described a shortcut to prove the results from [141] and [188]. We effectively use this in our proof of Ozawa's Theorem 24.20. This approach avoids passing through the nonseparability of the metric space (equipped with d_{cb}) of n-dimensional operator spaces (see §20.3 for more on this) and allows us to put forward the constant $C_0(n)$ defined in (24.13).

25

Open problems

Besides the main problem discussed in Chapters 12 and 13, many related questions remain open, and at least some of them are probably more accessible.

1. If a C^*-algebra has both WEP and LLP, is it nuclear?

Of course a positive answer will solve negatively the Connes–Kirchberg problem. So perhaps we should rephrase this as: Is there a nonnuclear C^*-algebra with both WEP and LLP? [Added in proof: while this book was at the printing stage, the author constructed such an example]. Is there any discrete group G for which $C^*(G)$ is such an example? Of course, A has both WEP and LLP if and only if the pair $(A, \mathscr{B} \oplus \mathscr{C})$ is nuclear.

2. If the pair $(A, \mathscr{Q})$ is nuclear where $\mathscr{Q}$ is the Calkin algebra, does it follow that A is nuclear?

It should be true if the Kirchberg conjecture is correct. See Remark 10.11.

There are many open questions involving discrete groups. It is natural to declare that a group G is WEP (resp. LLP) if $C^*(G)$ has the WEP (resp. LLP). (As for the WEP of $C^*_\lambda(G)$, it is equivalent to amenability by Corollary 9.29). By Proposition 3.5 and Remark 7.20 both properties pass to subgroups.

Clearly, amenable groups have both properties, since $C^*(G)$ is then nuclear.

Of course, groups with WEP are very poorly understood, since we do not even know whether free groups are WEP.

3. Is there any nonamenable WEP group?

Curiously, however, although it should be much easier since free groups are clearly LLP, there are very few known examples of nonamenable groups with LLP, besides free groups. In fact, until A. Thom's paper [244] no explicit example was known. For instance, the following very interesting question is still open:

4. Is the product of two free groups (say $\mathbb{F}_2 \times \mathbb{F}_2$) LLP?

This boils down, of course (by (4.11)), to the question whether $C^*(\mathbb{F}_2) \otimes_{\max} C^*(\mathbb{F}_2)$ has the LLP. Note that by (4.13) and Theorem 9.44 the LLP for groups is stable by free products.

The example of Thom [244] is a group with Kazhdan's property (T). We discuss this in more detail in §17. Thom's example is approximately linear (i.e. "hyperlinear"), with property (T) but *not* residually finite. It follows that for that group G, $C^*(G)$ fails the LLP.

Recall that the Kirchberg conjecture is equivalent to the assertion that *every* C^*-algebra is QWEP, but there is a candidate for a counterexample:

5. It seems to be open whether $C_\lambda^*(G)$ is QWEP when $G = SL_3(\mathbb{Z})$.

In sharp contrast, M_G is QWEP since $SL_3(\mathbb{Z})$ is RF (see Proposition 12.18). In connection with this, there is no example of discrete group G for which the inclusion $C_\lambda^*(G) \rightarrow M_G$ is not max-injective in the sense defined in Definition 7.18. This is true when G is a free group (see [39, p. 384]). In fact it follows from Remarks 7.32 and 10.22 (with Corollary 23.36) that if G is weakly amenable then $C_\lambda^*(G) \rightarrow M_G$ is max-injective. If $C_\lambda^*(G) \rightarrow M_G$ is max-injective, then M_G QWEP implies $C_\lambda^*(G)$ QWEP by Corollary 9.65.

Concerning exactness, a discrete group G is called exact if $C_\lambda^*(G)$ is exact. Ozawa's results [187] show that a certain group $\mathcal{G}$ called Gromov's monster, and which, as the name indicates, is extremely hard to construct, is not exact. See [14, 96, 182]. Until recently this was the only known example of nonexact group. However, a remarkable example of nonexact residually finite group was constructed more recently by Osajda in [183]. More examples, and simpler ones, would be most welcome.

Analogously, the exactness of the full C^*-algebra $C^*(G)$ is not better understood, as shown by the following longstanding open problem:

6. Does the exactness of $C^*(G)$ imply the amenability of G?

Recall (see Theorem 3.30) that G is amenable if and only if either $C^*(G)$ or $C_\lambda^*(G)$ is nuclear. Thus, if G is amenable $C^*(G)$ and $C_\lambda^*(G)$ are exact (and actually identical).

The WEP for $C_\lambda^*(G)$ is better understood, since as we showed in Corollary 9.29, $C_\lambda^*(G)$ has the WEP if and only if G is amenable. Analogously, if we assume that $C_\lambda^*(G)$ is QWEP or G approximately linear (in other words hyperlinear) -assumptions for which no counterexample is known-then $C_\lambda^*(G)$ has the LLP if and only if G is amenable. Indeed, by Corollary 9.41 if a QWEP C^*-algebra has the LLP then it has the WEP. Can this be proved without the a priori assumption that $C_\lambda^*(G)$ is QWEP? Equivalently:

7. Does LLP $\Rightarrow$ WEP hold for $C^*_\lambda(G)$?

We showed in §24.3 that $\mathscr{B} \otimes_{\min} \mathscr{B}$ fails WEP, and we know that $\mathscr{B} \otimes_{\max} \mathscr{B} \neq \mathscr{B} \otimes_{\min} \mathscr{B}$ (see Corollary 18.12) but the following seems to be open:

8. Show that $\mathscr{B} \otimes_{\max} \mathscr{B}$ fails WEP.

Concerning the question whether it is QWEP see Remark 13.2.

By Theorems 23.7 and 23.34, and similar results in §23.5, we know that a C^*-algebra $A \subset B(H)$ has the WEP if a certain kind of operators from A to a Hilbert space $\mathcal{H}$ (namely those associated to a positive definite state on $\overline{A} \otimes_{\max} A$) admit an extension of the same kind from $B(H)$ to $\mathcal{H}$. This line of thought naturally leads us to the following questions.

9. Let $A \subset B(H)$ be a unital C^*-subalgebra. Assume that any bounded linear $u : A \to \ell_2$ admits an extension $\tilde{u} : B(H) \to \ell_2$ with $\|\tilde{u}\| = \|u\|$. Does it follow that A has the WEP?

10. More generally, let $A \subset B$ be an inclusion of C^*-algebras. Assume that any $u : A \to \ell_2$ admits an extension $\tilde{u} : B \to \ell_2$ with $\|\tilde{u}\| = \|u\|$. Is there a contractive projection $P : B^{**} \to A^{**}$?

Here the assumption that the norm is preserved is essential. Indeed, in the setting of questions 9 and 10, by the author's version of the noncommutative Grothendieck theorem, there is always an extension $\tilde{u}$ with $\|\tilde{u}\| \leq C\|u\|$ where C is a universal constant. See [210] for more information and references in this direction. See also our memoir [205, ch. 5] for a general discussion of maps such as $u : A \to \ell_2$.

Concerning injectivity of von Neumann algebras and possible generalizations of Tomiyama's Theorem 1.45 the following is open:

11. Does the existence of a *bounded* linear projection from a von Neumann algebra $\mathcal{M}$ to a (von Neumann) subalgebra $M \subset \mathcal{M}$ imply the existence of a contractive one?

In particular, if G is a nonamenable group, although it seems likely to be true, there is no known proof of the absence of *bounded* projections from $B(\ell_2(G))$ onto M_G. However, if G contains $\mathbb{F}_2$ this was proved by Haagerup and the author in [118]. See also [53]. For a von Neumann subalgebra $M \subset B(H)$ let $\lambda(M) = \inf \|P\|$ where the infimum runs over all linear projections $P : B(H) \to M$ onto M. Note that $\lambda(M)$ remains the same for any completely isometric embedding of M in $B(H)$. It is proved in [201] that $\lambda(M \overline{\otimes} N) \geq \lambda(M)\lambda(N)$ for any pair M, N of von Neumann algebras.

Let $A/\mathcal{I}$ be the quotient of a C^*-algebra by a (self-adjoint closed) ideal $\mathcal{I}$ as usual. We saw that if X is a separable operator space and $A/\mathcal{I}$ is nuclear then any complete contraction $u : X \to A/\mathcal{I}$ admits a completely contractive lifting $\widehat{u} : X \to A$ (see Corollary 9.49). Using the theory of M-ideals, Ando and Choi–Effros independently proved (see [125, p. 59]) that if X is separable and $A/\mathcal{I}$ has the bounded approximation property then any *bounded* $u : X \to A/\mathcal{I}$ admits a *bounded* lifting $\widehat{u} : X \to A$ (note that for Banach spaces local reflexivity and hence local liftings are given "for free"). In particular when $A/\mathcal{I}$ is separable the identity of $A/\mathcal{I}$ admits a *bounded* lifting. Combined with Haagerup's subsequent result [104] that $C_\lambda^*(\mathbb{F}_\infty)$ has the metric approximation property, this shows that, in the case $A = C^*(\mathbb{F}_\infty)$ and $A/\mathcal{I} = C_\lambda^*(\mathbb{F}_\infty)$, the natural quotient map $C^*(\mathbb{F}_\infty) \to C_\lambda^*(\mathbb{F}_\infty)$ has a contractive lifting, and hence there is a *bounded* projection $P : C^*(\mathbb{F}_\infty) \to \mathcal{I}$, but no *completely bounded* one, since, by Proposition 7.34, there is no *completely bounded* lifting.

It is a longstanding open question whether the preceding Ando–Choi–Effros theorem holds without any assumption on $A/\mathcal{I}$, more precisely:

12. The following basic questions are open:

(i) If X is a separable Banach space is it true that any bounded
$u : X \to A/\mathcal{I}$ admits a bounded lifting $\widehat{u} : X \to A$?

(ii) Is there always a bounded linear projection $P : A \to \mathcal{I}$ when $A/\mathcal{I}$ is separable?

(iii) Is there always a bounded linear projection $P : A \to \mathcal{I}$ when A is separable?

These are three equivalent forms of the same question. Indeed, taking $X = A/\mathcal{I}$, the existence of a lifting for $u = Id_{A/\mathcal{I}}$ is equivalent to the existence of a bounded projection $P : A \to \mathcal{I}$. This shows that "yes" to (i) implies "yes" to (ii) and the latter trivially implies "yes" to (iii). Let X and $u : X \to A/\mathcal{I}$ be as in (i) and let $q : A \to A/\mathcal{I}$ be the quotient map. There is clearly a separable C^*-subalgebra $A_1 \subset A$ such that $q(A_1) \supset u(X)$. Therefore a "yes" to (iii) implies that $A_1 \cap \mathcal{I}$ is complemented in A_1 or equivalently that the identity of $q(A_1)$ admits a bounded lifting in A_1. This implies a fortiori that u is liftable, so that "yes" to (iii) implies "yes" to (i).

In Banach space theory, well-known work by Sobczyk shows that the space c_0 is *separably* injective (i.e. complemented in any *separable* superspace) but not injective (see [167, p. 106]). See Zippin's paper [265] for a proof that c_0 is actually the *only* (infinite-dimensional) separably injective Banach space (up to isomorphism). See [181] for a discussion of the analogous open questions for operator spaces, with $K(\ell_2)$ playing the role of c_0. See [266] for a broad survey of bounded linear projections in Banach space theory.

Appendix

Miscellaneous background

Our intention here is to help the reader remember why certain basic facts are true and how they are interlaced. We use deliberately a telegraphic style. We refer the reader to the many existing basic reference books for a more harmonious presentation of the various topics we survey.

A.1 Banach space tensor products

Let X, Y be Banach spaces. Any $t \in X \otimes Y$ (algebraic tensor product) can be written as a finite sum $t = \sum_1^N x_j \otimes y_j$ $(x_j \in X, y_j \in Y)$. The smallest possible N is called the rank of t.

Let $\mathrm{Bil}(X \times Y)$ denote the space of bounded bilinear forms on $X \times Y$.

We have a canonical linear *injective* map from $X \otimes Y$ to the space $\mathrm{Bil}(X^* \times Y^*)$. Namely this is the linear mapping taking $x \otimes y$ $(x \in X, y \in Y)$ to the form defined on $X^* \times Y^*$ by $(x', y') \mapsto x'(x)y'(y)$ $(x' \in X^*, y' \in Y^*)$. The latter form is separately weak* continuous on $X^* \times Y^*$.

Remark A.1 Similarly, we have a canonical linear *injective* map from $X^* \otimes Y^*$ to the space $\mathrm{Bil}(X \times Y)$. This is the linear mapping taking $x' \otimes y'$ $(x' \in X^*, y' \in Y^*)$ to the form defined on $X \times Y$ by $(x, y) \mapsto x'(x)y'(y)$ $(x \in X, y \in Y)$. In the case of biduals, it follows that if $t \in X^{**} \otimes Y^{**}$ vanishes on $X^* \otimes Y^*$ then $t = 0$.

The projective and injective tensor norms of t are defined by

$$\|t\|_\wedge = \inf\left\{ \sum_1^N \|x_j\| \|y_j\| \right\}$$

where the inf runs over all possible ways to write t as precedingly and

$$\|t\|_\vee = \sup\left\{ |\langle t, \xi \otimes \eta \rangle| \mid (\xi, \eta) \in B_{X^*} \times B_{Y^*} \right\}$$

$$= \sup\left\{ \left| \sum_1^N \xi(x_j)\eta(y_j) \right| \mid (\xi, \eta) \in B_{X^*} \times B_{Y^*} \right\}.$$

Note that by homogeneity, we also have

$$\|t\|_{\wedge} = \inf\left\{\left(\sum_{1}^{N}\|x_j\|^2\right)^{1/2}\left(\sum_{1}^{N}\|y_j\|^2\right)^{1/2}\middle| t = \sum_{1}^{N} x_j \otimes y_j\right\}.$$

Let α be either $\wedge$ or $\vee$. These norms satisfy both $\|x \otimes y\|_{\alpha} = \|x\|\|y\|$ for any $(x, y) \in X \times Y$ and similarly for the dual norm on $X^* \otimes Y^*$ we have $\|\xi \otimes \eta\|_{\alpha}^* = \|\xi\|\|\eta\|$ for any $(\xi, \eta) \in X^* \times Y^*$. Any norm $\|\ \|_{\alpha}$ on $X \otimes Y$ satisfying both conditions is called "reasonable." Then the projective ($\alpha = \wedge$) and injective ($\alpha = \vee$) norm are respectively the largest and smallest among the reasonable tensor norms. As is easy to check, they are dual to each other: if $\|\ \|_{\alpha} = \|\ \|_{\wedge}$ (resp. $\|\ \|_{\alpha} = \|\ \|_{\vee}$) then $\|\ \|_{\alpha}^* = \|\ \|_{\vee}$ (resp. $\|\ \|_{\alpha}^* = \|\ \|_{\wedge}$).

We denote by $X\overset{\wedge}{\otimes}Y$ (resp. $X\overset{\vee}{\otimes}Y$) the respective completions of $X \otimes Y$.

Note that we have canonical isometric identifications

$$X\overset{\wedge}{\otimes}Y = Y\overset{\wedge}{\otimes}X \quad \text{and} \quad X\overset{\vee}{\otimes}Y = Y\overset{\vee}{\otimes}X.$$

The dual of $X\overset{\wedge}{\otimes}Y$ can be canonically identified with the space $\mathrm{Bil}(X \times Y)$. The duality is the one obtained by extending the pairing

$$\langle F, t\rangle = \sum F(x_k, y_k)$$

for $F \in \mathrm{Bil}(X \times Y)$ and $t = \sum_{1}^{N} x_k \otimes y_k \in X \otimes Y$, for which we have $|\langle F, t\rangle| \leq \|F\|\|t\|_{\wedge}$.

The space $\mathrm{Bil}(X \times Y)$ can be naturally isometrically identified either with $B(X, Y^*)$ or equivalently with $B(Y, X^*)$. Thus we have

$$(X\overset{\wedge}{\otimes}Y)^* \simeq B(X, Y^*) \simeq B(Y, X^*). \tag{A.1}$$

When X or Y is an L_1-space, the projective tensor norm can be computed more explicitly: For any measure space (Ω, μ), we have (isometrically)

$$L_1(\mu)\overset{\wedge}{\otimes}Y \simeq L_1(\mu; Y),$$

where the latter space is meant in "Bochner's sense."

See [200] for a discussion of the pairs (X, Y) such that $X\overset{\wedge}{\otimes}Y = X\overset{\vee}{\otimes}Y$, that goes beyond the scope of the present volume. Note however that in the latter case the two norms are not equal as in the C^*-case, but only equivalent.

A.2 A criterion for an extension property

Let B, C be Banach spaces. The projective norm on the algebraic tensor product was just defined for any $t \in B \otimes C$ by

$$\|t\|_{\wedge} = \inf\left\{\sum_{1}^{N}\|b_j\|_B\|c_j\|_C\right\}$$

the infimum being on all possible representations of t as a *finite* sum $t = \sum_{1}^{N} b_j \otimes c_j$ $(b_j \in B, c_j \in C)$. Let $B \otimes_{\wedge} C$ denote the resulting normed space; for our present purpose we do not need to complete it. For the linear operator $T : C^* \to B$ corresponding to t, for any $\varepsilon > 0$, there are an integer N and a factorization

$$T : C^* \xrightarrow{T_1} \ell_\infty^N \xrightarrow{D} \ell_1^N \xrightarrow{T_2} B \qquad (A.2)$$

where D is diagonal, T_1 is weak* continuous, and

$$\|t\|_\wedge \leq \|T_1\|\|D\|\|T_2\| \leq (1+\varepsilon)\|t\|_\wedge.$$

The correspondence is like this: if $t \neq 0$ we may assume $\|b_j\|\|c_j\| \neq 0$, then for any $\xi \in C^*$ we set $T_1(\xi) = \sum \xi(c_j/\|c_j\|)e_j$, $T_2(e_j) = b_j/\|b_j\|$ and let D be the diagonal matrix with coefficients $(\|b_j\|\|c_j\|)$. This leads to

$$\|t\|_\wedge = N_\wedge(T) \qquad (A.3)$$

where $N_\wedge(T) = \inf\{\|T_1\|\|D\|\|T_2\|\}$ the infimum being over all integers $N \geq 1$ and all factorizations of T as in (A.2), with D diagonal and T_1 weak* continuous. (Warning: $N = \infty$ is not allowed here, as it leads to a smaller norm, namely the nuclear norm).

In the case $B = C^*$, there is a natural linear form on $C^* \otimes C$ denoted by $t \mapsto \mathrm{tr}(t)$ that takes $\xi \otimes x$ ($\xi \in C^*, x \in C$) to $\xi(x)$. We have clearly

$$\forall t \in C^* \otimes C \quad |\mathrm{tr}(t)| \leq \|t\|_\wedge. \qquad (A.4)$$

Let $T : C \to C$ be the finite rank linear map associated to t and let $E \subset C$ be any finite-dimensional subspace such that $E \supset T(C)$. Let $T_E : E \to E$ be the restriction of T to E. Then $\mathrm{tr}(t) = \mathrm{tr}(T_E)$, the latter being is of course the usual (linear algebraic) trace of T_E, which is independent of E. Thus it is natural to define the trace of a finite rank T on an infinite-dimensional C simply by setting $\mathrm{tr}(T) = \mathrm{tr}(t)$. Then (A.4) becomes

$$|\mathrm{tr}(T)| \leq N_\wedge(T).$$

We already mentioned the classical (and easy to see) fact that $[B \otimes_\wedge C]^* \simeq B(B, C^*)$ isometrically. Let $U \in B(B, C^*)$. Since $[B \otimes_\wedge C]^* \simeq B(B, C^*)$ we have for the corresponding duality

$$|\langle U, t\rangle| \leq \|t\|_\wedge \|U\| = N_\wedge(T)\|U\|.$$

The finite rank operators $UT : C^* \to C^*$ and $TU : B \to B$ have the same trace and in fact

$$\langle U, t\rangle = \mathrm{tr}(UT) = \mathrm{tr}(TU).$$

Therefore, we have

$$|\mathrm{tr}(UT)| \leq N_\wedge(T)\|U\|. \qquad (A.5)$$

Proposition A.2 (Dual criterion for extension) *Let B, C Banach spaces. Let $A \subset B$ be a closed subspace and let $j : A \to B$ be the inclusion mapping. Let $u : A \to C^*$ be a linear mapping. The following are equivalent:*

(i) *There is a linear map $\tilde{u} : B \to C^*$ extending u with norm ≤ 1.*
(ii) *For any weak* continuous $v : C^* \to A$ of finite rank we have $|\mathrm{tr}(uv)| \leq N_\wedge(jv)$.*

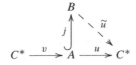

Proof Assume (i). Note $uv = \tilde{u}jv$. Then clearly (applying (A.5) with $U = \tilde{u}$ and $T = jv$)

$$|\text{tr}(uv)| = |\text{tr}(\tilde{u}jv)| \leq \|\tilde{u}\|N_\wedge(jv) \leq N_\wedge(jv).$$

The converse is a simple application of the Hahn–Banach theorem. Assume (ii). Let

$$S = C \otimes A \subset C \otimes_\wedge B.$$

Any $s \in C \otimes A$ corresponds to a weak* continuous finite rank map $v_s : C^* \to A$. We equip S with the norm induced by $C \otimes_\wedge B$. Observe that by (A.3) (ii) means that the linear form $f : S \to \mathbb{C}$ defined by $f(s) = \text{tr}(uv_s)$ has norm ≤ 1. Let $\tilde{f} \in [C \otimes_\wedge B]^*$ be its Hahn–Banach extension with $\|\tilde{f}\|_{[C\otimes_\wedge B]^*} \leq 1$. Let $\tilde{u} : B \to C^*$ be the associated operator. We have $\|\tilde{u}\| = \|\tilde{f}\|_{[C\otimes_\wedge B]^*} \leq 1$ and $\tilde{u}$ is the extension required in (i). Indeed, since $\tilde{f}_{|S} = f$ we have $\langle \tilde{u}(a), c \rangle = \tilde{f}(a \otimes c) = f(a \otimes c) = \langle u(a), c \rangle$ for any $a \in A, c \in C$. $\qquad\square$

A.3 Uniform convexity of Hilbert space

It is convenient to record here the following elementary fact expressing the uniform convexity of Hilbert space. The latter is usually formulated for pairs of unit vectors (i.e. $n = 2$ in (A.6)), but the generalization to n-tuples is straightforward:

Lemma A.3 *Let $x = (x_1, \ldots, x_n)$ be an n-tuple in the unit ball of a Hilbert space H. Then*

$$\forall \varepsilon > 0 \quad \left\| n^{-1} \sum_1^n x_k \right\|^2 > 1 - \varepsilon \Rightarrow \max_{1 \leq i \neq j \leq n} \|x_i - x_j\| < 2\sqrt{n\varepsilon}. \tag{A.6}$$

Proof Let $M(x) = n^{-1} \sum_1^n x_k$. A simple verification (this is a classical fact on the variance) shows that

$$n^{-1} \sup_k \|x_k - M(x)\|^2 \leq n^{-1} \sum_1^n \|x_k - M(x)\|^2 = n^{-1} \sum_1^n \|x_k\|^2 - \|M(x)\|^2,$$

which implies (A.6). $\qquad\square$

A.4 Ultrafilters

Most readers will surely know what is a free or nontrivial ultrafilter $\mathcal{U}$ on a set I. The following quick introduction is meant to allow those with less familiarity to grasp the minimum terminology necessary to follow the present notes. For our purposes, the notion that matters is "the limit along an ultrafilter," and this can be explained easily. Let $f : \ell_\infty(I) \to \mathbb{C}$ be a unital *-homomorphism (a fortiori f is positive). Then for any subset $\alpha \subset I$ we have $f(1_\alpha) \in \{0, 1\}$. The collection formed by the subsets $\alpha \subset I$ such that $f(1_\alpha) = 1$ is what is called an ultrafilter on I. Since indicators form a total set in $\ell_\infty(I)$, f is entirely determined by its values on the indicators of subsets, so that the correspondence $f \leftrightarrow \mathcal{U}$ is 1-1. So any ultrafilter $\mathcal{U}$ comes from a unique functional $f_\mathcal{U}$. The traditional notation is then to denote

$$\forall x \in \ell_\infty(I) \quad \lim_\mathcal{U} x_i = f_\mathcal{U}(x),$$

and to refer to $\lim_{\mathcal{U}} x_i$ as the limit of x_i along $\mathcal{U}$. In this viewpoint, for a subset $\alpha \subset I$, we have

$$\alpha \in \mathcal{U} \Leftrightarrow \lim_{\mathcal{U}} 1_\alpha = 1.$$

The trivial ultrafilters are those that are associated to the functionals δ_i ($i \in I$) defined by $\delta_i(x) = x_i$. The other ones are called *nontrivial* (or "free"). They are characterized by the property that $\lim_{\mathcal{U}} x_i = 0$ whenever $i \mapsto x_i$ is finitely supported.

The existence of nontrivial $\mathcal{U}$'s can be deduced easily from the pointwise compactness of the set of states on $\ell_\infty(I)$, which shows that the set of cluster points of $(\delta_i)_{i \in I}$ when "$i \to \infty$" is nonvoid, in other words that

$$\cap_{\alpha \subset I, |\alpha| < \infty} \overline{\{\delta_i \mid i \notin \alpha\}}$$

is nonvoid, which follows from the finite intersection property.

To illustrate the preceding terminology, let $(a_i)_{i \in I}$ be a bounded family in $\mathbb{R}$ and let $\ell = \lim_{\mathcal{U}} a_i$. Then for any $\varepsilon > 0$ the set $I_\varepsilon = \{i \in I \mid |a_i - \ell| < \varepsilon\}$ belongs to $\mathcal{U}$. Indeed, since $\varepsilon 1_{I \setminus I_\varepsilon} \le (|a_i - \ell|)$ and $f_{\mathcal{U}}$ is positive, we must have $\varepsilon \lim_{\mathcal{U}} 1_{I \setminus I_\varepsilon} \le \lim_{\mathcal{U}} |a_i - \ell| = 0$, and hence $\lim_{\mathcal{U}} 1_{I \setminus I_\varepsilon} = 0$ or equivalently $I_\varepsilon \in \mathcal{U}$.

The converse is also true: if for any $\varepsilon > 0$ the set $I_\varepsilon = \{i \in I \mid |a_i - \ell| < \varepsilon\}$ belongs to $\mathcal{U}$, then we must have $\ell = \lim_{\mathcal{U}} a_i$, because $(|a_i - \ell|) \le \varepsilon 1_{I_\varepsilon} + (\|a\|_\infty + |\ell|)1_{I \setminus I_\varepsilon}$ implies $\lim_{\mathcal{U}} |a_i - \ell| \le \varepsilon$.

More generally when (a_i) and ℓ are points in a topological space we say that $\ell = \lim_{\mathcal{U}} a_i$ if for any neighborhood V of ℓ the set $\{i \in I \mid a_i \in V\}$ belongs to $\mathcal{U}$.

Remark A.4 Let (a_n) be a bounded sequence of reals. Then (a_n) converges when $n \to \infty$ if and only if the limits $\lim_{\mathcal{U}} a_n$ are independent of $\mathcal{U}$ ($\mathcal{U}$ nontrivial ultrafilter on $\mathbb{N}$), and then the limit of (a_n) is their common value. Indeed, it is easy to show that the set formed of all the limits $\lim_{\mathcal{U}} a_n$ ($\mathcal{U}$ nontrivial ultrafilter on $\mathbb{N}$) coincides with the set of cluster points of (a_n).

Remark A.5 Let $\mathcal{U}$ and $\mathcal{V}$ be distinct ultrafilters on I. We claim that there is a disjoint partition $I = \alpha \cup \beta$ such that $\lim_{\mathcal{U}} 1_\alpha = 1$ and $\lim_{\mathcal{V}} 1_\beta = 1$. Indeed, by what precedes there must exist an infinite subset $\gamma \subset I$ such that $\lim_{\mathcal{U}} 1_\gamma \ne \lim_{\mathcal{V}} 1_\gamma$. Then either $\lim_{\mathcal{U}} 1_\gamma = 1$ and then we can take $\beta = I \setminus \gamma$ (and $\alpha = \gamma$), or $\lim_{\mathcal{U}} 1_\gamma = 0$ and then we can take $\alpha = I \setminus \gamma$ (and $\beta = \gamma$).

Remark A.6 Let I be a directed set, meaning that I is given with a partial order such that for any pair $i, j \in I$ there is $k \in I$ such that $i \le k$ and $j \le k$. Assuming I infinite, any function $x : I \to \mathbb{C}$ can be viewed as a "generalized sequence" (one also speaks of the net associated to the directed set I) and the meaning of $\lim x(i) = \ell$ is copied on the usual one: $\forall \varepsilon > 0 \exists j \in I$ such that $|x(i) - \ell| < \varepsilon \ \forall i \ge j$.

We claim that there is an ultrafilter $\mathcal{U}$ on I such that for any $x \in \ell_\infty(I)$ that admits a limit ℓ in the preceding sense we have $\lim_{\mathcal{U}} x(i) = \ell$. Equivalently, for any $i \in I$ the set $\alpha^i = \{j \in I \mid j \ge i\}$ satisfies $\lim_{\mathcal{U}} 1_{\alpha^i} = 1$. With terminology from the theory of filters, or nets and subnets one usually says that the ultrafilter $\mathcal{U}$ refines the filter or the net associated to the directed set I.

To prove that such a $\mathcal{U}$ exists, just observe that by the directedness of I and the compactness of the set of states on $\ell_\infty(I)$ the intersection $\cap_{i \in I} \overline{\{\delta_j \mid j \ge i\}}$ is nonvoid.

A.5 Ultraproducts of Banach spaces

Let $(H_i)_{i \in I}$ be a family of Banach spaces. We recall the usual definition of the ultraproduct of $(H_i)_{i \in I}$ with respect to an ultrafilter $\mathcal{U}$ on the set I. Let $X = \left(\oplus \sum_{i \in I} H_i \right)_\infty$. For any $x = (x_i) \in X$ we define a seminorm $\psi_{\mathcal{U}}$ on X by

$$\psi_{\mathcal{U}}(x) = \lim_{\mathcal{U}} \|x_i\|_{H_i}.$$

We will denote by

$$\mathcal{H}_{\mathcal{U}} = X / \ker(\psi_{\mathcal{U}})$$

the resulting Banach space. We call it the ultraproduct of $(H_i)_{i \in I}$ with respect to $\mathcal{U}$.

When all the H_i's are identical to a single space H we say that $\mathcal{H}_{\mathcal{U}}$ is an ultrapower of H, and we denote

$$\mathcal{H}_{\mathcal{U}} = H^{\mathcal{U}}.$$

Let $q : X \to X / \ker(\psi_{\mathcal{U}})$ be the quotient map. For any element $x = (x_i) \in X$, by convention we denote by $(x_i)_{\mathcal{U}}$ the corresponding element in $\mathcal{H}_{\mathcal{U}}$, i.e. we set

$$(x_i)_{\mathcal{U}} = q(x).$$

With this notation, we have

$$\|(x_i)_{\mathcal{U}}\|_{\mathcal{H}_{\mathcal{U}}} = \lim_{\mathcal{U}} \|x_i\|_{H_i}.$$

The most important case for us is the one when the H_i's are Hilbert spaces. In that case it is easy to check that we can equip $\mathcal{H}_{\mathcal{U}}$ with a scalar product defined by setting for any pair $x, y \in X$

$$\langle (x_i)_{\mathcal{U}}, (y_i)_{\mathcal{U}} \rangle = \lim_{\mathcal{U}} \langle x_i, y_i \rangle.$$

Clearly the right-hand side depends only on $q(x), q(y)$, so this is legitimate. The resulting space $\mathcal{H}_{\mathcal{U}}$ is a Hibert space.

A.6 Finite representability

A Banach space X is finitely representable in another one Y (X f.r. Y in short) if for any $\varepsilon > 0$ any finite-dimensional subspace of X is $(1 + \varepsilon)$-isomorphic to some subspace of Y.

Note that X f.r. Y and Y f.r. Z implies X f.r. Z.

Lemma A.7 *We have X f.r. Y if and only if X embeds isometrically in an ultrapower of Y (in the sense of §A.5).*

Proof Assume X f.r. Y. Let $\mathcal{S}$ be the set of all the finite-dimensional subspaces of X directed by inclusion. Let $I = \mathcal{S} \times \mathbb{N}$. Let $i = (E, m)$ and $i' = (E', m')$ be elements of I. We define a partial order on I by declaring that $i \le i'$ if $E \subset E'$ and $m \le m'$. Then I is a directed set. Let $(E, m) \in I$. Since X f.r. Y there is an operator $u_i : E \to Y$ such that $\|x\| \le \|u_i(x)\| \le (1 + 1/m)\|x\|$ for any $x \in E$. Let $\mathcal{U}$ be an ultrafilter on I refining the associated net (so that in particular $\lim_{\mathcal{U}} 1/m = 0$) as explained in Remark A.6. For any $x \in X$ we have $x \in E$ for all $i = (E, m)$ large enough in I. Thus $u_i(x)$ is well defined. Otherwise we set, say, $u_i(x) = 0$. Then the mapping $u : X \to Y^{\mathcal{U}}$ defined by $u(x) = (u_i(x))_{\mathcal{U}}$ is an isometric embedding.

Conversely assume $X \subset Y^{\mathcal{U}}$ isometrically. To show X f.r. Y it suffices to show that $Y^{\mathcal{U}}$ f.r. Y. Let $E \subset Y^{\mathcal{U}}$ be finite dimensional with basis $x_1, \ldots, x_n$. We may assume $x_k = (x_k(i))_{\mathcal{U}}$. For any scalar coefficients $a_1, \ldots, a_n$ we have $\left\| \sum a_k x_k \right\| = \lim_{\mathcal{U}} \left\| \sum a_k x_k(i) \right\|$. Fix $\varepsilon > 0$. Let $\mathcal{N}_\varepsilon$ be a finite ε-net in the unit ball of E. Choosing i large enough we can obtain

$$\forall a \in \mathcal{N}_\varepsilon \quad (1 - \varepsilon) \left\| \sum a_k x_k \right\|_E \leq \left\| \sum a_k x_k(i) \right\|_Y \leq (1 + \varepsilon) \left\| \sum a_k x_k \right\|_E . \quad \text{(A.7)}$$

Noting that any element a in the unit ball B_E of E can be written (by successive approximations) in the form $a = a^0 + \varepsilon a^1 + \varepsilon^2 a^2 + \cdots$ with $a^0, a^1, a^2, \ldots$ all in $\mathcal{N}_\varepsilon$, we obtain after a simple calculation

$$\forall a \in B_E \quad (1 - \delta_\varepsilon) \left\| \sum a_k x_k \right\| \leq \left\| \sum a_k x_k(i) \right\| \leq (1 + \varepsilon)(1 - \varepsilon)^{-1} \left\| \sum a_k x_k \right\|,$$
$$\text{(A.8)}$$

where $\delta_\varepsilon \to 0$ when $\varepsilon \to 0$. This shows that $Y^{\mathcal{U}}$ f.r. Y. □

Lemma A.8 *If X f.r. Y then there is a set I and a (metric) surjection $q : \ell_\infty(I; Y^*) \to X^*$ taking the closed unit ball of $\ell_\infty(I; Y^*)$ onto that of X^*.*

Proof Let $\mathcal{S}, I, \mathcal{U}$ and $u : X \to Y^{\mathcal{U}}$ be as in the first part of the preceding proof. We define $Q : \ell_\infty(I; Y^*) \to Y^{\mathcal{U}*}$ by setting for any $\xi = (\xi_i)_{i \in I}$ in $\ell_\infty(I; Y^*)$

$$\forall y = (y_i)_{\mathcal{U}} \in Y^{\mathcal{U}} \quad Q(\xi)(y) = \lim_{\mathcal{U}} \langle \xi_i, y_i \rangle.$$

Clearly $\|Q\| \leq 1$. Let

$$q = u^* Q : \ell_\infty(I; Y^*) \to X^*.$$

Clearly $\|q\| \leq 1$. Let $\eta \in B_{X^*}$. For any $i = (E, m)$, let $E_i = u_i(E) \subset Y$. By the preceding argument for all $i = (E, m)$ large enough u_i is an isomorphism from E to E_i. Let $f_i = \eta_{|E} u_i^{-1} \in E_i^*$. Then $\|f_i\|_{E_i^*} \leq 1$. Let $\xi_i \in Y^*$ denote a Hahn–Banach extension of f_i to Y, so that $\|\xi_i\|_{Y^*} \leq 1$ and hence $\|\xi\|_{\ell_\infty(I; Y^*)} \leq 1$. Let $x \in X$. Then

$$\langle u^* Q(\xi), x \rangle = \langle Q(\xi), u(x) \rangle = \lim_{\mathcal{U}} \langle \xi_i, u_i(x) \rangle = \lim_{\mathcal{U}} f_i(u_i(x)) = \eta(x).$$

Thus we conclude $q(\xi) = \eta$. □

A.7 Weak and weak* topologies: biduals of Banach spaces

Let X be a Banach space with (closed) unit ball B_X. As usual the weak topology on X is denoted by $\sigma(X, X^*)$, while the weak* topology on X^* is denoted by $\sigma(X^*, X)$. In general the latter is distinct from its weak topology $\sigma(X^*, X^{**})$. Of course both are weaker than the norm topology.

However, the following well-known result allows to conveniently pass from weak to strong convergence in many interesting cases.

Theorem A.9 (Mazur's theorem) *For any convex subset $C \subset X$, the weak (i.e. $\sigma(X, X^*)$) closure of C coincides with its norm closure.*

Remark A.10 Mazur's theorem is often used in the following form. Suppose we have a net (x_i) in X that converges weakly to a limit $x \in X$. Then we can form a net (x'_β) that converges in norm to x and that is such that each x'_β is a convex combination of elements of the original net (x_i). More precisely, we can arrange for the following supplementary property. Assume without loss of generality that the nets are with respect to directed sets of indices (sometimes called generalized sequences). Then we can make sure that for any i there is β such that for all $\eta \geq \beta$ the point x'_η is in the convex hull of $\{x_\xi, \xi \geq i\}$. Indeed, this is easy to check using the observation that $x \in \overline{\mathrm{conv}(\{x_\xi, \xi \geq i\})}^{\mathrm{weak}}$, and hence by Mazur's theorem for any $\varepsilon > 0$ there is $x'_{i,\varepsilon} \in \mathrm{conv}(\{x_\xi, \xi \geq i\})$ such that $\|x'_{i,\varepsilon} - x\| < \varepsilon$, so that we may use for the β's the set of pairs (i, ε) directed in the obvious way.

Now consider a set Γ, functions $u_i : \Gamma \to X$ and assume that the net (u_i) tends pointwise on Γ to a limit $u : \Gamma \to X$ with respect to the weak topology of X. We claim that there is a net (u'_β) tending to u pointwise with respect to the norm topology such that each u'_β is a convex combination of the u_i's. To check this fix a finite subset $F \subset \Gamma$. We will apply the first part with X replaced by X^F equipped with say (the choice is largely irrelevant) the norm of $\ell_\infty(F, X)$. Consider then the elements x_i in X^F defined by $x_i = (u_i(\gamma))_{\gamma \in F} \in X^F$, and let $x = (u(\gamma))_{\gamma \in F} \in X^F$. By the first part for any $\varepsilon > 0$ there is u in the convex hull of the u_i's such that $\|u(\gamma) - u_i(\gamma)\| < \varepsilon$ for any $\gamma \in F$. Using this, the claim follows, we leave the remaining details to the reader.

Remark A.11 Let $D \subset X$ be total, i.e. such that the linear span of D is norm dense in X. Then any *bounded* net (x_i) in X^* that is convergent (resp. Cauchy) with respect to pointwise convergence on D is convergent (resp. Cauchy) with respect to the weak* topology. The verification of this fact is entirely elementary.

In general, a bounded net in X does not converge weakly in X. However, since bounded subsets of X^{**} are relatively $\sigma(X^{**}, X^*)$-compact, there is a subnet that converges for $\sigma(X^{**}, X^*)$ to some point in X^{**}. More precisely, it is a well-known fact that

$$\overline{B_X}^{\sigma(X^{**}, X^*)} = B_{X^{**}}. \tag{A.9}$$

In these notes, we use on several occasions the following useful reformulation:

Proposition A.12 (Biduals as quotients of ℓ_∞-sums) *There is a set I such that if we set $X_i = X$ for any $i \in I$ and*

$$X_I = \left(\oplus \sum_{i \in I} X_i \right)_\infty,$$

there is a metric surjection $\varphi : X_I \to X^{**}$ such that

$$\varphi(B_{X_I}) = B_{X^{**}}.$$

In particular X^{**} is isometric to a quotient Banach space of X_I.

Proof Let I be a base of the set of neighborhoods of 0 in X^{**} for $\sigma(X^{**}, X^*)$. By (A.9) for any $x \in B_{X^{**}}$ there is $(x_i)_{i \in I}$ in the closed unit ball of X_I such that $x_i \in x + i$ for any $i \in I$. Observe that $\lim x_i = x$ with respect to $\sigma(X^{**}, X^*)$ the limit being relative to the directed net formed by the neighborhood base I. Let $\mathcal{U}$ be an utrafilter refining this net. Let

$$\forall y = (y_i)_{i \in I} \in X_I \quad \varphi(y) = \lim_{\mathcal{U}} y_i.$$

By the weak* compactness of $B_{X^{**}}$ the latter limit exists and $\|\varphi\| \le 1$. By the preceding observation we have $\varphi(B_{X_I}) = B_{X^{**}}$. Consequently, X^{**} is isometrically isomorphic to $X_I / \ker(\varphi)$. $\qquad\square$

Remark A.13 The following fact is one more well-known consequence of the Hahn–Banach theorem: let $u : X \to Y$ be a linear mapping between Banach spaces. Then $u : X \to Y$ is an isometry if and only if $u^{**} : X^{**} \to Y^{**}$ is also one.

A.8 The local reflexivity principle

In Banach space theory, the important "principle of local reflexivity" (from [170]) says that *every* Banach space X has the following property called "local reflexivity":

$$B(E, X)^{**} = B(E, X^{**}) \text{ (isometrically)} \tag{A.10}$$

for any finite-dimensional Banach space E.

Recall that $B(E, X)$ can be identified with the injective tensor product $E^* \overset{\vee}{\otimes} X$. The main point is that $B(E, X)^*$ can be identified isometrically with the projective tensor product $E \overset{\wedge}{\otimes} X^*$. From this (A.10) is immediate.

The typical application of (A.10) is this:

Lemma A.14 *Let X be any Banach space. Let $E \subset X^{**}$ be a finite-dimensional subspace. There is a net of maps $u_i : E \to X$ with $\|u_i\| \to 1$ such that*

$$\forall x \in E \cap X, \quad u_i(x) = x$$

and

$$\forall x \in E, \quad u_i(x) \to x \text{ for } \sigma(X^{**}, X^*).$$

Proof Let $u_E : E \to X^{**}$ denote the inclusion map and let $B = B(E, X)$. Note $\|u_E\| = 1$. By (A.10) the unit ball of $B(E, X)$ is $\sigma(B^{**}, B^*)$-dense in that of $B(E, X^{**})$. Therefore there is a net of maps $v_i : E \to X$ ($i \in I$) with $\|v_i\| \le 1$ such that $v_i(x) \to u_E(x)$ for any $x \in E$ with respect to $\sigma(X^{**}, X^*)$. For any $x \in E \cap X$, $v_i(x) - x$ lies in X and tends $\sigma(X, X^*)$ to 0. By Mazur's classical theorem A.9, 0 lies in the norm closure of $\text{conv}(\{v_i(x) - x \mid i \in I\})$, and also of $\text{conv}(\{v_j(x) - x \mid j \ge i\})$ for any choice of $i \in I$. Therefore (see Remark A.10) we can find a net formed of convex combinations of the v_i's that we denote (abusively) by $w_i : E \to X$ satisfying still that $w_i(x) \to u_E(x)$ for any $x \in E$ with respect to $\sigma(X^{**}, X^*)$ but in addition such that $\|w_i(x) - x\| \to 0$ for any $x \in E \cap X$. Note $\|w_i\| \le 1$. Fix $\varepsilon > 0$. Let $e_1, \dots e_d$ be a linear basis of $E \cap X$ (assuming $E \cap X \ne \{0\}$). Let $e_1^*, \dots e_d^*$ be biorthogonal linear functionals on $E \cap X$. By Hahn–Banach we may assume $e_k^* \in X^*$. We then set $u_i(x) = w_i(x) + \sum_k e_k^*(x)(e_k - w_i(e_k))$ for any $x \in E$. Then $u_i(x) = \sum_k e_k^*(x)e_k = x$ for any $x \in E \cap X$. Moreover $\|u_i\| \le \|w_i\| + \sum_k \|e_k^*\| \|e_k - w_i(e_k)\|$, so that for all i large enough we have $\|u_i\| \le 1 + \varepsilon$. Lastly $\|u_i(x) - w_i(x)\| \to 0$ for any $x \in E$, so that we still have $u_i(x) \to u_E(x)$ for any $x \in E$ with respect to $\sigma(X^{**}, X^*)$. $\qquad\square$

As a consequence of (A.10) we have:

Proposition A.15 *For any Banach space X we have X^{**} f.r. X.*

Proof Fix a finite-dimensional $E \subset X^{**}$. Let $B = B(E, X)$. Let $u : E \to X^{**}$ be the inclusion mapping. By (A.10) there is a net of mappings $u_i : E \to X$ with $\|u_i\| \leq 1$ tending in the sense of $\sigma(B^{**}, B^*)$ to u, i.e. such that $\xi(u_i(e)) \to \xi(e)$ for all $e \in E$ and all $\xi \in X^*$. Let $e \in E$. We have $\limsup \|u_i(e)\| \leq \|e\|$ and also $\liminf \|u_i(e)\| \geq \sup_{\xi \in B_{X^*}} \liminf |\xi(u_i(e))| = \|e\|$. Therefore $\|u_i(e)\| \to \|e\|$. Let $\varepsilon > 0$ and let $\mathcal{N}_\varepsilon$ be a finite ε-net in the unit ball of E. Choosing i large enough we can obtain

$$\forall e \in \mathcal{N}_\varepsilon \quad (1 - \varepsilon)\|e\|_E \leq \|u_i(e)\|_X \leq \|e\|_E.$$

Then we conclude by arguing as for the passage from (A.7) to (A.8) that X^{**} f.r. X. $\square$

A.9 A variant of Hahn–Banach theorem

Lemma A.16 *Let S be a set and let $\mathcal{F} \subset \ell_\infty(S)$ be a convex cone of real valued functions on S such that*

$$\forall f \in \mathcal{F} \quad \sup_{s \in S} f(s) \geq 0.$$

Then there is a net (λ_i) of finitely supported probability measures on S such that the limit of $\int f d\lambda_i$ exists for any $f \in \mathcal{F}$ and satisfies

$$\forall f \in \mathcal{F} \quad \lim \int f d\lambda_i \geq 0.$$

If S is a weak* compact convex subset of the dual X^* of a Banach space X, and $\mathcal{F}$ is formed of affine weak* continuous functions on S, then there is $s \in S$ such that

$$\forall f \in \mathcal{F} \quad f(s) \geq 0.$$

Proof Let $\ell_\infty(S, \mathbb{R})$ denote the space all bounded *real valued* functions on S with its usual norm. In $\ell_\infty(S, \mathbb{R})$ the set $\mathcal{F}$ is disjoint from the set $C_- = \{\varphi \in \ell_\infty(S, \mathbb{R}) \mid \sup \varphi < 0\}$. Hence by the Hahn–Banach theorem (we separate the convex set $\mathcal{F}$ and the convex open set C_-) there is a nonzero $\xi \in \ell_\infty(S, \mathbb{R})^*$ such that $\xi(f) \geq 0$ $\forall f \in \mathcal{F}$ and $\xi(f) \leq 0$ $\forall f \in C_-$. Let $M \subset \ell_\infty(S, \mathbb{R})^*$ be the cone of all finitely supported (nonnegative) measures on S viewed as functionals on $\ell_\infty(S, \mathbb{R})$. Since we have $\xi(f) \leq 0$ $\forall f \in C_-$, ξ must be in the bipolar of M for the duality of the pair $(\ell_\infty(S, \mathbb{R}), \ell_\infty(S, \mathbb{R})^*)$. Therefore, by the bipolar theorem, ξ is the limit for the topology $\sigma(\ell_\infty(S, \mathbb{R})^*, \ell_\infty(S, \mathbb{R}))$ of a net of finitely supported (nonnegative) measures ξ_i on S. We have for any f in $\ell_\infty(S, \mathbb{R})$, $\xi_i(f) \to \xi(f)$ and this holds in particular if $f = 1$, thus (since ξ is nonzero) we may assume $\xi_i(1) > 0$, hence if we set $\lambda_i(f) = \xi_i(f)/\xi_i(1)$ we obtain the first assertion.

If $\mathcal{F}$ is formed of affine functions on a weak* closed convex set $S \subset X^*$, let $s_i = \int s d\lambda_i(s)$ be the barycenter of λ_i. We have $\int f d\lambda_i = f(s_i)$ for all $f \in \mathcal{F}$. By the weak* compactness of S, there is a subnet for which s_i converges weak* to some point $s \in S$ and since f is assumed weak* continuous, we have

$$\lim \int f d\lambda_i = \lim f(s_i) = f(s),$$

and hence $f(s) \geq 0$ for any $f \in \mathcal{F}$. $\square$

A.10 The trace class

Let H, K be Hilbert spaces. We denote by $\overline{H}$ the complex conjugate Hilbert space (the same space but with complex conjugate scalar multiplication). Recall the canonical identifications

$$H^* = \overline{H} \quad \overline{H}^* = H \quad \text{and} \quad (\overline{K})^* = K.$$

With the latter, (A.1) implies the isometric identity

$$(\overline{K} \overset{\wedge}{\otimes} H)^* = B(H, K).$$

The space $\overline{K} \overset{\wedge}{\otimes} H$ can be identified with the space $S_1(K, H)$ of trace class operators, i.e. the operators $T : K \to H$ such that for some (or all) orthonormal bases (e_i) of K we have $\sum \langle e_i, |T|e_i \rangle < \infty$ (here $|T| = (T^*T)^{1/2}$). We equip this space with the norm

$$\|T\|_{S_1} = \sum \langle e_i, |T|e_i \rangle = \text{tr}(|T|).$$

Then $\overline{K} \overset{\wedge}{\otimes} H \simeq S_1(K, H)$ isometrically, for the correspondence that takes a tensor $t = \sum_1^n \overline{x_k} \otimes y_k$ to the operator T defined by $T(\xi) = \sum_1^n \langle x_k, \xi \rangle y_k$ ($\xi \in K$).

That same correspondence $t \mapsto T$ gives us an isometric identification

$$\overline{K} \otimes_2 H \simeq S_2(K, H),$$

where $S_2(K, H)$ denotes the space of Hilbert–Schmidt mappings $T : K \to H$, with norm defined by

$$\|T\|_{S_2} = \left(\sum \|T(e_i)\|^2 \right)^{1/2} = (\text{tr}(T^*T))^{1/2}.$$

One also defines the Schatten p-class $S_p(K, H)$ with norm $\|T\|_{S_p} = (\text{tr}|T|^p)^{1/p}$ for other values of $p \in [1, \infty)$ but we do not use them in this volume (see e.g. [215]).

In particular, when $K = H$, for $p = 1, 2$ we denote simply $S_p(H) = S_p(H, H)$.

The preceding shows that $B(H)$ admits as its predual the space $\overline{H} \overset{\wedge}{\otimes} H$ (or equivalently the space $S_1(H)$). We will sometimes denote that predual by $B(H)_*$, especially when we view it as a subspace of $B(H)^*$.

Remark A.17 For any $T \in S_1(H)$, we have clearly $T^* \in S_1(H)$ and, for any $a \in B(H)$, $aT \in S_1(H)$ and $Ta \in S_1(H)$. Moreover, $\|T^*\|_{S_1} = \|T\|_{S_1}$, $\|aT\|_{S_1} \le \|T\|_{S_1}\|a\|$ and $\|Ta\|_{S_1} \le \|T\|_{S_1}\|a\|$. It follows that the mappings $x \mapsto x^*$, $x \mapsto ax$ and $x \mapsto xa$ are all continuous from $B(H)$ to itself equipped with the weak* topology.

A.11 C^*-algebras: basic facts

A C^*-algebra A is a Banach $*$-algebra equipped with a norm $\| \ \|$ satisfying $\|x^*\| = \|x\|$, $\|xy\| \le \|x\|\|y\|$ for all $x, y \in A$, and more importantly

$$\|x^*x\| = \|x\|^2.$$

Gelfand's theory tells us that for any such algebra there is an isometric $*$-homomorphism $\pi : A \to B(H)$. Thus A is embedded (or "realized") in $B(H)$.

Moreover, if A is unital (i.e. has a unit element 1) then there is an embedding such that $\pi(1) = Id_H$. This is proved using the GNS construction that we describe in §A.13.

Any element x in a C^*-algebra A can be decomposed as $x = a + ib$ with $a, b \in A$ self-adjoint given by $a = (x + x^*)/2$ and $b = (x - x^*)/(2i)$. Furthermore, using the functional calculus in the commutative C^*-algebras generated respectively by a and b we can decompose x further as

$$x = a^+ - a^- + i(b^+ - b^-). \tag{A.11}$$

We often use the following simple operator analogue of the classical Cauchy–Schwarz inequality. Let $(a_i)_{1 \leq i \leq n}$ and $(b_i)_{1 \leq i \leq n}$ be finitely supported families of operators in $B(H)$. We have

$$\left\| \sum a_i b_i \right\| \leq \left\| \sum a_i a_i^* \right\|^{1/2} \left\| \sum b_i^* b_i \right\|^{1/2}. \tag{A.12}$$

This follows from

$$\left\| \sum a_i b_i \right\| = \sup \left| \left\langle \xi, \left(\sum a_i b_i \right) \eta \right\rangle \right|$$

$$= \sup \left| \sum \langle a_i^* \xi, b_i \eta \rangle \right| \leq \sup_\xi \left(\sum \| a_i^* \xi \|^2 \right)^{1/2} \sup_\eta \left(\sum \| b_i \eta \|^2 \right)^{1/2}$$

and

$$\left\| \sum a_i a_i^* \right\|^{1/2} = \sup_\xi \left(\sum \langle a_i^* \xi, a_i^* \xi \rangle \right)^{1/2}, \quad \left\| \sum b_i^* b_i \right\|^{1/2} = \sup_\eta \left(\sum \langle b_i \eta, b_i \eta \rangle \right)^{1/2}, \tag{A.13}$$

where the suprema run over all ξ, η in the unit ball of H.

Assuming A unital, an element $x \in A$ is called unitary if $x^* x = xx^* = 1$. When A is realized in $B(H)$ these correspond exactly to the unitary operators on H that are in A.

The following classical result is very useful when dealing with metric properties:

Theorem A.18 (Russo–Dye theorem) *The convex hull of the set of unitary elements of a unital C^*-algebra is dense in the unit ball.*

The proof appears in all major books such as [226, 240].

Kadison and Pedersen proved the following refinement improved later on by Haagerup (see [106, 112, 144]): if x in a unital C^*-algebra satisfies $\|x\| \leq 1 - 2/n$ then there are unitaries $u_1, \ldots, u_n$ such that $x = (u_1 + \cdots + u_n)/n$. Actually the first result in that direction had been proved a few years earlier by Popa [216].

Corollary A.19 *For any (bounded) positive linear map $u : A \to B$ between unital C^*-algebras, we have $\|u\| = \|u(1)\|$.*

Proof It suffices to show that $\|u(x)\| \leq \|u(1)\|$ for any unitary element $x \in A$. We may clearly replace A by the commutative C^*-algebra generated by x. Then the statement reduces to the case when $A = C(T)$ for some compact set T. The case when T is finite is easy. Indeed, we may assume $A = \ell_\infty^n$, and then denoting by (e_j) the canonical basis of ℓ_∞^n and observing that u positive means $u(e_j) \geq 0$ for all j, we have by (A.12) for any $z = \sum z_j e_j \in B_{\ell_\infty^n}$

$$\|u(z)\| = \left\| \sum z_j u(e_j) \right\| = \left\| \sum u(e_j)^{1/2} \times z_j u(e_j)^{1/2} \right\| \leq \left\| \sum u(e_j) \right\| = \|u(1)\|.$$

The proof can then be completed using Remark A.20. $\qquad\square$

A.12 Commutative C^*-algebras

By spectral theory any *commutative*, closed, self-adjoint and unital subalgebra $A \subset B(H)$ is isometrically $*$-isomorphic to an algebra of continuous functions on a compact set T.

By definition, the spectrum of A, denoted by σ_A, is the set of nonzero multiplicative $*$-homomorphic functionals $f : A \to \mathbb{C}$ (sometimes called "characters"). If A is unital this is a compact set for the pointwise topology. In the nonunital case it is a locally compact space, which we can compactify by adding a point at ∞. When A is commutative and unital, there is an isometric $*$-isomorphism from A to the C^*-algebra, denoted by $C(\sigma_A)$, formed of all the continuous functions on σ_A. Thus A can be identified with $C(T)$ for $T = \sigma_A$. When A is commutative and nonunital, A can be identified with the C^*-algebra, denoted by $C_0(T)$, formed of all the continuous functions on T that tend to 0 at ∞ (of course the latter condition becomes void when T is compact, so $C_0(T) = C(T)$ in that case).

Let $A \subset B(H)$ be a unital C^*-subalgebra. For any $x \in A$ let $\sigma_x \subset \mathbb{C}$ denote the spectrum of x (i.e. the set of $z \in \mathbb{C}$ such that $zI - x$ is noninvertible). If x is a normal operator, i.e. $x^*x = xx^*$ (in particular if x is self-adjoint) we have

$$\|x\| = \sup_{z \in \sigma_x} |z|. \tag{A.14}$$

Moreover, if $x \in A$ is invertible in $B(H)$ then its inverse is in A, so the spectra of x relative either to A or to $B(H)$ are the same. In particular, it is the same as the spectrum of x in the unital C^*-subalgebra $A_x \subset A$ generated by x. If x is normal (in particular if it is self-adjoint), A_x is isomorphic to $C(\sigma_x)$. This allows us to make sense of $f(x) \in A$ ("functional calculus") for any $f \in C(\sigma_x)$. Note that the spectrum of a continuous function in $C(T)$ (T compact) is just the closure of its range. Therefore, if x is normal

$$\forall f \in C(\sigma_x) \quad \sigma_{f(x)} = f(\sigma_x). \tag{A.15}$$

For convenience we record here a basic observation on the space $C(T)$, that gives a convenient way to reduce many questions to the case when T is finite. Recall that we denote by ℓ_∞^n the space $\mathbb{C}^n$ equipped with the sup-norm. For any n-element set T the space $C(T)$ can be identified to ℓ_∞^n.

Remark A.20 Let $A = C(T)$ with T compact. Then the identity of A is the pointwise limit of a net of maps $u_i : A \to A$ of the form

$$u_i : A \xrightarrow{v_i} \ell_\infty^{n(i)} \xrightarrow{w_i} A$$

where $n(i)$ are integers and v_i, w_i are unital positive contractions.

Indeed, for any $\varepsilon > 0$ and any finite set $x_1, \ldots, x_k \in A$, there is a finite open covering $(U_m)_{1 \le m \le n}$ of T such that for all $1 \le m \le n$ and $1 \le j \le k$ the oscillation of x_j on U_m is $\le \varepsilon$. Let $(\varphi_m)_{1 \le m \le n}$ be a partition of unity subordinated to this covering. We have $0 \le \varphi_m \le 1$, $1 = \sum_{1 \le m \le n} \varphi_m$ and $\operatorname{supp}(\varphi_m) \subset U_m$. Fix points $\omega_m \in U_m$ arbitrarily chosen. We define $v : A \to \ell_\infty^n$ and $w : \ell_\infty^n \to A$ by setting for $x \in A$ and $y = (y_m) \in \ell_\infty^n$

$$v(x) = (x(\omega_m))_{1 \le m \le n} \text{ and } w(y) = \sum_1^n y_m \varphi_m.$$

Then $\|wv(x_j) - x_j\| = \|\sum_{1 \le i \le n} \varphi_m[x_j(\omega_m) - x_j]\| \le \varepsilon$ for any $1 \le j \le k$. Moreover, v and w are unital positive maps with $\|v\| \le 1$ and $\|w\| \le 1$. This proves the assertion.

A.13 States and the GNS construction

Let A be a C^*-algebra. An element $f \in A_+^*$ with $\|f\| = 1$ is called a state. To any state we can associate an inner product on A by setting for any $a, b \in A$

$$\langle a, b \rangle = f(a^*b).$$

We then obtain a Hilbert space denoted by $L_2(f)$, after passing to the quotient by the kernel and completing. There is a natural mapping $A \to L_2(f)$. The left multiplication by $a \in A$ defines an element of $B(L_2(f))$ denoted by $\pi_f(a)$. Then $a \mapsto \pi_f(a)$ is a $*$-homomorphism from A to $B(L_2(f))$, and there is a unit vector $\xi_f \in L_2(f)$ such that $\overline{\pi_f(A)\xi_f} = L_2(f)$ (i.e. ξ_f is cyclic) and

$$\forall a \in A \quad f(a) = \langle \xi_f, \pi_f(a)\xi_f \rangle. \tag{A.16}$$

This is called the "GNS construction" for Gelfand–Naimark–Segal. We will say that (A.16) is the GNS factorization of f.

Remark A.21 In the converse direction let $\pi : A \to B(H)$ be cyclic with cyclic unit vector ξ. Then $f(\cdot) = \langle \xi, \pi(\cdot)\xi \rangle$ is a state and the associated π_f is unitarily equivalent to π. Indeed, the correspondence $\pi(\cdot)\xi \mapsto \pi_f(\cdot)\xi$ defines a unitary $u : H \to L_2(f)$ such that $\pi(\cdot) = u^*\pi_f(\cdot)u$.

Remark A.22 Note that we can perform this construction starting only from a functional f of norm 1 on a dense $*$-subalgebra $\mathcal{A} \subset A$ such that $f(x^*x) \ge 0$ for any $x \in \mathcal{A}$. Indeed, such an f extends to a state on A: by Hahn–Banach f extends to $f' \in A^*$ with $\|f'\| = 1$ and by the density of $\mathcal{A}$ we have $f'(x^*x) \ge 0$ for any $x \in A$.

If A admits a state f that is faithful, i.e., such that $f(x^*x) = 0$ implies $x = 0$, then the $*$-homomorphism $\pi_f : A \to B(L_2(f))$ is an embedding.

In general, the operation of right multiplication by $a \in A$ is not bounded on $L_2(f)$ for the norm we defined as $\|x\|_{L_2(f)} = f(x^*x)^{1/2}$ ($x \in A$). But it would be had we chosen to define it by $\|x\|_{L_2(f)} = f(xx^*)^{1/2}$. This difficulty disappears when both inner products coincide on A. This holds in particular when

$$\forall x, y \in A \quad f(xy) = f(yx).$$

In that case, the functional $f \in A_+^*$ is called *tracial*. We have then $\|x\|_{L_2(f)} = \|x^*\|_{L_2(f)}$ for any $x \in A$, so that $x \mapsto x^*$ defines an antilinear isometric involution J on $L_2(f)$ and the operation of right multiplication $b \mapsto ba$ defines a bounded linear map $R_f(a) \in B(L_2(f))$ such that

$$\forall a, b \in A \quad J(R_f(a)b) = \pi_f(b^*)J(a).$$

Moreover, we have $R_f(ab) = R_f(b)R_f(a)$ so R_f is a $*$-homomorphism on the algebra A^{op} that is the same as A but with the reversed product. It is important to observe that the ranges of π_f and R_f mutually commute.

Let T be any locally compact space. Then a state f on $C_0(T)$ can be identified with a probability measure μ_f on T, and $L_2(f)$ can be identified with $L_2(\mu_f)$. With this identification, the natural mapping $\pi_f : C_0(T) \to B(L_2(\mu_f))$ takes a function $x \in C_0(T)$ to the operator of multiplication by x on $L_2(\mu_f)$. Then f is faithful if and only if μ_f has full support on T. In that case π_f realizes $C_0(T)$ as multiplication operators acting on $L_2(\mu_f)$.

In any case, when A is commutative, there is a set I and an isometric $*$-homomorphism $\pi : A \to B(\ell_2(I))$ such that all the operators in the range of π are diagonal. Indeed, assuming $A = C_0(T)$ we can take $I = T$ viewed as a discrete set, and define $\pi(f) \in B(\ell_2(I))$ as the diagonal operator with coefficients $(f(t))_{t \in T}$.

Remark A.23 Let T be a compact space. It is well known that $C(T)^*$ is isometric to the space $M(T)$ formed of the complex measures μ on T equipped with the norm $\|\mu\|_{M(T)} = |\mu|(T)$. It is well known and elementary that a measure $\mu \in M(T)$ with $|\mu|(T) = 1$ is positive if and only if $\mu(T) = 1$. Moreover this holds if and only if $\Re(\mu)(T) = 1$. The generalization of this on a noncommutative unital C^*-algebra A is immediate: for any $f \in A^*$ with $\|f\|_{A^*} = 1$ the functional f is positive if and only if $f(1) = 1$ or if and only if $\Re(f)(1) = 1$. Indeed, to verify that $f(x) \geq 0$ for any $x \geq 0$ we obviously may restrict to the unital C^*-algebra generated by x, but the latter being commutative, the problem reduces to the preceding measure space case.

A.14 On $*$-homomorphisms

Let $\pi : A \to B(K)$ be a unital $*$-homomorphism on a C^*-algebra A. We claim that

$$\|\pi\| = 1. \tag{A.17}$$

Assume $x = x^* \in A$. Then $\sigma_{\pi(x)} \subset \sigma_x$ (because $zI - x$ is invertible implies $\pi(zI - x) = zI - \pi(x)$ invertible). Therefore by (A.14) $\|\pi(x)\| \leq \|x\|$ whenever $x^* = x$. But now for an arbitrary $x \in A$, we have $\|\pi(x)\| = \|\pi(x^*x)\|^{1/2} \leq \|x^*x\|^{1/2} = \|x\|$, proving the claim, since $\pi(1) = 1$ guarantees $\|\pi\| \geq 1$.

If π is injective then it is isometric. Indeed, using the preceding idea it suffices to show that injectivity forces $\sigma_{\pi(x)} = \sigma_x$ when $x = x^*$. Indeed, if $\sigma_{\pi(x)} \neq \sigma_x$ there is $z \in \sigma_x \setminus \sigma_{\pi(x)}$, so we can find a continuous function f on σ_x such that $f(z) = 1$ but $f_{|\sigma_{\pi(x)}} = 0$. Let $y = f(x)$. Recall this is defined using (A.14) and limits of polynomials, therefore $\pi(y) = f(\pi(x))$, and by (A.15) $\sigma_y = f(\sigma_x) \neq \{0\}$ but $\sigma_{\pi(y)} = f(\sigma_{\pi(x)}) = \{0\}$. Thus $y \neq 0$ but $\pi(y) = 0$, i.e. π is not injective.

If A is not unital, one can show that π extends to a unital $*$-homomorphism on the unitization of A (namely $\mathbb{C}I + A$), and the preceding results remain true. Recapitulating:

Proposition A.24 *For any $*$-homomorphism $\pi : A \to B$ between C^*-algebras the range $\pi(A)$ is closed and $\|\pi\| = 1$ (assuming $\pi \neq 0$). If π is injective it is isometric. If π is injective with dense range, it is automatically a surjective isometric isomorphism.*

Proof We may assume $B \subset B(K)$. We have a canonical factorization $A \to A/\ker(\pi) \to B$, with an injective $*$-homomorphism $A/\ker(\pi) \to B \subset B(K)$, which by what precedes must be isometric. Thus the range is isometric to $A/\ker(\pi)$. Since the latter is complete, the range is closed. The other assertions are now obvious. $\qquad\square$

Corollary A.25 *On a C^*-algebra the C^*-norm is unique.*

Proof We apply Proposition A.24 to the identity of A viewed as a ∗-homomorphism from $(A, \| \ \|_1)$ to $(A, \| \ \|_2)$, where $(\| \ \|_1, \| \ \|_2)$ are C^*-norms on A. $\qquad \square$

Corollary A.26 *On a ∗-algebra, if two C^*-norms are equivalent, they are equal.*

Proof After completion they become C^*-norms on the *same* C^*-algebra, so by the preceding Corollary they coincide. More explicitly, if $0 < \|x\|_1 / \|x\|_2 = \theta < 1$, then $0 < \|(x^*x)^n\|_1 / \|(x^*x)^n\|_2 = \theta^{2n} \to 0$. $\qquad \square$

Remark A.27 Let $\mathcal{A} \subset A$ be a dense (resp. unital) ∗-subalgebra $\mathcal{A} \subset A$ in a (resp. unital) C^*-algebra A. Let $\pi : \mathcal{A} \to B(H)$ be a (resp. unital) ∗-homomorphism admitting a cyclic unit vector $\xi \in H$. Let $f : \mathcal{A} \to \mathbb{C}$ be defined by $f(x) = \langle \xi, \pi(x)\xi \rangle$. If $\|f\| = 1$ then $\|\pi\| = 1$ and hence π extends to a (resp. unital) ∗-homomorphism on A. Indeed, f extends to a form $f' \in A_+^*$ (see Remark A.22). For any $x, y \in \mathcal{A}$ we have $y^*x^*xy \le \|x\|^2 y^*y$ in the order of A, and hence $f'(y^*x^*xy) \le \|x\|^2 f'(y^*y)$ which means $f(y^*x^*xy) \le \|x\|^2 f(y^*y)$ or equivalently $\langle \pi(x)\pi(y)\xi, \pi(x)\pi(y)\xi \rangle \le \|x\|^2 \langle \pi(y)\xi, \pi(y)\xi \rangle$. Since $\pi(\mathcal{A})\xi$ is dense in H (and since $1 = \|f\| \le \|\pi\|$), we conclude that $\|\pi\| = 1$. Actually, the mere continuity of f with respect to the norm induced by A implies that of π.

Recall the following classical notions in operator theory.

Definition A.28 (Weak and strong operator topology) A net (T_i) of operators in $B(H)$ converges in the weak operator topology (w.o.t. in short) to $T \in B(H)$ if $\langle \eta, T_i \xi \rangle \to \langle \eta, T\xi \rangle$ for all $\eta, \xi \in H$.

We say that $T_i \to T$ in the strong operator topology (s.o.t. in short) if $\|T_i \xi - T\xi\| \to 0$ for all $\xi \in H$. If both $T_i \to T$ and $T_i^* \to T^*$ in s.o.t. then we say that $T_i \to T$ in the strong* topology.

Remark A.29 (Universal representation of a C^*-algebra) Since the C^*-norm is the same in all representations of a (complete) C^*-algebra in $B(H)$ whatever H may be, it often makes no difference to us which representation we use. However, if we choose a representation with (apparently redundant) multiplicity, certain questions involving comparisons between the weak* (or the weak operator) topology of $B(H)$ and the weak topology can have a simpler answer. For instance, let $A \subset B(H)$, let H^∞ denote the direct sum of countably many copies of H and let $\pi : A \to B(H^\infty)$ be the direct sum of the embeddings $A \subset B(H)$, so that $\pi(a)$ ($a \in A$) acts "diagonally" on H^∞. Then the description of the trace class operators (see §A.10) shows that a net $(T_i) \in B(H)$ converges weak* in $B(H)$ if and only if $\pi(T_i)$ converges in the w.o.t. in $B(H^\infty)$.

This motivates consideration of the "universal" representation π_U of a C^*-algebra A, which is defined as follows. Let $S(A)$ be the set of states of A. For any $f \in S(A)$, let $H_f = L_2(f)$ and let $\pi_f : A \to B(H_f)$ be the cyclic representation (from the GNS construction) we then define $H_U = \oplus_{f \in S(A)} H_f$ and the universal representation π_U as the direct sum

$$A \ni a \mapsto \pi_U(a) = \bigoplus_{f \in S(A)} \pi_f(a) \in B(H_U).$$

Note that, up to unitary equivalence, any cyclic representation can be identified to some π_f (see Remark A.21) and any representation of A is a direct sum of cyclic representations (which explains the term universal). In particular, it is clear that π_U is an isometric representation of A.

Remark A.30 Let (a_i) be a net in A. Then (a_i) converges for $\sigma(A^{**}, A^*)$ to some $a'' \in A^{**}$ if and only if $\pi_U(a_i)$ converges in the w.o.t. in $B(H_U)$.

If $a'' = a \in A$ this happens if and only if $\pi_U(a_i) \to \pi_U(a)$ in the w.o.t. of $B(H_U)$.

Indeed, $a_i \to a''$ for some $a'' \in A^{**}$ with respect to $\sigma(A^{**}, A^*)$ if and only if $(f(a_i))$ is Cauchy for any $f \in A^*$ or equivalently for any $f \in S(A)$, and this is clearly implied by its w.o.t. counterpart in $B(H_U)$. The converse is obvious since $a \mapsto \langle \eta, \pi_U(a)\xi \rangle$ is in A^* for any $\eta, \xi \in H_U$.

Similarly $\pi_U(a_i) \to \pi_U(a)$ in the w.o.t. of $B(H_U)$ if and only if $a_i \to a$ in the weak topology of A.

A.15 Approximate units, ideals, and quotient C^*-algebras

Most of the C^*-algebraic questions we consider in these notes can be reduced to C^*-algebras A with a unit element. When A is not unital, the role of the unit element is played by an approximate unit, and fortunately all C^*-algebras have approximate units (see [240]). By a (bounded) approximate unit in a Banach algebra one usually means a bounded net of elements (x_i) such that for any $x \in A$, $\|x_i x - x\| \to 0$ along the net. As usual we implicitly assume that our nets (x_i) are indexed by a directed set I.

In the case of a C^*-algebra A, it turns out that the whole set $I = \{x \in A_+, \|x\| < 1\}$ is upward directed, meaning that for all $x, y \in I$ there is $z \in I$ such that $x \leq z$ and $y \leq z$, so that the whole collection $I = \{x \in A_+, \|x\| < 1\}$ forms a net (x_i) (with the rare feature that $i \mapsto x_i$ is the identity!) and it can be checked (see [240, p. 26]) that this net is an approximate unit for A.

We will need the more refined key notion of "quasi-central approximate unit" (see [13]) for an ideal $\mathcal{I} \subset A$ in a C^*-algebra A. Let $q : A \to A/\mathcal{I}$ be the quotient mapping. Then there is a non decreasing net (σ_i) in the unit ball of $\mathcal{I}$ with $\sigma_i \geq 0$ such that for any a in A and any b in $\mathcal{I}$

$$\|a\sigma_i - \sigma_i a\| \to 0 \quad \text{and} \quad \|\sigma_i b - b\| \to 0. \tag{A.18}$$

Such a net is called a "quasi-central approximate unit" for $\mathcal{I} \subset A$.

Here is a brief sketch of proof that they exist. Assume $A \subset B(H)$. Let (x_i) be an approximate unit for $\mathcal{I}$, with $x_i \geq 0$ and $\|x_i\| \leq 1$. We denote by σ the weak* topology in $B(H)$. Let $p \in \overline{\mathcal{I}}^\sigma$ be the weak* (or weak operator topology since the net is bounded) limit of (x_i). Clearly $px = xp = x$ for any $x \in \mathcal{I}$. Therefore, $px = xp = x$ for any $x \in \overline{\mathcal{I}}^\sigma$. Moreover, since $\mathcal{I}$ is an ideal, $x_i y$ and $y x_i$ are in $\mathcal{I}$ for any $y \in A$, from which we deduce that yp and py are in $\overline{\mathcal{I}}^\sigma$. This implies (take $x = yp$ or $x = py$) that $yp = pyp = py$, so that p commutes with A. Therefore for any $y \in A$, $yx_i - x_i y \to 0$ for σ. But if we choose for the embedding $A \subset B(H)$ the universal representation of A (see Remark A.29), the weak* topology σ coincides on A with the weak topology (see Remark A.30). Then $yx_i - x_i y \to 0$ weakly in A. Passing to convex combinations we can replace (cf. Mazur's theorem A.9) this weak limit by a norm limit, and we obtain the desired net (σ_i) with σ_i in the convex hull of $\{x_j \mid j \geq i\}$. See [13] or [67] for more details. The main properties we will use are summarized in the following.

Lemma A.31 *Let $A, \mathcal{I}, q$ and (σ_i) be as previously. Then, for any a in A, we have both*

$$\|\sigma_i^{1/2}a - a\sigma_i^{1/2}\| \to 0 \text{ and } \|(1 - \sigma_i)^{1/2}a - a(1 - \sigma_i)^{1/2}\| \to 0. \tag{A.19}$$

Moreover we have

$$\forall a \in A \quad \|q(a)\| = \lim \|a - \sigma_i a\|, \tag{A.20}$$

$$\forall a, b \in A \quad \limsup \|\sigma_i^{1/2}a\sigma_i^{1/2} + (1 - \sigma_i)^{1/2}b(1 - \sigma_i)^{1/2}\| \leq \max\{\|a\|, \|q(b)\|\}, \tag{A.21}$$

$$\lim \|\sigma_i^{1/2}a\sigma_i^{1/2} + (1 - \sigma_i)^{1/2}a(1 - \sigma_i)^{1/2} - a\| = 0. \tag{A.22}$$

Proof The first assertion (A.19) is immediate using a polynomial approximation of $t \to \sqrt{t}$, so we turn to (A.20). Fix $\varepsilon > 0$. Let $x \in A$ be such that $q(x) = q(a)$, $\|x\| < \|q(a)\| + \varepsilon$. Then $a - x \in \mathcal{I}$ implies $\|(1 - \sigma_i)(a - x)\| \to 0$ hence $\|(1 - \sigma_i)a\| \leq \|(1 - \sigma_i)x\| + \|(1 - \sigma_i)(a - x)\|$, therefore, since $\|1 - \sigma_i\| \leq 1$, we find $\limsup \|(1 - \sigma_i)a\| \leq \|x\| < \|q(a)\| + \varepsilon$. Also $\|q(a)\| \leq \|(1 - \sigma_i)a\|$ implies $\|q(a)\| \leq \liminf \|(1 - \sigma_i)a\|$. This proves (A.20).

To verify (A.21) note that by (A.12) we have

$$\|\sigma_i^{1/2}a\sigma_i^{1/2} + (1 - \sigma_i)^{1/2}b(1 - \sigma_i)^{1/2}\| \leq \|\sigma_i^{1/2}a^*a\sigma_i^{1/2}$$
$$+ (1 - \sigma_i)^{1/2}b^*b(1 - \sigma_i)^{1/2}\|^{1/2}$$

and hence

$$\|\sigma_i^{1/2}a\sigma_i^{1/2} + (1 - \sigma_i)^{1/2}b(1 - \sigma_i)^{1/2}\| \leq \max\{\|a\|, \|b\|\}.$$

Now if we replace b by b' such that $q(b') = q(b)$ we have by the second parts of (A.19) and (A.18)

$$\lim \|(1 - \sigma_i)^{1/2}(b - b')(1 - \sigma_i)^{1/2}\| = \lim \|(1 - \sigma_i)(b - b')\| = 0,$$

hence the left-hand side of (A.21) is $\leq \max\{\|a\|, \|b'\|\}$. Taking the infimum over b' we obtain (A.21). Finally, by (A.19) we have a fortiori for any a in A

$$\|\sigma_i a - \sigma_i^{1/2}a\sigma_i^{1/2}\| \to 0 \text{ and } \|(1 - \sigma_i)a - (1 - \sigma_i)^{1/2}a(1 - \sigma_i)^{1/2}\| \to 0, \tag{A.23}$$

from which (A.22) is immediate. □

Lemma A.32 *Let $\mathcal{I} \subset A$ be a (closed two-sided) ideal in a C^*-algebra. Let $q : A \to A/\mathcal{I}$ be the quotient map. Then $\forall x \in A \; \forall \varepsilon > 0 \; \exists x_1 \in A$ with $q(x_1) = q(x)$ such that $\|x_1\| < \|q(x)\| + \varepsilon$ and $\|x_1 - x\| \leq \|x\| - \|q(x)\|$.*

Proof Let (σ_i) be as before. We set $x_1 = \sigma_i(x\|x\|^{-1}\|q(x)\|) + (1 - \sigma_i)x$.

We will show that when i is large enough this choice of x_1 satisfies the announced properties. First we have clearly $q(x_1) = q(x)$ (since $\sigma_i \in \mathcal{I}$). We introduce

$$x_1' = \sigma_i^{1/2}(x\|x\|^{-1}\|q(x)\|)\sigma_i^{1/2} + (1 - \sigma_i)^{1/2}x(1 - \sigma_i)^{1/2}.$$

Choosing i large enough, by the first assertion of the preceding lemma we may on the one hand assume $\|x_1 - x_1'\| < \varepsilon/2$ and, by (A.21), also $\|x_1'\| \leq \|q(x)\| + \varepsilon/2$ whence $\|x_1\| < \|q(x)\| + \varepsilon$. On the other hand, we have $\|x - x_1\| = \|\sigma_i x\|x\|^{-1}(\|x\| - \|q(x)\|)\| \leq \|x\| - \|q(x)\|$. □

Thus we obtain the following well-known fact:

Lemma A.33 *For any x in A, there is $\tilde{x}$ in A such that $q(\tilde{x}) = q(x)$ and $\|\tilde{x}\|_A = \|q(x)\|_{A/\mathcal{I}}$.*

Proof Using Lemma A.32 we can select by induction a sequence $x, x_1, x_2, \ldots$ in A such that $q(x_n) = q(x)$, $\|x_n\| \leq \|q(x)\| + 2^{-n}$ and $\|x_n - x_{n-1}\| \leq \|x_{n-1}\| - \|q(x_{n-1})\| \leq 2^{-n+1}$. Since it is Cauchy, this sequence converges and its limit $\tilde{x}$ has the announced property.   □

Remark A.34 When A is a dual space and $\mathcal{I}$ a weak* closed ideal, the situation is much simpler. Indeed, the net (A.18) now has a subnet that converges weak* to a limit $p \in \mathcal{I}$. By (A.18) p is a self-adjoint projection in the center of A that is a unit element for $\mathcal{I}$. Therefore the mapping $x \mapsto px = xp$ is a projection from A to $\mathcal{I}$ that is also a weak* continuous *-homomorphism. We thus obtain a decomposition

$$A = (1 - p)A \oplus pA$$

that can be rewritten equivalently as

$$A \simeq (A/\mathcal{I}) \oplus \mathcal{I}. \tag{A.24}$$

In other words, if we denote by $Q : A \to A/\mathcal{I}$ the quotient map, the correspondence $(1 - p)x \mapsto Q(x)$ is a *-isomorphism from $(1 - p)A$ to $A/\mathcal{I}$.

Remark A.35 Actually, in the preceding remark, we can use any bounded approximate unit (x_i) of $\mathcal{I}$ to produce the projection p. Indeed, passing to a subnet we may assume that x_i tends weak* to $p \in \mathcal{I}$. Then since $x_i x \to x$ and (taking adjoints) $xx_i^* \to x$ in norm for any $x \in \mathcal{I}$, we derive $px = x$, $xp^* = x$ so that in particular $pp^* = p^* = p$, so that $p^2 = p$ (see Remark A.17 for clarification). For any $y \in A$ we have $yp \in \mathcal{I}$ and $py \in \mathcal{I}$ since $\mathcal{I}$ is an ideal, and hence $yp = p(yp) = (py)p = py$.

Remark A.36 More generally if the C^*-algebra A is a dual space and $\mathcal{I} \subset A$ a weak* closed *left* ideal (meaning $A\mathcal{I} \subset \mathcal{I}$), we claim that there is a (self-adjoint) projection $P \in A$ such that $\mathcal{I} = AP$.

To check this let $\mathcal{I}' = \{x \mid x^* \in \mathcal{I}\}$. Then $\mathcal{I}'$ is a weak* closed right ideal, and $\mathcal{I} \cap \mathcal{I}'$ a weak* closed C^*-subalgebra. As in the preceding remark, any net forming an approximate unit for the C^*-algebra $\mathcal{I} \cap \mathcal{I}'$ now has a subnet that converges weak* to a limit $P \in \mathcal{I} \cap \mathcal{I}'$ that is the unit of $\mathcal{I} \cap \mathcal{I}'$ and is a (self-adjoint) projection in A. Therefore we have $\mathcal{I} \cap \mathcal{I}' = (\mathcal{I} \cap \mathcal{I}')P = P(\mathcal{I} \cap \mathcal{I}')$. We claim that $\mathcal{I} = AP$. Indeed, let $x \in \mathcal{I}$. Then $x^*x \in \mathcal{I}$ and hence $x^*x \in \mathcal{I} \cap \mathcal{I}'$. This gives us $Px^*x = x^*xP = Px^*xP$ and hence $(x - xP)^*(x - xP) = 0$. Thus $x = xP$ which means $x \in AP$. Thus $\mathcal{I} \subset AP$. Since $P \in \mathcal{I}$ the converse is obvious, proving the claim. See also [226, p. 24], [240, p. 123], or [146, p. 443].

A.16 von Neumann algebras and their preduals

For the convenience of the reader, we recall now a few facts concerning von Neumann algebras. A von Neumann algebra on a Hilbert space H is a self-adjoint subalgebra of

$B(H)$ that is equal to its bicommutant. For $M \subset B(H)$ we denote by M' (resp. M'') its commutant (resp. bicommutant).

By a well-known result due to Sakai (see [240, p. 133]), a C^*-algebra A is C^*-isomorphic to a von Neumann algebra if and only if it is isometric to a dual Banach space, i.e. if and only if there is a closed subspace $X \subset A^{**}$ such that $X^* = A$ isometrically. For instance the subspace $X \subset A^{**}$ formed of all the weak* continuous linear forms on A^* clearly satisfies this. For a general dual Banach space A there may be several subspaces $X \subset A^{**}$ such that $X^* = A$ isometrically, however, when A is a dual C^*-algebra X is the *unique one*. Thus the predual is unique and is denoted by A_*. In the case $A = B(H)$, the predual $B(H)_*$ can be identified with the space of all trace class operators on H, equipped with the trace class norm, and it is easy to check that a von Neumann algebra $M \subset B(H)$ is automatically $\sigma(B(H), B(H)_*)$-closed. Thus, when it is a dual space, Sakai's theorem says that A can be realized as a von Neumann algebra on a Hilbert space H and its predual can be identified with the quotient of $B(H)_*$ by the preannihilator of A. Moreover, A is weak* separable if and only if it can be realized on a separable H.

In the early literature, a C^*-algebra that is C^*-isomorphic to a von Neumann algebra is called a W^*-algebra. The latter term is less often used nowadays.

Remark A.37 (Weak* continuous operations) Let $M \subset B(H)$ be a von Neumann subalgebra. Let (x_i) be a net in M. If $x_i \to x \in M$ in the weak* sense then $x_i^* \to x^*$ and, for any $a \in M$, $ax_i \to ax$, $x_i a \to xa$ all in the weak* sense. In other words the mappings $x \mapsto x^*$, $x \mapsto ax$, and $x \mapsto xa$ are weak* to weak* continuous. This is an easy consequence of Remark A.17.

Remark A.38 (On isomorphisms of von Neumann algebras) The unicity of the predual is essentially equivalent to the following useful fact. Let $\pi : M \to M^1$ be an isometric (linear) isomorphism between two C^*-algebras that are both dual spaces (and hence C^*-isomorphic to von Neumann algebras). Then π and π^{-1} are both continuous for the weak* topologies (i.e. "normal").

Indeed, $\pi^*(M_*^1) \subset M^*$ is automatically a predual of M, and hence must coincide with M_*, and similarly of course for π^{-1}.

In other words any C^*-isomorphism $u : M_1 \to M_2$ from a von Neumann algebra onto another one is automatically bicontinuous for the weak$-*$ topologies ($\sigma(M_1, M_{1*})$ and $\sigma(M_2, M_{2*})$) (see *e.g.* [240, vol. I p. 135] for more on this).

Remark A.39 (Commutative von Neumann algebras) If a commutative C^*-algebra A is isomorphic to a von Neumann algebra then there is a locally compact space (T, m) equipped with a positive Radon measure m such that A is isomorphic (as a C^*-algebra) to $L_\infty(T, m)$.

This is a classical structural result that summarizes "abstractly" the spectral theory of unitary operators. See [240, p. 109].

A concrete consequence is that any unitary U in A can be written as $U = \exp ix$ for some self-adjoint $x \in A$. This last fact (proved directly in [145, pp. 313–314]) does not hold in a general C^*-algebra, but some weaker version is available, see [145, p. 332].

Remark A.40 (The predual as a quotient of the trace class) The space $B(H)_*$ can be identified with the projective tensor product $\overline{H} \hat{\otimes} H$. This is a particular case of (A.1).

The duality is defined for all $t = \sum \overline{x_k} \otimes y_k \in \overline{H} \hat{\otimes} H$ with $\sum \|x_k\| \|y_k\| < \infty$ and $b \in B(H)$ by:

$$\langle b, t \rangle = \sum \langle x_k, b y_k \rangle. \tag{A.25}$$

We have then

$$\|t\|_{B(H)_*} = \inf \left\{ \sum_1^\infty \|x_k\| \|y_k\| \right\}$$

where the infimum is over all possible representations of t in the form $t = \sum_1^\infty \overline{x_k} \otimes y_k$ with $\sum \|x_k\| \|y_k\| < \infty$, and the duality is as in (A.25).

Let $M \subset B(H)$ be a weak* closed unital subalgebra (in other words a von Neumann algebra). Let $X = B(H)_*/M_\perp$ where $M_\perp$ is the pre-orthogonal of M. Standard functional analysis tells us that $X^* = M$ isometrically, so that $M_* \subset M^*$ can be identified isometrically with X. Thus any $f \in M_*$ can be represented by a (nonunique) element T of $B(H)_*$ (or equivalently of the trace class) and $\|f\|_{M_*} = \inf \|T\|_{B(H)_*}$ where the infimum runs over all possible such T's.

The next lemma is a useful refinement that is special to von Neumann algebras. The point is that we can obtain equality in (A.26) so the preceding infimum is attained.

Lemma A.41 *Let $f \in M_*$.*

(i) *There are $x_k, y_k \in H$ such that*

$$\sum_1^\infty \|x_k\| \|y_k\| = \|f\|_{M_*}, \tag{A.26}$$

and

$$\forall b \in M \quad f(b) = \sum \langle y_k, b x_k \rangle. \tag{A.27}$$

Moreover, there is a partial isometry u in M such that $f(u) = \|f\|_{M_*}$.

(ii) *Let $u \in M$ be a partial isometry such that $f(u) = \|f\|_{M_*}$. Then for any $b \in B(H)$ we have*

$$f(uu^* b) = f(bu^* u) = f(b). \tag{A.28}$$

Moreover there is a positive $g \in M_*$ such that for any $b \in M$

$$f(b) = g(u^* b). \tag{A.29}$$

(iii) *If f is a (normal) state on M, (i) holds with $(x_k) = (y_k)$ and there is a normal state $\tilde{f}$ on $B(H)$ extending f.*

Proof For the classical fact in (i) we refer to [240, Ex. 1, p. 156], [226, p. 78], or any other standard text for a proof. In [146, th. 7.1.12 p. 462] (i) is proved for positive f's. Then (i) follows by the polar decomposition of f (see [146, th. 7.3.2 p. 474]) to which (i) is closely related, and which is essentially the same as (A.29). Since the set $C_f = \{x \in B_M \mid f(x) = \|f\|_{M_*}\}$ is convex and weak* compact it has extreme points. Since it is a face of B_M its extreme points are extreme points in B_M, and it is well known (see [240, p. 48]) that the latter are partial isometries. This shows that C_f contains a partial isometry $u \in M$.

(ii) By spectral theory, it is known that the trace class operator associated to f as in (A.27) can be rewritten in the form

$$f(b) = \sum_{i \in I} \lambda_i \langle d_i, be_i \rangle, \tag{A.30}$$

where $\lambda_i > 0$ for any $i \in I$ (I is at most countable) satisfies $\sum_{i \in I} \lambda_i = \|f\|_{M_*}$ and $(e_i), (d_i)$ are orthonormal systems. After normalization, we may assume $f(u) = \|f\|_{M_*} = 1$. The reader should keep in mind that for vectors in Hilbert space $\|x\| = \|y\| = \langle y, x \rangle$ implies $x = y$. We have $\sum_{i \in I} \lambda_i \langle d_i, ue_i \rangle = 1 = \sum_{i \in I} \lambda_i$. This forces $\langle d_i, ue_i \rangle = 1$, and hence $d_i = ue_i$, for all $i \in I$. Therefore $1 = \|ue_i\|^2 = \langle e_i, u^*ue_i \rangle$ and hence $u^*ue_i = e_i$. Similarly $uu^*d_i = d_i$ for all $i \in I$. Using this, (A.28) is an immediate consequence of (A.30).

Let $g(x) = f(ux)$ ($x \in M$). Then $g(1) = \|f\|_{M_*} = \|g\|_{M_*}$ and hence $g \geq 0$. By (A.28) we have $f(b) = g(u^*b)$.

(iii) If f is a state, we have $1 = f(1) = \|f\|$ so that $1 = \sum_1^\infty \|x_k\| \|y_k\| = \sum_1^\infty \langle y_k, x_k \rangle$ and hence (assuming $\|x_k\| \|y_k\| \neq 0$) we must have $\|x_k\| \|y_k\| = \langle y_k, x_k \rangle$ and hence $x_k = y_k$ for all i. Then

$$\tilde{f}(b) = \sum_1^\infty \langle x_k, bx_k \rangle \quad (b \in B(H)) \tag{A.31}$$

is the desired extension. □

A linear map $u : M \to M^1$ between von Neumann algebras is called *normal* if it is continuous for the $\sigma(M, M_*)$ and $\sigma(M^1, M_*^1)$ topologies, or equivalently if there is a map $v : M_*^1 \to M_*$ of which u is the adjoint.

Remark A.42 In other words, $u : M \to M^1$ is normal if and only if the map $u^* : M^{1*} \to M^*$ satisfies $u^*(M_*^1) \subset M_*$. Since u^* is norm continuous, it suffices for this to know that $u^*(V) \subset M_*$ for some norm-total subset $V \subset M_*^1$. For instance, assuming $M^1 \subset B(H^1)$, we may take for V the set formed of the (normal) linear forms f on M^1 of the form $f(x) = \langle \xi', x\xi \rangle$ with ξ, ξ' running over a dense linear subspace of H^1. Using (A.27) one checks easily that the latter V is dense in M_*^1.

In particular, the normal linear forms on M are exactly those that are in the predual $M_* \subset M^*$. Furthermore:

Remark A.43 (GNS for normal forms) The GNS representation $\pi_f : M \to B(H_f)$ associated as in §A.13 to a normal form $f : M \to \mathbb{C}$ is normal. This is a particular case of the preceding remark. Just observe that $x \mapsto f(a^*xb) = \langle \pi_f(a)\xi_f, \pi_f(x)\pi_f(b)\xi_f \rangle$ is in M_* if $f \in M_*$.

The following fact is well known.

Theorem A.44 *A positive linear functional $\varphi \in M^*$ is normal if and only it is completely additive, meaning $\varphi(\sum p_i) = \sum \varphi(p_i)$ for any family (p_i) of mutually orthogonal projections in M.*

Proof By definition, "$\varphi \in M_*$" means that there are x, y in $\ell_2(H)$ such that $\varphi(a) = \sum \langle ax_n, y_n \rangle$ for all a in M. Clearly this implies the complete additivity of φ. The problem is to check the converse. To check this we will use the following facts.

Fact 1 Assume φ completely additive. If $(P_j)_{j \in I}$ is a family of mutually orthogonal projections with sum $P = \sum P_j$, such that $P_j \cdot \varphi \in M_*$ for all j, then $P \cdot \varphi \in M_*$. Here

$P \cdot \varphi$ (resp. $\varphi \cdot P$, resp. $P \cdot \varphi \cdot P$) is the linear form on M defined by $P \cdot \varphi(x) = \varphi(xP)$, (resp. $\varphi \cdot P(x) = \varphi(Px)$, resp. $P \cdot \varphi \cdot P(x) = \varphi(PxP)$). Indeed, fix $\varepsilon > 0$ and let $J \subset I$ be a finite subset such that (here we use complete additivity) $\varphi\left(P - \sum_J P_j\right) < \varepsilon$ and let $P_J = \sum_J P_j$. Then by Cauchy–Schwarz we have

$$|(P - P_J) \cdot \varphi(x)|^2 \leq \varphi(x^*x)\varepsilon \leq \varphi(1)\varepsilon\|x\|^2 \qquad \forall x \in M$$

and hence $\|P \cdot \varphi - P_J \cdot \varphi\|_{M^*} \leq (\varphi(1)\varepsilon)^{1/2}$. Clearly $P_J \cdot \varphi \in M_*$; therefore since M_* is norm closed in M^* we conclude that $P \cdot \varphi \in M_*$.

Fact 2 If $\varphi \in M_+^*$ and P is a projection in M such that $P \cdot \varphi \cdot P \in M_*$, then $P \cdot \varphi \in M_*$. Indeed, by (iii) in Lemma A.41 there is (x_n) is $\ell_2(H)$ such that $P \cdot \varphi \cdot P(a) = \sum \langle x_n, ax_n \rangle$. Then by Cauchy–Schwarz again

$$|P \cdot \varphi(a)| \leq \varphi(Pa^*aP)^{1/2}\varphi(1)^{1/2} = \left(\sum_n \|aPx_n\|^2\right)^{1/2} \varphi(1)^{1/2}$$

which implies that there is (y_n) in $\ell_2(H)$ (with norm $\leq \varphi(1)^{1/2}$) such that $P \cdot \varphi(a) = \sum \langle y_n, aPx_n \rangle$.

Fact 3 Given φ, ψ in M_+^* if $\varphi(q) \leq \psi(q)$ for any projection q in M then $\varphi \leq \psi$.

Indeed, given x in M_+, to show $\varphi(x) \leq \psi(x)$ it suffices to show that $\varphi_{|L} \leq \psi_{|L}$ where L is the commutative von Neumann algebra generated by x. Then this fact becomes obvious (e.g. because we can approximate x in norm by nonnegative step functions).

We now complete the proof of the theorem. Let $(P_j)_{j \in I}$ be a maximal family of mutually orthogonal (nonzero) projections such that $P_j \cdot \varphi$ is normal for all j. (This exists by Zorn's lemma or, say, transfinite induction). By Fact 1, it suffices to show that $\sum P_j = 1$. Assume to the contrary that $Q = 1 - \sum P_j \neq 0$. We will show that there is a projection $0 \neq Q' \leq Q$ such that $Q' \cdot \varphi \in M_*$; this will contradict maximality and thus prove that $Q = 0$.

Pick any h in H such that $Qh \neq 0$ and adjust its normalization so that $\varphi(Q) < \langle h, Qh \rangle$. Let $\omega_h \in M_*$ be defined by $\omega_h(a) = \langle h, ah \rangle$. Let (Q_β) be a maximal family of mutually orthogonal (nonzero) projections in M such that $Q_\beta \leq Q$ and $\omega_h(Q_\beta) \leq \varphi(Q_\beta)$. Then we must have $\sum Q_\beta \neq Q$, because otherwise (by complete additivity of ω_h) we find

$$\omega_h(Q) = \sum \omega_h(Q_\beta) \leq \sum \varphi(Q_\beta) \leq \varphi(Q),$$

which contradicts our choice of h. Let then $Q' = Q - \sum Q_\beta \neq 0$. By maximality of (Q_β) we must have $\omega_h(q) > \varphi(q)$ for any projection $0 < q \leq Q'$ in M. By Fact 3 this implies $Q' \cdot \omega_h \cdot Q' \geq Q' \cdot \varphi \cdot Q'$, so that $Q' \cdot \varphi \cdot Q' \in M_*$ and hence by Fact 2, $Q' \cdot \varphi \in M_*$.

As announced this contradicts the maximality of (Q_β), therefore we conclude $\sum Q_\beta = 1$ and $\varphi \in M_*$ by Fact 1. $\qquad\square$

Remark A.45 Theorem A.44 shows that if $\psi \in M_+^*$ is such that $0 \leq \psi \leq \varphi$ then $\varphi \in M_* \Rightarrow \psi \in M_*$. The latter fact can also be derived from part (iii) in Lemma A.41 and Lemma 23.11.

Theorem A.46 (von Neumann bicommutant theorem) *Let $\mathcal{A} \subset B(H)$ be a unital $*$-subalgebra. The closures of $\mathcal{A}$ either for the weak operator topology (w.o.t.), the* *strong operator topology (s.o.t.) (see Definition A.28) or for the weak* topology* $(\sigma(B(H), B(H)_*))$ *all coincide, and they are equal to the bicommutant* $\mathcal{A}''$.

Thus to check that $x \in B(H)$ belongs to a von Neumann subalgebra $M \subset B(H)$ it suffices to check that $xy = yx$ for any $y \in M'$. See e.g. [145, p. 326] or [240, p. 74] for a proof.

Theorem A.47 (Kaplansky's density theorem) *Let* $\mathcal{A} \subset B(H)$ *be a* **-subalgebra. The closures of* $B_{\mathcal{A}}$ *either for the weak operator topology (w.o.t.), the strong operator topology (s.o.t.) (see Definition A.28) or for the weak* topology* $(\sigma(B(H), B(H)_*))$ *all coincide, and if* $\mathcal{A}$ *is unital they are equal to the unit ball of* $M = \mathcal{A}''$.

See e.g. [145, p. 329] or [240, p. 100] for a proof.

Throughout what follows M denotes a von Neumann algebra.

Note that it is an elementary fact (in general topology) that the w.o.t. coincides with the weak* topology on any *bounded* subset of M. Indeed, since such a set is relatively weak* compact, $\sigma(M, M_*)$ coincides on it with $\sigma(M, D)$ for any total separating subset of M_*. Thus $B_{\mathcal{A}}$ clearly has the same closure in both. But it is nontrivial that whenever $\mathcal{A}$ is weak*-dense in M,

$$\overline{B_M \cap \mathcal{A}}^{\sigma(M, M_*)} = B_M.$$

Since $B_M \cap \mathcal{A}$ is a bounded, convex subset of M, its closure is the same for $\sigma(M, M_*)$, for the strong operator topology or even for the so-called strong* operator topology in which a net of operators $T_i \in B(H)$ converges to $T \in B(H)$ if both $T_i h \to Th$ and $T_i^* h \to T^* h$ for any $h \in H$.

Remark A.48 Let $\pi : M \to B(H)$ be an injective normal *-homomorphism. We will show that its range $\pi(M)$ is weak*-closed. Then by Remark A.38, $\pi^{-1} : \pi(M) \to M$ is normal and $\pi(M)$ is a von Neumann subalgebra in $B(H)$, isomorphic to M. To show that $\pi(M)$ is weak*-closed, by Kaplansky's theorem (A.47) it suffices to show that the unit ball of $\pi(M)$ is weak*-closed. But the latter, being the image of B_M which is $\sigma(M, M_*)$-compact, is itself weak*-compact.

Remark A.49 Let $\pi : M \to B(H)$ be a normal *-homomorphism. We claim that $\pi(M)$ is weak* closed and isomorphic to $M/\ker(\pi)$. Indeed, let $\mathcal{I} = \ker(\pi)$. Clearly $\mathcal{I}$ is a weak*-closed (two-sided, self-adjoint) ideal so that the quotient $M/\mathcal{I}$ is a dual with predual $(M/\mathcal{I})_* = \mathcal{I}_\perp = \{f \in M_* \mid f(x) = 0 \ \forall x \in \mathcal{I}\}$. The injective *-homomorphism $\pi_1 : M/\mathcal{I} \to \pi(M)$ canonically associated to π being clearly normal, the claim follows from the preceding Remark A.48. By Remark A.36 we have $M \simeq (M/\mathcal{I}) \oplus \mathcal{I}$.

A.17 Bitransposition: biduals of C^*-algebras

Let A be a C^*-algebra. The bidual of A can be equipped with a C^*-algebra structure as follows: let $\pi_U : A \to B(H)$ be the universal representation of A (i.e. the direct sum of all cyclic representations of A as in Remark A.29). Then the bicommutant $\pi_U(A)''$, which is a von Neumann algebra on H, is isometrically isomorphic (as a Banach space) to the bidual A^{**} (see the proof of Theorem A.55). Using this isomorphism as an identification, we will view A^{**} as a von Neumann algebra, so that the canonical inclusion $A \to A^{**}$ is a $*$-homomorphism. This inclusion possesses a fundamental extension property (see Theorem A.55), that we first discuss in a broader Banach space framework.

Let X, Y be Banach spaces. Then any linear map $u : X \to Y^*$ admits a (unique) weak$*$ to weak$*$ continuous extension $\ddot{u} : X^{**} \to Y^*$ such that $\|\ddot{u}\| = \|u\|$. Indeed, just consider $u^*_{|Y} : Y \to X^*$ and set

$$\ddot{u} = (u^*_{|Y})^*. \tag{A.32}$$

$$
\begin{array}{ccc}
X^{**} & & \\
\uparrow & \searrow & \ddot{u} = (u^*_{|Y})^* \\
\Big\uparrow & & \searrow \\
X & \xrightarrow{\;\;u\;\;} & Y^*
\end{array}
$$

The unicity of $\ddot{u}$ follows from the $\sigma(X^{**}, X^*)$-density of B_X in $B_{X^{**}}$. Since $\ddot{u} : X^{**} \to Y^*$ is weak$*$ to weak$*$ continuous, we can describe its value at any point $x'' \in X^{**}$ like this: for any net (x_i) in X with $\sup \|x_i\| < \infty$ tending weak$*$ to x'', we have (the limit being meant in $\sigma(Y^*, Y)$)

$$\ddot{u}(x'') = \lim u(x_i). \tag{A.33}$$

Indeed, for any $y \in Y$ we have $\langle \ddot{u}(x''), y \rangle = \langle x'', u^*_{|Y} y \rangle = \lim \langle x_i, u^*_{|Y} y \rangle = \lim \langle u(x_i), y \rangle$.

We record here two simple facts about maps such as $\ddot{u}$.

Proposition A.50 *Let X be a Banach space, $D \subset X$ a closed subspace, so that we have $D^{**} \subset X^{**}$ as usual. Let $i_D : D \to D^{**}$ denote the canonical injection. The following properties of a bounded linear map $T : X \to D^{**}$ are equivalent:*

(i) $T_{|D} = i_D$.
(ii) *The mapping $P = \ddot{T} : X^{**} \to D^{**}$ is a linear projection onto D^{**}.*

Moreover, when these hold the projection P is continuous with respect to the weak$*$ topologies of X^{**} and D^{**}, and $\|P\| = \|T\|$.

Proof Assume (i). Any $x'' \in B_{X^{**}}$ is the weak$*$ limit of a net (x_i) in B_X, and we have (with limit meant for $\sigma(D^{**}, D^*)$) $\ddot{T}(x'') = \lim \ddot{T}(x_i)$. If $x'' \in D^{**}$ we can choose $x_i \in B_D$ and then $\ddot{T}(x'') = \lim \ddot{T}(x_i) = \lim x_i = x''$. This shows (ii) with $\|P\| \leq \|T\|$. Conversely, if (ii) holds then $T = \ddot{T}_{|X}$ clearly satisfies (i) and $\|T\| \leq \|P\|$. □

Remark A.51 Let $u : X \to B$ be a bounded linear map between Banach spaces. Let $i_B : B \to B^{**}$ be the inclusion and let $v = i_B u : X \to B^{**}$. Then $\ddot{v} = u^{**}$. Indeed, with the notation in the preceding proof we have (with limits all meant for $\sigma(B^{**}, B^*)$) $\ddot{v}(x'') = \lim \ddot{v}(x_i) = \lim u(x_i) = u^{**}(x)$.

Remark A.52 When X is a dual space (say $X = Y^*$) there is a contractive projection $P : X^{**} \to X$. Indeed, if $u = Id_X$ then $P = \ddot{u}$ is such a projection.

Remark A.53 (The bidual as solution of a universal problem) Let $X \subset Z$ be an isometric inclusion of Banach spaces. Assume that Z is a dual space and that for any Y and any $u : X \to Y^*$ there is a unique weak$*$ continuous $\widehat{u} : Z \to Y^*$ extending u with

$\|\widehat{u}\| = \|u\|$. Then it is an easy exercise to see that Z is isometrically isomorphic to X^{**} via an isomorphism that transforms the inclusion $X \to Z$ into $i_X : X \to X^{**}$ (and $\widehat{u}$ into $\ddot{u}$).

Remark A.54 (Bidual of subspace or quotient) Let $\mathcal{I} \subset X$ be a closed subspace of a Banach space X. The space $\mathcal{I}^{**}$ can be naturally identified with the $\sigma(X^{**}, X^*)$-closure $\overline{\mathcal{I}}^{w*} \subset X^{**}$ of $\mathcal{I}$ in X^{**}. Indeed, if $v : \mathcal{I} \to X$ denotes the inclusion map, then $v^{**} : \mathcal{I}^{**} \to X^{**}$ is an isometric embedding with range $\overline{\mathcal{I}}^{w*}$. Similarly, the quotient space $X^{**}/\overline{\mathcal{I}}^{w*}$ can be naturally identified with $(X/\mathcal{I})^{**}$. More precisely, let $q : X \to X/\mathcal{I}$ be the quotient map, consider $u = i_{X/\mathcal{I}} q : X \to (X/\mathcal{I})^{**}$. Then $\ddot{u} : X^{**} \to (X/\mathcal{I})^{**}$ is a metric surjection such that $\ker(\ddot{u}) = \overline{\mathcal{I}}^{w*}$, and hence $\ddot{u}$ defines an isometric isomorphism $X^{**}/\overline{\mathcal{I}}^{w*} \to (X/\mathcal{I})^{**}$. These classical facts follow from the Hahn–Banach theorem.

Let A be a C^*-algebra and let S denote the set of states of A. For each $f \in S$, let $\pi_f : A \to B(H_f)$ denote the GNS representation associated to f, so that there is a unit vector $\xi_f \in H_f$ such that

$$\forall a \in A \quad f(a) = \langle \xi_f, \pi_f(a)\xi_f \rangle. \tag{A.34}$$

As before (see Remark A.29), we denote by $\pi_U : A \to \left(\oplus \sum_{f \in S} B(H_f) \right)_\infty \subset B(\oplus_{f \in S} H_f)$ the universal representation taking $a \in A$ to the block diagonal operator with coefficients $(\pi_f(a))_{f \in S}$. Let $\mathcal{H} = \oplus_{f \in S} H_f$ and let $\underline{\xi}_f$ denote the unit vector of $\mathcal{H}$ with coefficients equal to 0 except at the f-place where it is equal to ξ_f. Then we have

$$\forall a \in A \quad f(a) = \langle \underline{\xi}_f, \pi_U(a)\underline{\xi}_f \rangle. \tag{A.35}$$

Theorem A.55 (C^*-algebra structure on A^{**}) *Let A be a C^*-algebra. There is a (unique) C^*-algebra structure on A^{**} (with the same norm) satisfying the following:*

(i) *The canonical inclusion $i_A : A \to A^{**}$ is a $*$-homomorphism.*
(ii) *For any von Neumann algebra M and any $*$-homomorphism $\pi : A \to M$ the mapping $\ddot{\pi} : A^{**} \to M$ is a $*$-homomorphism.*

Proof Let

$$\mathcal{M} = \pi_U(A)'' = \overline{\pi_U(A)}^{\text{weak}*}.$$

Note that $\pi_U : A \to \pi_U(A)$ is isometric. Indeed, for any $a \in A$ we have

$$\|a\|^2 = \|a^*a\| = \sup_{f \in S} f(a^*a) = \sup\langle \underline{\xi}_f, \pi_U(a^*a)\underline{\xi}_f \rangle$$

$$\leq \|\pi_U(a^*a)\| = \|\pi_U(a)\|^2 \leq \|a\|^2.$$

From (A.35) one sees that the correspondence $a \mapsto \pi_U(a)$ is a homeomorphism from $(B_A, \sigma(A, A^*))$ to $(B_\mathcal{M}, \text{w.o.t.})$ or equivalently (by Remark A.11) $(B_\mathcal{M}, \sigma(\mathcal{M}, \mathcal{M}_*))$. More precisely, the correspondence $(x_i) \mapsto (\pi_U(x_i))$ is a bijection from the set of $\sigma(A, A^*)$ – Cauchy nets in B_A to that of $\sigma(\mathcal{M}, \mathcal{M}_*)$ – Cauchy nets in $B_{\pi_U(A)}$ (over the same index set). Taking (A.33) and Kaplansky's density theorem into account, this means that $\ddot{\pi}_U$ defines a bijection from $B_{A^{**}}$ to $B_\mathcal{M}$. In other words, $\ddot{\pi}_U : A^{**} \to \mathcal{M}$ is an isometric isomorphism. Thus we can equip A^{**} with a C^*-algebra structure by transplanting that of $\mathcal{M}$. This means that we define the product and involution in A^{**} as

$$\forall x'', y'' \in A^{**} \quad x'' \cdot y'' = \ddot{\pi}_U^{-1}(\ddot{\pi}_U(x'')\ddot{\pi}_U(y'')) \text{ and } x''^* = \ddot{\pi}_U^{-1}(\ddot{\pi}_U(x'')^*).$$

Since $\ddot{\pi}_U$ extends π_U the property (i) is immediate. To check the second one we observe that since any $\pi : A \to M$ decomposes as a direct sum of cyclic representation, it suffices to check it assuming that π has a cyclic vector. Then π is unitarily equivalent to π_f for some f (see Remark A.21), so that we are reduced to the case $\pi = \pi_f : A \to B(H_f)$, and $M = \pi_f(A)''$. Since the latter is weak* closed it suffices to prove that $\ddot{\pi}_f : A^{**} \to B(H_f)$ is a $*$-homomorphism, or equivalently that $\ddot{\pi}_f \ddot{\pi}_U^{-1} : \mathcal{M} \to B(H_f)$ is one. This turns out to be very easy: indeed, if we denote by $Q_f : \left(\oplus \sum_{f \in S} B(H_f) \right)_\infty \to B(H_f)$ the coordinate projection, which is clearly a weak* continuous $*$-homomorphism, we have $\pi_f = Q_f \pi_U$, and hence by (A.33) $\ddot{\pi}_f = Q_f \ddot{\pi}_U$, from which we see that $\ddot{\pi}_f \ddot{\pi}_U^{-1} \mid_{\mathcal{M}} = Q_f$.

Lastly the uniqueness follows from the observation that if A_{bis}^{**} is the same Banach space as A^{**} but with another C^*-algebra structure then (ii) with π equal to the inclusion $A \subset A_{\text{bis}}^{**}$ leads to a $*$-homomorphism $\ddot{\pi} : A^{**} \to A_{\text{bis}}^{**}$ for which the underlying linear map is the identity of A^{**}. This means A_{bis}^{**} is identical to A^{**}. $\quad\square$

Remark A.56 By Remark A.51, if $\pi : A \to B$ is a $*$-homomorphism between C^*-algebras so is $\pi^{**} : A^{**} \to B^{**}$, since $\pi^{**} = \ddot{v}$ with $v = i_B \pi$ and v is a $*$-homomorphism.

Remark A.57 For any C^*-algebra A, let $B_A^+ = B_A \cap A_+$. Let $\pi : A \to B(H)$ be a $*$-homomorphism. Let $M = \overline{\pi(A)}^{w*} \subset B(H)$ where the closure is meant in the weak* sense. We claim that

$$B_M^+ = \overline{B_{\pi(A)}^+}^{w*}.$$

The inclusion $\overline{B_{\pi(A)}^+}^{w*} \subset B_M^+$ is obvious. To show the converse, let $x \in B_M^+$, then $x = y^* y$ for some $y \in M$. By Kaplansky's theorem (A.47), there is a bounded net (y_i) in $\pi(A)$ such that $y_i \to y$ in s.o.t. Then $y_i^* y_i \to y^* y = x$ in w.o.t. and hence weak* (since the net is bounded), so that $x \in \overline{B_{\pi(A)}^+}^{w*}$. This proves the claim.

Applying this to the universal representation we obtain:

$$B_{A^{**}}^+ = \overline{B_A^+}^{\sigma(A^{**}, A^*)}. \tag{A.36}$$

We have a natural identification $M_n(A)^{**} \simeq M_n(A^{**})$ as vector spaces. A moment of thought shows that this is isometric:

Proposition A.58 *The identification $M_n(A)^{**} \simeq M_n(A^{**})$ is an (isometric) $*$-isomorphism.*

Proof We know that $i_A : A \to A^{**}$ is a $*$-homomorphism. It follows that $Id_{M_n} \otimes i_A : M_n(A) \to M_n(A^{**})$ is also one. Let $\sigma = Id_{M_n} \otimes i_A$. By the characteristic property of biduals $\ddot{\sigma} : M_n(A)^{**} \to M_n(A^{**})$ is a $*$-homomorphism. The latter must be an isomorphism since $M_n(A)^{**}$ and $M_n(A^{**})$ can both be identified as Banach spaces with the direct sum of n^2 copies of A^{**}. $\quad\square$

Remark A.59 Let $\mathcal{I} \subset A$ be an ideal as in §A.15. Then $\mathcal{I}^{**} \subset A^{**}$ is a weak* closed ideal and $A^{**}/\mathcal{I}^{**} = (A/\mathcal{I})^{**}$. By (A.24), we have a canonical identification

$$A^{**} \simeq A^{**}/\mathcal{I}^{**} \oplus \mathcal{I}^{**} \simeq (A/\mathcal{I})^{**} \oplus \mathcal{I}^{**}. \tag{A.37}$$

Indeed, this follows by taking π equal to the canonical map $\pi : A \to A/\mathcal{I} \subset (A/\mathcal{I})^{**}$. Then $\ddot{\pi} : A^{**} \to (A/\mathcal{I})^{**}$ is a *-homomorphism onto $(A/\mathcal{I})^{**}$ with kernel $\mathcal{I}^{**}$ (see Remark A.54). Then (A.37) follows from Remark A.34.

Remark A.60 (Important warning) When $M \subset B(H)$ is a von Neumann algebra, its bidual M^{**} (just like any C^*-algebra bidual) is itself isomorphic to a von Neumann algebra so that M^{**} can be realized as a weak* closed *-subalgebra $M^{**} \subset B(\mathcal{H})$. It is important to be aware that there are two distinct embeddings of M in M^{**}. The first one is of course the canonical inclusion $i_M : M \to M^{**}$. This is a unital faithful *-homomorphism that in general is *not normal*. Its range $i_M(M)$ is weak*-dense in M^{**}. In particular, being *not weak* closed* in general its range is *not* a von Neumann subalgebra of $B(\mathcal{H})$.

The second one appears when one considers the mapping $\ddot{u} : M^{**} \to M$ where u is the identity on M. We know this is a unital weak* continuous (i.e. normal) *-homomorphism, so that $\mathcal{I} = \ker(\ddot{u})$ is a weak* closed two-sided ideal. As observed in Remark A.34 applied to M^{**} there is a central projection $p \in M^{**}$ such that $\mathcal{I} = pM^{**}$ and the mapping $\psi_M : M \to M^{**}$ defined by $\psi_M(x) = (1-p)x$ is a normal embedding of M in M^{**}, so that its range $\psi_M(M)$ is a weak* closed subalgebra of M^{**}. However, if $\dim(M) = \infty$, ψ_M is *not unital* and the unit of $\psi_M(M)$ is $1 - p$. The confusion of these two embeddings can be a source of mistakes for beginners (as the author remembers!).

A.18 Isomorphisms between von Neumann algebras

We would like to describe more precisely the structure of isomorphisms for von Neumann algebras. Let $M \subset B(H)$ and $N \subset B(\mathcal{H})$ be von Neumann subalgebras. An isomorphism $\mathcal{W} : M \to N$ of the form $\mathcal{W}(x) = U^{-1}xU$ for some unitary $U : \mathcal{H} \to H$ will be called a spatial isomorphism. When $\widehat{H} = K \otimes_2 H$ and $N = \{Id_K \otimes x \mid x \in M\}$ the isomorphism $\mathcal{A} : M \to N$ defined by $\mathcal{A}(x) = Id_K \otimes x$ will be called an amplification. When $\mathcal{H} \subset H$ is invariant under M (this means $\mathcal{H} = p(H)$ for some $p \in M'$), any *-homomorphism $\mathcal{V} : M \to N \subset B(\mathcal{H})$ of the form $\mathcal{V}(x) = P_{\mathcal{H}}x_{|\mathcal{H}}$ ($x \in M$) where $P_{\mathcal{H}} : H \to \mathcal{H}$ denotes the orthogonal projection viewed as acting into $\mathcal{H}$ (or $\mathcal{V}(x) = pxp$ viewed as an element of $B(\mathcal{H})$) will be called a compression. The following is classical.

Theorem A.61 *Any weak* continuous (also called "normal") *-homomorphism $\pi : M \to N$ can be written as a composition $\pi = \mathcal{W}\mathcal{V}\mathcal{A}$ for some amplification $\mathcal{A} : M \to M^1$, compression $\mathcal{V} : M^1 \to N^1$ and spatial isomorphism $\mathcal{W} : N^1 \to N$, where $M^1 \subset B(H^1)$ and $N^1 \subset B(K^1)$ are von Neumann algebras.*

If π is an isomorphism, then $\mathcal{V}$ must be an isomorphism.

Sketch of proof Recall $M \subset B(H)$ and $N \subset B(\mathcal{H})$. By the usual decomposition of π as a direct sum of cyclic representations, it suffices to prove the statement in the cyclic case. Let $\xi \in \mathcal{H}$ be a cyclic unit vector for π. Then $x \mapsto f(x) = \langle \xi, \pi(x)\xi \rangle$ is a normal state on M. Let $\tilde{f} : B(H) \to \mathbb{C}$ be a *normal* state extending f (see Lemma A.41). By (A.31) there is a unit vector $\eta = \sum e_i \otimes x_i \in \ell_2 \otimes_2 H$ such that $\tilde{f}(b) = \langle \eta, [Id \otimes b]\eta \rangle$. Then a simple verification shows that the correspondence $\pi(x)\xi \mapsto [Id \otimes x]\eta$ $(x \in M)$ extends to an isometric embedding $S : H \subset \ell_2 \otimes_2 H$ such that $S(H)$ is invariant under $M^1 = [Id \otimes M]$, and moreover

$$\forall x \in M \quad \pi(x) = S^*[Id \otimes x]S.$$

Let $\mathcal{A}(x) = Id \otimes x$ $(x \in M)$. Let $U : H \to S(H)$ be the unitary that is the same operator as S but with range $S(H)$. We obtain $\pi(x) = U^{-1}(P_{S(H)}\mathcal{A}(x)_{|S(H)})U$ $(x \in M)$ or equivalently $\pi = \mathcal{W}\mathcal{V}\mathcal{A}$ with $\mathcal{W}(\cdot) = U^{-1} \cdot U$ and $\mathcal{V}(\cdot) = P_{S(H)} \cdot {}_{|S(H)}$. For a more detailed proof see [72] (p. 55 in the French edition, p. 61 in the English one). $\qquad\square$

A.19 Tensor product of von Neumann algebras

Let $M \subset B(H)$ and $N \subset B(K)$ be von Neumann algebras. We have a natural embedding $M \otimes N \subset B(H \otimes_2 K)$ of the algebraic tensor product into $B(H \otimes_2 K)$. We define the tensor product in the von Neumann sense $M \bar{\otimes} N$ as follows:

$$M \bar{\otimes} N = \overline{M \otimes N}^{\text{weak}*} \subset B(H \otimes_2 K),$$

and by the bicommutant Theorem A.46 we also have $M \bar{\otimes} N = (M \otimes N)''$.

In particular, with this definition we have $B(H) \bar{\otimes} B(K) = B(H \otimes_2 K)$.

A.20 On σ-finite (countably decomposable)
von Neumann algebras

A von Neumann algebra is called σ-finite or countably decomposable (the terminology is not unanimous) if it admits a normal faithful state. Equivalently, this means that any family of mutually orthogonal nonzero (self-adjoint) projections in M is at most countable. Any von Neumann algebra on a separable Hilbert space is σ-finite.

Recall that a vector $\xi \in H$ is separating for $M \subset B(H)$ if $m \in M, m\xi = 0 \Rightarrow m = 0$. The following basic facts are classical:

Lemma A.62 *Let $M \subset B(H)$ be a von Neumann algebra and let $\xi \in H$. Then ξ is cyclic for M if and only if it is separating for M'.*

Theorem A.63 *Any σ-finite von Neumann algebra can be realized for some H as a von Neumann subalgebra $M \subset B(H)$ in such a way that both M and M' have a cyclic vector.*

Remark A.64 (On the nonseparable case) In the commutative case, saying that M is σ-finite is the same as saying that M is (isomorphic to) the L_∞-space of a σ-finite measure space (Ω, μ). Equivalently, there is $f \in L_1(\mu)$ with $f > 0$ almost everywhere. If $M \subset B(H)$ with H separable then M is σ-finite, but this sufficient condition is not necessary, even in the commutative case (consider e.g. $\Omega = \{-1, 1\}^I$ with the usual product probability and I uncountable).

Many results on von Neumann algebras are easier to handle in the σ-finite (=countably decomposable) case where almost all the interesting examples lie. A classical way to reduce consideration to the latter case is via the following fundamental structural result:

Theorem A.65 (Fundamental reduction to σ-finite case) *Any von Neumann algebra M admits a decomposition as a direct sum*

$$M \simeq \left(\oplus \sum\nolimits_{i \in I} B(\mathcal{H}_i) \bar{\otimes} N_i \right)_\infty$$

where the N_i's are σ-finite (=countably decomposable) and the $\mathcal{H}_i$'s are Hilbert spaces.

See [72, ch. III, §1, lemma 7; p. 224 in the French edition and p. 291 in the English one] for a detailed proof.

Note that if M is σ-finite we can take for I a singleton with $N_i = M$ and $\mathcal{H}_i = \mathbb{C}$.

Corollary A.66 *For any $f \in M_*$ there is a (self-adjoint) projection $p \in M$ such that pMp is σ-finite and $f(x) = f(pxp)$ for any $x \in M$.*

Proof With the notation of Theorem A.65 we may assume $N_i \subset B(H_i)$ so that $M \subset B(H)$ with $H = (\oplus_{i \in I} \mathcal{H}_i \otimes_2 H_i)_2$. There are $x_k, y_k \in H$ such that (A.27) holds. We can clearly find a countable subset $J \subset I$ and separable subspaces $\mathcal{K}_i \subset \mathcal{H}_i$ such that $x_k, y_k \in K$ for all $k \geq 1$ with $K = (\oplus_{i \in J} \mathcal{K}_i \otimes_2 H_i)_2$ viewed as a subspace of H. Let $p = P_K = \sum_{i \in J} P_{\mathcal{K}_i} \otimes 1_{N_i}$. Then $pMp \simeq \left(\oplus \sum_{i \in J} B(\mathcal{K}_i) \bar{\otimes} N_i \right)_\infty$ is σ-finite and p has the required property. □

A.21 Schur's lemma

A subspace $E \subset H$ is called "invariant" under an operator $T \in B(H)$ if $T(E) \subset E$. It is called "reducing" if it is also invariant under T^*. In the latter case the orthogonal projection P_E commutes with T.

Let G be a discrete group. Consider an irreducible unitary representation $\pi : G \to B(H_\pi)$ with $\dim(H_\pi) < \infty$. By "irreducible" we mean that there is no nontrivial subspace of H_π that is left invariant by $\pi(G)$ (the range of π). Note that since $\pi(G)$ is a self-adjoint subset of $B(H_\pi)$, a subspace $E \subset H_\pi$ is invariant under $\pi(G)$ if and only if the orthogonal projection P_E commutes with the operators in $\pi(G)$. In that case, we have $\pi(t)P_E = P_E\pi(t)P_E = P_E\pi(t)$ and the mapping $t \mapsto \pi(t)_{|E}$ can be viewed as a unitary representation of G in $B(E)$ such that $\pi(t) = \pi(t)_{|E} \oplus \pi(t)_{|E^\perp}$. By definition π is irreducible if and only if there is no nontrivial decomposition of this type, nontrivial meaning that $\{0\} \neq E \neq H_\pi$.

Equivalently, the commutant of $\pi(G)$ (which is a C^*-algebra) is equal to $\mathbb{C}I$ (here we denote by I the identity on H_π). Indeed, the commutant is clearly linearly spanned by its self-adjoint elements; their spectral projections being orthogonal projections commuting with $\pi(G)$ must be equal to either 0 or the identity, and hence they must be in $\mathbb{C}I$. Thus π is irreducible if and only if the commutant of $\pi(G)$ consists of scalar multiples of the identity.

Remark A.67 Any finite-dimensional unitary representation can be decomposed as a direct sum of irreducible ones.

Indeed, if it is irreducible this is obvious, and if not then it is the sum of representations of lower dimensions so that we can obtain the result by induction on the dimension (starting with dimension 1 which is obviously irreducible).

Let $\mathbb{S}_N$ denote the symmetric group, i.e. the set of permutations of a set with N elements. Let $\chi = N^{-1/2} \sum_1^N e_k$. The following simple example is useful:

Lemma A.68 *The natural unitary representation $\pi : \mathbb{S}_N \to B(\ell_2^N)$ that acts on ℓ_2^N by permuting the basis vectors (i.e. $\pi(\sigma)(e_k) = e_{\sigma(k)}$) decomposes as the direct sum of the trivial representation on $\mathbb{C}\chi$ and an irreducible representation π^0 that is the restriction of π to $\chi^\perp$.*

More generally, if a group G acts by permutation on $\{1, 2, \ldots, N\}$, in a bitransitive way, meaning that for any $i \neq j$ and $i' \neq j'$ there is $\sigma \in G$ such that $\sigma(i) = i'$ and $\sigma(j) = j'$, then the corresponding unitary representation $\pi^0 = \pi_{|\chi^\perp} : G \to B(\chi^\perp)$ is irreducible on $\chi^\perp$.

Proof Let G be a group acting bitransitively on $\{1, 2, \ldots, N\}$ with associated unitary representation $\pi : G \to B(\ell_2^N)$. A fortiori the action is transitive, so that $\pi(\sigma)\chi = \chi$ for any $\sigma \in G$, and hence also $\pi(\sigma)\chi^\perp = \chi^\perp$. Therefore $\pi = \pi_{|\mathbb{C}\chi} \oplus \pi_{|\chi^\perp}$. The irreducibility of $\pi^0 = \pi_{|\chi^\perp}$ can be checked as follows. Using transitivity and bitransitivity one checks easily that the commutant of the range of π is formed of matrices $[a_{ij}]$ that are constant both on the diagonal and outside of it. These are matrices in the linear span of the identity and the orthogonal projection P_0 onto $\mathbb{C}\chi$, or equivalently $\mathrm{span}[P_0, I - P_0]$. From this one deduces easily that the commutant of the range of π^0 in $B(\chi^\perp)$ is formed of multiples of the identity on $\chi^\perp$. $\square$

We denote by $\widehat{G}_{fd}$ the set of finite-dimensional irreducible unitary representations of G with the convention that we identify two representations if they are unitarily equivalent.

Let $\pi, \sigma \in \widehat{G}_{fd}$. Consider the representation $\rho = \bar{\pi} \otimes \sigma : G \to \overline{H_\pi} \otimes_2 H_\sigma$ defined by $\rho(t) = \overline{\pi(t)} \otimes \sigma(t)$ $(t \in G)$. Using the identification $\overline{H_\pi} \otimes H_\sigma = S_2(H_\pi, H_\sigma)$ we may view $\overline{\pi(t)} \otimes \sigma(t)$ as an operator on $S_2(H_\pi, H_\sigma)$. More precisely, we have for any $\xi \in S_2(H_\pi, H_\sigma)$

$$\rho(t)\xi = \sigma(t)\xi\pi(t)^*. \tag{A.38}$$

Let I denote the identity operator on H_π. By (A.38), I is an invariant vector for $\bar{\pi} \otimes \pi$. The following classical result of Schur is very well known.

Lemma A.69 (Schur's lemma) *Let G be any group.*

(i) *For any $\pi \in \widehat{G}_{fd}$, the representation $[\bar{\pi} \otimes \pi]_{|I^\perp}$ has no nonzero invariant vector.*

(ii) *For any pair $\pi \neq \sigma \in \widehat{G}_{fd}$ the representation $\bar{\pi} \otimes \sigma$ has no nonzero invariant vector.*

Proof (i) Let $\xi \in S_2(H_\pi, H_\pi)$ be an invariant vector. Then $\xi = \pi(t)\xi\pi(t)^*$ for any $t \in G$. Therefore, ξ commutes with $\pi(G)$. By the preceding remarks, $\xi \in \mathbb{C}I$. Therefore any invariant $\xi \in I^\perp$ must be $= 0$.

(ii) Assume otherwise that $\bar{\pi} \otimes \sigma$ has an invariant vector $\xi \neq 0$. Viewing ξ as an element of $\overline{H_\pi} \otimes H_\sigma = S_2(H_\pi, H_\sigma)$ we would have $\sigma(t)\xi\pi(t)^* = \xi$ for any t in G (ξ is called an "intertwiner"). It follows that $\pi(t)\xi^*\xi\pi(t)^* = \xi^*\xi$ for any t in G. By the irreducibility of π, we must have $\xi^*\xi = I$. Arguing similarly with σ, we find $\xi\xi^* = I$ and hence ξ must be unitary. But then $\sigma(t)\xi\pi(t)^* = \xi$ implies $\sigma(t) = \xi\pi(t)\xi^*$, which would mean that $\sigma \simeq \pi$ which is excluded. $\qquad\square$

References

[1] C. Akemann, J. Anderson, and G. Pedersen, Triangle inequalities in operator algebras, *Lin. Multi. Alg.* **11** (1982), 167–178.

[2] C. Akemann and P. Ostrand, Computing norms in group C^*-algebras, *Amer. J. Math.* **98** (1976), 1015–1047.

[3] C. Anantharaman-Delaroche, Amenability and exactness for dynamical systems and their C^*-algebras, *Trans. Amer. Math. Soc.* **354** (2002), 4153–4178.

[4] C. Anantharaman and S. Popa, "An introduction to II_1-factors", Cambridge University Press, Cambridge, to appear.

[5] T. B. Andersen, Linear extensions, projections, and split faces, *J. Funct. Anal.* **17** (1974), 161–173.

[6] J. Anderson, Extreme points in sets of positive linear maps in $\mathcal{B}(\mathcal{H})$, *J. Funct. Anal.* **31** (1979), 195–217.

[7] G. Anderson, A. Guionnet, and O. Zeitouni, *An introduction to random matrices*, Cambridge University Press, Cambridge, 2010.

[8] H. Ando and U. Haagerup, Ultraproducts of von Neumann algebras, *J. Funct. Anal.* **266** (2014), 6842–6913.

[9] H. Ando, U. Haagerup, and C. Winsløw, Ultraproducts, QWEP von Neumann algebras, and the Effros–Maréchal topology, *J. Reine Angew. Math.* **715** (2016), 231–250.

[10] R. Archbold and C. Batty, C^*-tensor norms and slice maps, *J. Lond. Math. Soc.* **22** (1980), 127–138.

[11] W. Arveson, Analyticity in operator algebras, *Amer. J. Math.* **89** (1967), 578–642.

[12] W. Arveson, Subalgebras of C^*-algebras, *Acta Math.* **123** (1969), 141–224. Part II. *Acta Math.* **128** (1972), 271–308.

[13] W. Arveson, Notes on extensions of C^*-algebras, *Duke Math. J.* **44** (1977), 329–355.

[14] G. Arzhantseva and T. Delzant, Examples of random groups, unpublished preprint, 2008.

[15] D. Avitsour, Free products of C^*-algebras, *Trans. Amer. Math. Soc.* **271** (1982), 423–435.

[16] J. Bannon, A. Marrakchi, and N. Ozawa, Full factors and co-amenable inclusions, arXiv:1903.05395, 2019.

[17] B. Bekka, P. de la Harpe, and A. Valette, *Kazhdan's property (T)*, Cambridge University Press, Cambridge, 2008.

[18] A. Ben-Aroya and A. Ta-Shma, Quantum expanders and the quantum entropy difference problem, arXiv:quant-ph/0702129, no. 3, 2007.

[19] A. Ben-Aroya, O. Schwartz, and A. Ta-Shma, Quantum expanders: motivation and constructions, *Theory Comput.* **6** (2010), 47–79.

[20] C. A. Berger, L. A. Coburn, and A. Lebow, Representation and index theory for C^*-algebras generated by commuting isometries, *J. Funct. Anal.* **27**, no. 1 (1978), 51–99.

[21] J. Bergh, On the relation between the two complex methods of interpolation, *Indiana Univ. Math. J.* **28** (1979), 775–778.

[22] J. Bergh and J. Löfström, *Interpolation spaces: an introduction*, Springer-Verlag, Berlin, 1976.

[23] B. Blackadar, Weak expectations and injectivity in operator algebras, *Proc. Amer. Math. Soc.* **68** (1978), 49–53.

[24] B. Blackadar, Weak expectations and nuclear C^*-algebras, *Indiana Univ. Math. J.* **27** (1978), 1021–1026.

[25] B. Blackadar, *Operator algebras: theory of C^*-algebras and von Neumann algebras*, Encyclopaedia of mathematical sciences, 122, Springer-Verlag, Berlin, 2006.

[26] D. P. Blecher and L. Labuschagne, Outers for noncommutative H_p revisited, *Studia Math.* **217** (2013), 265–287.

[27] D. P. Blecher and C. Le Merdy, *Operator algebras and their modules: an operator space approach*, Oxford University Press, Oxford, 2004.

[28] D. Blecher and V. Paulsen, Explicit constructions of universal operator algebras and applications to polynomial factorization, *Proc. Amer. Math. Soc.* **112** (1991), 839–850.

[29] D. Blecher, Z. J. Ruan, and A. Sinclair, A characterization of operator algebras, *J. Funct. Anal.* **89** (1990), 188–201.

[30] F. Boca, Free products of completely positive maps and spectral sets, *J. Funct. Anal.* **97** (1991), 251–263.

[31] F. Boca, A note on full free product C^*-algebras, lifting and quasidiagonality, operator theory, operator algebras and related topics (Timişoara, 1996), 51–63, Theta Found., Bucharest, 1997.

[32] C. Bordenave and B. Collins, Eigenvalues of random lifts and polynomial of random permutations matrices, *Ann. of Math.* **190** (2019), 811–875.

[33] J. Bourgain, Real isomorphic complex Banach spaces need not be complex isomorphic, *Proc. Amer. Math. Soc.* **96** (1986), 221–226.

[34] M. Bożejko, Some aspects of harmonic analysis on free groups, *Colloq. Math.* **41** (1979), 265–271.

[35] M. Bożejko and G. Fendler, Herz–Schur multipliers and completely bounded multipliers of the Fourier algebra of a locally compact group, *Boll. Unione Mat. Ital.* (6) **3-A** (1984), 297–302.

[36] L. Brown, Ext of certain free product C^*-algebras, *J. Operator Theory* **6** (1981), 135–141.

[37] L. Brown, *Invariant means and finite representation theory of C^*-algebras*, Memoirs of the American Mathematical Society, **184**, American Mathematical Society, Providence, RI, 2006.

[38] L. Brown and K. Dykema, Popa algebras in free group factors, *J. Reine Angew. Math.* **573** (2004), 157–180.

[39] N. P. Brown and N. Ozawa, *C*-algebras and finite-dimensional approximations*, Graduate studies in mathematics, 88, American Mathematical Society, Providence, RI, 2008.

[40] S. Burgdorf, K. Dykema, I. Klep, and M. Schweighofer, Addendum to "Connes' embedding conjecture and sums of Hermitian squares" [*Adv. Math.* 217, no. 4 (2008) 1816–1837], *Adv. Math.* **252** (2014), 805–811.

[41] J. de Cannière and U. Haagerup, Multipliers of the Fourier algebras of some simple Lie groups and their discrete subgroups, *Amer. J. Math.* **107** (1985), 455–500.

[42] P.-A. Cherix, M. Cowling, P. Jolissaint, P. Julg, and A. Valette, *Groups with the Haagerup property. Gromov's a-T-menability*, Birkhäuser Verlag, Basel, 2001.

[43] W. M. Ching, Free products of von Neumann algebras, *Trans. Amer. Math. Soc.* **178** (1973), 147–163.

[44] M. D. Choi, A Schwarz inequality for positive linear maps on C^*-algebras, *Illinois J. Math.* **18** (1974), 565–574.

[45] M. D. Choi and E. Effros, Nuclear C^*-algebras and the approximation property, *Amer. J. Math.* **100** (1978), 61–79.

[46] M. D. Choi and E. Effros, Nuclear C*-algebras and injectivity. The general case, *Indiana Univ. Math. J.* **26** (1977), 443–446.

[47] M. D. Choi and E. Effros, Injectivity and operator spaces, *J. Funct. Anal.* **24** (1977), 156–209.

[48] M. D. Choi and E. Effros, Separable nuclear C^*-algebras and injectivity, *Duke Math. J.* **43** (1976), 309–322.

[49] M. D. Choi and E. Effros, The completely positive lifting problem for C^*-algebras, *Ann. Math.* **104** (1976), 585–609.

[50] E. Christensen, E. Effros, and A. Sinclair, Completely bounded multilinear maps and C^*-algebraic cohomology, *Invent. Math.* **90** (1987), 279–296.

[51] E. Christensen and A. Sinclair, On von Neumann algebras which are complemented subspaces of $B(H)$, *J. Funct. Anal.* **122** (1994), 91–102.

[52] E. Christensen and A. Sinclair, Module mappings into von Neumann algebras and injectivity, *Proc. Lond. Math. Soc.* **71** (1995), 618–640.

[53] E. Christensen and L. Wang, Von Neumann algebras as complemented subspaces of $B(\mathcal{H})$. *Internat. J. Math.* **25** (2014), 1450107, 9 pp.

[54] K. McClanahan, C^*-algebras generated by elements of a unitary matrix, *J. Funct. Anal.* **107** (1992), 439–457.

[55] P. M. Cohn, *Basic algebra*, Springer, London, 2003.

[56] B. Collins and C. Male, The strong asymptotic freeness of Haar and deterministic matrices, *Ann. Sci. Éc. Norm. Supér.* **47** (2014), 147–163.

[57] B. Collins and K. Dykema, A linearization of Connes' embedding problem, *New York J. Math.* **14** (2008), 617–641.

[58] W.W. Comfort, S. Negrepontis, *The theory of ultrafilters*, Springer, New York, 1974.

[59] A. Connes, Caractérisation des espaces vectoriels ordonnés sous-jacents aux algèbres de von Neumann, *Ann. Inst. Fourier (Grenoble)* **24** (1974), 121–155.

[60] A. Connes, A factor not anti-isomorphic to itself, *Bull. Lond. Math. Soc.* **7** (1975), 171–174.

[61] A. Connes, Classification of injective factors. Cases II_1, II_∞, III_λ, $\lambda \neq 1$, *Ann. Math. (2)* **104** (1976), 73–115.

[62] J. H. Conway and N. J. A. Sloane, *Sphere packings, lattices and groups*, Third edition. Springer-Verlag, New York, 1999.

[63] K. Courtney and D. Sherman, "The universal C^*-algebra of a contraction", arXiv:1811.04043, 2018, to appear.

[64] M. Cowling and U. Haagerup, Completely bounded multipliers of the Fourier algebra of a simple Lie group of real rank one, *Invent. Math.* **96** (1989), 507–549.

[65] H. S. M. Coxeter and W. O. J. Moser, *Generators and relations for discrete groups*, Springer-Verlag, Berlin-Göttingen-Heidelberg, 1957.

[66] M. Cwikel and S. Janson, Interpolation of analytic families of operators, *Studia Math.* **79** (1984), 61–71.

[67] K. Davidson. C^*-*algebras by example*, Fields Institute publication. Toronto, American Mathematical Society, Providence, RI, 1996.

[68] K. Davidson and E. Kakariadis, A proof of Boca's theorem, *Proc. Roy. Soc. Edinburgh Sect. A* **149** (2019), 869–876.

[69] A. M. Davie, Matrix norms related to Grothendieck's inequality, *Banach spaces (Columbia, MO, 1984), Lecture Notes in Mathematics*, **1166**, Springer, Berlin, 1985.

[70] A. Devinatz, The factorization of operator valued analytic functions, *Ann. Math.* **73** (1961), 458–495.

[71] J. Diestel, J. H. Fourie, and J. Swart, *The metric theory of tensor products. Grothendieck's résumé revisited*, American Mathematical Society, Providence, RI, 2008.

[72] J. Dixmier, *Les Algèbres d'Opérateurs dans l'Espace Hilbertien (Algèbres de von Neumann)*, Gauthier-Villars, Paris 1969. (In translation: *von Neumann algebras*, North-Holland, Amsterdam–New York 1981.)

[73] R. Douglas and R. Howe, On the C^*-algebra of Toeplitz operators on the quarterplane, *Trans. Amer. Math. Soc.* **158** (1971), 203–217.

[74] R. Douglas and C. Pearcy, Von Neumann algebras with a single generator, *Michigan Math. J.* **16** (1969), 21–26.

[75] K. Dykema and K. Juschenko, Matrices of unitary moments, *Math. Scand.* **109** (2011), 225–239.

[76] K. Dykema, V. Paulsen and J. Prakash, Non-closure of the set of quantum correlations via graphs, *Comm. Math. Phys.* **365** (2019), 1125–1142.

[77] E. Effros and U. Haagerup, Lifting problems and local reflexivity for C^*-algebras, *Duke Math. J.* **52** (1985), 103–128.

[78] E. Effros, M. Junge, and Z. J. Ruan, Integral mappings and the principle of local reflexivity for noncommutative L_1-spaces, *Ann. Math.* **151** (2000), 59–92.

[79] E. Effros and C. Lance, Tensor products of operator algebras, *Adv. Math.* **25** (1977), 1–34.

[80] E. Effros and Z. J. Ruan, *Operator Spaces*, Oxford University Press, Oxford, 2000.

[81] G. Elek and E. Szabó, Hyperlinearity, essentially free actions and L_2-invariants. The sofic property, *Math. Ann.* **332** (2005), 421–441.

[82] G. Elliott, On approximately finite-dimensional von Neumann algebras, *Math. Scand.* **39** (1976), 91–101.

[83] G. Elliott, On approximately finite-dimensional von Neumann algebras. II, *Canad. Math. Bull.* **21** (1978), 415–418.

[84] G. Elliott and E. Woods, The equivalence of various definitions for a properly infinite von Neumann algebra to be approximately finite dimensional, *Proc. Amer. Math. Soc.* **60** (1976), 175–178.

[85] P. Eymard, L'algèbre de Fourier d'un groupe localement compact, *Bull. Soc. Math. France* **92** (1964), 181–236.

[86] I. Farah, B. Hart, and D. Sherman, Model theory of operator algebras I: stability, *Bull. Lond. Math. Soc.* **45** (2013), 825–838.

[87] I. Farah, B. Hart, and D. Sherman, Model theory of operator algebras III: elementary equivalence and II1 factors, *Bull. Lond. Math. Soc.* **46** (2014), 609–628.

[88] I. Farah, B. Hart, and D. Sherman, Model theory of operator algebras II: model theory, *Israel J. Math.* **201** (2014), 477–505.

[89] I. Farah and S. Shelah, A dichotomy for the number of ultrapowers, *J. Math. Log.* **10** (2010), 45–81.

[90] D. Farenick, A. Kavruk, and V. Paulsen, C^*-algebras with the weak expectation property and a multivariable analogue of Ando's theorem on the numerical radius, *J. Operator Theory* **70** (2013), 573–590.

[91] D. Farenick, A. Kavruk, V. Paulsen, and I. Todorov, Characterisations of the weak expectation property, *New York J. Math.* **24A** (2018), 107–135.

[92] D. Farenick and V. Paulsen, Operator system quotients of matrix algebras and their tensor products, *Math. Scand.* **111** (2012), 210–243.

[93] J. Friedman, A proof of Alon's second eigenvalue conjecture and related problems, *Mem. Amer. Math. Soc.* **195**, 910 (2008).

[94] J. Friedman, A. Joux, Y. Roichman, J. Stern, and J.-P. Tillich, The action of a few permutations on r-tuples is quickly transitive, *Random Struct. Algo.* **12** (1998), 335–350.

[95] T. Fritz, Tsirelson's problem and Kirchberg's conjecture, *Rev. Math. Phys.* **24** (2012), 1250012, 67 pp.

[96] M. Gromov, Random walk in random groups, *Geom. Funct. Anal.* **13** (2003), 73–146.

[97] L. Gross, A non-commutative extension of the Perron–Frobenius theorem, *Bull. Amer. Math. Soc.* **77** (1971), 343–347.

[98] A. Grothendieck, Résumé de la théorie métrique des produits tensoriels topologiques, *Boll. Soc. Mat. São-Paulo* **8** (1953), 1–79. Reprinted in *Resenhas* **2** (1996), no. 4, 401–480.

[99] E. Guentner, N. Higson, and S. Weinberger, The Novikov conjecture for linear groups, *Publ. Math. Inst. Hautes Études Sci.* **101** (2005), 243–268.

[100] A. Guichardet, Tensor products of C^*-algebras, *Dokl. Akad. Nauk. SSSR* **160** (1965), 986–989.

[101] A. Guichardet, *Tensor products of C^*-algebras (Part I. Finite tensor products. Part II. Infinite tensor products)*, Lecture Notes Series **12** and **13**, Aarhus Universitet, 1969.

[102] U. Haagerup, The standard form of von Neumann algebras, *Math. Scand.* **37** (1975), 271–283.

[103] U. Haagerup, An example of a nonnuclear C^*-algebra, which has the metric approximation property, *Invent. Math.* **50** (1978–1979), 279–293.

[104] U. Haagerup, Injectivity and decomposition of completely bounded maps, *Operator algebras and their connections with topology and ergodic theory*, 170–222, Lecture Notes in Mathematics, **1132**, Springer, Berlin, Heidelberg, 1985.

[105] U. Haagerup, A new proof of the equivalence of injectivity and hyperfiniteness for factors on a separable Hilbert space, *J. Funct. Anal.* **62** (1985), 160–201.

[106] U. Haagerup, On convex combinations of unitary operators in C*-algebras, *Mappings of operator algebras (Philadelphia, PA, 1988)*, 1–13, Progr. Math., **84**, Birkhäuser Boston, Boston, MA, 1990.

[107] U. Haagerup, Self-polar forms, conditional expectations and the weak expectation property for C^*-algebras, Unpublished manuscript (1993).

[108] U. Haagerup, Group C^*-algebras without the completely bounded approximation property, *J. Lie Theory* **26** (2016), 861–887.

[109] U. Haagerup, S. Knudby, and T. de Laat, A complete characterization of connected Lie groups with the approximation property, *Ann. Sci. Éc. Norm. Supér.* **49** (2016), 927–946.

[110] U. Haagerup and J. Kraus, Approximation properties for group C^*-algebras and group von Neumann algebras, *Trans. Amer. Math. Soc.* **344** (1994), 667–699.

[111] U. Haagerup, M. Junge, and Q. Xu, A reduction method for noncommutative L_p-spaces and applications, *Trans. Amer. Math. Soc.* **362** (2010), 2125–2165.

[112] U. Haagerup, R. Kadison, and G. Pedersen, Means of unitary operators, revisited, *Math. Scand.* **100** (2007), 193–197.

[113] U. Haagerup and T. de Laat, Simple Lie groups without the approximation property, *Duke Math. J.* **162** (2013), 925–964.

[114] U. Haagerup and T. de Laat, Simple Lie groups without the approximation property II, *Trans. Amer. Math. Soc.* **368** (2016), 3777–3809.

[115] U. Haagerup and M. Musat, Factorization and dilation problems for completely positive maps on von Neumann algebras, *Comm. Math. Phys.* **303** (2011), 555–594.

[116] U. Haagerup and M. Musat, An asymptotic property of factorizable completely positive maps and the Connes embedding problem, *Comm. Math. Phys.* **338** (2015), 141–176.

[117] U. Haagerup and G. Pisier, Factorization of analytic functions with values in non-commutative L_1-spaces, *Canadian J. Math.* **41** (1989), 882–906.

[118] U. Haagerup and G. Pisier, Bounded linear operators between C^*-algebras, *Duke Math. J.* **71** (1993), 889–925.

[119] U. Haagerup and S. Thorbjoernsen, Random matrices and K-theory for exact C*-algebras, *Doc. Math.* **4** (1999), 341–450 (electronic).

[120] U. Haagerup and S. Thorbjørnsen, A new application of random matrices: $Ext(C^*_{red}(F_2))$ is not a group, *Ann. Math.* **162** (2005), 711–775.

[121] U. Haagerup and C. Winsløw, The Effros-Maréchal topology in the space of von Neumann algebras, *Amer. J. Math.* **120** (1998), 567–617.

[122] U. Haagerup and C. Winsløw, The Effros-Maréchal topology in the space of von Neumann algebras. II, *J. Funct. Anal.* **171** (2000), 401–431.

[123] H. Hanche-Olsen and E. Störmer, *Jordan operator algebras*, Pitman, Boston, 1984.

[124] A. Harcharras, On some "stability" properties of the full C^*-algebra associated to the free group F_∞, *Proc. Edinburgh Math. Soc.* **41** (1998), 93–116.

[125] P. Harmand, D. Werner, and W. Werner, *M-ideals in Banach spaces and Banach algebras*, Lecture Notes in Mathematics, **1547** Springer-Verlag, Berlin, 1993.

[126] P. de la Harpe, *Topics in geometric group theory*, The University of Chicago Press, Second printing, with corrections and updates, Chicago, 2003.

[127] P. de la Harpe and A. Valette, La propriété (T) de Kazhdan pour les groupes localement compacts (avec un appendice de Marc Burger). *Astérisque* **175** (1989), Soc. Math. France, Paris.

[128] S. Harris, A non-commutative unitary analogue of Kirchberg's conjecture, *Indiana Univ. Math. J.* **68** (2019), 503–536.

[129] S. Harris and V. Paulsen, Unitary correlation sets, *Integral Equations Operator Theory* **89** (2017), 125–149.

[130] A. Harrow, Quantum expanders from any classical Cayley graph expander, *Quantum Inf. Comput.* **8** (2008), 715–721.

[131] A. Harrow and M. Hastings, Classical and quantum tensor product expanders, *Quantum Inf. Comput.* **9** (2009), 336–360.

[132] M. Hastings, Random unitaries give quantum expanders, *Phys. Rev. A* (3) **76** no. 3 (2007), 032315, 11 pp.

[133] H. Helson, *Lectures on invariant subspaces*, Academic Press, New York, 1964.

[134] F. Hiai and Y. Nakamura, Distance between unitary orbits in von Neumann algebras, *Pacific J. Math.* **138** (1989), 259–294.

[135] S. Itoh, Conditional expectations in C^*-crossed products, *Trans. Amer. Math. Soc.* **267** (1981), 661–667.

[136] P. Jolissaint, A characterization of completely bounded multipliers of Fourier algebras, *Colloq. Math.* **63** (1992), 311–313.

[137] M. Junge and C. Le Merdy, Factorization through matrix spaces for finite rank operators between C^*-algebras, *Duke Math. J.* **100**, (1999), 299–319.

[138] M. Junge, M. Navascues, C. Palazuelos, D. Peréz-García, V.B. Scholz, and R.F. Werner, Connes' embedding problem and Tsirelson's problem, *J. Math. Phys.* **52** (2011), 012102, 12 pp.

[139] M. Junge, C. Palazuelos, D. Perez-García, I. Villanueva, and M. M. Wolf, Operator Space theory: a natural framework for Bell inequalities, *Phys. Rev. Lett.* **104**, 170405 (2010).

[140] M. Junge and C. Palazuelos, Large violation of Bell inequalities with low entanglement, *Comm. Math. Phys.* **306** (2011), 695–746.

[141] M. Junge and G. Pisier, Bilinear forms on exact operator spaces and $B(H) \otimes B(H)$, *Geom. Funct. Anal.* **5** (1995), 329–363.

[142] R. Kadison, Isometries of operator algebras, *Ann. Math.* **54** (1951), 325–338.

[143] R. Kadison, A generalized Schwarz inequality and algebraic invariants for operator algebras, *Ann. Math.* **56** (1952), 494–503.

[144] R. Kadison and G. Pedersen, Means and convex combinations of unitary operators, *Math. Scand.* **57** (1985), 249–266.

[145] R. Kadison and J. Ringrose, *Fundamentals of the theory of operator algebras*, Vol. I, Birkhäuser Boston, Inc., Boston, MA, 1983.

[146] R. Kadison and J. Ringrose, *Fundamentals of the theory of operator algebras*, Vol. II, Birkhäuser Boston, Inc., Boston, MA, 1992.

[147] R. Kadison and J. Ringrose, *Fundamentals of the theory of operator algebras*, Vol. IV, Birkhäuser Boston, Inc., Boston, MA, 1992.

[148] M. Kassabov, Symmetric groups and expanders, *Inv. Math.* **170** (2007), 327–354.

[149] A. Kavruk, Tensor products of operator systems and applications. Thesis (Ph.D.), University of Houston, 2011.

[150] A. Kavruk, V. Paulsen, I. Todorov, and M. Tomforde, Tensor products of operator systems, *J. Funct. Anal.* **261** (2011), 267–299.

[151] A. Kavruk, V. Paulsen, I. Todorov, and M. Tomforde, Quotients, exactness, and nuclearity in the operator system category, *Adv. Math.* **235** (2013), 321–360.

[152] A. Kavruk, The weak expectation property and Riesz interpolation, arXiv:1201.5414, 2012.

[153] A. Kavruk, Nuclearity related properties in operator systems, *J. Operator Theory* **71** (2014), 95–156.

[154] E. Kirchberg, C^*-nuclearity implies CPAP, *Math. Nachr.* **76** (1977), 203–212.

[155] E. Kirchberg, On nonsemisplit extensions, tensor products and exactness of group C^*-algebras, *Invent. Math.* **112** (1993), 449–489.

[156] E. Kirchberg, Commutants of unitaries in UHF algebras and functorial properties of exactness, *J. Reine Angew. Math.* **452** (1994), 39–77.

[157] E. Kirchberg, Discrete groups with Kazhdan's property T and factorization property are residually finite, *Math. Ann.* **299** (1994), 551–563.

[158] E. Kirchberg, Exact C*-algebras, tensor products, and the classification of purely infinite algebras, *Proceedings of the International Congress of Mathematicians, Vol. 1, 2 (Zürich, 1994)*, 943–954, Birkhäuser, Basel, 1995.

[159] E. Kirchberg, On subalgebras of the CAR-algebra, *J. Funct. Anal.* **129** (1995), 35–63.

[160] E. Kirchberg, On restricted perturbations in inverse images and a description of normalizer algebras in C^*-algebras, *J. Funct. Anal.* **129** (1995), 1–34.

[161] E. Kirchberg, Personal communication.

[162] E. Kirchberg and N. C. Phillips, Embedding of exact C^*-algebras in the Cuntz algebra O_2, *J. Reine Angew. Math.* **525** (2000), 17–53.

[163] I. Klep and M. Schweighofer, Connes' embedding conjecture and sums of Hermitian squares, *Adv. Math.* **217** (2008), 1816–1837.

[164] V. Lafforgue and M. De la Salle, Noncommutative L_p-spaces without the completely bounded approximation property, *Duke Math. J.* **160** (2011), 71–116.

[165] C. Lance, On nuclear C^*-algebras, *J. Funct. Anal.* **12** (1973), 157–176.

[166] F. Lehner, A characterization of the Leinert property, *Proc. Amer. Math. Soc.* **125** (1997), 3423–3431.

[167] J. Lindenstrauss and L. Tzafriri, *Classical Banach spaces, vol. I, Sequence spaces*, Springer Verlag, Berlin 1976.

[168] F. Lehner, Computing norms of free operators with matrix coefficients, *Amer. J. Math.* **121** (1999), 453–486.

[169] E. Lieb, Convex trace functions and the Wigner–Yanase–Dyson conjecture, *Adv. Math.* **11** (1973), 267–288.

[170] J. Lindenstrauss and H. P. Rosenthal, The $\mathcal{L}_p$ spaces, *Israel J. Math.* **7** (1969), 325–349.

[171] T. Loring, *Lifting solutions to perturbing problems in C^*-algebras*, Fields Institute Monographs, *American Mathematical Society*, Providence, RI, 1997.

[172] A. Lubotzky, *Discrete groups, expanding graphs and invariant measures*, Progress in Math, **125**. Birkhäuser, 1994.

[173] A. Lubotzky, What is Property (τ)? *Notices Amer. Math. Soc.* **52** (2005), 626–627.

[174] A. Lubotzky, Expander graphs in pure and applied mathematics, *Bull. Amer. Math. Soc.* **49** (2012), 113–162.

[175] A. Lubotzky, R. Phillips, and P. Sarnak, Hecke operators and distributing points on S^2, I, *Comm. Pure and Applied Math.* **39** (1986), 149–186.

[176] D. Mc Duff, Uncountably many II_1 factors, *Ann. Math.* **90** (1969), 372–377.

[177] A. I. Malcev, On isomorphic matrix representations of infinite groups of matrices (Russian), *Mat. Sb.* **8** (1940), 405–422 & *Amer. Math. Soc. Transl. (2)* **45** (1965), 1–18.

[178] N. Monod, Groups of piecewise projective homeomorphisms, *Proc. Natl. Acad. Sci. USA* **110** (2013), 4524–4527.

[179] A. Nica, Asymptotically free families of random unitaries in symmetric groups, *Pacific J. Math.* **157** (1993), 295–310.

[180] T. Oikhberg and É. Ricard, Operator spaces with few completely bounded maps, *Math. Ann.* **328** (2004), 229–259.

[181] T. Oikhberg and H. P. Rosenthal, Extension properties for the space of compact operators, *J. Funct. Anal.* **179** (2001), 251–308.

[182] D. Osajda, Small cancellation labellings of some infinite graphs and applications, arXiv:1406.5015, 2014.

[183] D. Osajda, Residually finite non-exact groups, *Geom. Funct. Anal.* **28** (2018), 509–517.

[184] N. Ozawa, On the set of finite-dimensional subspaces of preduals of von Neumann algebras, *C. R. Acad. Sci. Paris Sér. I Math.* **331** (2000), 309–312.

[185] A non-extendable bounded linear map between C^*-algebras, *Proc. Edinb. Math. Soc. (2)* **44** (2001), 241–248.

[186] N. Ozawa, On the lifting property for universal C^*-algebras of operator spaces, *J. Operator Theory* **46** no. 3, suppl. (2001), 579–591.

[187] N. Ozawa, Amenable actions and exactness for discrete groups, *C. R. Acad. Sci. Paris Sér. I Math.* **330** (2000), 691–695.

[188] N. Ozawa, An application of expanders to $\mathbb{B}(\ell_2) \otimes \mathbb{B}(\ell_2)$, *J. Funct. Anal.* **198** (2003), 499–510.

[189] N. Ozawa, About the QWEP conjecture, *Internat. J. Math.* **15** (2004), 501–530.

[190] N. Ozawa, Examples of groups which are not weakly amenable, *Kyoto J. Math.* **52** (2012), 333–344.

[191] N. Ozawa, About the Connes embedding conjecture: algebraic approaches, *Jpn. J. Math.* **8** (2013), 147–183.

[192] N. Ozawa, Tsirelson's problem and asymptotically commuting unitary matrices, *J. Math. Phys.* **54** (2013), 032202, 8 pp.

[193] N. Ozawa and G. Pisier, A continuum of C^*-norms on $B(H) \otimes B(H)$ and related tensor products, *Glasgow Math. J.* **58** (2016), 433–443.

[194] A. Paterson, *Amenability*, American Mathematical Society, Mathematical Surveys and Monographs, **29**, 1988.

[195] V. Paulsen, *Completely bounded maps and dilations*, Pitman Research Notes 146. Pitman Longman (Wiley) 1986.

[196] V. Paulsen, *Completely bounded maps and operator algebras*, Cambridge University Press, Cambridge, 2002.

[197] V. Paulsen and C.-Y. Suen, Commutant representations of completely bounded maps, *J. Operator Theory* **13** (1985), 87–101.

[198] V. Pestov, Operator spaces and residually finite-dimensional C^*-algebras, *J. Funct. Anal.* **123** (1994), 308–317.

[199] J. P. Pier, *Amenable locally compact groups*, Wiley Interscience, New York, 1984.

[200] G. Pisier, *Factorization of linear operators and the geometry of Banach spaces*, *CBMS (Regional Conferences of the A.M.S.)* no. 60 (1986), Reprinted with corrections 1987.

[201] G. Pisier, Remarks on complemented subspaces of von Neumann algebras, *Proc. Royal Soc. Edinburgh* **121** A (1992), 1–4.

[202] G. Pisier, Espace de Hilbert d'opérateurs et interpolation complexe, *Comptes Rendus Acad. Sci. Paris, Série I* **316** (1993), 47–52.

[203] G. Pisier, Projections from a von Neumann algebra onto a subalgebra, *Bull. Soc. Math. France* **123** (1995), 139–153.

[204] G. Pisier, A simple proof of a theorem of Kirchberg and related results on C^*-norms, *J. Operator Theory* **35** (1996), 317–335.

[205] G. Pisier, The operator Hilbert space OH, complex interpolation and tensor norms, *Memoirs Amer. Math. Soc.* **122** no. 585 (1996), 1–103.

[206] G. Pisier. Quadratic forms in unitary operators, *Linear Algebra Appl.* **267** (1997), 125–137.

[207] G. Pisier, *Similarity problems and completely bounded maps*. Second, Expanded Edition, Springer Lecture Notes, **1618** (2001).

[208] G. Pisier, *Introduction to operator space theory*, Cambridge University Press, Cambridge, 2003.

[209] G. Pisier, Remarks on $B(H) \otimes B(H)$, *Proc. Indian Acad. Sci. (Math. Sci.)* **116** (2006), 423–428.

[210] G. Pisier, Grothendieck's theorem, past and present, *Bull. Amer. Math. Soc.* **49** (2012), 237–323.

[211] G. Pisier, Random matrices and subexponential operator spaces, *Israel J. Math.* **203** (2014), 223–273.

[212] G. Pisier, Quantum expanders and geometry of operator spaces, *J. Europ. Math. Soc.* **16** (2014), 1183–1219.

[213] G. Pisier, On the metric entropy of the Banach–Mazur compactum, *Mathematika* **61** (2015), 179–198.

[214] G. Pisier, *Martingales in Banach spaces*, Cambridge University Press, Cambridge, 2016.

[215] G. Pisier and Q. Xu, Non-commutative L_p-spaces, *Handbook of the geometry of Banach spaces*, vol. II, North-Holland, Amsterdam, 2003.

[216] S. Popa, On the Russo–Dye theorem, *Michigan Math. J.* **28** (1981), 311–315.

[217] S. Popa, A short proof of "injectivity implies hyperfiniteness" for finite von Neumann algebras, *J. Operator Theory* **16** (1986), 261–272.

[218] S. Popa, Markov traces on universal Jones algebras and subfactors of finite index, *Invent. Math.* **111** (1993), 375–405.

[219] W. Pusz and S. L. Woronowicz, Form convex functions and the WYDL and other inequalities, *Lett. Math. Phys.* **2** (1977/78), 505–512.

[220] W. Pusz and S. L. Woronowicz, Functional calculus for sesquilinear forms and the purification map, *Rep. Mathematical Phys.* **8** (1975), 159–170.

[221] T. Pytlik and R. Szwarc, An analytic family of uniformly bounded representations of free groups, *Acta Math.* **157** (1986), 287–309.

[222] F. Rădulescu, A comparison between the max and min norms on $C^*(F_n) \otimes C^*(F_n)$, *J. Operator Theory* **51** (2004), 245–253.

[223] F. Rădulescu, Combinatorial aspects of Connes's embedding conjecture and asymptotic distribution of traces of products of unitaries, *Operator Theory* 20, 197–205, *Theta Ser. Adv. Math.* **6**, Theta, Bucharest, 2006.

[224] R. A. Rankin, *Modular forms and functions*, Cambridge University Press, Cambridge, 1977.

[225] M. Rieffel, Induced representations of C^*-algebras, *Adv. Math.* **13** (1974), 176–257.

[226] S. Sakai, C^*-*algebras and* W^*-*algebras*, Springer-Verlag, New York, 1971.

[227] P. Sarnak, What is an expander? *Notices Amer. Math. Soc.* **51** (2004), 762–763.

[228] A. Selberg, On the estimation of Fourier coefficients of modular forms, Proceedings of the Symposium Pure Mathematics, Vol. VIII, American Mathematical Society, Providence, RI, 1965, pp. 1–15.

[229] D. Sherman, On cardinal invariants and generators for von Neumann algebras, *Canad. J. Math.* **64** (2012), 455–480.

[230] W. Slofstra, The set of quantum correlations is not closed, *Forum Math. Pi* **7** (2019), e1, 41 pp.

[231] W. Slofstra, A group with at least subexponential hyperlinear profile, arXiv:1806.05267, 2018.

[232] R. R. Smith, Completely bounded module maps and the Haagerup tensor product, *J. Funct. Anal.* **102** (1991), 156–175.

[233] E. Størmer, On the Jordan structure of C^*-algebras, *Trans. Amer. Math. Soc.* **120** (1965), 438–447.

[234] E. Størmer, Multiplicative properties of positive maps, *Math. Scand.* **100** (2007), 184–192.

[235] E. Størmer, *Positive linear maps of operator algebras*, Springer, Heidelberg, 2013.

[236] C-Y. Suen, Completely bounded maps on C*-algebras, *Proc. Amer. Math. Soc.* **93** (1985), 81–87.

[237] R. Szwarc, An analytic series of irreducible representations of the free group, *Ann. Inst. Fourier* **38** (1988), 87–110.

[238] M. Takesaki, A note on the cross-norm of the direct product of C^*-algebras, *Kodai Math. Sem. Rep.* **10** (1958), 137–140.

[239] M. Takesaki, Duality for crossed products and the structure of von Neumann algebras of type III, *Acta Math.* **131** (1973), 249–310.

[240] M. Takesaki, *Theory of Operator algebras, vol. I*, Springer-Verlag, Berlin, Heidelberg, New York, 1979.

[241] M. Takesaki, *Theory of Operator algebras, vol. II*, Springer-Verlag, Berlin, Heidelberg, New York, 2003.

[242] M. Takesaki, *Theory of Operator algebras, vol. III*, Springer-Verlag, Berlin, Heidelberg, New York, 2003.

[243] T. Tao, *Expansion in finite simple groups of Lie type*, American Mathematical Society, Providence, RI, 2015.

[244] A. Thom, Examples of hyperlinear groups without factorization property, *Groups Geom. Dyn.* **4** (2010), 195–208.

[245] J. Tomiyama, Tensor products and projections of norm one in von Neumann algebras, Lecture Notes, University of Copenhagen, 1970.

[246] J. Tomiyama, Tensor products and approximation problems of C^*-algebras, *Publ. Res. Inst. Math. Sci.* **11** (1975/76), 163–183.

[247] J. Tomiyama, On the product projection of norm one in the direct product of operator algebras. *Tôhoku Math. J.* **11**, no. (1959) 305–313.

[248] J. Tomiyama, On the projection of norm one in W^*-algebras. III, *Tôhoku Math. J.* (2) **11** (1959) 125–129.

[249] A. Tonge, The complex Grothendieck inequality for 2×2 matrices, *Bull. Soc. Math. Grèce (N.S.)* **27** (1986), 133–136.

[250] S. Trott, A pair of generators for the unimodular group, *Canad. Math. Bull.* **5** (1962), 245–252.

[251] B.S. Tsirelson, Quantum generalizations of Bell's inequality, *Lett. Math. Phys.* **4** (1980), 93–100.

[252] A. Valette, Minimal projections, integrable representations and property (T), *Arch. Math.* (Basel) **43** (1984), 397–406.

[253] D. Voiculescu, Property T and approximation of operators, *Bull. London Math. Soc.* **22** (1990), 25–30.

[254] D. Voiculescu, K. Dykema, and A. Nica, *Free Random Variables*, American Mathematical Society, Providence, RI, 1992.

[255] S. Wassermann, On tensor products of certain group C^*-algebras, *J. Funct. Anal.* **23** (1976), 239–254.

[256] S. Wassermann, Injective W^*-algebras, *Proc. Cambridge Phil. Soc.* **82** (1977), 39–47.

[257] S. Wassermann, A pathology in the ideal space of $L(H) \otimes L(H)$, *Indiana Univ. Math. J.* **27** (1978), 1011–1020.

[258] S. Wassermann, *Exact C^*-algebras and related topics*, Lecture Notes Series, **19**. Seoul National University, Seoul, 1994.

[259] S. Wassermann, C^*-algebras associated with groups with Kazhdan's property T, *Ann. Math.* **134** (1991), 423–431.

[260] D. Werner, Some lifting theorems for bounded linear operators, *Functional analysis (Essen, 1991)*, 279–291, Lecture Notes in Pure and Applied Mathematics, **150**, Dekker, New York, 1994.

[261] P. Willig, On hyperfinite W^*-algebras, *Proc. Amer. Math. Soc.* **40** (1973), 120–122.

[262] J. S. Wilson, On characteristically simple groups, *Math. Proc. Cambridge Philos. Soc.* **80** (1976), 19–35.

[263] S. Woronowicz, Selfpolar forms and their applications to the C^*-algebra theory, *Rep. Mathematical Phys.* **6** (1974), 487–495.

[264] C. Zhang, *Representation and geometry of operator spaces*, Ph.D. thesis, University of Houston, 1995.

[265] M. Zippin, The separable extension problem, *Israel J. Math.* **26** (1977), 372–387.

[266] M. Zippin, Extension of bounded linear operators, *Handbook of the geometry of Banach spaces, Vol.* 2, 1703–1741, North-Holland, Amsterdam, 2003.

Index